Molecular Medicines for Cancer

Concepts and Applications of Nanotechnology

Molecular Medicines for Cancer

Concepts and Applications of Nanotechnology

Edited by
Deepak Chitkara
Anupama Mittal
Ram I. Mahato

CRC Press is an imprint of the
Taylor & Francis Group, an **informa** business

CRC Press
Taylor & Francis Group
6000 Broken Sound Parkway NW, Suite 300
Boca Raton, FL 33487-2742

First issued in paperback 2021

ISBN-13: 978-0-367-78083-8 (pbk)
ISBN-13: 978-1-138-03515-7 (hbk)

Library of Congress Cataloging-in-Publication Data

Names: Chitkara, Deepak, editor. | Mittal, Anupama, editor. | Mahato, Ram I., editor.
Title: Molecular medicines for cancer : concepts and applications of nanotechnology / [edited by] Deepak Chitkara, Anupama Mittal and Ram I. Mahato.
Description: Boca Raton : CRC Press, [2018] | Includes bibliographical references and index.
Identifiers: LCCN 2018015862| ISBN 9781138035157 (hardback : alk. paper) | ISBN 9781315269214 (ebook)
Subjects: | MESH: Neoplasms--drug therapy | Molecular Targeted Therapy--methods | Drug Delivery Systems | Nanostructures--therapeutic use | Theranostic Nanomedicine--methods | Molecular Medicine--methods
Classification: LCC RC271.C5 | NLM QZ 267 | DDC 616.99/4061--dc23
LC record available at https://lccn.loc.gov/2018015862

Visit the Taylor & Francis Web site at
http://www.taylorandfrancis.com

and the CRC Press Web site at
http://www.crcpress.com

I dedicate this book to my little daughters, Samaira (four years) and Aarohi (two years), whose smiles eradicate all the stresses; to my late mother for devoting her life to my upbringing; and to my teachers who always mentored me to grow into a better person.

Deepak Chitkara

I would like to dedicate this book to my parents Ved Prakash Mittal and Urmil Mittal for their unconditional love and support; to my family and to my teachers for showing the right path to achieve a better me.

Anupama Mittal

I dedicate this book to my wife Subhashini, my children Kalika and Vivek for their love and support; my mother Sarswati for believing in me; and to my students and mentors who have always helped me in my quest for learning and in achieving higher goals.

Ram I. Mahato

Contents

SECTION I Nanotechnology-Based Approaches to Target Cancer

SECTION II Imaging Technologies in Cancer

SECTION III Oligonucleotides and Gene-Based Therapies

Foreword

In today's era of information and communication technology, getting information about a particular subject has become increasingly fast and easy, but the readers are often seen struggling to precisely identify the best source from the plethora of information available. Here arises the need of expert opinion on the subject matter to guide the readers and that is what the editors have successfully accomplished in this book. Talking particularly about the "*molecular medicines*," although summed up in two words, the field is very broad, covering the mechanistic aspects of how a disease could be targeted at the molecular level either by using a novel molecule or employing novel tools for delivering established therapeutics. Further, applications of nanotechnology have generated enormous interest in the medical field in the recent past with many of these new technologies heading towards the clinic. To harness the translational potential of nanotechnology, it is essential to understand how these systems interact with biological targets, ultimately shaping the outcome of the therapy. As rightly said by physicist Richard Feynman, "*there is plenty of room at the bottom*." The idea of nanotechnology discussed in his famous lecture has now seen the light of the day with its application not being limited to a particular domain of physics but also extending its arms into almost all areas of current research. The field of molecular medicines has witnessed many of its major advancements due to nanotechnology, which is hence rightly covered in this book, while the research is still continuing to unleash the mysteries lying very deep at the bottom. The concept of "*magic bullets*" proposed by German Nobel laureate Paul Ehrlich in the 1900s has always been the driving force in research, particularly cancer research, to resolve the challenges associated with the conventional strategies and provide personalized therapy to patients; these endeavors are rendered fruitful by applying concepts of nanotechnology for precise delivery of newer therapies. In spite of significant achievements in the field, delivery science still needs a precise spacio-temporal control over the molecular medicines to hit the target accurately at the right concentration and at the right time; a nanotechnology-based approach would certainly enable this quantum leap.

This book discusses how nanotechnology based approaches precisely deliver small molecules as well as oligonucleotides and gene-based therapies for cancer treatment presenting all the essential aspects of molecular medicines. The editors have convened the leading experts and researchers to put forth their views on this perpetually advancing field in an easy to understand text for academicians, researchers, undergraduate, and graduate students. I wish them all the best for this book and look forward for further compilations in the future as well.

Kenneth H. Cowan, MD
Director, Fred & Pamela Buffett Cancer Center
University of Nebraska Medical Center

Preface

The term "nano" has paved its way into all the fields of science and technology and has shown promising improvements over the conventionally used methodologies. Talking about the field of medicine, it has emerged as a game-changing option by providing new ways of reaching out at the target disease and modifying the therapeutic outcome. With the merger of nanotechnology and medicine, a new term, "nanomedicine," has gained prominence and is applied to applications of nanotechnology in medicine encompassing both therapeutics as well as diagnostics. The reason behind the enormous interest in the field is not only due to the new properties that materials start exhibiting at "nano" scale but also the ability to deliver these therapeutics/diagnostics at the subcellular/molecular level. On the other hand, simultaneous advancements in the field of molecular medicines that deal with the medical interventions targeting molecular structures and mechanisms involved in disease progression require novel technologies to make their therapeutic targets achievable. Particularly in cancer, several molecular mechanisms have been shown to impact its progression, aggressiveness, and chemoresistance. Our growing understanding of the mechanistic association of aberrant cell signaling as well as genetic and epigenetic modifications with cancer has enabled the design of molecular therapeutics. Small targeted molecules, antibodies, and oligonucleotides have been shown to selectively target the molecular structures in the cell thereby influencing the signal transduction process. Also, RNA interference (RNAi), including siRNA and miRNA, has exhibited a marked progress over the past decade with several RNAi technologies in clinical trials including CALLA 01 and ALN-VSP02. The challenges of delivering these molecular medicines to the desired site at therapeutic concentrations have rationalized the use of nanotechnology approaches for achieving the therapeutic goal. There is an increasing body of evidence that demonstrates the role of nanotechnology in influencing the outcome of molecular therapy. Keeping this in mind, we have gathered an array of interrelated topics to apprise the readers with the fundamentals of nanotechnology vis-à-vis the recent advancements in delivering the molecular medicines.

The book has been divided into three sections with a total of 19 chapters that reflect the recent literature as well as the experience of the authors. Section 1 consists of seven chapters covering recent approaches for targeting cancer. Chapter 1 provides an introduction to nanomedicines used for cancer particularly focusing on the nano-based products that have reached the clinic or are under clinical trials. A brief account of different aspects of nanomedicines such as targeting mechanisms, *in vivo* transport as well as clinically relevant animal models required to assess these nanomedicines is also provided. Chapter 2 focuses on the intricate relationship between the physico-chemical properties of the nanoparticles and their *in vivo* journey through various intravascular and transvascular transport routes and biological processes. The significance of nanoparticle size, shape, and surface characteristics with respect to their biological properties such as particle transportation, pharmacokinetics, biodistribution, tumor penetration, cellular uptake, and particle clearance from blood and tissues is highlighted in this chapter. Once into the cell, nanomedicine should be able to target a particular subcellular organelle. The particle surface could be suitably modified for this purpose; this is thoroughly dealt with in Chapter 3 wherein, several strategies are outlined to target the sub-cellular structures including mitochondria, nucleus, lysosomes, and endoplasmic reticulum. Chapter 4 focuses on the reversal of chemoresistance wherein, several key mechanisms involved in the emergence of chemoresistance are discussed to provide the reader with a clear understanding of key pathways utilized by cancer cells to evade the drug effects. A brief account of the general concepts of nanomedicines is then provided for better understanding followed by a thorough description of nanomedicines utilized for the reversal of chemoresistance. Chapter 5 provides an in-depth understanding on dendrimers as a powerful multifunctional platform for delivering cancer therapeutics and discusses several biomedical applications of peptide-modified dendrimers. Chapter 6 explores an important component involved in cancer progression i.e. tumor microenvironment (TME) as several therapeutic strategies now aim to manipulate TME and to disrupt the cross-talk between tumor and stroma. A detailed account of TME is given in this chapter followed by nanotherapeutic approaches for TME-specific delivery. The last chapter of this section, Chapter 7 provides an insight into the recent nanotherapeutics containing active ingredients of traditional Chinese medicine (AITCM).

Section 2 comprises three chapters (8, 9, and 10) that deal with imaging technologies for cancer, recent trends in theranostic nanosystems and nanotechnologies for immunodiagnosis and immunotherapy, respectively. Herein, a detailed discussion of the imaging techniques in cancer is first provided to apprise the reader with the current state of the art followed by how nanotechnology could be utilized to improve the diagnostic capability. The emergence of theranostic systems that encompass both the diagnostic and therapeutic modalities together and enable personalized therapy is also focused upon in this section. Further, a thorough account on advancements in the fields of immunodiagnostics and immunotherapies vis-à-vis the emerging nanotechnologies that combine the sensitivity of nanomaterials with the specificity of the immunological interactions is provided.

Section 3 comprises nine chapters and provides a detailed account on the emerging gene-based therapies. Chapter 11 gives a brief introduction on the nucleic acids (DNA and RNA) and their use as cancer therapeutics. Further, various nanotechnology-based approaches utilized to deliver these nucleic acid-based therapeutics are discussed followed by the challenges in their *in-vivo* and intracellular delivery. The reader is then apprised of the non-coding RNAs (long non-coding RNAs [lncRNAs] and microRNAs [miRNAs] in Chapters 12 and 13, respectively). LncRNAs are 200–1,000 nucleotides long, non-coding transcripts involved in numerous essential cellular processes and epigenetic mechanisms including genomic imprinting, transcription, translation, chromatin modification, cell development, and differentiation and apoptosis. On the other hand, miRNAs are small oligonucleotides that regulate the expression of target mRNA by binding to 3'-untranslated regions (UTRs) resulting in translation repression. Molecular pathways involved in cancer including WNT/β-catenin, TGF-β/Smad, PI3K/AKT, and p53 signaling that could be targeted by miRNAs are discussed followed by current strategies and challenges in miRNA-based therapies. Further, Chapter 14 provides an introduction to siRNA therapeutics and their delivery strategies including chemical modifications, RNA-based nanostructures, and lipid systems. Fine tuning of these systems for on-demand release is then discussed. Chapters 15 and 16 provide a detailed account on the lipidic and polymeric carriers for delivering the RNAi-based therapeutics, respectively. Recently it has been shown that both RNA and DNA could be used as nanoscaffolds for delivering various functionalities for regulating cell function and gene expression. The same is dealt with in detail in Chapter 17. Genetic mutations and genomic instability have long been thought of as drivers of the tumorigenesis process. The damage caused by endogenous or exogenous agents needs to be repaired to avoid these mutations. DNA repair mechanisms involve identification of the alterations in DNA molecules followed by a correction to restore the integrity of the genome. On the other hand, cancers with low mutations have pointed at the role of epigenetics, which is defined as the inheritable changes in gene expression with no alterations in DNA sequences. A connection between disruptions of the epigenome and tumor progression has been demonstrated by several studies. A detailed account of both DNA repair mechanisms and epigenetics in cancer is provided in Chapter 18. Recently, CRISPR/Cas9 has been shown as a potential tool for editing the genome and thereby correcting the detrimental mutations. Chapter 19 provides an account on the CRISPR/Cas9 technology and its application in cancer for generation of tumor models as well as gene- and cell-based therapy.

This book presents an overview of the entire field of molecular medicines in cancer treatment and the use of nanotechnology as an efficient delivery approach. The content presented here has never before been compiled into a single book. The book brings together the leading experts and researchers to provide an account on the topic, such that the academicians, industrial researchers, higher-level undergraduate and graduate students may easily comprehend the molecular therapy concepts and the applications thereof and could integrate them with the ever-advancing field of nanotechnology. This will further enable the reader to understand and structure the concepts and applications of nanotechnology in a more comprehensive manner.

Acknowledgments

While editing this book, we have been constantly asking ourselves: What/who is driving this field of work? We recognize the tireless efforts made by leaders in this field that set forth a vision for the future. We take this opportunity to especially acknowledge the contributions of the pioneers in the field, professors Vincent H. L. Lee, Mitsura Hashida, and Sung Wan Kim whose research endeavors have largely shaped our current knowledge.

Editing a book takes an immense effort to put together and interlink the available information and make it a meaningful object. This could not have been achieved without the contribution of leading experts who graciously agreed to put forth their precious time in writing the chapters. We would like to thank them for entrusting us as an appropriate means for spreading their knowledge and views. Further, we would also like to acknowledge the efforts of the reviewers who provided their critical comments and suggestions on the chapters that helped us to improve the book. We also extend our gratitude to our students, mentors, and colleagues for sharing their thoughts in the overall designing of the book.

We especially thank the staff members of CRC Press, Taylor & Francis Group for lending their expertise in the planning, preparation, and production of this book. Particularly we would like to acknowledge Hilary Lafoe and Natasha Hallard for constantly providing the much necessary support in converting a draft into an assembled book.

Deepak Chitkara
Anupama Mittal
Ram I. Mahato

Editors

Deepak Chitkara is an Assistant Professor at the Department of Pharmacy, Birla Institute of Technology and Science (BITS)-Pilani, Vidya Vihar Campus, India. He obtained his Ph.D. in Pharmaceutical Sciences from the National Institute of Pharmaceutical Education and Research (NIPER), SAS Nagar, India. He was an exchange research scholar at University of Tennessee Health Science Center, Memphis, TN for one year. After that he did his post-doctoral training at University of Nebraska Medical Center, Omaha, NE in the area of nanomedicines for pancreatic cancer. He is the recipient of Ranbaxy Science Scholar Award-2011. His research interests include the nano-based delivery systems for small molecules, miRNAs, and CRISPR/Cas genome editing tools. He has been working in the area of nanotechnology since 2007.

Dr. Chitkara has developed and taught courses on advanced drug delivery systems and advanced pharmaceutical technology to students of the Department of Pharmacy, BITS-Pilani. The mechanisms, designing, delivery, and therapeutic applications of small molecules, proteins, and peptides and RNAi are extensively discussed in these courses.

Anupama Mittal is an Assistant Professor at the Department of Pharmacy, Birla Institute of Technology and Science (BITS)-Pilani, Vidya Vihar Campus, India. She graduated from the Department of Pharmaceutics at National Institute of Pharmaceutical Education and Research (NIPER), SAS Nagar, India. Thereafter, she worked as a post-doctoral research associate at the University of Tennessee Health Science Center, Memphis, TN and the University of Nebraska Medical Center, Omaha, NE.

She has been associated with several classroom courses entitled Advanced Physical Pharmaceutics, Physical Pharmacy, Instrumental Methods of Analysis, and Pharmaceutical Administration and Management. She has also been teaching these courses to the industry professionals of different pharmaceutical industries including Lupin, Wockhardt, Sun Pharma, etc.

Her research interests include nanomedicines and exosomes for the treatment of cancer and diabetes and regenerative medicine. Her research group is also actively engaged in developing self-assembling drug conjugates for disease treatment. Her work has been published in several high impact journals of international repute and she has filed three product patents also.

Ram I. Mahato is a Professor and Chairman of the Department of Pharmaceutical Sciences, University of Nebraska Medical Center, Omaha, NE. He was a professor at the University of Tennessee Health Science Center, Research Assistant Professor at the University of Utah, Senior Scientist at GeneMedicine, Inc., and as a postdoctoral fellow at the University of Southern California in Los Angeles, Washington University in St. Louis, and Kyoto University, Japan. He received a Ph.D. in Drug Delivery from the University of Strathclyde and a B.S. from China Pharmaceutical University.

Dr. Mahato has published 140 papers, 17 book chapters, holds two US patents, and has edited/written eight books and ten journal issues (total Google citations=9554 and h-Index=56). He was a feature editor of the Pharmaceutical Research (2006–2013) and an editorial board member of eight journals. He is a CRS and AAPS Fellow, a permanent member of BTSS/NIH study section, and an ASGCT scientific advisor. He applies sound principles in pharmaceutical sciences in the context of the latest advances in life and material sciences to solve challenging drug delivery problems in therapeutics.

Contributors

Kirill A. Afonin
Department of Chemistry
University of North Carolina
Charlotte, North Carolina

Narsireddy Amreddy
Department of Pathology
Stephenson Cancer Center
University of Oklahoma Health Sciences Center
Oklahoma City, Oklahoma

Anish Babu
Department of Pathology
Stephenson Cancer Center
University of Oklahoma Health Sciences Center
Oklahoma City, Oklahoma

Surinder K. Batra
Department of Biochemistry and Molecular Biology
Fred and Pamela Buffet Cancer Center
University of Nebraska Medical Center
Omaha, Nebraska

Claudia A. Benavente
Department of Pharmaceutical Sciences
Department of Developmental and Cell Biology
Chao Family Comprehensive Cancer Center
University of California
Irvine, California

Rakesh Bhatia
Department of Biochemistry and Molecular Biology
University of Nebraska Medical Center
Omaha, Nebraska

Rajan Sharma Bhattarai
Department of Pharmaceutical Sciences
University of Nebraska Medical Center
Omaha, Nebraska

Aniketh Bishnu
Imaging Cell Signaling and Therapeutics Lab
Advanced Centre for Treatment Research and Education in Cancer, Tata Memorial Centre
Homi Bhaba National Institute
Mumbai, India

Andrew Cannon
Department of Biochemistry and Molecular Biology
University of Nebraska Medical Center
Omaha, Nebraska

Morgan Chandler
Department of Chemistry
University of North Carolina
Charlotte, North Carolina

Rashmi Chaudhari
Department of Biosciences and Bioengineering
Indian Institute of Technology Bombay
Mumbai, India

Amit Kumar Chaudhary
Department of Pharmaceutical Sciences
University of Nebraska Medical Center
Omaha, Nebraska

Birendra Chaurasiya
Department of Pharmaceutics
China Pharmaceutical University
Nanjing, China

Meiwan Chen
State Key Laboratory of Quality Research in Chinese Medicine
Institute of Chinese Medical Sciences
University of Macau
Macau, China

Guofeng Cheng
Shanghai Veterinary Research Institute
Chinese Academy of Agricultural Sciences
Key Laboratory of Animal Parasitology
Ministry of Agriculture
Shanghai, China

Deepak Chitkara
Department of Pharmacy
Birla Institute of Technology and Science-Pilani [BITS]
Pilani Campus
Rajasthan, India

Abhilash Deo
Imaging Cell Signaling and Therapeutics Lab
Advanced Centre for Treatment Research and Education in Cancer, Tata Memorial Centre
Homi Bhaba National Institute
Mumbai, India

Ajit Dhadve
Imaging Cell Signaling and Therapeutics Lab
Advanced Centre for Treatment Research and Education in Cancer, Tata Memorial Centre
Homi Bhaba National Institute
Mumbai, India

Damjan Glavač
Department of Molecular Genetics
Institute of Pathology
Faculty of Medicine
University of Ljubljana
Ljubljana, Slovenia

Zhaopei Guo
State Key Laboratory of Quality Research in Chinese Medicine
Institute of Chinese Medical Sciences
University of Macau
Macau, China

Suprit Gupta
Department of Biochemistry and Molecular Biology
University of Nebraska Medical Center
Omaha, Nebraska

Vinod Kumar Gupta
Department of Pharmaceutical Chemistry
Mithibai College Campus
Mumbai, India

Channabasavaiah B. Gurumurthy
Mouse Genome Engineering Core Facility
Vice Chancellor for Research Office
University of Nebraska Medical Center
Omaha, Nebraska

Brad Hall
Department of General Surgery
University of Nebraska Medical Center
Omaha, Nebraska

Mateja M. Jelen
Department of Molecular Genetics
Institute of Pathology
Faculty of Medicine
University of Ljubljana
Ljubljana, Slovenia

Siddharth Jhunjhunwala
Centre for BioSystems Science and Engineering
Indian Institute of Science
Bengaluru, India

Abhijeet Joshi
INSPIRE Faculty
Centre for Biosciences and Biomedical Engineering
Indian Institute of Technology Indore
Indore, India

Chie Kojima
Department of Applied Chemistry
Graduate School of Engineering, Osaka Prefecture University
Osaka, Japan

Sushil Kumar
Department of Biochemistry and Molecular Biology
University of Nebraska Medical Center
Omaha, Nebraska

Shyh-Dar Li
Faculty of Pharmaceutical Sciences
The University of British Columbia
Vancouver, Canada

Yanan Li
Department of Pharmaceutics
China Pharmaceutical University
Nanjing, China

Ram I. Mahato
Department of Pharmaceutical Sciences
University of Nebraska Medical Center
Omaha, Nebraska

Reza Mahjub
Department of Pharmaceutics
School of Pharmacy
Hamadan University of Medical Sciences
Hamadan, Iran

and

Faculty of Pharmaceutical Sciences
The University of British Columbia
Vancouver, Canada

Meghna Mehta
Department of Radiation Oncology
Stephenson Cancer Center
University of Oklahoma Health Sciences Center
Oklahoma City, Oklahoma

Vinayak Sadashiv Mharugde
Department of Pharmacy
Birla Institute of Technology and Science-Pilani [BITS]
Rajasthan, India

Lucia Milanova
Institute of Biology and Ecology
Faculty of Science
Pavol Jozef Šafárik University
Košice, Slovakia

Anupama Mittal
Department of Pharmacy
Birla Institute of Technology and Science-Pilani [BITS]
Rajasthan, India

Souvik Mukherjee
Imaging Cell Signaling and Therapeutics Lab
Advanced Centre for Treatment Research and Education in Cancer, Tata Memorial Centre
Homi Bhaba National Institute
Mumbai, India

Anupama Munshi
Department of Radiation Oncology
Stephenson Cancer Center
University of Oklahoma Health Sciences Center
Oklahoma City, Oklahama

Ranganayaki Muralidharan
Department of Pathology
Stephenson Cancer Center
University of Oklahoma Health Sciences Center
Oklahoma City, Oklahoma

Jiayi Pan
Center for Pharmaceutical Biotechnology and Nanomedicine
Northeastern University
Boston, Massachusetts

Gaurav Pandey
Centre for Biosciences and Biomedical Engineering
Indian Institute of Technology Indore
Indore, India

Martin Panigaj
Institute of Biology and Ecology
Faculty of Science
Pavol Jozef Šafárik University
Košice, Slovakia

Janani Panneerselvam
Department of Pathology
Stephenson Cancer Center
University of Oklahoma Health Sciences Center
Oklahoma City, Oklahoma

Lorena Parlea
RNA Structure and Design Section
RNA Biology Laboratory
Center for Cancer Research, National Cancer Institute
Frederick, Maryland

Sudeep Pukale
Department of Pharmacy
Birla Institute of Technology and Science-Pilani [BITS]
Rajasthan, India

Anu Puri
RNA Structure and Design Section
RNA Biology Laboratory
Center for Cancer Research
National Cancer Institute
National Institutes of Health
Frederick, Maryland

Rajagopal Ramesh
Department of Pathology
Graduate Program in Biomedical Sciences
University of Oklahoma Health Sciences Center
Oklahoma City, Oklahoma

Eupa Ray
Institute of Nano Science and Technology [INST]
Habitat Centre
Mohali, India

Pritha Ray
Imaging Cell Signaling and Therapeutics Lab
Advanced Centre for Treatment Research and Education in Cancer, Tata Memorial Centre
Homi Bhaba National Institute
Mumbai, India

Brandon Roark
Department of Chemistry
University of North Carolina
Charlotte, North Carolina

Can Sarisozen
Center for Pharmaceutical Biotechnology and Nanomedicine
Northeastern University
Boston, Massachusetts

Bruce A. Shapiro
RNA Structure and Design Section
RNA Biology Laboratory
Center for Cancer Research
National Cancer Institute
National Institutes of Health
Frederick, Maryland

Ankur Sharma
Institute of Nano Science and Technology [INST]
Habitat Centre
Mohali, India

Saurabh Sharma
Department of Pharmacy
Birla Institute of Technology and Science-Pilani [BITS]
Rajasthan, India

Sourabh Shukla
Department of Biomedical Engineering
Case Western Reserve University
Cleveland, Ohio

Krishna Pal Singh
Department of Molecular Biology, Biotechnology, and Bioinformatics
College of Basic Science and Humanities
Chaudhary Charan Singh Haryana Agricultural University
Hisar, India

Nicole F. Steinmetz
Department of Biomedical Engineering
Department of Radiology
Department of Materials Science and Engineering
Department of Macromolecular Science and Engineering
Case Comprehensive Cancer Center
Department of Molecular Biology and Microbiology
Division of General Medical Sciences-Oncology
Case Western Reserve University
Cleveland, Ohio

Chalet Tan
Department of Pharmaceutics and Drug Delivery
University of Mississippi
Mississippi

Bhushan Thakur
Imaging Cell Signaling and Therapeutics Lab
Advanced Centre for Treatment Research and Education in Cancer, Tata Memorial Centre
Homi Bhaba National Institute
Mumbai, India

Christopher Thompson
Department of Biochemistry and Molecular Biology
University of Nebraska Medical Center
Omaha, Nebraska

Vladimir P. Torchilin
Center for Pharmaceutical Biotechnology and Nanomedicine
Northeastern University
Boston, Massachusetts

Jiasheng Tu
Department of Pharmaceutics
China Pharmaceutical University
Nanjing, China

Kalpesh Vaghasiya
Institute of Nano Science and Technology [INST]
Habitat Centre
Mohali, India

Rahul Kumar Verma
Institute of Nano Science and Technology [INST]
Habitat Centre
Mohali, India

Mathias Viard
RNA Structure and Design Section
RNA Biology Laboratory
Leidos Biomedical Research Inc.
Center for Cancer Research
National Cancer Institute
National Institutes of Health
Frederick, Maryland

Viktor Viglasky
Institute of Chemistry
Faculty of Science
Pavol Jozef Šafárik University
Košice, Slovakia

Faye Walker
Neuroscience Research Institute
University of California
Santa Barbara, California

Amy M. Wen
Department of Biomedical Engineering
Case Western Reserve University
Cleveland, Ohio

and

Wyss Institute of Biologically Inspired Engineering
Harvard University
Boston, Massachusetts

Paul Zakrevsky
RNA Structure and Design Section
RNA Biology Laboratory
Center for Cancer Research
National Cancer Institute
National Institutes of Health
Frederick, Maryland

Lihui Zhu
Shanghai Veterinary Research Institute
Chinese Academy of Agricultural Sciences
Key Laboratory of Animal Parasitology
Ministry of Agriculture
and
Shanghai Academy of Agricultural Sciences
National Research Center of Poultry Engineering and Technology
Shanghai, China

Loredana Zocchi
Department of Pharmaceutical Sciences
University of California
Irvine, California

Section I

Nanotechnology-Based Approaches to Target Cancer

1 Nanomedicines for Cancer

Jiasheng Tu, Birendra Chaurasiya, and Yanan Li

CONTENTS

1.1 INTRODUCTION

Recently, nanotechnologies have been gaining popularity in the medical field for the treatment as well as diagnosis of various diseases. Formulations prepared by nanotechnology are generally engineered below 100 nm size (Farokhzad & Langer, 2006). By definition, nanotechnology works on two principles: (i) nanoscale size of the whole system or its vital components; (ii) man-made nature and unique characteristics of new materials that arise due to their nanoscopic size (Godin et al., 2010). Nanotechnology is a convergent system of various research areas like chemistry, biology, physics, mathematics, and engineering. This multidisciplinary effort made nanotechnology a unique delivery system to be used in clinical application (Shvedova, Kagan, & Fadeel, 2010). These nanoplatforms have been proven to carry varieties of therapeutic and diagnostic agents such as drugs, genes, and imaging agents at targeted site in a safe and effective manner. Their unique attributes such as ultra-small size, large surface area-to-mass ratio, and high reactivity makes them deliver varieties of theranostics. With these multidisciplinary efforts, these nanocarriers loaded with various therapeutic agents have started to be used in clinical practice as nanomedicines (Liu, Miyoshi, & Nakamura, 2007). Nanocarriers play a major role in the improvement of solubility, bioavailability, and in decreasing the potential toxicity of chemotherapeutic drugs over conventional formulation strategies. This pivotal partnership between nanocarriers and theranostic agents represents a changing paradigm over the last two decades in the drug delivery system to provide nanomedicines for clinical use for many disease conditions like diabetes, asthma, allergies, infections, and so on, most notably in cancer treatment (Brannon-Peppas & Blanchette, 2004; Forrest & Kwon, 2008; Kawasaki & Player, 2005). For therapeutic applications, these nanomedicines precisely deliver the therapeutic agents to the targeted site in a controlled manner without significant systemic side effects. For diagnosis applications, these nanocarriers help to detect abnormalities on a molecular scale such as fragments of viruses, precancerous cells, and diseases markers that are not able to be identified with the traditional diagnosis system.

Nanotechnology, although recently applied to prepare medication for clinical use, was recognized as a drug delivery system long ago. The first nanotechnology based preparation was made of lipid vesicles in 1960s and was later described as liposomes in 1965 (Bangham, Standish, & Watkins, 1965). Similarly, several other nanotechnologies were established over the passage of time, for example, the first controlled-release polymer system of macromolecules was studied in 1976; the first long circulating stealth polymeric nanoparticle was described in 1994; the first quantum dot bioconjugate was described in 1998; and the first nanowire-based nanosensor described in 2001 (Farokhzad & Langer, 2006). History shows that the nanocarrier systems were explored more than 50 years ago but have got popularity in drug delivery date back about 40 years (Marty, Oppenheim, & Speiser, 1977). The first nanomedicine of anthracyclines was prepared in the form of nanosized phospholipid vesicles (liposomes) to reduce cardiotoxicity at the end of 1970s (Forssen & Tökès, 1981). The landmark of nanotechnology-based nanomedicines was harnessed in clinical practice after approval of Doxil®, the first doxorubicin-loaded liposome approved by the Food and Drug Administration (FDA) in 1995 (Barenholz, 2012). Most common nanoplatforms studied today are polymer-based nanoparticles, nanoshells, micelles, liposomes, dendrimers, quantum dots, magnetic nanoparticles, silicone oxide-based nanoparticles, and engineered viral-based nanoparticles (Ferrari, 2005). In this chapter we have focused on nanotechnology-based nanomedicines in clinical uses for various ailments and diagnoses and have included some nanomedicines that are under clinical trial.

1.2 NANOMEDICINES IN CLINICAL USE AND UNDER CLINICAL TRIALS

According to an earlier survey conducted by the European Science and Technology Observatory, in last two decades huge progress has been made in the development of nanotechnology-based therapeutics and diagnostics (Wagner et al., 2006). Based on survey data conducted in recent years, the FDA has approved 20 nanotechnology-based nanomedicines (Table 1.1) for cancer treatment and 67 nanodevices (not listed). A total of 122 therapeutics were under development and more than 795 nano-products were in ongoing clinical trials (Hare et al., 2017). Among these products, liposomal and polymer-based drugs are the dominant groups, which account for more than 85% of the total number. Recently ongoing therapeutic clinical trials are listed in Table 1.2. All the products listed in tables were obtained from www.fda.gov and https://clinicaltrials.gov database.

1.2.1 Liposome-Based Nanomedicines

Liposome is derived from the Greek words: *lipo* ("fat") and *soma* ("body"). It was first described by British hematologist Alec D. Bangham in 1961. It is a spherical vesicle composed of phospholipids especially phosphatidylcholine, but also includes other lipids like egg phosphatidylethanolamine (Sahoo, Parveen, & Panda, 2007). Liposomes are categorized as unilamellar, multilamellar, and cochleate vesicles. Unilamellar vesicles contain one lipid bilayer and generally have a diameter ranging from 50 to 250 nm. They contain a large aqueous core that is preferentially used to encapsulate water-soluble drugs. Multilamellar vesicles comprise several concentric lipid bilayers in an onion-skin arrangement and usually have diameters of 1–5 μm. Their high lipid content allows multilamellar vesicles to passively encapsulate hydrophobic drugs. Compared with the multilamellar vesicles above, cochleate vesicles comprise several lipid bilayers that are not concentric (Figure 1.1). Based on liposome preparation methods, the size distribution of liposomes varies from 25–1000 nm (Weissig, Pettinger, & Murdock, 2014). Liposomes have been widely used as pharmaceutical carriers in the past decades because of their unique abilities to (a) encapsulate both hydrophilic and hydrophobic therapeutic agents with high efficiency, (b) protect the encapsulated drugs from undesired effects of external conditions, (c) be functionalized with specific ligands that can target specific cells, tissues, and organs of interest, (d) be coated with inert and biocompatible polymers such as polyethylene glycol (PEG), in turn prolonging the liposome circulation half-life *in vivo*, and (e) form desired formulations with needed composition, size, surface charge, and other properties (Moghimi & Szebeni, 2003; Torchilin, 2005).

Liposome-based nanomedicines approved by the FDA for the treatment of cancer are listed in Table 1.1. Doxil® was the first liposome-based anticancer nanomedicine approved by the FDA in 1995 for the treatment of AIDS-associated Kaposi's sarcoma, ovarian cancer, and multiple myeloma, as well as for metastatic breast cancer (Zhang et al., 2008). Doxil® was prepared by encapsulating doxorubicin into stealth liposome carriers comprised of hydrogenated soy phosphatidylcholine, cholesterol, and PEGylated phosphoethanolamine (Figure 1.2). Doxil® has shown prolonged

TABLE 1.1
Nanotechnology-Based Nanomedicines Approved by the FDA for Clinical Use

Brand Name	Payloads	Route	Indications	Company	Approval Year
			Liposome Based Nanomedicines		
Doxil®/ Caelyx®	PEG-Liposome doxorubicin	i.m.	HIV-related Kaposi's sarcoma; ovarian cancer; multiple myelomas; breast cancer	Janseen-Cilag Pty, Ltd.	1995 2005
Myocet®	Liposomal doxorubicin	i.v.	HIV-related Kaposi's sarcoma; ovarian cancer; multiple myelomas; breast cancer	Zeneus	2008
DepoCyt®	Liposomal cytarabine	i.t.	Malignant lymphomatous meningitis	SkyePharma	1996
DaunoXome®	Liposomal daunorubicin	i.v.	HIV-related Kaposi's sarcoma	Gilead Sciences	1996
Marqibo®	Liposomal vincristine	i.v.	Acute Lymphoblastic Leukemia	Onco TCS	2012
Onco-TCS®	Liposomal vincristine	i.v.	Non-Hodgkin's lymphoma	Enzon & INEX	2004
Onivyde®	Liposomal Irinotecan	i.v.	Metastatic pancreatic cancer	Merrimack Pharmaceutical, Inc.	2015
Lipusu®	Liposomal paclitaxel	i.v.	NSCLC, breast cancer	Sike Pharmaceutical	2006
PICN®	Liposomal paclitaxel	i.v.	Metastatic breast cancer	Sun Pharmaceutical	2014
LEP-ETU®	Liposomal paclitaxel	i.v.	Ovarian cancer	Insys Therapeutics, Inc.	2015
Mepact®	Liposomal Mifamurtide	i.v.	Mononuclear phagocyte targeting for Osteosarcoma	IDM Pharma	2009

(*Continued*)

TABLE 1.1 (CONTINUED)
Nanotechnology-Based Nanomedicines Approved by the FDA for Clinical Use

Brand Name	Payloads	Route	Indications	Company	Approval Year
		Polymer-Based Nanomedicines			
Genexol-PM®	Methoxy-PEG-poly(D,L-lactide) Paclitaxel	i.v.	Metastatic breast cancer	Samyang	2005
Neulasta®	PEG–granulocyte colony-stimulating factor	s.c.	Neutropenia associated with cancer chemotherapy	Amgen	2002
Eligard®	Leuprolide acetate and polymer (PLGH [poly (DL-Lactide-co-glycolide])	i.v.	Prostate cancer	Tolmar	2002
		Protein-Based Nanomedicines			
Abraxane®	Albumin-bound paclitaxel Nanoparticles	i.v.	Breast cancer, NSCLC, pancreatic cancer	Celgene	2005 2012 2013
Ontak®	Engineered Protein combining IL-2 and diphtheria toxin	i.v.	Cutaneous T-Cell Lymphoma	Seragen, Inc	2008
Kadcyla®	ado-trastuzumab emtansine	i.v.	Metastatic breast cancer	Genentech	2013
		Nanotechnology-Based Miscellaneous Nanomedicines			
Nanotherm®	Iron oxide	Intra-dural	Glioblastoma	MagForce	2010
Ryanodex®	Dantrolene sodium	i.v.	Malignant hypothermia	Eagle Pharmaceutical	2014

TABLE 1.2
Nanotechnology-Based Nanomedicines in Clinical Trials

Brand Name	Payloads	Route	Indications	Company	Status Phase
		Liposome-Based Nanomedicines			
L-Annamycin	Liposomal annamycin	i.v.	Acute lymphocytic leukemia, acute myeloid leukemia	Callisto	I
Thermodox®	Liposomal doxorubicin	i.v.	Liver, breast cancers	Celsion	III
Lipolatin®	Liposomal cisplatin	i.v.	NSCLC	Regulon	III
9NC-LP®	Liposomal 9-nitrocamptothecin	aerosol	Hepatocellular carcinoma	ChemWerth	II/III
SPI-077®	Liposomal cisplatin	i.v.	Solid tumors	ALZA Pharmaceutical	I/II/III
Lipoxal®	Liposomal oxaliplatin	i.v.	Advanced cancers	BioCentury	II
EndoTAG-1®	Liposomal paclitaxel	i.v.	Breast, liver, pancreatic cancers	BioCentury	II
LE-DT®	Liposomal docetaxel	i.v.	Pancreatic, prostatic cancers	NeoPharma	I/II
TKM-080301®	Liposomal PLK1 siRNA	i.v.	Neuroendocrine tumors	Arbutus Biopharma	I/II
Atu027®	Liposomal PLK3 siRNA	i.v.	Pancreatic cancer	BioCentury	I/II
2B3-101®	Liposomal doxorubicin	i.v.	Solid tumors	Netherlands Cancer Institute	I/II
SLIT® cisplatin	Liposomal cisplatin	Aerosol	Osteogenic sarcoma metastatic to the lung	Transave	II
Sarcodoxome®	Liposomal doxorubicin	i.v.	Soft tissue sarcoma	GP-Pharm	I/II
OSI-211®	Liposomal lurtotecan	i.v.	Ovarian cancer	OSI Pharmaceuticals	II
OncoTCS®	Liposomal vincristine	i.v.	Non-Hodgkin's lymphoma	Inex, Enzon	II/III
NL CPT-11®	Liposomal Irinotecan	i.v.	Reccurent high grade glioma	Merrimack Pharmaceutical	I
MTL-CEBPA®	Liposomal CEBPA saran	i.v.	Liver cancer	MiNA Theraputics	I
TL1®	Liposomal topotecan	i.v.	Various solid tumors	Sagent Pharmaceutical	I
IHL-305®	Liposomal Irinotecan	i.v.	Advanced solid tumors	Taiwan Liposome Co.	I

(*Continued*)

TABLE 1.2 (CONTINUED)
Nanotechnology-Based Nanomedicines in Clinical Trials

Brand Name	Payloads	Route	Indications	Company	Status Phase
		Liposome-Based Nanomedicines			
ATI-1123	Liposomal vinorelbine	i.v.	Solid tumors	Azaya Theraputics	I
Alocrest®	Liposomal cisplatin	i.v.	Breast and lung cancers	Spectrum Pharmaceutical	I
LiPlaCis®	Liposomal cisplatin	i.v.	Advanced solid tumors	LiPlasome Pharma	I
MCC-465®	Liposomal targeted doxorubicin	i.v.	Metastatic stomach cancer	Mitsubishi Pharma Corporation	I
SGT-53®	Liposomal targeted p^{53} gene	i.v.	Various solid tumors	SynerGene Therapeutics	I
		Polymer-Based Nanomedicines			
ProLindac	HPMA copolymer–DACH Palatinate	i.v.	Ovarian cancers	Access Pharmaceuticals	II
Hepacid	PEG–arginine deaminase	i.v.	Hepatocellular carcinoma	Phoenix	I/II
Prothecan	PEG–camptothecin	i.v.	Various cancers	Enzon	I/II
SP1049C	Pluronic block-copolymer Doxorubicin	i.v.	Esophageal carcinoma	Supratek Pharma	II
Oncaspar®	PEG-Asparginase	i.v.	Leukemia		III
Xyotax® (Opaxio)	Poliglumex Paclitaxel	i.v.	NSCLC, glioma, breast, ovarian cancer	Cell Therapeutics, Inc.	III
CT-2106	Polyglutamate camptothecin	i.v.	Colorectal and ovarian cancers	Cell Therapeutics, Inc.	I/II
Xyotax®	Polyglutamate paclitaxel	i.v.	Non-small-cell lung cancer, ovarian cancer	Cell Therapeutics, Inc.	III

(Continued)

TABLE 1.2 (CONTINUED)
Nanotechnology-Based Nanomedicines in Clinical Trials

Brand Name	Payloads	Route	Indications	Company	Status Phase
Transdrug®	Poly(iso-hexyl-cyanoacrylate) Doxorubicin	i.a.	Hepatocellular carcinoma	BioAlliance Pharma	I/II
Taxoprexin®	Decosahexaenoic acid-paclitaxel	i.v.	Various solid tumors	Luitpold Pharmaceutical	II/III
PK1	Hydroxypropyl methacrylamide doxorubicin	i.v.	Breast, lung, colon cancer	Polymer Laboratories	II
PegAsys/ Pegintron®	PEG- IFN-α2a/b	i.v.	Melanoma/ Leukemia	Genetech	II
ProLindac®	PEG-Oxaliplatin		Various solid tumors	Access Pharmaceutical	II
DEP®	Dendrimer-docetaxel	i.v.	Advanced cancer	Starpharma	I
Paclical®	Micellar paclitaxel	i.v.	Ovarian cancer	Oasmias	III
NK105®	Micellar paclitaxel	i.v.	Gastric cancer	GSK	III
NK911®	Micellar doxorubicin	i.v.	Various solid tumors	GSK	II
NC-4016	Micellar Oxaliplatin	i.v.	Various solid tumors	Toudai Tlo Ltd	I
Nanotechnology-Based Miscellaneous Nanomedicines					
AI-850	Paclitaxel nanoparticles	i.v.	Solid tumors	Acusphere	I
BA-003	Doxorubicin nanoparticles	Oral, i.v.	Hepatocellular carcinoma	BioAlliance Pharma	III

(*Continued*)

TABLE 1.2 (CONTINUED)
Nanotechnology-Based Nanomedicines in Clinical Trials

Brand Name	Payloads	Route	Indications	Company	Status Phase
BIND-014	Docetaxel nanoparticles	i.v.	NSCLC	BIND Therapeutics	II
CriPec®	Docetaxel nanoparticles	i.v.	Metastatic cancer, Solid Tumors	Cristal Therapeutics	I
CRLX101	Camptothecin nanoparticles	i.v.	NSCLC	Cerulean Pharma Inc.	II
Rexin-G	dnG1Plasmid DNA	i.v.	Various solid tumors	Epeius Biotechnologies	I/II
ABI-008	Docetaxel nanoparticles	i.v.	Breast, prostate cancers	Celgene Corporation	I/II
ABI-009	Rapamycin nanoparticles	i.v.	Non-hematologic malignancies	Celgene Corporation	I/II
CYT-6091	rhTNF colloidal gold nanoparticles	i.v	Unspecified Adult Solid Tumor	CytImmune Sciences	I
Docetaxel-PNP	Docetaxel nanoparticles	i.v.	Various solid tumors	Samyang Biopharmaceutical	I
CALAA-01	Anti-RRRM2 siRNA	i.v.	Various solid tumors	Calando Pharmaceuticals	I
CIM331	Humanized monoclonal antibody nemolizumab	s.c.	Skin cancer	Chugai Pharmaceutical Co.	II
SP1049C	Doxorubicin	i.v.	Adenocarcinoma	Surpatek Pharma	I/II

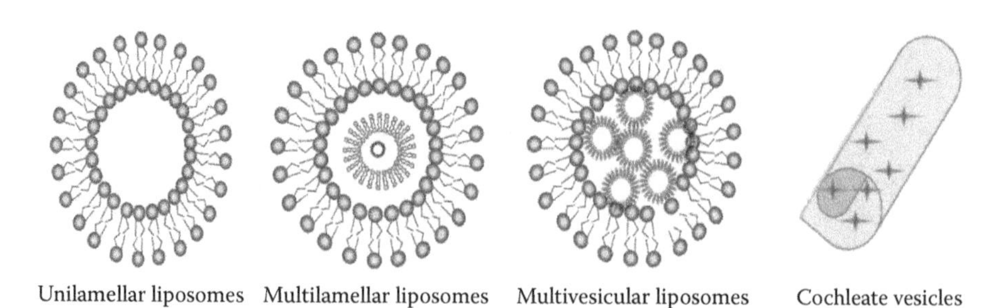

FIGURE 1.1 Basic structures of liposomes. (From Zununi Vahed, S. et al., *Materials Science & Engineering. C, Materials for Biological Applications, 71*, 1327–1341, doi: 10.1016/j.msec.2016.11.073, 2017.)

doxorubicin circulation half-life and enhanced drug deposition (about 16-fold) in the tumor tissue in comparison to free doxorubicin (Bobo et al., 2016). It was reported that 1 ml of commercial Doxil® dispersion contain 2.3×10^{14} liposomes and each liposome contains ~10,000 molecules of doxorubicin, above 95% of which was in the crystalline phase (Barenholz, 2012). Many investigations in animal models have reported that Doxil® extravasates and accumulates as intact liposomes in tumors through "leaky" vasculature and inside the tumor it moves by convection and distributes through the tumor, also called the EPR effect (Maeda, Bharate, & Daruwalla, 2009). A comprehensive review on Doxil® was summarized by Barenholz (2012).

Another liposome-based anticancer nanomedicine is DaunoXome® (daunorubicin) approved by FDA for the treatment of AIDS associated Kaposi's sarcoma. Similarly, Myocet®, is a PEGylated liposomal doxorubicin approved in Europe and Canada for metastatic breast cancers. In the last three years, FDA has approved three liposomal anticancer nanomedicines for clinical use, i.e. Lipusu® (Paclitaxel liposome, 2014); Onivyde® (Irinotecan liposome, 2015); and LEP-ETU® (Paclitaxel liposomal, 2015).

Lipusu® was approved by the State Food and Drug Administration of China in 2003 for clinical use in China for the treatment of ovarian cancer and non-small cell lung cancer by i.v. administration; later in 2014 it was approved by the FDA for global use. It is a cholesterol-based liposome prepared by film dispersion method (Ye et al., 2013). In animal studies in a rat model, it was found that Lipusu® displayed higher distribution in the liver, spleen, and lungs rather than in the kidney and heart, which was highly observed in cremophor-based paclitaxel preparation i.e. Taxol® (Zhang, Huang, & Gao, 2009). It was reported that the maximum tolerated dose of paclitaxel in liposomal formulation is 200 mg/kg, while the conventional paclitaxel dosage is 30 mg/kg (Zhang, Huang, & Gao, 2009). Similarly, Onivyde® is a topoisomerase-I inhibitor and was approved by FDA in 2015 for the treatment of advanced (metastatic) pancreatic cancer (Drugs.com, 2015). Patients treated with Onivyde® in addition to flurouracil and folinic acid were shown to have a 6.1-month median survival *vs* 4.2 months with the addition of the liposomal irinotecan formulation. Like all of the approved liposomal systems, the Onivyde® delivery mechanism is based on passive targeting. Likewise, LEP-ETU® was approved by the FDA in 2015 for the treatment of ovarian cancer.

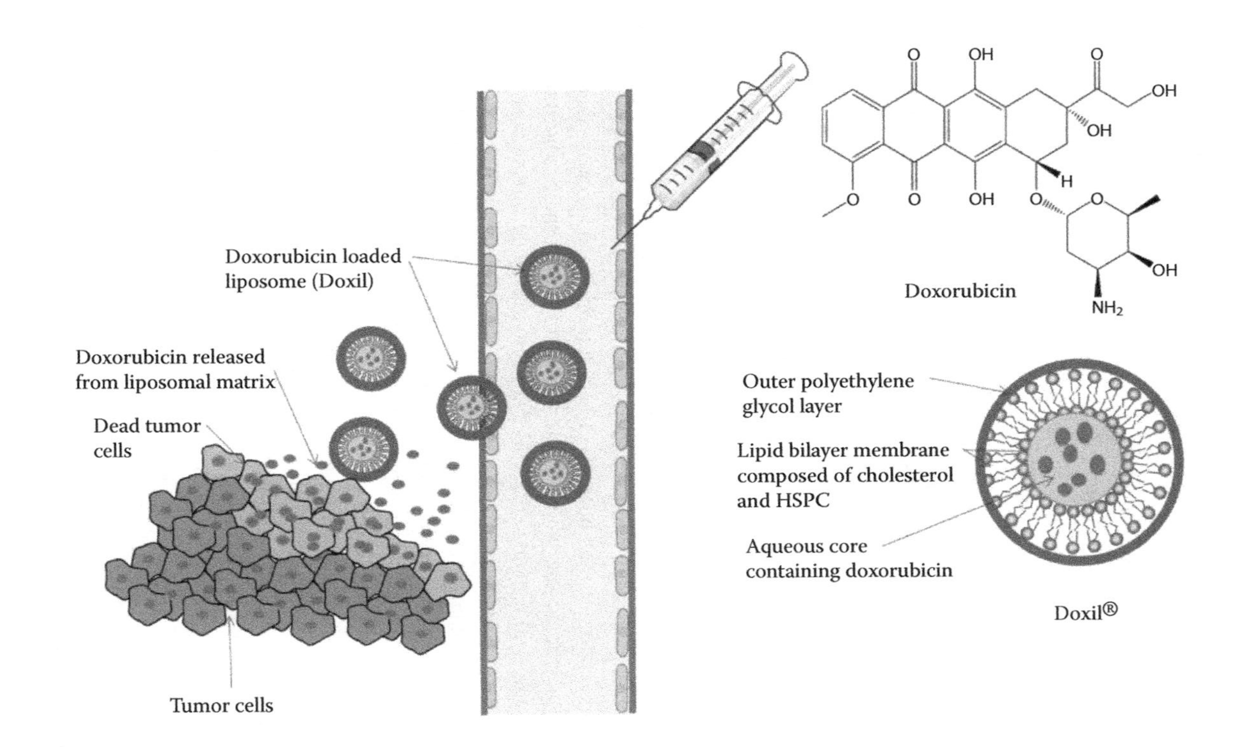

FIGURE 1.2 Schematic diagram of Doxil® and its mechanism of transmission. (From Barenholz, Y., *Journal of Controlled Release, 160*(2), 117–134, doi: http://dx.doi.org/10.1016/j .jconrel.2012.03.020, 2012.)

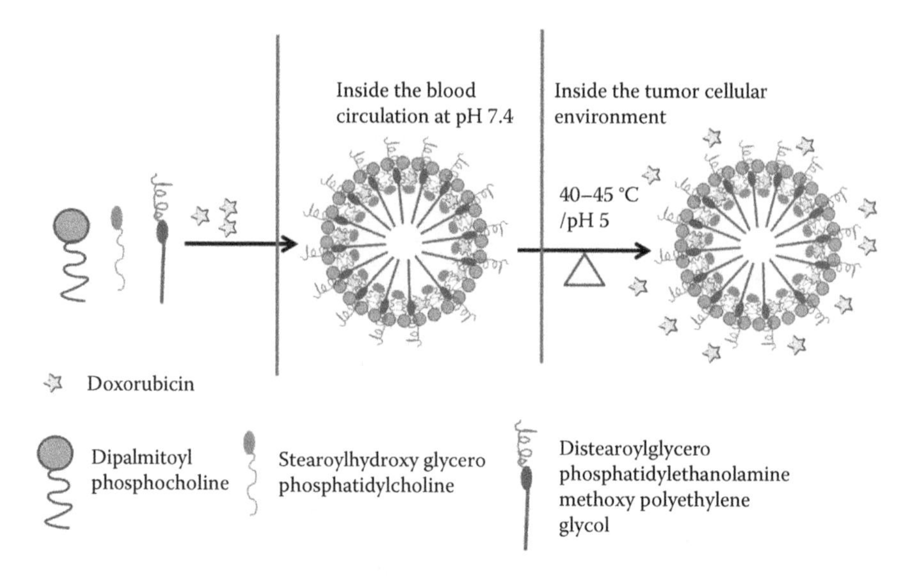

FIGURE 1.3 Schematic diagram of thermosensitive Thermodox liposome. (From Liu, D. et al., *Theranostics, 6*(9), 1306–1323, doi: 10.7150/thno.14858, 2016.)

There are many more liposomal anticancer preparations under clinical trials. Thermodox® a multifunctional doxorubicin preparation, is in Phase III. It is formulated in the same way as Doxil® but is formulated with thermosensitive lipids that degrade the bilayer when exposed to high heat (40–45°C). It is designed to release loaded drugs in the tumor at a specific site where it interacts with the thermal radiofrequency of tumor cells (Figure 1.3) (Saxena et al., 2015). There are many other liposome-based nanoformulations for cancer treatment under clinical trials listed in Table 1.2.

Liposomes are popular versatile carriers used to deliver verities of biologically active molecules in a controlled manner to a target site. These are relatively non-toxic systems have enough potential to entrap both hydrophilic and hydrophobic drugs. Their hydrophilic outer layers provide a good platform to decorate with multiple cell targeting ligands. Thus, they represent very effective carriers for delivering multiple active molecules for enhanced therapeutic effect. The final amount of the encapsulated drug is affected by a selection of an appropriate preparation method providing a preparation of liposomes of various size, lamellarity, and physico-chemical properties. The modification of liposomes permits a passive or active targeting of the tumor site. This effect enables an efficient drug payload into the malignant cells of the tumor, while the non-malignant cells become minimally impacted (Koudelka & Turanek, 2012).

1.2.2 Polymer-Based Nanomedicines

The polymeric drug delivery system is another well-studied nanocarrier system in drug delivery platforms currently in clinical practice. The polymeric system is further classified into polymeric-drug conjugates, polymeric micelles, and polymeric nanoparticles. The schematic diagram of these polymeric systems is shown in

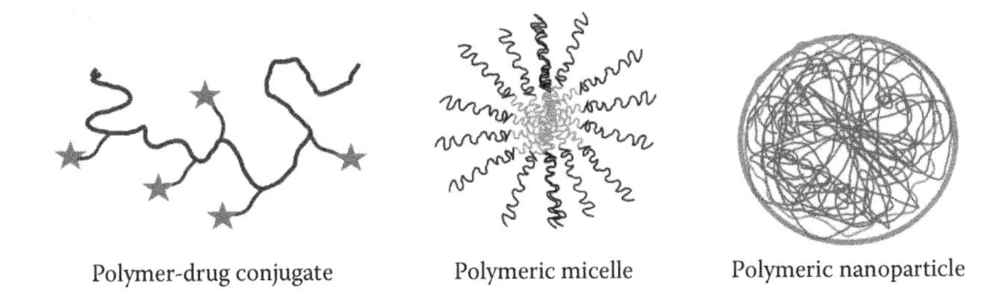

FIGURE 1.4 Schematic diagram of various polymeric drug delivery systems.

Figure 1.4. Small-molecule therapeutic agents, especially anticancer chemotherapeutic agents, usually have two limitations: a short circulation half-life, which leads to frequent administrations, and non-site-specific targeting, resulting in undesired systemic side effects. The conjugation of small-molecule drugs to polymeric nanocarriers can improve the undesirable adverse effects. Polymer-drug conjugates not only prolong the *in vivo* circulation time from several minutes to several hours but also reduce cellular uptake to the endocytic route (Zhang et al., 2008). Polymeric conjugates enhance the plasma stability and solubility of the payload while reducing its immunogenicity. Polymer nanomedicines usually fall into one of two categories: (a) polymer-drug conjugates for increased drug half-life and bioavailability, and (b) degradable polymer architectures for controlled-release applications.

To date, the FDA has approved three polymer-based nanomedicines for cancer treatment and many more are under clinical trials. In polymeric nanomedicines, drugs are combined with hydrophilic polymers to increase blood circulation for biocompatibility (Alconcel, Baas, & Maynard, 2011). The most studied polymer is poly (ethylene glycol) (PEG). Neulasta® is one of the most popular polymer-based nanomedicines approved by the FDA for the treatment of neutropenia. Neulasta® is a PEGylated granulocyte colony-stimulating factor, due to its PEGylation, the biological half-life in plasma was increased from 3–4 hours to 15–80 hours in comparison to its basic filgrastim (Benbrook, 2014). There are other two polymer-based anticancer nanomedicines approved by the FDA, Eligard®, and Genexol-PM®. Eligard® is a PEGylated leuprolide acetate suspension, a GnRH agonist, for subcutaneous injection for the management of advanced prostate cancer. It is designed to deliver leuprolide acetate at a controlled rate over a one-, three-, four-, or six-month therapeutic period.

Polymeric carriers not only extend the circulation time of established drugs but also establish stable release patterns for hydrophobic drugs for controlled release to maintain prolonged therapeutic index. This is achieved by using slowly degradable functionality that subsequently leads to a kinetically driven release of the drug. PLGA is a well-established degradable polymer that slowly decomposes into the constituent monomeric units over controlled time courses. Another chemotherapeutic, Camptothecin (a DNA topoisomerase I inhibitor), has been encapsulated in cyclodextrin-PEG copolymers to form nanoparticles ~20–50 nm in diameter (Svenson et al. 2011). These nanoparticles (CRLX101) are administered by intravenous injection and utilize the so-called enhanced permeability and retention effect (EPR) that relies on leaky vasculature in tumors to increase accumulation of the drug molecules at the target of interest. CRLX101 has achieved Phase II trials in patients with rectal, ovarian, tubal, and peritoneal cancer and is an example of classical nanomedicine therapeutics utilizing biocompatible polymeric nanoparticles. There are many other polymeric nanoformulations under clinical trials listed in Table 1.2.

1.2.3 Protein-Based Nanomedicines

Proteins used as a carrier are biocompatible and biodegradable. Protein nanoparticles cover a number of different nanomedicine classes, from drugs conjugated to endogenous protein carriers to engineered proteins where the active therapeutic is the protein itself, and to combined complex platforms that rely on protein motifs for targeted therapeutic delivery. Early protein nanoparticles sought to use the natural properties of protein circulating in serum, allowing dissolution and transport of drug compounds in blood during circulation. This approach consisted of natural protein combined with known drugs in order to reduce toxicity (Bobo et al., 2016).

Albumin has been the most common protein used as a carrier in a nano-drug delivery system for a decade. It can self-assemble and has a huge capacity to load both hydrophilic and hydrophobic drugs as well as genetic materials. Abraxane® is a prototype example of a protein-based, anticancer nanomedicine approved by the FDA in 2005 for the treatment of metastatic breast cancer. Preparation of Abraxane® was based on albumin-bound nanoparticles (NAB)-technology, in which paclitaxel was loaded in human serum albumin nanoparticles with a size of 130 nm. A basic preparation technique of albumin nanoparticles is schematically described in Figure 1.5. As paclitaxel is insoluble in aqueous media, it needs a surfactant and an organic solvent as co-solvents for solubility. This Abraxane® was designed to replace the toxic solvent Cremophor, used to increase the solubility of the paclitaxel (Trickler, Nagvekar, & Dash, 2008). Cremophor is a polyethoxylated castor oil that contains ricinolic acid,

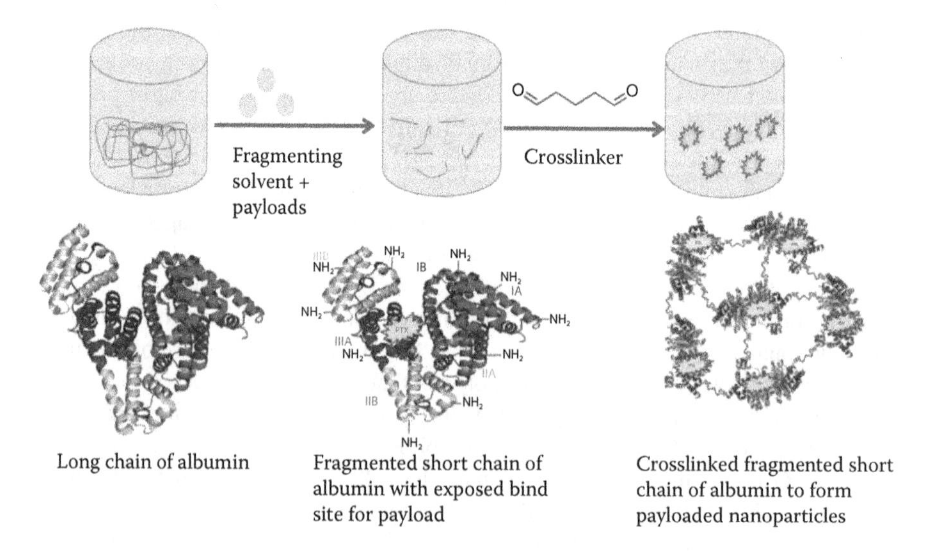

FIGURE 1.5 Schematic process for preparation of drug loaded albumin nanoparticles.

which has marked cytotoxicity effect. Abraxane® cannot only reduce the toxic effect of Cremophor in Taxol® but also enhances the pharmacokinetic efficiency of paclitaxel. After successful therapeutic results of Abraxane®, many other formulations were designed based on NAB-technology, for examples, NAB-docetaxel, NAB-rapamycin, and NAM-heat shock protein inhibitor. There are other two protein-based anticancer drugs approved by the FDA, Ontak®, and Kadcyla® and many are under clinical trials.

Ontak® is prepared by fusion of Denileukin Diftitox with an engineered protein combination of interleukin-2 and diphtheria toxin. It was designed to target the interleukin-2 receptor to introduce diphtheria toxin into the lymphoma's malignant cells to kill the cell. This interleukin-2 targeted therapy appears to be an effective treatment for peripheral T-cell lymphomas. It was reported that Ontak® in combination with other first line chemo drugs has increased survival rates up to 63.3% in comparison with therapy using first-line chemo drugs only, which is only 32–35%. Ontak® is not myelosuppressive, nor is it associated with significant organ toxicity. Ontak® represented one of the first actively targeted proteinaceous nanoparticles (Foss, 2006). Kadcyla® is another proteinaceous nanoparticle approved by the FDA in 2013 with payload ado-trastuzumab emtansine for metastatic breast cancer treatment (Figure 1.6). It is administered prior to trastuzumab and taxane. Kadcyla® is designed to selectively target epidermal growth factor receptor-2-expressing cells for the treatment of metastatic breast cancer (Weissig et al., 2014).

Rexin-G®, another good example of proteinaceous nanomedicines for targeting genes, is under clinical trials in Phase I/II. Its active targeting relies on a collagen-binding peptide from human von Willebrand factor (vWF). This protein normally induces platelet aggregation in the instance of vascular injury. In Rexin-G®, vWF serves to enhance particle deposition by guiding the whole particle into a tumor where exposed collagen is often found (Chawla et al., 2010). In contrast to previous protein

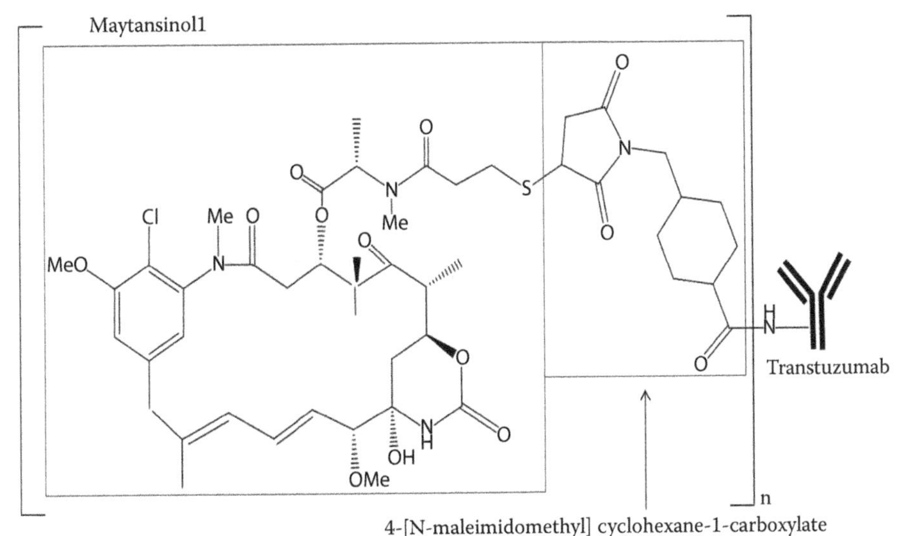

FIGURE 1.6 Molecular structure and sequence of Kadcyla®.

nanoparticles, Rexin-G® is a mixed system that is based on the murine leukemia virus. The von Willenbrand factor-derived binding motif is expressed in the modified viral envelope for particle delivery. The proteinaceous envelope is responsible for nanoparticle accumulation and ultimate transfection of a cytotoxic *cyclin G1* gene. As opposed to receptor-specific targeting, Rexin-G® is targeted against the general disease state characteristics found in tumor environments. Avoiding reliance on a specific receptor may avoid the confounding effects of mutation and adaptation. Rexin-G®'s proponents have stated this general targeting of invasive cancer characteristics has improved delivery of the genetic payload to where it is needed while reducing target selection of normal tissues and tumor adaptation (Gordon & Hall, 2010). There are many more proteinaceous nanomedicines under clinical trials listed in Table 1.2.

1.2.4 Micelles-Based Nanomedicines

Micelles are an emerging nanoplatform to deliver various molecules in a controlled manner to the biological target site. It consists of both hydrophilic as well as hydrophobic motifs to load hydrophobic drugs in the core of the micelles. It also provides excellent outer surface decorating property to bind with target-specific ligands for site-specific delivery of therapeutics. It is self-assembled system; therefore, during preparation it is required to balance the hydrophobic/hydrophilic ratio to obtain the desired size and morphology. The internal core is hydrophobic to load hydrophobic molecule and the outer surface is hydrophilic to enhance dissolution in the plasma and cellular matrix. The flexible and tunable properties of polymeric micelles have contributed to extensive investigations in this field. However, the relatively low physical stability and thermodynamically driven disassembly upon dilution in plasma are the main drawbacks of polymeric micelles. A promising approach to overcome these drawbacks and optimize the properties of single micelles as above, is the combination of two or more distinct amphiphilic polymers, which are called mixed micelles (Figure 1.7). Compared with the single micelles, the mixed micelle is able to achieve

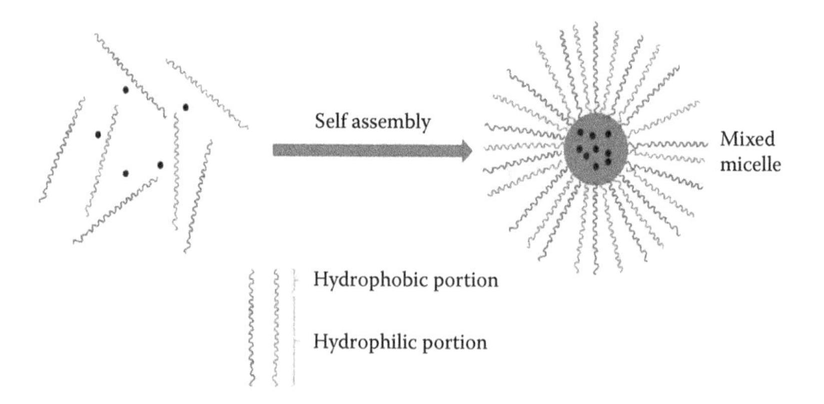

FIGURE 1.7 The schematic diagram of mixed micelles. (From Cagel, M. et al., *European Journal of Pharmaceutics and Biopharmaceutics, 113*, 211–228, doi: 10.1016/j.ejpb.2016.12.019, 2017.)

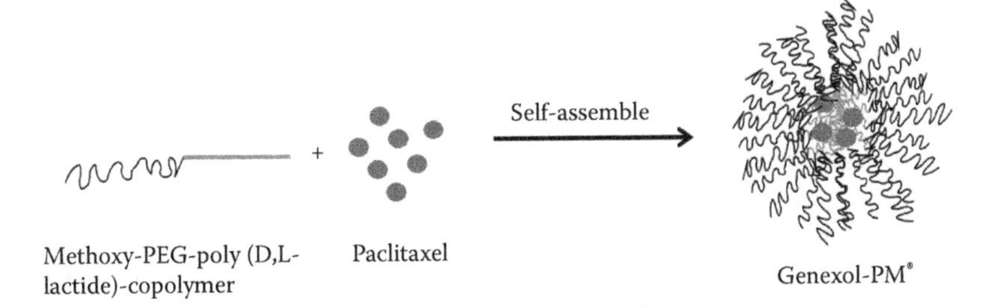

FIGURE 1.8 Schematic representation of the self-assemble phenomena of Genexol-PM®.

the following advantages such as lower CMC, improved stability, enhanced drug loading, and more accurate control for physicochemical property.

To date one polymeric micelles-based nanoformulation is approved by the FDA for cancer therapy, Genexol-PM®. It is a Methoxy-PEG-poly (D,L-lactide)-paclitaxel approved for the management of metastatic breast cancer. The preparation method is illustrated below in Figure 1.8. Genexol-PM® was found to have a three-fold high maximum tolerated dose in nude mice, and two- to three-folds higher levels of biodistribution in various tissues including the liver, spleen, kidney, and lung in comparison to Taxol®.

Many polymeric nanomedicines that are under clinical trials are listed in Table 1.2. BIND-014 is the first micellar nanomedicine for prostate cancer; it is in clinical trial Phase II. It is a docetaxel-loaded, core-shell polymeric micelle that contains a degradable and hydrophobic polymeric core and a hydrophilic PEG shell. It is designed to target prostate-specific antigen, which is a prominent protein marker on the surface of many prostate cancer cells (Kaittanis et al., 2014). BIND micelles are considered as next generation therapeutics for cancer treatments that contain kinase inhibitors as future drug therapeutics. Another similar micellar nanomedicine is CriPec®, in which docetaxel was loaded in lactate-based hydrophobic blocks and contains PEG-based hydrophilic blocks. In preclinical investigation, complete remission of breast cancer in rodent models was found, which shows favorable biodistribution for effective therapy (Rijcken et al., 2005). CALAA-01 is another micellar anticancer nanomedicine for various tumors that is under clinical investigation. Mixed micelles are another hybrid group of micelles that are more stable than normal micelles. SP1049C, an example of these mixed micelles, entered into clinical trial and has successfully passed Phase I/II. This novel Pgp-targeting micellar formulation is composed of the loaded doxorubicin and two non-ionic copolymers, Pluronic L-61, and Pluronic F-127. The targeting ability is achieved due to Pluronic L-61 which disrupts Pgp function in the cell membrane of doxorubicin-resistant cancer cells and selectively depletes cellular ATP (Valle et al., 2011). In addition, after releasing the loaded cargo, micelles can dissociate into block copolymers and be eliminated by filtration through kidneys, which would avoid the long-term side effect.

1.2.5 Nanotechnology-Based Miscellaneous Nanomedicines

Metallic nanoparticles are well-studied due to their effective therapeutic as well as imaging properties. There are various iron oxide-based nanomedicines under clinical trials and some are approved by the FDA only for iron deficiency treatment. To date there is only one metal based-nanomedicine approved by the FDA for cancer treatment, namely Nanotherm®. This consists of aminosilane-coated superparamagnetic iron oxide nanoparticles designed for the treatment of glioblastoma using local tumoral hyperthermia (Thiesen & Jordan, 2008). For the treatment, this preparation is directly injected into the tumor where it gets heated in the tumoral hyperthermic environment (40–45°C), then the tumor cells are programmed to die. In its clinical trials, the overall survival period of the patients treated with Nanotherm® was increased by 12 months (Maier-Hauff et al., 2011). Recently, gold has been utilized as a nanomedicine in clinical trials due to its unique combination of optical properties, thermal properties, and easy to modify size, shape, and surface structure (Yeager et al., 2014). Although having versatility in preparation and wide coverage in research fields, to date there is no gold nanomedicine approved by the FDA for cancer therapy and only a few are under clinical trials. CYT-6091 is a gold-based anticancer nanomedicine that has successfully passed its clinical trial Phase I (Libutti et al., 2010). It is a recombinant human tumor necrosis factor (rhTNF) bound to colloidal gold linked up with PEG linker to enhance the biocompatible antifouling layer. During preclinical study in animal models, it was found that the gold increased the immobilization of rhTNF three-fold more than the free rhTNF without any toxicity and the PEG linker aided the cellular uptake *via* EPR effect.

Crystalline nanoparticles are another well-studied platform; they are composed of 100% drug molecules. Due to their nanoscale size, the solution stability is maintained and leads to enhance the diffusion-based mass transfer through the biological membrane. To date there is only one FDA-approved crystalline nanomedicine for cancer treatment, Ryanodex®. It is a crystalline nanosuspension of Dantrolene sodium indicated for the treatment of malignant hyperthermia. During clinical trials, the bioavailability of Ryanodex® was compared with Dantrolene sodium, and it was found that C_{max} for Ryanodex® was 44% more than Dantrolene followed by intravenous injection administration (Kobayashi et al., 2003).

1.3 TARGETING MECHANISM OF NANOMEDICINES

1.3.1 Passive Targeting

The enhanced permeability and retention effect (EPR) effect was first proposed by Maeda in 1986. The incompletely hyperpermeable vasculature and impaired lymphatic drainage enhanced permeability of large particles in the tumor and avoided the clearance of particles that caused retention in solid tumor (Matsumura & Maeda, 1986). Generally, nanomedicines (sub-200 nm) are able to extravasate and accumulate within the interstitial space of tumors by circulating through the much larger endothelial pores (10 to 1000 nm) (Fang, Nakamura, & Maeda, 2011). The nanomedicines include nano-proteins, macromolecules, liposomes, micelles, conjugates, and other nanoparticles. The EPR effect is considered as the major underlying mechanism since the nanomedicine developed and accumulated inside the tumor (Figure 1.9) (Prabhakar et al., 2013). However, there is too much controversy on the efficiency of the EPR effect in humans through the clinical trials. In fact, the compromised effect of EPR effect for tumor in humans is attributable to the complex

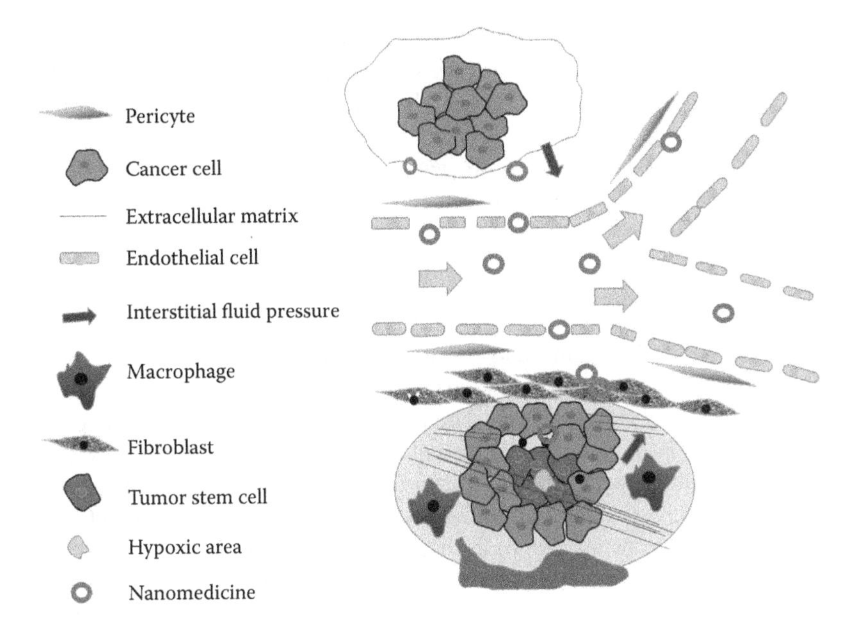

FIGURE 1.9 Schematic representation of EPR effects of passage of nanomedicine through vascular leaky aperture to approach the tumoral cells. (From Danhier, F., *Journal of Controlled Release, 244*, 108–121, doi: 10.1016/j.jconrel.2016.11.015, 2016.)

heterogeneity of the tumors, such as a hypoxic gradient, which can severely limit the efficacy of drugs delivered by passive targeting. Another factor that limits the retention of nanomedicine drugs in the tumor is increased interstitial fluid pressure that can result in reduced or abrogated transport of drug (Jain & Stylianopoulos, 2010). Furthermore, the extracellular matrix of certain malignancies, such as pancreatic cancer, limits drug penetration into the tumor (Provenzano et al., 2012).

While the number of researches citing "EPR" for nanoparticles has extremely increased, this creative flourish has largely failed in the clinical trials. Although the EPR effect has been well achieved in small animal models, the clinic data is not as apparent as in animal models. Nanomedicines have almost uniformly failed to produce improved efficacy in the clinic compared with the standard therapies, in spite of tremendously successful preclinical results. For instance, doxorubicin-loaded PEGylated liposomes (Doxil®) (Rose et al., 2006) was approved to treat the AIDS-associated Kaposi's sarcoma and ovarian cancers with pre-clinically greater therapeutic efficacy compared with standard therapies in 1995. The main benefit of Doxil® in treating solid tumors was the reduced cardiotoxicity gained by limiting free doxorubicin exposure to cardiac tissue. However, these appeared to aggravate notably skin toxicity in the form of palmar-plantar erythrodysesthesia (PPE)—hand and foot syndrome because of the extended circulation time of the PEGylated liposomes, which provides time for the kinetically slow process of extravasation into the skin (Sutton et al., 2005). In addition, Doxil® failed to achieve enhanced tumor accumulation over free doxorubicin, indicating that EPR is compromised in human tumors without as well understood as in pre-clinical trials. Another example is Abraxane®, an albumin-bound nanoparticle of paclitaxel, was approved to maintain the therapeutic benefits of paclitaxel but eliminate the toxicities associated with the emulsifier Cremophor®EL in the free paclitaxel formulation (Taxol®) (Bertrand et al., 2014). The maximum tolerated dose (MTD) of Abraxane® was approximately 50% higher than that for Taxol® (Morenoaspitia & Perez, 2005). Thus, the clinical efficiency of Abraxane® is probably not due to nanoparticles characteristics but to the removal of Cremopho®EL from the formulation that causes toxicities of its own (Sparreboom et al., 2005).

1.3.2 Active Targeting

Based on the passive targeting and the flexible surface of nanomedicine, specific ligands are modified on the surface of the nanocarrier that are able to bind to the over-expressed receptors at the target site. The first targeting system mainly concentrates on the antibodies (F [ab'], scFv) as targeting moieties because of the high specificity and wide availability. Since then, many other targeting moieties such as other proteins (transferrin, ankyrin repeat protein, affibodies), peptides (LyP-1, F3 peptide, iRGD, KLWVLPK, aptides), nucleic acid-based ligands (A10 aptamer, A9 CGA aptamer), and small molecules (folic acid, triphenylphosphine [TPP], nucleic acid aptamers [ACUPA]) have been investigated. The active targeting delivery system is able to achieve more efficiently specific target towards cancer tissue, which is classified as two types on cellular targets: (i) target the receptors of cancer cells, for instance, CD44, transferrin, folate, epidermal growth factors, and others; (ii). target the tumor microenvironment such as the overexpression of vascular endothelial growth factor (VEGF) integrins, HER2, and others. Most active targeting nanomedicines are internalized via receptor-mediated endocytosis that can recognize and enter into the specific targeted cell with higher specificity compared with the passive accumulation. Besides, receptor-mediated endocytosis can inhibit drug resistance by changing the internalized pathways to some extent. Despite a large number of publications in this area, only a few investigations with active targeting are under investigation in clinical trials (Table 1.3). Table 1.3 conclude the target to interact between the cells and the nanomedicine in clinic.

1.3.3 Stimuli-Responsive Nanomedicine

There are many investigations on the stimuli-responsive nanomedicine recently. The stimuli-responsive systems act in response to physical, chemical or biological triggers that promote release of drugs by interfering with the phase, structure or conformation of the nanocarrier. Triggers can be classified into internal (patho-physiological/pathochemical condition) and external (physical stimuli such as temperature, light, ultrasound, magnetic force, and electric fields) stimuli (Figure 1.10) (Holme et al., 2012; Jhaveri, Deshpande, & Torchilin, 2014; Tannock & Rotin, 1989).The advantage of stimuli-responsive systems is minimizing systemic exposure to the anti-drug. With the use of an external stimulus to improve efficacy of nanomedicines, two formulations have managed to reach clinical trials: the thermosensitive liposome containing doxorubicin, ThermoDox, for the treatment of breast cancers and hepatocellular carcinomas. Another formulation is the magnetic nanoparticle, NanoTherm, for the treatment of glioblastomas (Mura, Nicolas, & Couvreur, 2013).

TABLE 1.3
Actively Targeted Nanomedicines for Cancer in Clinical Trials (Phase I/II)

Name	System	Target	Payload	Ref
BIND-014	Polymeric NPs	Prostate specific membrane antigen (PSMA)	Docetaxel	(Hrkach et al., 2012)
MCC-465	Liposomes	Gastric cancer (Ab fragment)	Doxorubicin	(Matsumura et al., 2004)
MM-302	Liposomes	HER2 (Ab scFv)	Doxorubicin	(Wickham & Futch, 2012)
MBP-426	Liposomes	Tf-Receptor	Oxaliplatin	(Byrne, Betancourt, & Brannonpeppas, 2008)
SGT-53	Liposomes	Tf-Receptor (Ab scFv)	p53 plasmid DNA	(Senzer et al., 2013)
CALAA-01	Polymeric NPs	Tf-Receptor	siRNA M2 subunit of ribonucleotide reductase	(Davis et al., 2010)
Anti-EGFR-ILs-dox	Liposomes	EGFR (Ab, cetuximab)	Doxorubicin	(Mamot et al., 2012)
Rexin-G	Retroviral vector	Collagen (viral envelope peptide)	Human cyclin G1 gene	(Chawla et al., 2009; Chawla et al., 2010)

1.4 DESIGN ASPECT OF NANOMEDICINES

To achieve enhanced antitumoral effect at the tumor site, there are four major steps for the nanomedicine entering into the cancer cells: (i) the circulation time of the nanomedicines within the vascular network; (ii) cross the tumor vessel wall into the tumor interstitial space; (iii) the interstitial transport of the particles within the tumor; (iv) the internalization by cancer cells for some drugs.

1.4.1 Vascular Transport

In theory, the size for the pore of the normal vessel wall is between 6–12 nm, which suggests that nanomedicines should be >12 nm to avoid crossing the vessel wall of normal tissues. In addition, clearance from the blood circulation is performed by filtration in the kidneys or by the reticuloendothelial system in the liver and spleen. Renal clearance is very rapid for small particles with a hydrodynamic diameter smaller than 5–6 nm, while clearance by the liver and the spleen is rapid for particles with diameter above 200 nm. So, systemically administered nanomedicines which are biodegradable and not toxic should be designed at a range of 12–200 nm to have a higher chance of reaching the tumor vasculature and extravasating into the tumor tissue. The larger the nanomedicine diameter, the shorter the half-life time (Figure 1.11).

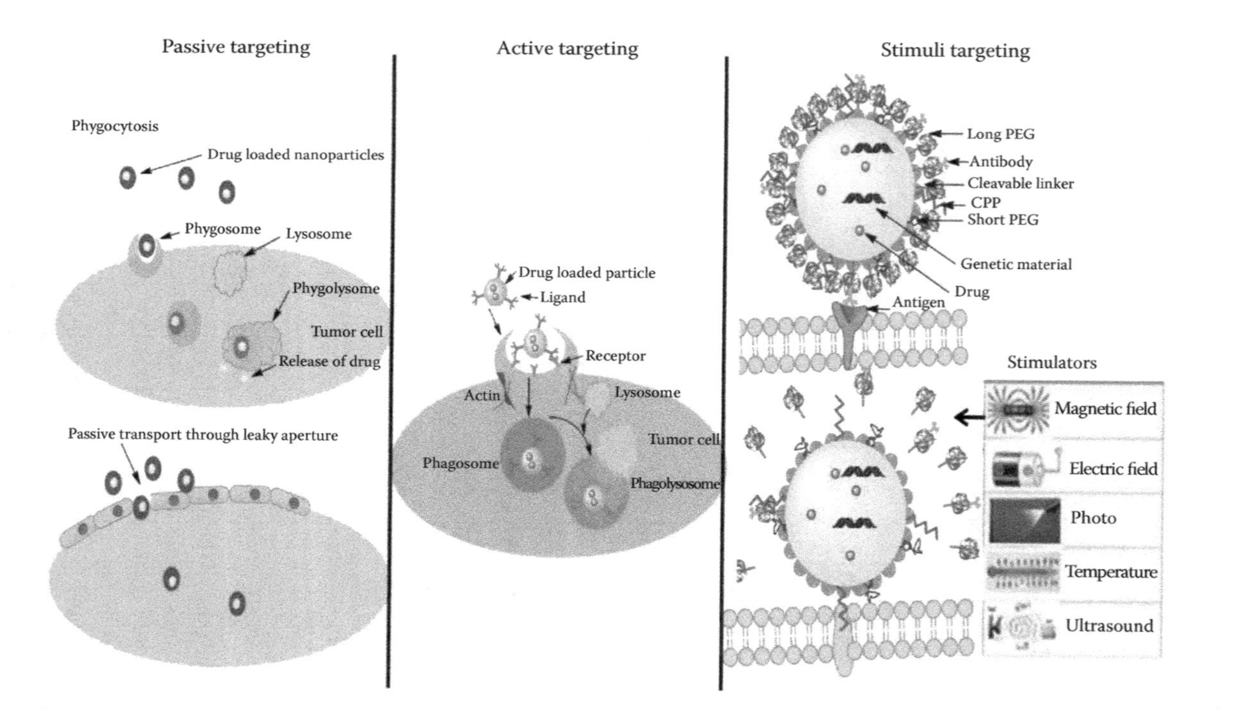

FIGURE 1.10 Schematic representation of targeting mechanism: passive targeting is generally achieved by an endocytosis (phagocytosis/pinocytosis) process in which the nanocarriers get attached with the endothelial membrane of the tumor cell and get internalized into the cell by the formation of phygosome which is further digested by the lysosome to rerelease the medications, diffusion is another mode of passive targeting in which the nanocarriers get internalized through the porous apertures on the epithelial membrane by the virtue of their size and surface charges; active targeting is achieved through ligand decoration on the nanopreparation, which is further recognized by the receptors on the epithelial surface. Similarly, stimuli targeting is achieved by the application of various physical stimuli (i.e. magnetic field, electric field, light, temperature and ultrasound) to activate the internalization and release of payloads inside the tumoral cells.

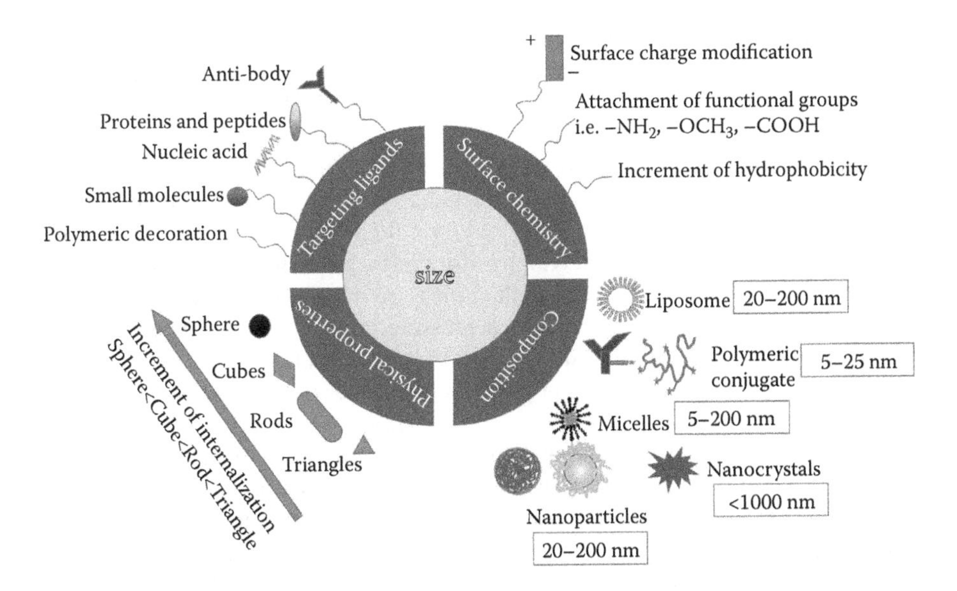

FIGURE 1.11 Schematic representation of various design aspects of nanomedicines to improve the tumoral internalization. (1) Decoration of various ligands can improve the active targeting and selective internalization, (2) modification on surface morphology can increase the passive targeting and cellular uptake by diffusion and phagocytosis process, (3) delivery of drugs using different nanocarriers can increase the stability and retention of drugs into the circulation system, and (4) improvement in physical properties like shape and size can also improve the cellular uptake into the tumoral cells.

Compared with spherical shape, filamentous particles processed longer circulation times, while nanotubes with diameters of less than 2 nm have rapid clearance from the kidneys, which drastically reduces their circulation times. Furthermore, the surface charges on the nanomedicine play a crucial role for the vascular transport of nanomedicines. The higher charged particles (anionic or cationic), the easier to be cleared from the reticuloendothelial system. And cationic particles induce binding of plasma lipoproteins to their surface that signals immune cells. Hence, it would be preferable for the nanoparticle formulations with neutral charge to transport across the vascular wall. A common practice of surface functionalization is PEGylated by adding a neutral layer of polyethylene glycol (PEG) to its surface. PEGylation is able to protect the nanomedicine to avoid opsonization and phagocytosis by the reticuloendothelial system (e.g. Kupffer cells, phagocytes, etc.), which is called "stealth" nanoparticles.

In conclusion, PEGylated nanoparticles with a hydrodynamic diameter above 12 nm but less than 200 nm could, in general, ensure sufficient circulation times and seem to selectively transport across the vascular wall in tumor area and reach to tumoral cell.

1.4.2 Transvascular Transport

The transvascular transport of nanomedicine passes through the pores of the leaky tumor vessels and delivers the drug into the interstitial space of tumor, which is the main mechanism of EPR effect (Gillies et al., 1999; Jain, 1987). Transvascular transport depends on two major points: (i) the difference between the microvascular and interstitial fluid pressure; (ii) the interactions between the nanomedicine and the vessel wall. The heterogeneity of tumor explains why the EPR effect failed in the clinic for some nanomedicine.

The interstitial fluid pressure is mainly controlled both by the permeability of the vessels and lymphatic dysfunction. Some tumors have large intercellular pores in the endothelial lining. Based on these, excessive fluid flows from the vascular to the interstitial space and insufficient lymphatic drainage causes the increased interstitial space pressure. When the interstitial fluid pressure in solid tumors is almost equal to microvascular pressure, transvascular pressure gradients reduce, which renders the diffused transport of molecules, so for very permeable tumors (e.g., openings larger than 200 nm) the transport of particles of all sizes will be hindered due to the elevated interstitial fluid pressure.

The interactions between the nanomedicine and the pores of the tumor vessel wall can be of three types: (i) steric due to collisions of the particle with the wall; (ii) hydrodynamic due to hydrodynamic forces induced by the motion of the particle within the fluid medium; (iii) electrostatic due to repulsion or attraction. All the above interactions are able to be controlled by the relative size of the particle to the openings of the wall. The research calculations indicate that neutral particles with a diameter larger than 60% of the openings of the vessel wall will be excluded due to steric interactions. Whereas if the opening size is too large for the particle, the interactions becomes not important and the transport would not be hindered.

Therefore, there is a size-dependent delivery of nanoparticles for less permeable tumors, that will benefit only small particles with sizes usually less than 60 nm. Note that MM-398 (ONIVYDE® [irinotecan liposome injection]) is of the same size as Doxil® at 100 nm, it has been clinically approved for the treatment of pancreatic ductal adenocarcinomas, which are considered to be poorly permeable.

As far as the effect of nanomedicine shape is concerned, transvascular flux of elongated particles, such as nanorods, is superior compared to spherical particles of equal hydrodynamic diameter. This could be explained by the fact that non-spherical particles traveling in the bloodstream interact more actively with the vessel wall as they exhibit tumbling and rolling motions. It is indicated that steric and hydrodynamic interactions may be lower for elongated particles. In conclusion, spherical nanoparticles with diameters in the range of 12–60 nm, elongated particles with similar hydrodynamic diameters and cationic particles are likely to have superior extravasation into the tumor tissue.

1.4.3 Interstitial Transport

Similar to transvascular transport, interstitial transport is controlled by the gradients of the interstitial fluid pressure within the tumor as well as by steric, hydrodynamic, and electrostatic interactions between the particles and the openings apertures (pores) of the interstitial space. However, the desmoplastic response observed in many tumors overproduces extracellular fibers and the pores can be on the order of 100 nm or smaller. As a result, nanomedicines with a hydrodynamic diameter larger than 50 nm might not be able to effectively and uniformly diffuse into the tumor interstitial space. Therefore, even if nanoparticles are able to cross the hyper-permeable tumor vessels, they might not be able to penetrate deep into the tumor interior, and thus, concentrate in the perivascular regions or end up in the surrounding normal tissue. In that case, the release rate of the nanomedicine will largely determine the distribution of the effective drug into the tumor (Figure 1.11).

In conclusion, particles with a diameter of 12–50 nm, elongated particles with similar hydrodynamic diameters and neutral particles are ideal to optimize nanoparticle penetration in the tumor interstitial space. Obviously, there is better penetration for smaller particles, but very small particles might be cleared rapidly from the tumor tissue due to limited retention. As discussed earlier for transvascular transport, the efficacy of particles larger than 50 nm will depend on the release kinetics of drugs from these particles.

1.4.4 Intracellular Transport

Some antitumor drugs require intracellular delivery while others are able to release the drug into the microenvironment. For the former drug, the nanomedicine needs a design accounting for effective size, surface, and shape on the cellular internalization (Figure 1.11).

A nanomedicine with a different size is able to be take up through various receptor-mediated endocytic processes. Additionally, it has been shown that nanomedicines with a size larger than 50 nm are more likely to be excluded from the cell nucleus. The cell membrane is covered with negatively charged, sulfated proteoglycans that are able to combine with cationic nanoparticles. As a result, *in vitro* experiments have shown that cationic nanoparticles were superior compared with the anionic counterparts, which indicated that surface charge might also determine the uptake mechanism.

Finally, as for the particle shape, there is research that has shown that for particles less than 100 nm, spherical nanoparticles are internalized more effectively than elongated particles. However, internalization is faster and more efficient for elongated particles with larger sizes. Additionally, the local geometry of the nanomedicine at the interaction area with the cell determines whether it will be internalized or not. Thus, the relevance of shape in cellular internalization of nanoparticles *in vivo* remains to be shown. Moreover, nanomedicines that release their therapeutic load to be effective once they enter the tumor microenvironment do not need to be internalized.

Except the above the design considerations, we need also to take into account the drug loading, release rate of the nanomedicine, and targeting capability. As this chapter focuses on the optimal delivery of nanoparticles, we propose three different strategies such as small size nanomedicine, elongated shape formulations, and multi-responsive nanomedicine systems that are responsive to the properties of the microenvironment.

1.5 ENHANCING NANOMEDICINE TRANSLATION THROUGH CLINICALLY RELEVANT MODELS

The high failure of translation from pre-clinic to clinic is partly due to the lack of predictive models or the inadequate use of existing preclinical test systems.

There is a critical requirement for the xenograft model that it should faithfully replicate the original patient tumor such as its histological appearance, invasiveness, and metastasis phenotype for pre-clinical studies to have clinical relevance that is able to as closely as possible achieve meaningful insight into the underlying molecular mechanisms of the tumor. However, there is growing evidence that the data obtained from subcutaneous xenograft models poorly translate into positive clinical responses for nanomedicine. Therefore, the *in vivo* models we constructed for developing new nanomedicines may only reflect a narrow spectrum of the human tumor pathophysiology. For example, early research with relevant models demonstrated the efficacy of liposomal doxorubicin in tumors with a high tumor cell density, low stromal content, and highly permeable vascularization. This efficacy has successfully been translated to the clinic, where Doxil® has shown its superior therapeutic efficacy in the treatment of multiple myeloma and AIDS-related Kaposi's sarcoma. Many predictive models for clinic uses have been developed as follows.

1.5.1 Standard Xenograft Models with Human Stroma Components

The discovery of incubating human cell lines to solid cancers intraperitoneally or subcutaneously in immunodeficient mice laid the foundation for studying tumor biology and nanomedicine response *in vivo*. The major drawback of this model is that many paracrine signals are assumed to be non-functional between xenografted human cells and the mouse host, due to cross-species incompatibility (Voskoglou-Nomikos, Pater, & Seymour, 2003). For instance, the granulocyte macrophage colony stimulating factor (GM-CSF) is the prominent example for this unsuitability. As murine HGF could not activate human c-Met, mouse models were developed, which express human HGF, to study HGF/c-Met-dependent human tumor progression and to evaluate c-Met inhibition in these xenograft models (Jeffers, Rong, & Vande Woude, 1996; Mazzone et al., 2004). Furthermore, interleukin-6 (IL-6) in mice is inactive on human cells, whereas human IL-6 can activate mouse IL-6R signaling. Based on that, the co-inoculation of human stromal cells and cancer cells could improve tumor growth, progression, angiogenesis, and invasion (Francone et al., 2007). Although the above models may not be directly predictive of the clinic, many studies have now been published using these relevant models to develop new perspectives on nanomedicines. Furthermore, further insight can be developed by using these models, where the tumor develops in situ.

1.5.2 Patient-Derived Tumor Xenograft (PDX) Model

Directly implanting freshly resected tumor pieces subcutaneously or orthotopically into immuno-compromised mice is called patient-derived xenograft (PDX) cancer models (Daniel et al., 2009; Jin et al., 2010). The vasculature in clinical tumors and PDX models are more mature and less permeable than in xenograft models, which develop over days, rather than weeks or months, and present properties that are less influenced by the tumor cell proliferation. The differential vessel distribution is more reliably demonstrated in PDX models than it is in solid tumors. Despite the great promise of PDX model as a preclinical model for predictive human cancer, there are several limitations for clinic uses. Firstly, tumor tissue and microenvironment are variable, leading to biases in testable material (Lum et al., 2012; Tentler et al., 2012). Then the tumor-host interaction is not fully established due to cross-species incompatibilities as pointed out above. In addition, as a prerequisite for further predictive potential of cancer, the constructed PDX has to be thoroughly characterized histopathologically, cytogenetically, and genomically by essential molecular marker expression as well as by global expression profiling and comparison must be made to the initial human tumor, which is time-consuming and expensive.

There are some other various pre-clinical models such as the humanized PDX models and human metastatic site models *in vivo* (Mcgonigle & Ruggeri, 2014).

1.5.3 Humanized PDX Model

To deal with the drawback that tumor immunity is lacking in the PDX models, the human hematopoietic system was reconstituted in the mice models. In the models, mice are transplanted with human hematopoietic stem and progenitor cells (HSPCs). After that, this should enable researchers to study human tumors in an environment with at least partially functional human immune system. In addition, HSPCs and the primary tumor should be from the same patient ideally. For instance, HSPCs and peripheral T-cells were transplanted into the same mice, which were also incubated with breast cancer cells.

1.5.4 Human Metastatic Site Model

Animal models of human breast cancer metastasis into functional human bone have been constructed in which orthotropic injection of metastatic human breast cancer cell resulted in metastatic dissemination to bone (Kuperwasser et al., 2005). Bone metastasis was directed to the implanted human bone fragment instead of the mouse skeleton, which indicates a species-specific osteotropism (Holzapfel et al., 2013). Such human bone metastasis models have been expanded to other cancers such as prostate cancer, and breast and prostate cancer colonization of human bone or tissue engineered human bone, which are in research now.

1.6 CHALLENGES AND CURRENT LIMITATIONS

Nanomedicine has become one of the most promising and advanced approaches in the forefront of cancer treatment. Decades of investigation suggest that nanomedicine therapeutics are effective in cancer treatment, both *in vitro* and *in vivo*. However, only very few nanocarrier-based cancer therapeutics have successfully entered clinical trials. Thus, it is essential to address the challenges in developing optimized nanomedicine products for clinical use.

In general, the physico-chemical characterization of nanomedicine include size distribution, surface charge, shape, porosity, drug loading, ligand characteristics, release behavior, biodegradability, stability, sterility, and batch-to-batch reproducibility (Figure 1.10) (Fubini, Ghiazza, & Fenoglio, 2010; Kettiger et al., 2013). Variability within these properties both before and after administration made

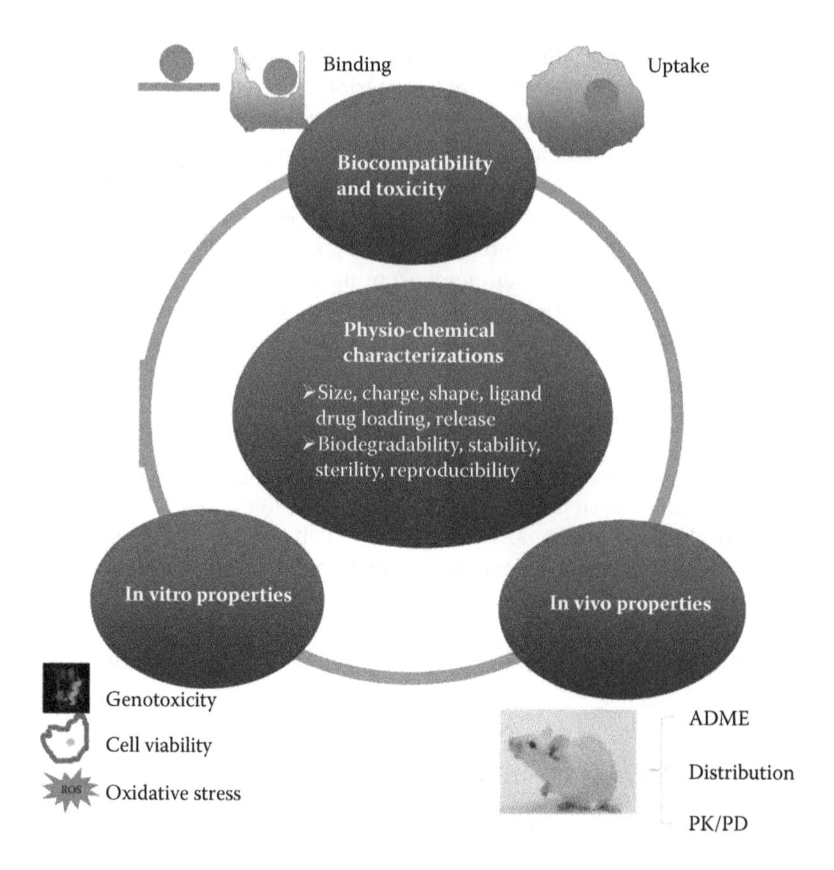

FIGURE 1.12 The physico-chemical properties of nanomedicines affect their pharmacokinetic and pharmacodynamic profiles. Detailed characterization of nanomedicine products is necessary to predict their performance in the clinical setting. (From Wicki, A. et al., *Journal of Controlled Release, 200*, 138–157, doi: 10.1016/j.jconrel.2014.12.030, 2015.)

characterizing these nanomedicine products problematic (Figure 1.12). Quantitative evaluative methods have to carry out the analysis of all necessary quality aspects of the nanomedicine.

Firstly, the size distribution (polydispersity) is essential for characterizing the heterogeneity of nanoparticles that play a major part in the changes *in vivo* such as the toxicity, pharmacokinetic behavior, biocompatibility, and *in vivo* efficiency. In particular, toxicity has been proposed to originate from nanomaterial size and surface area, composition, and shape. In addition, degradability of the nanocarrier is an important component of acute and long-term toxicity for clinical use (Aillon et al., 2009; Dobrovolskaia & Mcneil, 2007; Nel et al., 2009). Furthermore, chemical composition at the surface of nanomedicine has shown promising effects in many applications, which potentially interact with biological components, alter biological function, and achieve specificity of nanomedicines that would not normally be taken up by certain cells. Regarding formulations currently in clinical use, micellar polymeric particles (i.e. NC-6004, NC-4016, NC-6300, and Genexol-PM®) as well as Abraxane® and CRLX101 were at the size distribution range of 12–50 nm. Liposomal formulations usually have a larger size around 100 nm (e.g. Doxil®, MM-398, Thermodox®). Then, modification of nanoparticles (targeting ligands) would increase the size of the nanoparticles, leading to compromised entry of these as targeted nanomedicine. For instance, MCC-465 and SGT-53 are targeted liposomes with sizes of 100 nm, while BIND-014 and CALAA-01 are polymeric nanoparticles with sizes of 100 and 75 nm, respectively. Whether the modified binding affinity overcomes the enlarged size diameter is not only determined by the characteristics of nanoparticles, but also the tumor type.

Based on the complicated quality aspects, nanomedicine products should be characterized on a batch-to-batch basis with multiple methods. Characterization of stability and storage aspects of nanomedicine products is also challenging. After being administrated, nanomedicines may interact with other biological fluids (e.g. blood serum) or biomolecules (e.g. proteins), which may lead to instability or aggregation of nanoparticles. Such interactions can significantly alter the function of nanomedicine compounds in biological systems *in vivo.* To sum up, the final formulation of nanomedicine products should be fully characterized under clinically relevant conditions. There is a substantial need to improve quality assessment of nanomedicine by developing well-defined and reproducible standards. Moreover, *in vitro* and *in vivo* models accurately representing the clinical setting must be developed.

1.7 ADVANTAGES AND DISADVANTAGES OF NANOTECHNOLOGY

Nanotechnology is considered as boon in medical practice. Many advantages have been recognized for nanotechnology-based drug delivery system. Due to nanoscale size, these nano-drug delivery systems have large surface mass area to load large amount of theranostics in comparison to conventional drug delivery system (Emerich & Thanos, 2007; Groneberg et al., 2006). Nanotechnology has improved solubility, diffusivity, blood circulation half-life, drug release efficiency, and immunogenicity of conventional drug delivery system. It has also improved patient compliance by reducing potential therapeutic toxicity and lowering the therapeutic cost. Many drugs can be simultaneously delivered to suppress drug resistance through nanotechnology (Zhang et al., 2008). Many nanotechnology-based theranostics are in clinical practices for prevention, diagnosis, and treatment of diseases. Better imaging and diagnostic agents are prepared by nanotechnology, which helps in early diagnosis for early treatment for better therapeutic success rates.

For better bioavailability and successful therapeutic effect, an individualized drug delivery system is necessary. Nanotechnology helps to delivery drugs directly at the site of action for better therapeutic effect. The pathophysiological condition and anatomical changes of diseased or inflamed tissues provide many advantages for the delivery of various nanotechnological products. Drug targeting can be achieved by taking advantage of the distinct pathophysiological features of diseased tissues (Haley & Frenkel, 2008). Thus, the differences between the physiological status of normal and diseased tissues provide a ladder for nanomedicines for site-specific targeting. This difference between normal and diseased tissues recognition is lacked by conventional drug delivery systems. Various nanosystems, as a result of their size and surface structure get accumulated in the targeted organ at a higher concentration than normal drugs (Vasir, Reddy, & Labhasetwar, 2005). Thus, this pathophysiological opportunity allows extravasation of the nanodrugs and their selective localization in the inflamed tissues (Allen & Cullis, 2004).

Treatment of many neurological diseases with a conventional drug delivery system show poor therapeutic outcomes, due to poor permeability through the blood-brain barrier (Calvo et al., 2001). But the delivery of drugs in the nano size to treat neurological diseases have shown effective therapeutic outcome (Garcia-Garcia et al., 2005). Recently, there has been a move away from cytotoxic cancer drugs towards molecularly targeted agents that have had little attention in the nanomedicine field. While highly efficacious, these drugs are associated with serious toxicity profiles on their own, and therefore the combination with nanoparticles may help to reduce toxicity and allow higher dosing (Ashton et al., 2016).

In this modernized technology era, many researchers are trying to improve the quality of human life by utilization of modern technologies. Recently, nanotechnology is rocking in medical field. Some scientists believe that these nanotechnology-based medicines have no health hazards and are safe for therapeutic use. But, some scientists believe that there are potential risks on long-term use of these nanotechnology-based medicines. They also argue that if it is mishandled than it may bring unexpected environmental as well as health hazards. The following are some potential risks always associated with nanotechnologies, which are beyond the researcher's knowledge and understanding.

Although some of these nanotechnology-based preparations are speculative, researchers have valid concerns regarding short- and long-term threats to the human body and environment because of recent laboratory findings about the toxicity of fullerenes (Aschberger et al., 2010). Fullerenes are carbon allotropes with 60 carbon atoms with a diameter of approximately 0.7 nm and molecular weight of 720 g/mol, also known as radical sponge (Johnson et al., 2010). It was reported that fullerenes get absorbed into the brains of fish and cause extensive neurological damages and physiological alterations (Sergio et al., 2013). In another report related with fullerenes, mentioned that earthworms after absorption of fullerenes survive for long time on dry soil (Rushton et al., 2010). Quantum dots are used in diagnosis of various ailments, the poisoning of quantum dots is well reported by Kim et al. (Kim et al., 2004). The real or potential drawbacks associated with nanotechnology raises much concern; although there is no quantitative way to determine it.

Nanoparticles are very different than their larger counterparts, with distinct electronic, magnetic, chemical, and mechanical properties. Nanoparticles have an increased surface area, which offers more space for interaction with other substances. This increased interaction with their surroundings means that substances at the nanoscale are more reactive and have higher toxicity than they do at their normal size. Adding to the concern of increased toxicity, substances that are stable in larger forms (e.g. aluminium) can also become reactive or explosive in nanoparticle form, creating the potential for health effects that are not seen when the substance is in its larger form.

Size and structural differences allow nanomaterials to migrate to different tissues and organs than their larger counterparts. There is also evidence that nanoparticles can be more completely absorbed by the body, increasing the substance's bioavailability. Nano-engineered materials also have the potential to increase the bioavailability of those chemicals, such as toxins. One study found that micronized titanium dioxide in sunscreens increases the skin's absorption of several pesticides (Brand et al., 2003).

Due to their extremely small size, nanoparticles have the potential to bypass the blood-brain barrier. They also have the potential to pass the placental barrier. In one investigation it was found that nanoparticles can easily reach to brain through nasal passage-ways, similarly, in another study shows that gold nanoparticles can move across the placenta from mother to fetus (Schulte et al., 2013). Recently, one study published in *Nature* revealed that carbon nanotubes (CNT) have same potential to cause cancer as asbestos (Poland et al., 2008). Once nanoparticles enter into the bloodstream, circulate throughout the body and internalized into different organs and tissues. Based on the nature of materials, clearance of these nanomaterials varies from a week to a year. Multiple administrations of these nanotechnology-based nanomedicines accumulate in the body and are retained for longer time, which may cause unknown harmful effects.

1.8 CONCLUSION

In this chapter we have discussed the nanomedicines approved by the FDA for cancer treatment (Figure 1.13) and the most prominent nanomedicines under clinical trials. We have also discussed about the mode of targeting, different modalities, challenges, advantages, and disadvantages of nanomedicines and nanotechnologies.

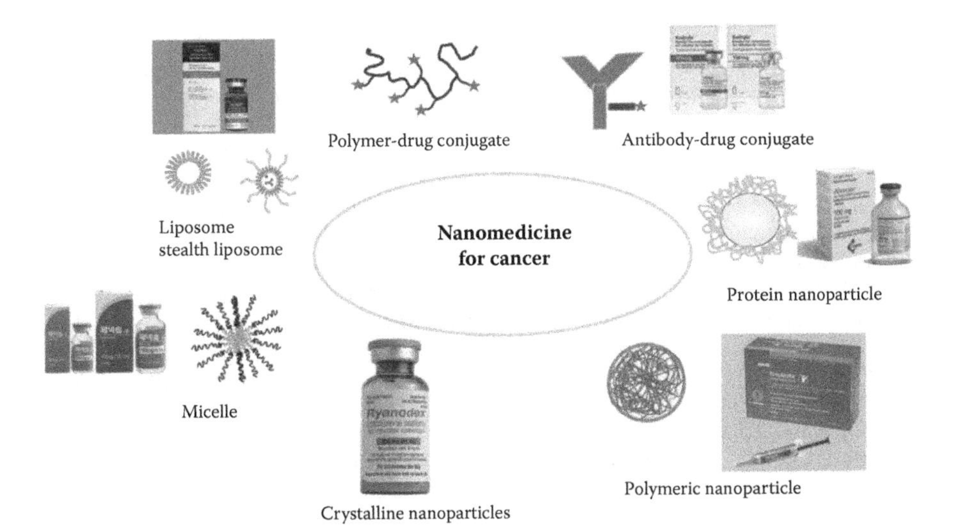

FIGURE 1.13 Most frequently used nanomedicine for cancer therapy.

It was shown that application of modern technologies and advancement in polymeric and non-polymeric materials resulted in many nanomedicines for cancer treatment. Also, this made a distinct trend that evolves beyond the initial goal for nanomedicines, i.e. enhanced bioavailability with reduced systemic toxicity, to engineered delivery systems that may allow for controlled release, active targeting of disease, and diversification of drug approaches beyond traditional chemotherapeutics. Most of the recently approved nanomedicines consist of relatively simple nanoparticles with PEG surface modification for long retention and enhanced bioavailability. In the future, more precise targeting approaches will be required to combat the increasing failure in chemotherapy.

REFERENCES

Aillon, K. L., Xie, Y. M., El-Gendy, N., et al. (2009). Effects of nanomaterial physicochemical properties on in vivo toxicity. *Advanced Drug Delivery Reviews, 61*(6), 457-466. doi: 10.1016/j.addr.2009.03.010

Alconcel, S. N. S., Baas, A. S., & Maynard, H. D. (2011). FDA-approved poly(ethylene glycol)–protein conjugate drugs. *Polymer Chemistry, 2*(7), 1442–1448.

Allen, T. M., & Cullis, P. R. (2004). Drug delivery systems: entering the mainstream. *Science, 303*(5665), 1818–1822.

Aschberger, K., Johnston, H. J., Stone, V., et al. (2010). Review of fullerene toxicity and exposure – appraisal of a human health risk assessment, based on open literature. *Regulatory Toxicology and Pharmacology, 58*(3), 455–473. doi: http://dx.doi.org/10.1016/j.yrtph.2010.08.017

Ashton, S., Song, Y. H., Nolan, J., et al. (2016). Aurora kinase inhibitor nanoparticles target tumors with favorable therapeutic index in vivo. *Science Translational Medicine, 8*(325).

Bangham, A., Standish, M. M., & Watkins, J. (1965). Diffusion of univalent ions across the lamellae of swollen phospholipids. *Journal of Molecular Biology, 13*(1), 238–IN227.

Barenholz, Y. (2012). Doxil® — The first FDA-approved nano-drug: lessons learned. *Journal of Controlled Release, 160*(2), 117–134. doi: http://dx.doi.org/10.1016/j.jconrel.2012.03.020

Benbrook, D. M. (2014). Biotechnology and biopharmaceuticals: transforming proteins and genes into drugs. *Clinical Infectious Diseases.*

Bertrand, N., Wu, J., Xu, X., et al. (2014). Cancer nanotechnology: the impact of passive and active targeting in the era of modern cancer biology. *Advanced Drug Delivery Reviews, 66*, 2–25.

Bobo, D. P., Robinson, K. J., Islam, J., et al. (2016). Nanoparticle-based medicines: a review of fda-approved materials and clinical trials to date. *Pharmaceutical Research, 33*(10), 2373–2387.

Brand, R. M., Pike, J., Wilson, R. M., et al. (2003). Sunscreens containing physical UV blockers can increase transdermal absorption of pesticides. *Toxicology and Industrial Health, 19*(1), 9–16.

Brannon-Peppas, L., & Blanchette, J. O. (2004). Nanoparticle and targeted systems for cancer therapy (vol 56, pg 1649, 2004). *Advanced Drug Delivery Reviews, 56*(11), 1649–1659. doi: 10.1016/j.addr.2004.02.014

Byrne, J. D., Betancourt, T., & Brannon-Peppas, L. (2008). Active targeting schemes for nanoparticle systems in cancer therapeutics. *Advanced Drug Delivery Reviews, 60*(15), 1615–1626.

Cagel, M., Tesan, F. C., Bernabeu, E., et al. (2017). Polymeric mixed micelles as nanomedicines: achievements and perspectives. *European Journal of Pharmaceutics and Biopharmaceutics, 113*, 211–228. doi: 10.1016/j.ejpb.2016.12.019

Calvo, P., Gouritin, B., Chacun, H., et al. (2001). Long-circulating PEGylated polycyanoacrylate nanoparticles as new drug carrier for brain delivery. *Pharmaceutical Research, 18*(8), 1157–1166.

Chawla, S. P., Chua, V. S., Fernandez, L. O., et al. (2009). Phase I/II and Phase II studies of targeted gene delivery in vivo: intravenous Rexin-G for chemotherapy-resistant sarcoma and osteosarcoma. *Molecular Therapy, 17*(9), 1651–1657.

Chawla, S. P., Chua, V. S., Fernandez, L. O., et al. (2010). Advanced Phase I/II studies of targeted gene delivery in vivo: intravenous Rexin-G for gemcitabine-resistant metastatic pancreatic cancer. *Molecular Therapy, 18*(2), 435–441.

Danhier, F. (2016). To exploit the tumor microenvironment: Since the EPR effect fails in the clinic, what is the future of nanomedicine? *Journal of Controlled Release, 244*, 108–121. doi: 10.1016/j.jconrel.2016.11.015

Daniel, V. C., Marchionni, L., Hierman, J. S., et al. (2009). A Primary xenograft model of small-cell lung cancer reveals irreversible changes in gene expression imposed by culture in vitro. *Cancer Research, 69*(8), 3364–3373.

Davis, M. E., Zuckerman, J. E., Choi, C. H. J., et al. (2010). Evidence of RNAi in humans from systemically administered siRNA via targeted nanoparticles. *Nature, 464*(7291), 1067–1070.

Dobrovolskaia, M. A., & Mcneil, S. E. (2007). Immunological properties of engineered nanomaterials. *Nature Nanotechnology, 2*(8), 469–478. doi: DOI 10.1038/nnano.2007.223

Drugs.com. (2015). FDA approves onivyde. Retrieved 3/16/2017, 2017, from www.drugs.com/newdrugs/fda-approves-onivyde-irinotecan-liposome-advanced-pancreatic-cancer-4283.html

Emerich, D. F., & Thanos, C. G. (2007). Targeted nanoparticle-based drug delivery and diagnosis. *Journal of Drug Targeting, 15*(3), 163–183.

Fang, J., Nakamura, H., & Maeda, H. (2011). The EPR effect: unique features of tumor blood vessels for drug delivery, factors involved, and limitations and augmentation of the effect. *Advanced Drug Delivery Reviews, 63*(3), 136–151.

Farokhzad, O. C., & Langer, R. (2006). Nanomedicine: developing smarter therapeutic and diagnostic modalities. *Advanced Drug Delivery Reviews, 58*(14), 1456–1459.

Ferrari, M. (2005). Cancer nanotechnology: opportunities and challenges. *Nature Reviews Cancer, 5*(3), 161–171.

Forrest, M. L., & Kwon, G. S. (2008). Clinical developments in drug delivery nanotechnology. *Advanced Drug Delivery Reviews, 60*(8), 861–862. doi: 10.1016/j.addr.2008.02.013

Forssen, E. A., & Tökès, Z. (1981). Use of anionic liposomes for the reduction of chronic doxorubicin-induced cardiotoxicity. *Proceedings of the National Academy of Sciences, 78*(3), 1873–1877.

Foss, F. (2006). Clinical experience with denileukin diftitox (ONTAK). *Seminars in Oncology, 33, Supplement 3*, 11–16. doi: http://dx.doi.org/10.1053/j.seminoncol.2005.12.017

Francone, T. D., Landmann, R. G., Chen, C., et al. (2007). Novel xenograft model expressing human hepatocyte growth factor shows ligand-dependent growth of c-Met–expressing tumors. *Molecular Cancer Therapeutics, 6*(4), 1460–1466.

Fubini, B., Ghiazza, M., & Fenoglio, I. (2010). Physico-chemical features of engineered nanoparticles relevant to their toxicity. *Nanotoxicology, 4*(4), 347–363.

Garcia-Garcia, E., Gil, S., Andrieux, K., et al. (2005). A relevant in vitro rat model for the evaluation of blood-brain barrier translocation of nanoparticles. *Cellular and Molecular Life Sciences, 62*(12), 1400–1408.

Gillies, R. J., Schornack, P. A., Secomb, T. W., et al. (1999). Causes and effects of heterogeneous perfusion in tumors. *Neoplasia, 1*(3), 197–207.

Godin, B., Sakamoto, J. H., Serda, R. E., et al. (2010). Emerging applications of nanomedicine for the diagnosis and treatment of cardiovascular diseases. *Trends in Pharmacological Sciences, 31*(5), 199–205.

Gordon, E. M., & Hall, F. L. (2010). Rexin-G, a targeted genetic medicine for cancer. *Expert Opinion on Biological Therapy, 10*(5), 819–832.

Groneberg, D. A., Giersig, M., Welte, T., et al. (2006). Nanoparticle-based diagnosis and therapy. *Current Drug Targets, 7*(6), 643–648.

Haley, B. B., & Frenkel, E. P. (2008). Nanoparticles for drug delivery in cancer treatment. *Urologic Oncology-seminars and Original Investigations, 26*(1), 57–64.

Hare, J. I., Lammers, T., Ashford, M. B., et al. (2017). Challenges and strategies in anti-cancer nanomedicine development: an industry perspective. *Advanced Drug Delivery Reviews, 108*, 25–38. doi: http://dx.doi.org/10.1016/j.addr.2016.04.025

Holme, M. N., Fedotenko, I. A., Abegg, D., et al. (2012). Shear-stress sensitive lenticular vesicles for targeted drug delivery. *Nature Nanotechnology, 7*(8), 536–543. doi: 10.1038 /nnano.2012.84

Holzapfel, B. M., Thibaudeau, L., Hesami, P., et al. (2013). Humanised xenograft models of bone metastasis revisited: novel insights into species-specific mechanisms of cancer cell osteotropism. *Cancer and Metastasis Reviews, 32*, 129–145.

Hrkach, J. S., Von Hoff, D. D., Ali, M. M., et al. (2012). Preclinical development and clinical translation of a PSMA-targeted docetaxel nanoparticle with a differentiated pharmacological profile. *Science Translational Medicine, 4*(128).

Jain, R. K. (1987). Transport of molecules in the tumor interstitium: a review. *Cancer Research, 47*(12), 3039–3051.

Jain, R. K., & Stylianopoulos, T. (2010). Delivering nanomedicine to solid tumors. *Nature Reviews Clinical Oncology, 7*(11), 653–664.

Jeffers, M., Rong, S., & Vande Woude, G. F. (1996). Hepatocyte growth factor/scatter factor—Met signaling in tumorigenicity and invasion/metastasis. *Journal of Molecular Medicine, 74*(9), 505–513. doi: 10.1007/bf00204976

Jhaveri, A., Deshpande, P., & Torchilin, V. (2014). Stimuli-sensitive nanopreparations for combination cancer therapy. *Journal of Controlled Release, 190*, 352–370. doi: 10.1016/j .jconrel.2014.05.002

Jin, K., Teng, L., Shen, Y., et al. (2010). Patient-derived human tumour tissue xenografts in immunodeficient mice: a systematic review. *Clinical & Translational Oncology, 12*(7), 473–480.

Johnson, D. R., Methner, M. M., Kennedy, A. J., et al. (2010). Potential for occupational exposure to engineered carbon-based nanomaterials in environmental laboratory studies. *Environmental Health Perspectives, 118*(1), 49–54.

Kaittanis, C., Shaffer, T. M., Ogirala, A., et al. (2014). Environment-responsive nanophores for therapy and treatment monitoring via molecular MRI quenching. *Nature Communications, 5*, 3384–3384.

Kawasaki, E. S., & Player, A. (2005). Nanotechnology, nanomedicine, and the development of new, effective therapies for cancer. *Nanomedicine-Nanotechnology Biology And Medicine, 1*(2), 101–109. doi: 10.1016/j.nano.2005.03.002

Kettiger, H., Schipanski, A., Wick, P., & Huwyler, J. (2013). Engineered nanomaterial uptake and tissue distribution: from cell to organism. *International Journal of Nanomedicine, 8*(1), 3255–3269.

Kim, S., Lim, Y. T., Soltesz, E. G., et al. V. (2004). Near-infrared fluorescent type II quantum dots for sentinel lymph node mapping. *Nat Biotech, 22*(1), 93–97. doi: www.nature .com/nbt/journal/v22/n1/suppinfo/nbt920_S1.html

Kobayashi, H., Kawamoto, S., Jo, S.-K., et al. (2003). Macromolecular MRI contrast agents with small dendrimers: pharmacokinetic differences between sizes and cores. *Bioconjugate Chemistry, 14*(2), 388–394.

Koudelka, S., & Turanek, J. (2012). Liposomal paclitaxel formulations. *Journal of Controlled Release, 163*(3), 322–334.

Kuperwasser, C., Dessain, S. K., Bierbaum, B. E., et al. (2005). A mouse model of human breast cancer metastasis to human bone. *Cancer Research, 65*(14), 6130–6138.

Libutti, S. K., Paciotti, G. F., Byrnes, A. A., et al. (2010). Phase I and pharmacokinetic studies of CYT-6091, a novel PEGylated colloidal gold-rhTNF nanomedicine. *Clinical Cancer Research, 16*(24), 6139–6149.

Liu, D., Yang, F., Xiong, F., et al. (2016). The smart drug delivery system and its clinical potential. *Theranostics, 6*(9), 1306–1323. doi: 10.7150/thno.14858

Liu, Y., Miyoshi, H., & Nakamura, M. (2007). Nanomedicine for drug delivery and imaging: a promising avenue for cancer therapy and diagnosis using targeted functional nanoparticles. *International Journal of Cancer, 120*(12), 2527–2537.

Lum, D. H., Matsen, C., Welm, A. L., et al. (2012). Overview of human primary tumorgraft models: comparisons with traditional oncology preclinical models and the clinical relevance and utility of primary tumorgrafts in basic and translational oncology research. *Current Protocols in Pharmacology, Chapter 14*, Unit 14 22. doi: 10.1002/0471141755.ph1422s59

Maeda, H., Bharate, G. Y., & Daruwalla, J. (2009). Polymeric drugs for efficient tumor-targeted drug delivery based on EPR-effect. *European Journal Of Pharmaceutics And Biopharmaceutics, 71*(3), 409–419. doi: 10.1016/j.ejpb.2008.11.010

Maier-Hauff, K., Ulrich, F., Nestler, D., et al. (2011). Efficacy and safety of intratumoral thermotherapy using magnetic iron-oxide nanoparticles combined with external beam radiotherapy on patients with recurrent glioblastoma multiforme. *Journal of Neuro-Oncology, 103*(2), 317–324. doi: 10.1007/s11060-010-0389-0

Mamot, C., Ritschard, R., Wicki, A., et al. (2012). Tolerability, safety, pharmacokinetics, and efficacy of doxorubicin-loaded anti-EGFR immunoliposomes in advanced solid tumours: a phase 1 dose-escalation study. *The Lancet Oncology, 13*(12), 1234–1241. doi: http://dx.doi.org/10.1016/S1470–2045(12)70476-X

Marty, J., Oppenheim, R., & Speiser, P. (1977). Nanoparticles—a new colloidal drug delivery system. *Pharmaceutica Acta Helvetiae, 53*(1), 17–23.

Matsumura, Y., Gotoh, M., Muro, K., et al. (2004). Phase I and pharmacokinetic study of MCC-465, a doxorubicin (DXR) encapsulated in PEG immunoliposome, in patients with metastatic stomach cancer. *Annals of Oncology, 15*(3), 517–525.

Matsumura, Y., & Maeda, H. (1986). A new concept for macromolecular therapeutics in cancer chemotherapy: mechanism of tumoritropic accumulation of proteins and the antitumor agent smancs. *Cancer Research, 46*(12), 6387-6392.

Mazzone, M., Basilico, C., Cavassa, S., et al. (2004). An uncleavable form of pro–scatter factor suppresses tumor growth and dissemination in mice. *Journal of Clinical Investigation, 114*(10), 1418-1432.

Mcgonigle, P., & Ruggeri, B. A. (2014). Animal models of human disease: challenges in enabling translation. *Biochemical Pharmacology, 87*(1), 162-171.

Moghimi, S. M., & Szebeni, J. (2003). Stealth liposomes and long circulating nanoparticles: critical issues in pharmacokinetics, opsonization and protein–binding properties. *Progress in Lipid Research, 42*(6), 463-478.

Morenoaspitia, A., & Perez, E. A. (2005). North central cancer treatment group N0531: Phase II Trial of Weekly Albumin–Bound Paclitaxel (ABI–007; Abraxane®) in Combination with Gemcitabine in Patients with Metastatic Breast Cancer. *Clinical Breast Cancer, 6*(4), 361-364.

Mura, S., Nicolas, J., & Couvreur, P. (2013). Stimuli–responsive nanocarriers for drug delivery. *Nature Materials, 12*(11), 991-1003. doi: 10.1038/NMAT3776

Nel, A. E., Madler, L., Velegol, D., et al. (2009). Understanding biophysicochemical interactions at the nano–bio interface. *Nature Materials, 8*(7), 543-557. doi: 10.1038/NMAT2442

Poland, C. A., Duffin, R., Kinloch, I. A., et al. (2008). Carbon nanotubes introduced into the abdominal cavity of mice show asbestos–like pathogenicity in a pilot study. *Nature Nanotechnology, 3*(7), 423-428.

Prabhakar, U., Maeda, H., Jain, R. K., et al. (2013). Challenges and key considerations of the enhanced permeability and retention effect for nanomedicine drug delivery in oncology. *Cancer Research, 73*(8), 2412–2417.

Provenzano, P. P., Cuevas, C., Chang, et al. (2012). Enzymatic targeting of the stroma ablates physical barriers to treatment of pancreatic ductal adenocarcinoma. *Cancer Cell, 21*(3), 418–429.

Rijcken, C. J. F., Veldhuis, T. F. J., Ramzi, A., et al. (2005). Novel fast degradable thermosensitive polymeric micelles based on PEG-block-poly(N-(2-hydroxyethyl)methacrylamide-oligolactates). *Biomacromolecules, 6*(4), 2343–2351.

Rose, P. G., Blessing, J. A., Lele, S., et al. (2006). Evaluation of PEGylated liposomal doxorubicin (Doxil) as second-line chemotherapy of squamous cell carcinoma of the cervix: a Phase II study of the Gynecologic Oncology Group. *Gynecologic Oncology, 102*(2), 210–213.

Rushton, E., Jiang, J., Leonard, S. S., et al. (2010). Concept of assessing nanoparticle hazards considering nanoparticle dosemetric and chemical/biological response metrics. *Journal of Toxicology and Environmental Health, 73*(5), 445–461.

Sahoo, S. K., Parveen, S., & Panda, J. J. (2007). The present and future of nanotechnology in human health care. *Nanomedicine: Nanotechnology, Biology and Medicine, 3*(1), 20–31.

Saxena, V., Johnson, C. G., Negussie, A. H., et al. (2015). Temperature-sensitive liposome-mediated delivery of thrombolytic agents. *International Journal of Hyperthermia, 31*(1), 67–73.

Schulte, P. A., Geraci, C. L., Hodson, L., et al. (2013). Overview of risk management for engineered nanomaterials. *Journal of Physics: Conference Series, 429*(1), 012062.

Senzer, N., Nemunaitis, J., Nemunaitis, D., et al. (2013). Phase I study of a systemically delivered p53 nanoparticle in advanced solid tumors. *Molecular Therapy, 21*(5), 1096–1103.

Sergio, M., Behzadi, H., Otto, A., et al. (2013). Fullerenes toxicity and electronic properties. *Environmental Chemistry Letters, 11*(2), 105–118. doi: 10.1007/s10311-012-0387-x

Shvedova, A. A., Kagan, V. E., & Fadeel, B. (2010). Close encounters of the small kind: adverse effects of man-made materials interfacing with the nano-cosmos of biological systems. *Annual Review of Pharmacology and Toxicology, 50*, 63–88.

Sparreboom, A., Scripture, C. D., Trieu, V., et al. (2005). Comparative preclinical and clinical pharmacokinetics of a cremophor-free, nanoparticle albumin-bound paclitaxel (ABI-007) and paclitaxel formulated in cremophor (taxol). *Clinical Cancer Research, 11*(11), 4136–4143. doi: 10.1158/1078-0432.ccr-04-2291

Sutton, G. P., Blessing, J. A., Hanjani, P., et al. (2005). Phase II evaluation of liposomal doxorubicin (Doxil) in recurrent or advanced leiomyosarcoma of the uterus: a Gynecologic Oncology Group study. *Gynecologic Oncology, 96*(3), 749–752.

Svenson, S., Wolfgang, M., Hwang, et al. (2011). Preclinical to clinical development of the novel camptothecin nanopharmaceutical CRLX101. *Journal of Controlled Release, 153*(1), 49–55.

Tannock, I. F., & Rotin, D. (1989). Acid pH in tumors and its potential for therapeutic exploitation. *Cancer Res, 49*(16), 4373–4384.

Tentler, J. J., Tan, A. C., Weekes, C. D., et al. (2012). Patient-derived tumour xenografts as models for oncology drug development. *Nature Reviews Clinical Oncology, 9*(6), 338–350. doi: 10.1038/nrclinonc.2012.61

Thiesen, B., & Jordan, A. (2008). Clinical applications of magnetic nanoparticles for hyperthermia. *International Journal of Hyperthermia, 24*(6), 467–474.

Torchilin, V. P. (2005). Recent advances with liposomes as pharmaceutical carriers. *Nature Reviews Drug Discovery, 4*(2), 145–160.

Trickler, W. J., Nagvekar, A. A., & Dash, A. K. (2008). A novel nanoparticle formulation for sustained paclitaxel delivery. *Aaps Pharmscitech, 9*(2), 486–493.

Valle, J. W., Armstrong, A., Newman, C., et al. (2011). A phase 2 study of SP1049C, doxorubicin in P-glycoprotein-targeting pluronics, in patients with advanced adenocarcinoma of the esophagus and gastroesophageal junction. *Investigational New Drugs, 29*(5), 1029–1037. doi: 10.1007/s10637-010-9399-1

Vasir, J. K., Reddy, M. K., & Labhasetwar, V. (2005). Nanosystems in drug targeting: opportunities and challenges. *Current Nanoscience, 1*(1), 47–64.

Voskoglou-Nomikos, T., Pater, J. L., & Seymour, L. (2003). Clinical predictive value of the in vitro cell line, human xenograft, and mouse allograft preclinical cancer models. *Clinical Cancer Research, 9*(11), 4227–4239.

Wagner, V., Dullaart, A., Bock, A., et al. (2006). The emerging nanomedicine landscape. *Nature Biotechnology, 24*(10), 1211–1217.

Weissig, V., Pettinger, T. K., & Murdock, N. (2014). Nanopharmaceuticals (part 1): products on the market. *International Journal of Nanomedicine, 9*(1), 4357–4373.

Wickham, T., & Futch, K. D. (2012). Abstract P5-18-09: A Phase I study of mm-302, a HER2-targeted liposomal doxorubicin, in patients with advanced, HER2- positive breast cancer. *Cancer Research, 72*.

Wicki, A., Witzigmann, D., Balasubramanian, V., et al. (2015). Nanomedicine in cancer therapy: Challenges, opportunities, and clinical applications. *Journal of Controlled Release, 200*, 138–157. doi: 10.1016/j.jconrel.2014.12.030

Ye, L., He, J., Hu, Z., et al. (2013). Antitumor effect and toxicity of Lipusu in rat ovarian cancer xenografts. *Food and Chemical Toxicology, 52*, 200–206.

Yeager, D., Chen, Y., Litovsky, S. H., et al. (2014). Intravascular Photoacoustics for image-guidance and temperature monitoring during plasmonic photothermal therapy of atherosclerotic plaques: a feasibility study. *Theranostics, 4*(1), 36–46.

Zhang, L., Gu, F. X., Chan, J., et al. (2008). Nanoparticles in Medicine: Therapeutic Applications and Developments. *Clinical Pharmacology & Therapeutics, 83*(5), 761–769.

Zhang, Q., Huang, X.-E., & Gao, L.-L. (2009). A clinical study on the premedication of paclitaxel liposome in the treatment of solid tumors. *Biomedicine & Pharmacotherapy, 63*(8), 603–607. doi: http://dx.doi.org/10.1016/j.biopha.2008.10.001

Zununi Vahed, S., Salehi, R., Davaran, S., et al. (2017). Liposome-based drug co-delivery systems in cancer cells. *Materials Science & Engineering. C, Materials for Biological Applications, 71*, 1327–1341. doi: 10.1016/j.msec.2016.11.073

2 Effect of Nanocarrier Size/Surface on Molecular Targeting in Cancer

Eupa Ray, Ankur Sharma, Kalpesh Vaghasiya, and Rahul Kumar Verma

CONTENTS

2.1 INTRODUCTION

The concept of the "magic bullet" proposed a century ago by the Nobel laureate Paul Ehrlich is turning into a reality with the approval of several nanoformulations for the treatment of cancer and other diseases. This approach has provided a platform for the development of various types of nanocarrier formulations, encapsulating therapeutics and delivering them to the site of action. Since, the first formulation in the form of liposome was proposed in 1974 by Gregoriadis et al., there has been tremendous increase in the number of nanocarrier formulations suitable for targeted drug delivery, which are either made of lipids or composed of polymers (Gregoriadis et al. 1974, Davis, Chen, and Shin 2008). These systems are exploited to incorporate drugs with poor biopharmaceutical profile or drugs associated with systemic toxicity issues. Nanocarrier systems help in protecting active constituents from degradation, augmenting bioavailability by enabling diffusion through the epithelium, enhancement of intracellular penetration and bio-distribution. Nanotechnology-based therapy encompasses elements with therapeutic entities, including proteins, peptides, DNA, RNA, small molecules and the constituents that are used to assemble the therapeutic entities, such as lipids and polymers. Such nanoparticles will have enhanced anticancer activity as compared to pure therapeutic entities (Davis, Chen, and Shin 2008).

Besides other physicochemical properties of the nanoparticles, their geometric parameters such as size and shape plays a significant role in the targeted cancer drug delivery. With precise control over size, shape and surface properties of nanoparticles, the circulation time and hence the pharmacokinetics and biodistribution of the encapsulated drugs can be modified. According to present thought, the diameter of nanoparticles should be in the range of 10–100 nm for optimum penetration and retention in cancerous tissues. The lower size limit is based on the measurements of sieving coefficients of the glomerular capillary-wall in the tumors. The typical transportation process in blood vasculature following the intravenous administration is shown in Figure 2.1.

Nanomedication is favored over conventional therapy because of its potential to facilitate the targeted delivery of drug to tumors owing enhanced permeability and retention (EPR) effect (Jain and Stylianopoulos 2010). The EPR effect helps the nanoformulation to concentrate in the desired tissue for an adequately longer time and facilitates the release of the drug in the vicinity of tissue. Experiments shown that, the nanoparticle with size <400 nm extravasates easily to the tumors (Peer et al. 2007). The pore cutoff size of the tumor depends on the site as well as tumor micro-environment, hence it is important to manipulate the size of the nanocarrier accordingly for enhanced efficacy of the drugs.

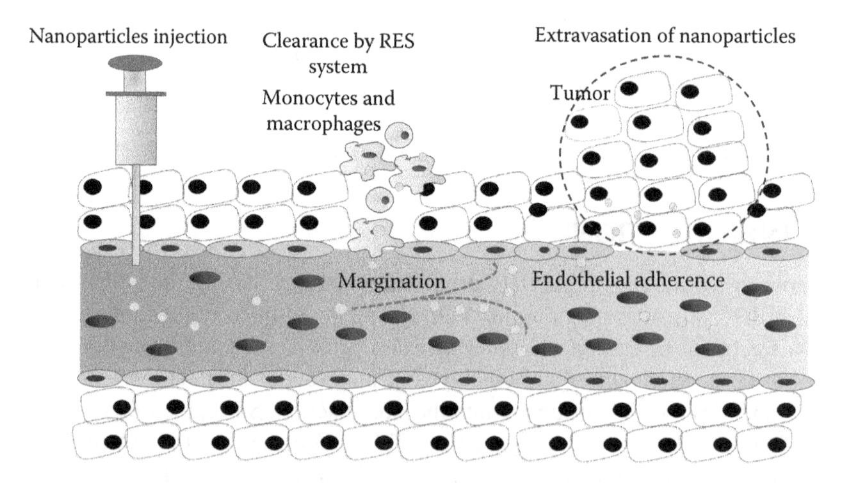

FIGURE 2.1 Transportation of nanoparticles in vasculature and various biological processes that influence its delivery.

2.2 PARTICLE SIZE IN TUMOR TARGETING

Particle size plays an important role in the vascular delivery of nanocarriers, which influences circulation half-life, attachment, rolling, adhesion phagocytosis, pharmacokinetics, biodistribution and tumor targeting (Champion, Walker, and Mitragotri 2008; Brown et al. 2010). Particles must be of adequately optimum size to carry a significant payload of bioactive material and be transported efficiently in the blood vessels. Several studies have been done to evaluate the effect of particle size on the transport, margination and adhesion of particles in a bloodstream inside the vessels (Champion, Walker, and Mitragotri 2008, Alexis et al. 2008, Vinogradov, Bronich, and Kabanov 2002, Owens and Peppas 2006). In cancer chemotherapy, targeting tumors with long circulating nanocarriers are a promising strategy. These nanomaterials accumulate in solid tumors through the Enhanced Permeation and Retention (EPR) effect (Matsumura and Maeda 1986) (Figure 2.2), which is characterized by leaky blood vessel and impaired lymphatic drainage in tumor tissues. This EPR in such tumors is solely due to the size of the nanocarriers. Generally, particles with diameters of 100–200 nm provides prolonged circulation half-lives, lesser RES uptake and are suitable for passive targeting of tumors by Enhanced Permeation and Retention (EPR) Effect (Champion, Walker, and Mitragotri 2008).

For intravenously administered nanoparticles, diameter is an important determinant of pharmacokinetics and biodistribution owing to the variable size of inter-endothelial pores lining the blood vessels. Nanoparticles with diameters smaller than 6 nm are quickly eliminated from the body as they can be excreted by glomerular filtration of kidneys (Choi et al. 2007). Nanoparticles with diameters larger than 200 nm accumulate in the spleen and liver where they are processed by the mononuclear phagocytic system (MPS) cells. To produce long-circulating nanoparticles that can accumulate inside tumor tissues a size-range of diameter between 30 nm and 200 nm is desired (Jain and Stylianopoulos 2010). Using this approach, researchers may induce passive accumulation of nanomaterials inside a tumor. In fact, it is possible to control the overall accumulation and penetration depth into the tumor by tuning the nanoparticle's diameter (Perrault et al. 2009). Researchers determined that the nanoparticle's capacity to navigate between the tumor interstitium after extravasation

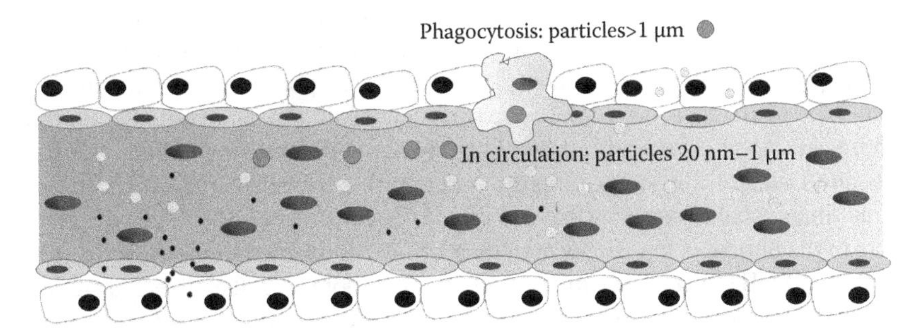

FIGURE 2.2 Impact of particle size on its transportation and its fate.

increased with decreasing size. In contrast, larger nanoparticles (>100 nm) do not extravasate far beyond the blood vessel because they remain trapped in the extracellular matrix between the cells. Thus, the smallest nanoparticles (~20 nm) penetrate deep into the tumor tissue, but they could not remain retained longer than 24 hours. Active targeting of nanoparticles change the intra-tissue localization of nanoparticles and enhance their internalization into cancerous cells (Choi, Alabi, et al. 2010, Lee et al. 2010). It is currently uncertain, how active targeting affects nanoparticle accumulation in the target tissue (Lee et al. 2010).

2.2.1 Tumor Type and Pore Cutoff Size

Previous quantitative studies have demonstrated that, the permeability of tumor vessels is usually higher than that of normal vessels (Jain 1987, Gerlowski and Jain 1986, Yuan et al. 1994). This increase in permeability is due to the defective basement membrane and fenestrated wall tissue. Further, it has been also well established that vascular permeability may depend on tumor type (Yuan et al. 1994, Sands et al. 1988), increase with tumor size (Zhang et al. 1992) and growth rate (Heuser and Miller 1986). It may be higher in the periphery than in the central region of tumors (Dvorak et al. 1988). However, the regulation of tumor vascular permeability is not well understood. It is generally accepted that the host-microenvironment plays a major role in determining the tumor morphology. There are inconsistencies in growth-rate, rate of angiogenesis, metastasis and the effectiveness of treatments among ectopic and orthotopic tumors (Fidler 1995, Fidler et al. 1994). Fukumura et al. showed that, the tumors grown in the liver had a significantly lower vessel density especially in the center, coincident with central necrosis, than the subcutaneous tumors (heterotopic tumor). The frequency distribution of vessel diameters of a liver tumor was slightly shifted towards smaller size compared to a subcutaneous tumor. The macromolecular vascular permeability in the liver tumor was significantly higher than in the subcutaneous tumor (Fukumura et al. 1997). In a similar study, Hobbs et al. investigated the practical limits of trans-vascular transport and its inflection by the microenvironment. In this study, five murine tumors comprising mammary carcinoma, colorectal carcinomas, hepatoma, glioma, and sarcoma were implanted in the dorsal skin-fold chamber and the pore cutoff size was calculated (Hobbs et al. 1998). The microenvironment of the tumor was moderated; spatially, by budding tumors in subcutaneous region; temporally, through induction of vascular regression in hormone-dependent tumors. Cancers grown subcutaneously showed a characteristic pore cutoff size ranging from 200 nm to 1.2 μm. This pore cutoff size was reduced in tumors grown after hormone withdrawal. Vessels induced in basic fibroblast growth-factor-containing gels had a pore cutoff size of 200 nm. Similarly, Yuan et al. performed studies to determine the maximum size of particles that can cross the tumor vessel wall. They did experiments in the human colon adenocarcinoma LS174T implanted in the dorsal skin fold chamber in SCID mice. To estimate the maximum pore size, they injected sterically stabilized liposomes of different sizes, and found that the maximum size of liposomes that can seep from the lumen of tumor vessels was in the range from 400 to 600 nm. The liposomes of 400 nm in diameter might enter into the tumor interstitium, while liposomes of 600 nm or more were excluded from extravascular space. The maximum pore size of the tumor vessel wall was assumed to be the maximum size of liposomes, which was around 500 nm (Yuan et al. 1995). These results may have major implications in the delivery of therapeutic agents in tumors. The delivery of nano-sized delivery systems may be reduced during tumor-regression induced by hormonal-ablation and permeability is independent of pore cutoff diameter providing the diameter of the delivery system is lesser than the pore diameter (Hobbs et al. 1998). In contrast to the above hypothesis of passive targeting due to hyper permeability of tumor tissue, Chauhan et al. reported that normalization of the tumor blood vessel leads to improvement in delivery of nanomedicines (Chauhan et al. 2012). This group demonstrated that, the elevated interstitial fluid pressure inside the tumors reduces the driving force for extravasation of fluid and macromolecules in tumors, which leads to a radially outward convection which opposes the inward diffusion (Jain and Baxter 1988). This phenomenon restricts the drug diffusion through the vessel-walls into tumors (Chauhan et al. 2011). Therapies against angiogenesis (Figure 2.1) may help to repair the tumor-vessel anomalies like large heterogeneous pores that enable leakiness by persuading vessel maturation (Tong et al. 2004, Nakahara et al. 2006). In order to determine the effect of vascular normalization on nanomedicine delivery, they studied the real-time delivery of particles ranging from 12 to 125 nm in diameter. Using the resulting data, they applied a novel physiologically based mathematical model of drug delivery to tumors and determined how anti-angiogenics affect pore size distributions. They also used this model to study how pore size distributions can be therapeutically modulated to optimize the delivery of different sizes of nanomedicines. Generally, the small size range nanoparticles (1–12 nm) show rapid tumor penetration as compared to larger particles (125–250 nm), which do not leave the vasculature to appreciable extent. The phenomenon of convection is dominant at small mean pore sizes. Increasing the mean pore size above a certain critical point (>140 nm) hinders nanoscale therapeutic delivery due to the rising interstitial fluid pressure (IFP). The combined effect of convection with increasing hydrodynamic and steric hindrance helps in transport of nanocarriers as pore sizes approach the therapeutic particle size. Hence, every size of nanocarrier has its own ideal mean pore size for optimal delivery into tumors. (Chauhan et al. 2012).

2.2.2 Circulation in Bloodstream

During intravascular movement, nanoparticles undergo various physiological processes including (i) margination to blood vessels, (ii) interaction with endothelial cells and (iii) cellular uptake. These processes are strongly influenced by various physicochemical properties of nanoparticles including the shape of the particles. Margination is the phenomenon which involves movement of particles in the bloodstream toward the vessel walls. Generally, nanoparticles loaded chemotherapeutic drugs are administered through systemic injection that circulates in the bloodstream before reaching the tumor site. In vascular targeting, nanocarriers are transported in bloodstream where they interact specifically with the endothelial cell-membrane by either non-selective van der Waals, electrostatic force or by ligand-receptor binding. It has been shown that the circulation and biodistribution of nanoparticles, like silica nanoparticles, silver nanoparticles, gold nanoparticles, quantum dots, and carbon nanotubes (CNT), are size-dependent after being systemically administered. For efficient tumor targeting, circulating nanoparticles in the bloodstream should evade macrophage uptake and spleen filtration (Jain 1987). The movement of nanoparticles in vessels includes motion, margination, adhesion and internalization. The size of particle plays a significant role in each of the phenomena and considerably influences particle clearance from the bloodstream and cellular uptake. The optimum size for the phagocytosis by macrophages is 1–5 µm and smaller particles are less expected to be taken up by macrophages (Yue et al. 2010). The particles of size between 20 nm–1 µm shows prolonged circulation with minimal clearance by macrophages (Choi, Liu, et al. 2010). Small nanoparticles (10–20 nm) offer good biodistribution after systemic administration by overpassing barrier membranes and tight endothelial junctions but are rapidly cleared after first pass glomerular filtration of the kidney (Choi et al. 2007). Drug delivery particles should be more than 20 nm in size to evade clearance through renal filtration. It has been found that particles of size ~10 µm (or more) tend to aggregate under physiological conditions and cause embolization in the blood vessels and several organs like liver, spleen and lungs. It is commonly accepted fact that nanoparticles of size between 20–200 nm demonstrate longer circulation half-lives and are appropriate for passive targeting to solid tumors through the EPR effect. Hence, nanocarriers for systemic administration should be larger than 20 nm to evade first pass filtration in the kidney and smaller than 200 nm to escape phagocytosis and sequestration reticuloendothelial system (RES). Decuzzi et al., studied the effect of particle size (nano- to micro-scale) on adhesion to cultured endothelial cells below a low shear flow setting (Figure 2.1). Spherical particles of nanoscale diameter (50, 100, 200, 500 and 750 nm) and micro-scale diameter (1, 6 and 10 µm) were studied. Results demonstrated that, movement of nanoparticles in the chamber was influenced by brownian motion while the transport of larger particles (microparticles) was guided by gravitation forces (Lee, Ferrari, and Decuzzi 2009).

In another study by Toy et al. who investigated the margination of liposomes (60–130 nm) in the wall of a microfluidic-chamber at low shear rates. They found that smaller liposomes marginates more rapidly than larger liposomes towards wall due rapid Brownian motion and greater diffusivity (Toy et al. 2013). The transport and targeting of nanocarriers to cancer tumors depends upon the size of tumor vasculature fenestrations, which are influenced by various factors like the type of tumor, aggravation of disease, stage of disease etc. Some vasodilator drugs that reduce hypertension cause vasodilation of tumor micro-vessels and hence facilitate the movement of slightly larger nanoparticles through them.

2.2.3 Nanoparticle Trafficking and Tumor Internalization

Tumors have immature and porous vasculature, which provides access to circulating nanoparticles (Matsumura and Maeda 1986, Maeda et al. 2003). Tumor accumulation is a function of (i) the rate of extravasation of the blood to the tumor space and (ii) the rate of clearance from tumor. The optimum particle size of gold nanoparticles, silica nanoparticles, single-walled carbon nanotubes and quantum dots for the cellular uptake and internalization in tumor cells was found to be 50 nm (Lu et al. 2009, Jin et al. 2009). Hobbs et al. showed that the rate of extravasation of bovine serum albumin (BSA) was independent of pore-size in a variety of tumor models. The results demonstrated that the extravasation of molecule of size ~7 nm, (much smaller than the trans-vascular pore) was independent of pore size but depended on the concentration gradient between blood and tumor (Hobbs et al. 1998). Several nanomedicines, including Doxil and Abraxane (diameters: 90 and 130 nm, respectively), showed significant antitumor activity in highly vascularized tumors such as Kaposi's sarcoma and breast cancer, and have been approved for clinical use (Northfelt et al. 1998, Gradishar et al. 2005). From the literature, nanomedicines in the sub-100 nm range are now regarded as being more important in the targeted delivery of drugs to the less permeable tumors (Dreher et al. 2006, Perrault et al. 2009). In support of the above results, Cabral et al. showed size-dependent penetration and efficacy of the nanocarrier formulation in poorly permeable pancreatic tumors. They examined the penetration and accumulation of a series of micellar nanocarriers with diameters less than 100 nm that encapsulated the tumurocidal agent 1,2-diaminocyclohexane-platinum(II) (DACHPt) (Cabral et al. 2011). In hypervascular tumors of highly permeable structure, sub-100 nm micellar nanomedicines showed no size-dependent restrictions on extravasation and penetration in tumors. It was observed that the nanomicelles that were smaller than 50 nm were only able to penetrate poorly permeable hypovascular tumors. Furthermore, increasing the permeability of hypovascular tumors using TGF-β signaling inhibitor improved the accumulation and distribution of the larger ~70 nm micelles, offering a new way to enhance the efficacy of larger nanomedicines. Perrault et al. systematically examined how particle design can be optimized for efficient tumor targeting. They measured size-dependent nanoparticle accumulation in tumors and correlated this phenomenon with blood pharmacokinetics. They demonstrated that the particle-design has significant importance for tumor targeting, which help in controlling delivery to tumors. It was observed that the blood half-life of the ~60 nm nanoparticles was 6.5 times longer than those of ~20 nm particles, owing to its mPEG layer. Half-lives of the nanocarriers can be improved by application of larger mPEG layers, but this also increases hydrodynamic diameter, which could influence their ability to extravagate and permeate into tumors (Perrault et al. 2009). From a particle design viewpoint, small hydrodynamic diameters and long half-lives appear to converge when particle size is below 50 nm and particles are protected with PEG layer of moderate molecular weight (MW 5kDa). The accumulation of particles and molecules in the tumor strives against uptake and removal by mononuclear phagocytic system (MPS) cells. Small particles (~20 nm) were cleared swiftly from the blood without a corresponding gathering in the liver and spleen. They are also unlikely to be cleared via the kidney, which has a hydrodynamic diameter cutoff of approximately 6 nm. These nanoparticles could have accumulated in lymphatic cells, subsequent extravasation into nonspecific tissues. A comprehensive study of how particle design influences nonspecific bio-distribution would be imperative and values further investigation.

2.2.4 Cellular Uptake and Cell-Particle Interactions

Following margination towards endothelial cells, a particle may adhere to the target cell, and then internalized to release therapeutic agents in the cytosol. In cancer therapy, the particles need to avoid uptake by macrophages in circulation whereas rendering improved uptake by diseased cells. In a study, the cellular uptake of spherical polystyrene particles with a size range (40 nm–15 µm) in dendritic cells were investigated. The results showed that the cellular uptake was more with smaller particles. The rate of internalization of particles was influenced by particle size, which further influenced subcellular distribution and their interaction with receptors and organelles (Foged et al. 2005). Cellular uptake of particles follows several mechanisms of internalization including phagocytosis, pinocytosis, clathrin-mediated endocytosis, caveolin-mediated endocytosis and clathrin/caveolin independent endocytosis. The pathway mechanism of cellular-internalization is subjective to the particle size of the nanocarrier. The internalization of smaller nanoparticles with a diameter lesser than 200 nm predominantly involves receptor-mediated endocytosis or clathrin-coated pits. This process requires small degree of rearrangement of the cell cytoskeleton. In particles with a size range of 200–500 nm, the mechanism of internalization shift to caveolae-mediated cellular uptake. Micron-sized particles (more than 1µm) are internalized via the phagocytosis process that needs a comprehensive rearrangement of the cell cytoskeleton. Particles with size range 500 nm to 1 µm are internalized through macro-pinocytosis of a mixed mode (Rejman et al. 2004). The particle size also influences its distribution inside the cells and organelles. It has been shown that larger quantum dots (~5.2 nm) were distributed all over the cytosol but do not enters the nucleoplasm whereas smaller one (~2.2 nm) effortlessly enters into nucleus (Lovric et al. 2005).

2.3 PARTICLE SHAPE ON TUMOR TARGETING

Beside the control over the particle-size, the shape of the nanoparticles plays a significant role in biological functions. Advanced fabrication approaches have allowed the construction of wide-ranging shapes of nanomaterials with great precision. These shapes include polyhedral, rods, ellipsoids, discs, cubes, urchins, dendrites, cylinders, hemispheres, cones, chains, binconcave discoids etc. Different shapes of nanoparticles show significant effect on their transport through the vasculature, their circulation half-lives, targeting proficiency, cellular uptake and following intracellular targeting. Several research reports on the significance of nanoparticle shape on various biological processes have driven attention toward studying and improving nanoparticle geometry for optimal targeting (Truong et al. 2015).

2.3.1 Circulation in Bloodstream

Reduction in the phagocytic clearance of nano drug delivery systems enhance blood circulation time in the vasculature (Yoo, Doshi, and Mitragotri 2010). Although, spherical structure is the most common geometry in use, asymmetrical shapes have also been found to be beneficial in several cases to enhance circulation time in the bloodstream with reduced collisions against the walls of the blood vessels (Muro et al. 2008). In different studies, the ability of nanomaterials of different aspect ratios was evaluated for their tendency to accumulate in the tumors. Asymmetrically shaped nanocarriers demonstrated a diverging hydrodynamic behaviour compared to the spheres that leads to align them with the blood flow, which further enhances circulation time. Nanoconstructs with a minimum of one stretched axis, such as oblate, ellipsoids, cones, binconcave, discoids etc. are found to be more appropriate than spherical nanoconstructs for nanomedicine applications. These systems have a positive orientation with the bloodstream and are less prone to phagocytic clearance by the reticuloendothelial systems (RES), which make them long circulation drug delivery systems (Sharma et al. 2010). Non-spherical nano drug delivery systems show diverse drug distribution profiles compared to their spherical equivalents, by providing a different mode for targeting specific organs and tissues, including liver, lung, breast, spleen and tumors. It was observed that rod-shaped micelles have a much longer circulation lifetime than that of spherical micelles, which was up to one week after IV injection (Geng et al. 2007) (Figure 2.3). The extent of phagocytosis of gold nanorods was found to be lesser than the spherical counterparts. This lead to lesser deposition of gold-rods in the liver, hence prolonged circulation time in the blood circulation and higher concentration in cancerous tumors (Dreaden et al. 2012). Studies demonstrated that the single-walled carbon nanotubes with an aspect ratio of 100:1 to 500:1 cleared effectively by glomerular filtration of kidney, which indicated that elongated nanocarriers could be eliminated by filtration process (Ruggiero et al. 2010). Rod-shaped (44 nm) nanomaterials were found to be more efficiently accumulated in tumors compared to nanospheres of similar size (35 nm). Both types of nanomaterials showed almost similar blood circulation kinetics but diffusion of rods into tumors was much faster and deeper (Durr et al. 2007). Similarly, Filo-micelles loaded with paclitaxel showed longer circulation time compared with spherical micelles, which lead to greater accumulation in tumors and higher efficacy and minimal non-specific accumulation (Geng et al. 2007). Elongated Dextran-coated nanochains (~50 nm) and spherical nanoparticles were evaluated for tumor targeting. Both the delivery systems showed significant accumulation into the tumor cells by extravasation, but it was observed that nanospheres reappeared in blood circulation after some time while nanochains were well retained inside the tumors

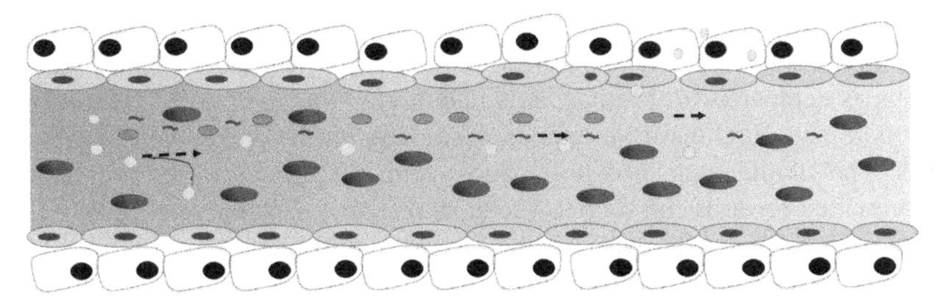

FIGURE 2.3 Effect of nanoparticle shape on its circulation and extravasation.

(Park et al. 2008). In another study, elongated viral-nanofilaments which present favorable EPR properties demonstrated augmented accumulation in tumors compared to the spherical viral nano construct (Wen and Steinmetz 2016).

2.3.2 Nanoparticle Transportation and Tumor Internalization

The effect of shape was shown to play an important role in the margination of nanoparticles towards the blood-vessel walls. Following the margination to the vicinity of blood vessels, the endothelial cell-interaction and eventually cellular internalization, were considerably dependent on the shape of the nanocarriers (Gentile et al. 2008). In contrast to spherical particles, non-spherical particles were exposed to more complex torques leading to tumbling and rotation that caused translational and rotational motions. As compared to nanospheres, elongated and rod shaped particles exhibited more contact surface area that lower the drag in blood vessels (Lee, Ferrari, and Decuzzi 2009). Tumbling of non-spherical particles in bloodstream help in adhesion and margination towards the endothelial walls. Hence, they were more likely to adhere to endothelial cells. These particles were exposed to larger areas of the endothelial cell-surface as compared to nanospheres which lead to strong adhesion. Discoid nanoparticles showed more lateral drift towards blood vessel walls and could exodus main vessels more easily as compared to conventional spherical particles.

2.3.3 Cellular Uptake and Cell-Particle Interactions

Recent studies showed that the shape of nanoparticles could affect the ability of cellular uptake by independently influencing attachment and phagocytosis processes. The shape and local geometry of nanoparticles at the interaction point between particle and cell-surface decides whether macrophages will uptake the particle or not. A wide range of nanocarriers in non-spherical forms competently internalized into cells after crossing the cell membranes. Several studies have showed the highest cellular uptake of nanorods, followed by spheres, cylinders and then cubes. Asymmetrical nanostructure may offer added level of control in presenting ligands towards targeted receptors. The longer axis interacts with more surface receptors on cells as compared to the short axis (Chithrani and Chan 2007). For example, spiky gold nano-urchins have ligands positioned on the spikes and between the spikes. The cellular uptake of this nanosystem was influenced by the way as it was presented towards the targeted cellular receptors (Hutter et al. 2010). In the case of ellipsoid particles, when the cell attached with the flat part of the particle there was lesser chance of internalization whereas, its attachment via the sharp part increased the probability of cellular uptake. The prolate-spheroids showed good attachment with the cellular surface but demonstrated lower cellular uptake, while oblate-spheroids displayed good phagocytosis after attachment (Champion, Walker, and Mitragotri 2008). These results revealed that both type of particles attach, internalize and accumulate with different orientations. Nanoparticles with a high aspect ratio had higher probability of attaching to the macrophage surface but lesser probability of being engulfed by the cells. Nanochains or wormlike particles exhibited a high aspect ratio and demonstrated reduced cellular uptake as compared with spherical counterparts (Geng et al. 2007). These particles were poorly taken by macrophages *in vivo* due to hydrodynamic shearing and showed extended half lives in the blood circulation. Another study showed that the phagocytosis of polystyrene ellipsoid nanoparticles in mouse macrophages cell lines was significantly lesser than nanospheres (Paul et al. 2013). Similarly, the uptake of rod-shaped nanoparticles developed by layer-by-layer technology into HeLa cells was lower as compared with spherical capsules (Shimoni et al. 2013). Wooley et al. demonstrated that spherical polymeric nanoparticles were more efficient in cellular uptake in Chinese Hamster Ovary (CHO) cells as compared to cylindrical nanoparticles. After functionalization with targeting moiety TAT-HIV1, spherical nanoparticles showed enhanced cellular uptake but there was insignificant change in the internalization of cylindrical particles (Zhang et al. 2008). Short nanorods were efficiently taken up by cells compared to long nanorods. Cellular uptake and intracellular trafficking of uncommon geometry such as disks, lamellae etc. were also studied. Staggered lamellae nanostructures showed prolonged blood circulation as well as fast cellular uptake by clathrin- or caveolae-independent endocytosis. Large-sized hydrogel nanodisks were found to be internalized into endothelial cells more competently than similar small-sized nanomaterials (Hu et al. 2013). Iron oxide nanorods and nanochains showed margination of seven-fold and 2.3-fold respectively as compared to nanospheres in same flow conditions. In another study, mesoporous rod and disc-like-particles were compared for margination in blood vessels. When the width of both particles was made constant, then nanodiscs marginated twice as compared to nanorods. Dissimilarities in adhesion capabilities of studied particles were probably one of the reasons for the different margination rates. Largely, the available research work showed that non-spherical nanostructures undergo cellular uptake more competently; nevertheless, the results are not consistent. With currently available contradictory evidence, it would be too early to draw conclusions on the impact of shape on the cellular uptake of nanoparticles. More comprehensive experimental work is required to assess the impact of the shape of nanoparticles on margination, cellular uptake and internalization. In addition, the effect of cell types and entry pathways also need to be elaborately studied.

2.4 INFLUENCE OF NANOCARRIER SURFACE PROPERTIES ON TUMOR TARGETING

The surface characteristics of the nanoparticle, such as surface charge, hydrophobicity influences the circulation life in blood vessels, tumor distribution, cellular interaction and uptake (normal/via opsonisation) and organelle targeting. The surface charge of nanoparticles is usually characterized by the zeta potential of a nanoparticle, which shows the electrical potential of particles and depends upon the solution in which they are distributed. For stable nanoformulations, the zeta potential over the particles should be above ± 30 mV, as the surface potential avoids aggregation of the nanoparticles. Surface properties significantly impact the opsonisation process that eventually commands immune response. Several surface modifications were applied to evade phagocytosis and to dodge the initiation of the immune response.

Functionalization of the particle surface with specific ligands help in enhanced cellular and organelle targeting.

2.4.1 Circulation in Bloodstream

Nanoparticle hydrophobicity, surface functional groups and surface charge are important parameters that regulate the level of blood components that attach to their surface, hence impacts the *in vivo* fate of nanoparticles. These properties of nanoparticles affect the opsonisation (binding of blood proteins), RES uptake and clearance. In order to increase the probability of drug targeting in cancer therapy, it is essential to reduce the opsonization and extend the circulation of nanoparticles *in vivo*. It is well recognized that positively charged nanocarriers have a higher degree of cellular uptake as compared to negative or neutral particles (Thorek and Tsourkas 2008, Slowing, Trewyn, and Lin 2006). It has been suggested that the cell membrane possesses a slight negative charge and cell uptake is driven by electrostatic attractions (Jin et al. 2009, Wang et al. 2010). A recent study demonstrated that this electrostatic attraction between membrane and positively charged nanoparticles favors adhesion onto the cell's surface, leading to cellular uptake. It has been shown that polystyrene particles functionalized with positively charged primary amine underwent considerably better phagocytosis compared to particles functionalized with other groups like hydroxyl, carboxyl or sulfate groups. Positively charged nanocarrier surfaces predictably have a high cellular uptake frequency but small blood circulation half-life, whereas negatively charged nanosystems shown a noticeable reduction in the rate of cellular uptake and elongated circulation time. Investigations showed that, fairly negative charged nanocarriers could diminish the undesirable clearance by RES and may help in delivering drugs effectively to the tumor sites. It is recognized that neutral or negatively charged nanoparticles showed less plasma protein adsorption and a slower rate of cellular uptake (Figure 2.4). The surface hydrophobicity of nanoparticles plays an important role in opsonisation and phagocytic uptake. Nanoparticles with hydrophobic surfaces are readily opsonised with plasma proteins and then favorably taken up by RES systems. Protein adsorption makes the particle surface more

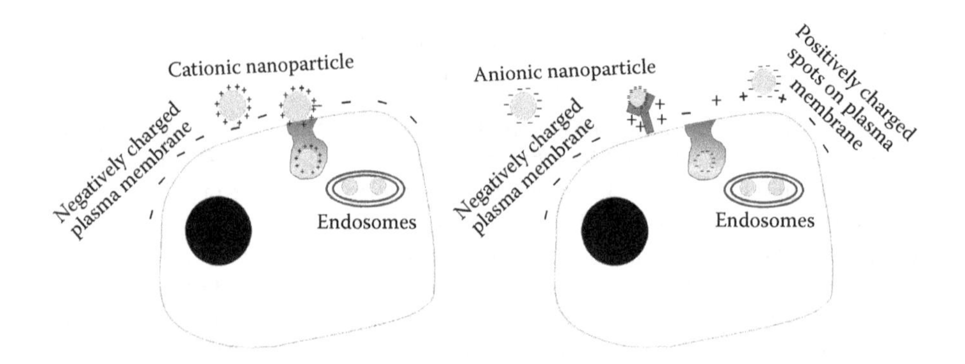

FIGURE 2.4 Effect of surface charge on nanoparticle uptake in cells.

hydrophobic. To enhance the possibility of tumor targeting, it is crucial to decrease the opsonisation phenomenon to extend the circulation. This can be attained by coating particles with hydrophilic polymers (or surfactants) like polyethylene glycol (PEG), poloxamer, poloxamine, and polysorbate 80 (Tween 80), polyethylene oxide etc. PEG showed substantial reduction in interactions with plasma proteins due its hydrophilicity and steric repulsion, hence decreasing opsonization and phagocytosis.

2.4.2 Nanoparticle Transportation and Tumor Internalization

Subsequently to the margination in blood vasculature, nanoparticle attachment with the endothelial wall and its internalization into tumor cells depends upon surface properties. The adherence of the nanoparticles to the cell surface is significantly influenced by the surface charge of nanoparticles. In the adhesion of nanoparticles with endothelial membrane, internalization of particles occurs through various mechanisms of endocytosis like clathrin, caveoli mediated, and clathin-caveoli independent pathways. Different endocytic mechanisms involved in the internalization of nanoparticles with a dissimilar surface charge and charge density. Positively charged nanoparticles, show higher rates of nonspecific cellular uptake through electrostatic interactions with negatively charged cellular membranes. In a study done by Thurston et al., it was found that positively charged liposomes were specially adhered and internalized by angiogenic endothelial cells of tumors compared with normal vasculature (Thurston et al. 1998). Noteworthy, positively charged nanoparticles escape from endosomes via 'proton sponge effect, minimizing the detrimental effects by lysosomes on drug. Negatively charged nanoparticles were efficient in transportation inside the blood vessels and delivering drugs into deep tumor tissues, while positively charge nanoparticles demonstrated higher cellular uptake. Ideally the nanocarrier for tumor targeting should have a neutral or slightly negative charge during systemic administration of formulation, which switches to a positive charge as it reaches to vicinity of tumor. Stimulus-guided, charge-switchable zwitter-ionic nanoparticles offer promise for an ideal drug targeting and maximum accumulation in tumors (Yuan et al. 2012). Hydrophobic particles are capable of moving by simple diffusion through the lipid bilayer membrane of cells. The particle surface can be improved to enhance interactions and cellular uptake. Hydrophilic PEG functionalized nanoparticles enhance circulation time but reduce the cellular uptake. Hence, nanoparticles necessitate a balance between hydrophilicity and hydrophobicity to sustain a movement in the blood and also achieve optimum cellular uptake. For example, Janus nanoparticles having hydrophobic/hydrophilic faces were developed by Bishop and coworker; they could penetrate the lipid bilayer, suggesting these are appropriate for drug delivery (Lee et al. 2013).

2.4.3 Surface Functionalization with Targeting Tumors

Many drug-loaded targeted delivery systems have shown promise in *in-vitro*, where significant increases in cellular uptake along with improved cytotoxicity was observed when compared to their non-targeted counterparts. Surface modification of nanocarriers might be helpful in achieving desirable properties and might make them suitable for controlled drug delivery and cellular targeting. In recent times, there has been a lot of research in active targeting of the tumor cells with ligand decorated nanocarriers to achieve enhanced efficacy and for mitigating the systemic side effects. Active targeting can be achieved by functionalization with surface-modifiers like stimuli sensitive polymer/lipids, soluble polymer and selective ligands (peptides, antibodies, folate, antibody, transferrin, lectins, epidermal growth factor [EGF], lipoprotein receptors, platelet-derived growth factor [PDGF]) (Holgado et al. 2012) (Figure 2.5). Surface modification provides functionality to nanocarriers for targeted drug delivery. Ligands attached on the nanocarrier surface may be any molecules that could selectively interact with the receptors of target cells. Ligand-coated nanocarriers can efficiently penetrate and be internalized by tumor cells by receptor recognition and interaction. The nature of the ligand used to coat the nanomaterial could also affect downstream biological responses. Further, the cellular uptake and cytotoxicity of nanoparticles could be significantly altered when the nanoparticles are coated with ligands targeting the receptor. Once bound to their receptor, nanoparticles will typically enter the cell via receptor-mediated endocytosis. The binding of ligand functionalized nanoparticles with the receptors cause reduction in the Gibbs free energy, which cause cell membrane to cloak over the nanoparticle to form a vesicle. Zhang et al., investigated and presented a detailed thermodynamic model for receptor-mediated endocytosis of ligand-coated nanoparticles. They identified an optimal radius of a ligand-coated nanocarrier at which the receptor mediated internalization reaches a maximum at physiologically relevant parameters and showed that the cellular uptake of nanoparticles is regulated by membrane tension and can be elaborately controlled by particle size. They concluded that the optimal nanoparticle radius for endocytosis is on the order of 25–30 nm, which is in good agreement with prior estimates (Zhang et al. 2009). Depending upon the surface and matrix properties of the nanoparticle, vesicles either fuse with lysosome for degradation of cargo or remain in cytoplasm. Larger ligand-coated nanoparticles (100–300 nm) do interaction with greater number of cell receptors to encourage cellular uptake as compared with smaller nanoparticles (5–10 nm). Thermodynamically, 40–60 nm is the optimum size of nanoparticles that were able to bind an adequate number of receptors to effectively form vesicles for internalization. Since different cells possess

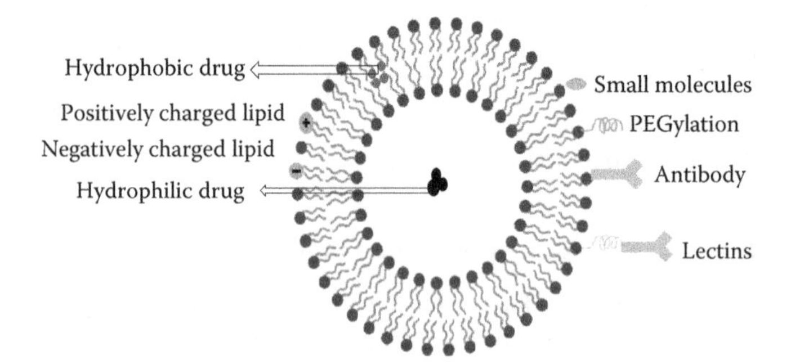

FIGURE 2.5 Surface modification of nanoparticles.

diverse and unique phenotypes, optimal nanoparticle uptake size may depend on the cell used in the study as each cell type expresses different levels of surface receptors. The different ligands that are frequently used in the cancer targeting therapies include antibodies, antibody fragments, peptides, proteins, siRNA, small molecules (folates, estrogen) and aptamers. This provides active targeting that offers extremely specific interactions between ligands and tissues and cells to stimulate the accumulation of nanoparticles in tumors.

Nanoparticles conjugated with an antibody against a specific tumor antigen have been developed for specific drug delivery. Monoclonal antibodies destroy cancer cells after binding to the specific antigens on cancer cells by various approaches including apoptosis induction, autophagy, obstructing growth factor receptors, activation of complement system etc. HER2 receptors, members of EGFR family, are overexpressed in several cancers including breast cancer. Trastuzumab (a HER2 receptor specific antibody) was conjugated to the surface of gelatin nanoparticles for targeting HER2-overexpressing cells (breast cancer cell line MCF7) (Wartlick et al. 2004). Rituximab (Fas agonist CH-11), a monoclonal antibody to the Fas receptor, conjugated with lipid nanoparticles demonstrated targeted complement-dependent cytotoxicity (CDC) and antibody-dependent cell-mediated cytotoxicity (ADCC) in mantle-cell lymphoma cell lines (Z138 and JVM2). This has been clinically approved for the treatment of non-Hodgkin's lymphoma. Several antibodies are FDA-approved in clinical applications include Avastin, which targets VEGF for the management of colorectal, lung, and breast cancer; Herceptin targeting, HER-2-overexpressing breast tumor cells); and Erbitux which binds to overexpressed epidermal growth factor receptor (EGFR) in colorectal cancer cells. The enhanced *in vivo* cell uptake and cell membrane binding of targeted micelles were found to delay their clearance from tumors that were overexpressing EGFR.

Aptamers are single-stranded DNA and RNA oligonucleotides/peptides that are capable of binding to particular target proteins. Drug encapsulated PLA- nanoparticles conjugated with an aptamer-targeting Prostate Specific Membrane Antigen (PSMA) selectively target biomarkers on prostate cancer cells. The attached PEG ensures enhanced circulation half-life of the nanoparticles and aptamer enhanced cellular uptake. Doxorubicin-loaded super paramagnetic iron oxide nanoparticles (SPION) conjugated with RNA aptamer were utilized to deliver drugs into prostate cancer cell lines. The resulting nanoparticles demonstrated enhanced selective killing efficiency for cancer cells compared to that of normal cells (Wang et al. 2008).

Folate is one of the most broadly used ligands for targeted drug delivery for cancer therapy. The folate receptors are overexpressed in several types of tumors such as ovarian cancer, uterine sarcomas, osteosarcomas, non-Hodgkin's lymphomas etc., which have a high binding avidity with folate. Nanocarriers conjugated with folate/folic acid bind to folate receptors and are internalized into the cell and internalized to the cytoplasm. Numerous studies have been done in devising and evaluating folate-conjugated liposomes, polymer conjugates, lipid nanoparticles and polymeric micelles for selective targeting. PLGA-PEG nanoparticles conjugated with folate offer improved cellular uptake and cytotoxicity of drug in MCF-7 breast cancer cells, and SKOV-3 ovarian cancer (Chen et al. 2011). Xiang and coworkers synthesized a folate derivative, folate-polyethylene glycol-cholesterol hemi-succinate (F-PEG-CHEMS), and investigated its folate-receptor targeting potential. F-PEG-CHEMS-conjugated liposomes displayed extraordinary targeting capabilities (Xiang et al. 2008).

Transferrin is another membrane glycoprotein that has been frequently utilized for tumor targeting applications. The transferrin receptor (TfR) is overexpressed to the magnitude of 10-fold on tumor cells compared to normal cells, which makes it promising choice for targeted delivery of bioactive material through nanocarrier systems. In a study, transferrin conjugated paclitaxel loaded PLGA-PVA nanoparticles showed three times more cellular uptake as compared to non-transferrin conjugated nanoparticles, consequently demonstrated better efficacy (Sahoo, Ma, and Labhasetwar 2004). There are several hormone receptors which are overexpressed in various type of cancers, are utilized as ligands for cancer targeting. Luteinizing hormone-releasing hormone (LHRH) is one of such targeting moiety against LHRH receptors, which is overexpressed on the surface of various cancer cells like breast cancer, ovarian cancer and prostate cancer. Song and group showed selective targeting using LHRH-decorated liposomes for enhanced cancer therapy. There are myriad studies to show ligand-based drug delivery systems for cancer therapy (He, Zhang, and Song 2010). Dendrimers, in this regards serve as potential nanocarriers for site-specific drug delivery (Duncan and Izzo 2005). Studies exploring PPI (DAB) dendrimers for the delivery of a ^{32}P-labeled triplex-forming oligonucleotide (ODN) showed that dendrimers could enhance uptake by MDA-MB-231 breast cancer cells by ~14 fold (Santhakumaran, Thomas, and Thomas 2004). Enhancement of cellular uptake was related to the concentration and dendrimer generation. The generation-4 dendrimer showed the most pronounced enhancement of cellular uptake.

2.5 CONCLUSION

Targeting is a multiphasic process that involves the regulation of specific particle design parameters for greater efficiency. The size, shape and surface properties of the nanoparticle contributes significantly to its transportation in the bloodstream, margination tendency, circulation half-life, adherence, tumor tissue and cellular uptake. The cellular uptake and cellular trafficking of nanoparticles were also influenced by their physicochemical properties. The engineering of these parameters in the nanoparticles significantly enhances the tumor targeting, efficacy of anticancer agents and reduces adverse effects on normal cells. Nanoparticles having size lower than 20 nm are excreted in the urine and those approximately 50–100 nm in size often exhibit optimal cellular uptake and intratumoral accumulation. Relatively large-sized nanoparticles, favorably accumulate at solid tumors because of enhanced permeability and retention effect. The shape of particles plays significant role in their *in vivo* behaviour and interaction with cells. Non-spherical nanoparticles show better properties than spherical particles in enhancing circulation time, endothelial adhesion, evading phagocytosis and cellular targeting. Oblate and elongated nanoparticles can be utilized to target tumors for enhanced accumulation in target tumor tissues compared to the spherical particles. Neutral or slightly negative-charged nanoparticles demonstrate better residence in blood vasculature and positive-charged particles better internalization into tumor cells. Polymer-functionalized (PEGylated) nanoparticles show long circulation times and can escape phagocytosis by RES and blood clearance. This chapter focuses on the complex relationships between physiological properties and biological response, during the optimization process to achieve optimal results. The basic knowledge provided here gives insight for the rationale design of tumor targeting nanoparticles.

REFERENCES

Alexis, F., E. Pridgen, L. K. Molnar, et al. 2008. "Factors affecting the clearance and biodistribution of polymeric nanoparticles." *Mol Pharm* 5(4):505–515. doi: 10.1021/mp800051m.

Brown, S. C., M. Palazuelos, P. Sharma, et al. 2010. "Nanoparticle characterization for cancer nanotechnology and other biological applications." *Methods Mol Biol* 624:39–65. doi: 10.1007/978-1-60761-609-2_4.

Cabral, H., Y. Matsumoto, K. Mizuno, et al. 2011. "Accumulation of sub-100 nm polymeric micelles in poorly permeable tumours depends on size." *Nat Nanotechnol* 6(12):815–823.

Champion, J. A., A. Walker, and S. Mitragotri. 2008. "Role of particle size in phagocytosis of polymeric microspheres." *Pharm Res* 25 (8):1815–1821. doi: 10.1007/s11095-008-9562-y.

Chauhan, V. P., T. Stylianopoulos, Y. Boucher, et al. 2011. "Delivery of molecular and nanoscale medicine to tumors: transport barriers and strategies." *Annu Rev Chem Biomol Eng* 2:281–298.

Chauhan, V. P., T. Stylianopoulos, J. D. Martin, et al. 2012. "Normalization of tumour blood vessels improves the delivery of nanomedicines in a size-dependent manner." *Nat Nanotechnol* 7(6):383–388.

Chen, J., S. Li, Q. Shen, et al. 2011. "Enhanced cellular uptake of folic acid-conjugated PLGA-PEG nanoparticles loaded with vincristine sulfate in human breast cancer." *Drug Dev Ind Pharm.* 37(11):1339–1346. doi: 10.3109/03639045.2011.575162.

Chithrani, B. D., and W. C. W. Chan. 2007. "Elucidating the mechanism of cellular uptake and removal of protein-coated gold nanoparticles of different sizes and shapes. *Nano Lett.* 7(6):1542–1550.

Choi, C. H. J., C. A. Alabi, P. Webster, et al. 2010. "Mechanism of active targeting in solid tumors with transferrin-containing gold nanoparticles." *Proc Natl Acad Sci U S A* 107(3):1235–1240.

Choi, H. S., W. Liu, F. Liu, et al. 2010. "Design considerations for tumour-targeted nanoparticles." *Nat Nanotechnol* 5(1):42–47. doi: 10.1038/nnano.2009.314.

Choi, H. S., W. Liu, P. Misra, et al. 2007. "Renal clearance of quantum dots." *Nat Biotechnol* 25(10):1165–1170.

Davis, M. E., Z. G. Chen, and D. M. Shin. 2008. "Nanoparticle therapeutics: an emerging treatment modality for cancer." *Nat Rev Drug Discov* 7(9):771–782. doi: 10.1038/nrd2614.

Dreaden, E. C., L. A. Austin, M. A. Mackey, et al. 2012. "Size matters: gold nanoparticles in targeted cancer drug delivery." *Ther Deliv* 3(4):457–478.

Dreher, M. R., W. Liu, C. R. Michelich, et al. 2006. "Tumor vascular permeability, accumulation, and penetration of macromolecular drug carriers." *J Natl Cancer Inst* 98(5):335–344.

Duncan, R., and L. Izzo. 2005. "Dendrimer biocompatibility and toxicity." *Adv Drug Deliv Rev* 57 (15):2215–2237.

Durr, N. J., T. Larson, D. K. Smith, et al. 2007. "Two-photon luminescence imaging of cancer cells using molecularly targeted gold nanorods." *Nano Lett* 7(4):941-5. doi: 10.1021/nl062962v.

Dvorak, H. F., J. A. Nagy, J. T. Dvorak, et al. 1988. "Identification and characterization of the blood vessels of solid tumors that are leaky to circulating macromolecules." *Am J Pathol* 133 (1):95.

Fidler, I. J. 1995. "Modulation of the organ microenvironment for treatment of cancer metastasis." *J Natl Cancer Inst* 87(21):1588–1592.

Fidler, I. J, C. Wilmanns, A. Staroselsky, et al. 1994. "Modulation of tumor cell response to chemotherapy by the organ environment." *Cancer Metastasis Rev* 13(2):209–222.

Foged, C., B. Brodin, S. Frokjaer, et al. 2005. "Particle size and surface charge affect particle uptake by human dendritic cells in an in vitro model." *Int J Pharm* 298(2):315–322. doi: 10.1016/j.ijpharm.2005.03.035.

Fukumura, D., F. Yuan, W. L. Monsky, et al. 1997. "Effect of host microenvironment on the microcirculation of human colon adenocarcinoma." *Am J Pathol* 151(3):679.

Geng, Y., P. Dalhaimer, S. Cai, et al. 2007. "Shape effects of filaments versus spherical particles in flow and drug delivery." *Nat Nanotechnol* 2(4):249–255.

Gentile, F., C. Chiappini, D. Fine, et al. 2008. "The effect of shape on the margination dynamics of non-neutrally buoyant particles in two-dimensional shear flows." *J Biomech* 41(10):2312–2318. doi: 10.1016/j.jbiomech.2008.03.021.

Gerlowski, L. E., and R. K. Jain. 1986. "Microvascular permeability of normal and neoplastic tissues." *Microvasc Res* 31(3):288–305.

Gradishar, W. J., S. Tjulandin, N. Davidson, et al. 2005. "Phase III trial of nanoparticle albumin-bound paclitaxel compared with polyethylated castor oil–based paclitaxel in women with breast cancer." *J Clin Oncol* 23(31):7794–7803.

Gregoriadis, G., E. J. Wills, C. P. Swain, et al. 1974. "Drug-carrier potential of liposomes in cancer chemotherapy." *Lancet* 1(7870):1313–1316.

He, Y., L. Zhang, and C. Song. 2010. "Luteinizing hormone-releasing hormone receptor-mediated delivery of mitoxantrone using LHRH analogs modified with PEGylated liposomes." *Int J Nanomedicine* 5:697–705.

Heuser, L. S., and F. N. Miller. 1986. "Differential macromolecular leakage from the vasculature of tumors." *Cancer*; 57(3):461–464.

Hobbs, S. K., W. L. Monsky, F. Yuan, et al. 1998. "Regulation of transport pathways in tumor vessels: role of tumor type and microenvironment." *Proc Natl Acad Sci U S A* 95(8):4607–4612.

Holgado, M. A., L. Martin-Banderas, J. Alvarez-Fuentes, et al. 2012. "Drug targeting to cancer by nanoparticles surface functionalized with special biomolecules." *Curr Med Chem* 19(19):3188–3195.

Hu, X., J. Hu, J. Tian, et al. 2013. "Polyprodrug amphiphiles: hierarchical assemblies for shape-regulated cellular internalization, trafficking, and drug delivery." *J Am Chem Soc* 135(46):17617–17629. doi: 10.1021/ja409686x.

Hutter, E., S. Boridy, S. Labrecque, et al. 2010. "Microglial response to gold nanoparticles." *ACS Nano* 4(5):2595–2606.

Jain, R. K. 1987. "Transport of molecules across tumor vasculature." *Cancer Metastasis Rev* 6(4):559–593.

Jain, R. K., and L. T. Baxter. 1988. "Mechanisms of heterogeneous distribution of monoclonal antibodies and other macromolecules in tumors: significance of elevated interstitial pressure." *Cancer Res* 48(24 Part 1):7022–7032.

Jin, H., D. A. Heller, R. Sharma, et al. 2009. "Size-dependent cellular uptake and expulsion of single-walled carbon nanotubes: single particle tracking and a generic uptake model for nanoparticles." *Acs Nano* 3(1):149–158.

Lee, H. Y., S. H. Shin, L. L. Abezgauz, et al. 2013. "Integration of gold nanoparticles into bilayer structures via adaptive surface chemistry." *J Am Chem Soc* 135(16):5950–5953. doi: 10.1021/ja400225n.

Lee, H., H. Fonge, B. Hoang, et al. 2010. "The effects of particle size and molecular targeting on the intratumoral and subcellular distribution of polymeric nanoparticles." *Mo Pharm.* 7(4):1195–1208.

Lee, S. Y., M. Ferrari, and P. Decuzzi. 2009. "Shaping nano-/micro-particles for enhanced vascular interaction in laminar flows." *Nanotechnology* 20(49):495101. doi: 10.1088/0957-4484/20/49/495101.

Lovric, J., H. S. Bazzi, Y. Cuie, et al. 2005. "Differences in subcellular distribution and toxicity of green and red emitting CdTe quantum dots." *J Mol Med (Berl)* 83(5):377–385. doi: 10.1007/s00109-004-0629-x.

Lu, F., S.-H. Wu, Y. Hung, et al. 2009. "Size effect on cell uptake in well-suspended, uniform mesoporous silica nanoparticles." *Small* 5(12):1408–1413.

Maeda, H., J. Fang, T. Inutsuka, et al. 2003. "Vascular permeability enhancement in solid tumor: various factors, mechanisms involved and its implications." *Int Immunopharmacol* 3(3):319–328.

Matsumura, Y., and H. Maeda. 1986. "A new concept for macromolecular therapeutics in cancer chemotherapy: mechanism of tumoritropic accumulation of proteins and the antitumor agent smancs." *Cancer Res* 46(12 Part 1):6387–6392.

Muro, S., C. Garnacho, J. A. Champion, et al. 2008. "Control of endothelial targeting and intracellular delivery of therapeutic enzymes by modulating the size and shape of ICAM-1-targeted carriers." *Mol Ther* 16(8):1450–1458. doi: 10.1038/mt.2008.127.

Nakahara, T., S. M. Norberg, D. R. Shalinsky, et al. 2006. "Effect of inhibition of vascular endothelial growth factor signaling on distribution of extravasated antibodies in tumors." *Cancer Res* 66(3):1434–1445.

Northfelt, D. W., B. J. Dezube, J. A. Thommes, et al. 1998. "Pegylated-liposomal doxorubicin versus doxorubicin, bleomycin, and vincristine in the treatment of AIDS-related Kaposi's sarcoma: results of a randomized phase III clinical trial." *J Clin Oncol* 16(7):2445–2451.

Owens, D. E., 3rd, and N. A. Peppas. 2006. "Opsonization, biodistribution, and pharmacokinetics of polymeric nanoparticles." *Int J Pharm* 307(1):93–102. doi: 10.1016/j.ijpharm.2005.10.010.

Park, J. H., G. von Maltzahn, L. Zhang, et al. 2008. "Magnetic iron oxide nanoworms for tumor targeting and imaging." *Adv Mater* 20(9):1630–1635. doi: 10.1002/adma.200800004.

Paul, D., S. Achouri, Y. Z. Yoon, et al. 2013. "Phagocytosis dynamics depends on target shape." *Biophys J* 105(5):1143–1150. doi: 10.1016/j.bpj.2013.07.036.

Peer, D., J. M. Karp, S. Hong, et al. 2007. "Nanocarriers as an emerging platform for cancer therapy." *Nat Nanotechnol*2(12):751–760. doi: 10.1038/nnano.2007.387.

Perrault, S. D., C. Walkey, T. Jennings, et al. 2009. "Mediating tumor targeting efficiency of nanoparticles through design." *Nano Lett* 9 (5):1909-1915.

Rejman, J., V. Oberle, I. S. Zuhorn, et al. 2004. "Size-dependent internalization of particles via the pathways of clathrin- and caveolae-mediated endocytosis." *Biochem J* 377(Pt 1):159–169. doi: 10.1042/BJ20031253.

Ruggiero, A., C. H. Villa, E. Bander, et al. 2010. "Paradoxical glomerular filtration of carbon nanotubes." *Proc Natl Acad Sci U S A* 107(27):12369–12374. doi: 10.1073/pnas.0913667107.

Sahoo, S. K., W. Ma, and V. Labhasetwar. 2004. "Efficacy of transferrin-conjugated paclitaxel-loaded nanoparticles in a murine model of prostate cancer." *Int J Cancer* 112(2):335–340. doi: 10.1002/ijc.20405.

Sands, H., P. L. Jones, S. A. Shah, et al. 1988. "Correlation of vascular permeability and blood flow with monoclonal antibody uptake by human Clouser and renal cell xenografts." *Cancer Res* 48(1):188–193.

Santhakumaran, L. M., T. Thomas, and T. J. Thomas. 2004. "Enhanced cellular uptake of a triplex-forming oligonucleotide by nanoparticle formation in the presence of polypropylenimine dendrimers." *Nucleic Acids Res* 32(7):2102–2112.

Sharma, G., D. T. Valenta, Y. Altman, et al. 2010. "Polymer particle shape independently influences binding and internalization by macrophages." *J Control Release* 147(3):408–412. doi: 10.1016/j.jconrel.2010.07.116.

Shimoni, O., Y. Yan, Y. Wang, et al. 2013. "Shape-dependent cellular processing of polyelectrolyte capsules." *ACS Nano* 7(1):522–530. doi: 10.1021/nn3046117.

Slowing, I., B. G .Trewyn, and V. S.-Y. Lin. 2006. "Effect of surface functionalization of MCM-41-type mesoporous silica nanoparticles on the endocytosis by human cancer cells." *J Am Chem Soc* 2006; 128(46):14792–14793.

Thorek, D. L. J., and A. Tsourkas. 2008. "Size, charge and concentration dependent uptake of iron oxide particles by non-phagocytic cells." *Biomaterials* 29 (26):3583-3590.

Thurston, G., J. W. McLean, M. Rizen, et al. 1998. "Cationic liposomes target angiogenic endothelial cells in tumors and chronic inflammation in mice." *J Clin Invest* 101(7):1401–1413. doi: 10.1172/JCI965.

Tong, R. T., Y. Boucher, S. V. Kozin, et al. 2004. "Vascular normalization by vascular endothelial growth factor receptor 2 blockade induces a pressure gradient across the vasculature and improves drug penetration in tumors. *Cancer Res* 64(11):3731–3736.

Toy, R., E. Hayden, A. Camann, et al. 2013. "Multimodal in vivo imaging exposes the voyage of nanoparticles in tumor microcirculation." *ACS Nano* 7(4):3118–3129. doi: 10.1021/nn3053439.

Truong, N. P., M. R. Whittaker, C. W. Mak, et al. 2015. "The importance of nanoparticle shape in cancer drug delivery." *Expert Opin Drug Deliv* 12(1):129–142. doi: 10.1517/17425247.2014.950564.

Vinogradov, S. V., T. K. Bronich, and A. V. Kabanov. 2002. "Nanosized cationic hydrogels for drug delivery: preparation, properties and interactions with cells." *Adv Drug Deliv Rev* 54(1):135–147.

Wang, A. Z., V. Bagalkot, C. C. Vasilliou, et al. 2008. "Superparamagnetic iron oxide nanoparticle-aptamer bioconjugates for combined prostate cancer imaging and therapy." *Chem Med Chem* 3(9):1311–1315. doi: 10.1002/cmdc.200800091.

Wang, J., S. Tian, R. A. Petros, et al. 2010. "The complex role of multivalency in nanoparticles targeting the transferrin receptor for cancer therapies." *J Am Chem Soc* 132(32):11306–11313.

Wartlick, H., K. Michaelis, S. Balthasar, et al. 2004. "Highly specific HER2-mediated cellular uptake of antibody-modified nanoparticles in tumour cells." *J Drug Target* 12(7):461–471. doi: 10.1080/10611860400010697.

Wen, A. M., and N. F. Steinmetz. 2016. "Design of virus-based nanomaterials for medicine, biotechnology, and energy." *Chem Soc Rev* 45(15):4074–4126. doi: 10.1039/c5cs00287g.

Xiang, G., J. Wu, Y. Lu, et al. 2008. "Synthesis and evaluation of a novel ligand for folate-mediated targeting liposomes." *Int J Pharm* 356(1–2):29–36. doi: 10.1016/j.ijpharm.2007.12.030.

Yoo, J. W., N. Doshi, and S. Mitragotri. 2010. "Endocytosis and intracellular distribution of PLGA particles in endothelial cells: effect of particle geometry." *Macromol Rapid Commun* 31(2):142–148. doi: 10.1002/marc.200900592.

Yuan, F., M. Dellian, D. Fukumura, et al. 1995. "Vascular permeability in a human tumor xenograft: molecular size dependence and cutoff size." *Cancer Res* 55(17):3752–3756.

Yuan, F., H. A. Salehi, Y. Boucher, et al. 1994. "Vascular permeability and microcirculation of gliomas and mammary carcinomas transplanted in rat and mouse cranial windows." *Cancer Res* 1994; 54(17):456–4568.

Yuan, Y. Y., C. Q. Mao, X. J. Du, et al. 2012. "Surface charge switchable nanoparticles based on zwitterionic polymer for enhanced drug delivery to tumor." *Adv Mater* 24(40):5476–5480. doi: 10.1002/adma.201202296.

Yue, H., W. Wei, Z. Yue, et al. 2010. "Particle size affects the cellular response in macrophages." *Eur J Pharm Sci* 41(5):650–657. doi: 10.1016/j.ejps.2010.09.006.

Zhang, K., H. Fang, Z. Chen, et al. 2008. "Shape effects of nanoparticles conjugated with cell-penetrating peptides (HIV Tat PTD) on CHO cell uptake." *Bioconjug Chem* 19(9):1880–1887. doi: 10.1021/bc800160b.

Zhang, R. D., J. E. Price, T. Fujimaki, et al. 1992. "Differential permeability of the blood-brain barrier in experimental brain metastases produced by human neoplasms implanted into nude mice." *Am J Pathol* 1992; 141 (5):1115.

Zhang, S., J. Li, G. Lykotrafitis, et al. 2009. "Size-dependent endocytosis of nanoparticles." *Adv Mater* 21 (4):419–424.

3 Nanocarrier Systems for Anticancer Drug Delivery at the Subcellular Level

Reza Mahjub and Shyh-Dar Li

CONTENTS

3.1 INTRODUCTION

Over the last decade, several nanodrug delivery systems have been studied for therapeutics, diagnosis, and imaging of cancer. It has been demonstrated that, following systemic administration, nanoparticles are rapidly removed from blood circulation by the mono-nuclear phagocytic system (MPS), also known as reticuloendothelial system (RES) (Roy and Li 2016). After bypassing the MPS clearance, nanocarriers must extravasate from blood circulation to tumor tissues. The vasculature system located in malignant tissues is structurally different from that located in normal ones and exhibits increased permeability for nanocarriers. Additionally, the lymphatic system responsible for removing substances from tissues is often immature in tumors, leading to accumulation of the extravasated nanoparticles. This feature regarding nanoparticle delivery to tumors is termed the Enhanced Permeation and Retention (EPR) Effect (Li and Huang 2008; Ernsting et al. 2013) and is the dominant mechanism of uptake in first generation of nanomedicines. For the second generation of nanomedicines, nanocarriers are surface modified with a ligand such as (monoclonal antibody, peptide, and folate) that targets a specific antigen at the target cell surface. After accumulating in tumors either as the first generation or second generation, nanocarriers must enter into the tumor cells to exert a pharmacological effect. Plasma membrane is a barrier for intracellular transport of nanoparticles. Endocytosis is the most dominant mechanism in cellular uptake of nanoparticles (Sahay et al. 2010). After taken up by the tumor cells, most of the anticancer drugs exhibit non-selective intracellular distribution that in most cases, may cause toxicity to other subcellular organelles. In addition, the efficacy of the anticancer treatment can be enhanced by specific intracellular drug localization. Therefore, the development of a drug delivery system targeting subcellular organelles (known as third generation nanomedicines) has been actively pursued. The main subcellular targets involved in anticancer therapy are schematically depicted in Figure 3.1.

The most straightforward approach for targeting a drug to its subcellular target is conjugating a targeting ligand with the drug (Sito et al. 2003; Whitmire et al. 2012). A number of factors for the ligand-drug conjugate need to be optimized for enhanced delivery. For example, the linker must be stable during the blood circulation and rapidly degrade intracellularly for efficient drug release. Further, the ligand needs to be highly specific to avoid off-target toxicity. Application of nanoplatforms is identified as the most efficient strategy for intracellular organelle targeting among all other approaches. Alternatively, multifunctional nanoparticles can be developed for subcellular drug targeting (Shi-Kam et al. 2005; Xiong et al. 2008; Ding et al. 2009; Shi et al. 2009). Various nanocarriers that were developed for organelle-specific drug delivery are summarized in Table 3.1. In this chapter, the role of subcellular organelles in cancer therapy and also various strategies for targeting of nanocarriers to these intracellular organelles including mitochondria, nucleus, lysosomes, and endoplasmic reticulum (ER) will be discussed.

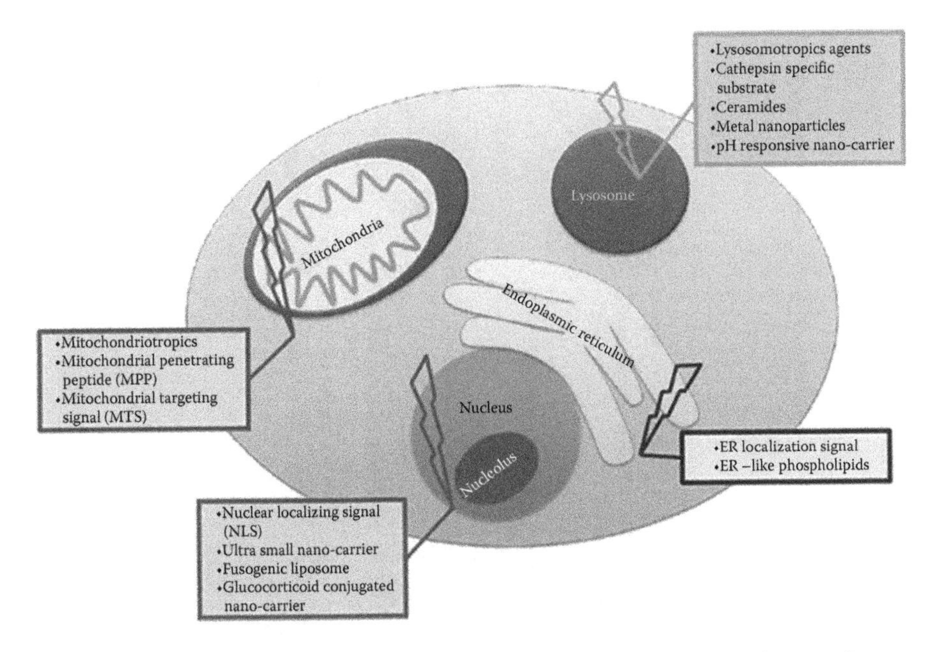

FIGURE 3.1 Main subcellular targets and appropriate strategies for anticancer therapy using nanocarriers.

3.2 MITOCHONDRIAL TARGETED DRUG DELIVERY

3.2.1 Mitochondria and Cancer Therapy

The mitochondrion is a double membrane, subcellular organelle with a size range of 0.75–3 μm. A mitochondrion is composed of an outer mitochondrial membrane, an inner mitochondrial membrane, inner membrane space, and the mitochondrial matrix. The mitochondrion is responsible for regulation of cellular energy by synthesizing ATP using oxidative phosphorylation (Hiendleder et al. 1999; Waterhouse et al. 2001; Gulbins et al. 2003). In 1999, it was suggested that the mitochondrion has a dominant role in cell death regulation (Bernardi et al. 1999). It was proposed that cell apoptosis is regulated through two major signaling pathways including extrinsic and intrinsic routes (Hengartner 2000). At the cell surface, binding of ligands to the death receptors activates the extrinsic pathway which leads to cytosolic activation of caspase-8. Caspase-8 transforms BH3-integrating-domain death agonist (BID), a pro-apoptotic member of the Bcl-2 protein family, to truncated BID (tBID) that is active and can translocate to the mitochondria. For the intrinsic pathway, translocation of tBID to the outer mitochondrial membrane promotes the release of apoptogenic factors from the mitochondrial matrix to the cytosol, such as cytochrome c (cyt c), second mitochondria-derived activator of caspases (smac), apoptosis inducing factor (AIF), and endonuclease G (Li et al. 2001; Barczyk et al. 2005; Adams and Cory 2007; Kroemer et al. 2007). Both intracellular and extracellular stress signals can induce the mitochondrial apoptosis pathways (Finkel 2001; Maddika et al. 2007). The extrinsic and intrinsic apoptosis pathways are schematically depicted in Figure 3.2.

TABLE 3.1
Nanocarriers for Targeting Various Subcellular Organelles for Anticancer Therapy

Subcellular Organelle	Nanostructure	Targeting Moiety	Cell Line	Active Compound	Reference
Mitochondria	MWCN	Rhodamine-123	MCF-7	Platinum derivatives	Yoong et al. 2014
	Polymeric micelle (PEG-PCL)	Triphenyl Phosphonium Bromide (TPP Br)	N 9	Co-Q 10	Sharma et al. 2012
	Pegylated liposomes	Rhodamine-123	HeLa	Paclitaxel	Biswas et al. 2011
	Surface-modified gold nanorods	Cetyltrimetyl Ammonium Bromide (CTAB)	HepG2, HeLa, LN Cap, A549	N/A	Wang et al. 2011
	Surface-modified gold nanoclusters	Triphenyl Phosphonium (TPP)	HepG2, HeLa	N/A	Zhuang et al. 2014
	DQasome (Liposome)	Dequalinium	Colo 205	Paclitaxel	D'Souza et al. 2008
	Mitosome	Dequalinium	K562-MDR K5262	Epirubicin	Men et al. 2011
	Liposome	Dequalinium	MCF-7	Daunorubicin	Zhang et al. 2012
	MITO-Porter liposome	DOPE, R8 Octa-arginine	HeLa	pDNA (GFP)	Yamada et al. 2008
	Surface-modified liposome	KLA Peptide	A549	Paclitaxel	Jiang et al. 2015
	Surface-modified CdSe -ZnSQuantum dots	Mito 8	Vero cells	N/A	Hoshino et al. 2004
	PAMAM dendrimers	TPSO	C6 Glioma	N/A	Denora et al. 2013
	Surface-modified MWCN	MTS peptide	Murine RAW 264	Oligonuclotide	Battigelli et al. 2013

(Continued)

TABLE 3.1 (CONTINUED)
Nanocarriers for Targeting Various Subcellular Organelles for Anticancer Therapy

Subcellular Organelle	Nanostructure	Targeting Moiety	Cell Line	Active Compound	Reference
Nucleus	CdSe -ZnSQuantum Dot	NLS (PKKKRKVKAA)	Hela	N/A	Maity et al. 2016
	Mesoporus silica nanoparticles	TAT Peptide	HeLa	Doxorubicin	Pan et al. 2012
	Liposome	$S4_{13}$-PV penetrating peptide	HeLa, TSA	luciferase encoded pDNA	Trabulo et al. 2008
	Polymeric micelles (PLA-PEG)	TAT pepptide	MCF-7	N/A	Sethuraman and Bae 2007
	PLGA nanoparticles	SV 40 (PKKRKV)	MCF-7	luciferase encoded pDNA	Jeon et al. 2007
	Gold nanoparticles	NLS Peptide (GGFSTSLRARKA)	HeLa	N/A	Nativo et al. 2008
	Supraparamagnetic (Fe_3O_4) nanoparticles	NLS Peptide (KKKRKV)	Hela	Rhodamine	Xu et al. 2008
	PLGA Nanoparticles	NLS peptide (DRQIKIWFQNRRMKWKK)	MCF-7	Doxorubicin	Misra and Sahoo 2010
	Surface-modified gold nanoparticles	NLS Peptide (CGGRKKRRQRRRAP)	HeLa	N/A	Yang et al. 2014
	Cationic PAMAM Dendrimer	Small Size (10 nm)	SKOV-3	Camptothecin	Hsu et al. 2011; Hsu et al. 2012
	Grafted gold nanoparticles	Nucleolin specific aptamers	MCF-7	N/A	Huo et al. 2014
	Surface-modified silver nanoparticles	NLS peptide (CGGGPKKRKVGG)	HSC-3	N/A	Austin et al. 2015
	Modified liposomes as T-MEND	Cardiolipin, DOPE	JAWSII	EGFP encoding pDNA	Akita et al. 2009
	Modified SLN	Dexamethsone	HepG2	EGFP encoding pDNA	Wang et al. 2012a

(Continued)

TABLE 3.1 (CONTINUED)
Nanocarriers for Targeting Various Subcellular Organelles for Anticancer Therapy

Subcellular Organelle	Nanostructure	Targeting Moiety	Cell Line	Active Compound	Reference
Lysosome	Surface-modified PEG-PLGA nanoparticles	Monoclonal antibody (mAb)	Chronic lymphocyte lukemia	Hydroxy chloroquine	Mansilla et al. 2012
	Surface-modified liposomes	Cathepsin B substrate (Gly-Leu-Phe-Gly)	HeLa	Doxorubicin	Maniganda et al. 2014
	Gold Zinc oxide nanoparticles	Cathepsin B substrate (Arg-Arg)	HepG2	N/A	Gao et al. 2014
	Modified liposomes	Transferrin, Exogenous ceramides	HeLa	N/A	Koshkaryev et al. 2014
	Iron oxide magnetic nanoparticles	EFG	HeLa	N/A	Domenech et al. 2013
	Cationic USPIO	Iron oxide	HCEC	N/A	Kenzaoui et al. 2012
	Polymeric micelle (PEG-4-(N)-Stearoyl gemicitabine	pH sensivity of Micelle	BxPC-3	Gemcitabine	Zhu et al. 2012
	Superparamagnetite (Fe_3O_4) Nanoparticles	MPEG-b-PMAA-b-PGMA	HeLa	Adriamycin	Guo et al. 2011
	Polymeric micelle (PEO-PMA)	Positive charge of PMA (pH Responsive)	H2009	N/A	Zhou 2011
Endoplasmic Reticulum (ER)	Polymerc micelle (PEG-PE)	Phosphatidylethanol amine	A 549	N/A	Wang et al. 2012b
	Polymeric micelle P(IPAAm-DMAAm)-*b*-PLA)	PLA	Bovine carotid endothelial cells (EC)	N/A	Akimoto et al. 2010
	Surface-modified gold nanoparticles	KDEL	C2C12	siRNA	Acharya and Hill 2014
	Modified Liposomes	L-α-phosphatidyl inositol, L-α-phosphatidylserine	Huh7.5	Calcein, HIV-1-Antivirals	Pollock et al. 2010

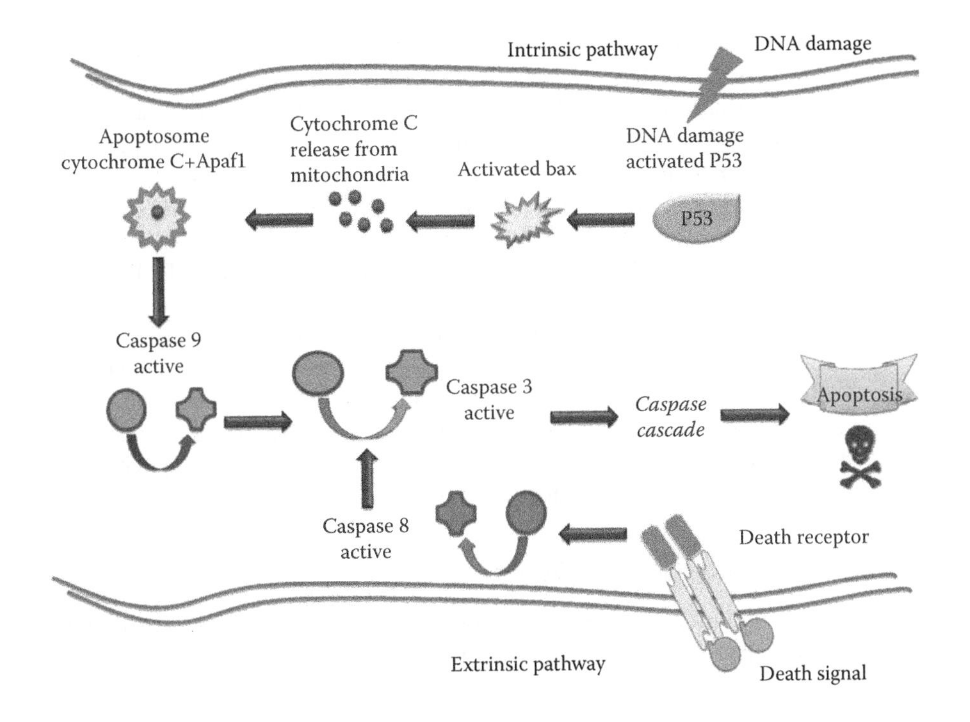

FIGURE 3.2 The extrinsic and intrinsic apoptotic pathways.

In addition, the significant role of mitochondria in bio-energetics and ATP synthesis, balancing of reduction-oxidation chain, regulation of calcium metabolism, and preparation of super oxides, makes the organelle a good target for cancer therapy (Serviddio et al. 2011). It has been shown that the outer and inner mitochondrial membranes can act as the barrier against permeation of nanoparticles. The high negative charge density of the mitochondrial matrix and impermeability of the inner membrane to ions including protons (H^+), are responsible for generation and maintenance of an inner mitochondrial trans-membrane potential ($\Delta\Psi m$) in the range of −160 mV to −180 mV, depending on the cell type. Although, the high negative value of $\Delta\Psi m$ is crucial for ATP synthesis by oxidative phosphorylation (Mitchell and Moyle 1965a; Mitchell and Moyle 1965b), it is reported to act as a barrier against drug permeation. The size of nanocarriers has a significant effect on their translocation into the mitochondria. It has been reported that while 6-nm gold nanoparticles failed to accumulate inside the mitochondria, similar gold particles with a diameter of 3 nm can permeate across the outer mitochondrial membrane (Salnikov et al. 2007). Strategies to overcome the mitochondrial barriers include disruption of the mitochondrial negative membrane potential, alteration in mitochondrial permeability, and destabilization of proteins in mitochondrial membrane.

There are some differences in the structure and function of mitochondria located in malignant cells as compared to normal cells, and these differences make mitochondria an attractive target for cancer therapy. In malignant cells, the rate of aerobic glycolysis is increased (Huang et al. 2011), electron and anion transport is increased, calcium uptake is enhanced, and activity of certain enzymes is decreased, such as cytochrome c oxidase (Agemy et al. 2011), adenine nucleotide translocase (Yamamada and Harashima 2013; Modica-Napolitano and Weissing 2015), and mitochondrial ATPase (Yousif et al. 2008). The mitochondrial trans-membrane potential ($\Delta\Psi m$) is shown to be higher in cancer cells (Pelicano et al. 2004; Ma et al. 2012). Tumor cells also pose higher levels of reactive oxygen species (ROS) and oxidative stresses compared to normal cells (Paine et al. 1975; Dam et al. 2012; Ma 2014).

Tumor-initiating cells (TIC) are a small subpopulation of tumor cells that play a significant role in initiation, growth, progression, and metastasis of cancers. Mitochondria located in TICs exhibit a higher level of reactive oxygen species (ROS) and more negative mitochondrial membrane potential ($\Delta\Psi m$) compared to mitochondria located in differentiated cancer cells. In TICs, enhanced expression of anti-apoptotic proteins results in increased resistance to apoptosis compared to differentiated cancer cells (Yang et al. 2016) and therefore, targeting mitochondria located in TICs is an attractive strategy to improve cancer therapy.

3.2.2 Mitochondrial Drug Delivery

Various strategies including surface modification of nanocarriers with different mitochondriotropics, mitochondrial-penetrating peptides (MPPs), and mitochondrial targeting signals (MTSs) have been applied for mitochondria-targeted drug delivery and are discussed in the following sections.

3.2.2.1 Mitochondriotropic Conjugated Nanocarriers

Mitochondriotropics are amphiphilic compounds that permeate across mitochondrial membrane and accumulate in the mitochondrial matrix, resulting in a reduction of mitochondrial membrane potential. They can be applied as a mitochondrial translocation enhancer for targeted delivery to the mitochondrial matrix (Weissig 2003; Weissig et al. 2004). Although various chemical structures were recognized as mitochondriotropics, lipophilic cations such as rhodamine and tetraphenylphosphonium (TPP) salts are the most studied mitochondriotropics. They can also facilitate permeation through phospholipid bilayer and enhance endosomal escape due to their high positive charge (Horobin et al. 2007). As mitochondria located in malignant cells pose a higher negative $\Delta\Psi m$ compared to that in normal cells, the interaction of mitochondriotropics with the inner mitochondrial membrane in malignant cells is stronger. Consequently, mitochondriotropics exhibit selectivity to mitochondria in tumor cells (Modica-Napolitano and Aprille 2001; Ito et al. 2009; Lim et al. 2009).

It was reported that multi-walled carbon nanotubes (MWCN) functionalized with rhodmine-110 as a mitochondriotropic displayed 4-fold increased mitochondrial uptake compared to the non-targeted MWCN. When carrying a platinum drug, the targeted MWCN exhibited increased potency against tumor cells (IC_{50} = 0.34 μM vs. 2.64 μM) (Yoong et al. 2014). Preparation of a block co-polymer micelle composed of PEG–poly (caprolactone) functionalized with triphenylphosphonium for mitochondrial delivery of co-enzyme Q10 has been reported in the literature (Sharma et al. 2012). Images obtained from confocal microscopy of FITC-labeled nanomicelles revealed mitochondrial localization of 30% of the intracellularly absorbed polymeric micelles. In another study, rodamine-123 (Rh 123) was conjugated to the distal end of a PEGylated lipid, which was used to prepare mitochondria-targeted liposomes. Paclitaxel delivery by the Rh-123-liposomes to the mitochondria in HeLa cells was increased by 5-fold compared to the control liposomes (Biswas et al. 2011).

It was demonstrated that surface modified gold nanorods using cetyltrimethyl ammonium bromide (CTAB) exhibited significant cytotoxicity against cancer cell lines including HepG2, HeLa, LNCaP, and A549 while slight cytotoxicity was reported in 16HBE and MSC normal cell lines (Wang et al. 2011). Images obtained by transmission electron microscopy (TEM) and confocal microscopy demonstrated accumulation of gold nanorods at the edges or around inner membrane of mitochondria in the tumor cells. They suggested that CTAB-gold nanorods accumulated in the mitochondria, decreased the membrane potential, and increased levels of reactive oxygen species (ROS), inducing cell death. Similarly, a study reported mitochondrial accumulation of gold nanoclusters with a surface modification of TPP in Hela and HepG2 cell lines (Zhuang et al. 2014). The preparation of liposomes containing dequalinium as a mitochondriotropic ligand (i.e. DQAsome) for mitochondrial delivery of paclitaxel has been reported in the literature (D'Souza et al. 2008). There was approximately a 5-fold increase in mitochondrial accumulation of the drug compared to plain liposomes. Similar results have been found in the literature (Men et al. 2011; Zhang et al. 2012).

MITO-Porters are liposomal formulations decorated with octa-arginine (R8) at the surface for mitochondrial delivery of macromolecules (Yamada et al. 2008). Confocal microscopy and flow cytometry results revealed that MITO-Porters exhibited three-times higher accumulation in mitochondria compared to cytoplasm.

3.2.2.2 Mitochondrial-Penetrating Peptide (MPP) Conjugated Nanocarriers

The development of paclitaxel-loaded liposomes containing a KLA peptide (i.e. D[KLAKLAK]2) as an MPP at the liposomal surface for mitochondria-targeted delivery was reported in the literature (Jiang et al. 2015). They demonstrated 4-fold higher mitochondrial accumulation of the KLA-liposomes compared to control liposomes in human lung adenocarcinoma A549 cells using confocal laser scattering microscopy (CLSM). The conjugation of Mito-8 (i.e. NH2-MSVLTPLLLRGLTGSARRLPVPRAKIHWLC-COOH), a peptide-based mitochondrial targeting agent, to the surface of quantum dots, and reported eight-fold increased mitochondrial accumulation compared to control quantum dots has been reported in the literature. (Hoshino et al. 2004). In another study, the translocator protein (TPSO), a 18kDa protein found in outer mitochondrial membrane, was used to decorate the surface of the PAMAM dendrimers, and 39% of the TPSO-dendrimers was found associated with mitochondria in C6 glioma cells (Denora et al. 2013).

Szeto-Schiller (SS) are small water-soluble peptides that can permeate the cell membrane and translocate in mitochondria. Due to their antioxidant capacity, these peptides protect mitochondria from oxidative stresses. One of these SS peptides, SS-31 with a sequence of [D-Arg-Lys-Phe-NH_2] (Elamipretide, Bendavia®), is currently under clinical trial for protection of mitochondrial injury during ischemia in patients suffering from acute coronary events (Breunig et al. 2008; Wongrakpanich et al. 2014).

3.2.2.3 Mitochondrial Targeting Signal (MTS) Conjugated Nanocarriers

Modification of nanocarriers with MTS is another approach for the development of mitochondria-targeted drug delivery systems. Although both MPP and MTS enhance mitochondrial localization there are some differences in structure, intracellular localization, and mechanisms of action between these two types of peptides. Mitochondrial-penetrating peptides (MPPs) are peptide-based mitochondrial transporters that are endogenously incorporated in the mitochondrial membrane. They act on mitochondrial membrane potential ($\Delta\Psi m$) and induce charge-driven mitochondrial uptake. Mitochondrial Targeting Signals (MTSs) including bacterial signal peptides and mitochondrial leader sequence peptides, are endogenous, naturally occurring N-terminal sequences of mitochondrial proteins that localize to mitochondria. MTS contain 2040 amino acids and are recognized by mitochondrial protein import machinery for mitochondrial localization (Stojanovski et al. 2003; Horton et al. 2008; Zhang et al. 2010). The application of MTS derived from Bak (i.e. bcl2 antagonist/killer) and Bax (i.e. bcl2- associated X protein) for mitochondrial targeted delivery of p53 has been reported in the literature (Matissek et al. 2014). They showed that p53-Bak MTS and p53-Bax MTS fusion proteins induced significant apoptosis through the mitochondrial pathway. In another study, MWCNs were surface conjugated with a MTS peptide, which exhibited increased mitochondrial accumulation compared to the non-modified MWCNs (Battigelli et al. 2013).

3.3 NUCLEAR TARGETED DRUG DELIVERY

3.3.1 The Nucleus: Role in Cancer Treatment

The nucleus is responsible for regulation of key cellular functions including replication and transcription of genes. It is also the major site of action for cellular growth factors that control cell reproduction, metabolism, and the cell cycle. Therefore, the nucleus plays a dominant role in regulation of cell death, and many cellular disorders including cancers originate from this subcellular organelle (Dam et al. 2012; Sakhrani and Padh 2013). Significant roles of nucleus in cancer development, progression, and metastasis, makes this organelle the main site of action for various anticancer drugs including doxorubicin, platinum drugs (e.g. cisplatin and carboplatin), and camptothecin, which exert their efficacy by DNA intercalation and reduction of nucleic acid replication by inhibition of topo isomerase 1 (topo-I), respectively (Wong et al. 1999; Yang et al. 2006; Frissen et al. 2008; Kim et al. 2011). Camptothecin binds at the interface between topo-I and DNA, establishing reversible topo-I-camptothecin-DNA complexes, and inhibits the rejoining step of the cleavage/relegation reaction of topo-I (Pommier et al. 1998). Photodynamic therapy for cancer is based on the generation of reactive oxygen species (ROS) from the photo-sensitizers such as haematoporphyrin derivatives, hydroxyethyl deuteroporphyrin, benzoporphyrin derivatives, and zinc phthalocyanine, and these ROS act on DNA located in the nucleus (Sobolev et al. 2000). Additionally, an exogenous pDNA must be delivered to the nucleus for expression of a therapeutic protein to perform gene therapy (Sui et al. 2011).

3.3.2 Nuclear Drug Delivery

Drug delivery to the nucleus is a complicated and challenging process with multiple barriers (Hakim and Usmani 2015). In tumor cells, overexpression of efflux pumps and increased cytoplasmic drug metabolism have been reported to decrease nuclear localization of many active compounds (Burrow et al. 2002; Duvvuri and Krise 2005). The nuclear envelope (NE) is identified as a double-membrane structure that surrounds the nucleus and separates the inner components from the cytoplasmic environment (Torchilin 2006). NE is considered as the most significant intracellular barrier for nuclear translocation of drugs. Structurally, the integrity of the nuclear envelope is disrupted by the nuclear pore complexes (NPCs), which are incorporated in the NE (Wente 2000) and can allow passive diffusion of small molecules with a molecular weight below 45 kDa and free transport of particles with a size smaller than 9 nm (Gasiorowski and Dean 2003; Wagstaff and Jans 2009), while the nuclear permeation of larger particles through NPCs is strongly restricted. Integral membrane proteins (IMPs) are a class of proteins that are embedded in biological membranes. In the nucleus, IMPs are located in the inner membrane of the nuclear envelope and mediate the formation of NPC and the nucleoskeleton (Wente 2000; Gasiorowski and Dean 2003). It was reported that due to the small size and precise selectivity of NPCs, only a small portion of cisplatin (approximately 11%) taken up by a cell was translocated into the nucleus (Perez 1998). Therefore, development of an effective nuclear targeted delivery system can improve the efficacy of cancer therapy.

Some viruses can naturally overcome the NPC barrier for nuclear localization via receptor mediated uptake. Modified viral vectors prepared from retroviruses (Culver et al. 1992; Somia et al. 1995; Paulus et al. 1996), adeno-associated viruses (Gregorevic et al. 2004; Maheshri et al. 2006), adenoviruses (Willson 1996; Curiel et al. 1991), and herpes simplex viruses (Palmer et al. 2000) have been developed for nuclear targeted delivery of genes and other active compounds. However, their wide use has been significantly limited due to the safety concerns including immunological reactions against viral vectors, probability of transformation of modified vectors to pathogenic phenotypes, and environmental biohazards (Baldo et al. 2013).

Nanocarriers can be designed for transport across the NPC barrier and therefore, in contrast to viral vectors (Hou et al. 2014), they pose high potential for providing safe and efficient nuclear targeted delivery (Li and Huang 2007). Different strategies including surface modification of nanocarriers with nuclear localization signals (NLS), aptamers, and glucocorticoids, preparation of ultra-small nanocarriers, and development of fusogenic liposomes can be applied for nuclear delivery.

3.3.2.1 Nuclear Localization Signal (NLS) Conjugated Nanocarriers

Nanoparticle surfaces modified with a peptide-based nuclear localization signal (NLS) have been widely studied (Belting et al. 2005; Fan et al. 2016). It is suggested that interaction between the NLS and the integral membrane protein (IMP) family receptors can facilitate the nuclear translocation of the nanocarriers (Jans et al. 1998). Nuclear targeting NLS-conjugated CdSe-ZnS quantum dots with a diameter of ~22 nm was studied (Maity and Stepensky 2016). Only 11.9% of the intracellularly accumulated non-modified quantum dots were localized in the nucleus, while 56.1% of the intracellular NLS-conjugated quantum dots were found in the nucleus. The effect of TAT peptide conjugation to mesoporous silica nanoparticles with varying diameters for nuclear drug delivery of doxorubicin was investigated (Pan et al. 2012). They demonstrated that the size of the nanoparticles was critical for nuclear translocation and the TAT conjugated nanoparticles with a diameter of 25–50 nm efficiently targeted the nucleus. SV40 large T-antigen derived from a tumor antigen of simian virus 40 has been extensively studied as a NLS. It consists of highly basic amino acids: Pro-Lys-Lys-Lys-Arg-Lys-Val (Kalderon et al. 1984). Preparation of complexes by mixing luciferase encoded plasmid DNA,$S4_{13}$-PV penetrating peptide derived from SV40 T-antigen and a cationic liposome composed of 1,2-dioleoyl-3-trimethylammonium-propane and 1,2-dioleoyl-sn-glycero-3-phosphoethanolamine was reported (Trabulo et al. 2008). The $S4_{13}$-PV containing lipoplex exhibited significantly higher nuclear translocation compared to other complexes prepared with a reverse sequence of NLS or a scrambled peptide in HeLa and TSA cell lines. The development of gold nanoparticles surface modified with two separate peptide ligands including either a sequence of TAT peptide (AGRKKRRQRRR) derived from HIV, or the Pntn peptide (GRQIKIWFQNRRMKWKK) as a cell-penetrating peptide (CPP) obtained from An-tennapedia protein of *Drosophila*, and also an NLS (GGFSTSLRARKA) was reported (Nativo et al. 2008). After cytosolic accumulation in HeLa cells, a large portion of the particles were detected near the nucleus due to NLS targeting. In another study, surface modification of superparamagnetic (Fe_3O_4) nanoparticles with an NLS peptide (KKKRKV) resulted in significant accumulation in the nucleus while the non-targeted nanoparticles failed to deliver the model drug to the nucleus (Xu et al. 2008).

Most studied NLSs are rich in lysine and arginine and are positively charged, resulting in non-specific cellular interactions. Therefore, many NLS-conjugated nanocarriers exhibited significant adverse effects that limited their *in vivo* applications (Schwarze et al. 1999; Sarko et al. 2010). Shielding the positive charge of NLS conjugated nanocarriers during the systemic circulation followed by re-exposure of the NLS in the tumorous region has been demonstrated as an effective strategy for overcoming this problem (Ma et al. 2012). Development of a pH sensitive, micellar drug delivery system composed of two co-polymers including poly (methacryloyl-sulfa dimethoxine) (PSD)-PEG as a pH-sensitive component and poly (l-lactic acid) (PLLA)-polyethylene glycol (PEG) conjugated with TAT as an NLS was investigated (Sethuraman and Bae 2007). They demonstrated that at pH 7.4, the positive charges of TAT were shielded by anionic PSD. The zeta potential of the micelles at pH 6.8–8.0 was ~0 mV, indicating complete charge shielding. In the acidic environment within a tumor (i.e. pH 6.0–6.6), the PSD-PEG segment (pKb = 7.0) became uncharged was detached from TAT and formed aggregates due to increased hydrophobicity. The zeta potential of the micelles increased to +6.0 mV, indicating de-shielding of the TAT. They demonstrated that this system was selective for tumor nuclear delivery compared to normal cells. Preparation of polymeric nanoparticles composed of NLS-conjugated PLGA for nuclear delivery of plasmid DNA encoding luciferase gene was reported in the literature (Jeon et al. 2007). The polymeric nanoparticles showed enhanced nuclear localization with sustained gene expression for >13 days.

The authors suggested that compared to nanoparticles prepared with poly (ethyleneimine), the NLS-PLGA nanoparticles were more effective for long-term gene expression.

Enhanced nuclear localization of doxorubicin delivered by nanoparticles composed of NLS-modified PLGA was demonstrated in the literature (Misra and Sahoo 2010). They reported increased cytotoxic potency for the NLS-conjugated nanoparticles (2.3 μM) compared to that for free drug (17.6 μM) in MCF-7 cells. The data were further supported by the confocal laser scattering microscopy (CLSM) showing increased nuclear accumulation of doxorubicin delivered by the nanoparticles compared to free drug. Similarly, the NLS-functionalized gold nanoparticles (AuNP) displayed enhanced nuclear accumulation in HeLa cells (Yang et al. 2014). It was shown that surface-modified silver nanoparticles (AgNP) with a NLS peptide (CGGGPKKKRKVGG) efficiently accumulated in the nucleus of human oral squamous cell carcinoma cells (HSC-3) and triggered caspase-independent apoptosis (Austin et al. 2015). Alternatively, NLS can be conjugated to a drug, followed by encapsulation into nanoparticles for improved nuclear delivery.

3.3.2.2 Other Approaches

Very recently, aptamers identified as oligonucleotides have been developed for specific binding to their target molecules and exhibit many advantages compared to antibodies. In contrast to antibodies, aptamers exhibit increased selectivity and affinity with the target molecule and enhanced stability during the storage without immunogenicity *in vivo* (Bruno 2013). Aptamers can be synthesized in a test tube using a polymerase chain reaction (PCR), and therefore, their mass production is much cheaper than antibodies.

Nucleolin is located mostly in fibrillar components around the fibrillar centers as well as in granular component of the nucleolus. Phosphorylation of the specific domains in nucleolin (i.e. RNA binding domains) enhances the cytoplasmic localization while dephosphorylation of the specified domains improves the nuclear translocation. In addition to the nucleus, nucleolin is also found to be overexpressed at the cell surface (Ginisty et al. 1990). The development of an aptamer known as anti-nucleolin AS1411 was studied (Soundararajan et al. 2008). They reported that the aptamer targeted and bound with nucleolin overexpressed in MCF-7 and MDA-MB-231 breast cancer cells. In this study, it was demonstrated that the cytotoxicity of AS1411 was four times greater in the tumor cells compared to normal MCF-10A mammary epithelial cells.

Gold nanoparticles modified with an aptamer that bound with nucleolin were demonstrated to localize in the nucleus, leading to increased deformations in the nuclear envelope and elevated activity of caspase-3 and caspase-7, which promoted cell apoptosis (Dam et al. 2012).

As mentioned previously, particulate systems with a diameter smaller than 9 nm can be easily transported across the NPC. In a study, high nuclear accumulation was reported with a 2-nm AuNP, while the 16-nm AuNP showed non-specific cytosolic delivery (Huo et al. 2014). Quantitative analysis using ICP-MS revealed intra-nuclear percentage of 40% for the 2-nm AuNP 24 h after incubation, while the 16-nm AuNP exhibited only 10% nuclear accumulation. Ultra-small PAMAM dendrimers with a diameter of 10 nm were shown to exhibit enhanced nuclear delivery

(Hsu et al. 2011; Hsu et al. 2012). Primary amines in PAMAM dendrimers were conjugated with 1, 2-dicarboxilic acid-cyclohexene anhydride (DCA) or folic acid via amide bonds. At pH 7.4, the amide bond is stable and the dendrimer was negatively charged. Following cellular uptake via folate receptor-mediated endocytosis, the acidic lysosomal environment triggered pH-dependent hydrolysis of the amides to release the cationic dendrimers, facilitating the lysosomal escape and the ultimate nuclear delivery via nuclear pores (Hsu et al. 2011; Hsu et al. 2012).

Fusogenic liposomes are a class of phospholipid vesicles, and their bilayer membrane exhibits enhanced ability to interact with biological membranes, promoting lipid mixing and therefore release of the content to the other site of the membrane. They can be produced by incorporation of either a fusogenic lipid or polymer that increases the fluidity of the vesicle and promotes destabilization of endosomal and nuclear membranes, facilitating endosomal escape and nuclear localization, respectively (Jhaveri and Torchilin 2016). In a study, a multifunctional, four-layered liposomal formulation, introduced as tetra-lamellar multifunctional envelope-type nanodevice (T-MEND) was developed for nuclear delivery of pDNA (Akita et al. 2009). The formulation was composed of a protamine-pDNA complex core surrounded by four lipid layers, including two endosomal fusogenic outer layers and two nuclear fusogenic inner layers. The endosomal fusogenic layers were composed of 2-dioleoyl-sn-glycero-3-phosphatidylethanolamine (i.e. DOPE) and phosphatidic acid modified with stearylated octarginine. These two layers fused with the endosomal membrane and mediated endosomal escape. The two nuclear fusogenic layers were prepared with cardiolipin and DOPE, which could fuse with nuclear envelope to enhance nuclear delivery. In this study, the stepwise membrane fusion was determined by Förster resonance energy transfer (FRET). Greater transfection efficiency by the four-layered liposomes was demonstrated compared to Lipofectamine 2000 and the two-layered liposomes prepared with either endosomal fusogenic layers or nuclear fusogenic layers (Akita et al. 2009).

Nuclear internalization of pDNA was enhanced by conjugation of glucocorticoid to nanoparticles. The glucocorticoid receptor (i.e. NR3C1) is a nuclear receptor that can bind to glucocorticoid and is responsible for regulation of gene transcription involved in cell metabolism, anti-inflammatory, and immune responses. Binding of the glucocorticoids to the nuclear glucocorticoid receptor (GR) induces dilation of NPC, leading to enhanced drug and gene delivery to the nucleus. Preparation of dexamethasone-conjugated solid lipid nanoparticles (SLN) carrying a pDNA encoded with EGFP was reported (Wang et al. 2012a). They showed enhanced *in vitro* and *in vivo* nuclear translocation of the glucocorticoid-conjugated SLN compared to the control nanoparticles (Wang et al. 2012a).

3.4 LYSOSOMAL TARGETED DRUG DELIVERY

3.4.1 Lysosomes

Lysosomes, found in all eukaryotic cells except erythrocytes, are intracellular cytoplasmic organelles separated by a single lipid bilayer (Saftig et al. 2010) that entraps various hydrolytic enzymes including proteases, lipases, glycosidases, nucleases, phospho-lipases, phosphatases, and sulfatases for degrading macromolecules and other cell components (Appelqvist et al. 2013). Lysosomes are responsible for catabolic functions involved in phagocytosis, endocytosis, autophagy (i.e. degradation of large subcellular organelles), cell signaling, tumor metastasis, and regulation of cell death and apoptosis (Shen and Mizushima 2014). The intra-lysosomal environment is acidic (pH 4.5) due to the activity of proton pumps located in lysosomal membrane (i.e. vacuolar H^+-ATPase), which transports cytosolic protons into the lysosomal compartment (Settembre et al. 2013). Some glycosylated proteins including Lysosome Associate Membrane Protein 1(Lamp-1) and 2 (Lamp-2) are incorporated in the lysosomal membrane, forming glycocalyx as a protective layer from intra-lysosomal acidic hydrolysis (Settembre et al. 2013; Wartosch et al. 2015). Lysosomal integral membrane proteins 1 and 2 are considered as other important lysosomal membrane proteins and play significant roles in membrane fusion, endocytosis, and exocytosis of macromolecules (Eskelinen et al. 2003; Saftig and Klumperman 2009).

3.4.2 Lysosomal Cell Death

Lysosomal cell death is mediated by lysosomal membrane permeabilization (LMP). Once the permeability of lysosomal membrane increases, intra-lysosomal contents such as digestive enzymes are released into the cytosol, inducing cell necrosis or apoptosis. Cathepsins are the most studied lysosomal proteases accumulating in the cytoplasm during LMP. Cathepsins mediate cytosolic proteolysis, which leads to cell death (Terman et al. 2006). According to the amino acid sequence at their active sites, cathepsins are categorized into cysteine (cathepsins B, C, F, H, K, L, O, S, V, W, and X), serine (cathepsins A and G), and aspartic cathepsins (cathepsin D and E) (Groth-Pedersen and Jaatela 2013). Although cathepsins pose maximum activity in the lysosomal acidic environment, they are also active in cytoplasm (Jhaveri and Torchilin 2016). In addition to their hydrolytic activity in cytosol, some studies have shown the correlation between the lysosomal apoptosis pathway and the mitochondrial apoptosis pathway. These studies revealed an increase in mitochondrial membrane permeability and activation of caspase dependent apoptosis after activation of the cathepsins-mediated lysosomal apoptotic pathways (Gomez-Sintes et al. 2016).

3.4.3 Lysosomal Drug Delivery

Lysosome-targeted approaches include preparation of nanocarriers containing lysosomotropic agents, employment of cathepsin specific substrates for synthesizing nanocarriers, incorporation of ceramides in nanoparticle formulation, and development of pH-responsive nanocarriers.

3.4.3.1 Nanocarriers Containing Lysosomotropic Agents

In recent years, the failure of caspase-dependent cell apoptosis pathways has been reported in malignant cells and is considered as one of the mechanisms for anticancer drug resistance. Lysosome-mediated apoptosis is an attractive alternative for cancer therapy. Lysosomotropic agents, including hydroxychloroquine, chloropromazine,

TABLE 3.2
Lysosomotropic Detergent

	Compounds
Lysosomotropic detergents	N-dodecyl difluroacetamide
	N-dodecyl trifluroacetamide
	N-dodecyl-2,2,2-trifluroethylamine
	N-octyl-2,2,2-trifluroethylamine
	N-hexadecyl-2,2,2-trifufluroethylamine
	N-octadecyl-2,2,2-trifulroethylamine
	N-dodecyl-2,2-difluroethylamine
	N-carbobenzoxy-glyceryl-N-dodecyl-phenylalanine amide
	N-dodecyl imidazole
	N-retinyl morpholine
	N-carbobenzoxy-glyceryl-N,N-bis(2-chlroethyl)-L-phenylalanine amide
	N-(t-butyloxy carbonyl) glyceryl-N,N-di92-chloroethyl)-L-phenylalanine amide
	O-methyl serine dodecylamine
	N-carbobenzoxyglycyl-N-[2-(perfluroooctyl) ethyl]-L-phenylalanineamide

Source: Raymond, A. et al., Lysosomotropic detergent therapeutic agent, United States Patent, Patent Number: 4,719, 312, 1986.

aripiprazole, and sphingosine are compounds that destabilize lysosomal membrane (Kagedal et al. 2001; Boya et al. 2003; Bechara et al. 2014). Because of their weak base characteristics, these compounds become protonated, positively charged, and eventually trapped in the lysosomal environment (Gomez-Sintes et al. 2016). This leads to preferential accumulation in the lysosomes (De Duve et al. 1974). Other lysosomotropic detergents are summarized in Table 3.2. Preparation of a PEG polyethylene glycol - PLGA [poly (lactic-co-glycolic acid] nanoparticle, which was surface-modified with a monoclonal antibody (mAb) and loaded with hydroxylchloroquine, a lysosomotropic agent was reported in the literature (Mansilla et al. 2010). Treatment of the drug-resistant chronic lymphocyte leukemia cells with the drug containing targeted nanoparticles induced ~four-fold increased cell apoptosis compared to the drug-free formulation.

3.4.3.2 Nanocarriers Modified with Cathepsin-Specific Substrates

The application of lysosomotropic agents has been limited due to their off-target toxicity in normal cells. Therefore, strategies that can differentiate the lysosomes in cancer cells from normal ones have been developed for cancer-specific and lysosome-targeted drug delivery. Among these strategies, including various substrates for cathepsin B (Cat B) in the targeted formulation is widely studied. Cat B, compared to other proteins in the cathepsin family, is the dominant hydrolase involved in malignant cells and is often overexpressed in cancer cells (Kirkegaard and Jaattela 2009). Encapsulation of doxorubicin into liposomes with surface conjugation of a tetra-peptide Gly-Leu-Phe-Gly, a substrate for Cat B was investigated (Maniganda et al. 2014). The conjugated liposomes posed higher lysosomal accumulation and higher cytotoxicity compared to both non-targeted liposomes and free doxorubicin in HeLa cells. This was due to the high affinity of the synthetic tetra peptide substrate with cathepsin B that was overexpressed in lysosomes of the malignant cells (Maniganda et al. 2014).

Conjugation of another Cat B substrate peptide with a sequence of Arg-Arg to the surface of a gold-zinc oxide nanoparticle led to increased lysosomal accumulation and activation of the LMP-dependent apoptosis in HepG2 cells (Gao et al. 2014). Resistance mechanisms against LMP-mediated cell apoptosis have been reported in cancer cells. Expression of heat shock protein 70 (HSP 70) and its translocation into the inner part of the lysosomal membrane in tumor cells can stabilize the membrane and inhibit LMP (Nylandsted et al. 2004; Kroemer and Jaattela 2005). Suppression of HSP 70 might be an efficient strategy to overcome resistance in tumor cells. Delivery of an anti-HSP 70 siRNA using chitosan-based nanoparticles was reported (Matokanovic et al. 2013), and the study showed that downregulation of HSP 70 induced significant toxicity in leukemia and glioblastoma cells.

3.4.3.3 Ceramide Containing Nanocarriers

In addition to nanocarriers modified with a cathepsin substrate, nanoplatforms containing exogenous ceramides can be used for triggering lysosomal apoptotic pathways in cells. It is believed that ceramides can destabilize lysosomes and enhance LMP (Hannun and Obeid 2008). Ceramides are types of sphingolipids located in plasma membrane and lysosomal membrane. Ceramides act as second messengers in cell signaling. The cytotoxic effects of transferrin-modified liposomes containing ceramides on HeLa cells was studied (Koshkaryev et al. 2012). They showed that the transferrin-liposomes exhibited increased selectivity for cancer cells and enhanced lysosomal uptake through the transferrin receptor (TfR) mediated endocytic pathway compared to plain liposomes. In this study, it is suggested that lysosomal delivery of ceramides increased the release of cathepsin D from lysosomes to the cytoplasm, triggering apoptosis in the cancer cells (Koshkaryev et al. 2012). It was demonstrated that ceramides could selectively interact with cathepsin D to enhance cathepsin dependent cell apoptosis (Heinrich et al. 1999).

3.4.3.4 Metal Nanoparticles

Induction of LMP using iron oxide magnetic nanoparticles targeted to the epidermal growth factor receptor (EGFR) that is overexpressed in cancer cells was reported (Domenech et al. 2013). After binding of the targeted magnetic nanoparticles to the EGFR, the ligand-receptor complexes were internalized into the endosomes. These targeted iron oxide nanoparticles then accumulated in the lysosome and were found mainly in the lipid bilayer of the lysosomes. Under an alternating magnetic field (AMF), the magnetic nanoparticles rotated and dissipated heat that mediated disruption of the membrane and consequently increased LMP, resulting in increased levels of Cat B in cytosol.

In a comprehensive study, the effects of different nanoparticles including uncoated ultra-small super paramagnetic iron oxide (USPIO), oleic acid coated USPIO, fluorescently labeled silica nanoparticles, TiO_2 nanoparticles, and polymeric nanoparticles prepared with PLGA-PEO [poly (ethylene oxide)] on induction of lysosomal autophagy were investigated in human cerebral endothelial cells (HCEC) (Kenzaoui et al. 2012). The MTT cytotoxicity assay revealed no cytotoxic effect for PLGA-PEO nanoparticles but maximum reduction in cell viability for uncoated USPIO. Other types of nanoparticles displayed medium cytotoxicity. Data obtained by Western blot showed overexpression of cathepsin B and D in cellular extracts, suggesting that the nanoparticles promoted lysosomal protease activity. TEM imaging data supported that the lysosomal accumulation and induction of autophagy could be the major mechanism for the cytotoxicity of these nanocarriers (Kenzaoui et al. 2012).

3.4.3.5 pH-Responsive Nanocarriers

Preparation of a series of pH-activatable co-polymeric micellar nanoparticles for targeting of endocytic organelles including lysosomes was reported (Zhou 2011). They investigated self-assembled micelles prepared with co-polymers of poly (ethylene oxide) (PEO)–poly methacrylate containing tertiary amines. It was shown that at a high pH, the tertiary amines were de-protonated and the micellar structure remained intact. At acidic pHs (5.0–5.5), the tertiary amines were protonated with increased solubility and charge repulsion, causing instability of the micelles and release of the entrapped drug in the lysosomes. Preparation of a drug delivery system consisting of superparamagnetic (Fe_3O_4) nanoparticles, coated with a tri-block co-polymer of MPEG [methoxy poly (ethylene glycol)] - P(MMA-co-nBMA) [poly (methacrilic acid-co-n-butyl methacrylate)]-PGMA [poly (glycerol monomethacrylate)] that contained adriamycin as the active ingredient was investigated (Guo et al. 2011). Entrapment of the drug into the micelles was mediated by pH-dependent ionic and hydrophobic interactions between adriamycin and the carboxylic groups, and therefore, the system exhibited pH-responsive drug release: 40% and 100% drug release at pH 7.4 and 3.5, respectively. In another study, an amphipathic conjugate was synthesized by conjugating PEG and 4-(N)-stearoyl gemcitabine via an acid-sensitive hydrozone bond. Multiple molecules of this conjugate then self-assembled into micelles (Zhu et al. 2012). It was demonstrated that the micelles exhibited pH-dependent drug release. At pH 5.5 almost all of the entrapped drug was released after four hours of incubation, but the nanocarriers were stable during eight hours of incubation at pH 6.8.

3.5 ENDOPLASMIC RETICULUM (ER) TARGETED DRUG DELIVERY

3.5.1 Endoplasmic Reticulum (ER) and Cell Apoptosis

The endoplasmic reticulum (ER), a eukaryotic subcellular organelle, is responsible for post translational modifications (e.g. glycosylation, lipidation, and formation of disulphide bonds) and appropriate folding of proteins. ER also plays an important role in intracellular calcium homeostasis (Healy et al. 2009). ER contains high concentrations of chaperones that enhance appropriate folding of newly synthesized proteins. These chaperons include calnexin (CNX), peptidyl prolyl isomerases (PPI), protein disulphide isomerase (PDI), glucose-regulated proteins (GRPs), and calreticulin (CRT) (Ruddon and Bedows 1997; Gething and Sambrook 1992). Chaperones also recognize correctly formed peptides and proteins and facilitate their translocation into the Golgi apparatus for further processing, while the inappropriately formed polypeptides are recognized for degradation in the lumen of the endoplasmic reticulum (Kuznetsov et al. 1997; Ellgaard et al. 1999). Some diseases such as cancer display ER dysfunctions, such as inappropriate folding of oncoproteins, a decreased energy level in ER caused by hypoxia, altered oxidizing capacity in ER, and a reduced calcium level in ER. The ER dysfunction in cancer cells leads to reduction in chaperone functionality and consequently increased accumulation of unfolded and/or misfolded proteins in the lumen of ER, initiating signals known as "ER stress". To respond to the ER stress, unfolded protein response (UPR) is induced inside the cell.

UPR is initiated by the activation of some ER trans-membrane stress sensors including inositol-requiring enzyme 1 (IRE1), activating transcription factor 6 (ATF 6), and pancreatic ER kinase-like ER kinase (PERK) (Schroder and Kaufman 2005). As the result of UPR, the expression of some proteins is increased, including chaperones, folding enzymes, and protein disulfide isomerase (Lee 2001), to enhance the folding capacity. In addition, a process known as ER-associated degradation (ERAD) that mediates the degradation of mis-folded proteins is also increased (Rutkowski and Kaufman 2004). Prolonged and excessive ER stress can cause cell apoptosis rather than UPR. This process is known as ER stress-induced apoptosis (Urano et al. 2004; Szegezdi et al. 2006). It is believed that activation of transcription factor C/EBP homologous protein (CHOP) due to excessive ER stress can mediate mitochondrial apoptotic pathways through activation of proapototic Bcl-2 family members (e.g. BH3) and caspase 12. CHOP is also involved in suppression of anti-apoptotic proteins (McCullough et al. 2001). Therefore, induction of a prolonged ER stress response in tumor cells is an attractive approach for anticancer therapy.

3.5.2 Drug Delivery to ER

ER-targeted drug delivery methods encompass surface modification of nanocarriers with an ER translocating signal and nanocarriers with phospholipids found in ER.

3.5.2.1 Surface Modification of Nanocarriers with ER Translocating Signals

A peptide bearing c-terminal sequence of Lys-Asp-Glu-Leu (KDEL) as an ER translocation signal was studied for surface modification of gold nanoparticles. KDEL is a trans-membrane endogenous peptide located in the lumen of endoplasmic reticulum and is recognized by the KDEL receptor (KDELR) in the ER-Golgi intermediate compartment. The modified gold nanoparticles were taken up by Sol 8 myogenic cells via clathrin-mediated endocytosis, and quickly localized to the ER 5–10 minutes post incubation by the coat protein I (COPI) mediated retrograde transport pathway through the Golgi apparatus. After 60 minutes post incubation, approximately 70% of the nanoparticles accumulated in the ER (Wang et al. 2013). It was demonstrated that the KDEL-modified gold nanoparticles delivered three times more siRNA to the ER compared to Lipofectamine 2000-modified gold nanoparticles in differentiated myotubes (C2C12) (Acharya and Hill 2014).

3.5.2.2 Nanocarriers with Phospholipids Found in ER

In a study, it was demonstrated that micelles prepared with poly (ethylene glycol)-phosphatidylethanolamine (PEG-PE) significantly accumulated in the lumen of ER in A549 tumor cells with 75.3% of the intracellular micelles localized in the ER (Wang et al. 2012b). The micelles later induced ER dependent apoptosis. The data revealed no significant difference in intracellular distribution and ER accumulation between the PEG-PE micelles and PE, suggesting that PE might mediate the ER targeting. ER is an important organelle for lipid metabolism, and PE as a phospholipid, exhibits a high affinity for ER. In contrast to tumor cells, normal cells exhibited better tolerance for the ER stress by the overexpression of UPR feedback proteins.

The ER-like liposomes composed of 1,2-dioleoyl-*sn*-glycero-3-phosphoethanolamine (DOPE), cholesteryl hemisuccinate (CH), and L-α-phosphatidylinositol (PI) displayed four-fold increased ER accumulation compared to control liposomes prepared with DOPE and CH only, indicating the critical role of PI for ER-targeted delivery (Pollock et al. 2010). They also showed that incorporation of 20 mol% L-α-phosphatidylserine (PS) to the control liposomes significantly increased their ER accumulation by three-fold.

3.6 CONCLUSION AND FUTURE PERSPECTIVES

Various nanocarrier systems have been developed to target diseased tissue, and they are often designed to release the payload in the extracellular environment in the tissue, followed by drug penetration through the cell membrane, reaching the subcellular target. Recently, these targeted delivery systems have been evolved to deliver the drug directly to the intracellular target for improved efficacy. A high amount of drug and multiple components can be incorporated into a single nanocarrier to overcome various delivery barriers and achieve effective delivery to the intracellular target, including mitochondria, nucleus, lysosomes, and endoplasmic reticulum (ER). In order to target subcellular organelles, nanoparticles are often modified with various targeting moieties. Mitochondriotropics, mitochondrial-penetrating peptides (MPPs), and mitochondrial targeting signals (MTSs) were used for mitochondrial translocation. Nuclear localization signals (NLSs), aptamers, and glucocorticoids were investigated for targeting nanocarriers to nucleus. Lysosomotropic agents, cathepsin-specific substrates, and ceramide were used for modifying nanoparticles to enhance their accumulation in lysosomes. Some lipids such as phosphatidylethanolamine (PE), L-α-phosphatidylinositol (PI), and L-α-phosphatidylserine (PS) as well as ER translocation peptide sequences such as KDEL were applied to target nanoparticles to ER.

As a majority of these targeted nanoparticles enter the cells by endocytosis, endosomal escape of the carriers is one of the limiting steps for delivery. Development of pH responsive or fusogenic nanocarriers can enhance the endosomal escape. Alternatively, caveolae-mediated endocytosis may bypass the lysosomal degradation. Development of a drug delivery system that can simultaneously target multiple subcellular organelles might be effective for drug therapeutics. One significant challenge for intracellularly targeted drug delivery is the correlation between *in vitro* findings and *in vivo* results. For drug delivery in an *in vivo* setting, the delivery system must overcome multiple barriers before reaching the target cell, including plasma stability, MPS clearance, renal clearance, and non-specific uptake by other tissues. Formulation components in these subcellularly targeted nanoparticles could induce non-specific uptake by other tissues and compromise the pharmacokinetics. As a result, the *in vivo* delivery may be poor while the *in vitro* cell culture results are promising. A robust *in vitro* system that can accurately predict the *in vivo* outcome for those nanoparticles must be developed. Moreover, these intracellularly targeted nanoparticles tend to incorporate multiple components in the formulation with complicated structures and procedures for fabrication. As such, the scale-up manufacturing of these nanocarriers may be difficult. Scientists from different disciplines must collaborate to develop robust and reproducible methods to manufacture these complicated nanosystems.

Development of efficient intracellularly targeted drug delivery systems also requires a further understanding of biological membranes, mechanisms involved in cellular uptake, and intracellular trafficking. To reduce off-target effects of the nanoparticles, we must improve our knowledge regarding to structural, functional, and pathophysiological differences of intracellular organelles in cancer cells and normal cells. Although the intracellular organelle targeting is still in its infancy, it is a rapidly growing field with significant potential to improve current nanoparticle drug delivery for cancer therapy.

ACKNOWLEDGMENTS

The authors would like to thank Ali Reza Pelarak at School of Pharmacy, Hamadan University of Medical Sciences, Hamadan, Iran, and Paniz Mahjoub, at Department of Chemical Engineering, Technical University of Hamadan, Iran for producing the figures in this chapter. Dr. Shyh-Dar Li would like to thank multiple funding agencies for supporting his research in targeted drug delivery, including Canadian Institutes of Health Research, National Institutes of Health, Natural Sciences and Engineering Research Council of Canada, Canada Foundation for Innovation, and Prostate Cancer Foundation. Dr. Li holds the CIHR New Investigator Award and the Angiotech professorship in Drug Delivery.

REFERENCES

Acharya, S., and R.A. Hill. 2014. High efficacy gold-KDEL peptide-siRNA nanoconstruct mediated transfection in C2C12 myoblasts and myotubes, *Nanomedicine. Nanotechnol. Biol. Med.* 10: 39–37.

Adams, J.M., and S. Cory. 2007. The Bcl-2 apoptotic switch in cancer development and therapy. *Oncogene.* 26: 1324–37.

Agemy, L., Friedmann-Morvinski D., Kotamraju, V.R., et al. 2011. Targeted nanoparticle enhanced proapoptotic peptide as potential therapy for glioblastoma. *Proc. Natl. Acad. Sci. U. S. A.* 108: 17450–5.

Akita, H., Kudo, A., Minoura, A., et al. 2009. Multi-layered nanoparticles for penetrating the endosome and nuclear membrane via a step-wise membrane fusion process. *Biomaterials.* 30: 940–9.

Appelqvist, H., Wäster, P., Kagedal, K., et al. 2013. The lysosome: from waste bag to potential therapeutic target. *J. Mol. Cell. Biol.* 5: 214–26.

Austin, L.A., Ahmad, S., Kang, B., et al. 2015. Cytotoxic effects of cytoplasmic-targeted and nuclear-targeted gold and silver nanoparticles in HSC-3 cells: a mechanistic study. *Toxicol. In Vitro.* 9: 694–705.

Baldo, A., Van den Akker, E., Bergmans, H.E., Lim, F., Pauwels, K. 2013. General considerations on the biosafety of virus-derived vectors used in gene therapy and vaccination. *Curr. Gene Ther.* 13: 385–94.

Barczyk, K., Kreuter, M., Pryjma, J. 2005. Serum cytochrome c indicates *in vivo*-apoptosis and it can serve as a prognostic marker during cancer therapy. *Int. J. Cancer.* 114: 167–73.

Battigelli, A., Russier, J., Venturelli, E., et al. 2013. Peptide-based carbon nanotubes for mitochondrial targeting. *Nanoscale* 5: 9110–17.

Bechara, A., Barbosa, C.M., Paredes-Gamero, E.J., et al. 2014. Palladacycle (BPC) antitumour activity against resistant and metastaticcell lines: the relationship with cytosolic calcium mobilisation and cathepsin Bactivity. *Eur. J. Med. Chem.* 79: 24–33.

Belting, M., Sandgren, S., Wittrup, A. 2005. Nuclear delivery of macromolecules: barriers and carriers. *Adv. Drug Deliv. Rev.* 57: 505–27.

Bernardi, P., Scorrano, L., Colonna, R., et al. 1999. Mitochondria and cell death Mechanistic aspects and methodological issues. *Eur. J. Biochem.* 264: 687–701.

Biswas, S., Dodwadkar, N.S., Sawant, R.R., et al. 2011. Surface modification of liposomes with rhodamine-123-conjugated polymer results in enhanced mitochondrial targeting. *J. Drug. Target.* 19: 552–61.

Boya, P., Gonzalez-Polo, R.A., Poncet, D., et al. 2003. Mitochondrial membrane permeabilizationis a critical step of lysosome-initiated apoptosis induced byhydroxychloroquine. *Oncogene.* 22: 3927–36.

Breunig, M., Bauer, S., Goepferich, A. 2008. Polymers and nanoparticles: intelligent tools for intracellular targeting. *Eur. J. Pharm. Biopharm.* 68: 112–28.

Bruno, J.G. 2013. A review of therapeutic aptamer conjugates with emphasis on new approach. *Pharmaceuticals.* 6: 340–57.

Burrow, S.M., Phoenix, D.A., Wainwright, M., et al. 2002. Intracellular localization studies of doxorubicin and Victoria Blue BO in EMT6-S and EMT6-R cells using confocal microscopy. *Cytotechnology.* 39: 15–25.

Culver, K.W., Ram, Z., Walbridge, S., et al. 1992. In vivo gene transfer with retroviral vector- producer cells for treatment of experimental brain tumors. *Science.* 256: 1550–2.

Curiel, D.T., Agarwal, S., Wagner, E., et al. 1991. Adenovirus enhancement of transferring-polylysine-mediated gene delivery. *Proc. Natl. Acad. Sci. U. S. A.* 88: 8850–4.

Dam, D.H., Lee, J.H., Sisco, P.N., et al. 2012. Direct observation of nanoparticle-cancer cell nucleus interactions. *ACS Nano.* 6: 3318–26.

De Duve, C., De Barsy, T., Poole, B., et al. 1974. Lysosomotropic agents. *Biochem. Pharmacol.* 23: 2511–31.

Denora, N., Laquintana, V., Lopalco, A., et al. 2013. In vitro targeting and imaging the translocator protein TSPO 18-kDa through G(4)-PAMAM-FITC labeled dendrimer. *J. Control. Release.* 172: 1111–25.

Ding, C., Gu, J., Qu, X., et al. 2009. Preparation of multifunctional drug carrier for tumor-specific uptake and enhanced intracellular delivery through the conjugation of weak acid labile linker. *Bioconjugate. Chem.* 20: 1163–70.

Domenech, M., Marrero-Berrios, I., Torres-Lugo, M., et al. 2013. Lysosomal membrane permeabilization by targeted magnetic nanoparticles in alternating magnetic fields. *ACS Nano.* 25: 5091–101.

D'Souza, G.G., Cheng, S.M., Bod Dapati, S.V., et al. 2008. Nanocarrier-assisted sub-cellular targeting to the site of mitochondria improves the pro-apoptotic activity of paclitaxel. *J. Drug. Target.* 16: 578–85.
Duvvuri, M., and J.P. Krise. 2005. Intracellular drug sequestration events associated with the emergence of multidrug resistance: a mechanistic review. *Front. Bio. Sci.* 10: 1499–1509.
Ellgaard, L., Molinari, M., Helenius, A. 1999. Setting the standards: quality control in the secretory pathway. *Science.* 286: 1882–8.
Ernsting, M.J., Murakami, M., Roy, A., et al. 2013. Factors controlling the pharmacokinetics, biodistribution and intratumoral penetration of nanoparticles. *J. Control. Release.* 127: 782–94.
Eskelinen, E.L., Tanaka, Y., Saftig, P. 2003. At the acidic edge: emerging functions for lysosomal membrane proteins. *Trends. Cell. Biol.* 13: 137–45.
Fan, Y., Li, C., Chen, D. 2016. pH-activated size reduction of large compound nanoparticles for in vivo nucleus-targeted drug delivery. *Biomaterials.* 85: 30–9.
Finkel, E. 2001. The mitochondrian: is it central to apoptosis? *Science.* 92: 624–6.
Gao, W., Cao, W., Zhang, H., et al. 2014. Targeting lysosomal membrane permeabilization to induce and image apoptosis in cancer cells by multifunctional Au-ZnO hybrid nanoparticles. *Chem. Commun (Camb).* 50: 8117–20.
Gasiorowski, J.Z., and D.A. Dean. 2003. Mechanisms of nuclear transport and interventions. *Adv. Drug Deliv. Rev.* 55: 703–16.
Gething, M.J., and J. Sambrook. 1992. Protein folding in the cell. *Nature.* 355: 33–45.
Ginisty, H., Sicard, H., Roger, B., et al. 1990. Structure and functions of nucleolin. *J. Cell. Sci.* 112:761–772.
Gomez-Sintes, R., Ledesma, M.D., Boya, P. 2016. Lysosomal cell death mechanism in aging. *Age. Res. Rev.* 32:150–68.
Gregorevic, P., Blakinship, M.J., Allen, J.M., et al. 2004. Systemic delivery of genes to striated muscles using adeno-associated viral vectors. *Nat. Med.* 10: 828–34.
Groth-Pedersen, L., and M. Jaatela. 2013. Combating apoptosis and multidrug resistant cancers by targeting lysosomes. *Cancer. Lett.* 332: 265–74.
Gulbins, E., Dreschers, S., Bock, J. 2003. Role of mitochondria in apoptosis. *Exp. Physiol.* 88: 85–90.
Guo, M., Que, C., Wang, C., et al. 2011. Multifunctional superparamagnetic nanocarriers with folate-mediated and pH-responsive targeting properties for anticancer drug delivery. *Biomaterials.* 32: 185–94.
Hakim, A., and O.S. Usmani. 2015. Enhancing nuclear translocation: perspectives in inhaled corticosteroid therapy. *Ther. Deliv.* 6: 443–51.
Hannun, Y.A., and L.M. Obeid. 2008. Principles of bioactive lipid signalling: lessons from sphingolipids, *Nat. Rev. Mol. Cell Biol.* 9: 139–50.
Healy, J.M.S., Gorman, A.M., Mousavi-Shafaei, P., et al. 2009. Targeting the endoplasmic reticulum-stress response as an anticancer strategy. *Eur. J. Pharmacol.* 625: 234–46.
Heinrich, M., Wickel, M., Schneider-Brachert,W., et al. 1999. Cathepsin D targeted by acid sphingomyelinase-derived ceramide. *The EMBO J.* 18: 5252–63.
Hengartner, M.O. 2000. The biochemistry of apoptosis. *Nature.* 407: 770–6.
Hiendleder, S., Schmutz, S.M., Erhartdt, G., et al. 1999. Trans mitochondrial differences and varying levels of heteroplasmy in nuclear transfer cloned cattle. *Mol. Reprod. Dev.* 54: 24–31.
Horobin, R.W., Trapp, S., Weissig, V. 2007. Mitochondriotropics: a review of their mode of action, and their applications for drug and DNA delivery to mammalian mitochondria. *J. Control. Release.* 121: 125–36.
Horton, K.L., Stewart, K.M., Fonseca, S.B., et al. 2008. Mitochondria-penetrating peptides. *Chem. Biol.* 15: 375–82.

Hoshino, A., Fujioka, K., Oku, T., et al. 2004. Quantum dots targeted to the assigned organelle in living cells. *Microbiol. Immunol.* 48: 985–94.

Hsu, J., Northrup, L., Bhowmick, T., et al. 2012. Enhanced delivery of alphaglucosidase for Pompe disease by ICAM-1-targeted nanocarriers: comparative performance of a strategy for three distinct lysosomal storage disorders. *Nanomedicine. Nanotechnol. Biol. Med.* 8: 731–9.

Hsu, J., Serrano, D., Bhowmick, T., et al. 2011. Enhanced endothelial delivery and biochemical effects of alpha-galactosidase by ICAM- 1-targeted nanocarriers for Fabry disease. *J. Control. Release.* 149: 323–31.

Huang, J.G., Leshuk, T., Gu, F.X. 2011. Emerging nanomaterials for targeting subcellular organelles. *Nano Today.* 6: 478–92.

Huo, S., Jin, S., Ma, X., et al. 2014. Ultra-small gold nanoparticles as carriers for nucleus-based gene therapy due to size dependent nuclear entry. *ACS Nano.* 8: 5852–62.

Ito, E., Yip, K.W., Katz, D., et al. 2009. Potential use of cetrimonium bromide as an apoptosis-promoting anticancer agent for head and neck cancer. *Mol. Pharmacol.* 76: 969–83.

Jans, D.A., Chan, S.K., Huebner, S. 1998. Signals mediating nuclear targeting and their regulation: application in drug delivery. *Med. Res. Rev.* 18: 189–223.

Jeon, O., Lin, H.W., Lee, M., et al. 2007. Poly(L-lactide-co-glycolide) nanospheres conjugated with a nuclear localization signal for delivery of plasmid DNA. *J. Drug. Target.* 15: 190–98.

Jhaveri, A., and V. Torchilin. 2016. Intracellular delivery of nanocarriers and targeting to subcellular organelles. *Expert. Opin. Drug. Deliv.* 13: 49–70.

Jiang, L., Li, L., He, X., et al. 2015. Overcoming drug-resistant lung cancer by paclitaxel loaded dual-functional liposomes with mitochondria targeting and pH-response. *Biomaterials.* 52: 126–39.

Kagedal, K., Zhao, M., Svensson, I., et al. 2001. Sphingosine-induced apoptosis is dependent on lysosomal proteases. *Biochem. J.* 359: 335–43.

Kalderon, D., Roberts, B.L., Richardson, W.D., et al. 1984. A short amino acid sequence able to specify nuclear location. *Cell.* 39: 499–509.

Kenzaoui, B.H., Bernasconi, C.C., Guney-Ayra, S., et al. 2012. Induction of oxidative stress, lysosome activation and autophagy by nanoparticles in human brain-derived endothelial cells. *Biochem. J.* 441: 813–21.

Kim, J.H., Chae, M., Kim, W.K., et al. 2011. Salinomycin sensitizes cancer cells to the effects of doxorubicin and etoposide treatment by increasing DNA damage and reducing p21 protein. *Br. J. Pharmacol.* 162: 773–84.

Kirkegaard, T., and M. Jaattela. 2009. Lysosomal involvement in cell death and cancer. *Biochim. Biophys. Acta.* 1793: 746–54.

Koshkaryev, A., Piroyan, A., Torchilin, V.P. 2012. Increased apoptosis in cancer cells in vitro and in vivo by ceramides in transferrin-modified liposomes. *Cancer. Biol. Ther.* 13: 50–60.

Kroemer, G., Galluzi, L., Brenner, C. 2007. Mitochondrial membrane permeabilization in cell death. *Physiol. Rev.* 87: 99–163.

Kroemer, G., and M. Jaattela. 2005. Lysosomes and autophagy in cell death control. *Nat. Rev. Cancer.* 5: 886–97.

Kuznetsov, G., Chen, L.B., Nigam, S.K. 1997. Multiple molecular chaperones complex with mis folded large oligomeric glycoproteins in the endoplasmic reticulum. *J. Biol. Chem.* 272: 3057–63.

Lee, A.S. 2001. The glucose-regulated proteins: stress induction and clinical applications. *Trends. Biochem. Sci.* 26: 504–10.

Li, L.Y., Luo, X., Wang, X. 2001. Endonuclease G is an apoptotic DNase when released from mitochondria. *Nature.* 412: 95–9.

Li, S.D., and L. Huang. 2007. Non-viral is superior to viral gene delivery. *J. Control. Release.* 123: 181–3.

Li, S.D., and L. Huang. 2008. Pharmacokinetics and biodistribution of nanoparticles. *Mol. Pharm.* 5: 496–504.

Lim, S.H., Wu, L., Burgess, K., Lee, H.B. 2009. New cytotoxic rosamine derivatives selectively accumulate in the mitochondria of cancer cells. *Anticancer Drugs.* 20: 461–8.

Ma, D. 2014. Enhancing endosomal escape for nanoparticle mediated siRNA delivery. *Nanoscale.* 6: 6415–25.

Ma, X., Zhang L.H., Wang, L.R., et al. 2012. Single-walled carbon nanotubes alter cytochrome c electron transfer and modulate mitochondrial function. *ACS Nano.* 6: 10486–96.

Maddika, S., Ande, S.R., Panigrahi, S. 2007. Cell survival, cell death and cell cycle pathways are interconnected: Implications for cancer therapy. *Drug. Resis. Updat.* 10: 13–9.

Maheshri, N., Koerber, J.T., Kaspar, B.K., et al. 2006. Directed evaluation of adeno-associated virus yields enhanced gene delivery vectors. *Nat. Biotechnol.* 24: 198–204.

Maity, A.R., and D. Stepensky. 2016. Efficient subcellular targeting to the cell nucleus of quantum dots densely decorated with a nuclear localization sequence peptide. *ACS. Appl. Mater. Interfaces.* 8: 2001–9.

Maniganda, S., Sankar, V., Nair, J.B., et al. 2014. A lysosome-targeted drug delivery system based on sorbitol backbone towards efficient cancer therapy. *Organic. Biomol. Chem.* 12: 6564–9.

Mansilla, E., Marin, G.H., Nunez, L., et al. 2010. The lysosomotropic agent, hydroxychloroquine, delivered in a biodegradable nanoparticles system, overcomes drug resistance of B-chronic lymphocytic leukemia cells *in vitro. Cancer. Biother. Radiopharm.* 25: 97–103.

Matissek, K.J., Okal. A., Mossalam. M., et al. 2014. Delivery of a monomeric p53 subdomain with mitochondrial targeting signals from pro-apoptotic Bak or Bax. *Pharm Res.* 31: 2503–15.

Matokanovic, M., Barisic, K., Filipovic-Grcic, J., et al. 2013. Hsp70 silencing with siRNA in nanocarriers enhances cancer cell death induced by the inhibitor of Hsp90. *Eur. J. Pharm. Sci.* 50: 149–58.

McCullough, K.D., Martindale, J.L., Klotz, L.O., et al. 2001. Gadd153 sensitizes cells to endoplasmic reticulum stress by down-regulating Bcl2 and perturbing the cellular redox state. *Mol. Cell. Biol.* 21: 1249–59.

Men, Y., Wang, X.X., Li, R.J., et al. 2011. The efficacy of mitochondrial targeting antiresistantepirubicin liposomes in treating resistant leukemia in animals. *Int. J. Nanomed.* 6: 3125–37.

Misra, R., and S.K. Sahoo. 2010. Intracellular trafficking of nuclear localization signal conjugated nanoparticles for cancer therapy. *Eur. J. Pharm. Sci.* 39: 152–63.

Mitchell, P., and J. Moyle. 1965a. Evidence discriminating between the chemical and the chemiosmotic mechanisms of electron transport phosphorylation. *Nature.* 208: 1205–6.

Mitchell, P., and J. Moyle. 1965b. Stoichiometry of proton translocation through the respiratory chain and adenosine triphosphatase systems of rat mitochondria. *Nature.* 208: 147–51.

Modica-Napolitano, J.S., and J.R. Aprille. 2001. Delocalized lipophilic cations selectively target the mitochondria of carcinoma cells. *Adv. Drug. Deliv. Rev.* 49: 63–70.

Modica-Napolitano, J.S., and V. Weissing. 2015. Treatment strategies that enhance the efficacy and selectivity of mitochondria-targeted anticancer agents. *Int. J. Mol. Sci.* 16: 17394–421.

Nativo, P., Prior, I.A., Burst, M. 2008. Uptake and intracellular fate of surface modified gold nanoparticles. *ACS Nano.* 2: 1639–44.

Nylandsted, J., Gyrd-Hansen, M., Danielewicz, A., et al. 2004. Heat shock protein 70 promotes cell survival by inhibiting lysosomal membrane permeabilization. *J. Exp. Med.* 200: 425–35.

Paine, P.L., Moore, L.C., Horowitz, S.B. 1975. Nuclear envelope permeability. *Nature.* 254: 109–14.

Palmer, J.A., Branston, R.H., Lilley, C.E., et al. 2000. Development and optimization of herpes simplex virus vectors for multiple long-term gene delivery to the peripheral nervous system. *J. Virol.* 74: 5604–18.

Pan, L., He, Q., Liu, J., et al. 2012. Nuclear-targeted drug delivery of tat peptide-conjugated monodisperse mesoporous silica nanoparticles. *J. Am. Chem. Soc.* 134: 5722–25.

Paulus, W., Baur, I., Boyce, F.M., et al. 1996. Self-contained, tetracycline-regulated retroviral vector system for gene delivery to mammalian cells. *J. Virol.* 70: 62–7.

Pelicano, H., Carney, D., Huang, P. 2004. ROS stress in cancer cells and therapeutic implications. *Drug Resist. Updates.* 7: 97–110.

Perez, R.P. 1998. Cellular and molecular determinants of cisplatin resistance. *Eur. J. Cancer.* 34: 1535–42.

Pollock, S., Antrobus, R., Newton, L., et al. 2010. Uptake and trafficking of liposomes to the endoplasmic reticulum. *FASEB J.* 24: 1866–78.

Pommier, Y., Proquier, P., Strumberg, D. 1998. Mechanism of action of eukaryotic topoisomerase I and drug targeted to enzyme. *Biochim. Biophys. Acta.* 1400: 83–106.

Raymond, A., Firestone, A., Fanwood, N.J. 1986. Lysosomotropic detergent therapeutic agent. United States Patent. Patent Number: 4,719, 312.

Roy, A., and S.D. Li. 2016. Modifying the tumor microenvironment using nanoparticle therapeutics. *Wiley. Interdiscip. Rev. Nanomed. Nanobiotechnol.* 8: 891–908.

Ruddon, R.W., and E. Bedows. 1997. Assisted protein folding. *J. Biol. Chem.* 272: 3125–28.

Rutkowski, D.T., and R.J. Kaufman. 2004. A trip to the ER: coping with stress. *Trends. Cell. Biol.* 14: 20–8.

Saftig, P., and J. Klumperman. 2009. Lysosome biogenesis and lysosomal membrane proteins: trafficking meets function. *Nat. Rev. Mol. Cell. Biol.* 10: 623–35.

Saftig, P., Schröder, B., Blanz, J. 2010. Lysosomal membrane proteins: life between acid and neutral conditions. *Biochem. Soc. Trans.* 38: 1420–3.

Sahay, D.L., Alakhova, D.Y., Kabanov, A.V. 2010. Endocytosis of nanomedicines. *J. Control. Release.* 145: 182–95.

Sakhrani, N.M., and H. Padh. 2013. Organelle targeting: third level of drug targeting. *Drug. Des. Dev. Ther.* 7: 585–99.

Salnikov, V., Lukyanenko, Y.O., Frederick C.A., et al. 2007. Probing the outer mitochondrial membrane in cardiac mitochondria with nanoparticles. *Biophys. J.* 92: 1058–71.

Sarko, D., Beijer, B., Boy, R.G., et al. 2010. The pharmacokinetics of cell-penetrating peptides. *Mol. Pharm.* 7: 2224–31.

Schroder, M., and R.J. Kaufman. 2005. The mammalian unfolded protein response. *Ann. Rev. Biochem.* 74: 739–89.

Schwarze, S.R., Ho, A., Vocero-Akbani, A., et al. 1999. In vivo protein transduction: delivery of a biologically active protein into the mouse. *Science.* 285: 1569–72.

Serviddio, G., Ronmano, A.D., Cassano, T., et al. 2011. Principles and therapeutic relevance for targeting mitochondria in aging and neurodegenerative diseases. *Curr. Pharm. Des.* 17: 2036–55.

Sethuraman, V.A., and Y.H. Bae. 2007. TAT peptide-based micelle system for potential active targeting of anti-cancer agents to acidic solid tumors. *J. Control. Release.* 118: 216–24.

Settembre, C., Fraldi, A., Medina, D.L., et al. 2013. Signals from the lysosome: a control centre for cellular clearance and energy metabolism. *Nat. Rev. Mol. Cell Biol.* 14: 283–96.

Sharma, A., Soliman, G.M., Al-Hajaj, N., et al. 2012. Design and evaluation of multifunctional nanocarriers for selective delivery of coenzyme Q10 to mitochondria. *Biomacromolecules.* 13: 239–52.

Shen, H.M., and N. Mizushima. 2014. At the end of the autophagic road: an emerging understanding of lysosomal functions in autophagy. *Trends. Biochem. Sci.* 39: 61–71.

Shi, M., Ho, K., Keating, A., et al. 2009. Doxorubicin-conjugated immuno-nanoparticles for intracellular anticancer drug delivery. *Adv. Funct. Mater.* 19: 1689–96.

Shi-Kam, N.W., Liu, Z., Dai, H. 2005. Functionalization of carbon nanotubes via cleavable disulfide bonds for efficient intracellular delivery of siRNA and potent gene silencing. *J. Am. Chem. Soc.* 127: 12492–93.

Sito, G., Swanson, J.A., Lee, K.D. 2003. Drug delivery strategy utilizing conjugation via reversible disulfide linkages: role and site of cellular reducing activities. *Adv. Drug. Deliv. Rev.* 55: 199–215.

Sobolev, A.S., Jans, D.A., Rosenkranz, A.A. 2000. Targeted intracellular delivery of photosensitizers. *Prog. Biophys. Mol. Biol.* 73: 51–90.

Somia, N.V., Zoppe, M., Verma, I.M. 1995. Generation of targeted retroviral vectors by using single-chain variable fragment: an approach to in vivo gene delivery. *Proc. Natl. Acad. Sci. U. S. A.* 92: 7570–4.

Soundararajan, S., Chen, W., Spicer, E.K., et al. 2008. The nucleolin targeting aptamer AS1411 destabilizes Bcl-2 messenger RNA in human breast cancer cells. *Cancer. Res.* 68: 2358–65.

Stojanovski, D., Johnston, A.J., Streimann, I., et al. 2003. Import of nuclear-encoded proteins into mitochondria. *Exp. Physiol.* 88: 57–64.

Sui, M., Liu, W., Shen, Y. 2011. Nuclear drug delivery for cancer hemotherapy. *J. Control. Release.* 155: 227–36.

Szegezdi, E., Logue, S.E., Gorman, A.M., et al. 2006. Mediators of endoplasmic reticulum stress-induced apoptosis. *EMBO Rep.* 7(9): 880–5.

Terman, A., Kurz, T., Gustafsson, B., Brunk, U.T., 2006. Lysosomal labilization. *IUBMB Life.* 58: 531–39.

Torchilin, V.P. 2006. Recent approaches to intracellular delivery of drugs and DNA and organelle targeting. *Annu. Rev. Biomed. Eng.* 8: 343–75.

Trabulo, S., Mano, M., Faneca, H., et al. 2008. S413-PV cell penetrating peptide and cationic liposomes act synergistically to mediate intracellular delivery of plasmid DNA. *J. Gene. Med.* 10: 1210–22.

Urano, F., Wang, X., Bertolotti, A., et al. 2000. Coupling of stress in the ER to activation of JNK protein kinases by transmembrane protein kinase IRE1. *Science.* 287: 664–6.

Wagstaff, K.M., and D.A. Jans. 2009. Nuclear drug delivery to target tumour cells. *Eur. J. Pharmacol.* 625: 174–80.

Wang, G., Norton, A.S., Pokharel, D., et al. 2013. KDEL peptide gold nanoconstructs: promising nanoplatforms for drug delivery. *Nanomedicine. Nanotechnol. Biol. Med.* 9: 366–74.

Wang, J., Fang, X., Liang, W. 2012a. Pegylated phospholipid micelles induce endoplasmic reticulum-dependent apoptosis of cancer cells but not normal cells. *ACS Nano.* 6 (6): 5018–30.

Wang, L., Liu, Y., Jiang, X., et al. 2011. Selective targeting of gold nanorods at the mitochondria of cancer cells: implications for cancer therapy. *Nano. Lett.* 11: 772–80.

Wang, W., Zhou, F., Ge, L., et al. 2012b. Transferrin-PEG-PE modified dexamethasone conjugated cationic lipid carrier mediated gene delivery system for tumor-targeted transfection. *Int. J. Nanomed.* 7: 2513–22.

Wartosch, L., Bright, N.A., Luzio, J.P., 2015. Lysosomes. *Curr. Biol.* 25: R315–16.

Waterhouse N.J., Goldstein, J.C., Kluck, R.M., et al. 2001. The (Holey) study of mitochondria in apoptosis., *Methods. Cell. Biol.* 66: 365–91.

Weissig, V. 2003. Mitochondrial-targeted drug and DNA delivery. *Crit. Rev. Ther. Drug. Carrier. Syst.* 20: 1–62.

Weissig V., Cheng, S.M., D'Souza, G.G. 2004. Mitochondrial pharmaceutics. *Mitochondrion.* 3(4): 229–44.

Wente, S.R. 2000. Gate keepers of the nucleus. *Science.* 288: 1374–7.

Whitmire, R.E., Wilson, D.S., Singh, A., et al. 2012. Self-assembling nanoparticles for intra-articular delivery of anti-inflammatory proteins. *Biomaterials.* 33: 7665–75.

Willson, J.M. 1996. Adenoviruses as gene delivery vehicles. *N. Eng. J. Med.* 334: 1185–7.

Wong, E., and C.M. Giandomenico. 1999. Current status of platinum-based antitumor drugs. *Chem. Rev.* 99: 2451–66.

Wongrakpanich, A., Geary, S.M., Joiner, M.L.A., et al. 2014. Mitochondria targeting particles. *Future. Med.* 9: 2531–43.

Xiong, X.B., Mahmud, A., Uludag, H., et al. 2008. Multifunctional polymeric micelles for enhanced intracellular delivery of doxorubicin to metastatic cancer cells. *Pharm. Res.* 25: 2555–66.

Xu, C., Xie, J., Kohler, N., et al. 2008. Monodisperse magnetite nanoparticles coupled with nuclear localization signal peptide for cell-nucleus targeting. *Chem Asian. J.* 3: 548–52.

Yamada, Y., Akita, H., Kamiya, H., et al. 2008. MITO-Porter: A liposome-based carrier system for delivery of macromolecules into mitochondria via membrane fusion. *Biochimica et Biophysica Acta.* 1778: 423–32.

Yamada, Y., and H. Harashima. 2013. Enhancement in selective mitochondrial association by direct modification of a mitochondrial targeting signal peptide on a liposomal based nanocarrier, *Mitochondrion.* 13: 526–32.

Yang, B., Dong, L., Neuzli, J. 2016. Mitochondria: an intriguing target for tumor initiating cells. *Mitochondrion.* 26: 89–93.

Yang, C., Uertz, J., Yohan, D., et al. 2014. Peptide modified gold nanoparticles for improved cellular uptake, nuclear transport, and intracellular retention. *Nanoscale.* 6: 12026–33.

Yang, S.H., Meta, M., Qiao, X., et al. 2006. A farnesyl transferase inhibitor improves disease phenotypes in mice with a Hutchinson–Gilford progeria syndrome mutation. *J. Clin. Invest.* 116: 2115–212.

Yoong, S.L., Wong, B.S., Zhou, Q.L., et al. 2014. Enhanced cytotoxicity to cancer cells by mitochondria-targeting MWCNTs containing platinum (IV) prodrug of cisplatin. *Biomaterials.* 35: 748–59.

Yousif L.F., Stewart K.M., Horton K.L., et al. 2008. Mitochondria-penetrating peptides: sequence effects and model cargo transport. *Chem. Biochem.* 10: 2081–8.

Zhang, L., Yao, H.J., Yu, Y., et al. 2012. Mitochondrial targeting liposomes incorporating daunorubicin and quinacerine for treatment of relapsed breast cancer arising from cancer stem cells. *Biomaterials.* 33: 565–82.

Zhou, K. 2011. Tunable, ultra-sensitive pH responsive nanoparticles targeting specific endocytic organelles in living cells. *Angew. Chem. Int. Ed Engl.* 50: 6109–14.

Zhu, S., Lansakara, D.S.P., Li, X., Cui, Z. 2012. Lysosomal delivery of a lipophilic gemcitabine prodrug using novel acid-sensitive micelles improved its antitumor activity. *Bioconjug. Chem.* 23: 966–80.

Zhuang, Q., Jia, H., Du, L., et al. 2014. Targeted surface functionalized gold nanoclusters for mitochondrial imaging. *Biosens. Bioelectron.* 55: 76–82.

4 Chemo-Resistance Reversal Using Nanomedicines

Can Sarisozen, Jiayi Pan, and Vladimir P. Torchilin

CONTENTS

4.1 INTRODUCTION

Nanomedicines have been approved for treatment of a large variety of diseases including systemic fungal infections, HIV-related Kaposi's sarcoma, age-related macular degeneration, multiple sclerosis, and cancer (Weissig et al. 2014). Since the oncology-based nanotechnology applications have received more than two-thirds of the recent research attention, this chapter focuses on cancer nanomedicines for the reversal of chemo-resistance (Etheridge et al. 2013). In the United States, over a million people are diagnosed with cancer each year. The monthly individual cost of treating cancer with a newly approved drug has exceeded $10,000 since 2010, which is ten times higher than the cost in 2000. The increased cost for cancer treatment not only reflects the increasing expenditures for new drug development, but also indicates a transformation in the types of new drugs used. Almost all the drugs approved by the Food and Drug Administration since 2010 are antibody-based cancer therapeutics rather than chemotherapeutics. The most critical hindrance to the development of cancer chemotherapy is the chemo-resistance of tumor cells. Just as in the development of resistance towards antibiotics by bacteria, after repeated exposure to a chemotherapy tumor cells develop defensive mechanisms and become invulnerable to the chemotherapy.

Many groups have investigated the mechanisms of chemo-resistance on a molecular level. Several key mechanisms have been identified (Table 4.1). These mechanisms involve change in cellular drug accumulation, detoxification of drug, inhibition of apoptosis, repair of DNA adducts, and tumor recovery via cancer stem cells (Figure 4.1). Many possible molecular targets are involved in the creation of

TABLE 4.1
Summary of Resistance Mechanisms to Some Common Chemotherapeutic Agents

Cytotoxic Agent	Resistance Mechanisms	Refs.
Methotrexate	Reduced cellular accumulation	(Dixon et al. 1994, Trippett et al. 1992)
	Increased activity of drug target	(Goker et al. 1995)
Cisplatin	Reduced cellular accumulation	(Veneroni et al. 1994)
	Inhibition of apoptosis	(Zhang et al. 2008)
	Increased DNA damage repair	(Dabholkar et al. 1994, Wiedemeyer et al. 2014)
Doxorubicin	Reduced cellular accumulation	(Goren et al. 2000)
	Inhibition of apoptosis	(Kelly et al. 2002)
Paclitaxel	Reduced cellular accumulation	(Jang et al. 2001)
	Inhibition of apoptosis	(Wang et al. 2005)
	Homologous recombination repair	(Blanchard et al. 2015)
5-fluorouracil	Decreased active metabolites	(Beck et al. 1994)
	Inhibition of apoptosis	(Liang et al. 2002)
Topoisomerase I inhibitors (Ex: Irinotecan)	Reduced cellular accumulation	(Thomas and Coley 2003)
	Inhibition of apoptosis	(Longley et al. 2006)
	Increased DNA damage repair	(Farmer et al. 2005)
	Decreased amount of drug targets	(Meijer et al. 1992)
Poly (ADP-ribose) polymerase inhibitors	Reduced cellular accumulation	(Oplustilova et al. 2012, Rottenberg et al. 2008)
	Decreased amount of drug targets	(Liu et al. 2009)
	Homologous recombination repair	(Ashworth 2008)

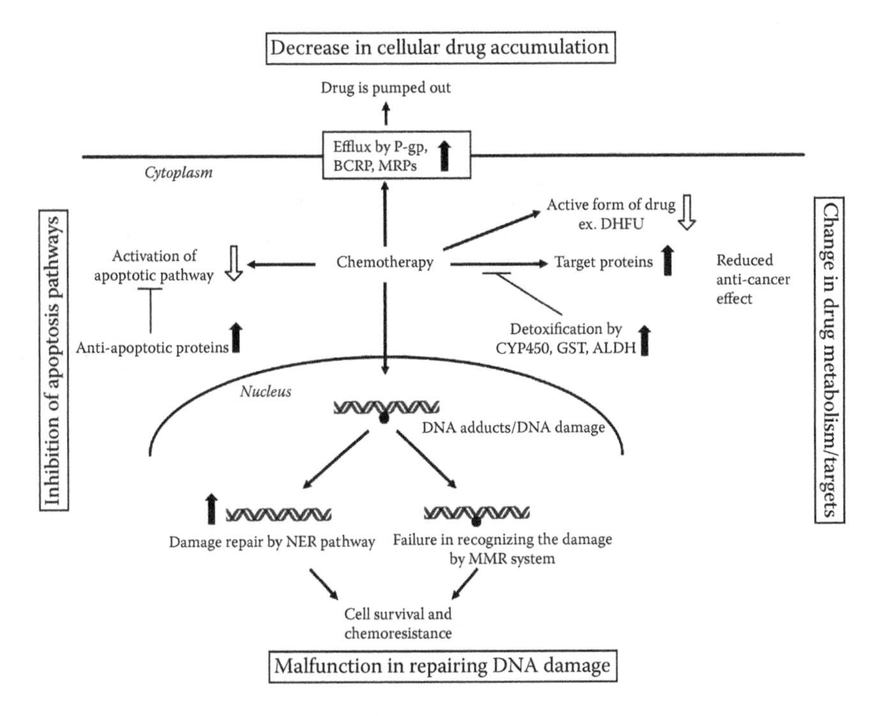

FIGURE 4.1 General chemoresistance mechanisms summarized in this chapter. The filled arrows indicate an upregulated/overexpressed pathways and proteins, while the hollow arrows indicate the downregulated/decreased pathways/events causing resistance.

multidrug chemoresistance using these pathways. Apart from these five main mechanisms, there are additional important molecules involved in chemo-resistance development such as PKC inhibitors and topoisomerase II. Based on the discovery of the molecular targets inducing chemo-resistance, researchers started to think about the possibility of re-validating existing and new chemotherapies with the reversal of chemo-resistance by either knocking out the molecules that contribute to chemoresistance or recovering the levels of components that facilitate apoptosis. In these scenarios, nanomedicines have turned out to be a suitable platform in fulfilling these aims.

4.2 CHEMO-RESISTANCE MECHANISMS IN CANCER

4.2.1 Change of Cellular Drug Accumulation

Fundamental to creation of an apoptotic effect, a drug molecule has to accumulate in cells. However, via the active transport within cells, these drug molecules are pumped out against their concentration gradients by membrane-bound transporters such as ATP-binding cassette transporters (ABC transporters). This situation is exaggerated especially in cancer cells where ABC transporters are overexpressed. Such a mechanism significantly decreases the amount of drug inside cells, suppressing the effectiveness of chemotherapy. Moreover, upregulation of transporters resulting from one type of chemotherapy can also contribute to pumping out other types of structurally non-related chemotherapeutics. Such a phenomenon, called the multi-drug resistant effect (MDR), limits the application of chemotherapy, and enhances tumor cell invulnerability to newly developed chemotherapies.

In 1979, a permeability-glycoprotein (P-glycoprotein) unique to drug-resistant cells was discovered that led to reduced colchicine permeability in Chinese hamster ovary cell mutants (Riordan and Ling 1979). This 17,000 Da protein encoded by ABCB1 was purified in 1985 and aroused great interest (Riordan et al. 1985). P-glycoprotein is one of the ABC transporters, which consists of 48 or 49 family members (Kimura et al. 2007). Besides P-glycoprotein, other ABC transporters including the multidrug resistant protein family (MRP1–MRP5) (Cole et al. 1992, Evers et al. 1998) and breast cancer resistance protein (BCRP) (Bates et al. 2001, Doyle et al. 1998) were found to transport molecules from cells against their concentration gradient.

In normal cells, ABC transporters function as cellular permeability modulators of diverse compounds. They are believed to be involved in preventing toxicity from xenobiotics by pumping them from cells. However, in malignancies, overexpressed ABC transporters enable cancer cells an enhanced pumping of xenobiotics. Among these xenobiotics, antitumor chemotherapeutics are a major class. Thus, overexpression of ABC transporters hinders the efficiency of various chemotherapeutics and results in multidrug resistance.

For better therapeutic design for treatment of multidrug resistance caused by ABC transporters, it is important to understand the molecular mechanisms of their upregulation. Among the ABC transporter family, MDR-1 transcribed P-glycoprotein (P-gp) is one that frequently gets mutated in tumor cells. Several pathways of P-gp amplification have been investigated (Kuo 2009). P-gp overexpression can be induced both internally and externally. It is straightforward to understand the external regulation of P-gp expression. For example, when a tumor-bearing patient is treated with a chemotherapeutic agent, the influx of the agent into cells triggers a series of MDR-1 gene promoters to produce more P-gp. These overexpressed P-gp transporters are then recruited to the cell membrane to promote a xenobiotic's removal. Additionally, upregulation of MDR-1 by external factors is also promoted by the presence of xenobiotics. Ultraviolet (UV), reactive oxygen species and ionizing radiation exposure also induce DNA changes that promote the transcription of MDR-1 (Hu et al. 2000, Ziemann et al. 1999).

Internally, stress conditions such as carcinogens and hypoxia have been reported to induce upregulation of P-gp. Constitutive activation by carcinogens, such as diethylnitrosamine (DEN), 2-acetylaminofluoren (2-AAF), and dimethylhydrazine, enhanced the activity of the MDR-1 promoter (Deng et al. 2001, Kuo et al. 2002). Under hypoxic conditions, it was reported that the c-Jun NH_2-terminal kinase (JNK) pathway was involved in the upregulation of MDR-1 expression. In this pathway, binding of hypoxia-inducible factor (HIF) to the hypoxia-response element located at the promoter of the MDR-1 gene upregulated P-gp expression (Liu et al. 2007, Comerford et al. 2002).

Clinically, most treatments targeting multidrug resistance have focused on P-gp because P-gp is highly expressed in colon, kidney, liver, and adrenocortical cancers

(Fojo et al. 1987, Goldstein et al. 1989). P-gp has a wide spectrum of substrates including methotrexate, cisplatin, vincristine, doxorubicin, and paclitaxel. Administration of a P-gp inhibitor increased the accumulation and efficiency of chemotherapy in multiple multidrug resistant cancer cells (Goren et al. 2000, Navarro et al. 2012, Tidefelt et al. 2000). Three generations of P-gp inhibitors have been developed in last two decades to overcome the multidrug resistance effect. The first generation of P-gp inhibitors was developed originally for other indications but later discovered to be P-gp substrates. For example, verapamil was first prescribed as a calcium channel blocker to treat hypertension. Later, it was found to have potential for inhibition of P-gp and was co-administered together with chemotherapies (Potschka et al. 2002, Bansal et al. 2009). The second generation of P-gp inhibitors was developed with a better clinical profile and fewer toxic side effects. However, the P-gp inhibitory dose of these two generations of inhibitors was still far above the tolerated dose, and the poor pharmacokinetic profile of these inhibitors brought difficulties in determining the optimum chemotherapeutic doses (Thomas and Coley 2003, Darby et al. 2011). The third generation of P-gp inhibitors was developed by high-throughput screening of combinational compounds. These compounds were highly specific toward P-gp, independent of P450 metabolism and much more potent than the previous two generations. Among the third generation of P-gp inhibitors, tariquidar, derived from anthranilamide, has drawn great attention. Tariquidar interacts with a distinct modulatory binding site and inhibits the ATPase of P-gp. This third generation of P-gp inhibitor, although most promising, was still hindered in applications in clinical settings by the existence of unfavorable toxicity resulting from a large tissue and organ distribution where P-gp activity is useful.

For a better antitumor performance, nanoformulation has been a good tool for reduction of toxicity and enhancement of the pharmacokinetic profile. Also, with nanoformulations, P-gp inhibitors can be delivered simultaneously with chemotherapy. By entering cells through endocytosis, vast amounts of a drug molecule can be delivered across the cell membrane as nanoformulations compared to its simple diffusion. It was also reported that degraded nanocarriers and polymers could block P-gp by direct interaction with P-gp and inhibit its function (Dabholkar et al. 2006, Kobayashi et al. 2007). An additional novel strategy in downregulating P-gp utilizes nucleic acid therapy. It is necessary to use such a nano-sized delivery system, to protect the nucleic acids from degradation. Usually, P-gp is involved in most of the cases by hurdling drug accumulation in cells, whereas other ABC transporters like MRPs and BCRP also contribute negatively to the performance of chemotherapy and remain as an obstacle to be solved.

4.2.2 Detoxification of the Drug

Most anticancer drugs need to be metabolized into their active form to be effective as anticancer agents. Also, anticancer drugs are made vulnerable to the *in vivo* detoxifying systems that make them more water soluble for excretion through urine or bile. These processes are associated with diverse enzymes, which can degrade, modify or complex the drug with detoxifying molecules. Corresponding genes or gene promoters of those enzymes are usually regulated in response to chemotherapies.

One consequence of this regulation is the decrease of enzymes that play a role in activation of anticancer pro-drugs. Another is the increased level of enzymes that metabolize anticancer drugs. Under these circumstances, the amount of active anticancer drug is reduced, resulting in treatment resistance.

Several enzymes, including glutathione transferase (GST), aldehyde dehydrogenase (ALDH), cytochrome P450 (CYP450), protect the *in vivo* environment from xenobiotics. However, in the presence of cancer, the existence of these defensive enzymes can be an obstacle. For example, more than 80% of 5-fluorouracil (5-FU) was metabolized by dihydropyrimidine dehydrogenase (DPD) to dihydrofluorouracil (DHFU) in the liver before reaching the tumor (Diasio and Harris 1989). Thus, the activity and pharmacological profiles of 5-FU were markedly influenced by the amount of DPD expressed in the liver. In addition, an elevated intratumoral DPD level resulting from consecutively administered 5-FU doses led to the resistance to 5-FU therapy (Beck et al. 1994, Diasio and Harris 1989, Diasio and Johnson 1999). In contrast, bioavailability of 5-FU was significantly increased when administrated together with DPD inhibitors in both preclinical and clinical studies (Adams et al. 1999, Diasio 1998). Increased level of GST has been identified in tumor cells resistant to platinum drugs (Kelland 1993). GST is widely expressed throughout the body and covalently binds cisplatin and oxaliplatin with glutathione (GSH), resulting in deactivation of these drugs (Ishikawa and Ali-Osman 1993). Tumor resistance due to the detoxification by GST has been observed in lung and ovarian cancers since the 1990s (Chang 2011, Raynaud et al. 1996). Additionally, irinotecan, used to treat colon cancer and small cell lung cancer, suffered from deactivation by CYP450 (Xu and Villalona-Calero 2002). A high level of ALDH is regarded as an identification marker for cancer stem cells (Vinogradov and Wei 2012) and is associated with resistance in various types of cancer cells (Cheung et al. 2007, Friedman et al. 1992, Honoki et al. 2010).

Importantly, some metabolized chemotherapies act on particular target proteins to trigger the downstream tumor-killing effect. These target proteins play a critical role in determining the efficiency of chemotherapy. For example, fluorodeoxyuridine monophosphate (FdUMP), the metabolite of 5-FU, is a potent inhibitor of thymidylate synthase (TS). The primary anticancer effect of 5-FU is initiated by the inhibition of TS. TS catalyzes the conversion of deoxyuridine monophosphate (dUMP) to deoxythymidine monophosphate (dTMP). With the inhibition of TS, the amount of dUMP rises, which damages DNA replication, leading to apoptosis. However, resistant cancer cell lines exhibit an upregulated TS compared to sensitive cancer cell lines (Copur et al. 1995, Saga et al. 2003). To achieve the tumor suppressing effect by inhibition of TS, a larger dose of 5-FU is required that results in resistance to 5-FU.

4.2.3 Inhibition of Apoptosis

A balance between cell survival and cell death is important in maintaining organism hemostasis. Most chemotherapies inhibit tumor growth by triggering cell apoptosis that is, programmed cell death. There are two major pathways involved in mediating apoptosis. One is the extrinsic pathway that is activated predominantly by engagement between death receptors and their respective ligands (Scaffidi et al. 1998). Death receptors are comprised of TNF, CD95, and TNF-related apoptosis-inducing ligand (TRAIL) receptors, which all belong to the tumor necrosis factor (TNF) superfamily. Typically, activation of these receptors by TNF, Fas (for CD95 receptors), and DR4/5 (for TRAIL receptors) recruits the adapter Fas-associated death domain (FADD), which in turn recruits death-inducing activated complex (DISC). The caspase-8 mediated apoptosis pathway is initiated after the cleavage of procaspase-8 and procaspase-10 by DISC to active caspases. The other pathway triggering apoptosis is the intrinsic pathway regulated by the mitochondrion. During chemotherapy, UV exposure, DNA damage or exposure to reactive oxygen species (ROS), the permeability of mitochondrial membrane is altered. A major consequence of the change in mitochondrial membrane permeability is the release of the pro-apoptotic proteins cytochrome c and apoptosis inducing factor (AIF) (Indran et al. 2011, Wilson et al. 2006). Released cytochrome c forms a complex, termed the apoptosome, with apoptotic protease activating factor-1 (APAF1), ATP, and the inactive initiator procaspase-9. Apoptosomes activate caspase-9 that initiates mitochondria-mediated apoptosis (Twiddy et al. 2004). Once caspase-8 and caspase-9 are activated, they cleave and activate caspase-3, caspase-6, and capase-7, which are regarded as "executioner" caspases. A group of cellular substrates, such as nuclear LAMINS, deoxyribonuclease, DNA fragmentation factor, and cytoskeletal proteins, are cleaved by these executioner caspases, followed by actual apoptosis with chromatin condensation, nuclear shrinkage, cell blebbing, and formation of apoptotic bodies.

In both pathways, vast numbers of proteins regulate the apoptotic self-destruction machinery by tight control of the process at different levels. In cancer cells, gene expression of these apoptosis-related protein malfunctions. Expression of the pro-apoptotic proteins is inhibited in some cases. Expression of anti-apoptotic proteins is exaggerated in other cases. Both of these situations can induce multidrug resistance to chemotherapeutics.

4.2.3.1 Overexpression of Anti-Apoptotic Proteins

In terms of overexpression of anti-apoptotic genes, the BCL-2 family proteins are the most prominent, despite the fact that BCL-2 family proteins are related both to the pro-apoptotic and to the anti-apoptotic pathway. Chromosomal translocation, when coupled with the BCL-2 gene, leads to amplified BCL-2 gene expression (McDonnell et al. 1989, Tsujimoto et al. 1985). BCL-2 contributes to tumorigenesis by cooperating with oncoproteins such as c-MYC (Vaux et al. 1988). In promyelocytic leukemia, BCL-2 cooperated with promyelocytic leukemia-retinoic-acid-receptor-α, resulting in tumor development (Kogan et al. 2001). Many different groups have shown that overexpression of BCL-2 confers a poor responsiveness to various kinds of chemotherapeutics both *in vivo* and *in vitro* (Findley et al. 1997, Miyashita and Reed 1992, Weller et al. 1995). In addition, cytotoxic drugs usually induced cell death through a pathway that influenced mitochondria outer membrane permeability (MOMP). BCL-2 blocks the MOMP apoptotic pathway.

Other than BCL-2, BCL-$X_{L,}$ and MCL1 (myeloid cell leukemia sequence 1) of the BCL-2 family are also involved in resistance of tumors to apoptosis.

BCL-X_L was found to result in resistance through multiple pathways. One of these pathways is similar to that of BCL-2. BCL-X_L is anchored to the outer membrane of the mitochondrion, suppressing apoptosis by preventing Bax and Bak from formation of pro-apoptotic homodimers (Hanada et al. 1995). Interestingly, Bax and Bak are also members of the BCL-2 family but perform pro-apoptotic functions. Oligomerization of Bax and Bak forms pores in the outer mitochondria membrane and releases cytochrome c into the plasma. However, BCL-2 and BCL-X_L have a pocket that binds Bax and Bak to bypass the pro-apoptotic effect.

Death-receptor mediated apoptosis is the extrinsic pathway for cell apoptosis. Inhibitors of death receptors can contribute to resistance to chemotherapy. Cellular FADD-like interleukin-1 β-converting enzyme-like protease-inhibitory proteins (c-FLIPs) interfere with the formation of DISC so that the cleavage of pro-caspase-8 to caspase-8 is inhibited (Micheau et al. 2002). Hence, the downstream apoptotic effect initiated by caspase-8 is also inhibited. It has also been found that substantial expression of FLIP was observed in diverse deadly cancer types involved in resistance to chemotherapies. Overexpression of c-FLIP inhibits the apoptosis induced by DNA damaging reagents such as 5-FU, irinotecan, and oxaliplatin in colorectal cancer (Longley et al. 2006) and reduces the sensitivity of breast cancer to both paclitaxel and docetaxel (Wang et al. 2005). Available data on c-FLIP supports the relationship between multi-drug resistance and its overexpression in lung cancer and ovarian cancer (Shivapurkar et al. 2002, Sun and Chao 2005). Thus, antisense oligonucleotide delivery approaches for downregulation of c-FLIP may develop as a new direction for research on MDR reversal.

Apart from the anti-apoptotic proteins mentioned above, another main class of anti-apoptotic proteins involves the inhibitors of the apoptosis (IAPs) family. Many oncoproteins, such as survivin, neuronal apoptosis inhibitor protein (NAIP), cellular IAP1/2, X-linked inhibitor of apoptosis (XIAP), and BIR-containing ubiquitin conjugating enzyme (BRUCE), belong to this family. IAPs inhibit apoptosis by binding to caspase-3, caspase-7, and caspase-9, the "executioner" caspases in the intrinsic apoptosis pathway. Increased expression of IAPs resulted in chemoresistance and a poor therapeutic effect in cancer patients (Hunter et al. 2007). Thus, downregulation of these proteins would provide an efficient method for cancer treatment.

4.2.3.2 Insufficient Expression of Pro-Apoptotic Proteins

In addition to overexpression of anti-apoptotic proteins, insufficient expression of pro-apoptotic proteins contributes to chemoresistance in cancer cells. As mentioned in Section 4.2.3, BCL-X_L inhibited apoptosis by interfering with the homodimerization of Bax, which triggers the intrinsic pathway of apoptosis. Without sufficient expression of Bax, however, the apoptosis signal pathway cannot be triggered either. Restoration of Bax by Bax mRNA delivery both *in vitro* and *in vivo* showed a promising effect by inducing apoptosis and suppressing tumorigenesis in various cancer cell types, including breast and prostate cancers (Bargou et al. 1996, Li et al. 2001, Yin et al. 1997).

Tumor suppressor protein p53 plays a central role in maintaining genomic integrity by regulating cell cycle arrest as well as cell death. Various chemotherapies inhibit tumor growth by causing DNA damage that consequently initiates the p53-mediated apoptosis pathway (Indran et al. 2011). Those signals of damage activate cellular checkpoint kinases such as Ataxia Telangiectasia Mutated (ATM) and ATM/Rad3-related (ATR), which phosphorylate p53 and modulate its interaction with mouse double minute 2 (MDM2). MDM2 sequesters p53 and negatively regulates p53 by promoting ubiquitin-mediated degradation.

p53 was also reported to regulate apoptosis in several additional ways. First, expression of the pro-apoptotic protein, Bax, in the intrinsic apoptotic pathway could be activated by p53, resulting in enhanced apoptotic events. Second, death receptors, i.e. DR5, could also be upregulated by p53. Increased levels of death receptors increased the chance of binding with ligands such as Fas and TNF, which would activate the extrinsic apoptotic pathway. In addition, the anti-apoptotic protein, survivin, could also be suppressed by p53 (Nakano et al. 2005). Mutations of p53 influence the sensitivity of tumor cells towards chemotherapies. Although the activity of p53 is also regulated by many other modulators, elevating p53 levels is one of the main strategies for overcoming tumor chemoresistance.

4.2.4 DNA Damage Repair

One of the major mechanisms that cause cancer cell death by chemotherapeutic drugs is DNA damage in cancer cells. This DNA damage promotes cell cycle arrest if left unrepaired and if recognized by DNA damaging response (DDR) factors (Bouwman and Jonkers 2012). Maintenance of DNA repair mechanisms is necessary for cells to stabilize DNA (Hoeijmakers 2001). However, in cancer cells where the DNA repair mechanism-related proteins are upregulated, chemotherapy resistance is developed (Chaney and Sancar 1996). For example, the nucleotide excision repair (NER) pathway is predominantly responsible for repairing DNA adducts in cellular DNA. In NER, the excision repair cross-complementation group 1 (ERCC1) cleaves the DNA sequence where there is a mismatch or adduct. The exposed DNA gap is reconstituted by the action of DNA polymerases and ligases. Different studies have shown that overexpression of ERCC1 mRNA confers increased resistance to platinum drugs. ERCC1 downregulation by antisense vectors sensitized tumor cells to platinum therapies both *in vitro* and *in vivo* (Dabholkar et al. 1994, Kohn et al. 1994, Selvakumaran et al. 2003).

One of the key mechanisms that recognizes genomic integrity and transduces apoptotic signals is the mismatch repair (MMR) system (Fink et al. 1998). Mutation in MMR genes such as MLH1 and MSH2 has been linked to the resistance to various genotoxic chemotherapies. Basically, a mismatched DNA sequence is recognized by the MMR system. The MMR system creates a gap by incising the thymine-containing strand and removing the thymine as well as adjacent bases. The gap created is filled via a repair synthesis mechanism. Under the circumstances where MMR genes were mutated, MMR related proteins, like MLH1 and MSH2, may no longer detect a mismatched DNA sequence. Failure to recognize a mismatched DNA sequence leads to failure to trigger MLH1/MSH2-mediated cell cycle arrest. Hence, loss of MMR function was regarded as one of the most important mechanisms of resistance to chemotherapeutics such as temozolomide (Zhang et al. 2012) and procarbazine (Armand et al. 2007), which were then currently being used in the clinic. The mechanism of action of these drugs involves large amounts of O^6-methyguanine adduct production on DNA. If not detected by MMR-related proteins, these alkylated guanines are not toxic to cells. MMR-related resistance to platinum drugs has been well studied in ovarian cancer, lung cancer, and gastric cancer. It was reported that a significant proportion of ovarian tumors had silenced MLH1 due to methylation. Such a silencing effect correlated with cisplatin resistance in some patients (Brown et al. 1997).

Additionally, BRCA1/2 was extensively involved in DNA double-strand break (DSB) repair by homologous recombination (HR) and conferred chemoresistance in cancer cells. Overexpression of BRCA1/2 followed by upregulation of cyclin D1 (Nakuci et al. 2006), inactivation of p53 (Chock, Allison, and Elshamy 2010) or by activation of survivin (Chock, Allison, Shimizu et al. 2010), desensitized cancer cells to DNA interference agents, such as paclitaxel, platinum drugs, and poly (ADP-ribose) polymerase (PARP) inhibitors (Bhattacharyya et al. 2000; Farmer et al. 2005). Regulation of BRCA gene expression can be achieved through interfering tyrosine receptor kinase (TRK), phophatidylinositide 3-kinase (PI3K), and KRAS signaling pathways (Wiedemeyer et al. 2014). However, it should be noted that interfering upstream of these pathways could affect different downstream results and limit therapeutic efficacy.

4.2.5 Cancer Stem Cells

Cancer stem cells (CSC), or cancer initiating cells, have drawn considerable attention in cancer research because of their tumor-initiating properties. Even when cancer stem cells occupy a very small portion of cells constituting the whole tumor, they maintain a tumorigenic ability. Failure to kill CSCs creates significant barriers to cancer treatment, including metastasis and relapse. Importantly, CSCs are more difficult to kill because they acquire chemoresistance through both the intrinsic and extrinsic mechanisms. Thus, chemotherapies may kill the bulk of cancer cells but not the CSC, which give rise to new tumors with increased invasiveness and malignancy.

CSCs have been characterized by special cell markers used in sorting CSCs for research. For example, CD34$^+$/CD38$^-$, and CD44$^+$/CD24$^-$ were used as markers for leukemia CSCs and solid tumor CSCs. In other tumors, CD133 was overexpressed by CSCs. These cell markers were closely related to chemoresistance in CSCs. Multiple transporters, such as Pgp, BCRP, and MRP, have been identified in CSCs (Shervington and Lu 2008). High levels of BCRP are related to CD133 overexpression in CSCs. However, the exact relation between them is not well known till date. Enhanced ALDH1 activity has been found to be associated with CSCs, decreasing the sensitivity of CSCs to treatment by detoxifying the chemotherapy agents. It is believed that inhibition of ALDH1 has the potential of sensitizing CSCs and inhibiting tumor growth (Rausch et al. 2010). CSCs also utilize pathways associated with proliferation in normal stem cells for their self-renewal and maintenance. It was demonstrated that the active Hedgehog (Hh) signaling pathway was involved in the rapid replication of leukemia stem cells (Kobune et al. 2009). Beta-catenin in the Wnt pathway and Jagged-1 in the Notch pathway were also found to contribute to the survival of CSCs in diverse types of cancer (Deng et al. 2010, Teng et al. 2010a,

Wang et al. 2009). Blocking these pathways and related proteins resulted in increased sensitivity to chemotherapies involving CSCs (Teng et al. 2010b).

In addition to protein-related chemoresistance mechanisms in CSCs, it has also been recognized that the particular microenvironment around CSCs protects and supports them for survival. Two major components of this microenvironment are the promoted angiogenesis and the hypoxic conditions. The vasculature around CSCs supports the proliferation of CSCs. This phenomenon indicates that disruption of the vasculature matrix in the CSC microenvironment may contribute to vulnerability of CSCs. The combination of antiangiogenic- and chemo-therapies has shown enhanced efficiency in reducing the number of CSCs in tissues as well as the CSCs in spheroids (Folkins et al. 2007). Reduction of oxygen in tumors has also been regarded as a major factor that leads to the tumorigenicity of CSCs. Under hypoxic conditions, hypoxia-inducible factors (HIFs) activate repair enzymes for double strand DNA breaks. In the presence of an increased ability to fix DNA damage, CSCs are rendered more resistant to DNA-damaging chemotherapy. Thus, another strategy to overcome drug resistance in CSCs is based on downregulation of HIFs or blockage of the HIF signaling pathway. Preclinical studies have shown that targeting HIFs with nucleic acid therapy and topoisomerase inhibitors was effective in overcoming drug resistance (Choi et al. 2009, Wirthner et al. 2008).

4.2.6 Strategies to Overcome Chemoresistance

Chemoresistance is a broadly defined term given to the number of possible mechanisms contributing to the loss or lack of response to a wide range of chemotherapeutics. Important efforts have been made in the antibacterial research area to clarify the definitions and harmonize the scientific language related to resistance (Magiorakos et al. 2012). While some of the terms have cross-discipline uses such as multidrug resistance (MDR), others such as extensive drug-resistance (XDR) or pan drug-resistance (PDR) have not found the common use in oncology. Although various mechanisms related to chemoresistance have been very well established and elucidated, circumventing them still requires clear definitions. Historically, the strategies that have been used to overcome or reverse the summarized chemoresistance mechanisms can be collected into the following main groups;

1. *Overcoming the drug efflux* by arresting the P-gp pump or other types of efflux proteins by small molecule modulators is the most direct and widely used approach to overcome multidrug resistance. Saturating the pumps by delivering high amounts of P-gp substrate drugs to the tumor microenvironment can also oversaturate the efflux mechanisms by acting as local drug depots.
2. *Provoking apoptosis* by initiating lysosomal membrane permeabilization (LMP) and release of the cathepsins into the cytoplasm or by mitochondrial targeting to initiate the intrinsic apoptosis pathway are the other approaches used to overcome resistance.
3. *Silencing the 'undruggable' proteins that contribute to chemoresistance.* Following the realization of wide applicability and selectivity of RNAi by

small interfering RNAs (siRNA), the downregulation of overexpressed proteins and simultaneous co-delivery of chemotherapeutics became one of the most efficient strategies for circumventing chemoresistance.

Some of these approaches require local drug delivery to the tumor microenvironment and some require intracellular organelle-targeted delivery. However, in the last four decades (DeVita 1975, DeVita et al. 1975) it has been established that, as long as the above strategies are pursued by using a single chemotherapeutic, the cancer therapy will fail due to either side effects caused by high doses of the chemotherapeutic or, more likely due to resistance development (Diaz et al. 2012). The fact that a small number of cancer cells resistant to any chemotherapeutic agent will always be present and remain in tumor after the single-agent therapy (monotherapy) makes the treatment of tumors with two or more anticancer drugs (combination therapy) targeting different pathways an essential strategy (Bozic et al. 2013). However, emerging evidence indicates that certain ratios of combined anticancer drugs can be synergistic and overcome drug resistance, while other ratios of the same drugs can be additive or even antagonistic (Abouzeid, Patel, Sarisozen et al. 2014, Abouzeid, Patel, and Torchilin 2014, Sarisozen et al. 2016). Thus, optimizing the ratios of the combination drugs is one of the most important aspects of resistance reversal (Mayer et al. 2006, Mayer and Janoff 2007).

Rational nanomedicine designs and nanotechnology-based approaches have the unique advantages of combining many required parameters in ways that can adapt and evolve, just like cancer itself. Thus, in Section 4.3, we will lay out some basics in the nanomedicine area and build on them for developing strategies to reverse chemoresistance.

4.3 GENERAL CONCEPTS IN CANCER NANOMEDICINE

Tumor heterogeneity and the complexity of chemo-resistance mechanisms, as summarized earlier, set a fundamental challenge for effective treatment of drug resistant cancers. The rationale of using nanotechnology-based medicine, more appropriately "nanomedicine", can easily be justified due to their well-established advantages. First, nano-sized drug carriers can overcome the problems associated with the low-water solubility of the cargo molecules. Low bioavailability, uptake, and delivery of such drugs can be enhanced by encapsulating them into nanoparticles as seen in paclitaxel carriers Genexol-PM® (Kim et al. 2004) and Abraxane/Celgene (Kratz 2014). Second, nanocarriers can influence the biodistribution of their cargos by protecting them from biodegradation or excretion. They can also improve the efficacy of the anticancer moieties by targeting them to the tumor areas, or in many cases specifically to the cancer cells. In recent years, rational nanoparticle designs have aimed to use tumor-associated stimuli to trigger the release of their payload only in the tumor microenvironment or inside the cancer cells. And more relevant to the context of this chapter, these advantages, individually or in combination with each other, can significantly improve the response of the chemo-resistant tumors to the anticancer treatments by targeting/overcoming the molecular mechanisms summarized in earlier sections. To better understand the reversal of resistance using nanomedicines, it is important to mention the fundamentals, advantages, and challenges of general concepts related to cancer nanotechnology.

4.3.1 EPR Effect, Particle Size, and Longevity in the Blood

In 1979, Maeda *et al.* demonstrated that styrene malic acid (SMA) conjugated anti-cancer protein neocarzinostatin (SMANCS) accumulated higher in the tumor tissue than its unconjugated (NCS) derivative (Maeda et al. 1979). The researchers showed that proteins bigger than 40 kDa could not only selectively accumulate in the tumor areas, but also stay there for prolonged time periods. After seven years of comprehensive research, they built up the theory of an Enhanced Permeation and Retention (EPR) Effect, which since then has become one of the golden standards in drug delivery (Matsumura and Maeda 1986, Maeda et al. 2000, Maeda 2012). The first part of this effect, the 'enhanced permeation,' is related to the differences in the endothelial lining of blood capillaries in the tumors (as well as infarcts and inflammation sites) compared to healthy blood vessel walls. The endothelial lining of the normal vasculature has a turnover rate of about one in 1,000 days while this time in tumors can be as short as 10 days (Hobson and Denekamp 1984). Thus, the newly formed vasculature around the tumors lacks a well-defined morphology and is more permeable due to the increase in the spacing between the endothelial cells, allowing large molecules with appropriate sizes to accumulate within the interstitial space in such areas. This accumulation works especially well with tumors because of the lack of lymphatic drainage, meaning that the molecules and particles that accumulate by vascular leakage will continue to stay at the tumor site (Maeda et al. 2000, Iyer et al. 2006, Maeda 2001, Maeda et al. 2001).

While there are examples of oral, pulmonary, nasal or transdermal administration of anticancer therapeutics (Dening et al. 2016, Garbuzenko et al. 2010, Liechty et al. 2011, Reddy et al. 2008), the majority of cancer nanomedicines are being administered systemically. Reliance on EPR effect has become the main approach for systemically administered nanomedicine formulations since it allows nano-sized drug formulations to accumulate selectively in the tumor tissues based on their sizes, which forms the basis of such passive targeting. To deal with the size requirements for nanomedicines, it is important to give some numbers here. Most normal vessel walls have a permeable size cut-off of around 6–12 nm (Sarin 2010), indicating that a given nanoparticle should be larger than this size to limit its distribution to healthy tissues. Moreover, the renal clearance via filtration is very fast for the particles and molecules with hydrodynamic diameters smaller than 6-8 nm (Choi et al. 2010). In contrast, the clearance by liver and spleen is more rapid for particles which have particle sizes larger than 200 nm (Longmire et al. 2008). In addition, the cut-off size in vasculature around tumor sites varies markedly from case to case (100–800 nm) but usually falls within the range of 100 and 500 nm. These numbers define a tentative particle size range of nanomedicines for enhanced tumor accumulation and minimized distribution to healthy tissues. In general, given their surface chemistry and physicochemical properties are kept the same, nanoparticles within the range of 6–200 nm have slower clearance (Popovic et al. 2010, Singh et al. 2006). Small nanomedicine systems, with sub-100 nm particle sizes, as well as low molecular weight macromolecules and proteins extravasate more readily than the larger systems (Cabral et al. 2011, Dreher et al. 2006, Wang et al. 2015).

While these numbers and ranges are important for nanomedicines in relation to the EPR effect, they also make it apparent that the nano-sized drug carriers have to stay in the circulation for prolonged time periods to generate a sufficient level of accumulation at the targeted tumor sites. Many *in vitro, in vivo* and *in-patient* studies have shown that the prolonged circulation times of the nanomedicines correlate with the higher accumulation at the leaky vasculature sites (Bakker-Woudenberg et al. 1992, Gabizon et al. 1994, Gabizon and Papahadjopoulos 1988). Thus, it is imperative to understand the nanomedicine fate after it enters the blood circulation. When nanoparticles are introduced into blood stream, different components, but mainly the proteins, rapidly cover their surface. This non-specific interaction with serum proteins alters the surface properties of the nanoparticles and can lead to opsonization leading to recognition by the mononuclear phagocytic system (MPS) and eventual clearance from the circulation. Given the fact that all the clinically approved nanomedicines (Table 4.2) claim to be subjected to EPR effect, preventing the opsonization and extending the blood circulation time have become the main goal of every successful nanomedicine-based approach. Among the various options to achieve this goal, coating nanoparticle surface with polyethylene glycol (PEG), i.e. PEGylation, is the first one that comes to mind, as was first suggested for liposomes (Klibanov et al. 1990). The PEGylation of nanomedicines prevents their interaction with opsonins by creating a steric hindrance and impedes their capture and clearance by MPS, hence makes them 'stealthy'. It should be noted that other types of biocompatible, soluble, and hydrophilic polymers for affording steric protection including poly(2-methyl-2-oxazoline) (Woodle et al. 1994), phosphatidyl polyglycerols (Maruyama et al. 1994), poly(acryloyl morpholine) (Monfardini et al. 1995), poly(acryl amide), poly(vinyl pyrrolidone) (Torchilin, Levchenko, Whiteman et al. 2001), and poly(vinyl alcohol) (Takeuchi et al. 1999) have been used to create stealth nanomedicines, but PEG remains the most popular choice.

The most important example of implementing the PEGylation and exploiting the EPR effect has been Doxil®, a stealth liposomal formulation of doxorubicin with a circulation half-life of more than two days (Gabizon et al. 2003, Gabizon et al. 2002). Doxil® is the first FDA-approved nanomedicine that underlined the advantages of a cancer nanomedicine which made it possible to reduce the severe side effects, mainly the cardiac toxicity, and increase the overall patient compliance (Barenholz 2012). Moreover, besides being the first FDA approved nano-drug formulation, Doxil® almost single handedly revived the scientific and financial interest in nanotechnology-based drug formulations and reassured its recovery after the largely skeptic environment of 1980s against liposomes and nanoformulations in general (Poste 1983). However, despite the significantly reduced toxicity profile of doxorubicin due to its improved pharmacokinetics and biodistribution (Northfelt et al. 1996, Gabizon et al. 2012, Gabizon et al. 1994), Doxil® failed to show significant increase in the rate of cancer progress free survival in the clinic (O'Brien et al. 2004). This lack of enhanced efficacy also underlines the fact that a nanomedicine's activity could not be tied solely to the enhanced accumulation/retention in the tumor site. In fact, EPR-driven drug accumulation in tumor sites was reported for only some tumor

TABLE 4.2
Examples of Cancer Nanomedicines in the Market and the Most Likely Candidates to Be Approved

Product Name (and company)	Anticancer Cargo	Indicated for	Current Phase
Doxil® (Janssen)	Doxorubicin HCl	Ovarian cancer (after failure of platinum-based therapy) AIDS related Kaposi's sarcoma (after failure of prior systemic chemotherapy) Multiple myeloma	Approved
Myocet™ (Teva UK)	Doxorubicin HCl	First-line treatment of metastatic breast cancer	Approved
Genexol-PM® (Samyang)	Paclitaxel	First-line treatment of metastatic or recurrent breast cancer First-line treatment of locally advanced or metastatic non-small cell lung cancer First-line treatment of ovarian cancer	Approved
Abraxane® (Celgene)	Paclitaxel	Metastatic breast cancer (after failure of combination chemotherapy) First-line treatment of locally advanced or metastatic non-small cell lung cancer in combination with carboplatin First-line treatment of metastatic pancreatic cancer	Approved
Onivyde® (Merrimack)	Irinotecan	Second-line treatment of metastatic pancreatic cancer	Approved
DaunoXome® (Galen)	Daunorubicin citrate	First-line treatment of HIV-associated Kaposi's sarcoma	Approved
Marqibo® (Spectrum)	Vincristine sulfate	Ph- acute lymphoblastic leukemia (ALL)	Approved
CPX-351 (*Vyxeos*, Celator)	Cyterabine: daunorubicin (5:1 fixed molar ratio)	Acute myeloid leukemia (AML)	Phase III (NCT01696084)
Opaxio™ (CTI Biopharm)	Paclitaxel	Maintenance therapy of ovarian cancer	Phase III (NCT00108745)
NKTR-102 (Nektar)	Etirinotecan pegol	Metastatic breast cancer	Phase III (NCT01492101)
NK105 (Nippon Kayaku)	Paclitaxel	Metastatic breast cancer	Phase III (NCT01644890)
NC-6004 Nanoplatin (NanoCarrier)	Cisplatin	Pancreatic cancer	Phase III (NCT02043288)
Lipoplatin™ (Regulon)	Cisplatin	Non-small cell lung cancer	Phase III (Europe)

types rather than for all solid tumors (Maeda 2015). The EPR effect is unlikely to be present and also be equal in all tumors. Due to the heterogeneity of a tumor and its stroma, one may find different endothelial lining openings in different parts of the same tumor. Tumors which have large endothelial openings eventually experience accumulation of large volumes of liquid in the tumor area as well. As a result, the interstitial fluid pressure can increase to a point matching the microvascular pressure and thus eliminate effective nanomedicine extravasation (Boucher et al. 1990, Jain and Stylianopoulos 2010). Moreover, the EPR effect is mainly responsible for the increase of drug concentrations in the tumor vicinity, not in the tumor cells. In addition, PEGylated nanomedicines may induce the production of antibodies that can accelerate the blood clearance of nanoparticles, particularly after repetitive administration (Ishida, Ichihara et al. 2006, Ishida et al. 2002, Ishida, Atobe et al. 2006). These drawbacks, which are associated with the lack of control, hinder the potential of nanomedicines and make it clear that the EPR effect is not the only means of effective nanomedicine activity in the tumors, including the chemo-resistant ones.

4.3.2 Active Targeting

Cancer cells or the cancer-associated tumor vasculature frequently show the presence of antigens or receptors, which are either uniquely expressed or overexpressed relative to normal cells of the surrounding normal tissues. Active targeting of nanomedicines can be achieved by attaching a targeting ligand such as monoclonal antibodies, transferrin, peptides, folate or certain sugar moieties onto their surface. In case of PEGylated nanomedicine formulations, the targeting ligand should be attached to the chemically activated distal end of the PEG chains to prevent stearic hindrance between the ligand and the PEG corona (Torchilin, Levchenko, Lukyanov et al. 2001). However, separation of active targeting from passive targeting is conditional and not always defines two distinct definitions. That is, the passive targeting will still be required for the nanomedicines to reach the target area before ligand-mediated interaction with the target cells occurs. Thus, EPR effect remains as a bottleneck for active targeted nanomedicines. Passive targeting and tumor accumulation via EPR effect work when creation of a drug "depot" in the vicinity of a tumor is sufficient to induce antitumoral effect. However, active targeting is specifically useful if the drugs are needed inside the individual tumor cells or intracellular organelles inside the cells.

Successful actively targeted nanomedicines require a balanced ligand content and surface exposure that prevents rapid clearance from the circulation by MPS, while achieving efficient binding to their targets. There are numerous examples of the use of active targeting for enhanced nanomedicine efficacy including PEGylated doxorubicin-loaded liposomes that have antibodies on their surface specific for human epidermal receptor 2 (HER2) (Gao, Zhong et al. 2009). Another antibody that has been successfully utilized on different types of nanomedicines is the antinuclear mAb 2C5, which possesses a nucleosome-restricted specificity. It has been shown that the doxorubicin-loaded 2C5-attached liposomes and micelles increased tumor cell cytotoxicity in *in vitro* and *in vivo* models (Elbayoumi and Torchilin 2009, 2007, Gupta and Torchilin 2007, Perche et al. 2012).

Certainly, antibodies are not the only targeting moieties that have been used to achieve active targeting of nanomedicines to chemo-resistant tumors. Folate, owing to its easy conjugation with nanomedicine formulations, its high affinity for folate receptors, and the lower expression of folate receptors in normal cells compared to cancer cells, has been used as a targeting ligand (Low et al. 2008). Numerous folate-targeted liposomal carriers have been described and found to be effective both *in vitro* and *in vivo* (Lee and Low 1995, Liu et al. 2011, Niu et al. 2011, Chaudhury et al. 2012). Another active targeting moiety that has been exploited due to its increased overexpression on the tumor cells is transferrin. Liposome-attached transferrin formulations have been used in mouse models to deliver different anticancer drugs such as doxorubicin (Li et al. 2009), boron compounds (Cheng et al. 2004), and ceramide (Koshkaryev et al. 2012).

The advantages of active targeting are also evident from the recent clinical trials of nanomedicines that have ligands on their surface for specific binding to cancer cells. MCC-465 is a PEGylated doxorubicin-loaded liposomal formulation targeted with the F(ab')2 fragment of the human monoclonal antibody GAH against cancerous stomach tissues (Matsumura et al. 2004) (Phase I in Japan). SGT-53 is another targeted liposomal formulation tagged with a single-chain antibody fragment against transferrin receptor (Pirollo et al. 2016) for delivery of the *p53* gene (Phase I, NCT00470613, NCT02354547; Phase II, NCT02340117, NCT02340156). BIND-014 is a polymeric nanoparticle formulation loaded with docetaxel (Hrkach et al. 2012, Von Hoff et al. 2016) that uses the tumor prostate-specific membrane antigen (PSMA) as its targeting moiety (Phase I, NCT01300533; Phase II, NCT02283320, NCT01792479, NCT01812746, NCT02479178). Finally, CALAA-01 contains human transferrin as the targeting ligand on the surface of a PEGylated cyclodextrin-based siRNA delivery system (Zuckerman et al. 2014) (Phase I, NCT00689065). It should also be noted that all the clinical trials involving targeted-nanomedicines were terminated in Phase I/II stages. Especially, discouraging results from the Phase II trials of BIND-014 showed that the nanomedicine candidate failed against cervical and head-and-neck cancers. Although it was found to be somewhat effective against one type of lung cancer, it was not clear whether the targeted nanomedicine exceeded the efficacy of conventional docetaxel treatment. At the end, BIND Therapeutics, Inc. had to file for bankruptcy, highlighting the challenges and high-risks associated with attempts to develop effective targeted nanomedicine therapies.

4.3.3 Tumor Penetration and Overcoming Physical Barriers

While the EPR effect has received much of the attention in nanomedicine research, it is becoming clear now that a deep and uniform tumor penetration by nanotherapeutics is equally important and a critical parameter for successful therapy of cancer including the chemo-resistant tumors. The challenges for tumor penetration and homogeneous drug distribution throughout the tumors include the dense interstitial matrix components such as collagen and other extracellular matrix proteins, elevated interstitial fluid pressure, and the stearic, size-dependent, and electrostatic interactions between the nanomedicines and openings in the tumors (Jain and Baxter 1988, Shi et al. 2017). Each of these parameters, as well as combinations of them, can significantly decrease the penetration of the nanomedicines reaching the tumor site. Moreover, the average distances between the tumor capillaries can be up to hundreds of micrometers, which poses another challenge for nanoparticle delivery to tumor cells distant from the vessels (Yoshii and Sugiyama 1988, West et al. 2001).

Particle size also plays an important role in the tumor penetration of nanomedicines. Generally, if particle size is smaller than the interstitial openings (pores), the stearic and size-dependent interactions can be considered insignificant. But these pore sizes can be as small as 100 nm in many tumors (Ramanujan et al. 2002). Thus, particle sizes smaller than this value may be required for homogeneous and efficient penetration and distribution in the tumor. For sub-100 nm nanoparticles, the penetration and tumor distribution increases with decreasing particle size (Goodman et al. 2007, Kim et al. 2010, Huo et al. 2013). Targeting is another factor related to tumor penetration/distribution of the nanomedicines. While it is generally regarded as a desirable option to be included in the nanomedicine design to ensure the cancer cell specificity, it does not guarantee sufficient and homogeneous drug or drug carrier distribution in tumor tissues. For example, nanomedicines targeted with high affinity antibodies to target antigens tend to bind efficiently to the target cells that are easily available on the tumor periphery. This high binding to the surface cells limits distribution in tumors by lowering penetration into deeper parts, while at the same time, non-targeted formulations can bypass this binding step and continue their journey to the deeper tumor layers. This phenomenon is called a 'binding-site/barrier effect' and can provide serious limitations for cancer cell targeted formulations.

The tumor penetration ability of the nanomedicine formulations holds even greater importance with regard to chemo-resistance. It has been reported that the heterogeneity of tumor tissues causes hypoxic and necrotic microregions as well as a decreased glucose gradient in the deeper layers, all of which can significantly contribute to the cancer cell differentiation. This differentiation among cancer cells can easily cause chemo-resistance related protein overexpression in some cells, as shown in three-dimensional cancer cell spheroid models (Hamilton 1998, Wartenberg et al. 2003, Comerford et al. 2002, Sutherland 1988, Kerbel et al. 1994).

4.4 A SPECIAL CASE: siRNA DELIVERY BY NANOMEDICINES

Gene silencing by RNA interference (RNAi) has undergone significant development in a relatively short time since its discovery in 1998 (Fire et al. 1998). This highly conserved process of post-transcriptional gene silencing is mediated by small interfering RNA (siRNA) molecules. siRNAs are usually 21-23 nucleotide-long double-stranded RNA molecules which are shorter than 10 nm in length (Guo et al. 2010, Hansen et al. 2001). The possibility of selectively downregulating chemo-resistance associated proteins in the cancer cells to overcome one of the greatest hurdles of effective clinical cancer therapy was a breakthrough in nanomedicine. However, it has also very rapidly become clear that successful *in vivo* siRNA or RNAi treatments come with serious challenges.

To exploit the RNAi *in vivo*, siRNA molecules need to reach their target cells' cytoplasm. Systemic administration of siRNAs must provide the following for successful therapy: (i) stability against serum nucleases, (ii) increased retention time in the circulation and reduced renal clearance, (iii) interaction with the target cells and reduced non-specific interactions, (iv) the ability to extravasate, (v) effective cellular entry and uptake into the RNA-induced silencing complex (RISC). Properties of siRNAs, such as their hydrophilicity, negative charge, small size, and exposure to degradation by serum nucleases, make every step a challenge.

Post-injection into the bloodstream, naked siRNA must navigate the circulation without being degraded by nucleases. Careful design and modification on the siRNA structure can overcome this problem (Guo et al. 2010) but the siRNAs need to stay in the circulation long enough to reach their target tissues and cells. With a pore size filtration barrier of 8 nm in the kidneys, naked siRNAs can quickly pass into the urine and be cleared from the circulatory system, in less than 10 minutes (Gao, Dagnaes-Hansen et al. 2009, Iversen et al. 2013, Soutschek et al. 2004). However, even when the siRNAs are formulated with the carriers for a larger size (generally >20 nm) to avoid the renal clearance, phagocytosis starts to play an important role. Prevention of opsonization is crucial to bypass the removal of the siRNA. This can be achieved by formulating a stearic barrier around the siRNA or its carrier, such as with polyethylene glycol (PEG) (Torchilin and Trubetskoy 1995). But then it is important to note that PEG also hinders the cellular interaction of the molecules (explained in more detail in the following sections). After siRNA leaves the bloodstream, it must overcome the diffusion barrier through the extracellular matrix which consists of polysaccharides and proteins that can slow down or even prevent siRNA and its carrier from reaching the target cells. Despite their size advantage, a net negative charge and hydrophilic character prevent siRNA from getting readily internalized by the target cells. To overcome this challenge, various strategies, such as complexation with positively charged polymers by electrostatic interaction, encapsulation into nanoparticles, and using different ligands (i.e. antibodies, cell-penetrating peptides, aptamers) can be implemented. However, effective internalization does not guarantee adequate protein silencing since the siRNA still needs to escape endosomal encapsulation and lysosomal degradation, be released into the cytoplasm and taken up by RISC (Oliveira et al. 2007). Endosomal escape remains the major barrier to efficient siRNA delivery since challenges up to that point have already been established over the last decades by using nanotechnology-based carriers and approaches.

Given the summarized multi-step challenges, the potential of siRNA therapy very much depends on innovative vehicles and strategies to overcome these challenges with high efficacy. This bottleneck for siRNA technology against chemo-resistance can only be overcome by nanomedicine-based delivery approaches.

4.5 NANOMEDICINES FOR THE REVERSAL OF CHEMO-RESISTANCE

Almost all the approved nanomedicines for cancer treatment rely heavily on the EPR effect and passive targeting. Key benefits of the EPR-based therapeutics are the accumulation of the drug in the tumor area, and limited biodistribution to healthy organs/tissues, and hence decreased side effects. This results in several advantages such as increased tolerance to the anticancer drugs, increase in the area under the curve for plasma and tumor, and extended drug exposure at the target site.

However, despite all these known advantages and the successful approval of Doxil® and other approved nanomedicines, they mostly failed to generate increased overall survival and progression-free survival times in patients (Prabhakar et al. 2013). Although the academy and industry focused on the EPR effect based on the assumption of a positive correlation between the EPR and efficacy, successful translation to the clinic was hard to achieve. A meta-study revealed that only 0.7% of the administered dose was delivered to the tumors in preclinical studies. What is more disappointing is the lack of improvement in the delivery efficiency over the last ten years of extensive research (Wilhelm et al. 2016). The overall conclusion may suggest that, after all, the achievements in the nanomedicine field have been limited to improved drug solubility, stability due to encapsulation, and longer circulation times that can result in tumor accumulation when compared to free drug. These achievements might have seemed groundbreaking a decade ago, however, they are no indication of actual tumor selectivity. For example, the Abraxane® extended life expectancy nearly ten weeks (Gradishar et al. 2005). While ten weeks seems insignificant, in the big picture for a disease such as cancer, it was more than enough to give hope to patients and to receive FDA approval. The demands from the regulatory agencies for more than incremental increases indicated the end of an era dependent solely on the EPR effect. In the light of the recent understanding of the pitfalls associated with passive targeting, the primary focus has shifted to actively-targeted nanomedicines, which have tended to outperform passive targeting in efficiency by approx. 30% (Wilhelm et al. 2016). However, it should be noted that even after overcoming several challenges that have been explained in this chapter many times, a given formulation may not get translated into the clinic and thus change the rules of the game. Chemotherapy often fails in the clinic not only because of insufficient drug accumulation and delivery into tumors but mostly due to the development of resistance against the chemotherapeutics.

The following sections of this chapter will focus some of the most common chemo-resistance mechanisms outlined earlier, and their reversal with nanomedicine-based approaches. To date, development and engineering of the drug delivery systems have been the priority. However, the complexity of chemo-resistance mechanisms, as well as the heterogeneity of cancer, have made it thus far impossible to develop a one-for-all nanosystem that can give all the desired advantages of nanomedicines and be able to reverse or prevent the resistance development in cancer. Instead of focusing on the nanomedicine types for resistance reversal, we have detailed examples of nanomedicines that are based on the tumor's biology. The nanomedicine-based approaches for this purpose can be grouped into four main categories:

1. Single-drug carriers that depend on the nanoencapsulation or formulation components' properties for reversal.
2. Two or more drug carriers for combination therapy-based reversal.
3. Oligonucleotide delivery systems for reversal.
4. Complex systems, eg. stimuli-sensitive carriers.

Overcoming chemo-resistance of tumors with nanomedicine is a broad and detailed subject. Due to the massive research data on this topic, we focused on selected strategies that used the different nanomedicine approaches listed.

4.5.1 Overcoming Drug Efflux-Mediated Resistance

The upregulation of drug efflux pumps is correlated with a poor diagnosis in cancer (Hornicek et al. 2000, Chan et al. 1997, Linn et al. 1995). P-gp, encoded by the human ABCB1/MDR1 gene was the first ABC family transporter to be identified (Chen et al. 1986), and since then it has become one of the most studied human transporter proteins (Ambudkar et al. 2003), ABCC1/MRP1 being another (Cole et al. 1992). As described earlier, chemo-resistance, or multidrug resistance (MDR) is a broad term that includes many different unique and complex molecular mechanisms. However, P-gp-mediated drug resistance is by far one of the most targeted mechanisms for chemo-resistance reversal by nanomedicine. P-gp is a glycoprotein with 170 kDa molecular weight, has six transmembrane helices called transmembrane domains (TMD) and cytoplasmic domains with ATP-binding sites (nucleotide-binding domain, NBD) (Aller et al. 2009). The structure of murine P-gp (Figure 4.2) was found to have 87% sequence identity with the human P-gp, thus its structure has been utilized in mechanistic studies (Aller et al. 2009, Zha et al. 2013).

P-gp has an unusually broad specificity for chemically unrelated substrates. Most of its substrates are hydrophobic and partition into the lipid bilayer. The P-gp affinity for actinomycin D and paclitaxel are 4000 and 100 times higher when the protein is in the lipid bilayer than inside the cell (Jin et al. 2012). This makes the protein act as a 'vacuum cleaner' for molecules as they pass through the bilayer into cells. In the case of cancer cells, their subsequent efflux decreases the intracellular concentrations of the chemotherapeutics and results in resistance.

The nature of the efflux action of P-gp also reveals opportunities to bypass it with nanomedicine. First, upon reaching the tumor site via EPR effect, a drug carrier could release its anticancer cargo and increase the drug concentration in the tumor

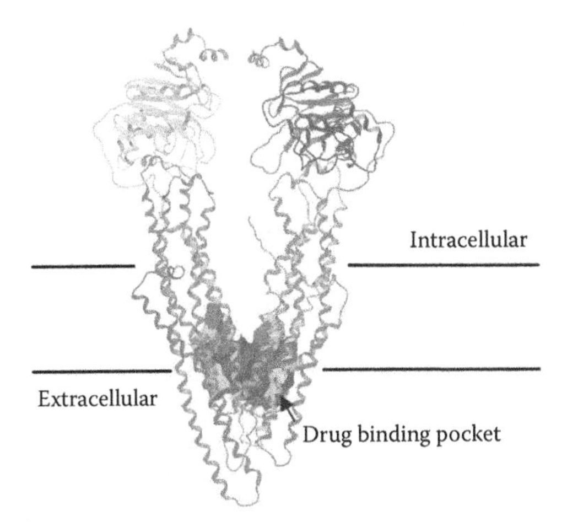

FIGURE 4.2 X-ray structure of P-gp. The area indicated by the arrow in the drug binding pocket represents hydrophobic and aromatic residues. Horizontal lines indicate approximate positioning of the lipid bilayer. (Modifed and reprinted from Zha, W. et al., *PLoS One*, 8 (1):e54349, 2013. Under Creative Commons CC0 license.)

microenvironment up to the point that the P-gp pumps get saturated and thus, allow increased diffusion of chemotherapeutics. Second, the very components used for preparing nanomedicine formulations can be used to alter the action of P-gp. Pluronics, A-B-A type block copolymers of poly(ethylene oxide), and poly(propylene oxide) are excellent examples for reversing chemo-resistance depending on the nanomedicine design. Surfactants, including Pluronic polymers, are amphiphilic molecules that can form micelles in aqueous environment when they are above their critical micelle concentration (CMC) (Torchilin 2005). Pluronics are important nanomedicines not only because of their ability to form micelles and carry the water-insoluble anticancer drugs to tumor site via EPR effect but also due to their capacity to inhibit P-gp pumps and sensitize the MDR cells (Alakhova and Kabanov 2014). While the sensitization mechanisms are multiple and complex, Pluronic P85 has been shown to be a strong inhibitor of P-gp by suppressing their ATPase activity by depleting the intracellular ATP levels (Batrakova et al. 2004, Batrakova et al. 2001). Pluronics also interact with the cell membrane and alter the lipid bilayer-P-gp interaction, which is important for efflux activity (Demina et al. 2005). Based on preclinical data, a Pluronic-based formulation SP1049C has entered the clinical trials (Valle et al. 2011). SP1049C consists of mixed micelles of Pluronic F127 ($(EO)_{100}$-$(PO)_{65}$-$(EO)_{100}$, HLB 22, MW 12,600 g/mol), and L61, encapsulating doxorubicin (Danson et al. 2004). In Phase II study with 21 patients, SP1049C demonstrated effective antitumor efficacy with 47% objective response rate after three weeks of therapy (Valle et al. 2011). It has been shown that Pluronic P85 was rapidly taken up by MDR cells following 15 minutes of exposure, associated with mitochondria and showed chemo-resistance reversal effects (Alakhova et al. 2010). Moreover, Pluronics also prevent the resistance development against anticancer drugs *in vitro* and *in vivo*. Human breast cancer cells, when exposed to doxorubicin behaved very differently in the presence and absence of Pluronic P85. While the cells cultured with P85/Dox nanomedicine formulation stayed sensitive to 10 ng/ml doxorubicin concentration, cells cultured with only doxorubicin eventually converted to MDR phenotype and tolerated significantly higher doses up to 10 μg/ml (Batrakova et al. 2006).

While drug molecules enter the cells by diffusion, and thus are exposed to P-gp efflux activity, nanomedicine formulations are usually taken up via endocytosis, which bypasses the membrane P-gp and are released into the cytoplasm. Dabholkar *et al.*, (Dabholkar et al. 2006) reported supportive data for bypassing of P-gp with nanomedicines consisting of long-circulating poly(ethylene glycol)2000–phosphatidylethanolamine conjugate (PEG_{2000}–PE) and d-α-tocopheryl polyethylene glycol 1000 succinate (TPGS) mixed micelles incorporating paclitaxel. The micelles themselves were not cytotoxic to P-gp overexpressing human cancer cell lines. Micelle formulations successfully solubilized paclitaxel without the need of other toxic ingredients. The rhodamine 123 uptake/efflux studies in the absence/presence of the P-gp inhibitor verapamil indicated that the nanomedicine internalization was not influenced by the inhibition of the P-gp pump, which assumes P-gp independent micelle internalization. Moreover, inclusion of TPGS in the formulation could also create P-gp inhibition and help to reverse the increased P-gp activity (Dintaman and Silverman 1999). Micellar nanomedicine systems, due to their favorable properties, have been one of the most studied nanomedicine systems for

MDR reversal. However, many other nanomedicine-based drug delivery systems, including polymeric nanoparticles, liposomes, and mesoporous silica nanoparticles are also reported to bypass P-gp-mediated drug efflux and create somewhat increased intracellular chemotherapeutic concentrations that depend on endocytosis and high locally released drug (Hu and Zhang 2012, Kirtane et al. 2013, Kapse-Mistry et al. 2014).

Nevertheless, arresting the P-gp activity by its modulators to prevent the drug efflux is another strategy which can exploit nanotechnology for chemo-resistance reversal. There are several generations of P-gp inhibitors. Generation one and two inhibitors (such as cyclosporine A or valspodar) inhibit the P-gp competitively. Valspodar (PSC-833), the most studied P-gp inhibitor, reached Phase II/III trials in combination with vincristine, doxorubicin, dexamethasone, and paclitaxel (Friedenberg et al. 2006, Lhomme et al. 2008). However, the trials failed to improve treatment outcomes and further introduced toxicity. The third generation of inhibitors (such as elacridar/GG918 and tariquidar/XR9051) inhibit P-gp noncompetitively, completely blocking the pump irreversibly at concentrations as low as 50 nM (Dale et al. 1998, Hyafil et al. 1993).

One of the reasons for using nanomedicine-based delivery of the P-gp inhibitors is selectivity. P-gp is distributed to a wide range of tissues and organs including liver, kidney, small intestine, colon, placenta, uterus, and brain, and has clear physiological functions including prevention of toxic compound absorption through the intestine or passing through the blood-brain barrier (BBB) (Krishna and Mayer 2000). Thus, systemic administration of P-gp modulators will have a body-wide distribution, inhibiting the normal function of P-gp and causing significant side effects. Thus, P-gp inhibitors should be delivered directly to their target tumor cells, just like anticancer compounds. These important points make the combination therapy with chemotherapeutics and P-gp inhibitors one of the most studied and effective approaches for chemo-resistance reversal in cancer. Sarisozen *et al.* encapsulated different generations of P-gp inhibitors into PEG-PE based micelles in combination with paclitaxel and investigated the MDR reversal properties of the formulations (Sarisozen et al. 2012b, a). Nanomedicine-based delivery of paclitaxel, when encapsulated alone into the micelles, increased the efficacy of the drug by bypassing the membrane-bound P-gp activity. Incorporation of the P-gp inhibitors further improved the cytotoxicity of paclitaxel in the MDR cells, up to the levels of sensitive cells. Moreover, modification of these micelles with active-targeting ligand increased their activity *in vitro* (Sarisozen et al. 2012b).

Zou *et al.* (Zou et al. 2017) used the same approach and encapsulated a third generation P-gp inhibitor, tariquidar, in combination with paclitaxel into transferrin-targeted long-circulating micelles. The nanoformulation was evaluated for targeting efficiency, cellular association, cellular internalization pathway, and cytotoxicity for reversal of PTX resistance on two multidrug resistant (MDR) ovarian carcinoma cell lines, SKOV-3TR, and A2780-Adr. P-gp inhibition with tariquidar increased the intracellular PTX levels and its cytotoxicity. Transferrin targeting increased the nanomedicine's ability to penetrate deeper layers of the model three-dimensional spheroid structures, and thus provided clear benefit in terms of effective anticancer drug distribution throughout the tumor masses. They also demonstrated that transferrin-targeting of these micellar nanoformulations can further enhance their efficacy against some MDR ovarian cancer cell lines. Specifically, the authors reported that the transferrin-targeted combination nanomedicine formulation completely reversed the chemo-resistance of human MDR ovarian cancer cells.

Along with micelles, liposomes have also been used extensively for combinatorial delivery of anticancer drugs with P-gp inhibitors. Patel *et al.* were first to co-encapsulate tariquidar with paclitaxel in a long-circulating liposomal combination nanomedicine formulation. They showed successful blocking of P-gp efflux activity in a rhodamine 123 uptake study on the drug-resistant human ovarian cancer cell line, SKOV-3TR. A simultaneous delivery of tariquidar and paclitaxel by long-circulating liposomes resulted in greater cytotoxicity in SKOV-3TR cells at a paclitaxel dose that was ineffective in the absence of tariquidar, demonstrating a significant reversal of MDR to paclitaxel. Following the successful results and proof of concept of combination delivery by liposomes, Zhang *et al.* investigated the same combination *in vitro* and *in vivo* using HeyA8-MDR and SKOV-3TR human chemo-resistant ovarian cancer cell lines (Zhang, Sriraman, et al. 2016). The combination treatment with liposomal nanomedicine formulations mediated apoptosis as shown with advanced microscopy techniques including label-free holomicroscopy and imaging cytometry, and resensitized the paclitaxel-resistant ovarian cancer cells to paclitaxel treatment. More importantly, *in vivo* results obtained with HeyA8-MDR tumor-bearing mice consistently showed that the liposomal combination treatment had no effect on angiogenesis and macrophage infiltration, but impaired proliferation and had a higher rate of apoptosis. The study addressed the main advantages of nanomedicines, that include effective solubilization of drugs and their co-encapsulation, EPR-mediated tumor inoculation, simultaneous delivery of the compounds selectively to the tumor site, successful blockage of P-gp, and an eventual resensitization of the drug-resistant tumors to chemotherapy.

Another successful combination treatment with P-gp inhibitors was achieved with polymeric nanoparticles encapsulating paclitaxel and tariquidar (Patil et al. 2009). Following the chemo-resistance reversal in *in vitro* MDR cell models by the co-loaded poly(D,L-lactide-co-glycolide) (PLGA) nanoparticles, the authors targeted the formulation with biotin for active targeting to tumors. Selective inhibition of P-gp efflux activity resulted in an eight-fold higher intracellular paclitaxel accumulation in drug-resistant JC cells. Biotin targeting improved the uptake of the particles by different MDR cells, up to 13-folds. *In vivo* tumor growth inhibition and survival studies also supported the MDR reversal by showing the highest tumor growth inhibition and longest survival with the targeted combination nanomedicine group. The authors tied the results to the successful inhibition of P-gp in the *in vitro* experiments and selective accumulation of the formulation in the tumor cells due to an EPR effect and targeting. However, it should be noted that, without determining the drug levels in the tumor and confirming the inhibition of P-gp activity *in vivo*, the results can only be considered as supportive.

One of the unique advantages of nanomedicine-mediated chemo-resistance reversal is the ability to downregulate proteins involved in the resistance mechanisms via an oligonucleotide delivery to tumors. As summarized in the earlier sections, successful *in vivo* oligonucleotide delivery depends on nanotechnology-based delivery strategies. Many groups have chosen to silence P-gp using siRNAs against the MDR1 gene (Figure 4.3). This is quite understandable. P-gp is one

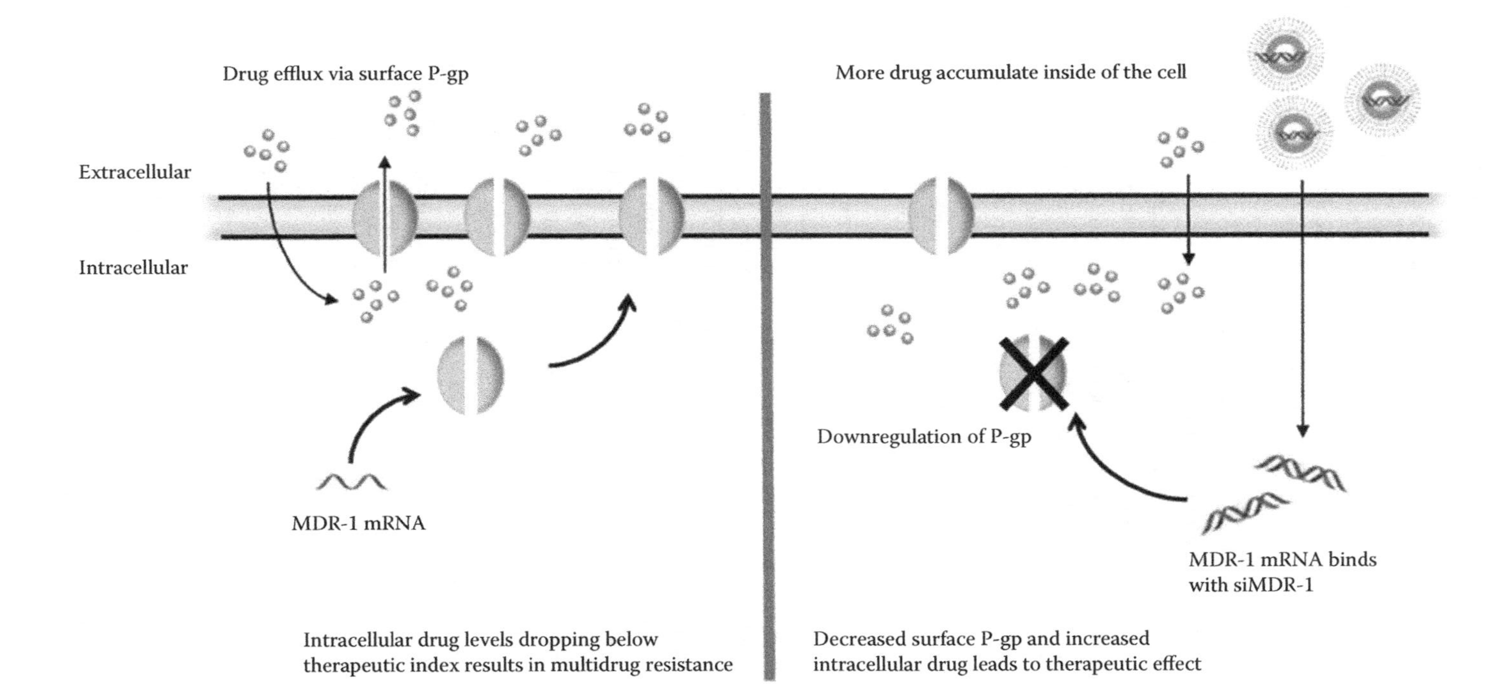

FIGURE 4.3 General scheme of MDR phenomena and anti-P-gp siRNA/drug combination treatment.

of the most studied pumps responsible for the MDR, and its downregulation can result in resensitization of cells against a wide range of chemotherapeutics. Also, being a transmembrane protein, its downregulation is easily followed by various techniques including flow cytometry (Swerts et al. 2004), immunohistochemistry (Schlaifer et al. 1990) or immunofluorescence (Meaden et al. 2002). Its downregulation by oligonucleotides (siRNA, mRNA or miRNA) can easily return the MDR cells to their sensitive stage and result in cell sensitivity to chemotherapeutics again.

Zhang *et al.* (Zhang, Zhu et al. 2016) synthesized a tri-block copolymer N-succinyl chitosan–poly-L-lysine–palmitic acid (NSC–PLL–PA) to co-deliver doxorubicin and siRNA against P-gp. The unique property of this micellar nanosystem was its responsiveness to changes in pH. The authors showed that the co-loaded micelle stability is lost under mildly acidic conditions compared to neutral pH, which corresponded to fast *in vitro* doxorubicin and siRNA release upon internalization by drug-resistant hepatocytes. Successful downregulation of the P-gp increased the doxorubicin efficacy. Patil *et al.* also used biotin-targeted PEI-PLGA based polymeric nanoparticle formulation to simultaneously deliver paclitaxel and P-gp siRNA to drug resistant cancer cells (Patil et al. 2010). They showed effective *in vivo* tumor growth inhibition by reversal of the chemo-resistance of tumor cells by downregulating P-gp.

Navarro *et al.* (Navarro et al. 2012) conjugated 1,2-distearoyl-sn-glycero-3-phosphoethanolamine (DOPE) to polyethyleneimine (PEI, 1.8 kDa, branched) and showed that the formulation of DOPE-PEI:siRNA complexes in MNP formulations did not significantly alter their GFP or P-gp silencing efficacy but considerably decreased their cytotoxicity compared to Lipofectamine 2000 or non-lipid-modified complexes. Treatment of multidrug resistant breast cancer cells (MCF-7/Adr) with MNPs loaded with siMDR-1 resulted in significant downregulation of P-gp. This decrease translated successfully into the sensitization of resistant cells to doxorubicin treatment 4, 8, 24, and 48h post siMDR-1 delivery. Regardless of the treatment regimen and time lag between siRNA and drug treatments, the effectiveness of doxorubicin was increased.

The overall advantages of lipid conjugates with low molecular weight PEI (1.8 kDa, branched) led Essex *et al.* to evaluate their *in vivo* potential using a multidrug resistant breast cancer model in mice (Essex et al. 2014). First, the tissue distribution of DOPE-PEI/siRNA complexes with a PEG coating, achieved by combining these complexes with micelle-forming PEGylated lipid PEG-PE, was evaluated. Four hours post-injection of the DOPE-PEI/PEG/siRNA micelles with an almost spherical shape, approx. 95 nm size and 13 mV surface charge, 22% of the injected dose (ID) of fluorescently-labeled siRNA was still in the circulation (vs. 9% ID of DOPE-PEI/siRNA complexes). The long-circulating ability introduced by PEG chains allowed significantly better tumor accumulation (8% ID) and lower siRNA doses in the lung and liver compared to DOPE-PEI/siRNA complexes alone. The high tumor accumulation of the delivery system should be noted as indicative of enhanced efficacy. In the second step, the therapeutic efficacy of the siMDR1-containing complexes in combination with doxorubicin was investigated in resistant breast cancer MCF-7/Adr tumors. One intravenous dose of siRNA (1.2 mg/kg, naked or in complexes) followed by one intravenous doxorubicin injection (2 mg/kg) after 48 hours was performed for 5 weeks. The MDR1 mRNA level in the excised tumors was reduced by half in the DOPE-PEI/siMDR1 group, while the DOPE-PEI/PEG/siMDR1 treatment resulted in almost complete downregulation of mRNA. The tumor volumes at the end of the study (day 40) were three-fold reduced as compared to the free siRNA + doxorubicin treatment group. Moreover, simultaneous delivery of siRNA and doxorubicin had no impact on the therapeutic outcome (Figure 4.4). The results indicate the versatility of DOPE-PEI/siRNA complexes and freedom from *in vivo* toxicity. These summarized results that related chemo-resistance reversal by modulating the intracellular drug concentrations using nanomedicine-based strategies give emphasis to the advantages of nanotechnology.

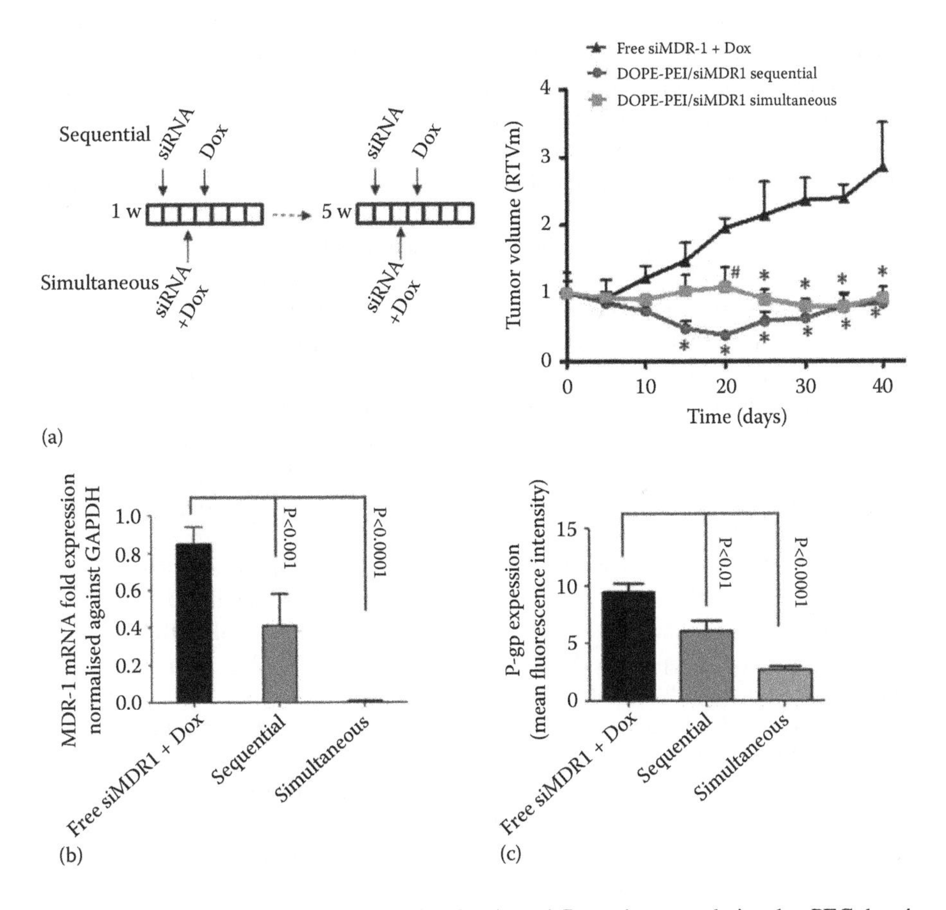

FIGURE 4.4 In vivo efficacy, MDR1 mRNA and P-gp downregulation by PEGylated DOPE-PEI-based nanocarriers. The treatment regimen is summarized in the upper left panel. (a) Mean relative tumor volume following the treatments. (b) and (c) are the MDR1 mRNA and P-gp protein levels at the final time point. Results are plotted as mean±SD. n = 5, performed in duplicates and triplicates for panels (b) and (c), respectively. (Reprinted from Essex, S. et al., *Gene Ther*, 22 (3):257–66, 2015. With permission.)

4.5.2 Nanomedicines Targeting the Enhanced DNA Repair System

Malfunction in DNA repair systems contributes chemoresistance to DNA-damaging drugs, such as alkylating agents and cisplatin. In the last two decades, numerous studies have focused on how to overcome the resistance to cisplatin, which is a widely used antitumor chemotherapeutic. It turned out that the application of nanotechnology in cisplatin therapy enhanced the tumor killing profiles based on several improvements. First, nanocarriers, such as liposomes and micelles, contribute to a much higher payload of cisplatin in the tumor site than the free drug. Drug resistance in tumors can be overwhelmed because of the significantly increased cisplatin internalized into the cells. Second, the protection provided by nanocarriers or PEGylated nanoconjugates of cisplatin prolong blood circulation time of cisplatin. More drug gets retained in the blood stream, hence more drug gets in touch with tumors and increases the tumor killing efficiency. Also, cisplatin in nanoformulations resulted in much lower systemic toxicity than free cisplatin due to its altered biodistribution. For example, hyaluronic-conjugated cisplatin nanoconjugates have been administered subcutaneously and distributed through the lymphatic system. Results showed that renal toxicity of cisplatin was significantly reduced in nanoconjugates compared to intravenously injected cisplatin (Cohen et al. 2009). Third, targeted delivery of cisplatin can be achieved by nanocarriers. Besides the passive targeting via EPR effect, various ligands and antibodies have been applied to the outer surface of cisplatin nanoparticles to deliver them to treat tumors in specific organs. Anti-EGFR antibody and prostate-specific membrane antigen (PSMA) were used as targeting moieties in nanomedicine formulations for cisplatin selective delivery to lung cancer and prostate cancer (Dhar et al. 2008, Peng et al. 2011). Chemoresistance to cisplatin could be decreased in tumors by application of traditional nanoformulated cisplatin. However, the chemoresistance was not fundamentally reversed and cells developed stronger chemoresistance by further effects on the MMR system as well as the NER system. This concern led to more advanced nanomedicines for treating chemoresistance caused by DNA damage repair.

The first strategy used the synergistic effect of two or more anticancer drugs. For example, cisplatin and doxorubicin were delivered together in polymer-based nanobins for a synergistic effect in killing cancer cells (Lee et al. 2010). In addition, mitomycin C, a potent DNA cross-linker, was co-loaded in polymer-lipid hybrid nanoparticles (PLN) with doxorubicin to achieve synergistic effect in treating DNA damage repair-induced multidrug resistant breast cancer cells (Shuhendler et al. 2010). One advantage of delivering two drugs at the same time is the increased therapeutic effect in killing cancer cells. The other one is that it makes it more difficult to develop chemoresistance against two different mechanisms of action. In this scenario, resistance development to one single drug will not completely deactivate the efficiency of the whole formulation.

The second strategy is interfering with multidrug resistance using nucleic acid therapy. Specific downregulation of multidrug resistance-related proteins with corresponding anti-sense nucleic acid has promise for reversal of resistance and re-sensitizing cancer cells to established chemotherapies. To protect nucleic acid molecules from degradation and digestion by enzymes, nanocarriers are essential.

As mentioned earlier, significantly elevated expression of ERCC1 in the NER pathway has been observed in multidrug resistant cancer cell lines. Downregulation of ERCC1 with ERCC1 siRNA was reported to disturb the NER pathway and inhibit the multidrug resistance that resulted from enhanced DNA damage repair. Downregulation of ERCC1 by its corresponding siRNA has been shown to be effective in sensitizing multidrug resistant ovarian and gastric cancer cells to cisplatin (Li et al. 2014, Selvakumaran et al. 2003). Mutation in the MMR system causes insufficient expression of MLH1 and MSH2, which trigger the downstream apoptotic events. However, few studies have investigated the restoration or repair of the mutated MMR-related genes. Other gene targets that can be modulated through nanotechnology are BRCA1 and PARP1. Downregulation of BRCA1 is regarded as a promising target for sensitizing tumors to chemotherapies, including paclitaxel and cisplatin. Lipidoid nanoparticles were used to deliver of PARP1 siRNA to treat BRCA1-deficient ovarian cancer and produced a tumor inhibitory effect both *in vivo* and *in vitro* (Goldberg et al. 2011). Based on these gene targets documented above, co-delivery of anti-sense nucleic acid therapy with chemotherapy can further enhance the therapeutic efficacy in treating multidrug resistant tumors.

4.5.3 Targeting Ceramide Metabolism with Nanomedicines for Resistance Reversal

Ceramide is the basic unit of sphingomyelin and can be produced *de novo* or by sphingomyelin hydrolysis. Ceramide-mediated signaling has been tied to cell cycle arrest and differentiation (Senchenkov et al. 2001). Ceramide lipids are natural pro-apoptotic mediators of cellular responses in cancer against chemotherapy, radiation or hypoxic stress conditions (Kolesnick and Kronke 1998). Ceramides are located in cell membranes as well as the outer mitochondrial membrane and are related to cytochrome c release to the cytoplasm and initiation of apoptosis. Neutralization of ceramide by overexpressed glucosylceramide synthase activity in cancer cells is linked to chemo-resistance (Barth et al. 2011).

While stimulating ceramide synthesis can be useful to overcome resistance, exogenous ceramide delivery, especially short-chain ceramide (ceramide C6), by nanomedicines has proven an effective way to tackle drug resistance (Barth et al. 2011). Several *in vitro* and *in vivo* studies have used various nanomedicine-based approaches to deliver ceramides, either alone or in combination with other chemotherapeutics to resistant cancer cells. Combination therapy of paclitaxel with ceramide C6 in poly(ethylene oxide)-modified poly(ε-caprolactone) nanoparticles resulted in more than a three-fold increase in tumor growth delay and doubling time in a SKOV-3TR drug resistant human ovarian cancer xenograft model (Devalapally et al. 2007). A polymer-blend nanoparticle system co-delivering paclitaxel and ceramide C6 showed higher paclitaxel accumulation in drug-resistant MCF-7TR tumors compared to free paclitaxel (van Vlerken et al. 2008) and reduced tumor volume by two-folds compared to paclitaxel monodelivery to tumors (van Vlerken et al. 2010).

Ceramide C6, a lipid, can easily locate itself in the lipid bilayers and thus, formulating it in liposomal formulations sounds logical. Koshkaryev *et al.* (Koshkaryev et al. 2012) developed PEGylated liposomes consisting of egg phosphatidylcholine and cholesterol, targeted with transferrin, and loaded with short-chain ceramides C6 and C16. By incorporating approx. 120 transferrin molecules on the outer surface of the liposomes, they showed that transferrin-targeted liposomes were internalized significantly more by the HeLa and A2780 ovarian carcinoma cells compared to non-targeted liposomes. The internalized liposomes eventually located themselves in the lysosomes, induced the lysosomal membrane permeabilization (LMP), and enhanced the apoptotic response of the cells. Ceramide-mediated apoptosis via the initiation of LMP translated into *in vivo* tumor growth inhibition in a A2780 xenograft mouse model and caused the greatest tumor volume growth inhibition and lowest tumor weights.

Sriraman et al. (Sriraman et al. 2016) investigated the cytotoxic potential of folic acid-targeted liposomal formulations co-encapsulating doxorubicin and ceramide C6 on different chemo-resistant cells including A2780-Adr and H69-AR. The results demonstrated that the co-delivery of ceramide C6 and doxorubicin using a liposomal platform correlates significantly with antiproliferative effect, due to cell cycle regulation, and subsequent induction of apoptosis. Furthermore, targeting the liposomal formulations with folic acid increased the cytotoxicity against monolayers as well as three-dimensional spheroids.

Nanomedicines give unique advantages for formulating ceramide to address drug resistance in cancer. They can also modulate the response of the resistant cancer cells to ceramide since it was shown that co-delivery of doxorubicin and ceramide exerts synergism only when formulated in liposomal carriers rather than free drug combinations (Fonseca et al. 2014).

4.5.4 Targeting Overexpressed Anti-Apoptotic Proteins by Nanomedicines

Proteins in the B-cell lymphoma 2 (BCL-2) family are key regulators of the apoptotic process and consist of prodeath (i.e. BAX, BAK, BIM) and prosurvival (i.e. BCL-2, BCL-X_L, MCL-1) members (Youle and Strasser 2008). BCL-2 was the first identified apoptosis regulator, and many cancers depend on BCL-2 mediated apoptosis escape to survive. Due to its significance and powerful anti-apoptotic regulatory function, inhibition or downregulation of this protein holds great importance in controlling the apoptotic pathways in cancer. Inhibiting the BCL-2 is a challenging task given that the blockage needs to be only in cancer cells, and any inhibitor needs to be delivered intracellularly. Here, nanomedicine-based approaches proved themselves very useful.

siRNA delivery by nanomedicines to cancer cells for selective downregulation of BCL-2 is one of the promising approaches for reversing drug resistance. Given the previously summarized challenges, it is not surprising that many different complex nanomedicine-based delivery systems fit under this heading.

Yu *et al.* developed cationic micellar nanocarriers from dual pH-responsive poly(2-(dimethyl amino)ethyl methacrylate)-block-poly(2-(diisopropyl amino)ethyl methacrylate) (PDMA-b-PDPA) di-block copolymers. The PDPA segment of the copolymer with a pKa around 6.3 can self-assemble to form hydrophobic cores for encapsulation of hydrophobic drugs (e.g. paclitaxel), whereas the PDMA block forms a positively charged shell allowing siRNA complexation to carrier (Yu et al. 2014). Upon cellular internalization and subsequent exposure to lower pH values in the endosomal compartments (pH 5.9–6.2), PDPA is protonated and the system dissociates. This dissociation causes the rapid release of paclitaxel and the protonated PDPA segment causes endosomal/lysosomal membrane disruption and intracellular siRNA release of BCL-2 siRNA. The nanomedicine system reversed the drug resistance of BCL-2-overexpressing A549 human non-small cell lung cancer cells by successful downregulation of BCL-2 via RNAi. Thus, the cells were sensitized to paclitaxel treatment *in vitro*.

Another nanomedicine formulation for targeting overexpressed pro-apoptotic proteins was reported by Wang et al. (Wang et al. 2016). They designed a delivery system consisting of hyaluronic acid (HA)-decorated PEI-PLGA for simultaneous delivery of doxorubicin along with the microRNA, MiR-542-3p. The authors showed that successful intracellular delivery of the miRNA induced p53 activation and downregulated survivin, an inhibitor of apoptosis (IAP) family member. The system protected and increased the intracellular concentrations of miRNA. This, in return, sensitized the triple negative breast cancer cells to the levels seen in drug-sensitive MCF-7 cells.

Very recently, an intriguing nanomedicinal formulation was reported for downregulation of survivin by another miRNA. Salzano *et al.* (Salzano et al. 2016) developed a novel dual stimuli-sensitive system for simultaneous delivery of miRNA-34a and doxorubicin. The novelty of the reported system resides in the presence of two stimuli-sensitive prodrugs, a matrix metalloproteinase 2 (MMP2)-sensitive doxorubicin conjugate and a reducing agent (glutathione, GSH)-sensitive miRNA-34a conjugate, in a single particle decorated with a polyethylene glycol corona and a cell-penetrating peptide (TATp). The combination system (Figure 4.5) can be considered as 'multi-headed missile' that drops (releases) its cargos selectively at the sites of action, and only when they are needed. The sensitivity of the system to MMP2 resulted in three-fold higher cytotoxicity in MMP2-overexpressing HT1080 cells compared to low MMP2-expressing MCF7 cells. In addition, cellular internalization of doxorubicin increased by more than 70% due to the inclusion of a cell-penetrating peptide (TATp) in the formulation. Treatment with the MMP2-sensitive nanomedicine also inhibited proliferation and migration of HT1080 cells. Moreover, GSH-sensitivity of the system allowed for an efficient downregulated expression of BCL-2, survivin, and notch1 (65%, 55%, and 46%, respectively). They also showed up to 50% decrease in cell viability and reduction of invasiveness in three-dimensional spheroid models. This kind of complex and multifunctional nanomedicine-based formulation that contains various desired components within themselves should lead the way to efficient treatment and reversal of chemoresistance in cancer cells.

4.6 CONCLUSION

Chemoresistance is a major factor for chemotherapy failure either for conventional small molecule treatments or nanomedicine-based approaches. As described in this chapter, the underlying molecular mechanisms of cancer resistance cover a broad range of factors. But resistance development remains *fait accompli* due to already

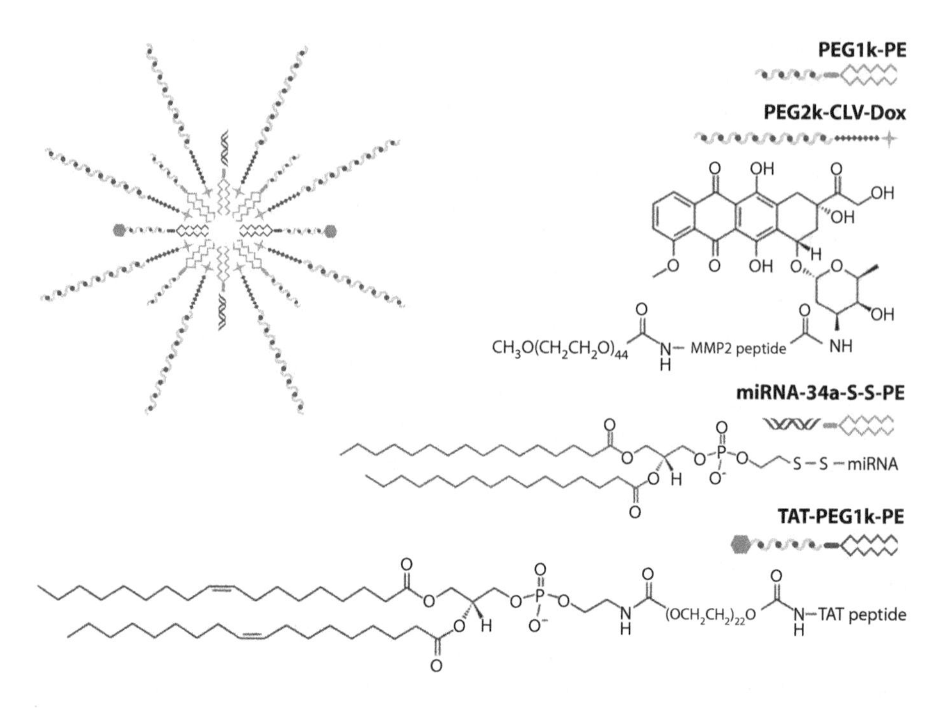

FIGURE 4.5 Schematic representation of the dual stimuli-sensitive system with the structural representation of the synthesized components. (Reprinted from Salzano, G. et al., *Small*, 12 (35):4837–48, 2016. With permission.)

existing drug resistant cells in the large solid tumors in addition to resistance acquirement during the course of the treatment with single agents. When combined with the heterogeneity of tumors, reversal of the chemoresistance sets a great challenge. However, increased number of identified resistance mechanisms in the last decades also increased the potential targets that can be exploited by different approaches to overcome chemoresistance. Significant advances and extensive knowledge that has been gathered in the nanomedicine field offered substantial advantages to tackle drug resistance by strategies summarized earlier.

While the vast number of nanomedicine-based approaches are being investigated and developed, relatively a few of promising candidates were able to reach the market. This indicates a necessity to shift the focus from the long-established paradigms and find new strategies that also internalizes the approaches to overcome hurdles related to chemoresistance. The future for successful nanomedicine-based therapies requires multifunctional approaches which include molecularly targeted agents, oligonucleotide delivery, peptide and protein drugs, and in most cases the combination of these. Relying on the formulation parameters or components of single-agent loaded nanomedicines for resistance reversal found some success, but it also made it clear that the combination therapies are the viable options for chemoresistance reversal. Many different chemoresistance reversal agents with different acting mechanisms such as P-gp inhibitors, pathway blockers, resistance related protein inhibitors can be delivered simultaneously with cytotoxic drugs to targeted tumor cells.

Nanotechnology-based drug delivery platforms also enable the use of resistance modulators that were thought to be unviable due to low solubility or stability issues, including siRNA/miRNAs and peptide/proteins. Many properties of the nanomedicine platforms can be fine-tuned depending on the type of tumor, location, resistance mechanism or subcellular target, the factors that are not allowing the personalization of traditional drug cocktail administration for resistance reversal.

Of course, there are many challenges involved with the nanomedicines concerning poor clinical translation, batch-to-batch inconsistencies, elevated immune responses in some cases, and even biocompatibility issues arising from the materials used in the formulations. However, with ever expanding knowledge in this field, the molecular mechanisms underlying the chemoresistance can be challenged and overcome by using nanomedicines. Utilization of nanomedicines in the clinic, growing number of successful clinical trials, even more promising preclinical data related to multifunctional tumor microenvironment sensitive approaches makes the possibility of reversing drug resistance a reality day by day.

REFERENCES

Abouzeid, A. H., N. R. Patel, C. Sarisozen, et al. 2014. "Transferrin-targeted polymeric micelles co-loaded with curcumin and paclitaxel: efficient killing of paclitaxel-resistant cancer cells." *Pharm Res* 31 (8):1938–45.

Abouzeid, A. H., N. R. Patel, and V. P. Torchilin. 2014. "Polyethylene glycol-phosphatidylethanolamine (PEG-PE)/vitamin E micelles for co-delivery of paclitaxel and curcumin to overcome multi-drug resistance in ovarian cancer." *Int J Pharm* 464 (1–2):178–84.

Adams, E. R., J. J. Leffert, D. J. Craig, et al. 1999. "In vivo effect of 5-ethynyluracil on 5-fluorouracil metabolism determined by 19F nuclear magnetic resonance spectroscopy." *Cancer Res* 59 (1):122–7.

Alakhova, D. Y., and A. V. Kabanov. 2014. "Pluronics and MDR reversal: an update." *Mol Pharm* 11 (8):2566–78.

Alakhova, D. Y., N. Y. Rapoport, E. V. Batrakova, et al. 2010. "Differential metabolic responses to pluronic in MDR and non-MDR cells: a novel pathway for chemosensitization of drug resistant cancers." *J Control Release* 142 (1):89–100.

Aller, S. G., J. Yu, A. Ward, et al. 2009. "Structure of P-glycoprotein reveals a molecular basis for poly-specific drug binding." *Science* 323 (5922):1718–22.

Ambudkar, S. V., C. Kimchi-Sarfaty, Z. E. Sauna, et al. 2003. "P-glycoprotein: from genomics to mechanism." *Oncogene* 22 (47):7468–85.

Armand, J. P., V. Ribrag, J. L. Harrousseau, et al. 2007. "Reappraisal of the use of procarbazine in the treatment of lymphomas and brain tumors." *Ther Clin Risk Manag* 3 (2):213–24.

Ashworth, A. 2008. "A synthetic lethal therapeutic approach: poly(ADP) ribose polymerase inhibitors for the treatment of cancers deficient in DNA double-strand break repair." *J Clin Oncol* 26 (22):3785–90.

Bakker-Woudenberg, I. A., A. F. Lokerse, M. T. ten Kate, et al. 1992. "Enhanced localization of liposomes with prolonged blood circulation time in infected lung tissue." *Biochim Biophys Acta* 1138 (4):318–26.

Bansal, T., G. Mishra, M. Jaggi, et al. 2009. "Effect of P-glycoprotein inhibitor, verapamil, on oral bioavailability and pharmacokinetics of irinotecan in rats." *Eur J Pharm Sci* 36 (4–5):580–90.

Barenholz, Y. 2012. "Doxil® --the first FDA-approved nano-drug: lessons learned." *J Control Release* 160 (2):117–34.

Bargou, R. C., C. Wagener, K. Bommert, et al. 1996. "Overexpression of the death-promoting gene bax-alpha which is downregulated in breast cancer restores sensitivity to different apoptotic stimuli and reduces tumor growth in SCID mice." *J Clin Invest* 97 (11):2651–9.

Barth, B. M., M. C. Cabot, and M. Kester. 2011. "Ceramide-based therapeutics for the treatment of cancer." *Anticancer Agents Med Chem* 11 (9):911–9.

Bates, S. E., R. Robey, K. Miyake, et al. 2001. "The role of half-transporters in multidrug resistance." *J Bioenerg Biomembr* 33 (6):503–11.

Batrakova, E. V., D. L. Kelly, S. Li, et al. 2006. "Alteration of genomic responses to doxorubicin and prevention of MDR in breast cancer cells by a polymer excipient: pluronic P85." *Mol Pharm* 3 (2):113–23.

Batrakova, E. V., S. Li, W. F. Elmquist, et al. 2001. "Mechanism of sensitization of MDR cancer cells by Pluronic block copolymers: selective energy depletion." *Br J Cancer* 85 (12):1987–97.

Batrakova, E. V., S. Li, Y. Li, et al. 2004. "Effect of pluronic P85 on ATPase activity of drug efflux transporters." *Pharm Res* 21 (12):2226–33.

Beck, A., M. C. Etienne, S. Cheradame, et al. 1994. "A role for dihydropyrimidine dehydrogenase and thymidylate synthase in tumour sensitivity to fluorouracil." *Eur J Cancer* 30A (10):1517–22.

Bhattacharyya, A., U. S. Ear, B. H. Koller, et al. 2000. "The breast cancer susceptibility gene BRCA1 is required for subnuclear assembly of Rad51 and survival following treatment with the DNA cross-linking agent cisplatin." *J Biol Chem* 275 (31):23899–903.

Blanchard, Z., B. T. Paul, B. Craft, et al. 2015. "BRCA1-IRIS inactivation overcomes paclitaxel resistance in triple negative breast cancers." *Breast Cancer Res* 17:5.

Boucher, Y., L. T. Baxter, and R. K. Jain. 1990. "Interstitial pressure gradients in tissue-isolated and subcutaneous tumors: implications for therapy." *Cancer Res* 50 (15):4478–84.

Bouwman, P., and J. Jonkers. 2012. "The effects of deregulated DNA damage signalling on cancer chemotherapy response and resistance." *Nat Rev Cancer* 12 (9):587–98.

Bozic, I., J. G. Reiter, B. Allen, et al. 2013. "Evolutionary dynamics of cancer in response to targeted combination therapy." *Elife* 2:e00747.

Brown, R., G. L. Hirst, W. M. Gallagher, et al. 1997. "hMLH1 expression and cellular responses of ovarian tumour cells to treatment with cytotoxic anticancer agents." *Oncogene* 15 (1):45–52.

Cabral, H., Y. Matsumoto, K. Mizuno, et al. 2011. "Accumulation of sub-100 nm polymeric micelles in poorly permeable tumours depends on size." *Nat Nanotechnol* 6 (12):815–23.

Chan, H. S., T. M. Grogan, G. Haddad, et al. 1997. "P-glycoprotein expression: critical determinant in the response to osteosarcoma chemotherapy." *J Natl Cancer Inst* 89 (22):1706–15.

Chaney, S. G., and A. Sancar. 1996. "DNA repair: enzymatic mechanisms and relevance to drug response." *J Natl Cancer Inst* 88 (19):1346–60.

Chang, A. 2011. "Chemotherapy, chemoresistance and the changing treatment landscape for NSCLC." *Lung Cancer* 71 (1):3–10.

Chaudhury, A., S. Das, R. M. Bunte, et al. 2012. "Potent therapeutic activity of folate receptor-targeted liposomal carboplatin in the localized treatment of intraperitoneally grown human ovarian tumor xenograft." *Int J Nanomedicine* 7:739–51.

Chen, C. J., J. E. Chin, K. Ueda, et al. 1986. "Internal duplication and homology with bacterial transport proteins in the mdrl (P-glycoprotein) gene from multidrug-resistant human cells." *Cell* 47 (3):381–9.

Cheng, K. T., P. C. Wang, and L. Shan. 2004. "Alexa Fluor 680-labeled transferrin-cationic (NBD-labeled DOPE-DOTAP) liposome-encapsulated gadopentetate dimeglumine complex." In *Molecular Imaging and Contrast Agent Database (MICAD)*. Bethesda (MD).

Cheung, A. M., T. S. Wan, J. C. Leung, et al. 2007. "Aldehyde dehydrogenase activity in leukemic blasts defines a subgroup of acute myeloid leukemia with adverse prognosis and superior NOD/SCID engrafting potential." *Leukemia* 21 (7):1423–30.

Chock, K., J. M. Allison, and W. M. Elshamy. 2010. "BRCA1-IRIS overexpression abrogates UV-induced p38MAPK/p53 and promotes proliferation of damaged cells." *Oncogene* 29 (38):5274–85.

Chock, K. L., J. M. Allison, Y. Shimizu, et al. 2010. "BRCA1-IRIS overexpression promotes cisplatin resistance in ovarian cancer cells." *Cancer Res* 70 (21):8782–91.

Choi, H. S., W. Liu, F. Liu, et al. 2010. "Design considerations for tumour-targeted nanoparticles." *Nat Nanotechnol* 5 (1):42–7.

Choi, Y. J., J. K. Rho, S. J. Lee, et al. 2009. "HIF-1alpha modulation by topoisomerase inhibitors in non-small cell lung cancer cell lines." *J Cancer Res Clin Oncol* 135 (8):1047–53.

Cohen, M. S., S. Cai, Y. Xie, et al. 2009. "A novel intralymphatic nanocarrier delivery system for cisplatin therapy in breast cancer with improved tumor efficacy and lower systemic toxicity in vivo." *Am J Surg* 198 (6):781–6.

Cole, S. P., G. Bhardwaj, J. H. Gerlach, et al. 1992. "Overexpression of a transporter gene in a multidrug-resistant human lung cancer cell line." *Science* 258 (5088):1650–4.

Comerford, K. M., T. J. Wallace, J. Karhausen, et al. 2002. "Hypoxia-inducible factor-1-dependent regulation of the multidrug resistance (MDR1) gene." *Cancer Res* 62 (12):3387–94.

Copur, S., K. Aiba, J. C. Drake, et al. 1995. "Thymidylate synthase gene amplification in human colon cancer cell lines resistant to 5-fluorouracil." *Biochem Pharmacol* 49 (10):1419–26.

Dabholkar, M., J. Vionnet, F. Bostick-Bruton, et al. 1994. "Messenger RNA levels of XPAC and ERCC1 in ovarian cancer tissue correlate with response to platinum-based chemotherapy." *J Clin Invest* 94 (2):703–8.

Dabholkar, R. D., R. M. Sawant, D. A. Mongayt, et al. 2006. "Polyethylene glycol-phosphatidylethanolamine conjugate (PEG-PE)-based mixed micelles: some properties, loading with paclitaxel, and modulation of P-glycoprotein-mediated efflux." *Int J Pharm* 315 (1–2):148–57.

Dale, I. L., W. Tuffley, R. Callaghan, et al. 1998. "Reversal of P-glycoprotein-mediated multidrug resistance by XR9051, a novel diketopiperazine derivative." *Br J Cancer* 78 (7):885–92.

Danson, S., D. Ferry, V. Alakhov, et al. 2004. "Phase I dose escalation and pharmacokinetic study of pluronic polymer-bound doxorubicin (SP1049C) in patients with advanced cancer." *Br J Cancer* 90 (11):2085–91.

Darby, R. A., R. Callaghan, and R. M. McMahon. 2011. "P-glycoprotein inhibition: the past, the present and the future." *Curr Drug Metab* 12 (8):722–31.

Demina, T., I. Grozdova, O. Krylova, et al. 2005. "Relationship between the structure of amphiphilic copolymers and their ability to disturb lipid bilayers." *Biochemistry* 44 (10):4042–54.

Deng, L., Y. C. Lin-Lee, F. X. Claret, et al. 2001. "2-acetylaminofluorene up-regulates rat mdr1b expression through generating reactive oxygen species that activate NF-kappa B pathway." *J Biol Chem* 276 (1):413–20.

Deng, Y. H., X. X. Pu, M. J. Huang, et al. 2010. "5-Fluorouracil upregulates the activity of Wnt signaling pathway in CD133-positive colon cancer stem-like cells." *Chin J Cancer* 29 (9):810–5.

Dening, T. J., S. Rao, N. Thomas, et al. 2016. "Oral nanomedicine approaches for the treatment of psychiatric illnesses." *J Control Release* 223:137–56.

Devalapally, H., Z. Duan, M. V. Seiden, et al. 2007. "Paclitaxel and ceramide co-administration in biodegradable polymeric nanoparticulate delivery system to overcome drug resistance in ovarian cancer." *Int J Cancer* 121 (8):1830–8.

DeVita, V. T., Jr. 1975. "Single agent versus combination chemotherapy." *CA Cancer J Clin* 25 (3):152–8.

DeVita, V. T., Jr., R. C. Young, and G. P. Canellos. 1975. "Combination versus single agent chemotherapy: a review of the basis for selection of drug treatment of cancer." *Cancer* 35 (1):98–110.

Dhar, S., F. X. Gu, R. Langer, et al. 2008. "Targeted delivery of cisplatin to prostate cancer cells by aptamer functionalized Pt(IV) prodrug-PLGA-PEG nanoparticles." *Proc Natl Acad Sci U S A* 105 (45):17356–61.

Diasio, R. B. 1998. "The role of dihydropyrimidine dehydrogenase (DPD) modulation in 5-FU pharmacology." *Oncology (Williston Park)* 12 (10 Suppl 7):23–7.

Diasio, R. B., and B. E. Harris. 1989. "Clinical pharmacology of 5-fluorouracil." *Clin Pharmacokinet* 16 (4):215–37.

Diasio, R. B., and M. R. Johnson. 1999. "Dihydropyrimidine dehydrogenase: its role in 5-fluorouracil clinical toxicity and tumor resistance." *Clin Cancer Res* 5 (10):2672–3.

Diaz, L. A., Jr., R. T. Williams, J. Wu, et al. 2012. "The molecular evolution of acquired resistance to targeted EGFR blockade in colorectal cancers." *Nature* 486 (7404):537–40.

Dintaman, J. M., and J. A. Silverman. 1999. "Inhibition of P-glycoprotein by D-alpha-tocopheryl polyethylene glycol 1000 succinate (TPGS)." *Pharm Res* 16 (10):1550–6.

Dixon, K. H., B. C. Lanpher, et al. 1994. "A novel cDNA restores reduced folate carrier activity and methotrexate sensitivity to transport deficient cells." *J Biol Chem* 269 (1):17–20.

Doyle, L. A., W. Yang, L. V. Abruzzo, et al.1998. "A multidrug resistance transporter from human MCF-7 breast cancer cells." *Proc Natl Acad Sci U S A* 95 (26):15665–70.

Dreher, M. R., W. Liu, C. R. Michelich, et al. 2006. "Tumor vascular permeability, accumulation, and penetration of macromolecular drug carriers." *J Natl Cancer Inst* 98 (5):335–44.

Elbayoumi, T. A., and V. P. Torchilin. 2007. "Enhanced cytotoxicity of monoclonal anticancer antibody 2C5-modified doxorubicin-loaded PEGylated liposomes against various tumor cell lines." *Eur J Pharm Sci* 32 (3):159–68.

Elbayoumi, T. A., and V. P. Torchilin. 2009. "Tumor-specific anti-nucleosome antibody improves therapeutic efficacy of doxorubicin-loaded long-circulating liposomes against primary and metastatic tumor in mice." *Mol Pharm* 6 (1):246–54.

Essex, S., G. Navarro, P. Sabhachandani, et al. 2015. "Phospholipid-modified PEI-based nanocarriers for in vivo siRNA therapeutics against multidrug-resistant tumors." *Gene Ther.* 22(3):257–66.

Etheridge, M. L., S. A. Campbell, A. G. Erdman, et al. 2013. "The big picture on nanomedicine: the state of investigational and approved nanomedicine products." *Nanomedicine* 9 (1):1–14.

Evers, R., M. Kool, L. van Deemter, et al. 1998. "Drug export activity of the human canalicular multispecific organic anion transporter in polarized kidney MDCK cells expressing cMOAT (MRP2) cDNA." *J Clin Invest* 101 (7):1310–9.

Farmer, H., N. McCabe, C. J. Lord, et al. 2005. "Targeting the DNA repair defect in BRCA mutant cells as a therapeutic strategy." *Nature* 434 (7035):917–21.

Findley, H. W., L. Gu, A. M. Yeager, et al. 1997. "Expression and regulation of Bcl-2, Bcl-xl, and Bax correlate with p53 status and sensitivity to apoptosis in childhood acute lymphoblastic leukemia." *Blood* 89 (8):2986–93.

Fink, D., S. Aebi, and S. B. Howell. 1998. "The role of DNA mismatch repair in drug resistance." *Clin Cancer Res* 4 (1):1–6.

Fire, A., S. Xu, M. K. Montgomery, et al. 1998. "Potent and specific genetic interference by double-stranded RNA in Caenorhabditis elegans." *Nature* 391 (6669):806–11.

Fojo, A. T., K. Ueda, D. J. Slamon, et al. 1987. "Expression of a multidrug-resistance gene in human tumors and tissues." *Proc Natl Acad Sci U S A* 84 (1):265–9.

Folkins, C., S. Man, P. Xu, et al. 2007. "Anticancer therapies combining antiangiogenic and tumor cell cytotoxic effects reduce the tumor stem-like cell fraction in glioma xenograft tumors." *Cancer Res* 67 (8):3560–4.

Fonseca, N. A., L. C. Gomes-da-Silva, V. Moura, et al. 2014. "Simultaneous active intracellular delivery of doxorubicin and C6-ceramide shifts the additive/antagonistic drug interaction of non-encapsulated combination." *J Control Release* 196:122–31.

Friedenberg, W. R., M. Rue, E. A. Blood, et al. 2006. "Phase III study of PSC-833 (valspodar) in combination with vincristine, doxorubicin, and dexamethasone (valspodar/VAD) versus VAD alone in patients with recurring or refractory multiple myeloma (E1A95): a trial of the Eastern Cooperative Oncology Group." *Cancer* 106 (4):830–8.

Friedman, H. S., O. M. Colvin, S. H. Kaufmann, et al. 1992. "Cyclophosphamide resistance in medulloblastoma." *Cancer Res* 52 (19):5373–8.

Gabizon, A., R. Catane, B. Uziely, et al. 1994. "Prolonged circulation time and enhanced accumulation in malignant exudates of doxorubicin encapsulated in polyethylene-glycol coated liposomes." *Cancer Res* 54 (4):987–92.

Gabizon, A., and D. Papahadjopoulos. 1988. "Liposome formulations with prolonged circulation time in blood and enhanced uptake by tumors." *Proc Natl Acad Sci U S A* 85 (18):6949–53.

Gabizon, A., H. Shmeeda, and Y. Barenholz. 2003. "Pharmacokinetics of pegylated liposomal Doxorubicin: review of animal and human studies." *Clin Pharmacokinet* 42 (5):419–36.

Gabizon, A., H. Shmeeda, and T. Grenader. 2012. "Pharmacological basis of pegylated liposomal doxorubicin: impact on cancer therapy." *Eur J Pharm Sci* 45 (4):388–98.

Gabizon, A., D. Tzemach, L. Mak, et al. 2002. "Dose dependency of pharmacokinetics and therapeutic efficacy of pegylated liposomal doxorubicin (DOXIL) in murine models." *J Drug Target* 10 (7):539–48.

Gao, J., W. Zhong, J. He, et al. 2009. "Tumor-targeted PE38KDEL delivery via PEGylated anti-HER2 immunoliposomes." *Int J Pharm* 374 (1–2):145–52.

Gao, S., F. Dagnaes-Hansen, E. J. Nielsen, et al. 2009. "The effect of chemical modification and nanoparticle formulation on stability and biodistribution of siRNA in mice." *Mol Ther* 17 (7):1225–33.

Garbuzenko, O. B., M. Saad, V. P. Pozharov, et al. 2010. "Inhibition of lung tumor growth by complex pulmonary delivery of drugs with oligonucleotides as suppressors of cellular resistance." *Proc Natl Acad Sci U S A* 107 (23):10737–42.

Goker, E., M. Waltham, A. Kheradpour, et al. 1995. "Amplification of the dihydrofolate reductase gene is a mechanism of acquired resistance to methotrexate in patients with acute lymphoblastic leukemia and is correlated with p53 gene mutations." *Blood* 86 (2):677–84.

Goldberg, M. S., D. Xing, Y. Ren, et al. 2011. "Nanoparticle-mediated delivery of siRNA targeting Parp1 extends survival of mice bearing tumors derived from Brca1-deficient ovarian cancer cells." *Proc Natl Acad Sci U S A* 108 (2):745–50.

Goldstein, L. J., H. Galski, A. Fojo, et al. 1989. "Expression of a multidrug resistance gene in human cancers." *J Natl Cancer Inst* 81 (2):116–24.

Goodman, T. T., P. L. Olive, and S. H. Pun. 2007. "Increased nanoparticle penetration in collagenase-treated multicellular spheroids." *Int J Nanomedicine* 2 (2):265–74.

Goren, D., A. T. Horowitz, D. Tzemach, et al. 2000. "Nuclear delivery of doxorubicin via folate-targeted liposomes with bypass of multidrug-resistance efflux pump." *Clin Cancer Res* 6 (5):1949–57.

Gradishar, W. J., S. Tjulandin, N. Davidson, et al. 2005. "Phase III trial of nanoparticle albumin-bound paclitaxel compared with polyethylated castor oil-based paclitaxel in women with breast cancer." *J Clin Oncol* 23 (31):7794–803.

Guo, P., O. Coban, N. M. Snead, et al. 2010. "Engineering RNA for targeted siRNA delivery and medical application." *Adv Drug Deliv Rev* 62 (6):650–66.

Gupta, B., and V. P. Torchilin. 2007. "Monoclonal antibody 2C5-modified doxorubicin-loaded liposomes with significantly enhanced therapeutic activity against intracranial human brain U-87 MG tumor xenografts in nude mice." *Cancer Immunol Immunother* 56 (8):1215–23.

Hamilton, G. 1998. "Multicellular spheroids as an in vitro tumor model." *Cancer Letters* 131 (1):29–34.

Hanada, M., C. Aime-Sempe, T. Sato, and J. C. Reed. 1995. "Structure-function analysis of Bcl-2 protein. Identification of conserved domains important for homodimerization with Bcl-2 and heterodimerization with Bax." *J Biol Chem* 270 (20):11962–9.

Hansen, K. M., H. F. Ji, G. Wu, et al. 2001. "Cantilever-based optical deflection assay for discrimination of DNA single-nucleotide mismatches." *Anal Chem* 73 (7):1567–71.

Hobson, B., and J. Denekamp. 1984. "Endothelial proliferation in tumours and normal tissues: continuous labelling studies." *Br J Cancer* 49 (4):405–13.

Hoeijmakers, J. H. 2001. "Genome maintenance mechanisms for preventing cancer." *Nature* 411 (6835):366–74.

Honoki, K., H. Fujii, A. Kubo, et al. 2010. "Possible involvement of stem-like populations with elevated ALDH1 in sarcomas for chemotherapeutic drug resistance." *Oncol Rep* 24 (2):501–5.

Hornicek, F. J., M. C. Gebhardt, M. W. Wolfe, et al. 2000. "P-glycoprotein levels predict poor outcome in patients with osteosarcoma." *Clin Orthop Relat Res* (373):11–7.

Hrkach, J., D. Von Hoff, M. Mukkaram Ali, et al. 2012. "Preclinical development and clinical translation of a PSMA-targeted docetaxel nanoparticle with a differentiated pharmacological profile." *Sci Transl Med* 4 (128):128ra39.

Hu, C. M., and L. Zhang. 2012. "Nanoparticle-based combination therapy toward overcoming drug resistance in cancer." *Biochem Pharmacol* 83 (8):1104–11.

Hu, Z., S. Jin, and K. W. Scotto. 2000. "Transcriptional activation of the MDR1 gene by UV irradiation. Role of NF-Y and Sp1." *J Biol Chem* 275 (4):2979–85.

Hunter, A. M., E. C. LaCasse, and R. G. Korneluk. 2007. "The inhibitors of apoptosis (IAPs) as cancer targets." *Apoptosis* 12 (9):1543–68.

Huo, S., H. Ma, K. Huang, et al. 2013. "Superior penetration and retention behavior of 50 nm gold nanoparticles in tumors." *Cancer Res* 73 (1):319–30.

Hyafil, F., C. Vergely, P. Du Vignaud, et al. 1993. "In vitro and in vivo reversal of multidrug resistance by GF120918, an acridonecarboxamide derivative." *Cancer Res* 53 (19):4595–602.

Indran, I. R., G. Tufo, S. Pervaiz, et al. 2011. "Recent advances in apoptosis, mitochondria and drug resistance in cancer cells." *Biochim Biophys Acta* 1807 (6):735–45.

Ishida, T., K. Atobe, X. Wang, et al. 2006. "Accelerated blood clearance of PEGylated liposomes upon repeated injections: effect of doxorubicin-encapsulation and high-dose first injection." *J Control Release* 115 (3):251–8.

Ishida, T., M. Ichihara, X. Wang, et al. 2006. "Injection of PEGylated liposomes in rats elicits PEG-specific IgM, which is responsible for rapid elimination of a second dose of PEGylated liposomes." *J Control Release* 112 (1):15–25.

Ishida, T., R. Maeda, M. Ichihara, et al. 2002. "The accelerated clearance on repeated injection of pegylated liposomes in rats: laboratory and histopathological study." *Cell Mol Biol Lett* 7 (2):286.

Ishikawa, T., and F. Ali-Osman. 1993. "Glutathione-associated cis-diamminedichloroplatinum(II) metabolism and ATP-dependent efflux from leukemia cells. Molecular characterization of glutathione-platinum complex and its biological significance." *J Biol Chem* 268 (27):20116–25.

Iversen, F., C. Yang, F. Dagnaes-Hansen, et al. 2013. "Optimized siRNA-PEG conjugates for extended blood circulation and reduced urine excretion in mice." *Theranostics* 3 (3):201–9.

Iyer, A. K., G. Khaled, J. Fang, et al. 2006. "Exploiting the enhanced permeability and retention effect for tumor targeting." *Drug Discov Today* 11 (17–18):812–8.

Jain, R. K., and L. T. Baxter. 1988. "Mechanisms of heterogeneous distribution of monoclonal antibodies and other macromolecules in tumors: significance of elevated interstitial pressure." *Cancer Res* 48 (24 Pt 1):7022–32.

Jain, R. K., and T. Stylianopoulos. 2010. "Delivering nanomedicine to solid tumors." *Nat Rev Clin Oncol* 7 (11):653–64.

Jang, S. H., M. G. Wientjes, and J. L. Au. 2001. "Kinetics of P-glycoprotein-mediated efflux of paclitaxel." *J Pharmacol Exp Ther* 298 (3):1236–42.

Jin, M. S., M. L. Oldham, Q. Zhang, et al. 2012. "Crystal structure of the multidrug transporter P-glycoprotein from Caenorhabditis elegans." *Nature* 490 (7421):566–9.

Kapse-Mistry, S., T. Govender, R. Srivastava, et al. 2014. "Nanodrug delivery in reversing multidrug resistance in cancer cells." *Front Pharmacol* 5:159.

Kelland, L. R. 1993. "New platinum antitumor complexes." *Crit Rev Oncol Hematol* 15 (3):191–219.

Kelly, M. M., B. D. Hoel, and C. Voelkel-Johnson. 2002. "Doxorubicin pretreatment sensitizes prostate cancer cell lines to TRAIL induced apoptosis which correlates with the loss of c-FLIP expression." *Cancer Biol Ther* 1 (5):520–7.

Kerbel, R. S., J. Rak, H. Kobayashi, et al. 1994. "Multicellular resistance: a new paradigm to explain aspects of acquired drug resistance of solid tumors." *Cold Spring Harb Symp Quant Biol* 59:661–72.

Kim, T. H., C. W. Mount, W. R. Gombotz, et al. 2010. "The delivery of doxorubicin to 3-D multicellular spheroids and tumors in a murine xenograft model using tumor-penetrating triblock polymeric micelles." *Biomaterials* 31 (28):7386–97.

Kim, T. Y., D. W. Kim, J. Y. Chung, et al. 2004. "Phase I and pharmacokinetic study of Genexol-PM, a cremophor-free, polymeric micelle-formulated paclitaxel, in patients with advanced malignancies." *Clin Cancer Res* 10 (11):3708–16.

Kimura, Y., S. Y. Morita, M. Matsuo, et al. 2007. "Mechanism of multidrug recognition by MDR1/ABCB1." *Cancer Sci* 98 (9):1303–10.

Kirtane, A. R., S. M. Kalscheuer, and J. Panyam. 2013. "Exploiting nanotechnology to overcome tumor drug resistance: Challenges and opportunities." *Adv Drug Deliv Rev* 65 (13–14):1731–47.

Klibanov, A. L., K. Maruyama, V. P. Torchilin, et al. 1990. "Amphipathic polyethyleneglycols effectively prolong the circulation time of liposomes." *FEBS Lett* 268 (1):235–7.

Kobayashi, T., T. Ishida, Y. Okada, et al. 2007. "Effect of transferrin receptor-targeted liposomal doxorubicin in P-glycoprotein-mediated drug resistant tumor cells." *Int J Pharm* 329 (1–2):94–102.

Kobune, M., R. Takimoto, K. Murase, et al. 2009. "Drug resistance is dramatically restored by hedgehog inhibitors in CD34+ leukemic cells." *Cancer Sci* 100 (5):948–55.

Kogan, S. C., D. E. Brown, D. B. Shultz, et al. 2001. "BCL-2 cooperates with promyelocytic leukemia retinoic acid receptor alpha chimeric protein (PMLRARalpha) to block neutrophil differentiation and initiate acute leukemia." *J Exp Med* 193 (4):531–43.

Kohn, E. C., G. Sarosy, A. Bicher, et al. 1994. "Dose-intense taxol: high response rate in patients with platinum-resistant recurrent ovarian cancer." *J Natl Cancer Inst* 86 (1):18–24.

Kolesnick, R. N., and M. Kronke. 1998. "Regulation of ceramide production and apoptosis." *Annu Rev Physiol* 60:643–65.

Koshkaryev, A., A. Piroyan, and V. P. Torchilin. 2012. "Increased apoptosis in cancer cells in vitro and in vivo by ceramides in transferrin-modified liposomes." *Cancer Biol Ther* 13 (1):50–60.

Kratz, Felix. 2014. "A clinical update of using albumin as a drug vehicle—A commentary." *Journal of Controlled Release* 190:331–336.

Krishna, R., and L. D. Mayer. 2000. "Multidrug resistance (MDR) in cancer. Mechanisms, reversal using modulators of MDR and the role of MDR modulators in influencing the pharmacokinetics of anticancer drugs." *Eur J Pharm Sci* 11 (4):265–83.

Kuo, M. T. 2009. "Redox regulation of multidrug resistance in cancer chemotherapy: molecular mechanisms and therapeutic opportunities." *Antioxid Redox Signal* 11 (1):99–133.

Kuo, M. T., Z. Liu, Y. Wei, et al. 2002. "Induction of human MDR1 gene expression by 2-acetylaminofluorene is mediated by effectors of the phosphoinositide 3-kinase pathway that activate NF-kappaB signaling." *Oncogene* 21 (13):1945–54.

Lee, R. J., and P. S. Low. 1995. "Folate-mediated tumor cell targeting of liposome-entrapped doxorubicin in vitro." *Biochim Biophys Acta* 1233 (2):134–44.

Lee, S. M., T. V. O'Halloran, and S. T. Nguyen. 2010. "Polymer-caged nanobins for synergistic cisplatin-doxorubicin combination chemotherapy." *J Am Chem Soc* 132 (48):17130–8.

Lhomme, C., F. Joly, J. L. Walker, et al. 2008. "Phase III study of valspodar (PSC 833) combined with paclitaxel and carboplatin compared with paclitaxel and carboplatin alone in patients with stage IV or suboptimally debulked stage III epithelial ovarian cancer or primary peritoneal cancer." *J Clin Oncol* 26 (16):2674–82.

Li, W., Z. Jie, Z. Li, et al. 2014. "ERCC1 siRNA ameliorates drug resistance to cisplatin in gastric carcinoma cell lines." *Mol Med Rep* 9 (6):2423–8.

Li, X. M., L. Y. Ding, Y. Xu, et al. 2009. "Targeted delivery of doxorubicin using stealth liposomes modified with transferrin." *Int J Pharm* 373 (1–2):116–123.

Li, X., M. Marani, J. Yu, et al. 2001. "Adenovirus-mediated Bax overexpression for the induction of therapeutic apoptosis in prostate cancer." *Cancer Res* 61 (1):186–91.

Liang, J. T., K. C. Huang, Y. M. Cheng, et al. 2002. "P53 overexpression predicts poor chemosensitivity to high-dose 5-fluorouracil plus leucovorin chemotherapy for stage IV colorectal cancers after palliative bowel resection." *Int J Cancer* 97 (4):451–7.

Liechty, W. B., M. Caldorera-Moore, M. A. Phillips, et al. 2011. "Advanced molecular design of biopolymers for transmucosal and intracellular delivery of chemotherapeutic agents and biological therapeutics." *J Control Release* 155 (2):119–27.

Linn, S. C., G. Giaccone, P. J. van Diest, et al. 1995. "Prognostic relevance of P-glycoprotein expression in breast cancer." *Ann Oncol* 6 (7):679–85.

Liu, M., D. Li, R. Aneja, et al. 2007. "PO(2)-dependent differential regulation of multidrug resistance 1 gene expression by the c-Jun NH2-terminal kinase pathway." *J Biol Chem* 282 (24):17581–6.

Liu, X., E. K. Han, M. Anderson, et al. 2009. "Acquired resistance to combination treatment with temozolomide and ABT-888 is mediated by both base excision repair and homologous recombination DNA repair pathways." *Mol Cancer Res* 7 (10):1686–92.

Liu, Y., S. Xu, L. Teng, et al. 2011. "Synthesis and evaluation of a novel lipophilic folate receptor targeting ligand." *Anticancer Res* 31 (5):1521–5.

Longley, D. B., T. R. Wilson, M. McEwan, et al. 2006. "c-FLIP inhibits chemotherapy-induced colorectal cancer cell death." *Oncogene* 25 (6):838–48.

Longmire, M., P. L. Choyke, and H. Kobayashi. 2008. "Clearance properties of nano-sized particles and molecules as imaging agents: considerations and caveats." *Nanomedicine (Lond)* 3 (5):703–17.

Low, P. S., W. A. Henne, and D. D. Doorneweerd. 2008. "Discovery and development of folic-acid-based receptor targeting for imaging and therapy of cancer and inflammatory diseases." *Acc Chem Res* 41 (1):120–9.

Maeda, H. 2001. "The enhanced permeability and retention (EPR) effect in tumor vasculature: the key role of tumor-selective macromolecular drug targeting." *Adv Enzyme Regul* 41:189–207.

Maeda, H. 2012. "Vascular permeability in cancer and infection as related to macromolecular drug delivery, with emphasis on the EPR effect for tumor-selective drug targeting." *Proc Jpn Acad Ser B Phys Biol Sci* 88 (3):53–71.

Maeda, H. 2015. "Toward a full understanding of the EPR effect in primary and metastatic tumors as well as issues related to its heterogeneity." *Adv Drug Deliv Rev* 91:3–6.

Maeda, H., T. Sawa, and T. Konno. 2001. "Mechanism of tumor-targeted delivery of macromolecular drugs, including the EPR effect in solid tumor and clinical overview of the prototype polymeric drug SMANCS." *J Control Release* 74 (1–3):47–61.

Maeda, H., J. Takeshita, and R. Kanamaru. 1979. "A lipophilic derivative of neocarzinostatin. A polymer conjugation of an antitumor protein antibiotic." *Int J Pept Protein Res* 14 (2):81–7.

Maeda, H., J. Wu, T. Sawa, et al. 2000. "Tumor vascular permeability and the EPR effect in macromolecular therapeutics: a review." *J Control Release* 65 (1–2):271–84.

Magiorakos, A. P., A. Srinivasan, R. B. Carey, et al. 2012. "Multidrug-resistant, extensively drug-resistant and pandrug-resistant bacteria: an international expert proposal for interim standard definitions for acquired resistance." *Clin Microbiol Infect* 18 (3):268–81.

Maruyama, K., S. Okuizumi, O. Ishida, et al. 1994. "Phosphatidyl Polyglycerols Prolong Liposome Circulation in-Vivo." *Int J Pharm* 111 (1):103–107.

Matsumura, Y., M. Gotoh, K. Muro, et al. 2004. "Phase I and pharmacokinetic study of MCC-465, a doxorubicin (DXR) encapsulated in PEG immunoliposome, in patients with metastatic stomach cancer." *Ann Oncol* 15 (3):517–25.

Matsumura, Y., and H. Maeda. 1986. "A new concept for macromolecular therapeutics in cancer chemotherapy: mechanism of tumoritropic accumulation of proteins and the antitumor agent smancs." *Cancer Res* 46 (12 Pt 1):6387–92.

Mayer, L. D., T. O. Harasym, P. G. Tardi, et al. 2006. "Ratiometric dosing of anticancer drug combinations: controlling drug ratios after systemic administration regulates therapeutic activity in tumor-bearing mice." *Mol Cancer Ther* 5 (7):1854–63.

Mayer, L. D., and A. S. Janoff. 2007. "Optimizing combination chemotherapy by controlling drug ratios." *Mol Interv* 7 (4):216–23.

McDonnell, T. J., N. Deane, F. M. Platt, et al. 1989. "bcl-2-immunoglobulin transgenic mice demonstrate extended B cell survival and follicular lymphoproliferation." *Cell* 57 (1):79–88.

Meaden, E. R., P. G. Hoggard, S. H. Khoo, et al. 2002. "Determination of P-gp and MRP1 expression and function in peripheral blood mononuclear cells in vivo." *J Immunol Methods* 262 (1–2):159–65.

Meijer, C., N. H. Mulder, H. Timmer-Bosscha, et al. 1992. "Relationship of cellular glutathione to the cytotoxicity and resistance of seven platinum compounds." *Cancer Res* 52 (24):6885–9.

Micheau, O., M. Thome, P. Schneider, et al. 2002. "The long form of FLIP is an activator of caspase-8 at the Fas death-inducing signaling complex." *J Biol Chem* 277 (47):45162–71.

Miyashita, T., and J. C. Reed. 1992. "bcl-2 gene transfer increases relative resistance of S49.1 and WEHI7.2 lymphoid cells to cell death and DNA fragmentation induced by glucocorticoids and multiple chemotherapeutic drugs." *Cancer Res* 52 (19):5407–11.

Monfardini, C., O. Schiavon, P. Caliceti, et al. 1995. "A Branched Monomethoxypoly(Ethylene Glycol) for Protein Modification." *Bioconjug Chem* 6 (1):62–69.

Nakano, J., C. L. Huang, D. Liu, et al. 2005. "Survivin gene expression is negatively regulated by the p53 tumor suppressor gene in non-small cell lung cancer." *Int J Oncol* 27 (5):1215–21.

Nakuci, E., S. Mahner, J. Direnzo, et al. 2006. "BRCA1-IRIS regulates cyclin D1 expression in breast cancer cells." *Exp Cell Res* 312 (16):3120–31.

Navarro, G., R. R. Sawant, S. Biswas, et al. 2012. "P-glycoprotein silencing with siRNA delivered by DOPE-modified PEI overcomes doxorubicin resistance in breast cancer cells." *Nanomedicine (Lond)* 7 (1):65–78.

Niu, R., P. Zhao, H. Wang, et al. 2011. "Preparation, characterization, and antitumor activity of paclitaxel-loaded folic acid modified and TAT peptide conjugated PEGylated polymeric liposomes." *J Drug Target* 19 (5):373–81.

Northfelt, D. W., F. J. Martin, P. Working, et al. 1996. "Doxorubicin encapsulated in liposomes containing surface-bound polyethylene glycol: pharmacokinetics, tumor localization, and safety in patients with AIDS-related Kaposi's sarcoma." *J Clin Pharmacol* 36 (1):55–63.

O'Brien, M. E., N. Wigler, M. Inbar, et al. 2004. "Reduced cardiotoxicity and comparable efficacy in a phase III trial of pegylated liposomal doxorubicin HCl (CAELYX/Doxil) versus conventional doxorubicin for first-line treatment of metastatic breast cancer." *Ann Oncol* 15 (3):440–9.

Oliveira, S., I. van Rooy, O. Kranenburg, et al. 2007. "Fusogenic peptides enhance endosomal escape improving siRNA-induced silencing of oncogenes." *Int. J. Pharm.* 331 (2):211–4.

Oplustilova, L., K. Wolanin, M. Mistrik, et al. 2012. "Evaluation of candidate biomarkers to predict cancer cell sensitivity or resistance to PARP-1 inhibitor treatment." *Cell Cycle* 11 (20):3837–50.

Patel, N. R., A. Rathi, D. Mongayt, et al. 2011. "Reversal of multidrug resistance by co-delivery of tariquidar (XR9576) and paclitaxel using long-circulating liposomes." *Int J Pharm.* 416 (1):296–9. doi: 10.1016/j.ijpharm.2011.05.082.

Patil, Y., T. Sadhukha, L. Ma, et al. 2009. "Nanoparticle-mediated simultaneous and targeted delivery of paclitaxel and tariquidar overcomes tumor drug resistance." *J Control Release* 136 (1):21–9.

Patil, Y. B., S. K. Swaminathan, T. Sadhukha, et al. 2010. "The use of nanoparticle-mediated targeted gene silencing and drug delivery to overcome tumor drug resistance." *Biomaterials* 31 (2):358–65.

Peng, X. H., Y. Wang, D. Huang, et al. 2011. "Targeted delivery of cisplatin to lung cancer using ScFvEGFR-heparin-cisplatin nanoparticles." *ACS Nano* 5 (12):9480–93.

Perche, F., N. R. Patel, and V. P. Torchilin. 2012. "Accumulation and toxicity of antibody-targeted doxorubicin-loaded PEG-PE micelles in ovarian cancer cell spheroid model." *J Control Release* 164 (1):95–102.

Pirollo, K. F., J. Nemunaitis, P. K. Leung, et al. 2016. "Safety and efficacy in advanced solid tumors of a targeted nanocomplex carrying the p53 gene used in combination with docetaxel: A phase 1b study." *Mol Ther* 24 (9):1697–706.

Popovic, Z., W. Liu, V. P. Chauhan, et al. 2010. "A nanoparticle size series for in vivo fluorescence imaging." *Angew Chem Int Ed Engl* 49 (46):8649–52.

Poste, G. 1983. "Liposome targeting in vivo: Problems and opportunities." *Biology of the Cell* 47 (1):19–37.

Potschka, H., M. Fedrowitz, and W. Loscher. 2002. "P-Glycoprotein-mediated efflux of phenobarbital, lamotrigine, and felbamate at the blood-brain barrier: evidence from microdialysis experiments in rats." *Neurosci Lett* 327 (3):173–6.

Prabhakar, U., H. Maeda, R. K. Jain, et al. 2013. "Challenges and key considerations of the enhanced permeability and retention effect for nanomedicine drug delivery in oncology." *Cancer Res* 73 (8):2412–7.

Ramanujan, S., A. Pluen, T. D. McKee, et al. 2002. "Diffusion and convection in collagen gels: implications for transport in the tumor interstitium." *Biophys J* 83 (3):1650–60.

Rausch, V., L. Liu, G. Kallifatidis, et al. 2010. "Synergistic activity of sorafenib and sulforaphane abolishes pancreatic cancer stem cell characteristics." *Cancer Res* 70 (12):5004–13.

Raynaud, F. I., D. E. Odell, and L. R. Kelland. 1996. "Intracellular metabolism of the orally active platinum drug JM216: influence of glutathione levels." *Br J Cancer* 74 (3):380–6.

Reddy, L. H., H. Ferreira, C. Dubernet, et al. 2008. "Squalenoyl nanomedicine of gemcitabine is more potent after oral administration in leukemia-bearing rats: study of mechanisms." *Anticancer Drugs* 19 (10):999–1006.

Riordan, J. R., K. Deuchars, N. Kartner, et al. 1985. "Amplification of P-glycoprotein genes in multidrug-resistant mammalian cell lines." *Nature* 316 (6031):817–9.

Riordan, J. R., and V. Ling. 1979. "Purification of P-glycoprotein from plasma membrane vesicles of Chinese hamster ovary cell mutants with reduced colchicine permeability." *J Biol Chem* 254 (24):12701–5.

Rottenberg, S., J. E. Jaspers, A. Kersbergen, et al. 2008. "High sensitivity of BRCA1-deficient mammary tumors to the PARP inhibitor AZD2281 alone and in combination with platinum drugs." *Proc Natl Acad Sci U S A* 105 (44):17079–84.

Saga, Y., M. Suzuki, H. Mizukami, et al. 2003. "Overexpression of thymidylate synthase mediates desensitization for 5-fluorouracil of tumor cells." *Int J Cancer* 106 (3):324–6.

Salzano, G., D. F. Costa, C. Sarisozen, et al. 2016. "Mixed nanosized polymeric micelles as promoter of doxorubicin and miRNA-34a co-delivery triggered by dual stimuli in tumor tissue." *Small* 12 (35):4837–4848.

Sarin, H. 2010. "Physiologic upper limits of pore size of different blood capillary types and another perspective on the dual pore theory of microvascular permeability." *J Angiogenes Res* 2:14.

Sarisozen, C., S. Dhokai, E. G. Tsikudo, et al. 2016. "Nanomedicine based curcumin and doxorubicin combination treatment of glioblastoma with scFv-targeted micelles: In vitro evaluation on 2D and 3D tumor models." *Eur J Pharm Biopharm* 108:54–67.

Sarisozen, C., I. Vural, T. Levchenko, et al. 2012a. "Long-circulating PEG-PE micelles co-loaded with paclitaxel and elacridar (GG918) overcome multidrug resistance." *Drug Deliv* 19 (8):363–70.

Sarisozen, C., I. Vural, T. Levchenko, et al. 2012b. "PEG-PE-based micelles co-loaded with paclitaxel and cyclosporine A or loaded with paclitaxel and targeted by anticancer antibody overcome drug resistance in cancer cells." *Drug Deliv* 19 (4):169–76.

Scaffidi, C., S. Fulda, A. Srinivasan, et al. 1998. "Two CD95 (APO-1/Fas) signaling pathways." *EMBO J* 17 (6):1675–87.

Schlaifer, D., G. Laurent, S. Chittal, et al. 1990. "Immunohistochemical detection of multidrug resistance associated P-glycoprotein in tumour and stromal cells of human cancers." *Br J Cancer* 62 (2):177–82.

Selvakumaran, M., D. A. Pisarcik, R. Bao, et al. 2003. "Enhanced cisplatin cytotoxicity by disturbing the nucleotide excision repair pathway in ovarian cancer cell lines." *Cancer Res* 63 (6):1311–6.

Senchenkov, A., D. A. Litvak, and M. C. Cabot. 2001. "Targeting ceramide metabolism--a strategy for overcoming drug resistance." *J Natl Cancer Inst* 93 (5):347–57.

Shervington, A., and C. Lu. 2008. "Expression of multidrug resistance genes in normal and cancer stem cells." *Cancer Invest* 26 (5):535–42.

Shi, J., P. W. Kantoff, R. Wooster, and O. C. Farokhzad. 2017. "Cancer nanomedicine: progress, challenges and opportunities." *Nat Rev Cancer* 17 (1):20–37.

Shivapurkar, N., J. Reddy, H. Matta, et al. 2002. "Loss of expression of death-inducing signaling complex (DISC) components in lung cancer cell lines and the influence of MYC amplification." *Oncogene* 21 (55):8510–4.

Shuhendler, A. J., R. Y. Cheung, J. Manias, et al. 2010. "A novel doxorubicin-mitomycin C co-encapsulated nanoparticle formulation exhibits anti-cancer synergy in multidrug resistant human breast cancer cells." *Breast Cancer Res Treat* 119 (2):255–69.

Singh, R., D. Pantarotto, L. Lacerda, et al. 2006. "Tissue biodistribution and blood clearance rates of intravenously administered carbon nanotube radiotracers." *Proc Natl Acad Sci U S A* 103 (9):3357–62.

Soutschek, J., A. Akinc, B. Bramlage, et al. 2004. "Therapeutic silencing of an endogenous gene by systemic administration of modified siRNAs." *Nature* 432 (7014):173–8.

Sriraman, S. K., J. Pan, C. Sarisozen, et al. 2016. "Enhanced cytotoxicity of folic acid-targeted liposomes co-loaded with C6 ceramide and doxorubicin: In vitro evaluation on HeLa, A2780-ADR, and H69-AR cells." *Mol Pharm* 13 (2):428–37.

Sun, C. L., and C. C. Chao. 2005. "Cross-resistance to death ligand-induced apoptosis in cisplatin-selected HeLa cells associated with overexpression of DDB2 and subsequent induction of cFLIP." *Mol Pharmacol* 67 (4):1307–14.

Sutherland, R. M. 1988. "Cell and environment interactions in tumor microregions: the multicell spheroid model." *Science* 240 (4849):177–84.

Swerts, K., B. de Moerloose, C. Dhooge, et al. 2004. "Comparison of two functional flow cytometric assays to assess P-gp activity in acute leukemia." *Leuk Lymphoma* 45 (11):2221–8.

Takeuchi, H., H. Kojima, T. Toyoda, et al. 1999. "Prolonged circulation time of doxorubicin-loaded liposomes coated with a modified polyvinyl alcohol after intravenous injection in rats." *Eur J Pharm Biopharm* 48 (2):123–129.

Teng, Y., X. Wang, Y. Wang, et al. 2010a. "Wnt/beta-catenin signaling regulates cancer stem cells in lung cancer A549 cells." *Biochem Biophys Res Commun* 392 (3):373–9.

Teng, Y., X. W. Wang, Y. W. Wang, et al. 2010b. "[Effect of siRNA-mediated beta-catenin gene on Wnt signal pathway in lung adenocarcinoma A549 cell]." *Zhonghua Yi Xue Za Zhi* 90 (14):988–92.

Thomas, H., and H. M. Coley. 2003. "Overcoming multidrug resistance in cancer: an update on the clinical strategy of inhibiting p-glycoprotein." *Cancer Control* 10 (2):159–65.

Tidefelt, U., J. Liliemark, A. Gruber, et al. 2000. "P-Glycoprotein inhibitor valspodar (PSC 833) increases the intracellular concentrations of daunorubicin in vivo in patients with P-glycoprotein-positive acute myeloid leukemia." *J Clin Oncol* 18 (9):1837–44.

Torchilin, V. P. 2005. "Lipid-core micelles for targeted drug delivery." *Curr Drug Deliv* 2 (4):319–27.

Torchilin, V. P., T. S. Levchenko, A. N. Lukyanov, et al. 2001. "p-Nitrophenylcarbonyl-PEG-PE-liposomes: fast and simple attachment of specific ligands, including monoclonal antibodies, to distal ends of PEG chains via p-nitrophenylcarbonyl groups." *Biochim Biophys Acta* 1511 (2):397–411.

Torchilin, V. P., T. S. Levchenko, K. R. Whiteman, et al. 2001. "Amphiphilic poly-N-vinylpyrrolidones: synthesis, properties and liposome surface modification." *Biomaterials* 22 (22):3035–44.

Torchilin, V. P., and V. S. Trubetskoy. 1995. "Which polymers can make nanoparticulate drug carriers long-circulating?" *Adv Drug Deliv Rev* 16 (2–3):141–55.

Trippett, T., S. Schlemmer, Y. Elisseyeff, et al. 1992. "Defective transport as a mechanism of acquired resistance to methotrexate in patients with acute lymphocytic leukemia." *Blood* 80 (5):1158–62.

Tsujimoto, Y., J. Cossman, E. Jaffe, et al. 1985. "Involvement of the bcl-2 gene in human follicular lymphoma." *Science* 228 (4706):1440–3.

Twiddy, D., D. G. Brown, C. Adrain, et al. 2004. "Pro-apoptotic proteins released from the mitochondria regulate the protein composition and caspase-processing activity of the native Apaf-1/caspase-9 apoptosome complex." *J Biol Chem* 279 (19):19665–82.

Valle, J. W., A. Armstrong, C. Newman, et al. 2011. "A phase 2 study of SP1049C, doxorubicin in P-glycoprotein-targeting pluronics, in patients with advanced adenocarcinoma of the esophagus and gastroesophageal junction." *Invest New Drugs* 29 (5):1029–37.

van Vlerken, L. E., Z. Duan, S. R. Little, et al. 2008. "Biodistribution and pharmacokinetic analysis of Paclitaxel and ceramide administered in multifunctional polymer-blend nanoparticles in drug resistant breast cancer model." *Mol Pharm* 5 (4):516–26.

van Vlerken, L. E., Z. Duan, S. R. Little, et al. 2010. "Augmentation of therapeutic efficacy in drug-resistant tumor models using ceramide coadministration in temporal-controlled polymer-blend nanoparticle delivery systems." *AAPS J* 12 (2):171–80.

Vaux, D. L., S. Cory, and J. M. Adams. 1988. "Bcl-2 gene promotes haemopoietic cell survival and cooperates with c-myc to immortalize pre-B cells." *Nature* 335 (6189):440–2.

Veneroni, S., N. Zaffaroni, M. G. Daidone, et al. 1994. "Expression of P-glycoprotein and in vitro or in vivo resistance to doxorubicin and cisplatin in breast and ovarian cancers." *Eur J Cancer* 30A (7):1002–7.

Vinogradov, S., and X. Wei. 2012. "Cancer stem cells and drug resistance: the potential of nanomedicine." *Nanomedicine (Lond)* 7 (4):597–615.

Von Hoff, D. D., M. M. Mita, R. K. Ramanathan, et al. 2016. "Phase I Study of PSMA-Targeted Docetaxel-Containing Nanoparticle BIND-014 in Patients with Advanced Solid Tumors." *Clin Cancer Res* 22 (13):3157–63.

Wang, J., W. Mao, L. L. Lock, et al. 2015. "The role of micelle size in tumor accumulation, penetration, and treatment." *ACS Nano* 9 (7):7195–206.

Wang, S., J. Zhang, Y. Wang, et al. 2016. "Hyaluronic acid-coated PEI-PLGA nanoparticles mediated co-delivery of doxorubicin and miR-542-3p for triple negative breast cancer therapy." *Nanomedicine* 12 (2):411–20.

Wang, Z., R. Goulet, 3rd, K. J. Stanton, et al. 2005. "Differential effect of anti-apoptotic genes Bcl-xL and c-FLIP on sensitivity of MCF-7 breast cancer cells to paclitaxel and docetaxel." *Anticancer Res* 25 (3c):2367–79.

Wang, Z., Y. Li, D. Kong, S. Banerjee, et al. 2009. "Acquisition of epithelial-mesenchymal transition phenotype of gemcitabine-resistant pancreatic cancer cells is linked with activation of the notch signaling pathway." *Cancer Res* 69 (6):2400–7.

Wartenberg, M., F. C. Ling, M. Muschen, et al. 2003. "Regulation of the multidrug resistance transporter P-glycoprotein in multicellular tumor spheroids by hypoxia-inducible factor (HIF-1) and reactive oxygen species." *FASEB J* 17 (3):503–5.

Weissig, V., T. K. Pettinger, and N. Murdock. 2014. "Nanopharmaceuticals (part 1): products on the market." *Int J Nanomedicine* 9:4357–73.

Weller, M., U. Malipiero, A. Aguzzi, J. et al. 1995. "Protooncogene bcl-2 gene transfer abrogates Fas/APO-1 antibody-mediated apoptosis of human malignant glioma cells and confers resistance to chemotherapeutic drugs and therapeutic irradiation." *J Clin Invest* 95 (6):2633–43.

West, C. M., R. A. Cooper, J. A. Loncaster, et al. 2001. "Tumor vascularity: a histological measure of angiogenesis and hypoxia." *Cancer Res* 61 (7):2907–10.

Wiedemeyer, W. R., J. A. Beach, and B. Y. Karlan. 2014. "Reversing platinum resistance in high-grade serous ovarian carcinoma: targeting BRCA and the homologous recombination system." *Front Oncol* 4:34.

Wilhelm, S., A. J. Tavares, Q. Dai, et al. 2016. "Analysis of nanoparticle delivery to tumours." *Nature Rev. Mat.*1:16014.

Wilson, T. R., D. B. Longley, and P. G. Johnston. 2006. "Chemoresistance in solid tumours." *Ann Oncol* 17 Suppl 10:x315–24.

Wirthner, R., S. Wrann, K. Balamurugan, et al. 2008. "Impaired DNA double-strand break repair contributes to chemoresistance in HIF-1 alpha-deficient mouse embryonic fibroblasts." *Carcinogenesis* 29 (12):2306–16.

Woodle, M. C., C. M. Engbers, and S. Zalipsky. 1994. "New amphipatic polymer lipid conjugates forming long-circulating reticuloendothelial system-evading liposomes." *Bioconjug Chem* 5 (6):493–96.

Xu, Y., and M. A. Villalona-Calero. 2002. "Irinotecan: mechanisms of tumor resistance and novel strategies for modulating its activity." *Ann Oncol* 13 (12):1841–51.

Yin, C., C. M. Knudson, S. J. Korsmeyer, et al. 1997. "Bax suppresses tumorigenesis and stimulates apoptosis in vivo." *Nature* 385 (6617):637–40.

Yoshii, Y., and K. Sugiyama. 1988. "Intercapillary distance in the proliferating area of human glioma." *Cancer Res* 48 (10):2938–41.

Youle, R. J., and A. Strasser. 2008. "The BCL-2 protein family: opposing activities that mediate cell death." *Nat Rev Mol Cell Biol* 9 (1):47–59.

Yu, H., Z. Xu, X. Chen, et al. 2014. "Reversal of lung cancer multidrug resistance by pH-responsive micelleplexes mediating co-delivery of siRNA and paclitaxel." *Macromol Biosci* 14 (1):100–9.

Zha, W., G. Wang, W. Xu, et al. 2013. "Inhibition of P-glycoprotein by HIV protease inhibitors increases intracellular accumulation of berberine in murine and human macrophages." *PLoS One* 8 (1):e54349.

Zhang, C. G., W. J. Zhu, Y. Liu, et al. 2016. "Novel polymer micelle mediated co-delivery of doxorubicin and P-glycoprotein siRNA for reversal of multidrug resistance and synergistic tumor therapy." *Sci Rep* 6:23859.

Zhang, J., M. F. Stevens, and T. D. Bradshaw. 2012. "Temozolomide: mechanisms of action, repair and resistance." *Curr Mol Pharmacol* 5 (1):102–14.

Zhang, L. J., Y. Z. Hao, C. S. Hu, et al. 2008. "Inhibition of apoptosis facilitates necrosis induced by cisplatin in gastric cancer cells." *Anticancer Drugs* 19 (2):159–66.

Zhang, Y., S. K. Sriraman, H. A. Kenny, et al. 2016. "Reversal of chemoresistance in ovarian cancer by co-delivery of a P-glycoprotein inhibitor and paclitaxel in a liposomal platform." *Mol Cancer Ther* 15 (10):2282–93.

Ziemann, C., A. Burkle, G. F. Kahl, et al. 1999. "Reactive oxygen species participate in mdrlb mRNA and P-glycoprotein overexpression in primary rat hepatocyte cultures." *Carcinogenesis* 20 (3):407–14.

Zou, W., C. Sarisozen, and V. P. Torchilin. 2017. "The reversal of multidrug resistance in ovarian carcinoma cells by co-application of tariquidar and paclitaxel in transferrin-targeted polymeric micelles." *J Drug Target* 25 (3):225–234.

Zuckerman, J. E., I. Gritli, A. Tolcher, et al. 2014. "Correlating animal and human phase Ia/Ib clinical data with CALAA-01, a targeted, polymer-based nanoparticle containing siRNA." *Proc Natl Acad Sci U S A* 111 (31):11449–54.

5 Multifunctional Dendrimers as Cancer Nanomedicines

Peptide-Based Targeting

Chie Kojima

CONTENTS

5.1 INTRODUCTION

Nanomedicine is a medical application of nanotechnology, which covers drug delivery, diagnosis, and tissue engineering (Wagner et al., 2006; Riehemann et al., 2009). Nanomedicine research requires the combined efforts of interdisciplinary teams with backgrounds in medicine, pharmaceutics, material science, and mechanical engineering. Researchers in materials science have prepared functional nanomaterials for biomedical applications with the aim of developing efficient delivery systems that target specific tissues. Numerous nanomaterials, such as inorganic nanomaterials, metal nanoparticles, liposomes, micelles, and polymers, have been applied to the aforementioned applications (Wagner et al., 2006; Riehemann et al., 2009). Multifunctional nanoparticles can be prepared by adding a host of compounds such as drug molecules, imaging probes and stimuli-sensitive materials. Furthermore, tailored preparation processes allow for controlled size and surface properties of the nanoparticles, which largely affect the biodistribution. As toxicity is a principal factor to consider in medication, biocompatible organic nanoparticles are often used in clinical trials. Liposomes and micelles are self-assembled nanoparticles whose size ranges from micrometer to submicron orders. Polymers are smaller and similar in size to proteins. Some compounds can be conjugated to the side chain of polymers as pendants (Haag & Kratz, 2006; Duncan 2012). Synthetic polymers with poly-dispersed molecular weight and uncontrollable conformation are quite different from proteins.

Dendrimers are synthetic polymers with highly branched and well-defined structures, whose molecular weight is uniform. Dendrimers are synthesized in a stepwise organic synthesis, whereas most polymers are typically prepared by the polymerization of the corresponding monomers (Tomalia et al., 1985; Tomalia et al., 1990; Tomalia 2005). The controllable and uniform molecular weight and chemical structure can lead to their chemical and biological reproducibility. Additionally, dendrimers possess a high volume of terminal functional groups and an inner core available for the conjugation and encapsulation of small molecules, respectively, which is a unique multifunctionalization pathway. A variety of functional materials, such as drug molecules, imaging agents and targeting compounds, have been conjugated to and/or encapsulated within a single dendrimer molecule to produce multifunctional dendrimers. Multifunctional dendrimers are powerful unimolecular multifunctional nanoparticles, which are smaller and more stable than self-assembled organic nanoparticles such as liposomes and micelles. The differences of liposomes, micelles, polymers, and dendrimers are listed in Table 5.1. There are numerous review articles on the biomedical applications of dendrimers (Lee et al., 2005; Svenson & Tomalia, 2005; Gajbhiye et al., 2007; Wolinsky & Grinstaff, 2008; Medina & El-Sayed, 2009; Kojima 2010; Astruc et al., 2010; Kannan et al., 2014; Kesharwani & Iyer, 2015; Luong et al., 2016). Theranostic applications are a recent hot topic and involve applications of a single dendrimer to both therapy and diagnosis (Paleos et al., 2008; Khandare et al., 2012, Sk & Kojima, 2015).

Peptides are another useful class of compounds in nanomedicine, whose activities can be controlled by their sequence. Some peptides, obtained from functional proteins, are powerful therapeutic agents as well as powerful ligands to the target molecules (Vlieghe et al., 2010). There are peptides derived from theoretical and experimental surveys. The targeting peptides are named cell-targeting peptides (CTPs) (Dissanayake et al., 2017). Some peptides are efficiently internalized into cells, which are named cell-penetrating peptides (CPPs) (Zorko & Langel, 2005;

TABLE 5.1
Comparison of Liposomes, Micelles, Polymers, and Dendrimers

Name	Components	Size	Properties
Liposome	Phospholipids	From submicron to several micron	Hydrophilic and hydrophobic drugs are loaded into the liposome and membrane, respectively. Lipid-assembled nanoparticle
Micelles	Detergents	10~20 nm	Hydrophobic drugs are loaded into the micelle. Detergent-assembled nanoparticle Micelles do not form at lower concentration than cmc (critical micelle concentration).
Polymeric micelles (Polymersome: polymer-based vesicle)	Block polymers	50~200 nm (submicron)	Drugs are loaded into the micelle. Polymer-self-assembled nanoparticle
Polymer	Polymers	<10 nm	Drugs are conjugated to polymer side chain as a pendant. Linear structure Molecular weight is not uniform.
Dendrimer	Dendrimers	<10 nm (without modification)	Drugs are conjugated at the termini and encapsulated into the interior. Unimolecular nanoparticle with globular structure Molecular weight is uniform.

Heitz et al., 2009; Koren & Torchilin, 2012). This chapter focuses on peptide-conjugated dendrimers in nanomedicine. Typically, the molecular weight of peptides is relatively small, which is beneficial to dendrimer conjugation. These peptide-conjugated dendrimers are associated with the target molecules and can further be internalized into the targeted cells. Detailed examples are shown in this chapter.

5.2 DENDRIMERS

Dendrimers were first reported by Dr. Tomalia in 1985 (Tomalia et al., 1985), whose name reflects their tree-like structure (Tomalia et al., 1985; Tomalia et al., 1990; Tomalia 2005). Their properties can be controlled by changing the inner core structure, building blocks, or terminal groups. Dr. Tomalia developed polyamidoamine (PAMAM) dendrimers, which have been widely studied and are commercially available. The dendrimer is synthesized by repeated reaction cycles using methyl acrylate and ethylenediamine, initiated from a core amine compound. This synthetic method is called a divergent method. A reaction cycle is referred to as a generation (G), and the molecular weight and the number of terminal groups are determined by generation (Figure 5.1). The number of terminal groups doubles

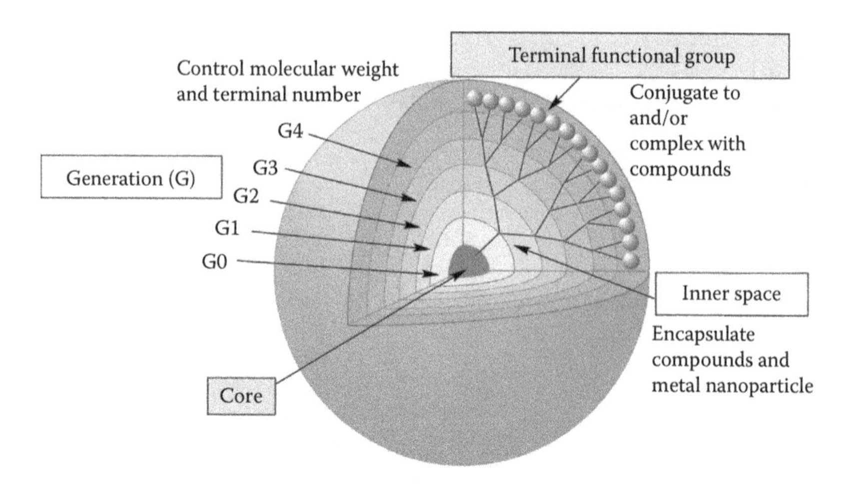

FIGURE 5.1 Dendrimer overview.

with each generation, and the molecular weight increases stepwise. The dendrimer diameter ranges from 2.2 nm to 13.5 nm (Tomalia et al., 1990; Svenson & Tomalia, 2005). The sizes of insulin, cytochrome c, and hemoglobin are 3.0, 4.5, and 5.5 nm, respectively, and correlate to G3, G4, and G5 dendrimers (Tomalia et al., 1990; Svenson & Tomalia, 2005).

The dendrimer inner core is thought to provide additional function. Furthermore, the inner cores of large generation dendrimers are isolated; however, there are reports of some compounds and metal nanoparticles being encapsulated in dendrimers (Lee et al., 2005; Svenson & Tomalia, 2005; Gajbhiye et al., 2007; Wolinsky & Grinstaff, 2008; Medina & El-Sayed, 2009; Kojima 2010; Astruc et al., 2010; Kannan et al., 2014; Kesharwani & Iyer, 2015; Sk & Kojima, 2015; Luong et al., 2016). The terminal functional groups of dendrimers have been widely used for conjugation and complex formation (Figure 5.1). Various compounds including drug agents, imaging probes, sugars, and peptides can be linked to the dendrimer periphery. In some cases, the modification is beyond the addition of compounds to the dendrimer. The modified moieties exhibit clustering effects. For example, the binding affinity of the sugar- and peptide-modified dendrimers is significantly larger than the corresponding original molecules (Crespo et al., 2005; Sadler & Tam, 2002; Kojima 2010; Astruc et al., 2010; Liu et al., 2012). Peptides linked to the surface of a dendrimer can also induce higher-order structures (Kojima 2016).

Multiple dendrimer types have been synthesized, in addition to PAMAM dendrimers, such as poly(propylene imine) (PPI) dendrimers, biodegradable polyester dendrimers and poly L-lysine dendrimers. PAMAM dendrimers, PPI dendrimers, 2,2-bis(hydroxymethyl)propionic acid (bis-MPA)-based dendrimers and phosphorous dendrimers were obtained from Sigma-Aldrich. bis-MPA-based dendrimers and dendritic polylysines are also available from the Polymer Factory (Sweden) and COLCOM (France), respectively. Clinical studies of poly L-lysine dendrimers as a drug carrier are performed by Starpharma (Australia) (Kesharwani & Iyer, 2015).

5.3 TARGETING STRATEGY

5.3.1 Active Targeting vs. Passive Targeting

Delivery to the target tissue is one of the most crucial issues in drug carrier development. Generally, carriers are modified with targeting ligands—themselves recognized by the target cells. This is named active targeting. Another targeting mechanism, named passive targeting, has been reported for cancer targeting (Figure 5.2). Passive targeting is based on the enhanced permeability and retention (EPR) effect, which derives from permeable neovessels surrounding tumor tissues and the lack of lymphatics in tumor tissues (Maeda & Matsumura, 1989; Maeda et al., 2000). Carriers with suitable size and blood retention properties can be passively accumulated to tumor tissues because of the EPR effect. Polyethylene glycol (PEG) is a biocompatible material, which can provide an inert surface because of the significant excluded volume effect. Thus, there are many studies showing PEG-modified (PEGylation) nanocarriers that lead to long blood circulation and passive targeting to tumor tissues (Ferrari, 2005; Alexis et al., 2010). Size-controllable dendrimers, modified with PEG chains, are a major design factor for nanocarriers having passive targeting ability to tumor tissues (Gajbhiye et al., 2007). Consequently, PEG- and

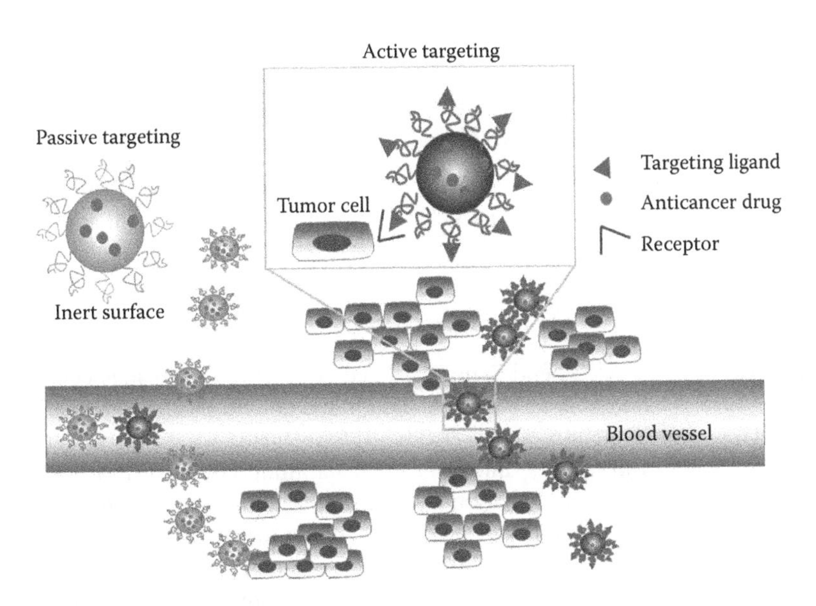

FIGURE 5.2 Active targeting and passive targeting.

ligand-modified dendrimers can possess both ligand-based active targeting properties and EPR effect-based passive targeting properties.

5.3.2 Ligands for Active Targeting

Antibodies are often used as targeting compounds because of their high and specific binding affinity to the target molecule. There are numerous reports on antibody-modified nanoparticles for drug delivery systems (DDSs) (Ferrari, 2005; Alexis et al., 2010). The molecular weight of the immunoglobulin antibody is 150 kDa, which is similar to, or larger than, dendrimers. Therefore, immunoglobulin is not a suitable targeting compound for dendrimers. Small ligand molecules have been used as targeting molecules. Folic acid is a ligand for folate receptors highly expressed at the surface of specific cancer cells, which have been used for cancer treatments (Ledermann et al., 2015). Sugar compounds, for example galactose, mannose, and sialic acid, are also able to target hepatic cells, immune cells, and specific cancer cells. Many sugar compounds can be conjugated to the periphery of a single dendrimer to induce clustering effects. Consequently, the binding affinity of sugar-conjugated dendrimers is increased over that of the corresponding sugar compound (Jain et al., 2012). Peptides are also a powerful ligand for active targeting. Peptide functionality can be controlled by its sequence, and synthetic procedures of peptides are well-established. Thus, peptides are attractive materials in DDSs (Dissanayake et al., 2017). Additionally, the multivalency of dendrimers is attributed to the clustering effects at the periphery as well as the multifunctionality. This chapter focuses on the delivery system employing various peptide-conjugated dendrimers, especially for cancer treatment applications.

5.4 PEPTIDE-CONJUGATED DENDRIMERS FOR ACTIVE TARGETING

Peptides used in DDSs are divided into two categories: CTPs and CPPs. Various CTP types have been conjugated to dendrimers in DDS studies, which are listed in Table 5.2.

5.4.1 Arginyl-Glycyl-Aspartic Acid (RGD)-Conjugated Dendrimers

RGD is a commonly used peptide in DDSs and is observed in various cell matrix proteins, such as fibronectin, collagen, laminin, and vitronectin (Hersel et al., 2003). RGD is a minimum-sequence that binds to integrin, which is a cell adhesion molecule (Hersel et al., 2003). Integrins are heterodimeric cell-surface receptors, which are formed by the combination of 18 α-subunit types and 8 β-subunit types. Different cells express different integrin heterodimer types, which decide cell properties (Desgrosellier & Cheresh, 2010). αv integrins and integrin α5β1 recognize the RGD sequence. Importantly, the αvβ3 integrin is expressed at low levels in most adult epithelia but is overexpressed on both the luminal surface of endothelial cells only during angiogenesis, and in numerous solid tumors (Desgrosellier & Cheresh, 2010, Hersel et al., 2003). Therefore, RGD peptides are often used as a DDS ligand

TABLE 5.2
Cell-Targeting Peptides (CTPs) Conjugated to Dendrimers for Drug Delivery Systems (DDSs)

Name	Sequence	Origin	Target: Molecule	Target: Tissue	Representative References
RGD	RGD	fibronectin, collagen, laminin, etc	integrins (αv integrins and integrin a5β1)	many	Zhong et al., 2014
RGD	cyclo(-RDGfK)	synthetic	αvβ3 integrin	cancer angiogenetic endothelial cells	Hersel et al., 2003
Angiopep-2	TFFYGGSRGKRN NFKTEEYC	aprotinin	low-density lipoprotein receptor-related protein	blood-brain barrier (BBB)	Demeule et al., 2008
GX-1	CGNSNPKSC	synthetic		human gastric cancer vasculature	Zhi et al., 2004
T7	HAIYPRH	synthetic	transferrin receptor	cancer	Lee et al., 2001
LHRH	Glp-HWSYGLRPG	luteinizing hormone-releasing hormone (LHRH)	LHRH receptor	cancer	Dharap et al., 2005
Bombesin	pEQRLGNQWAVGHLM	bombesin	gastrin-releasing peptide receptor	cancer	Cescato et al., 2008
CLT1	CGLIIQKNEC	synthetic	fribronectin-fibrin complexes	cancer	Pilch et al., 2006
FSH33	YTRDLVYKDPAR PKIQKTCTF	follicle stimulating hormone (FSH)	FSH receptor	ovarian cancers	Agris et al., 1992
Somatostatin	AGCKNFFWKTFTSC	somatostatin	somatostatin receptors	neuroendocrine tumors	Wängberg et al., 1997
13mer peptide	VHFFKNIVTPRTP	myelin basic protein (p-MBP)		B-cell leukemic cells	Gellerman et al., 2013
PADRE	aKXVAAWTLKAAaZC	Pan human leukocyte antigen-antigen related epitope		antigen-presenting cells	Daftarian et al., 2011
HA110–120	SFERFEIFPKEC	influenza hemoagglutinin		antigen-presenting cells	Daftarian et al., 2011

(Continued)

TABLE 5.2 (CONTINUED)
Cell-Targeting Peptides (CTPs) Conjugated to Dendrimers for Drug Delivery Systems (DDSs)

Name	Sequence	Origin	Target		Representative References
			Molecule	Tissue	
CREKA	CREKA	synthetic	fibrin (clotted plasma protein)	cancer glioblastoma multiforme	Simberg et al., 2007
Lyp-1	CGNKRTRGC	synthetic	mitochondrial/cell surface protein, p32/gC1qR	breast cancer cells tumor-associated lymphatic vessels atherosclerotic plaque	Laakkonen et al., 2002 Fogal et al., 2008 Seo et al., 2014
Pep-1	CGEMGWVRC	synthetic	IL-13Rα2	cancer	Pandya et al., 2012
LTP-1	CTSGTHPRC	synthetic		lung alveolar epithelial primary cells	Morris et al., 2011
H6	YLFFVFER	synthetic	human epidermal receptor 2 (HER2)	breast cancer	Wang et al., 2015
PTP	KTLLPTP	synthetic	plectin-1	pancreatic cancer	Kelly et al., 2008
SB105-A10	ASLRVRIKK			virus	Donalisio et al., 2010
Ac-TZ14011	Rr-Nal-CY-Cit-RkPYR-Cit-CR	designed from self-defense peptides of horseshoe crabs	chemokine receptor 4 (CXCR4)	metastatic tumors	Hanaoka et al., 2006
Poly(aspartic acid)	$D_{4\text{-}6}$	synthetic	hydroxyapatite	bone	Ouyang et al., 2009
LYRAG	LYRAG			immune cells	Gaertner et al., 2011
SP94	SFSIIHTPILPL	synthetic		hepatic cells	Lo et al., 2008

for cancer treatment (Zhong, et al., 2014; Hersel et al., 2003). Although linear RGD peptides were used classically, Kessler's group developed alternative αvβ3 integrin-specific bound peptides such as cyclo-(RDGfK). The cyclic peptides enhanced the binding affinity to the target integrin, compared with the corresponding linear peptides. Furthermore, the non-natural amino acid, D-Phe (f), prevents peptide degradation by proteinase. Hence, cyclic-RGD peptides have been used as a target ligand within DDS carriers (Hersel et al., 2003).

Saraswathy et al. and He et al. reported that the anticancer drug, doxorubicin (DOX), was encapsulated in the RGD-terminated PEG-conjugated PAMAM dendrimers. These dendrimers showed efficient cytotoxicity against U87MG glioma cells (Saraswathy et al., 2015; He et al., 2015). Yan et al. prepared dual targeting dendrimers to penetrate the blood-brain barrier (BBB). They conjugated RGD and the Angiopep-2 peptide to a single dendrimer. Angiopep-2 is a low-density ligand from the lipoprotein receptor-related protein and induces BBB transcytosis. Thus, this dendrimer can accumulate into glioma in the brain *via* BBB penetration. They demonstrated that the two-peptide-conjugated dendrimer exhibited the efficient accumulation in brain tumors (Yan et al., 2012). Delivery into the brain is the most difficult issue in DDS studies, because of BBB. This research provides a solution to solve it. Mekuria et al. prepared DOX-encapsulated dendrimers with RGD peptide or IL-6 antibody to compare their targeting properties. In this case, the RGD peptide did not function well. They directly conjugated the peptide to the dendrimer, suggesting the importance of the conjugation procedure (Mekuria et al., 2016). Jiang et al. reported RGD- and 5-fluorouracil (5FU)-conjugated dendrimers for bone cancer delivery (Jiang et al., 2013). Hill et al. attempted to prepare a tooth delivery system by employing an RGD-conjugated dendrimer (Hill et al., 2007).

RGD-conjugated dendrimers function as a wide range of delivery vehicles for various bioactive materials including nucleic acids and photo-responsive materials. Waite et al. reported the use of RGD-conjugated dendrimers for siRNA delivery to glioblastoma (Waite & Roth, 2009). Liu et al. succeeded in delivering siRNA to prostate cancer cells *in vivo* using complexes of amino-terminal PAMAM dendrimers, siRNA and RGD-containing poly(glutamic acid) peptide. They also achieved tumor growth suppression by silencing the heat shock protein 27 (Hsp27) in prostate cancer (Liu et al., 2014). Yuan et al. used an RGD-conjugated dendrimer for delivery of the photosensitizer, Ce6. Photosensitizers generate reactive oxygen species by photo-irradiation to kill affected cells, which are used for photodynamic therapy. It was possible to induce photocytotoxicity against A375 cells, a non-pigmented melanoma cell line, by using RGD- and Ce6-conjugated dendrimers (Yuan et al., 2015). Metal nanoparticles such as gold nanorods/nanoparticles have been encompassed within photothermal therapy, because they generate heat by photo-irradiation to kill cancer cells. Li et al. reported the delivery of gold nanorods modified with RGD-conjugated dendrimers to melanoma *via* photothermal therapy (Li et al., 2010). Furthermore, tumor-selective photothermal effects of other metal nanoparticles-encapsulated RGD-conjugated dendrimers have also been observed (Zhou et al., 2016).

RGD-conjugated dendrimers have also been applied to the imaging of tumor tissues. Esophageal squamous cell carcinoma was visualized by RGD-conjugated dendrimers labeled with the near infrared dye, Cy5.5 (Li, Q. et al., 2016). Gold nanoparticles were encapsulated in RGD-conjugated dendrimers, which were applied to x-ray computed tomography (CT) imaging of tumors (Li et al., 2015). Chen et al. conjugated gadolinium chelates to an RGD-conjugated dendrimer, followed by the encapsulation of gold nanoparticles. The dendrimers could act as a dual imaging probe for both CT imaging and magnetic resonance imaging (MRI) to detect tumors (Chen et al., 2015). Iron oxide nanoparticles are also used as an MRI imaging probe. Yang et al. reported an MRI image of a tumor in the presence of an RGD-conjugated dendrimer modified with iron oxide nanoparticles (Yang et al., 2015). Zhu et al. prepared a gold nanoparticle-entrapped dendrimer conjugated to RGD and α-tocopheryl succinate (α-TOC). α-TOC conjugated to dendrimers generates reactive oxygen species to kill targeted cancer cells, and gold nanoparticles can detect the targeted cancer cells by CT imaging. Thus, dendrimer-nanoparticle materials play specific roles in cancer diagnosis and cancer therapy—an example of a theranostic application (Zhu et al., 2015). Shen et al. also reported an example of theranostic nanoparticles using a dendritic polymer (dendri-graft polylysine [DGL]). An iron oxide nanoparticle (MNP) was coated initially with DGL. RGD-linked GX-1 (CGNSNPKSC) was then conjugated and DOX was loaded. DOX and iron oxide nanoparticles showed an anticancer effect and MRI-detectable properties, respectively. RGD-linked GX-1 was able to specifically mediate $\alpha_v\beta_3$-integrin/vasculature endothelium dual receptors. Thus, MNP–DGL–RGD-GX1–DOX nanoparticles showed accelerated anticancer effects and efficient tumor tissue detection (Shen et al., 2017).

5.4.2 T7-Conjugated Dendrimers

Han et al. reported the T7 peptide (HAIYPRH)-conjugated dendrimer for targeting the transferrin receptor (Han et al., 2010a; Han et al., 2011a; Han et al., 2011b). It was reported that the transferrin receptor is overexpressed on numerous tumor cells, because tumor cells require higher levels of iron to maintain cellular survival. Thus, DDSs targeting transferrin receptors have been classically used against cancers (Daniels et al., 2012). Transferrin is a ligand of the transferrin reporter, whose molecular weight is 80 kDa. Thus, transferrin is not useful as a dendrimer ligand because of size limitations. Lee et al. discovered the T7 peptide, obtained by the phage display method (Lee et al., 2001). As the binding affinity is relatively high (K_d = ~10 nM), endogenous transferrin would not inhibit the uptake of T7-based DDSs (Oh et al., 2009). Han et al. studied T7-conjugated dendrimers encapsulating DOX for hepatic carcinoma (Han et al., 2010a). They also applied the dendrimers to the co-delivery of the therapeutic gene encoding human tumor necrosis factor-related apoptosis-inducing ligand (pORF-hTRAIL) and DOX (Han et al., 2011b). Furthermore, imaging probes were also added to the dendrimer for cancer cell imaging (Han et al., 2011a).

5.4.3 Luteinizing Hormone-Releasing Hormone (LHRH)-Conjugated Dendrimers

LHRH (GIp-HWSYGLRPG) works as a releasing hormone in the brain. The synthetic peptide analog has been used as a ligand to numerous cancer cells such as breast, ovarian, and prostate cancer cells, because the LHRH receptors are overexpressed in these cancer cells, but not detectably expressed in most visceral organs (Dharap et al., 2005). Baker's group first published an LHRH-conjugated dendrimer (Bi et al., 2008). Minko's group synthesized an LHRH-conjugated dendrimer with the anticancer drug, paclitaxel (TAX). They also prepared linear polymer and liposomes, which contain both LHRH and TAX, and compared their targeting ability and antitumor activity *in vitro* and *in vivo*. All nanoparticles delivered the anticancer drug to the target lung carcinoma with similar efficiency (Saad et al., 2008). The same group also applied the siRNA-dendrimer to deliver siRNA to ovarian carcinoma (Shah et al., 2013). Taratula et al. reported a phthalocyanine-loaded LHRH-conjugated dendrimer for photodynamic therapy against the A2780/AD cells, which is drug-resistance human ovarian carcinoma (Taratula et al., 2013). Furthermore, they reported the addition of graphene to the phthalocyanine-loaded LHRH-conjugated dendrimer across multiple applications to include photodynamic therapy, photothermal therapy, and near infrared imaging (Taratula et al., 2015).

5.4.4 Bombesin-Conjugated Dendrimers

Bombesin (pEQRLGNQWAVGHLM) is a neuropeptide bound to the gastrin-releasing peptide receptor. It is known that the receptor is highly expressed in various tumors such as prostate, lung, and breast cancers. Hence, bombesin has been employed in DDSs as a ligand to target these specific cancers (Cescato et al., 2008). Lindner et al. reported PEG-modified bombesin (PEPSIN)-conjugated dendrimers having positron emission tomography (PET) detectable cores. They synthesized numerous dendrimers with different PEPSIN numbers and various PEG linkers to compare their targeting ability against prostate carcinoma. The bivalent dendrimer with long linkers showed the highest targeting activities (Lindner et al., 2014). Mendoza-Nava et al. prepared dual ligand (folic acid and bombesin)-conjugated dendrimers and nuclear imaging probes to target breast cancer cells (Mendoza-Nava et al., 2016).

5.4.5 CLT1-Conjugated Dendrimers

CLT1 (CGLIIQKNEC) is a cyclic peptide obtained from a phage library screening, which specifically binds to the fibronectin-fibrin complexes in the extravascular compartment of tumors (Pilch et al., 2006). Tan et al. synthesized polylysine dendrimers with a cubic silsesquioxane core conjugating to Gd(III) chelates and CLT1 peptides, for the MRI imaging of breast cancer cells (Tan et al., 2010). Furthermore, the group reported dual imaging of prostate cancer cells by employing dendrimers conjugated to Gd(III) chelates, CLT1 peptides and the fluorescence dye (Cy5) (Tan et al., 2014).

5.4.6 Other Peptide-Conjugated Dendrimers

The follicle-stimulating hormone (FSH) plays a significant role in the ovaries, and acts as a ligand that binds to the FSH receptor (FSHR). As primordial follicles do not express FSHR, FSHR-targeted drug carriers potentially preserve the fertility in younger women with ovarian cancers. The FSH33 peptide—having the FSH binding domain sequence (33–53)—exhibited the strongest binding affinity to FSHR. This peptide has been employed as a ligand to target ovarian tumor tissues (Agris et al., 1992). Modi et al. reported a FSH33-conjugated PAMAM G5 dendrimer, which successfully targeted SKOV-3 ovarian cancer cells *in vitro* and *in vivo* (Modi et al., 2014). Somatostatin is a peptide ligand of the somatostatin receptors (SR) overexpressed in neuroendocrine tumors (Wängberg et al., 1997). Orocio-Rodríguez et al. prepared a radioactive somatostatin (99m Tc-Tyr3-Octreotide peptide)-modified PAMAM G3.5 dendrimer, which was effective in visualizing SR-positive AR42J cancer cells *in vivo*. Furthermore, the group also prepared the same peptide-modified gold nanoparticle and compared the biodistribution. The accumulated nanoparticle concentrations at the tumor tissues were almost identical (Orocio-Rodríguez et al., 2015). Gellerman et al. employed the myelin basic protein peptide (87–99) as a ligand to target B-cell leukemic cells. They prepared dendrimers conjugated to the peptide at the core with the anticancer drug, chlorambucil, at the termini. The higher generation dendrimer exhibited more efficient anticancer activity to murine B-cell leukemic cells (Gellerman et al., 2013). The pan human leukocyte antigen-antigen related epitope (PADRE) and the influenza hemagglutinin peptide (110–120) are active in targeting the immune cells. These major histocompatibility complex (MHC) class II-targeting peptide-conjugated dendrimers are able to deliver DNA to antigen-presenting cells (Daftarian et al., 2011).

Some targeting peptides are obtained from peptide libraries and phage display methods. CREKA is a linear pentapeptide obtained from *in vivo* phage display. As CREKA binds to fibrin (clotted plasma protein) in the blood vessels and stroma of tumors, this pentapeptide is a targeting ligand for tumor tissues (Simberg et al., 2007). LyP-1 is a cyclic 9-amino-acid peptide obtained from a phage-displayed peptide library, which binds to breast cancer cells and tumor-associated lymphatic vessels (Laakkonen et al., 2002). It has been reported that the receptor for Lyp-1 is the mitochondrial, cell-surface protein, p32/gC1qR (Fogal et al., 2008). Lempens et al. synthesized a CREKA-conjugated dendrimer together with the Lyp-1-conjugated analog and compared their targeting ability to the individual peptides themselves. The binding affinity was enhanced after conjugation to the dendrimer (Lempens et al., 2011). Zhao et al. reported delivery to glioblastoma multiforme *via* a CREKA-conjugated dendrimer vehicle (Zhao et al., 2015). Seo et al. reported atherosclerotic plaque imaging using Lyp-1-conjugated dendrimers on the basis that p32 is overexpressed on plaque-associated macrophages (Seo et al., 2014). Pep-1 is a ligand of IL-13Rα2 that targets glioma *via* IL-13Rα2-mediated endocytosis (Pandya et al., 2012). Jiang et al. added Pep-1 to a PEGylated dendrimer to investigate the targeting effect of Pep-1. The Pep-1-conjugated PEGylated dendrimer was accumulated to IL-13Rα2-positive U87MG cells in the brain more efficiently than the PEGylated dendrimer (Jiang et al., 2016). This dendrimer could overcome BBB and achieved the brain delivery. Morris et al. obtained LTP-1 as a targeting ligand for pulmonary epithelial transport from a cyclic 7-mer peptide phage display library against primary rat lung alveolar epithelial primary cells. They conjugated LTP-1 to a PAMAM dendrimer, which provided the efficient lung adsorption properties *in vitro* and *ex vivo* (Morris et al., 2011). The H6 peptide was obtained as a ligand to the human epidermal receptor 2 (HER2) in breast cancers (Wang et al., 2015). Rostami et al. prepared H6-conjugated PAMAM dendrimers encapsulating DOX, for the treatment of HER2-positive cancer cells (Rostami et al., 2016). The plectin-1 targeted peptide (PTP) is a novel biomarker of pancreatic cancer (Kelly et al., 2008). Li et al. used PTP for co-delivery of TAX and siRNA to pancreatic cancer. They previously prepared a branched poly(ethylene glycol) modified with dendrimers through disulfide linkages (PSPG), in which TAX and therapeutic siRNA were loaded. The addition of PTP to their system improved accumulation of the dendrimers in tumors and the therapeutic effects *in vivo* (Li et al., 2017). Antiviral dendrimers have been developed using the SB105-A10 peptide, which are conjugated to the four-terminal polylysine dendrimer and exhibit antiviral effects against human papillomaviruses (HPVs), the herpes simplex virus (HSV), the human respiratory syncytial virus (RSV), and the human immunodeficiency virus (HIV) (Donalisio et al., 2010; Donalisio et al., 2012; Luganini et al., 2011; Bon et al., 2013).

Ac-TZ14011 is a peptide ligand toward the chemokine receptor 4 (CXCR4) and was developed based on the structure-activity relationships of self-defense peptides of horseshoe crabs. CXCR4 is a member of G-protein-coupled membrane receptors overexpressed in specific cancers, which is a potential target for metastatic tumors. Hence, Ac-TZ14011 has been used as an imaging probe toward metastatic tumors (Hanaoka et al., 2006). Kuil et al. synthesized dendrimers with imaging probes at the core and Ac-TZ14011 at the termini for cancer imaging. The group synthesized monomeric, dimeric, and tetrameric forms of Ac-TZ14011 conjugates to compare their targeting activities. The dimeric conjugate showed the best targeting activities *in vitro* and *in vivo* (Kuil et al., 2011). Oligo (aspartic acid) was used as a delivery vehicle to cancer cells in bone tissue. Ouyang et al. synthesized dendrimers with one (monomeric), two (dimeric), and three (trimeric) chains of Asp4–6 to compare their hydroxyapatite binding properties. The dimeric dendrimer exhibited the most efficient binding properties (Ouyang et al., 2009). These results suggest that steric hindrance can occur during multiple peptide conjugation to diminish the peptide effects. Furthermore, Pan et al. prepared an oligo(Asp)-conjugated dendrimer with the nonsteroidal anti-inflammatory drug, naproxen (Pan et al., 2012). Gaertner et al. and Medina et al. prepared both sugar- and peptide-conjugated dendrimers for dual targeting. Sugar moieties functioned well; however, the peptide analogs functioned inadequately in both cases. The results suggest that clustering effects are effective for sugar-based targeting, but ineffective for peptide-based targeting in this case (Gaertner et al., 2011; Medina et al., 2013).

Protein-conjugated dendrimers have been reported by Shah et al. and Breurken et al. The former report related to antibody-conjugated dendrimers, whereas the latter report focused on the collagen binding protein, (CNA35), conjugated to dendrimers. In both cases, the dendrimers were used as a "knot" of proteins to produce protein clusters. Binding affinities were improved after dendrimer conjugation (Shah et al., 2014; Breurken et al., 2011).

TABLE 5.3
Typical Cell-Penetrating Peptides (CPPs)

CPP	Sequence	Origin
Tat	YGRKKRPQRRR	HIV 1 Tat
gp41 fusion seq.	GALFLGWLGAAGSTMGA	HIV 1 env. gp41
Penetratin	RQIKIWFQNRRMKWKK	PAntp (43-58)
Transportan	GWTLNSAGYLLGKINKALAALAKKIL	Synthetic peptide
MAP	KLALKLALKALKAALKLA	Model amphipathic peptide
SynB1	RGGRLSYSRRRFSTSTGR	Synthetic peptide
Prion	MANLGYWLLALFVTMWTDVGLCKKRPKP	N-term (1-28) of prion protein
SV40	PKKKRKV	SV40NLS
Oligoarginine	$R_{4\text{-}16}$	synthetic peptide

5.5 PEPTIDE-CONJUGATED DENDRIMERS FOR EFFICIENT CELL INTERNALIZATION

Numerous CPPs have been reported in DDS studies. Representative examples are listed in Table 5.3 and further CPP examples have been reviewed elsewhere (Zorko & Langel, 2005; Heitz et al., 2009; Koren & Torchilin, 2012). The TAT peptide is the sequence of 48–60 HIV TAT proteins, which is the most classical and the most common CPP (Brooks et al., 2005). The TAT peptide has been conjugated to multiple dendrimers to enhance the cell internalization of nanoparticles such as metal nanoparticles and to complex with DNA or siRNA (Zhou et al., 2016; Yang et al., 2009; Han et al., 2010b; Han et al., 2011c). However, these CPPs do not have sufficient targeting activity. Sun's group reported the conjugation of both CTPs and CPPs to a single dendrimer, which could reach the target cells as well as internalize the cells efficiently (Ma et al., 2017; Li, J. et al., 2016). Two types of dendrimers were prepared using TAT and RGD peptides or a RGDTAT peptide. TAT and RGD peptides were separately conjugated to the dendrimer termini (Ma et al., 2017). In the other case, the RGDTAT peptide was conjugated to the dendrimer (Li, J. et al., 2016). Both dendrimers enhanced the RGD targeting effects. Notably, the TAT- and RGD-conjugated dendrimer penetrated deeply into the spheroid (Li, J. et al., 2016).

5.6 RECENT ADVANCES OF PEPTIDE-CONJUGATED DENDRIMERS IN DDS FOR CANCER TREATMENT

Kondo et al. discovered tumor-homing, cell-penetrating peptides using mRNA display technology, listed in Table 5.4 (Kondo et al., 2012). These peptides were selectively bound to specific cancer cells and thereafter internalized into the targeted cells. For example, CPP44 bound to M160 receptors on the cell surface of hepatic tumor cells and acute myelogenous leukemia (AML) cells and were then internalized. Recent advances in molecular biological studies have shown that certain peptides regulate the key molecule in specific diseases. For example, it was reported that particular peptides recovered tumor suppressor proteins such as $p16^{INK4a}$ and

TABLE 5.4
Tumor-Homing Cell-Penetrating Peptides

Name	Sequence	Target Cell*	Origin (Histological Type)
CPP2	DSLKSYWYLQKFSWR	Lovo	Colon (adenocarcinoma)
CPP28	RLIMRIYAPTTRRYG	U2OS	Bone (osteosarcoma)
CPP30	RLYMRYYSPTTRRYG	MCF-7	Breast (adenocarcinoma)
CPP33	RLWMRWYSPRTRAYG	A549	Lung (adenocarcinoma)
CPP44	KRPTMRFRYTWNPMK	HepG2	Liver (hepatoblastoma)
		K562	Haematopoietic (myelogenous leukaemia)

Source: Kondo, E. et al., *Nat. Commun.*, 3:951–63, 2012.

*Tested using HeLa [uterus (squamous cell carcinoma)], Lovo, A549, MCF-7, MKN45 [stomach (adenocarcinoma)], HepG2, LNCap [prostate (adenocarcinoma)], KPK [kidney (renal cell carcinoma)], U2OS, RC-15 [skeletal muscle (rhabdomyosarcoma)], RD-ES [undefined (Ewing's sarcoma)], H28 [pleura (mesothelioma)], K562, U251 [brain (glioblastoma)], NHDF [non-neoplastic, normal human dermal fibroblast] cells.

p14^{ARF}, deficient in cancer cells, and inhibited tumor growth (Kondo et al., 2004; Saito et al., 2013). Kondo et al. reported that the CPP44-linked p16^{INK4a} peptide had selective antitumor effects against hepatic tumor cells and AML cells *in vitro* and *in vivo* (Kondo et al., 2012).

We developed multifunctional dendrimers with tumor-targeting activity, cell-penetrating activity and antitumor activity, by using the tumor-homing CPP44 and the antitumor peptide, p16^{INK4a} (Kojima et al., 2018). Two types of multifunctional dendrimers were designed (Figure 5.3). One dendrimer design was to separately

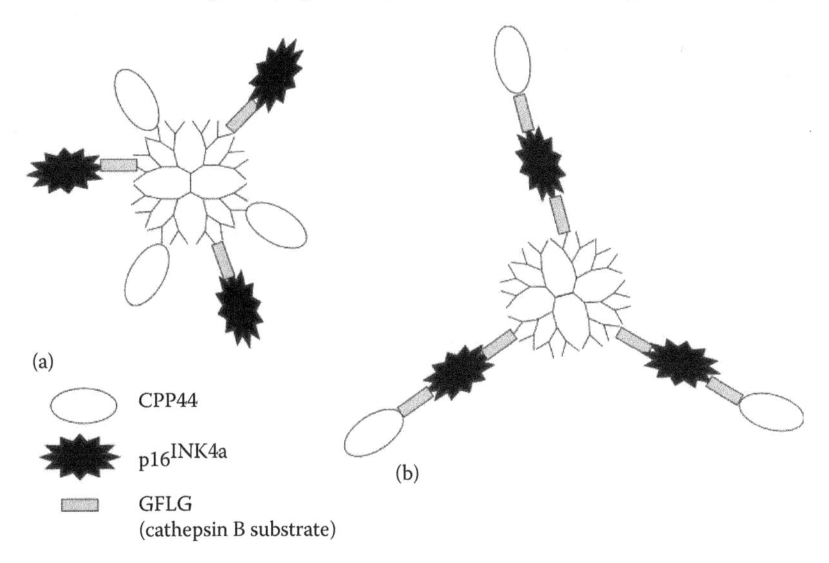

FIGURE 5.3 Parallel (a) and tandem (b) peptide-conjugated dendrimers with tumor-homing cell-penetrating activity and antitumor activity.

conjugate CPP44 and $p16^{INK4a}$ peptides (parallel-linked dendrimer). The cathepsin B substrate—a Gly-Phe-Leu-Gly (GFLG) sequence—was inserted at the C-terminus of the antitumor $p16^{INK4a}$ peptide. Cathepsin B is a proteinase, generally activated in lysosomes and upregulated in cancer (Palermo & Joyce, 2008). Thus, the GFLG peptide has been widely used as a linker between drugs and polymers in DDSs for cancer therapy, enabling antitumor agents to be released from the carriers by cathepsin B (Palermo & Joyce, 2008; de la Rica et al., 2012). As the antitumor peptide was released from the carriers, the GFLG linker-containing peptide exhibited enhanced antitumor activity (Saito et al., 2016). The second dendrimer design was to conjugate to a CPP44-linked $p16^{INK4a}$ peptide, referred to as a tandem-linked dendrimer, in which two GFLG linkers were inserted at the connecting interface of CPP44 and $p16^{INK4a}$ and the C-terminus. These tandem and parallel peptide-conjugated dendrimers were selectively associated with AML cells and showed antitumor activity, and the antitumor effects were observed to be higher than the CPP44-linked $p16^{INK4a}$ peptide alone. Thus, these dendrimers showed tri-functionality, as expected. Interestingly, the cell association and the antitumor activity were different. The tandem peptide-conjugated dendrimer showed greater association with cells than the parallel peptide-conjugated dendrimer. It is possible that the conjugated antitumor peptide interfered with the targeting functions of CPP44 because of steric hindrance. Conversely, the parallel peptide-conjugated dendrimer exhibited higher antitumor effects than the tandem peptide-conjugated dendrimer. These results may derive from $p16^{INK4a}$ peptide cleavage from the dendrimer. The $p16^{INK4a}$ peptide is released from the parallel peptide-conjugated dendrimer in a single digestion step. In contrast, to release the $p16^{INK4a}$ peptide from the tandem peptide-conjugated dendrimer a two-step digestion is necessary. Thus, higher $p16^{INK4a}$ peptide release may be available from the parallel peptide-conjugated dendrimer to cells than the tandem peptide-conjugated dendrimer. These results suggest that the conjugation strategy is important for designing multifunctional dendrimers (Kojima et al., 2018).

5.7 PEPTIDE CONJUGATION TO DENDRIMERS

There are three important aspects of the peptide conjugation to dendrimer: conjugation strategy, the linker and binding number. Peptide conjugation methods are summarized in Figure 5.4. As most dendrimers contain reactive groups such as amino

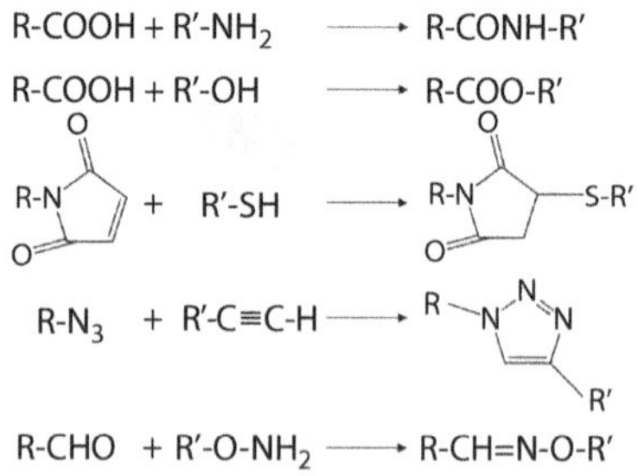

$$\text{R-COOH} + \text{R'-NH}_2 \longrightarrow \text{R-CONH-R'}$$

$$\text{R-COOH} + \text{R'-OH} \longrightarrow \text{R-COO-R'}$$

$$\text{R-N(maleimide)} + \text{R'-SH} \longrightarrow \text{R-N(succinimide)-S-R'}$$

$$\text{R-N}_3 + \text{R'-C}{\equiv}\text{C-H} \longrightarrow \text{R-(1,2,3-triazole)-R'}$$

$$\text{R-CHO} + \text{R'-O-NH}_2 \longrightarrow \text{R-CH=N-O-R'}$$

FIGURE 5.4 Peptide conjugation chemistry used in dendrimer studies.

groups, hydroxyl groups and/or carboxylic groups at the core and/or the termini, there are several strategies for peptide-dendrimer conjugation. Most peptides conjugate to the dendrimers *via* amide or ester bonding, where condensation between the amino/hydroxyl groups and the carboxylic groups undergoes. However, the condensation is not applicable in the presence of conjugating peptides having reactive side groups. The second reaction pathway involves the Michael-addition reaction between a thiol and maleimide. In this case, maleimide is necessary to link the dendrimer, and the conjugating peptides require cysteine functionality having a thiol group in the side chain. A click reaction between an alkyne and azide can also result in peptide conjugation to a dendrimer. The click reaction is fast and quantitative under mild conditions; however, a metal catalyst is necessary (Gauthier & Klok, 2008). Oxime ligation between an aminooxy group and a carbonyl group from a ketone/aldehyde is another method to conjugation a peptide to a dendrimer. Aminooxy groups were added to the dendrimer, and levulinic acid possessing carbonyl moieties was added to the conjugating peptide. Oxime ligation is useful for peptide conjugation because the reaction occurs selectively in the absence of metal catalysis (Gaertner et al., 2011; Lempens et al., 2011; Seo et al., 2014). The strategic design behind multifunctional dendrimers should follow either the conjugation of a single multifunctional peptide to a dendrimer, or conjugation of different peptides to a single dendrimer.

As multi-point binding can increase the binding affinity to the targeted molecule, conjugation of several peptides to a single dendrimer can increase the targeting ability. However, steric hindrance should also be considered in designing peptide-conjugated dendrimers. Targeting peptides should be exposed on the dendrimer and be recognized by the targeted molecules. Inaccessible peptides or high-density peptide packing do not function efficiently. To circumvent this issue, a linker can be inserted between the peptide and the dendrimer, or by controlling the peptide binding number to the dendrimer. Thus, optimizing the peptide-conjugated dendrimer structure is necessary to obtain efficient targeting activities.

5.8 OTHER PEPTIDE DENDRIMERS AS CANCER NANOMEDICINES

In this chapter, I have focused on peptide-conjugated dendrimers, which are a type of peptide dendrimer. Amino acid- and peptide-based dendrimers (Type Ia and Type Ib) are also categorized into peptide dendrimers (Crespo et al., 2005). A typical example of a Type Ia dendrimer is the polylysine dendrimer. As previously described, Tan et al. used polylysine dendrimers with a cubic silsesquioxane core for imaging of breast cancer cells (Tan et al., 2010; Tan et al., 2014). Additionally, Yang et al. reported the conjugation of the TAT peptide and nuclear localization signal (NLS) peptide to polylysine dendrimers for gene delivery (Yang et al., 2009). Shen et al. also reported an iron oxide nanoparticle coated with dendri-graft polylysine, which was linked to RGD-linked GX-1 and encapsulated DOX for theranostic application (Shen et al., 2017). Type Ib peptide-based dendrimers have a peptide substrate as a building block and the dendrimer is degraded by an enzyme to induce drug release (Darbre & Reymond, 2006). This kind of system is also useful as cancer nanomedicine.

5.9 CONCLUSIONS

This chapter focuses on peptide-modified dendrimer nanomedicines. CTPs, CPPs, tumor-homing cell-penetrating peptides, and antitumor peptides were conjugated to dendrimers. Multifunctional peptides can be produced by linking several peptides. Furthermore, dendrimers can play a role as a host of various peptide of the multivalency. Thus, peptide-based multifunctional nanoparticles are easily designed and prepared using dendrimers. Efficient targeting ability and multifunctional properties are observed when tuning the structure of peptide-conjugated dendrimers.

5.10 FUTURE PERSPECTIVES

A goal of cancer nanomedicines is their use in clinical applications and the ultimate goal is that research products are marketed. However, studies of multifunctional dendrimers are still at preclinical stages. A reason for this is that toxicity is hampering the use of multifunctional dendrimers in clinical studies. Amino-terminated dendrimers are toxic, whereas acetylated, carboxyl-terminated and PEGylated dendrimers show reduced toxicity (Malik et al., 2000; Kolhatkar et al., 2007). The most popular PAMAM dendrimers are not approved by the Food and Drug Administration (FDA). In contrast, polylysine dendrimers have been used in a clinical setting. Clinical studies of an antiviral polylysine dendrimer (VivaGel®, Starpharma (Australia)) started in 2004 and it is currently in Phase III (Kesharwani & Iyer, 2015). This dendrimer is at the forefront of potential clinical applications. The same company started clinical applications of polylysine dendrimers for docetaxel (DTX) delivery into solid tumors, which is the first clinical application of dendrimers as cancer nanomedicines. Liposomes are classically studied as cancer nanomedicine and some liposomes are already marketed. Liposomes are categorized into four types. First-generation liposomes (conventional liposomes) are constructed using only lipids. Second-generation liposomes are PEG-modified liposomes, which prolongs the drug's circulation in the bloodstream. Third-generation liposomes contain ligands or antibodies for targeting. Fourth-generation liposomes are multifunctional liposomes, which can target specific tissue, and control drug release and detect them. Some conventional and PEGylated liposomes are on the market as cancer nanomedicines, but others are not (Sercombe et al., 2015). This may be because the selection of targeted molecules is difficult. Cancer cells are heterogeneous in tumor tissues and their properties change rapidly. Thus, it is difficult to target all cancer cells even in the same tumor tissue and overcoming this drawback for clinical application of targeted nanomedicines is important.

ACKNOWLEDGMENT

Mariko Mitsui is thanked for assisting in the preparation of figures and tables.

REFERENCES

Agris, P., R. Guenther, H. Sierzputowska-Gracz, et al. 1992. Solution structure of a synthetic peptide corresponding to a receptor binding region of FSH (hFSH-β33–53). *J. Protein Chem.* 11:495–507.

Alexis, F., E. M. Pridgen, R. Langer, et al. 2010. Nanoparticle technologies for cancer therapy. *Handb. Exp. Pharmacol.* 197:55–86.

Astruc, D., E. Boisselier, and C. Ornelas. 2010. Dendrimers designed for functions: from physical, photophysical, and supramolecular properties to applications in sensing, catalysis, molecular electronics, photonics, and nanomedicine. *Chem. Rev.* 110:1857–959.

Bi, X., X. Shi, and J. R. Baker Jr. 2008. Synthesis, characterization and stability of a luteinizing hormone-releasing hormone (LHRH)-functionalized poly(amidoamine) dendrimer conjugate. *J. Biomater. Sci. Polym. Ed.* 19:131–42.

Bon, I., L. David, R. Marco, et al. 2013. Peptide-derivatized SB105-A10 dendrimer inhibits the infectivity of R5 and X4 HIV-1 strains in primary PBMCs and cervicovaginal histocultures. *PLoS One.* 8:e76482.

Breurken, M., E. H. M. Lempens, R. P. Temming, et al. 2011. Collagen targeting using multivalent protein-functionalized dendrimers. *Bioorg. Med. Chem.* 19:1062–71.

Brooks, H., B. Lebleu, E. Vivès. 2005. Tat peptide-mediated cellular delivery: back to basics. *Adv. Drug Deliv. Rev.* 57:559–77.

Cescato, R., T. Maina, and B. Nock, et al. 2008. Bombesin receptor antagonists may be preferable to agonists for tumor targeting. *J. Nucl. Med.* 49:318–26.

Chen, Q., H. Wang, H. Liu, et al. 2015. Multifunctional dendrimer-entrapped gold nanoparticles modified with RGD peptide for targeted computed tomography/magnetic resonance dual-modal imaging of tumors. *Anal. Chem.* 87:3949–56.

Crespo, L., G. Sanclimens, M. Pons, et al. 2005. Peptide and amide bond-containing dendrimers. *Chem. Rev.* 105:1663–82.

Daftarian, P., A. E. Kaifer, W. Li, et al. 2011. Peptide-conjugated PAMAM dendrimer as a universal DNA vaccine platform to target antigen-presenting cells. *Cancer Res.* 71:7452–62.

Daniels, T. R., E. Bernabeu, J. A. Rodríguez, et al. 2012. The transferrin receptor and the targeted delivery of therapeutic agents against cancer. *Biochim. Biophys. Acta.* 1820:291–317.

Darbre, T., and J.-L. Reymond. 2006. Peptide dendrimers as artificial enzymes, receptors, and drug-delivery agents. *Acc. Chem. Res.* 39:925–34.

de la Rica, R., D. Aili, and M. M. Stevens. 2012. Enzyme-responsive nanoparticles for drug release and diagnostics. *Adv. Drug Deliv. Rev.* 64:967–78.

Demeule, M., A. Regina, C. Che, et al. 2008. Identification and design of peptides as a new drug delivery system for the brain. *J. Pharmacol. Exp. Ther.* 324:1064–72.

Desgrosellier, J. S. and D. A. Cheresh. 2010. Integrins in cancer: biological implications and therapeutic opportunities. *Nat. Rev. Cancer* 10:9–22.

Dharap, S. S., Y. Wang, P. Chandna, et al. 2005. Tumor-specific targeting of an anticancer drug delivery system by LHRH peptide. *Proc. Natl. Acad. Sci. USA* 102:12962–7.

Dissanayake, S., W. A. Denny, S. Gamage, et al. 2017. Recent developments in anticancer drug delivery using cell penetrating and tumor targeting peptides. *J. Control. Release* 250:62–76.

Donalisio, M., M. Rusnati, A. Civra, et al. 2010. Identification of a dendrimeric heparan sulfate-binding peptide that inhibits infectivity of genital types of human papillomaviruses. Antimicrob. *Agents Chemother.* 54:4290–9.

Donalisio, M., M. Rusnati, V. Cagno, et al. 2012. Inhibition of human respiratory syncytial virus infectivity by a dendrimeric heparan sulfate-binding peptide. *Antimicrob. Agents Chemother.* 56:5278–88.

Duncan, R. 2012. Polymer conjugates as anticancer nanomedicines. *Nat. Rev. Cancer* 6:688–701.

Ferrari, M. 2005. Cancer nanotechnology: opportunities and challenges. *Nat. Rev. Cancer* 5:161–71.

Fogal, V., L. Zhang, S. Krajewski, et al. 2008. Mitochondrial/cell surface protein p32/gC1qR as a molecular target in tumor cells and tumor stroma. *Cancer Res.* 68:7210–8.

Gaertner, H. F., F. Cerini, A. Kamath, et al. 2011. Efficient orthogonal bioconjugation of dendrimers for synthesis of bioactive nanoparticles. *Bioconjug. Chem.* 22:1103–14.

Gajbhiye, V., P. V. Kumar, R. K. Tekade, et al. 2007. Pharmaceutical and biomedical potential of PEGylated dendrimers. *Curr. Pharm. Des.* 13:415–29.

Gauthier, M. A. and H.-A. Klok. 2008. Peptide/protein–polymer conjugates: synthetic strategies and design concepts. *Chem. Commun.* 2008:2591–611.

Gellerman, G., S. Baskin, L. Galia, et al. 2013. Drug resistance to chlorambucil in murine B-cell leukemic cells is overcome by its conjugation to a targeting peptide. *Anti-Cancer Drugs* 24:112–9.

Haag, R., and F. Kratz. 2006. Polymer therapeutics: concepts and applications. *Angew. Chem. Int. Ed.* 45:1198–215.

Han, L., R. Huang, S. Liu, et al. 2010a. Peptide-conjugated PAMAM for targeted doxorubicin delivery to transferrin receptor overexpressed tumors. *Mol. Pharm.* 7:2156–65.

Han, L., A. Zhang, H. Wang, et al. 2010b. Tat-BMPs-PAMAM conjugates enhance therapeutic effect of small interference RNA on U251 glioma cells in vitro and in vivo. *Hum. Gene Ther.* 21:417–26.

Han, L., J. Li, S. Huang, et al. 2011a. Peptide-conjugated polyamidoamine dendrimer as a nanoscale tumor-targeted T1 magnetic resonance imaging contrast agent. *Biomaterials* 32:2989–98.

Han, L., R. Huang, J. Li, et al. 2011b. Plasmid pORF-hTRAIL and doxorubicin co-delivery targeting to tumor using peptide-conjugated polyamidoamine dendrimer. *Biomaterials* 32:1242–52.

Han, L., A. Zhang, H. Wang, et al. 2011c. Construction of novel brain-targeting gene delivery system by natural magnetic nanoparticles. *J. Appl. Poly. Sci.* 121:3446–54.

Hanaoka, H., T. Mukai, H. Tamamura, et al. 2006. Development of a ^{111}In-labeled peptide derivative targeting a chemokine receptor, CXCR4, for imaging tumors. *Nucl. Med. Biol.* 33:489–94.

He, X., C. S. Alves, N. Oliveira, et al. 2015. RGD peptide-modified multifunctional dendrimer platform for drug encapsulation and targeted inhibition of cancer cells. *Colloids Surf. B* 125:82–9.

Heitz, F., M. C. Morris, and G. Divita. 2009. Twenty years of cell-penetrating peptides: from molecular mechanisms to therapeutics. *Br. J. Pharmacol.* 157:195–206.

Hersel, U., C. Dahmen, and H. Kessler. 2003. RGD modified polymers: biomaterials for stimulated cell adhesion and beyond. *Biomaterials* 24:4385–415.

Hill, E., R. Shukla, S. S. Park, et al. 2007. Synthetic PAMAM-RGD conjugates target and bind to odontoblast-like MDPC 23 cells and the predentin in tooth organ cultures. *Bioconjug. Chem.* 18:1756–62.

Jain, K., P. Kesharwani, U. Gupta, et al. 2012. A review of glycosylated carriers for drug delivery. *Biomaterials* 33:4166–86.

Jiang, B., J. Zhao, Y. Li, et al. 2013. Dual-targeting Janus dendrimer-based peptides for bone cancer: synthesis and preliminary biological evaluation. *Lett. Org. Chem.* 10:594–601.

Jiang, Y., L. Lv, H. Shi, et al. 2016. PEGylated Polyamidoamine dendrimer conjugated with tumor homing peptide as a potential targeted delivery system for glioma. *Colloids Surf. B* 147:242–9.

Kannan, R. M., E. Nance, S. Kannan, et al. 2014. Emerging concepts in dendrimer-based nanomedicine: from design principles to clinical applications. *J. Intern. Med.* 276:579–617.

Kelly, K. A., N. Bardeesy, R. Anbazhagan, et al. 2008. Targeted nanoparticles for imaging incipient pancreatic ductal adenocarcinoma. *PLoS Med.* 5:e85.

Kesharwani, P., and A. K. Iyer. 2015. Recent advances in dendrimer-based nanovectors for tumor-targeted drug and gene delivery. *Drug Discov. Today.* 20:536–47.

Khandare, J., M. Calderon, N. M. Dagia, et al. 2012. Multifunctional dendritic polymers in nanomedicine: opportunities and challenges. *Chem. Soc. Rev.* 41:2824–48.

Kojima, C. 2010. Design of stimuli-responsive dendrimers. *Expert Opin. Drug Deliv.* 7:307–19.

Kojima, C. 2016. Design of biomimetic interfaces at the dendrimer periphery and their applications. In *Stimuli-Responsive Interfaces Fabrication and Application*, ed. T. Kawai, and M. Hashizume, 209–28. Springer.

Kojima, C., K. Saito, and E. Kondo. 2018. Design of peptide–dendrimer conjugates with tumor homing and antitumor effects. *Res. Chem. Intermed.* https://doi.org/10.1007/s11164-018-3280-9.

Kolhatkar, R. B., K. M. Kitchens, P. W. Swaan, et al. 2007. Surface acetylation of polyamidoamine (PAMAM) dendrimers decreases cytotoxicity while maintaining membrane permeability. *Bioconjug. Chem.* 18:2054–60.

Kondo, E., K. Saito, Y. Tashiro, et al. 2012. Tumour lineage-homing cell-penetrating peptides as anticancer molecular delivery systems. *Nat. Commun.* 3:951–63.

Kondo, E., M. Seto, K. Yoshikawa, et al. 2004. Highly efficient delivery of p16 antitumor peptide into aggressive leukemia/lymphoma cells using a novel transporter system. *Mol. Cancer Ther.* 3:1623–30.

Koren, E., and V. P. Torchilin. 2012. Cell-penetrating peptides: breaking through to the other side. *Trends Mol. Med.* 18:385–93.

Kuil, J., T. Buckle, J. Oldenburg, et al. 2011. Hybrid peptide dendrimers for imaging of chemokine receptor 4 (CXCR4) expression. *Mol. Pharm.* 8:2444–53.

Laakkonen, P., K. Porkka, J. A. Hoffman, et al. 2002. A tumor-homing peptide with a targeting specificity related to lymphatic vessels. *Nat. Med.* 8:751–5.

Ledermann, J. A., S. Canevari, and T. Thigpen. 2015. Targeting the folate receptor: diagnostic and therapeutic approaches to personalize cancer treatments. *Ann. Oncol.* 26:2034–43.

Lee, C. C., J. A. MacKay, J. M. J. Fréchet et al. 2005. Designing dendrimers for biological applications. *Nat. Biotechnol.* 23:1517–26.

Lee, J. H, J. A. Engler, J. F. Collawn, et al. 2001. Receptor mediated uptake of peptides that bind the human transferrin receptor. *Eur. J. Biochem.* 268:2004–12.

Lempens, E. H., M. Merkx, M. Tirrell, et al. 2011. Dendrimer display of tumor-homing peptides. *Bioconjug. Chem.* 22:397–405.

Li, J., X. Zhang, M. Wang, et al. 2016. Synthesis of a bi-functional dendrimer-based nanovehicle co-modified with RGDyC and TAT peptides for neovascular targeting and penetration. *Int. J. Pharm.* 501:112–23.

Li, K., Z. Zhang, L. Zheng, et al. 2015. Arg-Gly-ASP-D-Phe-Lys peptide-modified PEGylated dendrimer-entrapped gold nanoparticles for targeted computed tomography imaging of breast carcinoma. *Nanomedicine* 10:2185–97.

Li, Q., W. Gu, K. Liu, et al. 2016. RGD conjugated, Cy5.5 labeled polyamidoamine dendrimers for targeted near-infrared fluorescence imaging of esophageal squamous cell carcinoma. *RSC Adv.* 6:74560–6.

Li, Y., H. Wang, K. Wang, et al. 2017. Targeted co-delivery of PTX and TR3 siRNA by PTP peptide modified dendrimer for the treatment of pancreatic cancer. *Small* 13:1602697.

Li, Z., P. Huang, X. Zhang, et al. 2010. RGD-conjugated dendrimer-modified gold nanorods for in vivo tumor targeting and photothermal therapy. *Mol. Pharm.* 7:94–104.

Lindner, S., C. Michler, B. Wängler, et al. 2014. PESIN multimerization improves receptor avidities and in vivo tumor targeting properties to GRPR-overexpressing tumors. *Bioconjug. Chem.* 25:489–500.

Liu, J., W. D. Gray, M. E. Davis, et al. 2012. Peptide- and saccharide-conjugated dendrimers for targeted drug delivery: a concise review. *Interface Focus* 2:307–24.

Liu, X., C. Liu, C. Chen, et al. 2014. Targeted delivery of Dicer-substrate siRNAs using a dual targeting peptide decorated dendrimer delivery system. *Nanomedicine* 10:1627–36.

Lo, A., C. T. Lin, and H. C. Wu. 2008. Hepatocellular carcinoma cell-specific peptide ligand for targeted drug delivery. *Mol. Cancer Ther.* 7:579–89.

Luganini, A., S. F. Nicoletto, L. Pizzuto, et al. 2011. Inhibition of herpes simplex virus type 1 and type 2 infections by peptide-derivatized dendrimers. *Antimicrob. Agents Chemother.* 55:3231–9.

Luong, D., P. Kesharwani, R. Deshmukh, et al. 2016. PEGylated PAMAM dendrimers: enhancing efficacy and mitigating toxicity for effective anticancer drug and gene delivery. *Acta Biomaterialia* 43:14–29.

Ma, P., H. Yu, X. Zhang, et al. 2017. Increased active tumor targeting by an αvβ3-targeting and cell-penetrating bifunctional peptide-mediated dendrimer-based conjugate. *Pharm. Res.* 34:121–35.

Maeda, H., and Y. Matsumura. 1989. Tumoritropic and lymphotropic principles of macromolecular drugs. *Crit. Rev. Ther. Drug Carrier. Syst.* 6:193–210.

Maeda, H., J. Wu, T. Sawa, et al. 2000. Tumor vascular permeability and the EPR effect in macromolecular therapeutics: a review. *J. Control. Release* 65:271–84.

Malik, N., R. Wiwattanapatapee, R. Klopsch, et al. 2000. Dendrimers: Relationship between structure and biocompatibility in vitro, and preliminary studies on the biodistribution of ^{125}I-labelled polyamidoamine dendrimers in vivo. *J. Control. Release* 65:133–48.

Medina, S. H., and M. E. H. El-Sayed. 2009. Dendrimers as carriers for delivery of chemotherapeutic agents. *Chem. Rev.* 109:3141–57.

Medina, S. H., G. Tiruchinapally, M. V. Chevliakov, et al. 2013. Targeting hepatic cancer cells with pegylated dendrimers displaying N-acetylgalactosamine and SP94 peptide ligands. *Adv. Healthcare Mater.* 2:1337–50.

Mekuria, S. L., T. A. Debele, H. Y. Chou, et al. 2016. IL-6 antibody and RGD peptide conjugated poly(amidoamine) dendrimer for targeted drug delivery of HeLa cells. *J. Phys. Chem. B* 120:123–30.

Mendoza-Nava, H., G. Ferro-Flores, F. D. M. Ramírez, et al. 2016. ^{177}Lu-Dendrimer conjugated to folate and bombesin with gold nanoparticles in the dendritic cavity: a potential theranostic radiopharmaceutical. *J. Nanomater.* Article ID1039258.

Modi, D. A., S. Sunoqrot, J. Bugno, et al. 2014. Targeting of follicle stimulating hormone peptide-conjugated dendrimers to ovarian cancer cells. *Nanoscale* 6:2812–20.

Morris, C. J., M. W. Smith, P. C. Griffiths, et al. 2011. Enhanced pulmonary absorption of a macromolecule through coupling to a sequence-specific phage display-derived peptide. *J. Control. Release* 151:83–94.

Oh, S., B. J. Kim, N. P. Singh, et al. 2009. Synthesis and anti-cancer activity of covalent conjugates of artemisinin and a transferrin-receptor targeting peptide. *Cancer Lett.* 274:33–9.

Orocio-Rodríguez, E., G. Ferro-Flores, C. L. Santos-Cuevas, et al. 2015. Two novel nanosized radiolabeled analogues of somatostatin for neuroendocrine tumor imaging. *J. Nanosci. Nanotech.* 15:4159–69.

Ouyang, L., W. Huang, G. He, et al. 2009. Bone targeting prodrugs based on peptide dendrimers, synthesis and hydroxyapatite binding in vitro. *Lett. Org. Chem.* 6:272–7.

Paleos, C. M., D. Tsiourvas, Z. Sideratou, et al. 2008. Multifunctional dendritic drug delivery systems: design, synthesis, controlled and triggered release. *Curr. Top. Med. Chem.* 8:1204–24.

Palermo, C., and J. A. Joyce. 2008. Cysteine cathepsin proteases as pharmacological targets in cancer. *Trends Pharmacol. Sci.* 29:22–8.

Pan, J., L. Ma, B. Li, et al. 2012. Novel dendritic naproxen prodrugs with poly(aspartic acid) oligopeptide: synthesis and hydroxyapatite binding in vitro. *Synth. Commun.* 42:3441–54.

Pandya, H., D. M. Gibo, S. Garg, et al. 2012. An interleukin 13 receptor α 2-specific peptide homes to human glioblastoma multiforme xenografts. *Neuro. Oncol.* 14:6–18.

Pilch, J., D. M. Brown, M. Komatsu, et al. 2006. Peptides selected for binding to clotted plasma accumulate in tumor stroma and wounds. *Proc. Natl. Acad. Sci. USA* 103:2800–4.

Riehemann, K., S. W. Scheider, T. A. Luger, et al. 2009. Nanomedicine-challenge and perspectives. *Angew. Chem. Int. Ed.* 48:872–97.

Rostami, I., Z. Zhao, Z. Wang, et al. 2016. Peptide-conjugated PEGylated PAMAM as a highly affinitive nanocarrier towards HER2-overexpressing cancer cells. *RSC Adv.* 6:107337–43.

Saad, M., O. B. Garbuzenko, E. Ber, et al. 2008. Receptor targeted polymers, dendrimers, liposomes: which nanocarrier is the most efficient for tumor-specific treatment and imaging? *J. Control. Release* 130:107–14.

Sadler, K., and J. P. Tam. 2002. Peptide dendrimers: applications and synthesis. *J. Biotechnol.* 90:195–229.

Saito, K., H. Iioka, C. Kojima, et al. 2016. Peptide-based tumor inhibitor encoding mitochondrial p14^{ARF} is highly efficacious to diverse tumors. *Cancer Sci.* 107:1290–301.

Saito, K., N. Takigawa, N. Ohtani, et al. 2013. Antitumor impact of p14^{ARF} on gefitinib-resistant non–small cell lung cancers. *Mol. Cancer Ther.* 12:1616–28.

Saraswathy, M., G. T. Knight, S. Pilla, et al. 2015. Multifunctional drug nanocarriers formed by cRGD-conjugated βCD-PAMAM-PEG for targeted cancer therapy. *Colloids Surf. B* 126:590–7.

Seo, J. W., H. Baek, L. M. Mahakian, et al. 2014. ^{64}Cu-labeled LyP-1-dendrimer for PET-CT imaging of atherosclerotic plaque. *Bioconjug. Chem.* 25:231–9.

Sercombe, L., T. Veerati, F. Moheimani, et al. 2015. Advances and challenges of liposome assisted drug delivery. *Front Pharmacol.* 6:286.

Shah, N. D., H. S. Parekh, and R. J. Steptoe. 2014. Asymmetric peptide dendrimers are effective linkers for antibody-mediated delivery of diverse payloads to b cells in vitro and in vivo. *Pharm. Res.* 31:3150–60.

Shah, V., O. Taratula, O. B. Garbuzenko, et al. 2013. Targeted nanomedicine for suppression of CD44 and simultaneous cell death induction in ovarian cancer: an optimal delivery of siRNA and anticancer drug. *Clin. Cancer Res.* 19:6193–204.

Shen, J. M., X. X. Li, L. L. Fan, et al. 2017. Heterogeneous dimer peptide-conjugated polylysine dendrimer-Fe_3O_4 composite as a novel nanoscale molecular probe for early diagnosis and therapy in hepatocellular carcinoma. *Int. J. Nanomedicine* 12:1183–200.

Simberg, D., T. Duza, J. H. Park, et al. 2007. Biomimetic amplification of nanoparticle homing to tumors. *Proc. Natl. Acad. Sci. USA* 104:932–6.

Sk, U. H., and C. Kojima. 2015. Dendrimes for theranostic applications. *Biomol. Concepts.* 6:205–17.

Svenson, S., and D. A. Tomalia. 2005. Dendrimers in biomedical applications—reflections on the field. *Adv. Drug Deliv. Rev.* 57:2106–29.

Tan, M., X. Wu, E. K. Jeong, et al. 2010. Peptide-targeted nanoglobular Gd-DOTA monoamide conjugates for magnetic resonance cancer molecular imaging. *Biomacromolecules* 11:754–61.

Tan, M., Z. Ye, D. Lindner, et al. 2014. Synthesis and evaluation of a targeted nanoglobular dual-modal imaging agent for MR imaging and image-guided surgery of prostate cancer. *Pharm. Res.* 31:1469–76.

Taratula, O., M. Patel, C. Schumann, et al. 2015. Phthalocyanine-loaded graphene nanoplatform for imaging-guided combinatorial phototherapy. *Int. J. Nanomedicine* 10:2347–62.

Taratula, O., C. Schumann, M. A. Naleway, et al. 2013. A multifunctional theranostic platform based on phthalocyanine-loaded dendrimer for image-guided drug delivery and photodynamic therapy. *Mol. Pharm.* 10:3946–58.

Tomalia, D. A., 2005. Birth of a new macromolecular architecture: dendrimers as quantized building blocks for nanoscale synthetic polymer chemistry. *Prog. Polym. Sci.* 30:294–324.

Tomalia, D. A., H. Baker, J. Dewald, et al. 1985. A new class of polymers: starburst-dendritic macromolecules. *Polymer J.* 17:117–32.

Tomalia, D. A., A. M. Naylor, and W. A. Goddard III. 1990. Starburst dendrimers: molecular-level control of size, shape, surface chemistry, topology, and flexibility from atoms to macroscopic matter. *Angew. Chem. Int. Ed. Eng.* 29:138–75.

Vlieghe, P., V. Lisowski, J. Martinez, et al. 2010. Synthetic therapeutic peptides: science and market. *Drug Discov. Today* 15:40–56.

Wagner, V., A. Dullaart, A.-K. Bock, et al. 2006. The emerging nanomedicine landscape. *Nature Biotech.* 24:1211–7.

Waite, C. L., and C. M. Roth. 2009. PAMAM-RGD conjugates enhance siRNA delivery through a multicellular spheroid model of malignant glioma. *Bioconjug. Chem.* 20:1908–16.

Wang, Z., W. Wang, X. Bu, et al. 2015. Microarray based screening of peptide nano probes for HER2 positive tumor. *Anal. Chem.* 87:8367–72.

Wängberg, B., O. Nilsson, V. Johanson, et al. 1997. Somatostatin receptors in the diagnosis and therapy of neuroendocrine tumors. *Oncologist.* 2:50–8.

Wolinsky, J. B., and M. W. Grinstaff. 2008. Therapeutic and diagnostic applications of dendrimers for cancer treatment. *Adv. Drug Deliv. Rev.* 60:1037–55.

Yan, H., L. Wang, J. Wang, et al. 2012. Two-order targeted brain tumor imaging by using an optical/paramagnetic nanoprobe across the blood brain barrier. *ACS Nano* 6:410–20.

Yang, J., Y. Luo, Y. Xu, et al. 2015. Conjugation of iron oxide nanoparticles with RGD-modified dendrimers for targeted tumor MR imaging. *ACS Appl. Mater. Interfaces* 7:5420–8.

Yang, S., D. J. Coles, A. Esposito et, al. 2009. Cellular uptake of self-assembled cationic peptide-DNA complexes: multifunctional role of the enhancer chloroquine. *J. Control. Release* 135:159–65.

Yuan, A., B. Yang, J. Wu, et al. 2015. Dendritic nanoconjugates of photosensitizer for targeted photodynamic therapy. *Acta Biomater.* 21:63–73.

Zhao, J., B. Zhang, S. Shen, et al. 2015. CREKA peptide-conjugated dendrimer nanoparticles for glioblastoma multiforme delivery. *J. Colloid Interface Sci.* 450:396–403.

Zhi, M., K. C. Wu, L. Dong, et al. 2004. Characterization of a specific phage-displayed Peptide binding to vasculature of human gastric cancer. *Cancer Biol. Ther.* 3:1232–5.

Zhong, Y., F. Meng, C. Deng, et al. 2014. Ligand-directed active tumor-targeting polymeric nanoparticles for cancer chemotherapy. *Biomacromolecules* 15:1955–69.

Zhou, Z., Y. Wang, Y. Yan, et al. 2016. Dendrimer-templated ultrasmall and multifunctional photothermal agents for efficient tumor ablation. *ACS Nano* 10:4863–72.

Zhu, J., F. Fu, Z. Xiong, M. Shen, et al. 2015. Dendrimer-entrapped gold nanoparticles modified with RGD peptide and alpha-tocopheryl succinate enable targeted theranostics of cancer cells. *Colloids Surf. B* 133:36–42.

Zorko, M., and U. Langel. 2005. Cell-penetrating peptides: mechanism and kinetics of cargo delivery. *Adv. Drug Deliv. Rev.* 57:529–45.

6 Targeting the Tumor Microenvironment

Andrew Cannon, Suprit Gupta, Rakesh Bhatia, Christopher Thompson, Brad Hall, Surinder K. Batra, and Sushil Kumar

CONTENTS

6.1 INTRODUCTION

Recently, focus in the field of oncology has shifted from cell autologous to heterotypic interactions between tumor cells and the surrounding tumor microenvironment (TME) indicating that cancer is a disease of tissue rather than of cells. These interactions include crosstalk between malignant cells and various components of stroma including cancer-associated fibroblasts (CAF), endothelium, infiltrating immune cells, and acellular extracellular matrix (ECM) components. Because of the vital role of these interactions in cancer progression and metastasis, significant efforts have been devoted to design therapies to target these interactions in the TME. Nanomedicine-based approaches are particularly promising for targeting mediators of tumor-stroma interaction and the downstream effector molecules of this crosstalk. Nanoparticles (NP) can be designed to specifically accumulate, modify, and deliver drugs and biomolecules to the TME thereby increasing the therapeutic efficacy and limiting toxicity associated with systemic administration of chemotherapeutic agents. In this chapter, we briefly highlight the general features of the TME that contribute to cancer development and progression, followed by in-depth discussion of the mechanistic basis on which nanoparticles have been formulated for TME-targeted delivery of therapeutic agents. Notably, we emphasize the advantages and limitations of each nanomedicine targeting approach. Next, we highlight critical molecules involved in tumor-stroma crosstalk that are amenable to nanomedicine-based therapeutic approaches. Finally, we conclude with a perspective on the future development of TME-based nanomedicines.

6.2 THE TUMOR MICROENVIRONMENT

The term "TME" refers to the abnormal cellular and biochemical environment surrounding malignant cells. This environment includes CAFs, endothelium, immune cells, and the milieu of cytokines, growth factors, and small molecules (Figure 6.1). Pathologically, many tumors including carcinomas of the pancreas, breast, stomach, and colon are frequently composed of as little as 5–10% neoplastic cells while the remainder of the tumor, frequently referred to as tumor stroma, is a complex mixture of both cellular and acellular components. These stromal components establish a complex network of reciprocal interactions between cancer and non-neoplastic cells. Importantly, both the abnormal biochemical environment and the milieu of signaling molecules occur concomitantly in tumors and are suggested to act cooperatively in promoting cancer progression. However, for the purpose of clarity, we will address the biochemical environment of the TME and heterotypic interactions within the TME separately.

6.2.1 CELLULAR COMPONENTS OF THE TME

The cellular compartment of the tumor stroma is composed of a variety of cell types including CAFs, myofibroblasts, endothelial cells, pericytes, smooth muscle, and resident or infiltrating immune cells (including macrophages, myeloid-derived suppressor cells, and T-cells). Interactions with neoplastic cells modulate the secretome

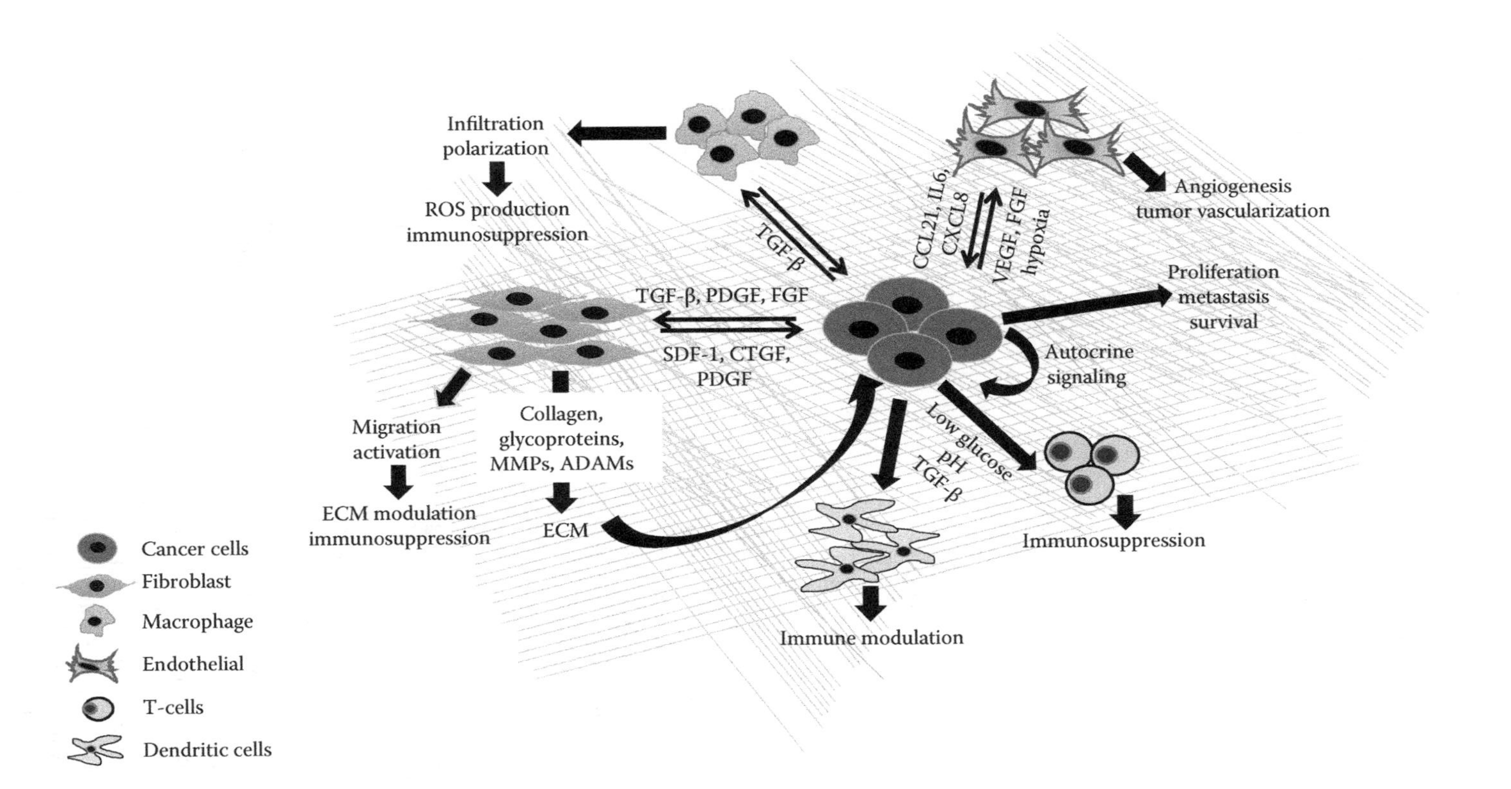

FIGURE 6.1 A schematic representation of the reciprocal interactions between TME components and cancer cells and resultant cellular processes. Each arrow between cells represents a heterotypic interaction in the TME whereas arrows pointing to text indicate cellular processes that result from those intercellular interactions.

of stromal cells which, leads to the distinct milieu of cytokines, growth factors, and small signaling molecules that activate signaling pathways shown to contribute to the malignant phenotype of cancer cells. Similarly, the ECM, consisting of collagens, proteoglycans, and glycoproteins, is largely produced by stromal cells and has the capacity to enhance oncogenic signaling in neoplastic cells. The contribution of each of these cellular stromal components towards cancer progression is discussed in this section.

6.2.1.1 Cancer-Associated Fibroblasts

While the origin and characteristic features of fibroblasts remain an area of active research, it is becoming increasingly clear that CAFs influence the behavior of cancer cells, especially in tumors with extensive desmoplasia. Under normal conditions, fibroblasts remain quiescent; however, in the case of tissue injury or chronic inflammation, fibroblasts are activated leading to their proliferation, production of ECM proteins and matrix degrading enzymes, and acquisition of a migratory phenotype (Kalluri et al. 2016). After resolution of inflammation and tissue repair, fibroblasts return to a quiescent state. However, during cancer progression, the chronic healing response is augmented by persistent tissue injury, secretion of transforming growth factor beta (TGF-β), platelet-derived growth factor (PDGF), and fibroblast growth factors (FGF) by cancer cells as well as dysregulation of p53 and Notch signaling in CAFs, resulting in continuous fibroblast activation and proliferation (Figure 6.1) (Kalluri et al. 2016). The function of CAFs in the TME is complex, as several autochthonous murine models of multiple malignancies have demonstrated conflicting functions of CAFs with regard to tumor promotion or tumor restraint. This contradiction suggests the existence of heterogeneity among CAFs. The expression of multiple markers with limited co-expression further supports the notion that CAFs entail both tumor promoting and restraining populations that differ depending on spatial relationships with other components of the TME and possibly stage of tumor progression (Kalluri et al. 2016).

Activated CAFs synthesize ECM proteins like collagen and fibronectin that constitute the predominant acellular components in tumors. ECM deposition is a dynamic process due to the production of ECM-degrading enzymes, such as matrix metalloproteinases (MMPs) and A disintegrin and metalloproteinase domain (ADAM) containing protein family members (Kalluri et al. 2016). The ECM proteins especially collagen cross-linking, orientation, and thickness increase the stiffness of the tumor that facilitates cancer cell proliferation, oncogenic signaling, invasion, EMT, and metastasis (Gilkes et al. 2014). Furthermore, collagen orientation has been shown to play a critical role in cancer cell migration, directionality, and immune cell recruitment (Gilkes et al. 2014). However, extensive deposition and cross-linking of the ECM impedes tumor perfusion, limiting the delivery of oxygen and nutrients to the tumor, and at the same time, posing a unique challenge to the delivery of chemotherapeutics (Gilkes et al. 2014). Another aspect of cancer biology that is significantly affected by CAFs is that of the immune response to the developing tumor. In general, CAFs are thought to promote an immunosuppressive microenvironment through the secretion of various cytokines including interleukins (IL)-4, -6, -8, -10; TGF-β, as well as CXC and CC chemokines (Kalluri et al. 2016). These cytokines suppress the activity of effector T-cells, promote T-cell differentiation to immunosuppressive phenotypes, and augment the immunosuppressive activity of cells of the myeloid lineage.

6.2.1.2 Cells of the Tumor Vasculature

Blood and lymphatic vessels are also an important component of the TME. These vessels are themselves composed of several cell types including endothelial cells, pericytes, and vascular smooth muscle cells, as well as supporting ECM. While the contributions of pericytes, vascular smooth muscle, and vascular ECM to tumor progression remain unknown, the endothelium plays significant roles during tumor progression. The endothelium is a specialized epithelium that constitutes the innermost layer of blood and lymphatic vessels and plays important roles in vascular response to intrinsic and extrinsic stimuli. In the TME, cancer cells secrete vascular endothelial growth factors (VEGFs), FGFs, and angiopoietins that recruit endothelial cells and their precursors and drive their proliferation (Figure 6.1) (Gomes et al. 2013). Interestingly, stimulation of endothelial cells by cancer cells further increases the production of VEGF as well as pro-inflammatory cytokines like IL-6 and CXCL8, which augment cancer cell proliferation in multiple malignancies. Similarly, cancer cells mediate lymphatic endothelial cell recruitment and proliferation through the secretion of VEGF-C, -D, and Cis. Stimulation of lymphatic endothelium results in increased secretion of CCL21, a CCR7 ligand, which acts as a chemoattractant for cancer cells expressing CCR7, suggesting a role of lymphatic endothelial cells in cancer cell chemotaxis, invasion, and metastasis (Gomes et al. 2013). In addition to its interaction with tumor cells, the endothelium also influences other cells in the tumor stroma. While an understanding of these interactions is developing, it is becoming clear that endothelial cells influence the recruitment of immune cells and fibroblasts to the tumor bed, thereby potentiating the role of these cell types in the TME.

6.2.1.3 Immune Cells

Oncogene-mediated chronic inflammation leads to recruitment and accumulation of immune cells in the tumor bed. Hematopoiesis gives rise to cells belonging to myeloid or lymphoid lineages. Cells of the myeloid lineage including macrophages, neutrophils, and myeloid-derived suppressor cells (MDSCs) are recruited to the TME by the CCL family of cytokines as well as by GM-CSF and M-CSF. Macrophages represent terminally differentiated monocytes and are classified as M1 or M2 polarized based on the expression of cell surface markers. M1 macrophages, marked by surface expression of CD68 and 86, stimulate T-cell response thus driving specific anti-tumor immune response (Elliott et al. 2017). In contrast, M2 macrophages demonstrate surface expression of CD68, 163, and 206 and are thought to express cytokines that suppress T-cell recruitment and cytotoxic activity (Elliott et al. 2017). Similar to macrophages, tumor associated neutrophils (TANs) are classified as being of N1 (antitumor response) or N2 (immune suppressive) polarization. Interestingly, the presence of TGF-β in the TME seems to drive the polarization of neutrophils towards the immunosuppressive N2 type (Elliott et al. 2017). Despite these observations, there are relatively few studies addressing the causative nature of TGF-β in neutrophil polarization in the TME and the overall functioning of neutrophils subtypes in the TME. MDSCs are immature cells of myeloid lineage that fail to achieve terminal differentiation before infiltrating an inflamed tissue. MDSCs are classified as monocytic-like MDSC (M-MDSC) or polymorphonuclear-like MDSC (PMN-MDSC) (Elliott et al. 2017). Similar to other immune cell classifications, MDSCs are classified based on immunophenotype. However, the ability to suppress T-cell function is a common feature of both subsets of MDSCs (Elliott et al. 2017).

Immune cells derived from the lymphoid lineage are also present within the TME. The lymphoid lineage can be divided into T-cells, NK cells, B-cells, and plasma cells. Of these, the role of T-cells in the TME has been investigated extensively, so, for the purpose of brevity, we will focus our discussion on T-cells. Several functionally distinct populations of T-cell subsets exist in the TME including, CD8+ cytotoxic T-cells, CD4+ Th1 and Th2 helper T-cells, and CD4+ CD25+ FoxP3+ T-regulatory cells. Cytotoxic (CD8+) T-cells are class restricted to MHC I, are involved in immune surveillance, and eliminate cells based on the tumor-associated antigen or tumor neoantigens derived from the intracellular compartment (Figure 6.1). Upon recognition of antigen presented on MHC I by T-cell receptor (TCR) and adequate co-stimulation through the interaction of CD28 with CD80/CD86 or binding of IL-2, a naïve CD8 T-cell is activated and undergoes clonal expansion (Anderson et al. 2017). Once activated, CD8+ T-cells bind an antigen on target cells and release perforin and granzyme that permeabilize the cytoplasmic membrane and activate caspase 8 thereby initiating the apoptotic cascade in the cancer cells displaying MHC I bound antigen (Anderson et al. 2017). However, several factors within the TME limit the activity of T-cells and allow cancer cells to evade the host immune system. For instance, multiple cells within the TME express PD-L1, which functions as an immune checkpoint causing effector T-cell anergy or death and limiting the ability of T-cells to execute antitumor immune reactions (Anderson et al. 2017). Additionally, the metabolic environment of the TME limits the antitumor immune response of T-cells through depletion of glucose and arginine by cancer cells and the immune-suppressive populations (Anderson et al. 2017).

Regarding CD4+ T-helper cells, Th1 subtype functions to coordinate the antitumor immune response of CD8+ T-cells. In contrast, Th2 cells are thought to promote humoral immune response over a CD8+ T-cell response; humoral immune response appears to be ineffectual in tumor immune surveillance (Accolla et al. 2014). Th17 represents yet another subgroup of CD4+ T-cells. Their function with respect to tumor immunity appears to be complex and dependent upon the interaction of these cells with CD8+ T-cells and T-regulatory cells. The critical subset of CD4+ T-cells are the T-regulatory cells, which suppress effector T-cells, macrophages, dendritic cells, NK cells, and B-cells through a variety of mechanisms including scavenging IL-2 and expression of CTLA-4, an immune checkpoint molecule thereby suppressing tumor immunity (Tanaka et al. 2017).

6.2.2 Biochemical Milieu in the Tumor

The biochemical environment within a tumor differs significantly from that found in healthy tissues. In general, tumors are hypoxic, acidic, and demonstrate elevated levels of reactive oxygen species (ROS). The underlying causes of these biochemical changes are multifactorial and often differ between tumor types. However, much of the hypoxia and acidity of the TME can be attributed to poor tumor perfusion in combination with the higher metabolic activity of neoplastic cells. Further, tumor hypoxia contributes to mitochondrial dysfunction resulting in a paradoxical increase in the elaboration of ROS. In certain cases, each of these biochemical features of the TME can contribute to the malignant behavior of neoplastic cells. This section reviews the evidence regarding the role of each of these TME features in cancer progression.

6.2.2.1 Tumor Hypoxia

Over the last 60 years, studies have established the critical role of hypoxia in the aggressiveness of solid tumors. Hypoxic conditions are evident by the necrotic centers of solid tumors and a strikingly lower respiratory quotient (Michiels et al. 2016). The factors leading to tumor hypoxia are multifaceted including the rapid rate of cancer cell proliferation and poor tumor perfusion. Tumors compensate for this hypoxia by releasing VEGF and FGF family members. These growth factors stimulate the recruitment and proliferation of cells required for the development of new blood vessels including endothelial cells, pericytes, and vascular smooth muscle, referred to as angiogenesis. Despite these compensatory mechanisms, signaling events required for angiogenesis are not sufficient to fully correct the hypoxic state within the tumor (Michiels et al. 2016). Furthermore, the rapid development of blood vessels in the TME results in the outgrowth of immature vessels that are unable to function normally leading to vascular dysfunction and intratumoral hemorrhage both of which contribute to the hypoxic TME (Michiels et al. 2016). Additionally, the metabolic activities of cancer cells and stromal cells rapidly consume the scant oxygen present in the TME (Michiels et al. 2016). Of note, the picture of tumor hypoxia is complicated by a growing body of evidence that suggests that low oxygen tension in the TME is not constant but cycles in a complex spatiotemporal pattern as a function of both distance of cells from a vascular bed as well as the presence of intermittent vascular dysfunction (Michiels et al. 2016). This cycling hypoxia is suggested to amplify the effect of hypoxia on cancer cells and the TME (Michiels et al. 2016). Cumulatively, tumor hypoxia is contributed by lack of vasculature relative to the amount of tissue, the vascular dysfunction that results from the rapid formation of immature blood vessels, and the metabolic rate of cells dwelling in the TME.

Counterintuitively, tumor hypoxia augments rather than limiting the cancer progression. As previously discussed, hypoxia in the TME stimulates angiogenesis, a process that is critical for metastasis by providing cancer cells access to the blood vessels (Ellis et al. 2004). Furthermore, hypoxia has been shown to stimulate the production of cytokines and growth factors, including TGF-β, CXCL12 (SDF-1), and CXCL13, by multiple cell types within the TME (Michiels et al. 2016). These soluble factors potently induce an invasive phenotype in cancer cells further augmenting the metastatic potential of malignant cells under hypoxia. Finally, hypoxia plays a major role in the polarization of macrophages towards an M2 phenotype and recruitment of T-regulatory cells to the TME (Elliott et al. 2017). Both of these functions of hypoxia result in a functionally immunosuppressive TME.

6.2.2.2 Tumor pH

The hypoxic TME leads to a lower extracellular pH compared to other tissues in the body. Briefly, the high metabolic rate, specifically glycolytic rate, of cancer cells coupled with rapid tumor growth causes tumors to outgrow their vascular supply leading to the production of organic acids such as lactic acid (Michiels et al. 2016). Further, poor perfusion in solid tumors limits the clearance of these acidic species resulting in acidification of the TME. In order to survive in this environment, cancer cells adopt several means by which they can effectively export these acids to the extracellular compartment. Cancer cells demonstrate increased carbonic anhydrase activity, which catalyzes the conversion of carbonic acid into bicarbonate (a base) and H+, compared to normal cells (Parks et al. 2011). The activity of carbonic anhydrases is coupled to the function of several transporters, including Na+/H+ exchangers (NHE), H+- ATPases, (HNE), NA+ dependent CL-/HCO3- exchangers, and Na+/HCO3- cotransporters (Parks et al. 2011). In general, the cumulative activities of these transports augment the efflux of H+ and influx/retention of HCO3- in cancer cells. The result of these adaptive mechanisms is the exportation of acidic species to the extracellular space. In fact, through these mechanisms, cancer cells are able to maintain a more basic intracellular pH than other cells despite the highly acidic extracellular milieu.

The acidity of the TME is not merely a challenge with which cancer cells must contend in order to survive, but rather contributes to cancer aggression by exerting a selective pressure that leads to the evolution of subclones with a more invasive/metastatic phenotype. For instance, NHE isoform-1 is a key component of tumor cell response to the acidic TME and is induced by the acidic conditions of the TME. However, NHE-1 also plays a role in the formation of focal adhesions (Parks et al. 2011). This is significant, as the turnover of focal adhesions is thought to be requisite for tissue invasion and metastasis. Other pH-regulating molecules are reported to aid in the metastatic dissemination of cancer cells, but the involvement, and mechanism of these molecules remain a topic of debate.

6.2.2.3 Reactive Oxygen Species

ROS are defined as chemical species containing a highly reactive or thermodynamically unstable atom of oxygen. The generation of ROS by cancer cells is closely related to high metabolic/glycolytic activity of cancer cells in combination with a hypoxic TME that increases mitochondrial dysfunction and thus the generation of ROS (Costa et al. 2014). This mechanism of ROS generation predominately contributes intracellular ROS; however, due to their reactive nature, ROS molecules are interconvertible between chemical species through enzymatic action or spontaneous dismutation. Further, infiltrating immune cells of the myeloid lineage express high levels of NADPH oxidase, an enzyme that specifically generates superoxide. Studies show that the activity of NADPH oxidase in myeloid cells contributes significantly to extracellular ROS in tumors. In the context of the TME, H_2O_2 plays a particularly important role in the generation of an oxidative environment as H_2O_2 is a highly reactive species that diffuses readily through the TME (Costa et al. 2014). The second determining factor of ROS concentration within a tissue is the amount and activity of antioxidants. Enzymatic antioxidants including the superoxide dismutases, peroxidases, glutathione peroxidase, catalase, and reductase catalyze the reactions of ROS with other molecules that finally result in degradation of hydrogen peroxide to water and molecular oxygen (Costa et al. 2014). Whereas, non-enzymatic antioxidants including vitamin A, C, and E function to scavenge macromolecular ROS adducts thereby limiting the damage caused by ROS and terminating free radical chain reactions (Costa et al. 2014).

In the context of cancer, elevated levels of ROS play a variety of important roles that contribute to the aggressive behavior of cancer cells. Within cancer cells, ROS promotes signaling through receptor tyrosine kinases leading to augmented proliferative, migratory, and survival signaling (Costa et al. 2014). These signaling events may work in conjunction with ROS mediated activation of NF-kB that results in epithelial to mesenchymal transition (EMT) causing epithelium-derived cancer cells to adopt a migratory/metastatic phenotype. Further, ROS increases the stemness of cancer stem cells thereby theoretically affecting both metastatic potential and chemotherapy resistance (Costa et al. 2014). In addition to its direct effects on cancer cells, ROS also affects the tumor stroma. For instance, elevated levels of ROS recruit and activate CAFs leading to the production of MMPs that degrade ECM and facilitate cancer cell invasion (Costa et al. 2014).

In conclusion, the TME represents a unique biochemical environment consisting of hypoxia, low pH, and increased levels of ROS. Each of these biochemical features of the TME contributes significantly to cancer progression, but at the same time presents an opportunity for targeting nanomedicines to the TME allowing for site-specific delivery of a therapeutic payload. Thus, the TME can be leveraged for the development of nanomedicines that would maximize therapeutic benefit and decrease dose-related toxicities associated with the administration of high dose cytotoxic therapies.

6.3 NANOPARTICLE APPROACHES TO TME-SPECIFIC DRUG DELIVERY

Early in the development of nanomedicines, researchers noticed that NPs, because of their physicochemical properties, preferentially accumulate in the TME. This effect, termed Enhanced Permeation and Retention (EPR), formed the basis of NP-based drug formulations for the treatment of cancer. Despite a natural propensity to accumulate in tumors, it was quickly understood that the intrinsic EPR properties of nanoparticles (NPs) were not sufficient to substantially improve therapeutic efficacy while minimizing the therapy-related adverse events. Perhaps even before the role of the TME in cancer progression was discovered, studies had highlighted that the features of the TME provided an opportunity for the development of TME specific nanomedicine delivery systems. The acidity of the TME was the first characteristic to be leveraged for the purpose of nanomedicine delivery, but soon, systems were designed to utilize the redox state of the TME, the elevated activity of proteases/peptidases in the TME, the presence of specific target molecules, and immune cell infiltration in the TME. This section describes the systems that have been or are being developed, which utilize features of the TME to enhance the accumulation of nanomedicines in the tumor (Table 6.1 and Figure 6.2).

TABLE 6.1
A Summary of the Features of the TME Used for Targeted Delivery of Nanomedicines to the TME

TME Feature	Functionalization	Mechanisms
Acidic pH	Charge switching surface conjugates	pH-mediated shift from anionic to cationic surface charge allow endocytosis
	pH sensitive internalizing peptide	pH-mediated peptide insertion into cell membrane, and conformational change causing cellular NP internalization
Oxidative Stress	Variable	Reducing intracellular environment mediates release of NP and delivery of the drug intracellularly
Proteinase	Variable	Proteinase-based cleavage of NP core mediates drug release
	Proteinase substrate and drug containing surface conjugate	Proteinase based cleavage of surface conjugates liberates drug from surface of NP
		Proteinase based cleavage of surface conjugates unmasks positive surface charge leading to endocytosis of the NP
Molecular Interaction	Receptor ligand Binding Partner Inhibitor Monoclonal antibody	Surface functionalization binds to a specific molecular target in TME mediating sequestration in the TME or mediating cellular endocytosis
Cellular	Variable	NPs targeted for uptake by a specific cell type within or near the tumor; cell traffics NP into the tumor as NP loaded cell is recruited to the TME

Note: Here functionalization refers to the molecules added to the nanoparticle core that allow for targeted NP delivery on the basis of a given TME feature; "Variable" in this column indicates that surface functionalization is not an intrinsic feature of the targeting mechanism. The "Mechanisms" column contains a description of how the NPs interact with the TME allowing for TME-targeted delivery.

6.3.1 pH Base Targeting of Nanomedicines to the TME

As described earlier, the TME is acidic in comparison to the majority of non-pathologic tissue in the body. In principle, the NPs designed to take advantage of the acidic TME work through their rapid dissociation or change in surface charge upon exposure to an acidic environment thereby allowing drug release or endocytosis of NPs by cells in the TME, respectively. Given that majority of the solid tumors have an acidic environment; this approach to develop tumor-specific nanomedicine delivery systems is appealing due to its broad applicability.

One of the first NP-based delivery systems to capitalize on the acidic TME was formulated from the biodegradable, cationic polymer poly-(beta-amino ester) (PbAE) modified by the physical adsorption of poly (ethylene oxide) (PEO) onto the surface of the NP (Shenoy et al. 2005). PbAE was selected for this drug delivery system due to its increased solubility at lower pH and low toxicity compared to

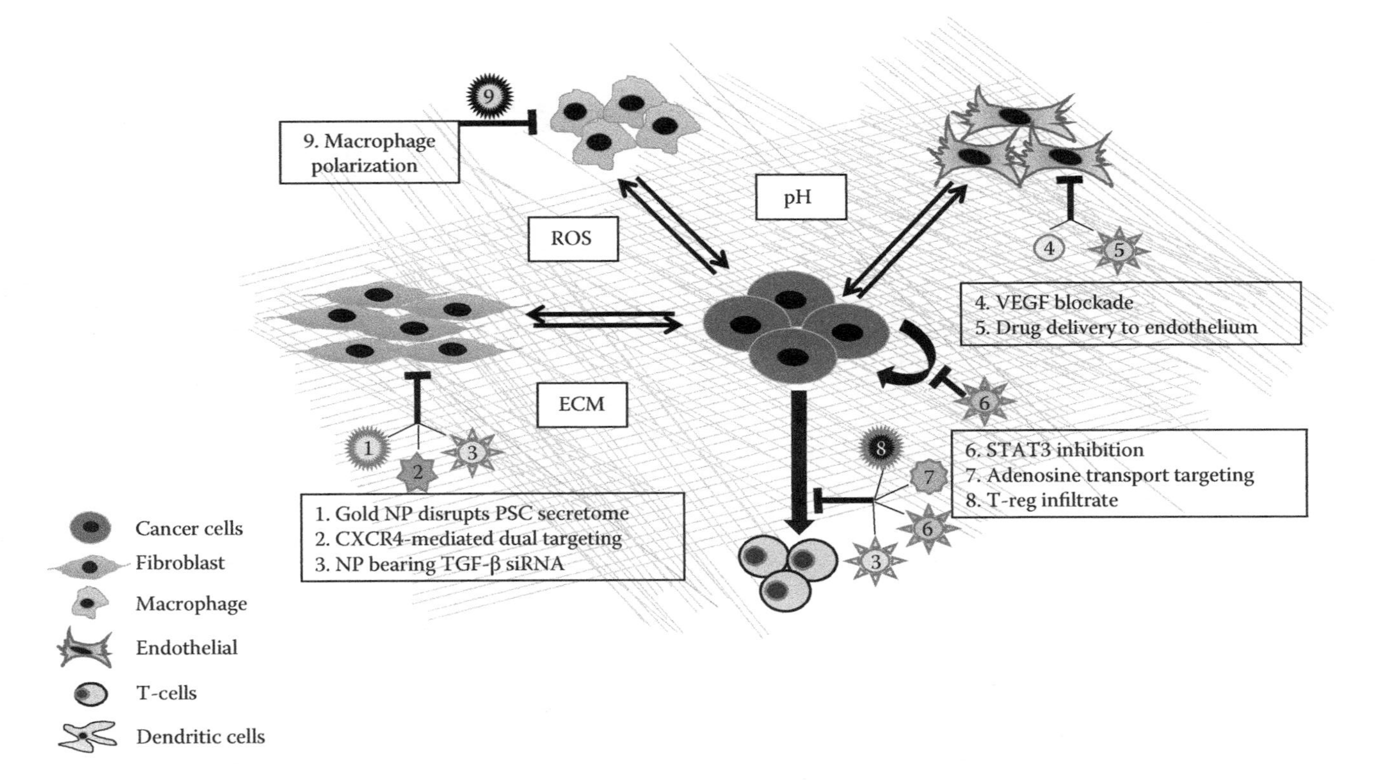

FIGURE 6.2 An illustrated overview of the nanomedicine-based approaches for targeting the tumor supportive roles of the TME and corresponding summary of the nanomedicine mechanism. Each arrow between cells represents a heterotypic interaction in the TME. Each inhibition symbol is associated with nanoparticles that function to block or decrease the associated heterotypic interaction. Note that nanomedicine approaches with multifaceted effects on the TME appear more than once.

other cationic polymers, which make it suitable for systemic administration. The surface adsorption of PEO onto a PbAE NP increased the circulating time of NPs and augmented passive accumulation of the NP in the tumor. PbAE is insoluble at physiologic pH allowing for the formation of NPs, however, as pH falls to 6.5, PbAE rapidly becomes soluble in aqueous solution, and the NP releases the therapeutic payload (Shenoy et al. 2005). In this study, NPs were used to encapsulate paclitaxel, an effective, yet toxic chemotherapeutic agent. Importantly, the administration of taxol drugs is complicated by the fact that taxanes are poorly soluble in water necessitating cremophore-based administration that considerably increases the toxicity of taxanes. Thus, the purpose of these PbAE-based NPs is two-fold, first, circumvention of cremophore-based administration of taxanes and second, targeting of therapeutic payload to the TME. Functionally, these NPs increased the delivery of paclitaxel to breast cancer and T-cell leukemia cells grown in culture and increased the cytotoxicity compared to pH-insensitive NP and traditional administration of paclitaxel (Shenoy et al. 2005; Lundberg et al. 2011). In spite of these benefits, the practicality of the use of this method clinically or even in preclinical models remains questionable. For instance, in the TME, pH below 6.8 has not been documented thus the release of drug from these NPs is likely not to occur specifically within the TME but rather within the endocytic vesicle. Therefore, any cell that endocytoses these NPs would release paclitaxel due to low pH of the endocytic vesicle. Thus, the ability of these NPs to release chemotherapeutic agent specifically and rapidly in the TME may be limited.

To circumvent these issues, one study used a pullulan-based, charge reversible (CAPL) shell on the modified PbAE core (Zhang et al. 2016). At physiologic pH, the CAPL shell carries a negative charge, which limits the endocytosis of this NP. However, when exposed to an acidic environment (pH 6.2–6.9), the charge of the CAPL shell becomes positive allowing endocytosis of the NP. In this way, endocytosis of the NPs is restricted to the cells residing in an acidic environment such as the TME. Upon endocytosis, the PbAE core is dissolved, and the drug is released. In nude mice bearing tumors derived from HepG2, a hepatocellular carcinoma (HCC) cell line, the use of CAPL shell PbAE NP decreased the tumor volume due to tumor-targeted cytotoxicity (Zhang et al. 2016). More importantly, there was a significant decrease in the accumulation of NP payload, paclitaxel, and CA4, in the liver and spleen compared to standard PbAE NP or administration of free paclitaxel and CA4 (Zhang et al. 2016).

Iron oxide is another material that can be used to develop pH-sensitive NPs for the purpose of both drug delivery systems and imaging contrast that exploit the acidic pH of the TME. The principle underlying iron oxide-based NP accumulation in the TME is similar to that of CAPL NPs in that a pH-sensitive polymer is conjugated to the surface of the iron oxide NP (Crayton et al. 2011). At physiologic pH, these surface polymers bear negative charges and remain in circulation for a longer duration while avoiding retention in the normal tissues. In the acidic pH of the TME however, the pH-sensitive polymer switches charges and NPs are internalized. One study used these NPs to develop a pH-sensitive contrast for magnetic resonance imaging (MRI). As a contrast agent, the super paramagnetic iron oxide (SPIO) accelerates the dephasing/relaxation of bulk water in the proximity of the NP thereby enhancing the signal from areas that have higher concentrations of the NP (Crayton et al. 2011). To this SPIO NP, glycol-chitosan (a glycosamine polymer) was added to the surface of the NP, which confers the NPs with the ability to switch surface charge in an acidic environment. These NPs significantly enhanced the signal from subcutaneously implanted tumors in nude mice compared to NPs of the same size with a negative or neutral surface charge at both physiologic and acidic pH (Crayton et al. 2011). Similar formulations were used in another study for both imaging and drug delivery using extremely small iron oxide NPs modified by the addition of pH-sensitive derivatives of poly (ethylene-glycol) poly (beta-benzyl-L-aspartate) (Ling et al. 2014). Once the surface charge of the NP becomes positive at low pH, cells endocytose the particles with greater efficiency and the additional drop in pH in endocytic vesicles completely dissociates the NPs. As an imaging contrast agent, these NPs enhanced T1- and T2-weighted MR images of the tumor (Ling et al. 2014). Additionally, when exposed to light, the iron core of the NP generates singlet oxygen which, is toxic to cells. At physiologic pH, these NPs are capable of self-quenching the light-mediated singlet oxygen due to Förster resonance energy transfer (FRET) between the components of the compact NP. However, upon swelling and dissociation in the endosome, the FRET is lost, and cytotoxic singlet oxygen remains unquenched. These NPs, as pH-sensitive photosensitizers, showed enhanced tumor accumulation and tumor cell killing as compared to administration of free Ce6 dye and Ce6 dye in pH-insensitive NPs in mice bearing xenograft tumors (Ling et al. 2014). Interestingly, these NPs swell (~70 nm at pH 7.4 to >100 nm at pH<6.5) upon exposure to low pH; this feature may enhance the EPR properties of the NP but may also limit the diffusion of the NP through the TME.

Smaller NPs (<60 nm) are expected to penetrate the TME with greater efficiency; however, smaller NPs tend to have decreased circulating times and less EPR effect than larger NPs (~70–100 nm) (Jain et al. 2010). A nanoparticle that decreases size upon exposure to acidic conditions could theoretically circumvent the limitations of both small and large NPs. This concept led to the development of a pH-sensitive NP of poly (aminoamide) dendrimers conjugated to both a platinum drug derivative and the pH-sensitive poly (2-azepane-ethyl-methylacrylate) (PAEMA) (Li et al. 2016b). At neutral pH, PAEMA is hydrophobic and insoluble resulting in the formation of large (80–100 nm) NPs around the centrally located PAMAM drug conjugate. However, at acidic pH (~6.5), PAEMA is protonated making it hydrophilic and triggering disintegration of the large NPs to around ~10 nm size with the PAEMA on the outside and an individual PAMAM drug conjugate on the inside of the NP. At low pH, these pH-sensitive NPs demonstrated increased diffusion of NPs and a more uniform distribution of therapeutic payload through tumor spheroids compared to the pH insensitive NPs. Under *in vitro* conditions, the pH-sensitive size switching of NPs resulted in better tumor accumulation and distribution of drug through tumor spheroids compared to pH-insensitive controls. *In vivo*, this delivery system inhibited tumor growth to a greater extent than control NPs (Li et al. 2016b).

Zwitterionic mesoporous silica (Z-MSN) is yet another biomaterial that can be functionalized in a pH-sensitive manner for the delivery of drug to an acidic environment. These NPs work in much the same way as those discussed previously. Upon entering an acidic environment, the surface of these NPs becomes positively charged allowing better tumor retention and endocytosis of the NPs. In addition, a pH-sensitive gatekeeper is added to these NPs that prevents drug diffusion out of the Z-MSN at physiologic pH (Khatoon et al. 2016). Maximal drug release was reported at a pH of 5.0 indicating that drug release occurs in the endosome. Z-MSN has been used to deliver doxorubicin (DOX) to the TME, and *in vitro* cell killing by the DOX loaded NPs was entirely pH-dependent (Khatoon et al. 2016). The *in vivo* studies demonstrated that the pH-sensitive zwitterionic NP formulation showed higher inhibition of tumor growth compared to the pH-insensitive formulation, indicating the feasibility of such an approach in clinic (Khatoon et al. 2016). Another study used MSN NPs for pH-sensitive drug delivery system, but rather than using a charge-swapping mechanism, they used a peptide (pH low insertion peptide, pHLIPss) to target the NP to the TME (Zhao et al. 2013). The peptides, upon exposure to low pH, are inserted into lipid bilayers and undergo a conformational change that transports the NP, which is conjugated to the peptide near the C-terminus, into the cell. This bypasses the requirement of endocytic pathways for the delivery of the drug to the cells present in the TME. Once inside the cell, intracellular glutathione is believed to reduce the disulfide bonds linking pHLIPss to the NP allowing diffusion of the drug from the MSN (Zhao et al. 2013). *In vitro* studies demonstrated efficient uptake of these NPs by breast cancer cells grown in cell culture conditions (Zhao et al. 2013).

Overall, the underlying principle of pH-sensitive NPs is that changes of surface charges from neutral or negative to positive modulate the way NP interacts with the cells in the acidic environment as compared to physiologic pH. To this basic concept of a pH-sensitive NP, additional features are added to improve the spatiotemporal activity of NPs including multilayered pH sensitivity, changes in size to allow better tumor penetration, and light sensitivity for diagnostic as well as therapeutic utility. In contrast to most pH-sensitive NPs, mechanisms that do not rely specifically on the change of surface charge for targeted delivery of drug to the acidic TME are being developed. These methods are of particular interest, as they do not require endocytosis by cells for further activation and therapeutic effect. This could potentially reduce the opportunity for malignant cells to develop resistance to these NPs. Finally, while many of these methodologies have been tested in animal models, it is difficult to determine how well will these function in the setting of the much more complex human diseases. It is likely that the complexity of human tumors will necessitate further modification of existing methods to improve the clinical utility of pH-targeted nanomedicines.

6.3.2 Redox-Based Targeting of Nanomedicines to the TME

In addition to low pH, the TME also contains higher concentrations of reactive oxygen species (ROS). While ROS contribute to the progression of malignancy, they represent another feature of the TME that nanomedicine formulations can exploit for tumor-specific delivery of drugs. These NPs are formulated using lipid-modified, disulfide-linked cysteine that alters the physicochemical properties of the NP including the size and redox responsiveness. The disulfide linkages are the redox-sensitive elements of the NPs and upon endocytosis by cells within the TME, intracellular glutathione rapidly reduces the disulfide linkages allowing disintegration of NPs and drug release (Wu et al. 2015). While these particles improved cell killing in reducing conditions *in vitro*, these studies are difficult to generalize to *in vivo* function (Wu et al. 2015). Additionally, there is no real mechanism for targeting these particles specifically to the TME. Thus, the oxidative environment of the TME remains largely unexploited in terms of directing the accumulation and activation of nanomedicines.

6.3.3 Proteinase/Peptidase-Based Targeting of Nanomedicines to the TME

Within the TME, different cell types including cancer cells, fibroblasts, and immune cells contribute enzymes that cleave a variety of extracellular proteins and maintain ECM homeostasis. It is suggested that the activity of these enzymes may be critical for the progression of cancer especially carcinomas of the breast and pancreas, which demonstrate rich desmoplasia and higher rates of ECM turnover. This proteolytic milieu of TME can be utilized to cleave a substrate from the surface of NPs thereby rendering them active. This mechanism, in general, allows for targeting NPs to the TME and can decrease the diameter of the NPs, which aids in the diffusion of the NP through the tumor.

The earliest studies to utilize the activity of matrix-degrading enzymes within the TME for the purpose of targeting a nanomedicine to the TME used gelatin as the core of the NP with a surface modified by the addition of quantum dots (a model of a 10 nm drug nanocarrier) (Wong et al. 2011). The underlying principle of this formulation is the presence of high concentrations of gelatinases/collagenases matrix metalloproteinase (MMP) 2 and 9 in the TME rapidly degrade the NPs to release the smaller quantum dots. Further, the perivascular areas near the invasive front of the tumor are particularly enriched in MMPs thus, upon extravasation, the NP would be rapidly degraded by MMPs releasing the payload for diffusion through the tumor. This study showed that, within 12 hours, the vast majority of the NPs are degraded both under *in vitro* and *in vivo* conditions (Wong et al. 2011). However, the methodologies used prevent the assessment of the overall accumulation of these NPs in the tumor following systemic administration and a practical analysis of their functionality.

Another study formulated a gelatin-based NP for the delivery of doxorubicin (DOX) in complex with DNA to the TME. Here a cationic gelatin NP encapsulated the DOX-DNA complex and was coated with human serum albumin through electrostatic interactions (Zhu et al. 2014). Upon entering the TME, albumin dissociates from the NP leaving the gelatin exposed to the MMPs within the TME. The subsequent degradation of the NP by MMPs liberates the drug. Importantly, the study demonstrated that NP formulation of DOX had significantly higher circulating time, increased tumor concentrations of DOX, and decreased accumulation in non-targeted organs including the heart (a major site of doxorubicin toxicity) (Zhu et al. 2014). This bio-distribution of DOX correlated to a decrease in tumor size and reduced DOX-related toxicity in mice (Zhu et al. 2014). However, this study was poorly controlled as all comparisons were made between the formulated NP and free DOX instead of a more appropriate set of controls including an MMP-insensitive NP with similar physicochemical properties.

Functionalized gold NPs rather than gelatin have also been used to design proteinase-sensitive NPs. In this formulation, DOX was conjugated to an MMP cleavable peptide, which was subsequently added to the surface of the gold NPs through a thiol linkage (Chen et al. 2013). As with many nanomedicines, the surface of these gold NPs was also modified by the addition of poly (ethylene-glycol) to maximize circulating times. The release of the drug from these NPs for therapeutic efficacy is dependent on the activity of MMP-mediated cleavage of the peptide linker or the reduction of the thiol linkage between the gold NP and the peptide linker (Chen et al. 2013). When cancer cells were treated with these NPs *in vitro*, DOX was efficiently released from the surface of the NP, and there was a decrease in the cell viability in a dose-dependent manner. The *in vivo* administration of the gold NP showed that DOX release was dependent on the function of MMP and was inhibited by concomitant administration of matrix metalloproteinase-2 inhibitor (TIMP-2) (Chen et al. 2013). These NPs, when used for the treatment of subcutaneous tumors, failed to improve the therapeutic efficacy over free DOX, but significantly decreased toxicity as indicated by decreased weight loss in mice receiving NP-DOX compared to free DOX (Chen et al. 2013).

While this targeting mechanism appears to be specific for a TME enriched in the proteinases/peptidases, other tissues, especially granulation tissues, express significant amounts of MMPs. Thus, healing tissue could also be targeted by these NPs. Conversely, not all tumors express high levels of the requisite enzymes. Thus, these nanomedicines may be ineffective for those patients with low MMP expression. Further, these approaches are dependent on enzyme activity rather than enzyme expression, and tumors frequently express or overexpress endogenous inhibitors of MMPs such as the TIMP family of proteins. Because of this fact, administration of these nanomedicines may require an assessment of MMP activity in the tumor, which would be difficult in a clinical setting. Overall, these studies show a great deal of promise for increasing the specific delivery of cytotoxic drugs to the TME. However, further characterization is required to determine the clinical feasibility of these approaches.

6.3.4 Molecular Interaction-Based Targeting of Nanomedicines to the TME

Thus far, we have discussed the features of the extracellular milieu of the TME that can be utilized for the formulation of nanomedicines. In this section, we will discuss specific molecular interactions in the TME that similarly aid in the accumulation of nanomedicine in the TME. It is important to note that unlike enzyme-based targeting mechanisms, the expression of the molecules being targeted is the critical factor, which can be assessed easily using standard pathological practices.

Among the first molecules to be used successfully for targeting NPs to the TME was secreted protein acidic and rich in cysteine (SPARC). Nano-albumin bound paclitaxel (Nab-paclitaxel) NPs were originally formulated to circumvent cremophore-based taxane administration (Hawkins et al. 2008). However, upon characterizing the nanomedicine, this formulation was found to interact specifically with SPARC, which is expressed in the TME of breast tumors (Desai et al. 2008).

This nanomedicine outperformed other NP formulations of taxanes in subsets of breast cancer patients expressing significantly higher levels of SPARC (Von Hoff et al. 2011). However, subsequent studies in pancreatic cancer demonstrated a poor correlation of Nab-paclitaxel response with SPARC expression. Nab-paclitaxel is now FDA approved for the first line treatment of metastatic pancreatic cancer and non-small cell lung cancer, as well as second line therapy for metastatic breast cancer following failure of anthracycline-based combination therapy.

In contrast to the previous studies where targeting of NPs to TME was achieved through specific interactions of binding proteins; inhibitors can also be used for molecular interaction-based targeting. One study used lipid-based NPs with their surface decorated with legumain inhibitor. Legumain is an endopepitdase expressed on breast cancer cells under hypoxic conditions and tumor-associated macrophages (Liao et al. 2011). Tumor bearing mice treated with legumain targeted NPs had a modest reduction in tumor burden compared to mice that received the same drug in an un-targeted NP formulation. While this study was largely unsuccessful in its approach to target delivery of NP to the TME, it suggested that synthetic ligands or substrates can be potentially used for creating TME-targeted NPs.

Along similar lines, nucleolin is another protein that is markedly overexpressed in multiple cancer types including breast cancer, gliomas, as well as tumor-associated blood vessels. As the name suggests, this protein is localized to the nucleus, but it functions to transport cargo from the cell surface to the nucleus making it accessible from the extracellular space. Because of its cell surface localization, and expression in cancer cells and tumor-associated vasculature, nucleolin is a candidate for the targeting of NPs to the TME. One study used the F3 peptide, a specific ligand of nucleolin for the delivery of GFP (green fluorescent protein) siRNA. The incubation of NPs with breast cancer cells grown in culture significantly down-regulated the expression of GFP compared to untargeted liposomes (Gomes-da-Silva et al. 2012). Later, the F3 peptide was used to target PEG-PLA NPs loaded with paclitaxel to intracranial glioma tumors in mice, which significantly improved the survival of mice that received the nanoformulation (Hu et al. 2013). Galectin-1 is yet another cell surface protein that can be used to target NP to the tumor. Importantly, expression of Galectin-1 on endothelial cells increased upon exposure to radiation, which was exploited for the administration of Anginex®-NPs decorated with platinum or arsenic in mice bearing orthotopic breast tumors. Anginex® is an anti-angiogenic peptide and a specific ligand of Galectin-1 (Sethi et al. 2015). These tumors were irradiated and treated with NP formulations of arsenic or cisplatin. Mice that received radiation showed a marked increase in intratumoral accumulation of Anginex® NPs. This study demonstrates the potential use of NPs to target or capitalize on adaptive changes in response to radiation or other therapeutic modalities. However, the fact that radiation-induced changes were necessary to induce Galectin-1 mediated targeting indicates that these NPs would be ineffective at targeting systemic disease.

Receptor-ligand interaction can also be used to target NP formulations to the TME. Insulin-like growth factor (IGF-1) and its receptor IGF-1R constitute an important signaling axis in multiple cancers. As a peptide growth factor receptor, IGF-1R is localized at the cell surface and is exposed to the extracellular milieu. A nanoformulation composed of DOX loaded into an iron oxide NP core with recombinant

IGF-1 conjugated to the surface was administered to mice bearing patient-derived pancreatic cancer xenografts. The IGF-1 conjugated NP demonstrated tumor-specific accumulation and conferred enhanced MRI contrast in xenograft tumors (Zhou et al. 2015). While this formulation did not reduce the side effects associated with DOX administration in terms of weight loss, it did reduce the overall tumor burden in these mice as compared to the albumin conjugated NP (Zhou et al. 2015). Although, this drug delivery system was shown to be more effective, one concern regarding this approach is the use of a natural ligand, which has the potential to activate oncogenic signaling pathways. Therefore, a more appealing approach would be targeting NPs to IGF receptor using an inhibitor or selecting a ligand that does not activate pro-tumorigenic pathways in cancer cells.

The clinical use of monoclonal antibodies has allowed yet another means by which nanomedicines can be targeted to specific molecules expressed in the TME. Epidermal growth factor receptor (EGFR) family members have been identified as key signaling receptors in numerous cancers and contribute to the malignant phenotype of cancer cells through a multitude of mechanisms. Underscoring the importance of EGF receptors in human malignancy, a plethora of therapeutics have been designed to inhibit the function of EGFR in cancer cells. Because of its high surface expression on many cancers, EGFR represents an excellent molecule for targeting nanomedicines to the TME. However, use of epidermal growth factor as the targeting moiety would be concerning due to its potential to activate EGFR signaling pathways that would undermine treatment efficacy. To circumvent this issue, several studies have used antibodies or antibody fragments to target chemotherapeutic carrying nanoformulations to the TME. One study used cetuximab (an FDA approved anti-EGFR inhibitory antibody) to target gemcitabine containing gold NPs to EGFR expressing pancreatic cancer cells. NP uptake correlated positively with EGFR expression in pancreatic cancer cell lines. Further, *in vivo* studies showed a significantly higher antitumor response from cetuximab-targeted NPs compared to isotype control antibody NPs. This response correlated with the amount of gold accumulated in the tumor, suggesting the increased effectiveness of the targeted NP compared to the combined effect of cetuximab and gemcitabine (Patra et al. 2008).

Another study took a similar approach for the treatment of HER2-positive breast cancer. HER2 is an orphan receptor (i.e., no known natural ligand) in the EGFR family. Trastuzumab, an FDA-approved anti-HER2 antibody, was conjugated to streptavidin, which was subsequently reacted with a biotinylated polymeric NP with endosomal disrupting properties. As a therapeutic mechanism, the NPs were loaded with siRNA against GAPDH, BCL-XL, and STAT3. Characterization of the anti-HER2 antibody conjugated NPs demonstrated increased cellular uptake compared to NPs conjugated to an antibody against bovine Herpes 1 virus (Palanca-Wessels et al. 2016). Additionally, treatment of cells with the anti-HER2 antibody conjugated NPs downregulated target genes in cell lines compared to control NPs (Palanca-Wessels et al. 2016). Further studies using these NPs in animal models are required to establish their potential clinical utility. Nonetheless this appears to be a particularly promising approach for the delivery of siRNA to the cancer cells.

It is important to point out that the targeting of nanomedicine to TME via specific molecular interactions has advantages and disadvantages in comparison to other methods. For instance, specific molecular interactions are diverse thereby providing flexibility when devising targeting strategies. These strategies can include the use of natural ligands of receptors, binding partners, inhibitors, and monoclonal antibodies. However, there are very few proteins that are expressed exclusively in the tumor tissue. Thus, extra-tumoral expression of the target molecules may contribute to decreased specificity of these delivery systems. At the same time, not all tumors express a given target molecule; therefore, these NPs cannot categorically be used to target a wide range of tumor types and may require gene expression analysis before their administration to ensure adequate expression of the target molecule.

6.3.5 Cell-Based Targeting of Nanomedicines to the TME

During tumor progression, multiple cell types and their precursors are recruited to the tumor bed. This provides an opportunity for targeting of delivery of nanomedicines to the TME through these cells. The general scheme for the delivery of nanomedicines includes targeting and uptake of NPs by the cell type of interest followed by the recruitment of nanomedicine-laden cells to the TME where the nanomedicine is ultimately delivered. Among the various cell types that are targeted to the TME, macrophages have been studied extensively. One study used macrophage-based targeting of NPs to liver metastases in breast cancer. Characteristically, liver metastasis derived from breast cancer respond poorly to the systemic therapy, in part due to poor vascularization that limits delivery of standard NPs or free drug-based therapeutics. To circumvent this, Tenai *et al.* developed a drug delivery NP that was taken up by resident or tumor associated macrophages in liver tissue, which further deliver the drug to cancer cells. Here, mesoporous silica NPs (MSNs), which appear to have an intrinsic affinity for macrophages were loaded with Nab-paclitaxel. Macrophages efficiently uptake the MSNs, and release the Nab-paclitaxel at an acidic pH (4.5) inside the macrophage endocytic compartment (Tanei et al. 2016). Under *in vivo* conditions, these NPs demonstrated increased therapeutic efficacy and reduced tumor burden compared to "free" Nab-paclitaxel (Tanei et al. 2016). To determine the precise role of macrophages in the trafficking of NP-bound drug, another study created a three-dimensional *in vitro* model and confirmed that macrophages actively trafficked Nab-paclitaxel into organoids and metastatic lesions in liver (Leonard et al. 2016). However, these NPs do not appear to be targeted specifically to macrophages residing in the TME, at least theoretically, therefore, they deliver chemotherapeutics to all macrophages in the body.

A unique cell-based drug delivery platform uses autonomous bacterial "nanoswimmers" to deliver drugs to the tumor. In principle, this delivery platform loads viable bacteria with NP encapsulated drug followed by intravenous administration of the nanoswimmers, which sense and move towards nutrients present in the TME (Zoaby et al. 2016). The cancer cells engulfed these nanoswimmers and the DOX is released from the NP thereby killing the bacterium, the engulfing cell, and the surrounding cancer cells. In cell culture, this approach was shown to be functional, however, *in vivo* validation is required to exploit its potential (Zoaby et al. 2016). This innovative method of drug delivery involves active homing rather than the passive EPR effect used by other nanomedicines, but it remains to be seen if this is a pragmatic method of active drug delivery on a nanoscale.

6.3.6 Combinational Approaches for Nanoparticle Delivery to the TME

Thus far, we have discussed the NP formulations that exploit a single attribute of the TME for specific targeting. Many of these nanomedicines demonstrate improved effectiveness and specificity of drug delivery. However, the combination of targeting mechanisms represents a promising modality to further enhance the accumulation and specificity of drug delivery to the TME. While there are numerous combinations of TME features that can be exploited for the delivery of NPs, we will discuss only a few possible combinations that appear to be effective.

One of the most well-researched combinations involve TME pH and molecular targeting, in part due to the flexibility of each of these targeting approaches. One study used an iron oxide NP core with surface modification through the addition of poly (ethyleneimine) bearing a positive charge blocked by pH-sensitive citraconic acid anhydride. The second targeting modality is chlorotoxin tethered to the surface of the NP. Currently, it is unclear as to how chlorotoxin binds to glioma cells with high affinity but reports of chlorotoxin's affinity for glioma cells have been well demonstrated. As a therapeutic model, siRNA for GFP was also added to the surface of the NPs. Under *in vitro* conditions, these NPs showed increased cancer cell targeting compared to acid-insensitive NPs bearing the same siRNAs. However, it is difficult to ascertain the effect of the chlorotoxin targeting mechanism (Mok et al. 2010). A similar study in pancreatic cancer focused on the dual targeting of NPs to pancreatic tumors. This platform used MSN core NP with both chitosan to confer pH sensitivity and urokinase plasminogen activator (UPA) to target the NPs to urokinase plasminogen activator receptor (UPAR) expressing cells (Gurka et al. 2016). The study demonstrated that addition of chitosan and UPA to the surface of the NP increased its accumulation in the tumor compared to NPs without UPA. Additional, control NP bearing chitosan or UPA decoration alone would help delineate the relative contribution of each targeting mechanism.

pH and enzymatic cleavage have also been combined as dual NP targeting modalities. In one study pH sensitivity was conferred by the use of charge switching functional groups on the surface of NPs and an MMP cleavable segment led to the unmasking of these functional groups. This combined approach allows the accumulation of NPs in the TME, higher diffusion through the tumor due to the small size and NP digestion mediated release of the chemotherapeutics. Another unique method of achieving dual sensitivity involves the use of cell-penetrating peptides (CPP) linked by an MMP cleavable linker to an anionic peptide quencher of CPP activity. Upon exposure to the acidic TME, the quenching peptide charge is neutralized allowing conformational change and MMP-mediated cleavage of the linker thereby exposing the active CPP and mediating NP internalization into cancer cells (Huang et al. 2013). Under *in vivo* conditions, the dual-targeted NPs more effectively target intracranial glioma cells in mice compared to control NPs and CPP-DNA conjugates (Huang et al. 2013). Further, this targeting modality increased the delivery of anti-VEGF siRNA to the tumor and increased the survival of mice while decreasing treatment associated weight loss compared to control NPs and CPP-DNA conjugates. However, due to the integrated nature of the targeting modalities, it is difficult to assess the relative contribution of the individual targeting modalities independently.

Combination of molecular targeting with enzymatic cleavage/activation is another effective method to improve tumor-specific administration of drugs. One study used this modality to bypass the blood-brain barrier for the targeting of NPs to the TME of glioma. Angiopep-2, a ligand of the low-density lipoprotein receptor that is expressed abundantly on an inflamed blood-brain barrier, was conjugated to the surface of a lipid NP. In addition to angiopep-2, the surface of this NP was decorated with a PEGylated, MMP-cleavable lipopeptide (Bruun et al. 2015). Upon cleavage of this lipopeptide, PEG is shed along with four glutamic acids leading to a charge switch from negative to positive that mediates an increase in endocytosis of the NP (Bruun et al. 2015). These NPs with the cleavable lipopeptide demonstrated increased cellular uptake and cytotoxicity mediated by the delivery of siRNAs (Bruun et al. 2015). Previous work has demonstrated that angiopep-2 mediates NP trafficking across the blood-brain barrier, but it remains difficult to interpret if the dual targeting mechanism used here is more efficient compared to previously reported NPs that used angiopep-2 as the sole targeting mechanism (Huang et al. 2011).

6.4 NANOMEDICINES IN MULTIMODALITY CANCER THERAPY

Despite the fact that many nanomedicine formulations have been designed to promote their accumulation in the tumor by exploiting features specific to TME, the presence of abundant ECM proteins in solid tumors represents a barrier to the diffusion of nanomedicines through the tumor. This limits the effective delivery of therapeutics. Several studies have noted that NPs administered in tumor-bearing mice accumulate in areas near tumor vasculature leaving large segments of poorly perfused tumor area untreated. To circumvent this problem, several groups have designed multistage NPs that exist as large NPs in circulation to improve their half-life and passive accumulation in the tumor. However, in TME, these NPs change or break down to release smaller nanocarriers capable of diffusion through the tumor ECM (Wong et al. 2011; Li et al. 2016a). Another approach is to combine administration of nanomedicines with other treatment modalities. For example, a combination of nanomedicine with concomitant radiation therapy improves the NP accumulation in, and diffusion through the TME. In terms of NP accumulation, the administration of radiotherapy was shown to induce vascular permeability (Appelbe et al. 2016; Giustini et al. 2012). Additionally, it was shown that irradiated tumors in mice had less interstitial fluid pressure allowing greater diffusion of the NPs through the tumor (Appelbe et al. 2016). The second treatment modality to increase the diffusion of nanomedicine through the desmoplastic tumor is the exogenous administration of PEGylated hyaluronidase. Hyaluronan is a glycosaminoglycan that makes up a large fraction of tumor ECM and has been reported to be a major cause of the elevated interstitial fluid pressure in tumors (Jacobetz et al. 2013). Treatment of tumor-bearing mice with hyaluronidase prior to the administration of NPs increased the uptake of the NP in the tumor by over two-fold and facilitate the diffusion of the NP through the tumor (Gong et al. 2016). These studies suggest that NP administration in combination with other therapeutic modalities potentiates the effects of both treatments. Therefore, understanding of the cooperation between NPs with other therapies will be critical to maximizing the benefits of combination therapies while minimizing the interference of each therapeutic modality.

6.5 THERAPEUTIC TARGETS IN THE TME

The heterotypic interactions within the TME are critical components of several of the hallmarks of cancer as delineated by Hanahan and Weinberg in 2000 and 2011 (Hanahan et al. 2000; Hanahan et al. 2011). Because of the critical nature of the TME in cancer progression, the interactions of cancer cells with the TME and the interactions between components of the TME have emerged as promising therapeutic targets. In this section, we will highlight various TME-associated molecules that have been targeted by nanomedicines for the purpose of undermining the tumor supportive role of the stroma.

6.5.1 Therapeutic Targeting of CAFs through Nanomedicines

Recently, CAFs have received significant attention due to their role in cancer progression. CAFs, because of their symbiotic relationship with cancer cells and their pro-tumorigenic role, can be targeted for therapeutic benefit. Interestingly, administration of naked gold NPs limits the proliferation of CAFs and modulates the secretome of both cancer cells and CAFs in pancreatic cancer including key signaling molecules like TGF-β, FGF, and IL8. Apparently, these responses are mediated by the ER stress response protein, IRE-1 (Saha et al. 2016). Further, the administration of naked gold NPs inhibited the growth of tumors derived from orthotopic implantation of CAFs and cancer cells together. The therapeutic efficacy of these NPs is based on their ability to disrupt the normal function of CAFs including the production secretory signaling molecules that alter the behavior of cancer cells and other cells in the TME.

One of the most important pathways involved in the communication between CAFs and cancer cells is mediated through stromal derived factor-1 (SDF-1, CXCL12)/CXCR4 signaling axis. SDF-1 is largely produced by CAFs; which binds to CXCR4 expressed on cancer cells and activates pro-survival, migratory, and angiogenic pathways in cancer. This signaling pathway appears to be fundamental to the crosstalk between cancer cells and CAFs in multiple malignancies, making it an attractive candidate for therapeutic targeting in combination with other therapies. One study developed a NP that used AMD3100, a CXCR4 inhibitor, to deliver sorafenib, a VEGFR inhibitor (Gao et al. 2015), or VEGF siRNA (Liu et al. 2015) to HCC. Both of these therapeutic strategies were effective in treating HCC *in vivo*. These NPs decreased the tumor vascularization and increased the survival of tumor bearing mice (Gao et al. 2015; Liu et al. 2015).

Another important mediator of tumor-CAF crosstalk is TGF-β. This growth factor can be secreted by a variety of cell types in the TME and elicits EMT in cancer cells, stimulates CAF proliferation and ECM synthesis, and contributes to the immunosuppression by promoting differentiation of T-cells into T-regulatory cells. To target the TGF-β pathway, NPs were designed to carry TGF-β siRNA (Xu et al. 2014). When administered in mice, these NPs effectively downregulated the expression of TGF-β and relieved some of the immune suppression associated with TGF-β overexpression (Xu et al. 2014).

6.5.2 Therapeutic Targeting of Tumor Vasculature through Nanomedicines

Angiogenesis, the formation of new blood vessels in a tumor is critical for its growth. In principle, inhibiting the process of angiogenesis would limit the supply of oxygen and nutrients to tumor cells and reduce the hematogenous spread due to decreased vascular density in the tumor. Many FDA-approved drugs are used as anti-angiogenic agents, but these have limited efficacy in the treatment of cancer. As previously discussed, NPs have been designed to combine VEGF as well as a CXCR4 blockade. The dual blockade prevents cancer cells from circumventing the VEGF blockade through the CXCR4-mediated angiogenic pathways (Gao et al. 2015; Liu et al. 2015). Additionally, the use of NPs has allowed for the successful administration of siRNAs *in vivo*. While siRNA-based therapies have not been approved for clinical use, NPs offer the best currently available approach for administration of these therapies (Liu et al. 2015). The final therapeutic option for targeting angiogenesis through NPs is the delivery of non-targeted cytotoxic agents specifically to tumor-associated endothelial cells. A study in breast cancer showed that targeting NPs loaded with DOX specifically to the tumor endothelium through the use of the nucleolin ligand F3 peptide increased the cytotoxicity of DOX in endothelial cells by 162-fold, which resulted in a significant decrease in tumor growth (Moura et al. 2012).

6.5.3 Therapeutic Modulation of the Tumor Immune Environment through Nanomedicines

The aspect of the TME that has received the most attention in terms of the therapeutic targeting is immune environment. Within the TME several factors conspire to suppress immune surveillance by cytotoxic T-cells. It is this immunosuppression that antibody-based therapies, such as nivolumab, are designed to overcome to boost an antitumor immune response. The strategy of harnessing the power of the host immune system to treat cancer has recently shown great promise. Nanomedicines have been used as a method of administration of immune modulating agents to TME to change the immunosuppressive milieu. The NP-based targeted delivery of a broad spectrum anti-inflammatory triterpenoid (CDDO-Me) to the TME in melanoma decrease the number of MDSCs and T-regulatory cells present in the TME and enhanced the efficacy of vaccine targeted to a tumor-associated antigen (Zhao et al. 2015). Similarly, NP-mediated targeted delivery of Signal Activator and Transduce-3 (STAT-3) inhibitors have been tested in melanoma and breast cancer, which decreased the expression of immunosuppressive cytokines and increased the infiltration of immune cells with anti-tumor activity (Liao et al. 2011; Huo et al. 2017).

Adenosine is another immunosuppressive factor and its high intratumoral concentration acts pleiotropically to enhance the immunosuppressive TME through activation of adenosine receptors (Young et al. 2014). Importantly, adenosine must be transported into the extracellular space before it can exert immunosuppressive effects. Adenosine transporters CD39 and CD73 can be targeted to prevent the export of adenosine and hence immunosuppression. A study used NPs to deliver siRNA for CD73 in mice bearing 4T1 cell-derived breast tumors. The administration of these NPs suppressed the expression of CD73. In theory, these studies represent a modality that can be used in combination with cancer vaccines or other immune-stimulating agents to augment antitumor immune response (Jadidi-Niaragh et al. 2016). As discussed previously, certain subsets of macrophages have been implicated in the immunosuppressive microenvironment. Cells within the TME secrete factors that recruit immunosuppressive macrophage populations to the TME, and these can be used as drug delivery systems (Tanei et al. 2016; Leonard et al. 2016). In addition, tumor-associated macrophages and the mechanisms that recruit them can also be targeted therapeutically to limit their contribution to the immunosuppressive TME. One approach is to down-regulate the expression of cytokines that mediate macrophage accumulation. In breast cancer, macrophage migration inhibitory factor (MIF) is a major mediator of macrophage recruitment to the tumor (Zhang et al. 2015). To target this signaling pathway, NPs have been formulated to deliver siRNA for MIF, which when administered in mice bearing orthotopic tumors, demonstrate the reduction in CD206 immunoreactivity in tissues, indicating decreased presence of M2 macrophages (Zhang et al. 2015). Mice treated with these NPs had decreased tumor growth and metastasis to the contralateral mammary tissue (Zhang et al. 2015).

6.6 CONCLUSION

Tumors are composed of a complex mixture of cellular and acellular components, which give rise to a unique biochemical environment. This environment contributes to the progression and chemotherapy resistance in cancer, but also provides features, which can be utilized for the purpose of both targeted delivery of therapeutics and identification of novel therapeutic targets. The targeted delivery of nanomedicines can be accomplished by exploiting the features of the TME for site-specific accumulation and activation for the release of therapeutics. NPs consisting of charge-switching polymers at their surface are particularly useful for NP homing to the acidic pH of the TME. While those with redox sensitive designs can, in theory, be targeted to the oxidative environment of the extracellular milieu or the reducing environment within cells present in the TME. Further mechanisms include proteinase cleavage-dependent activation of NPs, targeted delivery through specific molecular interactions within the TME, and cell-mediated trafficking of NPs to the TME. Additionally, combinations of targeting mechanisms have the potential to reduce intrinsic disadvantages of individual mechanism of targeted drug delivery. Overall, the use of TME-targeted nanomedicines in clinical practice promises to reduce the systemic toxicity associated with cancer therapies and increase the efficacy of chemotherapies through increasing intratumoral drug concentrations.

In addition to providing targeting modalities, the TME is an important source of emerging therapeutic targets for the treatment of cancer. The fact that the TME plays such a critical role in the progression of cancer supports the idea that the interaction between cancer, ECM, and other cellular stromal compartments can be effectively targeted to slow the rate of cancer progression or cause tumor regression. The interaction of cancer cells with CAFs, endothelial, and immune cells have been tried as potential therapeutic targets in multiple cancers. Nanomedicine presents an opportunity to develop novel treatment modalities for targeting specific tumor-stromal interactions, which otherwise would not be viable treatment options. Notably, nanoformulation allows for the administration of gene-based therapies under *in vivo* conditions. While these treatment methods are still in their infancy, they provide hope that they will one day be useful clinical tools.

Moving forward, nanomedicines face a variety of challenges. In particular, the TME prevents the uniform diffusion of NPs through the tumor. The resultant residual tumors may lead to the recurrence of potentially drug-resistant and aggressive tumors. To face these challenges, several strategies are already being employed. One such strategy involves the use of multistage NPs, which can change size upon entering the TME to allow for diffusion of smaller particles through the entire tumor mass. The other is the combination of nanomedicine with additional treatment modalities, such as radiation or heparanase that decrease the interstitial pressure within the tumor thereby facilitating entry of NPs deeper into the tumor core. Future work will undoubtedly reveal new capabilities of nanomedicines while at the same time improve the efficacy of already existing drug delivery systems for the treatment of human malignancy.

ACKNOWLEDGMENTS AND GRANT SUPPORT

The authors on this chapter, in parts, were supported by the grants from the Fred W. Upson Grant-in-Aid, NIH (SPORE P50 CA127297, R01 CA183459, R21 AA026428, F30 CA225117, RO1 CA195586 and RO1 GM113166).

REFERENCES

Accolla, R. S., L. Lombardo, R. Abdallah, et al. "Boosting the Mhc Class Ii-Restricted Tumor Antigen Presentation to Cd4+ T Helper Cells: A Critical Issue for Triggering Protective Immunity and Re-Orienting the Tumor Microenvironment toward an Anti-Tumor State." *Front Oncol* 4 (2014): 32.

Anderson, K. G., I. M. Stromnes, and P. D. Greenberg. "Obstacles Posed by the Tumor Microenvironment to T Cell Activity: A Case for Synergistic Therapies." *Cancer Cell* 31, no. 3 (Mar 13 2017): 311–25.

Appelbe, O. K., Q. Zhang, C. A. Pelizzari, et al. "Image-Guided Radiotherapy Targets Macromolecules through Altering the Tumor Microenvironment." *Mol Pharm* 13, no. 10 (Oct 03 2016): 3457–67.

Bruun, J., T. B. Larsen, R. I. Jolck, et al. "Investigation of Enzyme-Sensitive Lipid Nanoparticles for Delivery of SiRNA to Blood-Brain Barrier and Glioma Cells." *Int J Nanomedicine* 10 (2015): 5995–6008.

Chen, W. H., X. D. Xu, H. Z. Jia, et al. "Therapeutic Nanomedicine Based on Dual-Intelligent Functionalized Gold Nanoparticles for Cancer Imaging and Therapy in Vivo." *Biomaterials* 34, no. 34 (Nov 2013): 8798–807.

Costa, A., A. Scholer-Dahirel, and F. Mechta-Grigoriou. "The Role of Reactive Oxygen Species and Metabolism on Cancer Cells and Their Microenvironment." *Semin Cancer Biol* 25 (Apr 2014): 23–32.

Crayton, S. H., and A. Tsourkas. "Ph-Titratable Superparamagnetic Iron Oxide for Improved Nanoparticle Accumulation in Acidic Tumor Microenvironments." *ACS Nano* 5, no. 12 (Dec 27 2011): 9592–601.

Desai, N. P., V. Trieu, L. Y. Hwang, et al. "Improved Effectiveness of Nanoparticle Albumin-Bound (Nab) Paclitaxel Versus Polysorbate-Based Docetaxel in Multiple Xenografts as a Function of Her2 and Sparc Status." *Anticancer Drugs* 19, no. 9 (Oct 2008): 899–909.

Elliott, L. A., G. A. Doherty, K. Sheahan, et al. "Human Tumor-Infiltrating Myeloid Cells: Phenotypic and Functional Diversity." *Front Immunol* 8 (2017): 86.

Ellis, L. M. "Angiogenesis and Its Role in Colorectal Tumor and Metastasis Formation." *Semin Oncol* 31, no. 6 Suppl 17 (Dec 2004): 3–9.

Gao, D. Y., T. Lin Ts, Y. C. Sung, et al. "Cxcr4-Targeted Lipid-Coated Plga Nanoparticles Deliver Sorafenib and Overcome Acquired Drug Resistance in Liver Cancer." *Biomaterials* 67 (Oct 2015): 194–203.

Gilkes, D. M., G. L. Semenza, and D. Wirtz. "Hypoxia and the Extracellular Matrix: Drivers of Tumour Metastasis." *Nat Rev Cancer* 14, no. 6 (Jun 2014): 430–9.

Giustini, A. J., A. A. Petryk, and P. J. Hoopes. "Ionizing Radiation Increases Systemic Nanoparticle Tumor Accumulation." *Nanomedicine* 8, no. 6 (Aug 2012): 818–21.

Gomes, F. G., F. Nedel, A. M. Alves, et al. "Tumor Angiogenesis and Lymphangiogenesis: Tumor/Endothelial Crosstalk and Cellular/Microenvironmental Signaling Mechanisms." *Life Sci* 92, no. 2 (Feb 07 2013): 101–7.

Gomes-da-Silva, L. C., A. O. Santos, L. M. Bimbo, et al. "Toward a SiRNA-Containing Nanoparticle Targeted to Breast Cancer Cells and the Tumor Microenvironment." *Int J Pharm* 434, no. 1–2 (Sep 15 2012): 9–19.

Gong, H., Y. Chao, J. Xiang, et al. "Hyaluronidase to Enhance Nanoparticle-Based Photodynamic Tumor Therapy." *Nano Lett* 16, no. 4 (Apr 13 2016): 2512–21.

Gurka, M. K., D. Pender, P. Chuong, et al. "Identification of Pancreatic Tumors in Vivo with Ligand-Targeted, Ph Responsive Mesoporous Silica Nanoparticles by Multispectral Optoacoustic Tomography." *J Control Release* 231 (Jun 10 2016): 60–7.

Hanahan, D., and R. A. Weinberg. "The Hallmarks of Cancer." *Cell* 100, no. 1 (Jan 07 2000): 57–70.

Hanahan, D., and R. A. Weinberg. "Hallmarks of Cancer: The Next Generation." *Cell* 144, no. 5 (Mar 04 2011): 646–74.

Hawkins, M. J., P. Soon-Shiong, and N. Desai. "Protein Nanoparticles as Drug Carriers in Clinical Medicine." *Adv Drug Deliv Rev* 60, no. 8 (May 22 2008): 876–85.

Hu, Q., G. Gu, Z. Liu, et al. "F3 Peptide-Functionalized Peg-Pla Nanoparticles Co-Administrated with Tlyp-1 Peptide for Anti-Glioma Drug Delivery." *Biomaterials* 34, no. 4 (Jan 2013): 1135–45.

Huang, S., J. Li, L. Han, et al. "Dual Targeting Effect of Angiopep-2-Modified, DNA-Loaded Nanoparticles for Glioma." *Biomaterials* 32, no. 28 (Oct 2011): 6832–8.

Huang, S., K. Shao, Y. Liu, et al. "Tumor-Targeting and Microenvironment-Responsive Smart Nanoparticles for Combination Therapy of Antiangiogenesis and Apoptosis." *ACS Nano* 7, no. 3 (Mar 26 2013): 2860–71.

Huo, M., Y. Zhao, A. B. Satterlee, et al. "Tumor-Targeted Delivery of Sunitinib Base Enhances Vaccine Therapy for Advanced Melanoma by Remodeling the Tumor Microenvironment." *J Control Release* 245 (Jan 10 2017): 81–94.

Jacobetz, M. A., D. S. Chan, A. Neesse, et al. "Hyaluronan Impairs Vascular Function and Drug Delivery in a Mouse Model of Pancreatic Cancer." *Gut* 62, no. 1 (Jan 2013): 112–20.

Jadidi-Niaragh, F., F. Atyabi, A. Rastegari, et al. "Downregulation of Cd73 in 4t1 Breast Cancer Cells through SiRNA-Loaded Chitosan-Lactate Nanoparticles." *Tumour Biol* 37, no. 6 (Jun 2016): 8403–12.

Jain, R. K., and T. Stylianopoulos. "Delivering Nanomedicine to Solid Tumors." *Nat Rev Clin Oncol* 7, no. 11 (Nov 2010): 653–64.

Kalluri, R. "The Biology and Function of Fibroblasts in Cancer." *Nat Rev Cancer* 16, no. 9 (Aug 23 2016): 582–98.

Khatoon, S., H. S. Han, M. Lee, et al. "Zwitterionic Mesoporous Nanoparticles with a Bioresponsive Gatekeeper for Cancer Therapy." *Acta Biomater* 40 (Aug 2016): 282–92.

Leonard, F., L. T. Curtis, P. Yesantharao, et al. "Enhanced Performance of Macrophage-Encapsulated Nanoparticle Albumin-Bound-Paclitaxel in Hypo-Perfused Cancer Lesions." *Nanoscale* 8, no. 25 (Jul 07 2016): 12544–52.

Li, H. J., J. Z. Du, X. J. Du, et al. "Stimuli-Responsive Clustered Nanoparticles for Improved Tumor Penetration and Therapeutic Efficacy." *Proc Natl Acad Sci U S A* 113, no. 15 (Apr 12 2016): 4164–9.

Li, H. J., J. Z. Du, J. Liu, et al. "Smart Superstructures with Ultrahigh Ph-Sensitivity for Targeting Acidic Tumor Microenvironment: Instantaneous Size Switching and Improved Tumor Penetration." *ACS Nano* 10, no. 7 (Jul 26 2016): 6753–61.

Liao, D., Z. Liu, W. J. Wrasidlo, et al. "Targeted Therapeutic Remodeling of the Tumor Microenvironment Improves an Her-2 DNA Vaccine and Prevents Recurrence in a Murine Breast Cancer Model." *Cancer Res* 71, no. 17 (Sep 01 2011): 5688–96.

Ling, D., W. Park, S. J. Park, et al. "Multifunctional Tumor Ph-Sensitive Self-Assembled Nanoparticles for Bimodal Imaging and Treatment of Resistant Heterogeneous Tumors." *J Am Chem Soc* 136, no. 15 (Apr 16 2014): 5647–55.

Liu, J. Y., T. Chiang, C. H. Liu, et al. "Delivery of SiRNA Using Cxcr4-Targeted Nanoparticles Modulates Tumor Microenvironment and Achieves a Potent Antitumor Response in Liver Cancer." *Mol Ther* 23, no. 11 (Nov 2015): 1772–82.

Lundberg, B. B. "Preparation and Characterization of Polymeric Ph-Sensitive Stealth(R) Nanoparticles for Tumor Delivery of a Lipophilic Prodrug of Paclitaxel." *Int J Pharm* 408, no. 1-2 (Apr 15 2011): 208–12.

Michiels, C., C. Tellier, and O. Feron. "Cycling Hypoxia: A Key Feature of the Tumor Microenvironment." *Biochim Biophys Acta* 1866, no. 1 (Aug 2016): 76–86.

Mok, H., O. Veiseh, C. Fang, et al. "Ph-Sensitive SiRNA Nanovector for Targeted Gene Silencing and Cytotoxic Effect in Cancer Cells." *Mol Pharm* 7, no. 6 (Dec 06 2010): 1930–9.

Moura, V., M. Lacerda, P. Figueiredo, et al. "Targeted and Intracellular Triggered Delivery of Therapeutics to Cancer Cells and the Tumor Microenvironment: Impact on the Treatment of Breast Cancer." *Breast Cancer Res Treat* 133, no. 1 (May 2012): 61–73.

Palanca-Wessels, M. C., G. C. Booth, A. J. Convertine, et al. "Antibody Targeting Facilitates Effective Intratumoral SiRNA Nanoparticle Delivery to Her2-Overexpressing Cancer Cells." *Oncotarget* 7, no. 8 (Feb 23 2016): 9561–75.

Parks, S. K., J. Chiche, and J. Pouyssegur. "Ph Control Mechanisms of Tumor Survival and Growth." *J Cell Physiol* 226, no. 2 (Feb 2011): 299–308.

Patra, C. R., R. Bhattacharya, E. Wang, et al. "Targeted Delivery of Gemcitabine to Pancreatic Adenocarcinoma Using Cetuximab as a Targeting Agent." *Cancer Res* 68, no. 6 (Mar 15 2008): 1970–8.

Saha, S., X. Xiong, P. K. Chakraborty, et al. "Gold Nanoparticle Reprograms Pancreatic Tumor Microenvironment and Inhibits Tumor Growth." *ACS Nano* 10, no. 12 (Dec 27 2016): 10636–51.

Sethi, P., A. Jyoti, E. P. Swindell, et al. "3d Tumor Tissue Analogs and Their Orthotopic Implants for Understanding Tumor-Targeting of Microenvironment-Responsive Nanosized Chemotherapy and Radiation." *Nanomedicine* 11, no. 8 (Nov 2015): 2013–23.

Shenoy, D., S. Little, R. Langer, et al. "Poly(Ethylene Oxide)-Modified Poly(Beta-Amino Ester) Nanoparticles as a Ph-Sensitive System for Tumor-Targeted Delivery of Hydrophobic Drugs. 1. In Vitro Evaluations." *Mol Pharm* 2, no. 5 (Sep–Oct 2005): 357–66.

Tanaka, A., and S. Sakaguchi. "Regulatory T Cells in Cancer Immunotherapy." *Cell Res* 27, no. 1 (Jan 2017): 109–18.

Tanei, T., F. Leonard, X. Liu, et al. "Redirecting Transport of Nanoparticle Albumin-Bound Paclitaxel to Macrophages Enhances Therapeutic Efficacy against Liver Metastases." *Cancer Res* 76, no. 2 (Jan 15 2016): 429–39.

Von Hoff, D. D., R. K. Ramanathan, M. J. Borad, et al. "Gemcitabine Plus Nab-Paclitaxel Is an Active Regimen in Patients with Advanced Pancreatic Cancer: A Phase I/Ii Trial." *J Clin Oncol* 29, no. 34 (Dec 01 2011): 4548–54.

Wong, C., T. Stylianopoulos, J. Cui, et al. "Multistage Nanoparticle Delivery System for Deep Penetration into Tumor Tissue." *Proc Natl Acad Sci U S A* 108, no. 6 (Feb 08 2011): 2426–31.

Wu, J., L. Zhao, X. Xu, et al. "Hydrophobic Cysteine Poly(Disulfide)-Based Redox-Hypersensitive Nanoparticle Platform for Cancer Theranostics." *Angew Chem Int Ed Engl* 54, no. 32 (Aug 03 2015): 9218–23.

Xu, Z., Y. Wang, L. Zhang, et al. "Nanoparticle-Delivered Transforming Growth Factor-Beta SiRNA Enhances Vaccination against Advanced Melanoma by Modifying Tumor Microenvironment." *ACS Nano* 8, no. 4 (Apr 22 2014): 3636–45.

Young, A., D. Mittal, J. Stagg, et al. "Targeting Cancer-Derived Adenosine: New Therapeutic Approaches." *Cancer Discov* 4, no. 8 (Aug 2014): 879–88.

Zhang, C., T. An, D. Wang, et al. "Stepwise Ph-Responsive Nanoparticles Containing Charge-Reversible Pullulan-Based Shells and Poly(Beta-Amino Ester)/Poly(Lactic-Co-Glycolic Acid) Cores as Carriers of Anticancer Drugs for Combination Therapy on Hepatocellular Carcinoma." *J Control Release* 226 (Mar 28 2016): 193–204.

Zhang, M., L. Yan, and J. A. Kim. "Modulating Mammary Tumor Growth, Metastasis and Immunosuppression by SiRNA-Induced Mif Reduction in Tumor Microenvironment." *Cancer Gene Ther* 22, no. 10 (Oct 2015): 463–74.

Zhao, Y., M. Huo, Z. Xu, et al. "Nanoparticle Delivery of Cddo-Me Remodels the Tumor Microenvironment and Enhances Vaccine Therapy for Melanoma." *Biomaterials* 68 (Nov 2015): 54–66.

Zhao, Z., H. Meng, N. Wang, et al. "A Controlled-Release Nanocarrier with Extracellular Ph Value Driven Tumor Targeting and Translocation for Drug Delivery." *Angew Chem Int Ed Engl* 52, no. 29 (Jul 15 2013): 7487–91.

Zhou, H., W. Qian, F. M. Uckun, et al. "Igf1 Receptor Targeted Theranostic Nanoparticles for Targeted and Image-Guided Therapy of Pancreatic Cancer." *ACS Nano* 9, no. 8 (Aug 25 2015): 7976–91.

Zhu, Q., L. Jia, Z. Gao, et al. "A Tumor Environment Responsive Doxorubicin-Loaded Nanoparticle for Targeted Cancer Therapy." *Mol Pharm* 11, no. 10 (Oct 06 2014): 3269–78.

Zoaby, N., J. Shainsky-Roitman, S. Badarneh, H., et al. "Autonomous Bacterial Nanoswimmers Target Cancer." *J Control Release* (Oct 12 2016).

7 Novel Nanoparticulate Drug Delivery Systems for Active Ingredients of Traditional Chinese Medicine in Cancer Therapy

Meiwan Chen and Zhaopei Guo

CONTENTS

7.1 INTRODUCTION

Cancer ranks as the second leading cause of death in the world, thus cancer therapy remains one of the research hotspots in medicine (Torre et al. 2015). Until now, surgery, chemotherapy, and radiotherapy have been the three main strategies in cancer treatment. Chemotherapy is efficient in killing cancer cells and prolonging patient's life in most cancer therapies, but it is hindered by side effects and high costs (Feng and Chien 2003). To find new prognosis and develop more effective treatments, traditional Chinese medicine (TCM) as a major strategy of complementary and alternative medicine, has captured the attention of researchers for its multiple biological activities, including antimicrobial infection (Seneviratne, Wong, and Samaranayake 2008, Tan and Vanitha 2004), anti-inflammation (Saw, Saw, and Chew 2013), anti-antioxidant (Chou et al. 2012), hepatoprotective activity (Kumar et al. 2011), and gastrointestinal protective function (Cho et al. 2000). With the development of modern technological approaches to phytochemistry and phytopharmacology, more and more active ingredients of traditional Chinese medicine (AITCM) are employed in cancer therapies (Khushnud and Mousa 2013). Increasing evidence shows that the combination of AITCM and chemotherapy or radiotherapy can enhance the antitumor effect, overcome drug resistance, reduce side effects, alleviate tumor-induced pains, and prolong survival time of advanced-stage cancer patients (Lo et al. 2012, Qi et al. 2010). AITCM has demonstrated the ability to be an adjuvant treatment for cancer. However, the application of AITCM is limited by several disadvantages, such as low bioavailability, poor solubility, poor stability, short biological half-life, ease of metabolism, and rapid elimination (Bonifacio et al. 2014, Liu and Feng 2015). For example, when curcumin was taken orally by patients, even at a high dose of 3.6 g/day the serum concentration could only reach 11.1 nmol/L (Anand et al. 2007), such a high dose of drug will cause damage to normal organs including the liver, heart, and kidneys. Using nanotechnology has proven an effective method in improving delivery of AITCM and has shown encouraging results. Benefiting from the nanoscale structure, nanocarriers could encapsulate hydrophobic AITCM to improve its solubility and stability. Although the enhanced permeability and retention (EPR) effect (Musthaba et al. 2009, Ansari and Islam 2012) in humans is currently under debate, it is likely that the nano-delivery systems can deliver more AITCM to tumor sites for certain late-stage and aggressive cancers where the vasculature is particularly leaky. In addition, nanocarriers with active targeting ability will further promote the accumulation of AITCM in tumor sites, while stimuli-responsive nanocarriers triggered by the tumor environment can quickly release AITCM, thereby improving the therapeutic effects of AITCM. This chapter gives an overview of the benefits of nanoparticles and highlights the application of different polymeric nanocarriers of AITCM including polymeric-based nanoparticles, stimuli-responsive delivery systems, and lipid delivery systems in cancer therapy.

7.2 USING NANOPARTICLES TO IMPROVE AITCM BIOAVAILABILITY

Generally, using nanoparticles to deliver AITCM shows a number of advantages, such as increasing drug solubility, enhancing targeting (including tissue and cell targeting etc.), reducing systemic toxicity (reduce the harm of drugs to normal tissue), and achieving controlled release of the drugs, resulting in improved bioavailability.

7.2.1 Nanoparticles Can Avoid Hydrophobic Drug's Aggregation and Enhance Their Solubility

The majority of AITCM including curcumin, resveratrol, camptothecin, and paclitaxel are hydrophobic compounds and difficult to administer because of their poor aqueous solubility. Given as drug crystals they are not stable in the bloodstream, easily assembling into large particles that induce clearance by the reticuloendothelial system, leading to very low bioavailability. Consequently, large amount of drugs are needed to compensate for the lower bioavailability, thus leading to acute toxicity and low patient compliance (Muqbil et al. 2011). To improve that, the hydrophobic drugs are encapsulated or grafted to polymer nanoparticles. On one hand, the drugs can be uniformly loaded into the hydrophobic core of the nanoparticles to avoid aggregation. On the other hand, the drug-loaded nanoparticles can be effectively dispersed in the aqueous solution, improving their bioavailability and thereby reducing the therapeutic dose to lower toxicity. For example, Ravindran et al. loaded the bioactive phytochemical, *Nigella sativa* thymoquinone in poly(lactide-co-glycolide) (PLGA) to obtain drug-loaded nanoparticles, with the particle sizes distributed from 150 and 200 nm, and drug encapsulation efficiency about 94%. The drug loaded nanoparticles showed a highly efficient anticancer effect against HCT116 cells (Ravindran et al. 2010). Curcumin is another hydrophobic drug with its utility in the clinic limited by poor bioavailability (Gupta, Patchva, and Aggarwal 2013). Takahashi et al. constructed drug-loaded nanoparticles by encapsulating curcumin into liposomes, and using it in Sprague Dawley mouse, achieving nearly five times the plasma curcumin levels than using only free curcumin (Takahashi et al. 2009). Similar improvement in bioavailability has also been obtained using other liposome nanoparticles and polymer nanoparticles (Bisht et al. 2010, Chun et al. 2012). In addition, we conjugated curcumin to pluronic F68 by pH-sensitive linker, which would self-assemble into nanoparticles about 100 nm in size. These nanoparticles showed pH-dependent drug release and improved anticancer effect over free curcumin (Fang et al. 2016). In other research, a novel method was developed to conjugate CPT to polymer in two steps: firstly, the cyclic monomer γ-camptothecin-glutamate N-car-boxyanhydride (Glu (CPT)-NCA) was synthesized, then ring opening polymerization (ROP) of the Glu (CPT)-NCA to obtain CPT-conjugated polymer for cancer therapy (Tai et al. 2014).

7.2.2 Nanoparticles Can Help Accumulation and Penetration in Tumor Tissue

For intravenous administration, nanoparticles must navigate multiple barriers to reach the target site, the first being the reticuloendothelial system (RES), a system of macrophages in liver, spleen, and bone marrow. During the blood circulation, particles below 50 nm and above 300 nm are easily taken up by the liver, and particles larger than 400 nm are cleared by the spleen. Therefore, particle sizes near

100 nm are regarded optimal for tumor accumulation by the enhanced permeability and retention (EPR) effect, and simultaneously, with the minimum clearance (Li and Huang 2008). Penetration of tumor stroma to reach the cancer cells plays an important role in improving cancer therapy efficacy. While nanoparticles loaded with AITCM may accumulate in the tumor tissue, the tumor microenvironment, such as abnormal and heterogeneous vasculature, stromal density, interstitial fluid pressure, and tumor associated macrophages (Ernsting et al. 2013), would severely hinder their deep penetration. To overcome these limitations, the properties of nanoparticles with suitable size, shape, surface charge, or target ligand should be properly designed to achieve deep penetration into tumor (Ernsting et al. 2013, Jain 2013).

In a recent work by J. Wang and co-workers, the correlations between the surface charge of nanoparticles with their pharmacokinetics, tumor accumulation, penetration and ultimate therapeutic effect, by constructing positively, neutral, or negatively charged docetaxel-loaded polyethylene glycol-b-poly lactide (PEG-b-PLA) nanoparticles with size about 100 nm were systematically studied. The result showed that although positively charged nanoparticles were slightly inferior in blood circulation time and tumor accumulation, their tumor growth inhibition ratio (cationic nanoparticles effectively suppressed tumor growth, with the inhibition ratio of 90%, compared to 60% of anion or neutral counterparts), tumor penetration capacity, and cellular uptake efficiency (2.5-fold higher than anion or neutral counterparts) were better than the anion or neutral counterparts (Wang, Zuo et al. 2016). In another study, H.L. Gao etc. designed size-shrinkable nanoparticles, which were constructed with Gelatin-AuNPs-DOX-PEG. The nanocarriers could be degraded by the matrix metalloprotease-2 (MMP-2) protein with shrinkage from 186.5 nm to 59.3 nm, which would help in penetrating deep into the tumor. Gelatin-AuNPs-DOX-PEG showed high intensity distribution in 4T1 and B16F10 tumor, resulting in a better antitumor effect (Ruan et al. 2015). The size-shrinkable nanoparticles could be regarded as the platform for delivering other anticancer drugs, such as camptothecin, paclitaxel, and curcumin.

7.2.3 Ligand-Decorated Nanoparticles Can Help Tumor Accumulation

Targeting can help AITCM loaded in nanoparticles reach specific organs and tissues more efficiently. Targeted drug delivery systems are usually divided into passive targeting and active targeting. Passive targeting is caused by the EPR effect, which is mainly associated with the particle intrinsic properties, such as size and surface properties (Kanapathipillai, Brock, and Ingber 2014, Panyam and Labhasetwar 2003). If nanoparticles can achieve long circulation, the EPR effect will be significant, and the drug-loaded carrier can accumulate in tumor up to 50-fold higher than in normal tissue (Iyer et al. 2006). Three properties would be desirable for carriers loaded with AITCM to achieve favorable tumor accumulation: (1) a particle size between 80 to 100 nm; (2) a particle surface charge that is neutral or negative; and (3) a particle design that can avoid recognition by RES (Gullotti and Yeo 2009). For instance, K.H. Min etc., using hydrophobically modified glycol chitosan nanoparticles to encapsulate camptothecin (CPT) and treat MDA-MB231 human breast cancer xenografts subcutaneously implanted in nude mice, found that the nanocarrier could both prolong blood circulation and achieve high accumulation in tumor, resulting in significant antitumor effect (Min et al. 2008). For better therapeutic efficacy, active targeting delivery systems constructed by conjugating peptides, antibodies, proteins or small molecules to the carrier have in general shown some positive effect. Among these, the peptide arginine-glycine-aspartate (RGD) is a commonly used ligand molecules for targeting, which is targeted to $\alpha_v\beta_3/\alpha_v\beta_5$ integrins. RGD can be applied to bind the integrins, which are usually overexpressed in angiogenic sites and tumors. Z.H. Tang etc. prepared amphiphilic copolymers by grafting α-tocopherol (VE) and polyethylene glycol (PEG) to poly(L-glutamic acid) (PLG), in order to co-deliver docetaxel (DTX) and cisplatin (CDDP) to treat melanoma (B16F1) tumor. Besides the remarkably long circulation, the drug-loaded nanoparticles also achieved excellent antitumor and anti-metastasis efficacy. More importantly, the drug delivery system showed lower side effects, which was attributed to the active targeting guiding the nanoparticles directly to tumor site (Song, Tang et al. 2014).

7.2.4 Nanoparticles Can Overcome Multidrug Resistance

Multidrug resistance (MDR) is a common cause for the failure of malignant tumor therapy, which is influenced by various factors including decreased drug influx, increased drug efflux, DNA repair interference, altered drug metabolism, and defective apoptotic machinery (Baguley 2010). Usually, MDR results from the combination of the aforementioned factors, suggesting that a single inhibitor is insufficient to overcome MDR (Iyer et al. 2013, Patel et al. 2013). Delivery by nanosystems is one promising strategy to overcome MDR due to several cellular or physiological factors. First of all, nanocarriers can carry high concentration of anticancer drugs, MDR inhibitors and genes, or combination drugs simultaneously to tumor sites, thereby increasing drug accumulation in the tumor and overcoming MDR (Iyer et al. 2013, Kunjachan et al. 2013). In addition, nanocarriers are able to enter cells by endocytosis to break through lysosomal trafficking, thus effectively avoiding drug efflux, leading to reversal of MDR (Kunjachan et al. 2013). Vincristine, a water-soluble alkaloid isolated from *Catharanthus roseus*, has been used to treat various tumors such as leukemia and breast cancer. However, toxicity and P-gp-mediated efflux largely limits its therapeutic efficacy and clinical application. With the aid of nanocarriers, Wang et al. constructed a multifunctional drug delivery system, containing targeted, pH-sensitive polymer PLGA-PEG-folate and cell penetrating functional polymer PLGA-PEG-R7. The results showed that using vincristine-Fol/R7 nanoparticles could significantly enhance cellular uptake efficiency, and improved killing effects on MCF-7 and MCF-7/Adr (an MDR variant with P-gp overexpression) cells, in comparison with folate- or R7-modified nanoparticles. Besides, folate-mediated endocytosis and R7-mediated strong intracellular penetration help vincristine-Fol/R7 nanoparticles escape P-glycoprotein-mediated drug efflux and largely accumulate in tumor cells (Wang, Dou et al. 2014). Some of the AITCM, like tetrandrine, berberine, ginsenoside Rb3, and curcumin, also have the capacity to inhibit MDR (Xu, Tian, and Shen 2013). With respect to reversal of MDR, these original inhibitors show advantages in lower side effects, enhanced antitumor effects, improved immune function, inhibited angiogenesis, suppressed metastasis and so on (Saha et al. 2012). Overall, combining AITCM MDR inhibitors with nanocarriers show great potential in cancer treatment for the integrated function of MDR inhibition and tumor therapy (Abouzeid et al. 2014).

7.2.5 Nanoparticles Can Control Drug Release

The designed nanoparticles are expected to control drug release in specific cells to get better efficacy and lower toxicity; the controlled release can be divided into two types. The first type is that the polymer particles can achieve stable and sustained release of drugs. This kind of drug release is affected by a variety of factors, such as particle size, drug loading level, properties of the drug, and microenvironment (Yallapu, Jaggi, and Chauhan 2013). S. Rocha et al. encapsulated green tea polyphenol epigallocatechin-3-gallate (EGCG) in polysaccharide nanoparticles for prostate cancer chemoprevention. It showed a burst release of 46% of EGCG in the first ten minutes, and 100% drug released in three hours (Rocha et al. 2011). A. Kumari and co-workers loaded quercetin to poly-lactide (PLA) nanoparticles, which showed two phases of drug release behavior, a burst release of 40%–45% in 30 minutes followed by a slower release of reaching 87.6% in 96 hours (Kumari et al. 2010). The second type of controlled release is externally triggered or sensitive to the environment. The latter is often termed a "smart" delivery system. External triggers can include light, heat, ultrasound, magnetism, and irradiation. Environmental factors can include pH, temperature, redox condition, and enzyme concentration. This kind of controlled drug release behavior will be discussed in greater details in Section 7.3.2. Different kinds of delivery system for AITCM delivering are listed in Table 7.1.

7.3 VARIOUS POLYMERIC NANOCARRIERS

7.3.1 Polymeric-Based Nanoparticles

Amphiphilic block copolymers composed of hydrophilic and hydrophobic parts are emerging as a class of drug carriers for cancer therapy. The most commonly used hydrophilic chains are polyethylene glycol (PEG), poly (vinyl alcohol) (PVA), or poly (vinyl pyrrolidone) (PVP), and the hydrophobic chains are composed of polycaprolactone (PCL), poly (lactic acid) (PLA), PLGA. Benefits of the amphiphilic property include fabrication into different nanostructures such as liposomes, micelles, and vesicles. Among them, micelles are the most commonly studied drug delivery system, because the unique core-shell structure can self-assemble into small and narrow-sized distribution particles in aqueous media. The hydrophobic inner core is an ideal space for lipophilic drug encapsulation, simultaneously improving drug solubility and protecting drugs against degradation (Wang, Shen et al. 2013, Chen, Lu et al. 2012). The hydrophilic shell can help enhance biocompatibility, stealth properties, and also prolongs blood circulation (Lv et al. 2014, Zhang, Chan, and Leong 2013).

Representative amphiphilic block copolymers used in drug delivery are PEG-PCL, PEG-PLA, and PEG-PLGA, which are widely used as amphiphilic polyether-polyester drug carriers. **(1)** PEG-PCL is prepared by ring-opening polymerization of ε-caprolactone using PEG as the initiator and stannous octoate ($Sn(Oct)_2$) as the

TABLE 7.1
Novel Nano-Delivery Systems Used to Deliver AIT

Active Ingredients	Nanocarrier	Tumor Category	Performance	Ref.
Curcumin	Arabinogalactan	Breast cancer	Inhibits tumor growth via overexpression of p53	(Moghtaderi, Sepehri, and Attari 2017)
Docetaxel	Lipids/PEG-b-PLA	Breast cancer	Enhances tumor penetration	(Wang, Zuo, et al. 2016)
Curcumin	Hyaluronic acid-vitamin E	4T1 tumor	Codelivered with Dox to overcome MDR	(Ma et al. 2017)
Curcumin	Lipids	Glioblastoma	Changes the polarity of tumor-associated microglia and eliminates glioblastoma	(Mukherjee et al. 2016)
Curcumin	Magnetic PLGA nanoparticles	Brain tumor	Combined with paclitaxel enhances treatment efficiency and reduces adverse effects	(Cui et al. 2016)
Paclitaxel	Folic acid-modified bovine serum albumin (FB)-coated lipid nanoparticle	Breast cancer	Lipoprotein-mimicking nanocomplex, enhancing antitumor activity	(Chen et al. 2015a)
Paclitaxel	pH-sensitive lipoprotein-mimic nanocarrier	Breast cancer	pH-sensitive lipoprotein-mimic, enhancing antitumor activity	(Chen et al. 2015b)
Paclitaxel	Polymeric micelles	Breast cancer	Inhibits P-gp activity	(Li et al. 2010)
Camptothecin	Hyaluronic acid	4T1 tumor	Camptothecin-conjugated HA for targeting and reduction-triggered release	(Chen et al. 2016)
Camptothecin	pH-sensitive zwitterionic polymer poly(carboxybetaine)	HeLa tumor	pH and esterase dual sensitive	(Li et al. 2014)
EGCG	Polysaccharide nanoparticles	Prostate cancer	Fast release time	(Rocha et al. 2011)
Quercetin	Lecithin-based cationic nanocarriers	B16F10 melanoma cell tumor	Interactions with nanoparticle and natural products to get higher entrapment efficiency	(Date et al. 2011)
Quercetin	Folate-modified lipid nanocapsules	Potential carrier	Enhances solubility and enhances antitumor activity	(Ding et al. 2014)
Oxymatrine	Polymersomes	Hepatic stellate cell	Superior in anti-fibrosis activity	(Yang et al. 2014)

catalyst (Zhou, Deng, and Yang 2003). PEG-PCL has good biodegradability and low cytotoxicity, which makes it suitable for *in vivo* usage. PEG-PCL-based delivery systems have been widely used in loading and controlled release of AITCM, such as curcumin (Feng et al. 2016, Song, Zhu et al. 2014) and paclitaxel (Loverde, Klein, and Discher 2012, Wang, Tang et al. 2016). Recently, the study of hydrophilic and hydrophobic ratios of PEG-PCL in building different nanostructures has received more attention (Wang, Xu, and Zhang 2009). For example, Zhang et al. studied the hydrophilic/hydrophobic ratios of PEG-PCL micelles (the ratios of PEG/PCL are 2:8, 3:7, 4:6, 5:5), under the condition of keeping their particles sizes, zeta potential, and morphology consistent; the result showed that the highest cellular internalization occurs when the hydrophilic/hydrophobic ratio is 5:5 (Zhang et al. 2013). Li et al. found that the morphology of PEG-PCL can be adjusted to spherical and rod-like shapes by changing the salt concentration (Li, Tang et al. 2016). More interestingly, the rod-like micelle showed higher cell uptake efficiency, higher drug loading efficiency, and faster drug release in acidic pH value compared to the spherical micelle. Rod-like micelles also showed some advantages *in vivo*, such as longer circulation, higher tumor accumulation, and more efficiency in tumor inhibition. **(2)** PEG-PLA is another kind of polyester copolymer used in drug delivery. PLA has three stereochemical forms: they are poly (L-lactide) (PLLA), poly (D-lactide) (PDLA), and their racemic poly (D, L-lactide) (PDLLA) (Ulery, Nair, and Laurencin 2011). PDLA has a faster degradation rate, while PLLA has more mechanical strength. Therefore, the degradation rate and mechanical strength can be adjusted by turning the ratio of PLLA and PDLA to form suitable PDLLA. More importantly, PLA-based formulations have been approved by the Food and Drug Administration (FDA) because of their nontoxic hydrolyzation *in vivo* in an aqueous environment (PLA first hydrolyzes into nontoxic lactic acid through the breakdown of the ester bond, then, is further metabolized into water and carbon dioxide through a citric acid cycle [Biswas et al. 2016]). PLA can be widely used in biomedical and tissue engineering. PEG-PLA is also prepared by ring-opening polymerization. Adjusting the PEG/PLA ratio can also reduce particle size, increase drug loading efficiency, and enhance circulation in blood (Xiao et al. 2010). Micelle-based PEG-PLA can be prepared no larger than 100 nm in diameter (Letchford and Burt 2007), and PTX-loaded mPEG-PLA has been used clinically, named Genexol-PM. **(3)** PEG-PLGA is one of the most popular copolymers, because after hydrolysis of PLGA, two endogenous metabolite monomers, lactic acid and glycolic acid, are released and readily metabolized by the body (Lu et al. 2009) G.X. Zhai et al. prepared the amphiphilic copolymer PLGA-PEG-PLGA to load curcumin. The drug-loaded micelle was smaller in size, about 26.3 nm, with higher entrapment efficiency at nearly 70%. More importantly, the plasma AUC $_{(0-\infty)}$ (area under the drug concentration-time curve values), $t_{1/2}$, $t_{1/2\beta}$ biological half-life and enhanced mean residence time (MRT) of curcumin micelles were largely increase compared to the curcumin solution (Song et al. 2011). Wang et al. used mPEG-PLGA to co-deliver doxorubicin (DOX) and hydrophobic paclitaxel. The drug-loaded nanoparticles (NPs) possessed better polydispersity, and the co-delivery system showed good drug release and cellular uptake properties. Furthermore, the co-delivery nanocarrier suppressed tumor cell growth more efficiently than the delivery of either DOX or paclitaxel alone at the same concentrations

(Wang et al. 2011). Compared to amphiphilic polymer micelles, the other kind of polymeric nanoparticles were prepared only using hydrophobic PLA or PLGA, using the emulsification method to encapsulate drugs into the core formed by polymeric nanoparticles or adsorb drugs onto the particles. The advantages of these nanoparticles are sustained drug release to maintain plasma drug concentrations, enhanced stability in the physiological environment, improved antitumor efficiency, and fewer side effects (Jain, Thanki, and Jain 2013, Desai, Date, and Patravale 2012, Mittal et al. 2007). Using the single emulsion solvent method, Snima et al. loaded silymarin into PLGA to get 60% of encapsulation efficiency. *In vitro* cell tests showed these particles had a preferential toxicity to prostate cancer cells, which indicates potential usage in prostate cancer therapy (Snima et al. 2014). Mukerjee et al. used a solid/oil/water emulsion solvent evaporation method to encapsulate curcumin into PLGA nanospheres for prostate cancer therapy. Its encapsulation efficiency was more than 90%, and their sizes were only 45 nm. An *in vitro* MTT assay showed that the IC 50 of encapsulated curcumin in PLGA was reduced to about 21 μM compared to free curcumin which was about 33 Mm (Mukerjee and Vishwanatha 2009). Xu et al. conjugated RGD to PLA to prepare oridonin-loaded polymeric nanoparticles by a spontaneous emulsification solvent diffusion method. The results showed that the antitumor efficacy was effectively increased by the nanoparticles targeting hepatocarcinoma 22 (H22)-derived tumors (Xu et al. 2012). Therefore, amphiphilic block copolymers are one type of widely used carrier with a hydrophilic shell and hydrophobic core to deliver AITCM, which could allow researchers to develop more intelligent delivery systems.

7.3.2 Stimuli-Responsive Delivery System

A stimuli-responsive delivery system is another important drug delivery system, which was prepared by using various kinds of polymers. It is sensitive to modest changes of the physiological and external stimulus, leading to obvious transitions of corresponding physiochemical properties (Alarcon, Pennadam, and Alexander 2005). Stimuli-responsive delivery systems can be divided into internal and external responses. The internal stimuli are often from biological systems, such as pH, temperature, redox, and enzyme environment. On the other hand, the external stimuli are regarded as light, heat, ultrasound, magnetism, and irradiation. Therefore, considerable efforts have been devoted to design different kinds of polymers to respond to one or more internal or external stimuli, triggering rapidly and controlling release of the loaded drugs. Considering the recent progress made, pH, thermo, and reduction-responsive polymer delivery systems have been widely investigated and utilized as AITCM delivery systems.

7.3.2.1 pH-Responsive Delivery System

pH responsive carriers are widely used to control release drugs at a special position. For tumors, because of their rapid growth, plenty of oxygen is consumed, leading to an acidic extracellular environment (pH 6.5–7.2) (Martin and Jain 1994, Mura, Nicolas, and Couvreur 2013). In addition, early endosomes (pH 5–6) and late lysosomes (pH 4–5) also support a more acidic environment for triggering drug release from pH responsive nanocarriers (Fleige, Quadir, and Haag 2012). The pH-responsive drug delivery system can be designed using two approaches: one is attaching drugs to the polymers through an acid-sensitive linker. In acidic surroundings, the linker will be cleaved and the drug will be rapidly released from the nanocarriers (Pang et al. 2016). The other is that the carrier itself has a pH-responsive property. These nanocarriers can change their hydrophilic-hydrophobic states through protonation-deprotonation in different pH conditions. With the transformation of hydrophilic-hydrophobic states, polymers will self-assemble into micelles to load drugs or break down their structures to rapidly release drugs (Ge and Liu 2013, Kamaly et al. 2016).

Acid-labile chemical bonds are often used to conjugate drug molecules to polymers. There are multiple, alternative pH-sensitive chemical bonds that can be used in the delivery system. Commonly pH-sensitive chemical bonds include acetal, ketal, hydrazone, and imine. These chemical bonds are stable in the physiological environment, while in acidic media they are rapidly degraded or hydrolyzed. Acetal and ketal linkers have a similar structure; they are two, single-bonded oxygen atoms attached to the same carbon atom. In acetal bonds, there is one carbon-bonded group and in ketal two carbon-bonded groups (Knorr et al. 2008, Murthy et al. 2003). A hydrazone bond contains a carbon-nitrogen double bond, formed through the reaction of ketones or aldehydes with hydrazine. The imine bond is similar to hydrazone, which is that the nitrogen is attached to another organic group or a hydrogen atom, rather than nitrogen (Ding et al. 2013, Gurski et al. 2010). Between the two, the hydrazone bond is more widely studied for its acuteness in drug release. Alani et al. prepared poly (ethylene glycol)-block-poly (aspartate-hydrazide) (PEG-P (Asp-Hyd)) as the backbone, then conjugated PTX, which was modified with levulinic acid (LEV), via hydrazone bonds. This polymer prodrug showed faster PTX release at pH 5.0 than at pH 7.4. Cytotoxicity experiment suggested that the polymer prodrug had similar cell growth inhibition in MCF7 and SK-OV-3 cell lines. More importantly, the polymer prodrug was more efficient in prolonging blood circulation, enhancing tumor accumulation, and pH dependent release (Alani et al. 2010). Other acid cleavable linkers, such as acetal, ketal, and imine bonds, are also used to prepared polymer-drug. For acetal and ketal bonds, their first-order hydrolysis rate is ten times faster with each unit pH decrease, but the hydrolysis rate can be controlled by altering their chemical structure. Frechet's group prepared various acetal-based drugs and polymers, and their hydrolysis half-lives exhibited a wide range from less than one minute to several days at pH 5.0; otherwise they are stable at pH 7.4 (Gillies, Goodwin, and Fréchet 2004). Recently, Li et al. prepared acetal-linked prodrug micelles by conjugating curcumin to mPEG-PLA. This prodrug micelle showed pH-dependent drug release behavior, with very low IC50, 51.7 ± 9.0 (μM), compared to their control 103.0 ± 17.8 (μM) in HepG2 cells (Li, Gao, et al. 2016).

Different from the acid-labile chemical bond-conjugated drug delivery system, pH-sensitive polymers are often a class of ionizable polyelectrolytes, for example, containing carboxylic, sulfonic acid, or ammonium salts. With the changes of pH, polymers are ionized and dramatically change their conformation. The conformation change of the polymers associated with the pH transformation is usually present as dissociation, destabilization, and changes of the partition coefficient between the drug and vehicle. The nanoscale inorganic materials such as calcium phosphate

(CaP) and calcium carbonate ($CaCO_3$) are also pH sensitive. They are stable at physiological pH, but in acidic surroundings, such as endo/lysosomes and solid tumors, they dissolve and release loaded drugs (Banerjee, Roy, and Bose 2011). pH-sensitive polymers can be classified into cationic and anionic polymers, depending on their protonated and deprotonated property. Anion polymers contain a carboxylic group. In acidic conditions, they are protonated and become hydrophobic, such as polyglutamic acid, polyacrylic acid, polyaspartic acid, and so on. Compared to anionic polymers, cationic polymers have a positively charged surface, which could be easily taken up by cells. Cationic polymers have an amine group, especially tertiary amines, which play a crucial role in pH-responsive property. The amine group binds protons in an acidic condition, to show positive charge and hydrophilic properties. In a basic condition, it will lose protons, and show hydrophobicity. Poly (2-(diethylamino) ethyl methacrylate) (PDEAEMA) is one of the typical pH-sensitive cationic polymers, with side chains containing an ionizable tertiary amine. Variety of PDEAEMA copolymers have been used for drug loading (Sun, Hong, and Pan 2010). For example, when drugs are loaded into PDEAEMA copolymer-based hydrogels, there is a more rapid release at pH 3–5 than at pH 7.4. The poly (beta-amino ester) (PBAE) copolymer is another pH-sensitive, polymer-based tertiary amine group. PBAE is synthesized by the reaction between bis(secondary amines) or primary amines and bis(acrylate ester) groups. In the physiological environment, PBAE is usually hydrophobic, because of tertiary amine groups that lose protons, into which hydrophobic drugs can be loaded. In the acidic tumor microenvironment, PBAE copolymer-based nanocarriers can dissolve rapidly and release the loaded contents. More importantly, there are plenty of ester bonds in PBAE, which makes PBAE-based copolymers degradable, and this will help to improve cytotoxicity of the polymers. Lots of studies have focused on degradation, non-cytotoxicity, pH sensitivity, high drug delivery efficacy, and the excellent tumor inhibition capacity of PBAE based nano-delivery systems (Fang et al. 2012, Green, Langer, and Anderson 2008, Song et al. 2012, Wang, Wang, and Hsiue 2005).

7.3.2.2 Reduction-Sensitive Delivery System

Reduction-sensitive polymers are usually designed to contain a disulfide bond in backbones or side chains. In the reduction environment, the disulfide bond can be cleaved, and this will lead to the polymer disruption. Glutathione (c-glutamyl-cysteinyl-glycine, GSH), is a reducing tri-peptide, and it is in much higher concentration intracellularly (2–10 mM) than extracellularly (2–20 μM) (Schafer and Buettner 2001). Intracellular GSH concentration is enough to break the disulfide bond, while extracellular GSH concentration has little effect on disulfide bond. For tumors, the GSH level is at least two- to four-fold higher than normal tissues or cells (Saito, Swanson, and Lee 2003, Huo et al. 2014, Manickam et al. 2010). The intracellular GSH can effectively cleave the disulfide bond, which results in carrier disruption and rapid release of loaded drugs in cells. Due to this benefit, many reduction-responsive nanocarriers based on disulfide bond have been designed for tumor therapy (Tian, Bian, and Yang 2016, Zhan et al. 2015, Holm et al. 2015, Liu et al. 2011, Klaikherd, Nagamani, and Thayumanavan 2009). Wang et al. reported a redox-responsive drug delivery system based on amphiphilic PEG-based ether-anhydride copolymer, mPEG-ss-1, 3-bis(carboxyphenoxy) propane-sebacic acid (mPEG-ss-CPP-SA). mPEG-ss-CPP-SA contained disulfide bonds between mPEG and CPP/SA segments, and self-assembled into micelles. The lipophilic curcumin can be effectively loaded into the micelles, and curcumin-loaded, redox-responsive micelles are more effective in inhibiting the growth of tumor cells due to the rapid release of therapeutic agents in the tumor microenvironment in contrast to redox-insensitive micelles (Wang, Yang et al. 2014). Up until now, a variety of disulfide bond-based polymers have been designed as drug carriers such as the reduction-triggered, shell-shedding delivery system, with PEG as the shell and the core made of PCL, PLA, poly (propylene sulfide) (PPS), or poly (γ-benzyl-L-aspartate) (PBLA) (Lale et al. 2015, Yang et al. 2016, Chu et al. 2014, Cerritelli, Velluto, and Hubbell 2007). A disulfide bond can also be used as a reduction-sensitive linker to conjugate anticancer drugs as polymeric prodrugs, such as camptothecin (CPT) (Khan et al. 2014, Cao et al. 2016), paclitaxel (PTX) (Yin, Wu, et al. 2015, Yin, Wang et al. 2015), etc. Liu et al. Designed a polymer-drug system by the ring opening of CPT-based prodrug (Liu et al. 2015). This design overcame poor drug loading and complex steps of conjugating prodrugs to polymer.

7.3.2.3 Other Stimuli-Responsive Delivery Systems

Other stimuli responsive delivery systems including thermo- and light-responsive systems are often used for drug delivery. 1) Poly (N-isopropylacrylamide) (PNIPAM) containing polymers are the most used thermo-responsive nanocarriers. These kinds of nanocarriers usually have a low critical solution temperature (LCST). When the nanocarriers accumulate in tumor at a temperature above LCST, the nanocarriers will be damaged and the loaded drug can be rapidly released (Kono, Yoshino, and Takagishi 2002). N. Sanoj Rejinold et al. reported on the chitosan-*g*-PNIPAM nanocarrier for curcumin delivery. The amount of PNIPAM grafting largely affected the LCST, and the different ratios of chitosan to PNIPAM, such as 1:9, 2:8, 3:7, and 4:6, led to different LCSTs, such as 44°C, 42°C, 40°C, 38°C, respectively. The drug release was severely affected by the LCST; when the temperature was below LCST (38°C), there was only 10% of curcumin released within 36 hours. In comparison, when the temperature was above LCST (38°C), there was 100% curcumin released within the same timeframe (Rejinold et al. 2011). Wang et al. prepared a thermally responsive nanogel based on chitosan-poly (N-isopropylacrylamide-*co*-acrylamide) CTS-poly (NIPAAm-*co*-AAm) for paclitaxel delivery. By adjusting the content of AAm, the critical aggregation concentration, such as CTS-poly (NIPAAm-*co*-AAm5.5), of the nanogels was reduced to 1.11 μg/mL, much smaller than CTS-poly (NIPAAm) nanogels (5.00 μg/mL), and the loading efficiency of PTX in CTS-poly (NIPAAm-*co*-AAm) was about $9.06 \pm 0.195\%$. The drug release was also affected by the temperature. Furthermore, the half maximal inhibitory concentration of PTX-loaded nanogels on SMMC 7721 cells was about 2.025 nmol/L, which was a ten-fold improvement relative to free PTX solutions. *In vivo*, PTX-loaded nanogels presented remarkably higher antitumor efficacy against human colon carcinoma cells HT-29 in a xenograft nude mice model after intravenous administration (Wang, Xu et al. 2014). 2) The light-responsive process usually involves chemical bonds cleavage, isomers interconversion, and chemical reactions rearrangement (Kamaly et al. 2016). Light-sensitive molecules, like spiropyran, azobenzene, and salicylideneaniline are common used to tune the polymer light responsive properties

(Dai, Ravi, and Tam 2009). For example, to design polymers contains hydrophobic pyrenyl-methyl esters in the side chains, and the side chains (pyrenyl-methyl) are easily cleaved from the hydrophilic backbone when irradiated by UV, leading to self-assembling aggregates dissociation (Dorresteijn et al. 2014). But UV is easily absorbed by the skin, which will limit their application. Recent studies showed that light-responsive groups can also be cleaved by NIR light (Yan et al. 2011). Light-triggered processes are clean and convenient, which has potential in drug delivery usage (Zhou, Wang, and Chang 2016, Liu et al. 2016). S.O. Poelma et al. developed a new classes of photochromic materials donor-acceptor Stenhouse adducts (DASA), that undergo a hydrophobic-to-hydrophilic polarity change triggered by visible light between 530 and 570 nm. By conjugation of an N, N-di-n-octyl-substituted DASA derivative with poly (ethylene glycol) (PEG, Mn 3kDa, PDI 1.1) through copper-mediated azide alkyne cycloaddition (CuAAC), the reaction leads to a visible light-responsive amphiphilic system. The DASA system demonstrated controlled delivery of small molecules, such as the chemotherapeutic agent (paclitaxel), to human breast cancer cells, triggered by micellar switching with low intensity and visible light (that the photoisomerization of DASA systems triggered by the very low light intensities is only ~1 $mWcm^{-2}$, and can effectively induce hydrophobic-to-hydrophilic polarity change) (Poelma et al. 2016).

7.4 LIPID-BASED DELIVERY SYSTEM

A lipid-based delivery system is used to enhance drug bioavailability and mainly includes liposomes and solid lipid nanoparticles (SLNs). On one hand, liposomes are composed of lipid bilayers and showed high biocompatibility. Liposomes can improve solubility and stability of hydrophobic anticancer drugs, such as curcumin, glycyrrhetinic acid, vinorelbine, and resveratrol (Li et al. 2011, Cadena et al. 2013, Guo et al. 2012), by encapsulating them into the lipid bilayer. In particular, liposomes have demonstrated advantages of improved pharmacokinetics, biodistribution, and decreased toxicity. For example, brucine can effectively cause tumor cell apoptosis, but high doses of brucine will cause severe central nervous system toxicity. When brucine was loaded into liposomes, antitumor activity was enhanced and side effects were largely decreased (Li et al. 2013, Chen, Yan, et al. 2012). In addition, P. Jourghanian et al. prepared curcumin loaded SLNs, which had 112 and 163 nm particle sizes before and after freeze drying, respectively. The prepared SLNs had more than 70% drug loading efficiency, and showed sustained curcumin release and within 48 hours, 90% of loaded curcumin was released (Jourghanian et al. 2016). J.B. Sun et al. reported on curcumin-loaded SLNs (C-SLNs), which were prepared by high-pressure homogenization with liquid lipid Sefsol-218®. Their mean particle size was about 150 nm, with 90% entrapment efficiency. C-SLNs exhibited prolonged inhibitory activity in cancer cells, as well as time-dependent increases in intracellular uptake. After intravenous administration to rats, the bioavailability of curcumin was increased by 1.25-fold (Sun et al. 2013).

On the other hand, various newly developed SLNs have been used as immune, thermo-, and pH-responsive antitumor therapy (Catania et al. 2013, Chen et al. 2013, Zhou et al. 2012). SLNs are lipids containing a colloidal carrier system, with higher physicochemical stability, and remain solid at room and body temperature. SLNs can support better protection for labile drugs. They can be prepared from

50–1000 nm in size and their production can be easily scaled larger (Pardeike, Hommoss, and Muller 2009, dos Santos et al. 2013). SLNs also have the advantages of enhancing drug solubility, improving bioavailability, controlling drug release, and reducing toxicity (Neves et al. 2013, Aboutaleb et al. 2014). In order to get better control of the drug release profile and better stability, researchers developed a new kind of nanostructured lipid carrier (NLC) from SLNs, which is regarded as the second generation of lipid nanoparticle. NLCs are a mixture of solid lipids and liquid lipids, which makes them more irregular in the matrix and leads to enhanced drug loading capacity (Das, Ng, and Tan 2012). AITCM including Docetaxel (Wang, Xie, et al. 2016), Berberine (Wang, Li, et al. 2014), oridonin (Wang, Wang, et al. 2014), and aloe-emodin (Chen, Wang, et al. 2015) have been loaded into SLNs for the treatment of a variety of cancers, such as lung cancer, hepatic cancer, and breast cancer.

7.5 OTHER DELIVERY SYSTEMS

Besides those discussed above, there are many other carriers, each with unique characteristics for drug loading and delivery. These nanocarriers include dendrimers, natural biopolymers, and inorganic nanoparticles.

Dendrimers are "tree like" polymers with a central core and branched units. Compared to other polymers, dendrimers have a unique structure and molecular weight (different generations have different molecular weight) and their tree like structures increase inner space, in which more drugs can be loaded (Oliveira et al. 2010). Dendrimers have been used as carriers for gene and drug delivery, especially DNA, antigens or antibodies, to improve the therapeutic effects in cancer treatment. Many kinds of dendrimers have been used as drug delivery systems including poly (amidoamine) (PAMAM), glycodendrimers, and polypropyleneimine (PPI) for AITCMs. Previous studies have exhibited that podophyllotoxin, puerarin, curcumin, and resveratrol can be loaded in dendrimers in cancer therapy. For example, the solubility of curcumin loaded in PAMAM dendrimers was increased more than 190 times, and the inhibitory effect on tumor cells was improved as seen in an *in vitro* cell assay (Wang, Xu et al. 2013). Now, a variety of novel dendrimers with special modification have been investigated to enhance bioavailability and therapeutic efficacy, and to decrease side effects; all signs of potential in biomedical application. For instance, P. Kesharwani et al. prepared a target dendritic nanocarrier by conjugate hyaluronic acid (HA) to 4.0G PAMAM, which was loaded with 3, 4-difluorobenzylidene curcumin (CDF) (HA-PAMAM-CDF). A HA-PAMAM-CDF nano-formulation with 9.3 nm particle size resulted in 1.71-fold increase in the IC_{50} value compared to non-targeted formulation (PAMAM-CDF) in pancreatic cancer cells, which showed prospect in clinical translation (Kesharwani et al. 2015). M. Alibolandi et al. modified AS1411 aptamer to 5.0G PAMAM by PEG linker. The encapsulation efficiency of camptothecin-loaded AS1411-targeted PEGylated dendrimers was more than 93% and the cellular uptake was largely enhanced. More importantly, camptothecin-loaded AS1411-targeted PEGylated dendrimers demonstrated site-specific abilities to effectively inhibit C26 tumor growth *in vivo* and significantly decrease systemic toxicity (Alibolandi et al. 2017).

Natural biopolymers have unique features compared to synthesized polymers, and also have the potential to be developed as drug delivery systems. They have more biocompatibility, biodegradability, biological recognition, and ease with processing into gels (Shelke et al. 2014). These kinds of biopolymers include hyaluronan, albumin, dextran, cellulose, chitosan, and so on, which have been successfully prepared into nanoparticles, hydrogels, and drug conjugates (Bielska et al. 2013, Vittorio et al. 2014, Manju and Sreenivasan 2011, Bu et al. 2013). For example, chitosan-based biopolymer nanoparticles have been used to encapsulate curcumin to improve their pharmacological function (Akhtar, Rizvi, and Kar 2012). Li et al. used chitosan and alginate derivatives construct a hydrogel system, incorporating nano-curcumin to achieve more stability and controlled release of curcumin (Li et al. 2012).

Inorganic nanocarriers also have unique features as potential drug carriers. Inorganic nanocarriers including mesoporous silica, gold nanoparticles, and magnetic nanoparticles, etc. have been used as multifunction delivery systems. Gold nanoparticles have the feature of surface plasmon resonance and can be used to release drugs in a controlled manner—drugs such as curcumin, silymarin, and ginkgolide A, by photothermal therapy (Pissuwan, Niidome, and Cortie 2011, Kabir et al. 2014, Weakley et al. 2011, Manju and Sreenivasan 2012). Gold nanoparticles embedded into multifunctional liposomes are reported to overcome MDR, enhance antitumor efficacy, and reduce side effects. Magnetic nanoparticles are a kind of Fe_3O_4-based nanoparticles, which have magnetic targeting function. Drug-loaded Fe_3O_4 nanoparticles can target tumors and cause heat-trigged drug release with a magnetic field. Wang et al. loaded gambogic acid into Fe_3O_4 nanoparticles, effectively enhancing solubility and inhibiting Panc-1 pancreatic cell proliferation and migration (Wang et al. 2012). The attractive feature of silica nanoparticles is their mesoporousness, which makes their specific surface area larger, and promises more drug storage. Pore diameter of silica nanoparticles can be easily adjusted, and their surface can be easily functionalized. For example, silybin can be loaded into mesoporous silica nanoparticles with high efficiency as there is nearly 60% drug loading capacity, and simultaneously, achieve sustained drug release (Cao et al. 2012). Different type of polymeric nanocarriers associated above was shown in Figure 7.1.

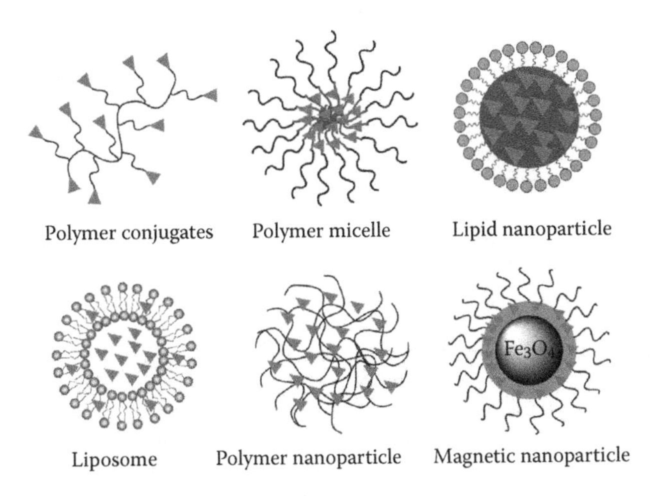

FIGURE 7.1 Various types of polymeric nanocarriers for AITCM delivery.

7.6 CONCLUSION

In summary, novel nanoparticulate drug delivery systems have been extensively utilized to deliver AITCM to improve therapeutic efficacy and reduce the side effects in cancer therapy. Despite much recent progress reported in this field, AITCM-loaded nanocarriers still have several problems to be addressed. First, it remains challenging to design multifunctional AITCM-loaded nanocarriers with superior delivery properties, and simultaneously evaluate their characteristics as well as therapeutic effects. Second, different nanoscale carriers exhibit different properties due to various bulk materials, leading to varied absorption, digestion, metabolism, and excretion in the body. Third, the potential toxicology of nanoparticles is unknown, more attention to the safety of nanoparticles is needed. Considering the fact that Chinese medicine possesses advantages in combination therapy, multiple-unit nanotechnology-based AITCM delivery systems may bring in next revolution in cancer therapy. According to the theory of Chinese medicine, a way to load different ingredients into multiple-unit nanocarriers to combine different type of herbs can further improve drugs' efficacy for clinical application. More in-depth research on novel AITCM delivery systems is important to push forward the future clinical translation.

REFERENCES

Aboutaleb, E., F. Atyabi, M. R. Khoshayand, et al. 2014. "Improved brain delivery of vincristine using dextran sulfate complex solid lipid nanoparticles: optimization and in vivo evaluation." *Journal of Biomedical Materials Research Part A* no. 102 (7):2125–2136. doi: 10.1002/jbm.a.34890.

Abouzeid, A. H., N. R. Patel, C. Sarisozen, and V. P. Torchilin. 2014. "Transferrin-targeted polymeric micelles co-loaded with curcumin and paclitaxel: efficient killing of paclitaxel-resistant cancer cells." *Pharmaceutical Research* no. 31 (8):1938–1945. doi: 10.1007/s11095-013-1295-x.

Akhtar, F., M. M. A. Rizvi, and S. K. Kar. 2012. "Oral delivery of curcumin bound to chitosan nanoparticles cured Plasmodium yoelii infected mice." *Biotechnology Advances* no. 30 (1):310–320.

Alani, A. W. G., Y. Bae, D. A. Rao, and G. S. Kwon. 2010. "Polymeric micelles for the pH-dependent controlled, continuous low dose release of paclitaxel." *Biomaterials* no. 31 (7):1765-1772. doi: 10.1016/j.biomaterials.2009.11.038.

Alarcon, C. D. H., S. Pennadam, and C. Alexander. 2005. "Stimuli responsive polymers for biomedical applications." *Chemical Society Reviews* no. 34 (3):276–285. doi: 10.1039/b406727d.

Alibolandi, M., S. M. Taghdisi, P. Ramezani, et al. 2017. "Smart AS1411-aptamer conjugated PEGylated PAMAM dendrimer for the superior delivery of camptothecin to colon adenocarcinoma in vitro and in vivo." *International Journal of Pharmaceutics* no. 519 (1–2):352–364. doi: 10.1016/j.ijpharm.2017.01.044.

Anand, P., A. B. Kunnumakkara, R. A. Newman, and B. B. Aggrawal. 2007. "Bioavailability of curcumin: problems and promises." *Molecular Pharmaceutics* no. 4 (6):807–818. doi: 10.1021/mp700113r.

Ansari, S. H., and M. F. Islam. 2012. "Influence of nanotechnology on herbal drugs: a review." *Journal of Advanced Pharmaceutical Technology & Research* no. 3 (3):142.

Baguley, B. C. 2010. "Multidrug resistance in cancer." *Multi-Drug Resistance in Cancer* no. 596:1–14. doi: 10.1007/978-1-60761-416-6_1.

Banerjee, S. S., M. Roy, and S. Bose. 2011. "pH tunable fluorescent calcium phosphate nanocomposite for sensing and controlled drug delivery." *Advanced Engineering Materials* no. 13 (1–2):B10–B17. doi: 10.1002/adem.201080036.

Bielska, D., A. Karewicz, K. Kaminski, et al. 2013. "Self-organized thermo-responsive hydroxypropyl cellulose nanoparticles for curcumin delivery." *European Polymer Journal* no. 49 (9):2485–2494. doi: 10.1016/j.eurpolymj.2013.02.012.

Bisht, S., M. Mizuma, G. Feldmann, et al. 2010. "Systemic administration of polymeric nanoparticle-encapsulated curcumin (nanocurc) blocks tumor growth and metastases in preclinical models of pancreatic cancer." *Molecular Cancer Therapeutics* no. 9 (8):2255–2264. doi: 10.1158/1535-7163.MCT-10-0172.

Biswas, S., P. Kumari, P. M. Lakhani, and B. Ghosh. 2016. "Recent advances in polymeric micelles for anti-cancer drug delivery." *European Journal of Pharmaceutical Sciences* no. 83:184–202. doi: 10.1016/j.ejps.2015.12.031.

Bonifacio, B. V., P. B. da Silva, M. A. D. Ramos, et al. 2014. "Nanotechnology-based drug delivery systems and herbal medicines: a review." *International Journal of Nanomedicine* no. 9:1–15. doi: 10.2147/Ijn.S52634.

Bu, L., L.-C. Gan, X.-Q. Guo, et al. 2013. "Trans-resveratrol loaded chitosan nanoparticles modified with biotin and avidin to target hepatic carcinoma." *International Journal of Pharmaceutics* no. 452 (1):355–362.

Cadena, P. G., M. A. Pereira, R. B. S. Cordeiro, et al. 2013. "Nanoencapsulation of quercetin and resveratrol into elastic liposomes." *Biochimica Et Biophysica Acta-Biomembranes* no. 1828 (2):309–316. doi: 10.1016/j.bbamem.2012.10.022.

Cao, D., J. He, J. Xu, et al. 2016. "Polymeric prodrugs conjugated with reduction-sensitive dextran–camptothecin and pH-responsive dextran–doxorubicin: an effective combinatorial drug delivery platform for cancer therapy." *Polymer Chemistry* no. 7 (25):4198–4212.

Cao, X., W.-W. Deng, M. Fu, et al. 2012. "In vitro release and in vitro-in vivo correlation for silybin meglumine incorporated into hollow-type mesoporous silica nanoparticles." *International Journal of Nanomedicine* no. 7:753–762.

Catania, A., E. Barrajon-Catalan, S. Nicolosi, F. Cicirata, and V. Micol. 2013. "Immunoliposome encapsulation increases cytotoxic activity and selectivity of curcumin and resveratrol against HER2 overexpressing human breast cancer cells." *Breast Cancer Research and Treatment* no. 141 (1):55–65. doi: 10.1007/s10549-013-2667-y.

Cerritelli, S., D. Velluto, and J. A. Hubbell. 2007. "PEG-SS-PPS: reduction-sensitive disulfide block copolymer vesicles for intracellular drug delivery." *Biomacromolecules* no. 8 (6):1966–1972.

Chen, C. H., H. Y. Hu, M. X. Qiao, et al. 2015a. "Anti-tumor activity of paclitaxel through dual-targeting lipoprotein-mimicking nanocarrier." *Journal of Drug Targeting* no. 23 (4):311–322. doi: 10.3109/1061186X.2014.994182.

Chen, C. H., H. Y. Hu, M. X. Qiao, et al. 2015b. "Tumor-targeting and pH-sensitive lipoprotein-mimic nanocarrier for targeted intracellular delivery of paclitaxel." *International Journal of Pharmaceutics* no. 480 (1–2):116–127. doi: 10.1016/j.ijpharm.2015.01.036.

Chen, D. Q., H. Y. Yu, H. J. Mu, et al. 2013. "Novel chitosan derivative for temperature and ultrasound dual-sensitive liposomal microbubble gel." *Carbohydrate Polymers* no. 94 (1):17–23. doi: 10.1016/j.carbpol.2012.12.069.

Chen, J., G. J. Yan, R. R. Hu, et al. 2012. "Improved pharmacokinetics and reduced toxicity of brucine after encapsulation into stealth liposomes: role of phosphatidylcholine." *International Journal of Nanomedicine* no. 7:3567–3577. doi: 10.2147/Ijn.S32860.

Chen, R., S. Wang, J. Zhang, M. Chen, and Y. Wang. 2015. "Aloe-emodin loaded solid lipid nanoparticles: formulation design and in vitro anti-cancer study." *Drug Delivery* no. 22 (5):666–674.

Chen, S. F., W. F. Lu, Z. Y. Wen, Q. Li, and J. H. Chen. 2012. "Preparation, characterization and anticancer activity of norcantharidin-loaded poly(ethylene glycol)-poly(caprolactone) amphiphilic block copolymer micelles." *Pharmazie* no. 67 (9):781–788. doi: 10.1691/ph.2012.1151.

Chen, Z. J., N. He, M. H. Chen, L. Zhao, and X. H. Li. 2016. "Tunable conjugation densities of camptothecin on hyaluronic acid for tumor targeting and reduction-triggered release." *Acta Biomaterialia* no. 43:195–207. doi: 10.1016/j.actbio.2016.07.020.

Cho, C.H., Q.B. Mei, P. Shang, et al. 2000. "Study of the gastrointestinal protective effects of polysaccharides from Angelica sinensis in rats." *Planta Medica* no. 66 (04):348–351.

Chou, S.-T., C.-P. Lai, C.-C. Lin, and Y. Shih 2012. "Study of the chemical composition, antioxidant activity and anti-inflammatory activity of essential oil from Vetiveria zizanioides." *Food Chemistry* no. 134 (1):262–268.

Chu, Y., H. Yu, Y. Ma, et al. 2014. "Synthesis and characterization of biodegradable pH and reduction dual-sensitive polymeric micelles for doxorubicin delivery." *Journal of Polymer Science Part A: Polymer Chemistry* no. 52 (13):1771–1780.

Chun, Y. S., S. Bisht, V. Chenna, et al. 2012. "Intraductal administration of a polymeric nanoparticle formulation of curcumin (NanoCurc) significantly attenuates incidence of mammary tumors in a rodent chemical carcinogenesis model: Implications for breast cancer chemoprevention in at-risk populations." *Carcinogenesis* no. 33 (11):2242–2249. doi: 10.1093/carcin/bgs248.

Cui, Y., M. Zhang, F. Zeng, et al. 2016. "Dual-targeting magnetic PLGA nanoparticles for codelivery of paclitaxel and curcumin for brain tumor therapy." *Acs Applied Materials & Interfaces* no. 8 (47):32159–32169. doi: 10.1021/acsami.6b10175.

Dai, S., P. Ravi, and K. C. Tam. 2009. "Thermo-and photo-responsive polymeric systems." *Soft Matter* no. 5 (13):2513–2533.

Das, S., W. K. Ng, and R. B. H. Tan. 2012. "Are nanostructured lipid carriers (NLCs) better than solid lipid nanoparticles (SLNs): development, characterizations and comparative evaluations of clotrimazole-loaded SLNs and NLCs?" *European Journal of Pharmaceutical Sciences* no. 47 (1):139–151. doi: 10.1016/j.ejps.2012.05.010.

Date, A. A., M. S. Nagarsenker, S. Patere, et al. 2011. "Lecithin-based novel cationic nanocarriers (Leciplex) II: improving therapeutic efficacy of quercetin on oral administration." *Molecular Pharmaceutics* no. 8 (3):716–726. doi: 10.1021/mp100305h.

Desai, P. P., A. A. Date, and V. B. Patravale. 2012. "Overcoming poor oral bioavailability using nanoparticle formulations–opportunities and limitations." *Drug Discovery Today: Technologies* no. 9 (2):e87–e95.

Ding, B. Y., P. Chen, Y. Kong, et al. 2014. "Preparation and evaluation of folate-modified lipid nanocapsules for quercetin delivery." *Journal of Drug Targeting* no. 22 (1):67–75. doi: 10.3109/1061186X.2013.839685.

Ding, M. M., N. J. Song, X. L. He, et al. 2013. "Toward the next-generation nanomedicines: design of multifunctional multiblock polyurethanes for effective cancer treatment." *Acs Nano* no. 7 (3):1918–1928. doi: 10.1021/nn4002769.

Dorresteijn, R., N. Billecke, S. H. Parekh, et al. 2014. "Polarity reversal of nanoparticle surfaces by the use of light-sensitive polymeric emulsifiers." *Journal of Polymer Science Part A Polymer Chemistry* no. 53 (2):200–205.

dos Santos, F. K., M. H. Oyafuso, C. P. Kiill, et al. 2013. "Nanotechnology-based drug delivery systems for treatment of hyperproliferative skin diseases - a review." *Current Nanoscience* no. 9 (1):159–167.

Ernsting, M. J., M. Murakami, A. Roy, and S. D. Li. 2013. "Factors controlling the pharmacokinetics, biodistribution and intratumoral penetration of nanoparticles." *Journal of Controlled Release* no. 172 (3):782–794. doi: 10.1016/j.jconrel.2013.09.013.

Fang, C., F. M. Kievit, O. Veiseh, et al. 2012. "Fabrication of magnetic nanoparticles with controllable drug loading and release through a simple assembly approach." *Journal of Controlled Release* no. 162 (1):233–241. doi: 10.1016/j.jconrel.2012.06.028.

Fang, X.-B., J.-M. Zhang, X. Xie, et al. 2016. "pH-sensitive micelles based on acid-labile pluronic F68–curcumin conjugates for improved tumor intracellular drug delivery." *International Journal of Pharmaceutics* no. 502 (1):28–37.

Feng, R., P. Deng, Z. Song, et al. 2016. "Glycyrrhetinic acid-modified PEG-PCL copolymeric micelles for the delivery of curcumin." *Reactive and Functional Polymers.*

Feng, S.-S., and S. Chien. 2003. "Chemotherapeutic engineering: application and further development of chemical engineering principles for chemotherapy of cancer and other diseases." *Chemical Engineering Science* no. 58 (18):4087–4114.

Fleige, E., M. A. Quadir, and R. Haag. 2012. "Stimuli-responsive polymeric nanocarriers for the controlled transport of active compounds: concepts and applications." *Advanced Drug Delivery Reviews* no. 64 (9):866–884.

Ge, Z. S., and S. Y. Liu. 2013. "Functional block copolymer assemblies responsive to tumor and intracellular microenvironments for site-specific drug delivery and enhanced imaging performance." *Chemical Society Reviews* no. 42 (17):7289–7325. doi: 10.1039/c3cs60048c.

Gillies, E. R., A. P. Goodwin, and J. M. J. Fréchet. 2004. "Acetals as pH-sensitive linkages for drug delivery." *Bioconjugate Chemistry* no. 15 (6):1254–1263.

Green, J. J., R. Langer, and D. G. Anderson. 2008. "A combinatorial polymer library approach yields insight into nonviral gene delivery." *Accounts of Chemical Research* no. 41 (6):749–759. doi: Doi 10.1021/Ar7002336.

Gullotti, E., and Y. Yeo. 2009. "Extracellularly activated nanocarriers: a new paradigm of tumor targeted drug delivery." *Molecular Pharmaceutics* no. 6 (4):1041–1051. doi: 10.1021/mp900090z.

Guo, B. H., Y. Cheng, W. Wu, L. P. Lin, and D. H. Lin. 2012. "HPLC assay and pharmacokinetics and tissue distribution study of glycyrrhetinic acid liposomes modified with galactosylated lipid." *Journal of Liposome Research* no. 22 (2):120–127. doi: 10.3109/08982104.2011.627515.

Gupta, S. C., S. Patchva, and B. B. Aggarwal. 2013. "Therapeutic roles of curcumin: lessons learned from clinical trials." *Aaps Journal* no. 15 (1):195–218. doi: 10.1208/s12248-012-9432-8.

Gurski, L. A., A. K. Jha, C. Zhang, X. Q. Jia, and M. C. Farach-Carson. 2010. "Hyaluronic acid-based hydrogels as 3D matrices for in vitro evaluation of chemotherapeutic drugs using poorly adherent prostate cancer cells (vol 30, pg 6076, 2009)." *Biomaterials* no. 31 (14):4248–4248. doi: 10.1016/j.biomaterials.2010.01.141.

Holm, R., K. Klinker, B. Weber, and M. Barz. 2015. "Synthesis of amphiphilic block copolypept (o) ides by bifunctional initiators: making peptomicelles redox sensitive." *Macromolecular rapid communications* no. 36 (23):2083–2091.

Huo, M., J. Yuan, L. Tao, and Y. Wei. 2014. "Redox-responsive polymers for drug delivery: from molecular design to applications." *Polymer Chemistry* no. 5 (5):1519–1528.

Iyer, A. K., G. Khaled, J. Fang, and H. Maeda. 2006. "Exploiting the enhanced permeability and retention effect for tumor targeting." *Drug Discovery Today* no. 11 (17–18):812–818. doi: 10.1016/j.drudis.2006.07.005.

Iyer, A. K., A. Singh, S. Ganta, and M. M. Amiji. 2013. "Role of integrated cancer nanomedicine in overcoming drug resistance." *Advanced Drug Delivery Reviews* no. 65 (13–14):1784–1802. doi: 10.1016/j.addr.2013.07.012.

Jain, A. K., K. Thanki, and S. Jain. 2013. "Co-encapsulation of tamoxifen and quercetin in polymeric nanoparticles: Implications on oral bioavailability, antitumor efficacy, and drug-induced toxicity." *Molecular Pharmaceutics* no. 10 (9):3459–3474. doi: 10.1021/mp400311j.

Jain, R. K. 2013. "Normalizing tumor microenvironment to treat cancer: bench to bedside to biomarkers." *Journal of Clinical Oncology* no. 31 (17):2205–U210. doi: 10.1200/Jco.2012.46.3653.

Jourghanian, P., S. Ghaffari, M. Ardjmand, S. Haghighat, and M. Mohammadnejad. 2016. "Sustained release curcumin loaded solid lipid nanoparticles." *Advanced Pharmaceutical Bulletin* no. 6 (1):17–21. doi: 10.15171/apb.2016.004.

Kabir, N., H. Ali, M. Ateeq, et al. 2014. "Silymarin coated gold nanoparticles ameliorates CCl 4-induced hepatic injury and cirrhosis through down regulation of hepatic stellate cells and attenuation of Kupffer cells." *RSC Advances* no. 4 (18):9012–9020.

Kamaly, N., B. Yameen, J. Wu, and O. C. Farokhzad. 2016. "Degradable controlled-release polymers and polymeric nanoparticles: mechanisms of controlling drug release." *Chemical Reviews* no. 116 (4):2602–2663. doi: 10.1021/acs.chemrev.5b00346.

Kanapathipillai, M., A. Brock, and D. E. Ingber. 2014. "Nanoparticle targeting of anti-cancer drugs that alter intracellular signaling or influence the tumor microenvironment." *Advanced Drug Delivery Reviews* no. 79–80:107–118. doi: 10.1016/j.addr.2014.05.005.

Kesharwani, P., L. X. Xie, S. Banerjee, et al. 2015. "Hyaluronic acid-conjugated polyamidoamine dendrimers for targeted delivery of 3,4-difluorobenzylidene curcumin to CD44 overexpressing pancreatic cancer cells." *Colloids and Surfaces B-Biointerfaces* no. 136:413–423. doi: 10.1016/j.colsurfb.2015.09.043.

Khan, A. R., J. P. Magnusson, S. Watson, et al. 2014. "Camptothecin prodrug block copolymer micelles with high drug loading and target specificity." *Polymer Chemistry* no. 5 (18):5320–5329.

Khushnud, T., and S. A. Mousa. 2013. "Potential role of naturally derived polyphenols and their nanotechnology delivery in cancer." *Molecular Biotechnology* no. 55 (1):78–86. doi: 10.1007/s12033-012-9623-7.

Klaikherd, A., C. Nagamani, and S. Thayumanavan. 2009. "Multi-stimuli sensitive amphiphilic block copolymer assemblies." *Journal of the American Chemical Society* no. 131 (13):4830–4838.

Knorr, V., V. Russ, L. Allmendinger, M. Ogris, and E. Wagner. 2008. "Acetal linked oligoethylenimines for use as pH-sensitive gene carriers." *Bioconjugate Chemistry* no. 19 (8):1625–1634. doi: 10.1021/bc8001858.

Kono, K., K. Yoshino, and T. Takagishi. 2002. "Effect of poly(ethylene glycol) grafts on temperature-sensitivity of thermosensitive polymer-modified liposomes." *Journal of Controlled Release* no. 80 (1–3):321–332. doi: Doi 10.1016/S0168-3659(02)00018-4.

Kumar, C. H., A. Ramesh, J. N. S. Kumar, and B. M. Ishaq. 2011. "A review on hepatoprotective activity of medicinal plants." *International Journal of Pharmaceutical Sciences and Research* no. 2 (3):501.

Kumari, A., S. K. Yadav, Y. B. Pakade, B. Singh, and S. C. Yadav. 2010. "Development of biodegradable nanoparticles for delivery of quercetin." *Colloids and Surfaces B-Biointerfaces* no. 80 (2):184–192. doi: 10.1016/j.colsurfb.2010.06.002.

Kunjachan, S., B. Rychlik, G. Storm, F. Kiessling, and T. Lammers. 2013. "Multidrug resistance: physiological principles and nanomedical solutions." *Advanced Drug Delivery Reviews* no. 65 (13–14):1852–1865. doi: 10.1016/j.addr.2013.09.018.

Lale, S. V., A. Kumar, S. Prasad, A. C. Bharti, and V. Koul. 2015. "Folic acid and trastuzumab functionalized redox responsive polymersomes for intracellular doxorubicin delivery in breast cancer." *Biomacromolecules* no. 16 (6):1736–1752.

Letchford, K., and H. Burt. 2007. "A review of the formation and classification of amphiphilic block copolymer nanoparticulate structures: micelles, nanospheres, nanocapsules and polymersomes." *European Journal of Pharmaceutics and Biopharmaceutics* no. 65 (3):259–269.

Li, C. L., J. X. Cui, C. X. Wang, et al. 2011. "Encapsulation of vinorelbine into cholesterol-polyethylene glycol coated vesicles: drug loading and pharmacokinetic studies." *Journal of Pharmacy and Pharmacology* no. 63 (3):376–384. doi: 10.1111/j.2042-7158.2010.01227.x.

Li, D., Z. Tang, Y. Gao, H. Sun, and S. Zhou. 2016. "A bio-inspired rod-shaped nanoplatform for strongly infecting tumor cells and enhancing the delivery efficiency of anticancer drugs." *Advanced Functional Materials* no. 26 (1):66–79.

Li, J., J. Chen, B. C. Cai, and T. Yang. 2013. "Preparation, characterization and tissue distribution of brucine stealth liposomes with different lipid composition." *Pharmaceutical Development and Technology* no. 18 (4):772–778. doi: 10.3109/10837450.2011.598165.

Li, M., M. Gao, Y. Fu, et al. 2016. "Acetal-linked polymeric prodrug micelles for enhanced curcumin delivery." *Colloids and Surfaces B: Biointerfaces* no. 140:11–18.

Li, S. D., and L. Huang. 2008. "Pharmacokinetics and biodistribution of nanoparticles." *Molecular Pharmaceutics* no. 5 (4):496–504. doi: 10.1021/mp800049w.

Li, X., S. Chen. B. Zhang, et al. 2012. "In situ injectable nano-composite hydrogel composed of curcumin, N,O-carboxymethyl chitosan and oxidized alginate for wound healing application." 2012 *International Journal of Pharmaceutics*. no. 437 (1–2):110–119.

Li, X., P. Li, Y. Zhang, et al. 2010. "Novel mixed polymeric micelles for enhancing delivery of anticancer drug and overcoming multidrug resistance in tumor cell lines simultaneously." *Pharmaceutical Research* no. 27 (8):1498–1511.

Li, Y., R. Y. Liu, J. Yang, et al. 2014. "Dual sensitive and temporally controlled camptothecin prodrug liposomes codelivery of siRNA for high efficiency tumor therapy." *Biomaterials* no. 35 (36):9731–9745. doi: 10.1016/j.biomaterials.2014.08.022.

Liu, B., C. Li, Z. Cheng, et al. 2016. "Functional nanomaterials for near-infrared-triggered cancer therapy." *Biomaterials science* no. 4 (6):890–909.

Liu, J., W. Huang, Y. Pang, et al. 2011. "Molecular self-assembly of a homopolymer: an alternative to fabricate drug-delivery platforms for cancer therapy." *Angewandte Chemie International Edition* no. 50 (39):9162–9166.

Liu, J., W. Liu, I. Weitzhandler, et al. 2015. "Ring-opening polymerization of prodrugs: a versatile approach to prepare well-defined drug-loaded nanoparticles." *Angewandte Chemie International Edition* no. 54 (3):1002–1006.

Liu, Y., and N. P. Feng. 2015. "Nanocarriers for the delivery of active ingredients and fractions extracted from natural products used in traditional Chinese medicine (TCM)." *Advances in Colloid and Interface Science* no. 221:60–76. doi: 10.1016/j.cis.2015.04.006.

Lo, L.-C., C.-Y. Chen, S.-T. Chen, et al. 2012. "Therapeutic efficacy of traditional Chinese medicine, Shen-Mai San, in cancer patients undergoing chemotherapy or radiotherapy: study protocol for a randomized, double-blind, placebo-controlled trial." *Trials* no. 13 (1):232.

Loverde, S. M., M. L. Klein, and D. E. Discher. 2012. "Nanoparticle shape improves delivery: rational coarse grain molecular dynamics (rCG-MD) of taxol in worm-like PEG-PCL micelles." *Advanced Materials* no. 24 (28):3823–3830.

Lu, J. M., X. W. Wang, C. Marin-Muller, et al. 2009. "Current advances in research and clinical applications of PLGA-based nanotechnology." *Expert Review of Molecular Diagnostics* no. 9 (4):325–341. doi: 10.1586/Erm.09.15.

Lv, L., Y. Y. Shen, M. Li, et al. 2014. "Preparation and in vitro evaluation of novel poly(anhydride-ester)-based amphiphilic copolymer curcumin-loaded micelles." *Journal of Biomedical Nanotechnology* no. 10 (2):324–335. doi: 10.1166/jbn.2014.1789.

Ma, W. Z., Q. Guo, Y. Li, et al. 2017. "Co-assembly of doxorubicin and curcumin targeted micelles for synergistic delivery and improving anti-tumor efficacy." *European Journal of Pharmaceutics and Biopharmaceutics* no. 112:209–223. doi: 10.1016/j.ejpb.2016.11.033.

Manickam, D. S., J. Li, D. A. Putt, et al. 2010. "Effect of innate glutathione levels on activity of redox-responsive gene delivery vectors." *Journal of Controlled Release* no. 141 (1):77–84.

Manju, S., and K. Sreenivasan. 2011. "Conjugation of curcumin onto hyaluronic acid enhances its aqueous solubility and stability." *Journal of Colloid and Interface Science* no. 359 (1):318–325.

Manju, S., and K. Sreenivasan. 2012. "Gold nanoparticles generated and stabilized by water soluble curcumin–polymer conjugate: blood compatibility evaluation and targeted drug delivery onto cancer cells." *Journal of Colloid and Interface Science* no. 368 (1):144–151.

Martin, G. R., and R. K. Jain. 1994. "Noninvasive measurement of interstitial ph profiles in normal and neoplastic tissue using fluorescence ratio imaging microscopy." *Cancer Research* no. 54 (21):5670–5674.

Min, K. H., K. Park, Y. S. Kim, et al. 2008. "Hydrophobically modified glycol chitosan nanoparticles-encapsulated camptothecin enhance the drug stability and tumor targeting in cancer therapy." *Journal of Controlled Release* no. 127 (3):208–218. doi: 10.1016/j.jconrel.2008.01.013.

Mittal, G., D. K. Sahana, V. Bhardwaj, M. N. V. R. Kumar. 2007. "Estradiol loaded PLGA nanoparticles for oral administration: effect of polymer molecular weight and copolymer composition on release behavior in vitro and in vivo." *Journal of Controlled Release* no. 119 (1):77–85. doi: 10.1016/j.jconrel.2007.01.016.

Moghtaderi, H., H. Sepehri, and F. Attari. 2017. "Combination of arabinogalactan and curcumin induces apoptosis in breast cancer cells in vitro and inhibits tumor growth via overexpression of p53 level in vivo." *Biomedicine & Pharmacotherapy* no. 88:582–594. doi: 10.1016/j.biopha.2017.01.072.

Mukerjee, A., and J. K. Vishwanatha. 2009. "Formulation, characterization and evaluation of curcumin-loaded PLGA nanospheres for cancer therapy." *Anticancer Research* no. 29 (10):3867–3875.

Mukherjee, S., J. Baidoo, A. Fried, et al. 2016. "Curcumin changes the polarity of tumor-associated microglia and eliminates glioblastoma." *International Journal of Cancer* no. 139 (12):2838–2849. doi: 10.1002/ijc.30398.

Muqbil, I., A. Masood, F. H. Sarkar, et al. 2011. "Progress in nanotechnology based approaches to enhance the potential of chemopreventive agents." *Cancers* no. 3 (1):428–445.

Mura, S., J. Nicolas, and P. Couvreur. 2013. "Stimuli-responsive nanocarriers for drug delivery." *Nature Materials* no. 12 (11):991–1003.

Murthy, N., J. Campbell, N. Fausto, A. S. Hoffman, and P. S. Stayton. 2003. "Design and synthesis of pH-responsive polymeric carriers that target uptake and enhance the intracellular delivery of oligonucleotides." *Journal of Controlled Release* no. 89 (3):365–374. doi: 10.1016/S0168-3659(03)00099-3.

Musthaba, S.M., S. Ahmad, A. Ahuja, J. Ali, and S. Baboota. 2009. "Nano approaches to enhance pharmacokinetic and pharmacodynamic activity of plant origin drugs." *Current Nanoscience* no. 5 (3):344–352.

Neves, A. R., M. Lucio, S. Martins, J. L. C. Lima, and S. Reis. 2013. "Novel resveratrol nanodelivery systems based on lipid nanoparticles to enhance its oral bioavailability." *International Journal of Nanomedicine* no. 8:177–187. doi: 10.2147/Ijn.S37840.

Oliveira, J. M., A. J. Salgado, N. Sousa, J. F. Mano, and R. L. Reis. 2010. "Dendrimers and derivatives as a potential therapeutic tool in regenerative medicine strategies - a review." *Progress in Polymer Science* no. 35 (9):1163–1194. doi: 10.1016/j.progpolymsci.2010.04.006.

Pang, X., Y. Jiang, Q. C. Xiao, et al. 2016. "pH-responsive polymer-drug conjugates: design and progress." *Journal of Controlled Release* no. 222:116–129. doi: 10.1016/j.jconrel.2015.12.024.

Panyam, J., and V. Labhasetwar. 2003. "Biodegradable nanoparticles for drug and gene delivery to cells and tissue." *Advanced Drug Delivery Reviews* no. 55 (3):329–347. doi: 10.1016/S0169-409x(02)00228-4.

Pardeike, J., A. Hommoss, and R. H. Muller. 2009. "Lipid nanoparticles (SLN, NLC) in cosmetic and pharmaceutical dermal products." *International Journal of Pharmaceutics* no. 366 (1–2):170–184. doi: 10.1016/j.ijpharm.2008.10.003.

Patel, N. R., B. S. Pattni, A. H. Abouzeid, and V. P. Torchilin. 2013. "Nanopreparations to overcome multidrug resistance in cancer." *Advanced Drug Delivery Reviews* no. 65 (13-14):1748–1762. doi: 10.1016/j.addr.2013.08.004.

Pissuwan, D., T. Niidome, and M. B. Cortie. 2011. "The forthcoming applications of gold nanoparticles in drug and gene delivery systems." *Journal of Controlled Release* no. 149 (1):65–71.

Poelma, S. O., S. S. Oh, S. Helmy, et al. 2016. "Controlled drug release to cancer cells from modular one-photon visible light-responsive micellar system." *Chemical Communications* no. 52 (69):10525–10528.

Qi, F., A. Li, Y. Inagaki, et al. 2010. "Chinese herbal medicines as adjuvant treatment during chemoor radio-therapy for cancer." *Bioscience Trends* no. 4 (6).

Ravindran, J., H. B. Nair, B. Y. Sung, et al. 2010. "Thymoquinone poly (lactide-co-glycolide) nanoparticles exhibit enhanced anti-proliferative, anti-inflammatory, and chemosensitization potential (Retracted article. See vol. 102, pg. 146, 2016)." *Biochemical Pharmacology* no. 79 (11):1640–1647. doi: 10.1016/j.bcp.2010.01.023.

Rejinold, N. S., P. R. Sreerekha, K. P. Chennazhi, S. V. Nair, and R. Jayakumar. 2011. "Biocompatible, biodegradable and thermo-sensitive chitosan-g-poly (N-isopropylacrylamide) nanocarrier for curcumin drug delivery." *International Journal of Biological Macromolecules* no. 49 (2):161–172. doi: 10.1016/j.ijbiomac.2011.04.008.

Rocha, S., R. Generalov, M. D. Pereira, S. V. Nair, and R. Jayakumar. 2011. "Epigallocatechin gallate-loaded polysaccharide nanoparticles for prostate cancer chemoprevention." *Nanomedicine* no. 6 (1):79–87. doi: 10.2217/Nnm.10.101.

Ruan, S. B., X. Cao, X. L. Cun, et al. 2015. "Matrix metalloproteinase-sensitive size-shrinkable nanoparticles for deep tumor penetration and pH triggered doxorubicin release." *Biomaterials* no. 60:100–110. doi: 10.1016/j.biomaterials.2015.05.006.

Saha, S., A. Adhikary, P. Bhattacharyya, T. Das, and G. Sa. 2012. "Death by design: where curcumin sensitizes drug-resistant tumours." *Anticancer Research* no. 32 (7): 2567–2584.

Saito, G., J. A. Swanson, and K.-D. Lee. 2003. "Drug delivery strategy utilizing conjugation via reversible disulfide linkages: role and site of cellular reducing activities." *Advanced Drug Delivery Reviews* no. 55 (2):199–215.

Saw, C. L.L., C. L. Saw, and L. S. T. Chew. 2013. "Traditional Chinese medicine: anti-inflammation for cancer prevention." In *Traditional Chinese Medicine*, 163–185.

Schafer, F. Q., and G. R. Buettner. 2001. "Redox environment of the cell as viewed through the redox state of the glutathione disulfide/glutathione couple." *Free Radical Biology and Medicine* no. 30 (11):1191–1212. doi: 10.1016/S0891-5849(01)00480-4.

Seneviratne, C., R. Jayampath, R. W. K. Wong, and L. P. Samaranayake. 2008. "Potent anti-microbial activity of traditional Chinese medicine herbs against Candida species." *Mycoses* no. 51 (1):30–34.

Shelke, N. B., R. James, C. T. Laurencin, and S. G. Kumbar. 2014. "Polysaccharide biomaterials for drug delivery and regenerative engineering." *Polymers for Advanced Technologies* no. 25 (5):448–460. doi: 10.1002/pat.3266.

Snima, K. S., P. Arunkumar, R. Jayakumar, and V. -K. Lakshmanan. 2014. "Silymarin encapsulated poly (D, L-lactic-co-glycolic acid) nanoparticles: a prospective candidate for prostate cancer therapy." *Journal of Biomedical Nanotechnology* no. 10 (4):559–570.

Song, W. T., Z. H. Tang, D. W. Zhang, et al. 2014. "Anti-tumor efficacy of c(RGDfK)-decorated polypeptide-based micelles co-loaded with docetaxel and cisplatin." *Biomaterials* no. 35 (9):3005–3014. doi: 10.1016/j.biomaterials.2013.12.018.
Song, W., Z. Tang, M. Li, et al. 2012. "Tunable pH-sensitive poly (β-amino ester) s synthesized from primary amines and diacrylates for intracellular drug delivery." *Macromolecular Bioscience* no. 12 (10):1375–1383.
Song, Z. M., R. L. Feng, M. Sun, et al. 2011. "Curcumin-loaded PLGA-PEG-PLGA triblock copolymeric micelles: preparation, pharmacokinetics and distribution in vivo." *Journal of Colloid and Interface Science* no. 354 (1):116–123. doi: 10.1016/j.jcis.2010.10.024.
Song, Z. M., W. X. Zhu, N. Liu, et al. 2014. "Linolenic acid-modified PEG-PCL micelles for curcumin delivery." *International Journal of Pharmaceutics* no. 471 (1-2):312–321. doi: 10.1016/j.ijpharm.2014.05.059.
Sun, J. B., C. Bi, H. M. Chan, et al. 2013. "Curcumin-loaded solid lipid nanoparticles have prolonged in vitro antitumour activity, cellular uptake and improved in vivo bioavailability." *Colloids and Surfaces B-Biointerfaces* no. 111:367–375. doi: 10.1016/j.colsurfb.2013.06.032.
Sun, J.-T., C.-Y. Hong, and C.-Y. Pan. 2010. "Fabrication of PDEAEMA-coated mesoporous silica nanoparticles and pH-responsive controlled release." *The Journal of Physical Chemistry C* no. 114 (29):12481–12486.
Tai, W., R. Mo, Y. Lu, T. Jiang, and Z. Gu. 2014. "Folding graft copolymer with pendant drug segments for co-delivery of anticancer drugs." *Biomaterials* no. 35 (25):7194–7203.
Takahashi, M., S. Uechi, K. Takara, Y. Asikin, and K. Wada. 2009. "Evaluation of an oral carrier system in rats: bioavailability and antioxidant properties of liposome-encapsulated curcumin." *Journal of Agricultural and Food Chemistry* no. 57 (19):9141–9146. doi: 10.1021/jf9013923.
Tan, B. K. H., and J. Vanitha. 2004. "Immunomodulatory and antimicrobial effects of some traditional Chinese medicinal herbs: a review." *Current Medicinal Chemistry* no. 11 (11):1423–1430.
Tian, Y., S. Bian, and W. Yang. 2016. "A redox-labile poly (oligo (ethylene glycol) methacrylate)-based nanogel with tunable thermosensitivity for drug delivery." *Polymer Chemistry* no. 7 (10):1913–1921.
Torre, L. A., F. Bray, R. L. Siegel, et al. 2015. "Global cancer statistics, 2012." *CA: A Cancer Journal for Clinicians* no. 65 (2):87–108.
Ulery, B. D., L. S. Nair, and C. T. Laurencin. 2011. "Biomedical applications of biodegradable polymers." *Journal of Polymer Science Part B: Polymer Physics* no. 49 (12):832–864.
Vittorio, O., V. Voliani, P. Faraci, et al. 2014. "Magnetic catechin–dextran conjugate as targeted therapeutic for pancreatic tumour cells." *Journal of Drug Targeting* no. 22 (5):408–415.
Wang, B.-L., Y. M. Shen, Q. W. Zhang, et al. 2013. "Codelivery of curcumin and doxorubicin by MPEG-PCL results in improved efficacy of systemically administered chemotherapy in mice with lung cancer." *International Journal of Nanomedicine* no. 8:3521–3531.
Wang, C. H., C. H. Wang, and G. H. Hsiue. 2005. "Polymeric micelles with a pH-responsive structure as intracellular drug carriers." *Journal of Controlled Release* no. 108 (1):140–149. doi: 10.1016/j.jconrel.2005.07.017.
Wang, C., H. Zhang, Y. Chen, F. Shi, and B. Chen. 2012. "Gambogic acid-loaded magnetic Fe." *International Journal of Nanomedicine* no. 7:781–787.
Wang, H., Y. Zhao, Y. Wu, et al. 2011. "Enhanced anti-tumor efficacy by co-delivery of doxorubicin and paclitaxel with amphiphilic methoxy PEG-PLGA copolymer nanoparticles." *Biomaterials* no. 32 (32):8281–8290. doi: 10.1016/j.biomaterials.2011.07.032.
Wang, H. X., Z. Q. Zuo, J. Z. Du, et al. 2016. "Surface charge critically affects tumor penetration and therapeutic efficacy of cancer nanomedicines." *Nano Today* no. 11 (2):133–144. doi: 10.1016/j.nantod.2016.04.008.

Wang, J., G. Yang, X. Guo, et al. 2014. "Redox-responsive polyanhydride micelles for cancer therapy." *Biomaterials* no. 35 (9):3080–3090.

Wang, L., X. P. Xu, Y. Zhang, et al. 2013. "Encapsulation of curcumin within poly (amidoamine) dendrimers for delivery to cancer cells." *Journal of Materials Science-Materials in Medicine* no. 24 (9):2137–2144. doi: 10.1007/s10856-013-4969-3.

Wang, L., H. Li, S. Wang, et al. 2014. "Enhancing the antitumor activity of berberine hydrochloride by solid lipid nanoparticle encapsulation." *AAPS PharmSciTech* no. 15 (4):834–844.

Wang, L., S. Wang, R. Chen, et al. 2014. "Oridonin loaded solid lipid nanoparticles enhanced antitumor activity in MCF-7 cells." *Journal of Nanomaterials* no. 2014.

Wang, L., X. Xie, D. Liu, et al. 2016. "iRGD-mediated reduction-responsive DSPE–PEG/LA–PLGA–TPGS mixed micelles used in the targeted delivery and triggered release of docetaxel in cancer." *RSC Advances* no. 6 (34):28331–28342.

Wang, T., X. Tang, J. Han, et al. 2016. "Biodegradable self-assembled nanoparticles of galactose-containing amphiphilic triblock copolymers for targeted delivery of paclitaxel to HepG2 cells." *Macromolecular Bioscience*.

Wang, Y., L. M. Dou, H. J. He, Y. Zhang, and Q. Shen. 2014. "Multifunctional nanoparticles as nanocarrier for vincristine sulfate delivery to overcome tumor multidrug resistance." *Molecular Pharmaceutics* no. 11 (3):885–894. doi: 10.1021/mp400547u.

Wang, Y. J., H. J. Xu, J. Wang, L. Ge, and J. B. Zhu. 2014. "Development of a thermally responsive nanogel based on chitosan-poly(n-isopropylacrylamide-co-acrylamide) for paclitaxel delivery." *Journal of Pharmaceutical Sciences* no. 103 (7):2012-2021. doi: 10.1002/jps.23995.

Wang, Y., H. Xu, and X. Zhang. 2009. "Tuning the amphiphilicity of building blocks: controlled self-assembly and disassembly for functional supramolecular materials." *Advanced Materials* no. 21 (28):2849–2864. doi: 10.1002/adma.200803276.

Weakley, S. M., X. Wang, H. Mu, et al. 2011. "Ginkgolide A-gold nanoparticles inhibit vascular smooth muscle proliferation and migration in vitro and reduce neointimal hyperplasia in a mouse model." *Journal of Surgical Research* no. 171 (1):31–39.

Xiao, R. Z., Z. W. Zeng, G. L. Zhou, et al. 2010. "Recent advances in PEG-PLA block copolymer nanoparticles." *International Journal of Nanomedicine* no. 5 (1):1057–1065.

Xu, D., W. Tian, and H. Shen. 2013. "P-gp upregulation may be blocked by natural curcuminoids, a novel class of chemoresistance-preventing agent." *Molecular Medicine Reports* no. 7 (1):115-121. doi: 10.3892/mmr.2012.1106.

Xu, J., J. H. Zhao, Y. Liu, N. P. Feng, and Y. T. Zhang. 2012. "RGD-modified poly(D,L-lactic acid) nanoparticles enhance tumor targeting of oridonin." *International Journal of Nanomedicine* no. 7:211–219. doi: 10.2147/Ijn.S27581.

Yallapu, M. M., M. Jaggi, and S. C. Chauhan. 2013. "Curcumin nanomedicine: a road to cancer therapeutics." *Current Pharmaceutical Design* no. 19 (11):1994–2010.

Yan, B., J.-C. Boyer, N. R. Branda, and Y. Zhao. 2011. "Near-infrared light-triggered dissociation of block copolymer micelles using upconverting nanoparticles." *Journal of the American Chemical Society* no. 133 (49):19714–19717.

Yang, J., Y. Hou, G. Ji, et al. 2014. "Targeted delivery of the RGD-labeled biodegradable polymersomes loaded with the hydrophilic drug oxymatrine on cultured hepatic stellate cells and liver fibrosis in rats." *European Journal of Pharmaceutical Sciences* no. 52:180–190.

Yang, Q., C. He, Z. Zhang, et al. 2016. "Redox-responsive flower-like micelles of poly (L-lactic acid)-b-poly (ethylene glycol)-b-poly (L-lactic acid) for intracellular drug delivery." *Polymer* no. 90:351–362.

Yin, T., J. Wang, L. Yin, et al. 2015. "Redox-sensitive hyaluronic acid–paclitaxel conjugate micelles with high physical drug loading for efficient tumor therapy." *Polymer Chemistry* no. 6 (46):8047–8059.

Yin, T., Q. Wu, L. Wang, et al. 2015. "Well-defined redox-sensitive polyethene glycol–paclitaxel prodrug conjugate for tumor-specific delivery of paclitaxel using octreotide for tumor targeting." *Molecular Pharmaceutics* no. 12 (8):3020–3031.

Zhan, Y., M. Gonçalves, P. Yi, et al. 2015. "Thermo/redox/pH-triple sensitive poly (N-isopropylacrylamide-co-acrylic acid) nanogels for anticancer drug delivery." *Journal of Materials Chemistry B* no. 3 (20):4221–4230.

Zhang, Y., H. F. Chan, and K. W. Leong. 2013. "Advanced materials and processing for drug delivery: The past and the future." *Advanced Drug Delivery Reviews* no. 65 (1):104–120. doi: 10.1016/j.addr.2012.10.003.

Zhang, Z., Q. Q. Qu, J. R. Li, and S. B. Zhou. 2013. "The effect of the hydrophilic/hydrophobic ratio of polymeric micelles on their endocytosis pathways into cells." *Macromolecular Bioscience* no. 13 (6):789–798. doi: 10.1002/mabi.201300037.

Zhou, F., H. Wang, and J. Chang. 2016. "Progress in the field of constructing near-infrared light-responsive drug delivery platforms." *Journal of Nanoscience and Nanotechnology* no. 16 (3):2111–2125.

Zhou, S., X. Deng, and H. Yang. 2003. "Biodegradable poly (ε-caprolactone)-poly (ethylene glycol) block copolymers: characterization and their use as drug carriers for a controlled delivery system." *Biomaterials* no. 24 (20):3563–3570.

Zhou, W. T., X. Q. An, J. Z. Wang, et al. 2012. "Characteristics, phase behavior and control release for copolymer-liposome with both pH and temperature sensitivities." *Colloids and Surfaces a-Physicochemical and Engineering Aspects* no. 395:225–232. doi: 10.1016/j.colsurfa.2011.12.034.

Section II

Imaging Technologies in Cancer

8 Noninvasive Imaging in Clinical Oncology

A Testimony of Current Modalities and a Glimpse into the Future

Aniketh Bishnu, Abhilash Deo, Ajit Dhadve, Bhushan Thakur, Souvik Mukherjee, and Pritha Ray

CONTENTS

8.1 INTRODUCTION

"Seeing is believing" has become the mantra of modern biomedical imaging science, which faithfully diagnoses, guides the course of treatment, monitors outcome of various diseases, and plays a critical role in the development of next generation scientific endeavor. Noninvasive imaging modalities, which utilize properties of the complete spectrum of electromagnetic radiation (from high energy gamma radiation to infra-red and sound) have been integrated in preclinical studies and clinical practices for the last three decades. These modalities primarily comprise radionuclide imaging, X-ray/computed tomography (X-ray/CT) imaging, optical and spectroscopic imaging, ultrasound (US) imaging, and magnetic resonance imaging (MRI). Each of these modalities uses a specific range of energy that can penetrate the tissue/organ of interest up to a certain depth and capture functional or structural alteration (if any) at a specific resolution. The sensitivity needed to find any alteration depends upon the nature of the electromagnetic wave used and the instrumental capability. The radionuclide imaging techniques exhibit the highest sensitivity and are able to detect small functional and biochemical changes. Contrarily, MRI suffers from a low sensitivity of 10^{-3}–10^{-5} mole/L (moles/liter) that is the minimum concentration of probe that must be present at the target site for successful MR imaging (Chen and Wu 2011, Massoud and Gambhir 2003). This low sensitivity is due to the small difference between atoms in a high energy state and a low energy state. However, the very high resolution of MRI counteracts the limitation of sensitivity and thus MRI has become the choice of imaging for various types of pathogenesis. In addition, the specificity of these imaging techniques relies upon the specific probe and the signature tagged. High specificity is an important criterion for accurate disease diagnosis especially for omitting false positive results. Both specificity and sensitivity are crucial to detect functional alteration (radionuclide and optical imaging), structural alteration (CT and US) or even both (MRI) in tissue/organ. Protein- or metabolism-specific probes tagged with radionuclide or MR contrast agents are used for Functional Imaging while dye-based contrast agents are utilized for X-ray/CT and MRI are useful for anatomical imaging. US imaging classically does not need any signature probe or contrast agent; however, modern US techniques often employ microbubble-based detection. Therefore, it is now possible to detect and diagnose finer tissue alteration and blood flow in certain diseased conditions with higher sensitivity by US imaging. The introduction of dual-modality positron emission tomography with computed tomography (PET-CT) and single photon emission computed tomography (SPECT-CT) systems to the clinical environment in the late 1990s is regarded as a revolutionary advance in modern diagnostic imaging, bringing precise anatomical localization to conventional PET and SPECT imaging techniques, and enhancing the quantitation capabilities of these modalities. The great success of PET-CT prompted development of combination of PET and MR scanners, leading to commercially available clinical positron emission tomography with magnetic resonance imaging (PET-MR) systems that again broadened the scope of precise detection and evaluation of therapeutic effect.

At present, the most extensive utilization of these powerful imaging technologies has been in cancer and neurodegenerative pathogenesis. Application of these techniques in regular practice in neurodegenerative diseases has been restricted to MRI and radionuclide imaging. On the other hand, each of these imaging modalities singly or in conjunction plays tremendous role in treatment of cancer patients. CANCER, this single term actually encompasses a multitude of diseases occurring in different organs that are biologically, functionally, and structurally quite distinct (Hanahan and Weinberg 2000). Thus, the diseases are obviously heterogeneous in their genetic, metabolic, and structural constituents except they share certain common disease-related features. Employment of different imaging modalities with different probes for diagnosis and monitoring therefore generate specific information relevant for treatment. Classical biochemical and histological tests often cannot describe longitudinal changes occurring due to treatment. Imaging technologies not only provide longitudinal information but also reveal the functional status of the disease in a noninvasive manner. For majority of the cancer types, CT, MRI or radionuclide imaging are routine practices in clinics. In addition, these technologies are also used in the evaluation of new therapies, especially targeted therapy in preclinical models and clinical trials.

Unfortunately, in spite of all these improvements, cancer is still a devastating disease. Although progression-free survival and overall survival for majority of cancers have significantly been improved due to superior therapy, better patient management, and medical service in the last two decades, a few cancers are still beyond control and overall survival is extremely poor. These cancers, namely, glioblastoma, pancreatic cancer, lung cancer, and ovarian cancer, need early diagnosis, targeted therapeutics, and better monitoring to gain control over these pestilent diseases. Noninvasive imaging techniques with standard and improved ability have been instrumental in achieving those goals for these cancers. In this chapter we discuss the detailed role of imaging in diagnosis and monitoring therapeutic response for glioblastoma, pancreatic, lung, and ovarian cancers. The first section describes the basic principle of these techniques followed by specific examples and outcomes of imaging for each disease and scope for the future generation imaging techniques.

8.2 CURRENT CANCER IMAGING TECHNIQUES

8.2.1 Magnetic Resonance Imaging (MRI)

Principle: When a human body or part of the body is placed in a magnetic field and a radio frequency pulse is applied, the gyromagnetic spin of hydrogen nuclei align themselves with the direction of the magnetic field (Larmor precession). Once this radio frequency is removed, these nuclei realign themselves to the magnetic field by releasing a weak radio frequency. This 'spin-relaxing' signal is captured by the conductive coils surrounding the human body and is used to obtain three-dimensional, gray scale MR images. MR image contrast also depends on two other tissue-specific parameters, the longitudinal relaxation time T1 and the transverse relaxation time T2. T1 measures the time required for the magnetic moment of the displaced nuclei to return to equilibrium (i.e., realign itself with magnetic field). T2 indicates the time required for the signal from a given tissue type to decay (Maravilla and Sory 1986, Araki et al. 1984, Mullins et al. 2005, Stall et al. 2010). Since hydrogen is the most abundant atom in every biomolecule in our body, subtle changes in composition and distribution of any of the biomolecules due to pathogenesis can be reliably measured by MRI.

8.2.1.1 Different Variants of MRI Imaging

Conventional MRI is an anatomical diagnostic imaging modality that detects anatomical abnormalities and changes in the normal physiology of the body. However, these anatomical changes are not sufficient to diagnose certain pathological conditions like distinguishing between therapy-induced necrosis, relapse, and tumor stages. Under these circumstances, specialized functional MR imaging (fMRI) sequences such as diffusion, perfusion, and spectroscopy are applied to improve the diagnostic potential (Rees 2003, Provenzale, Mukundan, and Barboriak 2006).

8.2.1.1.1 Diffusion MRI

Principle: In MR diffusion imaging, three opposing magnetic fields are applied to establish a gradient magnetic field. These magnetic field gradients establish voxel gradients that cause a loss of phase coherence of diffusing protons. Voxel is a three-dimensional point used for the construction of a computer-based model or graphic simulation. The loss of signal is in a linear relationship to the distance covered by the diffusion of 'un-restricted' protons at a given time period. Protons bound to macromolecules or retained in physical barriers will have 'restricted' diffusion that does not affect their net phases. Apparent diffusion coefficients (ADCs) can be calculated by ratios of intravoxel signal intensities between two sequences acquired by differing gradient strengths. Different image sequence acquisitions such as stronger and faster gradients, multi-shot echo-planar acquisitions, and phased array head coils parallel acquisition techniques are used to improve the specificity of diffusion MRI. ADC complements the conventional MRI to improve specificity of the diagnosis and characterization of the malignant tissues (Rees 2003, Provenzale, Mukundan, and Barboriak 2006, Al-Okaili et al. 2006).

8.2.1.1.2 Diffusion Tensor Imaging

Principle: In diffusion tensor imaging (DTI), directional maps of the cerebral white matter are generated by applying more than six magnetic field gradients in the same echo-planar sequences. For example, the proton diffusion in white matter is reliant on the orientation of the white matter bundle. Diffusion of the particles is greatest along the length of the myelinated neuronal axis and least in its perpendicular direction due to physical constraint of the membrane. As a result, particle diffusion measurements can be used to generate the three-dimensional directional maps of white matter tracts *in vivo*. A dedicated algorithm detects these tracts ('tensors') and aligns them on a three-dimensional map ('ellipsoids'). In ellipsoids, each imaging voxel is represented by eigenvalue (a vector representation for each voxel) that reflects the primary orientation and maximal diffusive area of the white matter in a specific imaging voxel. The fractional anisotropy (FA) is a measure to express the difference between diffusivity of two eigenvalues used to generate the diffusion maps that are represented as color FA. Similar to whole brain map, these DTI acquisition sequences can be applied over a defined area of a brain to generate temporal maps of white matter tract (also known as tractography) (Ferda et al. 2010).

8.2.2 Multi-Detector or Multi-Slice Computed Tomography

Principle: In CT, a part of the body or whole body is subjected to low-power X-ray radiation and attenuation of these X-rays are recorded on the detector. A dedicated algorithm reconstructs this X-ray attenuation pattern into CT scan. Three-dimensional X-ray attenuation patterns from various angles are reconstructed computationally to create multi-axial slices of the whole body. Multi-slice or multi-detector CT (MDCT) denotes the ability of a scanner to simultaneously acquire multiple slices. The MDCT era started in 1992 with a dual slice scanner and currently a scanner with 16-slice systems is used in the clinics. Another advancement has been the increased rotational speed of the detectors (from 1.0 to about 0.375 s/rotation) leading to reduced scan time for a given portion of a body, decreasing usage of X-ray tube power, and minimizing exposure (Ulzheimer and Flohr 2009).

MDCT enables anatomic evaluation of the whole body in an infinite number of planes and projections while maintaining image resolution and quality. Several image processing techniques and algorithms have been developed for analysis of the scans. Multiplanar reformation (MPR) is an algorithm used to stitch orthogonal or oblique sagittal, axial, and coronal virtual sections to reconstruct three-dimensional images for MDCT (Yitta et al. 2009, Lell et al. 2006) . It is extremely useful for detecting lesion/s, particularly if there is more than one lesion in the same area. Maximum intensity projection (MIP) is another post-processing technique in which single layer of brightest voxels along a line (or projection) at a specified angle (orthogonal or oblique) is displayed. A major drawback is that it lacks in-depth information (i.e., displays only the density of objects and not spatial information). This technique better suits the studies involving contrast agents as the post-processed images display the contrast enhanced structures with partial suppression of the background (e.g., vessels). Shaded surface display (SSD), another processing technique combines depth information as well as tissue density based on the preset thresholds; the first layer of voxels within defined density thresholds is used for display, leading to the visualization of the surface of all structures that fulfill the threshold conditions. Here depth information is preserved but the attenuation information is scaled proportionately. Finally, volume rendering (VR) utilizes all the information available in the volume data set and then groups of voxels are selected within a series of defined threshold densities each of which is color coded with appropriate depth shading/opacity. This is considered one of the best three-dimensional techniques, particularly for intraoperative navigation (Benvenuti et al. 2005).

CT Angiography (CTA) and Intracranial Perfusion (CTP) are sequence of CT imaging technique performed via rapid data acquisition cued to the arrival of an arterial bolus of contrast agent administered through intravenous injection. In CTA, high-resolution axial images are acquired longitudinally (i.e., in the direction of flow) and then evaluated using MIP or VR post processing with thresholds set for contrast enhanced vessels (Lell et al. 2006). Similar to CTA, CTP is another dynamic imaging technique used after contrast agent administration, in which a pixel-by-pixel time-density curve is created by rapid data acquisition over a stationary area of interest (Huang et al. 2014).

8.2.3 Radionuclide Imaging

Principle: The basic principle of radionuclide imaging revolves around systemic administration of radionuclide in a subject followed by detection and quantification of radiation produced due to the decay of the radionuclide. This decay leads to generation of radiation like X-rays, γ rays, and β rays. These radiations are either detected by scintillation counter or through ionization detector. These techniques have a lower spatial resolution than MRI (typically several mm for PET compared with approx. 1 mm for MRI) but can detect picomolar concentrations of isotopes with no depth limit and with high sensitivity (Zanzonico 2012).

8.2.3.1 Different Variants of Radionuclide Imaging

8.2.3.1.1 Positron Emission Tomography

PET is a functional imaging modality that provides metabolic information for tumors. Accumulation of a specific probe (targeted towards a biologically active molecule) labeled with positron or β^+ emitters such as fluorine-18, carbon-11 occurs after systemic administration in specific tissues based on the concentration of the targeted biomolecules. The positrons then undergo annihilation after colliding with nearby electrons in the tissue thus producing two 180^0 apart high energy photons of 511 KeV that are detected by coincidental hits on the PET detectors within nanoseconds of each other. These PET detectors are primarily composed of scintillation crystals like Bismuth germanite (BGO) (Kapoor, McCook, and Torok 2004). Depending on the radiotracer used, various molecular processes can be visualized, most of them relating to increased cell proliferation and metabolism in tumors. 18F-fluorodeoxyglucose (^{18}F-FDG), a glucose analog, is the most widely used tracer molecule in PET imaging for the detection of metabolically active tumors (Avril 2004). However, some tumors show less or equal intake of ^{18}F-FDG PET due to hypo- or isometabolism of glucose and thus more specific radiotracers such as methyl-[11C]-L-Methionine (MET) and 3'-deoxy-3'-[18F] fluoro-L-thymidine (FLT) are required to image increased activity of membrane transporters for amino acids and nucleosides, respectively (Bergstrom et al. 1987).

8.2.3.1.2 Single-Photon Emission Computed Tomography

In SPECT, gamma emitter radio-isotopes such as Technetium-99m, Iodine-131, Iodine-123 or Indium-111 are administered to patients and are then scanned with gamma camera for single photons emitted by the gamma emitter. Depending on the depth and position of the radiotracer inside the body, the emitted photons interact either directly with the gamma camera or after number of interactions with the neighboring tissue. Interaction with biological molecules in the surrounding tissues leads to scattering and energy loss by the photons. Thus, only those photons that interact directly and have the highest energy are used for reconstruction of the image while deflected and low energy photons are neglected. Images are collected through 360° rotation of the single or two gamma camera heads around the subject and then reconstituted to form a three-dimensional image. Technetium-99m is one of the radioisotopes used in SPECT (Livieratos 2012). A single strike by a direct photon originated at the source or from a scattered photon from the surrounding area on the detectors leads to significant noise generation in SPECT imaging. To reduce the noise, lead collimeters are introduced in SPECT, which significantly compromise the sensitivity of the SPECT imaging.

8.2.4 Ultrasound Imaging

Principle: The basic principle of US imaging is when pulses of sound waves are transmitted to human tissue with the help of piezoelectric crystals, 1% of these sound waves are reflected back due to different tissue elements that vary in their density thus leading to acoustic mismatch. These mismatches act as reflectors of sound waves that are again detected by same piezoelectric crystals thus leading to the formation of an image. The resolution of a US image depends upon the intensity of the waves used for generation of the image.

8.2.4.1 Different Variants of US Imaging

8.2.4.1.1 Doppler or Three-Dimensional Color Doppler Ultrasound Imaging

A standard US B-mode imaging is used for the localization of tumors but does not provide any ideas of vascularity around the tumor mass, which is dealt with Doppler technology. The Doppler technique is quite different from normal B-mode US imaging where sound waves are imparted at an angle to the blood vessel. These sound waves after reflecting from the red blood cells either get augmented or dampened in intensity based on the direction and velocity of the blood. With the advancement in image processing technology, it is now possible to construct a three-dimensional tumor image along with its information of vasculature (Fleischer et al. 2010).

8.2.4.1.2 Contrast Enhanced Ultrasound

The use of contrast agents is very common with the modalities such as MRI and CT, which are aimed to enhance the visualization properties of various organs, vessels, and cavities. Currently, microbubble-based contrast agents are used in diagnostic imaging to enhance the signal intensities in US imaging. This advancement is also called as contrast enhanced ultrasound (CEUS) and it has been used as a diagnostic tool for many organs as well as for diagnosis of neoplastic lesions. Furthermore, CEUS provides better understanding of blood perfusion through tumor tissue due to its ability to highlight microcirculation. The microbubbles consist of air or inert gas encapsulated in a layer of protein or polymers. Microbubbles are typically 5 μm in diameter similar to the size of red blood cells and can therefore be transported into the smallest capillaries and across the lungs, thus allowing the visualization of the arterial system after venous injection. Considering CEUS characteristics, it can be used for real-time monitoring of the vascularization dynamics, blood circulation patterns, and tissue architecture (Fleischer et al. 2010). Currently CEUS is also been used intraoperatively to guide the surgeon during tumor dissection (Mattei et al. 2016).

8.3 IMAGING OF CLINICALLY CHALLENGING CANCER

8.3.1 Gliomas

8.3.1.1 Introduction to Glioma

The preponderance of tumors in central nervous system (CNS), also known as the intracranial (related to occurrence within the cranium involving the brain or other structures such as cranial nerves, meninges) tumors, has reached a significant number so far. Malignant gliomas which happen to be the cancer of glial cells is, by far, the most prevalent primary malignancy of CNS with an annual incidence of 5/100,000 persons (Wen and Kesari 2008). Glial cells comprise of roughly six different cell types, namely oligodendrocytes, astrocytes, ependymal cells, Schwann cells, microglia, and satellite cells. Earlier it was thought that these glial cells outnumbered the neurons but the recent isotropic fractionator technique of counting implies that the ratio of glial cells to neuron is approximately 1:1 or even less (von Bartheld, Bahney, and Herculano-Houzel 2016). Malignancy in the former wreaks havoc in the nervous system simply by its presence in equal number as compared to neurons and its ability to divide in addition to the invasiveness of the cancer itself. Malignancy of brain could be of two types: a primary type that consists of malignancy arising in the brain parenchyma (e.g., gliomas, medulloblastomas, ependymomas) or in extraneural structures (for instance meningiomas, acoustic neuromas, and other schwannomas) and a secondary type that originates in tissues outside the brain and spreads to brain. Among the primary tumors, the low grade (grade I & II) tumors tend to have average proliferative potential and lesser chances of recurrence but can transform into anaplastic astrocytoma and glioblastoma. The World Health Organization (WHO) grade IV designation is assigned to cytologically malignant, highly mitotically active, necrosis-prone neoplasms with rapid invasive properties, and a fatal outcome. Grade IV malignancies include glioblastoma (GBM), most embryonal neoplasms, and many sarcomas as well. Invasion of tumor to all parts of CNS and a tendency for cranio-spinal dissemination are telltale properties of some grade IV gliomas (Cohen et al. 2005). Rare forms of malignant gliomas are GBM variants include gliosarcomas, which contain a prominent sarcomatous element; giant cell glioblastomas, which have multinucleated giant cells; small-cell glioblastomas, which are associated with amplification of the epidermal growth factor receptor (EGFR); and glioblastomas with oligodendroglial features, which can give a better prognosis than standard GBM (Smith et al. 2001).

8.3.1.2 Current Imaging Modalities for Glioma

The clinical history of high-grade gliomas is short (less than three months in more than 50% of cases) compare to most of the low-grade gliomas such as astrocytoma (Wick et al. 2009, Khan et al. 2016, Sizoo et al. 2014, Burger and Green 1987). Treatment approaches and strategies are primarily dependent on the early diagnosis, precise evaluation of 'true' tumor extension and its relationship with surrounding anatomic structures (Wang and Jiang 2013). The standard clinical care is based upon the diagnostic capabilities of the invasive stereotactic or surgical biopsy of suspicious tissue and its histopathological analysis. These invasive histopathological analyses of clinically suspicious masses are the gold standard for diagnosis of malignancy, tumor characterization, and its grading. However, discontinuous and milder symptoms of neoplasm are confused with the other pathologies compromising early diagnosis. Moreover, these stereotactic biopsy procedures may not always be possible and involve substantial mortality risk (McGirt et al. 2003, Jackson et al. 2001). Often, sampling errors and mixed tissue architecture may limit the diagnostic accuracy of the survey. This appeals to the necessity for a noninvasive technique that can be used for the diagnosis with comparable accuracy as histologic diagnosis without any risk to the patient.

Noninvasive imaging modalities play a crucial role in management of CNS neoplasms specifically for diagnosis, therapy planning, and assessing therapy response (Eskandary et al. 2005). Currently neuro-imaging modalities have evolved from an anatomy-driven discipline to the functional multimodal assessment of CNS lesions, incorporating biochemical properties (e.g., indicators of cell membrane synthesis) as well as physiological properties (e.g., hemodynamic variables) (Mabray, Barajas, and Cha 2015).

Current CNS imaging approaches including MRI, CT, and PET-MRI are practiced for the preliminary evaluation, preoperative management, and routine longitudinal follow-up of patients with high-grade gliomas (Watanabe, Tanaka, and Takeda 1992). A brief comparison of current imaging modalities has been summarized in Table 8.1. Early trial in 1993 on 56 children with pediatric brain stem gliomas had shown that MR scans are more efficient for diagnosing tumor occurrence compared to biopsy procedures and also this study reported postoperative complications in 11% of the children due to biopsy procedure (Albright et al. 1993). In certain cases, MRI serves as a useful tool for early detection as well as detection of recurrent brain tumors. Sixteen independent studies with a combined cohort of 19,559 normal participants have reported detection of CNS-neoplasm in 135 individuals by MRI and CT that would have otherwise remained undiagnosed (Morris et al. 2009). CT is also a traditional diagnostic modality for imaging CNS neoplasms.

8.3.1.2.1 Magnetic Resonance Imaging in Glioma Management

In normal brain MR images, T1-weighted cerebro spinal fluid (CSF) appears hypointense (dark) while a T2-weighted image appears hyperintense (bright) and a dynamic image contrast can be obtained by varying sequence capture parameters (proton density, T1, and T2). All these variable signal intensities on T1, T2, and proton density-weighted images are interpreted by dedicated MRI-algorithm (Maravilla and Sory 1986, Araki et al. 1984).

Pathological conditions in the brain including neoplasm show differential signal characteristics compared to normal tissues. Standard management for high-grade gliomas involves surgical resection followed by adjuvant radiation and chemotherapy necessities the assessment of 'true' tumor mass versus adjacent vital normal tissue. Most of these pathologic lesions can be distinguished into five major

TABLE 8.1
Comparative Look at the Various Imaging Modalities Associated with Glioma Management

	Magnetic Resonance Imaging	Computed Tomography	Positron Emission Tomography	Ultrasound
Role in Cancer Management	Primary imaging modalities for preliminary evaluation of glioma, tumor staging, routine longitudinal follow-up and evaluation of therapy response	Preferred imaging modality for patients with calcification, hemorrhage and bone changes related to tumors	Discriminates between tumor relapse and therapy-induced necrosis	Intraoperative detection of tumor and normal tissue boundaries. Ultra-sonography-guided surgical resection is used to reduce unnecessary damage to brain tissue
Sensitivity	96%	85%	85%	80%
Specificity	90%	85%	89%	85%
Cost	More expensive than both US and CT	More expensive than US	Most expensive among the four	Least expensive
Duration of the Procedure	30–45 min	1–5 min	30 min	15–20 min
Contrast Agent and Radiation Exposure	Gadolinium-based and no exposure	Iodine-based contrast agent and 10–20 mSv radiation exposure	18-F and iodine based and 20–30 mSv radiation exposure.	No side effects
Associated Hazard or Toxicity	Claustrophobia	Renal toxicity, iodine allergy	Renal toxicity, iodine allergy	No side effects

anatomical groups such as solid mass, cysts, sub-acute blood injuries, acute/chronic blood injuries, and fat accumulation based on their specific signal characteristics on the three basic images: T2-weighted, proton density-weighted (PD)/FLAIR, and T1-weighted. A comparative study conducted on 40 high-grade glioma patients by Stall et al. (2010) demonstrated that tumor volumes interpreted from T2, FLAIR, and T1 acquisition sequences are different and not interchangeable. T2 and FLAIR showed maximum overlap between 63.6% clinical tumor volume and 82.1% planning tumor volume. 'Planning tumor volume' is a critical clinical aspect that marks the tumor margin to define the uncertainty during surgery or radiotherapy and hence predicts the risk of damage to surrounding vital organs. Strikingly, both T2 and FLAIR sequences did not show any difference in the predicted toxicity towards normal brain structures (Stall et al. 2010).

Neoplastic tissue possesses differential contrast over normal tissue in a magnetic resonance image acquirement. Conventional MR images can predict the differences in tumor physiology that are important predictors for assessment of tumor grade, therapy response, and disease prognosis. In general, a solid region of high-grade neoplasms appears hypointense in T1-weighted sequences and hyperintense in T2, with a higher signal in areas of greater cellularity. Necrotic tissue can be differentiated from solid mass by its hyperintense appearance in T2-weighted sequences, which may appear as hypo-, iso- or hyperintense in T1-weighted sequences. A study on 27 patients with high-grade glioma undergoing proton beam radiation therapy showed that combination of two MR sequences accurately identified therapy-induced necrosis in 59.3% of patients and for rest of the cases early tumor recurrences were detected. The hyperintense appearance of necrotic tissue depends on the degree of protein degradation primarily involving hemoglobin (Mullins et al. 2005).

In most cases, T2-weighted images are the preferred modality for the diagnosis of brain neoplasms. Patients with a suspected neoplasm are subjected to a T2-weighted diagnostic sequence (spin-echo and FLAIR) to acquire the images. If an abnormality is found, T1-weighted images or contrast-enhanced scans are recommended for characterization of the lesion. The paramagnetic gadolinium-based tracers are used as contrast agents due to their superior biologic tolerance. Under normal circumstances, gadolinium-tracers do not cross the intact blood-brain barrier (BBB), which is permeable during tumor growth due to neo-vascularization (Rees 2003). Contrast agents enhance the intensity of irregular tumor margins and help in identification of proliferative space of the tumor. Performing 565 MRI scans in 67 patients, Galanis et al. (2006) showed that gadolinium-enhanced scans are significantly more predictive for patient survival compared to T1- or T2-weighted images (Galanis et al. 2006).

Noninvasive imaging modalities such as MRI or CT have the role of documenting the anatomic condition, iatrogenic changes, and normal physiology of the brain pre- and post-surgery. However, morphologic imaging modalities are not always sufficient in such sequelae as well as under the medically complicated scenario. Also, these modalities fail to detect cellular and subcellular changes. This obstacle can be overcome by applying improved fMRI techniques such as diffusion weighted imaging (DWI), DTI, perfusion, and spectroscopy that empower the diagnostic potential of MRI imaging and in many cases facilitates early diagnosis (Rees 2003, Provenzale, Mukundan, and Barboriak 2006).

As mentioned earlier, the revised WHO classification subdivides gliomas into four grades (I–IV) based on specific histologic features of the tumor such as cellularity, nuclear atypia, mitotic activity, pleomorphism, vascular hyperplasia, and necrosis. DWI utilizes cellularity as a target to distinguish between different grades of gliomas. On a DWI-MR scan, a high-grade tumor characterized with compact structure (high cellularity) appears least bright due to low ADC. The use of echo-planar imaging (EPI) reduces the overall timing of acquisition, which reduces the artifacts due to physiological motion (normal brain pulsation). A solid intracranial tumor causes a decrease in the ADC, which is increased in surrounding necrotic tissue, edema, and cyst. The physical mass of a tumor tissue restricts the proton diffusion and thus provides a distinct contrast over a normal brain tissue (Al-Okaili et al. 2006). A study by Kono et al. (2001) with 56 patients showed that patients with grade IV glioblastoma (n = 9) had lower ADC values than grade II tumor (n = 8) (P value 0.0008). Metastatic tumors (n = 21) did not show any significant difference compared to primary glioblastoma tumors. Therefore, quantitation of the ADC values can be used as a measure for distinguishing tumor grades (Kono et al. 2001).

As explained in the principle, DTI is a specialized MR modality that utilizes directional maps generated form diffusion anisotropy of tissue for detection of variety of intracranial pathologies, including patients with glioma. Mean diffusivity (MD) is used to distinguish between low- vs high-grade gliomas, tissue edema vs infiltrative neoplasm, and solitary vs multifocal primary neoplasm. DTI acquisition is also implemented in preoperative planning for patients with high-grade glioma. These scans help surgeons to resect the maximum extent of the neoplastic mass without damaging the neighboring tissue. DTI helps to discover the major white matter tracts (e.g., corticospinal tracts) and discriminate between displaced, edematous, infiltrated or destroyed tracks. A retrospective study with 24 patients by Ferda et al. (2010) showed that sensitivity and specificity of FA map to discriminate between low- and high-grade gliomas is 81% and 87% respectively. Grade II tumors demonstrated a uniform tumor structure as compared to grade III infiltrating tumor structure. However, grade IV tumors have a variable margin on the FA maps. The combination of the contrast enhancement pattern along with FA map evaluation can improve the discrimination between low- and high-grade glial tumors (Ferda et al. 2010).

Magnetic Resonance Perfusion (MRP) imaging has also been utilized in the diagnosis and management of gliomas. A hyper-intense region on a conventional MRI post gadolinium administration essentially depends on the degree of angiogenesis and compromised BBB. In MR perfusion imaging, cerebral blood flow (CBF), and cerebral blood volume (CBV) are quantified to map virtual 'angiograph' of gliomas. Though debatable, MRP is also used to discriminate high-grade gliomas from low-grade gliomas based on elevated cerebral blood flow and cerebral blood volume. In these studies, MRP provides relatively higher spatial resolution even in low-grade neoplasm over conventional PET and SPECT. Two different MRP scans can be performed as arterial spin labeling and the first pass bolus methods (Jackson et al. 2008, Jeong et al. 2015).

Advantages MR diffusion imaging can be used preoperatively to distinguish between a solitary metastasis from a primary glioma, which appears hyper-intense on peripheral T2 and FLAIR. It can be used to monitor the therapy response such as alteration in histological and fluid quantity in the intracellular and extracellular spaces, which discriminates between therapy-induced necrosis from recurrent neoplasm. Though relapse and therapy-induced necrosis have distinct histology, these are indistinguishable on a conventional MRI due to tumor proximity, adjoining vasogenic edema, localized mass effect, and a compromised blood-brain barrier (BBB) (primes gadolinium enhancement) (Castillo et al. 2001).

MR diffusion imaging can also assist in identifying radiation-induced necrotic cellular paucity compared to densely packed cells from recurrent neoplastic glioma. One study demonstrated significantly lower ADC value in relapse tissues compared to therapy-induced necrosis in patients presented with WHO grade III and IV gliomas. These studies set a mean ADC ratio threshold of 1.62 and above for a therapy-induced necrotic tissue while values below 1.62 are associated with tumor recurrence (Hein et al. 2004, Tsui et al. 2001).

Disadvantages Despite the superior detection of brain tumors with MRI, certain limitations exist. Enhancement of comparison between low-grade and high-grade gliomas cannot be generalized as not all the high-grade gliomas show contrast enhancement after gadolinium administration. Conversely, low-grade gliomas like pilocytic astrocytoma in children and young adults are hyperintense post gadolinium administration. This limits the sensitivity of conventional MR imaging to 55–83% for histologic grading of the glial neoplasms. In clinics, the preliminary function of MR imaging is to evaluate disease progression and its impact on the adjacent normal brain tissues. Limited conventional MRI features in conjunction with demographic data and the clinical presentation can help to distinguish the different neoplasms and to identify the more aggressive masses.

8.3.1.2.2 Computed Tomography in Glioma Management

MRI is the imaging modality of choice for brain (neoplasm) imaging. However, CT is superior in evaluating patient physiology such as calcification, hemorrhage, and bone changes related to tumors (Whelan et al. 1988, Morris et al. 2009). Also, CT plays a crucial role in the management of patients for whom MRI is not suitable, such as patients with pacemakers or metallic devices, as well as critically ill or unstable patients. Recent advancements in the instrumentation such as the multichannel, spiral CT scanner have rejuvenated its role in preoperative localization, intraoperative navigation, and radiation therapy targeting. These advancements in CT facilitate a sophisticated multiplanar three-dimensional imaging as well as dynamic imaging (i.e., CTP and CTA). With the help of CT as an imaging modality, Eskandary et al. (2005) diagnosed eight cases with tumors bearing incidental abnormalities from 3,000 asymptomatic participants who were not showing any signs of brain abnormalities in initial clinical evaluation (Eskandary et al. 2005). However, limited data are currently available regarding CT perfusion imaging in brain tumor imaging.

Advantage The major application for microvascular assessment using dynamic, contrast-enhanced CT is understanding the extent of differentiation of the most malignant region of the tumor before conducting stereotactic biopsy. This often turns out to be valuable information for differentiating between radiation necrosis, post-surgical scar tissue, and a recurrent tumor.

Disadvantage The major limitation of CTP is that only a limited volume of the brain can be scanned.

8.3.1.2.3 Radio-Nuclide Imaging in Glioma Management

PET and SPECT provide the opportunity to image multiple dynamic biological processes in situ in brain tumors. Despite the inherent background of ^{18}F-FDG in a normal brain scan, it is the most widely used tracer for nuclear imaging of gliomas. Technetium-99m-labeled compounds, such as 99mTc-sesta-methoxyisobutylisonitrile (99mTc-MIBI), 99mTc-hexamethyl-propyleneamine-oxime (99mTc-HMPAO), and 99mTc-tetrofosmin (99mTc-TF) are most widely used SPECT probes for glioma imaging (Mabray, Barajas, and Cha 2015, Watanabe, Tanaka, and Takeda 1992). This field is rapidly advancing and other metabolic tracers are being investigated.

^{18F}-U is routinely used to discriminate between tumor relapse from radiation-induced necrosis in patients with clinical suspicion of recurrent glioma. However, use of ^{18}F-FDG in brain tumor imaging is limited, as a normal brain is isometobolic or hypermetabolic compared to tumor tissue. With the exception of aggressive high-grade tumors (glioblastoma multiforme), normal parenchyma shows a high background due to higher glucose metabolism. Hence, it is often difficult to differentiate it from normal brain parenchyma. To overcome this limitation, various amino acid-based PET tracers have been developed as their uptake is higher in the tumor tissues compared to normal brain tissue (Alexiou et al. 2012). 11C-MET is the best-studied amino acid tracer for evaluation of recurrent glioma. In spite of convincing clinical results, the use of 11C-MET remains restricted to a few centers due to absolute requirement of an on-site cyclotron because of short half-life of 11C (T1/2 20 min vs 110 min of ^{18}F-FDG). This limitation has led to the development of various 18F-labeled aromatic amino acid analogues such as 3, 4-Dihydroxy-6-[18F] fluorophenylalanine (18F-FDOPA). Overall sensitivity of these tracer uptakes varies from 85% to 100% and specificity varies from 89% to 100%. The accuracy of 18F-FDOPA PET in evaluating low-grade and high-grade gliomas has been found to be superior to that of ^{18}F-FDG PET (Karunanithi et al. 2013, Chen et al. 2006, Becherer et al. 2003, Fueger et al. 2010). A study by Kato et al. (2008) with 95 primary glioma patients reported a significant correlation between ^{18}F-FDG, MET, and 11C-choline uptake with Ki-67 index. However, MET has been proven to be superior for visual tumor evaluation as it is able to produce a better tumor to background delineation (Kato et al. 2008). Other probes include O-(2-[18F]-fluoroethyl)-L-tyrosine (18F-FET), 39-deoxy-18F-FLT, and 18F-fluoromisonidazole.

Various SPECT radiotracers also have been evaluated for glioma imaging. These include Technetium-99m-labeled compounds, such as 99mTc-MIBI, 99mTc-HMPAO, and 99mTc-TF. Vos et al. (2007), demonstrated that 201Tl SPECT is superior to conventional MRI in differentiating between recurrence and therapy induced necrosis. In this study 201Tl SPECT sensitivity in detecting recurrence was found to be 43% to 100% with specificity ranging from 25% to 100% (Vos et al. 2007). Bleichner-Perez, et al. (2007) with 201 patient 99mTc-MIBI SPECT scans reported

89% sensitivity, 83% specificity, and 87% accuracy for the detection of tumor recurrence in high-grade gliomas (Bleichner-Perez et al. 2007).

Advantages PET and SPECT scans are the most accurate imaging modality for detection of micro-metastasis within the brain as well as in the whole body. These scans can differentiate between recurrent tumors from radiation necrosis.

Disadvantage Inherent background in normal brain scan limits the sensitivity of radio-nucleotide scan.

8.3.1.2.4 Multimodal Imaging of Glioma

PET-CT Diagnostic imaging with single functional (PET or SPECT) or anatomical (CT or MRI) imaging modality constrains the clinical investigation. The combination of two imaging modalities provides greater confidence in image registration, accurate diagnosis eliminating the need of multiple scans. PET-CT or SPECT-CT are useful tools for diagnosis and management of a variety of neurological diseases and cancers. Increased tumor uptake of 99mTc-TF in SPECT is correlated with aggressive behavior and serves as an independent prognostic factor in patients with malignant glioma (Karunanithi et al. 2013).

PET-MR Integrated PET-MRI systems offers the ability to perform brain anatomical MR imaging simultaneously with physiologic PET imaging. In PET-MR imaging, MR scans are used to acquire anatomical tumor volume followed by radio tracer 18F-flouromisoidazole (FMISO) administration to detect physiological tissue hypoxia using PET images. A preliminary study of 22 participants with glioblastoma demonstrated an association between both the pre-radiation tumor volume and degree of tumor hypoxia measured by FMISO PET and a shorter time to tumor progression and survival (Gerstner et al. 2016).

8.3.1.2.5 Intraoperative Imaging Modalities

Intraoperative imaging modalities have been used by many neurosurgeons to localize and distinguish tumor tissue from normal tissue. These modalities help to detect brain shift (a random brain movement during surgery) and tissue deformation during surgery, which improves the extent of resection and reduces neurologic morbidity. Recent studies have confirmed that the overall survival of patient with gliomas is highly dependent on the success of the surgical dissection of the tumor tissue. In 90% of the glioma cases, treatment fails due to recurrence from un-resected part of the tumor post-surgery. In either low- or high-grade gliomas, surgery is commonly the initial therapeutic approach followed by radiation treatment. Maximum/complete resection of the tumor tissue favors the comprehensive treatment along with extended tumor recurrence time and patient survival. Under these circumstances, micro-neurosurgical techniques play a very crucial role to identify tumor vs normal tissue during surgeries and enable the greatest degree of tumor resection. An integrated application of neuro-navigation, intraoperative MRI, intraoperative fluorescence labeling, and intraoperative sonography also enhances these factors.

In recent years, 'neuro-navigation' has been developed to assist neurosurgeons in performing surgery more safely and efficaciously. Neuro-navigation detects functional sectors of brain tissue and distinguishes 'true' tumor boundaries. However, it cannot provide surgeons with information about intraoperative dynamic changes such as brain shifting due to loss of cerebrospinal fluid (CSF), tumor debulking, and brain deformation caused by patient positioning (Gerard et al. 2017).

Intraoperative MRI could be a promising technology for detecting residual tumor mass during surgery. Unfortunately, the use of metallic surgical instruments poses a serious limitation. These problems can be overcome by intraoperative sonography, which is a real-time monitoring modality and is less costly than other methods. Several reports in last few years have demonstrated the benefits of sonography for intraoperative imaging and guidance in brain surgery. Sonographically guided resection of cerebral gliomas exhibits better curative effect than conventional surgery and significantly delays tumor recurrence and increases survival. However, many factors can influence the sonographic image definition, such as edema, air, pads, the hole formed by removing the lesion, and the shape of the removed bone window. These factors may decrease the sensitivity and specificity for evaluation of tumor remnants.

8.3.1.3 Future Directions of Glioma Imaging

MRI scans remain an important tool for diagnosis, staging, and evaluating therapy response to glioma. Yet early detection of the glioma is a major clinical concern. Sensitivity of CT and PET scans limits the early diagnostic capacity. Development of new fMRI methodologies and sequence algorithms are improving diagnosis accuracy. In future, improvement in DCE, DWI, and DTI MRI may serve as an early diagnostic tool for glioma.

The recent advancements in intraoperative imaging modalities using ultrasound has improved the accuracy of tumor resection and patient survival rates. Currently, the future lies in the amalgamation of more sensitive technologies such as MRI modalities. Moreover, progress has been made in making ultra-surgical manipulations for better tumor resection, which will increase the mean overall survival rate of this deadly disease.

8.3.2 Lung Cancer

8.3.2.1 Introduction to Lung Cancer

Lung cancer, with the highest incidence rate among all malignancies across the globe remains a deadly disease due to a high mortality rate both in men and women (Ferlay et al. 2015). There has been a significant increase in incidence (approximately half of the total cases occurring globally) of lung cancer in developing countries. Approximately half of the cases now occur in developing countries. Incidence of lung cancer has increased dramatically in women during last decade and is linked to increased smoking and occupational hazards (Dela Cruz, Tanoue, and Matthay 2011, Ridge, McErlean, and Ginsberg 2013). The two major forms of lung cancer are non-small cell lung cancer/NSCLC (about 85% of all lung cancers) and small-cell lung cancer/SCLC. NSCLC is further divided into three subtypes; adenocarcinoma, squamous cell carcinoma, and large cell lung carcinoma. Adenocarcinoma and squamous cell carcinoma constitute 90% of the NSCLC and remain the major cancer subtypes observed among lung cancer patients. While smoking is the major cause for increased lung cancer incidence, it is strongly associated with squamous cell carcinoma and SCLC, while adenocarcinoma is more common among non-smokers (Sun, Schiller, and Gazdar 2007). These subtypes are further sub-divided by histopathological characteristics and the molecular signatures of the disease. Standard treatment for lung cancer is based on TNM (T-Tumor, N-Node, and M-Metastasis) staging, which determines the course of lung treatment that includes surgery, platinum-based chemotherapy, and radiation as the first line of treatment (Mirsadraee et al. 2012). Surgical resection is performed when the disease is confined to the primary site without involvement of mediastinal nodes. Pleural effusion, node involvement, and distal metastasis are treated by chemotherapy and radiation. Understanding the underlying molecular mechanism of lung carcinogenesis led to discovery of several targetable molecules like Epidermal Growth Factor Receptor (EGFR), c-MET, KRAS, and LKB1 (Shackelford et al. 2013, An et al. 2012). Similar to other cancers, early detection remains a key for better treatment outcome and survival.

8.3.2.2 Diagnosis of Lung Cancer

Accurate diagnosis and staging of lung cancer are essential to determine the best possible treatment for the patient but the process is often very complex. Fitness of the patient augments the complexity and itself may influence both diagnostic and treatment procedures. Sufficient tissue sampling is required to enable a pathologist to further classify the tumor into different subtypes. Additional tissue samples are often required to determine the presence of specific biomarkers to predict the choice and efficacy of targeted therapies. Diagnostic tools and methods used for lung cancer diagnosis largely affect the morbidity and mortality rate during diagnosis. Both noninvasive and minimally invasive methods are used for lung cancer diagnosis.

8.3.2.3 Noninvasive Methods

Noninvasive visualization of cancer has played an important role in both screening and diagnosis of the disease. It has also been an important tool for following the disease progression during and after the treatment. Recent developments in noninvasive imaging has played important role in TNM staging and improved treatment of lung cancer. The comprehensive and comparative glance at the various imaging modalities associated with lung cancer management is listed in Table 8.2.

8.3.2.3.1 Computed Tomography in Lung Cancer Management

The chest radiograph, an imaging modality was used to detect abnormal mass in lungs in earlier days. Though it could detect lesions in the lung, initial clinical trials with chest radiograph for lung cancer screening in 1980s showed no significant beneficial outcome for lung cancer mortality (Fontana et al. 1986, Muhm et al. 1983). The major drawback for chest radiography is that the technique is unable to detect the disease at early stage. In fact, investigators of a recent large-scale, randomized cancer screening (known as the Prostate, Lung, Colorectal, and Ovarian [PLCO]) trial, consisting of 1,54,901 participants also concluded that annual screening with chest radiographs over a four-year period did not significantly decrease lung cancer

TABLE 8.2
Comparative Look at the Various Imaging Modalities Associated with Lung Cancer Management

Technique	Computed Tomography	Positron Emission Tomography	Magnetic Resonance Imaging	Ultrasound
Role in lung cancer management	Primary and an important imaging modality for lung cancer diagnosis and T-staging of the disease	PET is an important modality for lung cancer for nodal and metastasis imaging	Mainly used as an additional tool for identifying extent of disease mainly for patients with superior sulcus tumors. Though not commonly used, it is a promising tool for cancer diagnosis	An imaging modality though not used for lung cancer diagnosis but has emerged as an important tool for image-guided thoracic biopsy removing need of more invasive techniques and reducing morbidity
Sensitivity	94%	96%	88%	86%
Specificity	73%	79%	97%	100%
Cost	Expensive than US	Most expensive among the four	More expensive than both US and CT	Least expensive
Duration of the procedure	Considering the imaging involves respiration hold multi-detector CT with 5–10 seconds is possible	30 min	Current scan time for single scan is 20 seconds, which is still a challenge for patients to hold breath. However, efforts are made to reduce it to 5 seconds	15–20 minutes
Contrast agent and radiation exposure	Iodine based contrast agent and radiation exposure of 10 mSv for regular CT and 1.5 mSV for low-dose CT scan	18FDG and iodine based and 20–30 mSv radiation exposure	Gadolinium based and no exposure	No side effects
Associated hazard or toxicity	Renal toxicity, iodine allergy	Renal toxicity, iodine allergy	Claustrophobia	No side effects
Availability	Easily available	In many hospitals	In few advanced hospitals and centers	In few advanced hospitals and centers
Special preparation before imaging	Fasting for 4 hours is required	Same as CT with 1 additional hour of rest	Antiperistaltic agents	None

mortality when compared with standard clinical tests (Oken et al. 2011). CT is superior to radiography in sensitivity and resolution and has emerged as an important diagnostic tool for lung cancer screening. In 1999, a small trial called Early Lung Cancer Action Project (ELCAP) consisting of 1,000 high-risk individuals undergoing CT scans reported that low-dose CT at baseline compared to chest radiography was better for detecting the disease at earlier stages. In this study, non-calcified pulmonary nodules (NCN) were detected three times higher (23% vs 7%) and stage I malignancies were detected six times higher (2.3% vs 0.4%) in low-dose CT compared to the chest radiograph. Out of 27 NCNs detected by low dose CT, 23 were of stage I and 19 out of these 23 nodules were non-detectable on a chest radiograph (Henschke et al.). Later an extended and bigger trial consists of 25,000 high-risk individuals (International ELCAP (I-ELCAP)) reported detection of 382 patients with tumors, a vast majority of which were at stage I (International Early Lung Cancer Action Program Investigators 2007). These initial clinical trials showed potential of CT for lung cancer screening and early detection diagnosis. Later large screening programs to understand the influence of smoking in lung cancer utilizing both chest radiography and CT were carried out. In the National Lung Screening Trial (NLST) conducted in the US, around 53,439 asymptomatic participants ranging from 55 to 74 years of age with a history of smoking at least 30 packs per year were enrolled and randomly assigned to undergo annual screenings for three years using either low-dose CT or chest radiography. It was seen that, the participants who were subjected to low-dose CT scan showed 27.3% positive screening results compared to just 9.2% of the participants who were subjected to only a chest x-ray. Lung cancer was diagnosed in 292 participants (1.1%) in the low-dose CT group versus 190 (0.7%) in the radiography group. It was found that, the number of lung cancers diagnosed at stage I were 158 with the CT scan compared to 70 with the chest X-ray. It was also noted that there was no significant difference in the number of patients diagnosed at late stage by both modalities (stage IIB to IV in 120 vs 112). The sensitivity and specificity of lung cancer detection with the CT scan were found to be 94% and 73% respectively. By contrast, the sensitivity and specificity of a chest x-ray in the detection of lung cancer were found to be 74% and 91%. Positive predictive value for CT and x-ray was 5.7% and 5.2% (National Lung Screening Trial Research et al. 2013). Currently, contrast enhanced CT has emerged as routine practice for lung cancer diagnosis and T-staging of lung cancer.

Advantages CT has proven an invaluable tool for early detection of lung cancer. CT can detect calcified nodes in lung increasing its predictive value.

Disadvantages However, CT has certain inherent limitations. It relies on its ability to resolve and distinguish different anatomical features such as lymph node size, pleural effusion etc. to detect abnormalities. Lymph node staging by CT involves measuring lymph node size and hence, disease may be over staged if enlarged benign lymph nodes are measured, or disease may be under staged if the microscopic nodes are classified as benign (Knoepp and Ravenel 2006). Pleural effusion is relatively common in patients with lung cancer. To determine resectability of the tumor or the use of chemo-radio therapy it is necessary to differentiate between benign and malignant effusion. CT has limited specificity for mediastinal metastasis and differentiating ability for benign and malignant pleural effusion.

8.3.2.3.2 Positron Emission Tomography in Lung Cancer Management

PET is highly sensitive and widely used technique in radio-nucleotide imaging of cancer. ^{18}F-FDG PET has become an excellent tool for accurate detection of mediastinal lymph node metastases as well as extra-thoracic metastases and is often been recommended to be performed after CT for confirmation of node and extra-thoracic metastasis. This technique has been accurate in differentiating benign from malignant lesions as small as 1 cm (Marom et al. 2002). PET has an overall sensitivity of 96% (range 83%–100%), a specificity of 79% (range 52%–100%), and an accuracy of 91% (range 86%–100%) (Gould et al. 2001, Zhuang et al. 2001). Since CT scan is unable to differentiate benign and malignant pleural effusions, thoracic biopsy remains as an option that has low predictive value. It may not predict malignancy in 30%–40% of patients with truly malignant pleural effusion (Gupta et al. 2002). PET imaging, however can differentiate between benign and malignant pleural effusions. In one such study, PET^{18}F-FDG imaging correctly detected the presence of malignant pleural effusion and malignant pleural involvement in 16 of 18 patients and excluded malignant effusion or pleural metastatic involvement in 16 of 17 patients (sensitivity, specificity, and accuracy of 88.8%, 94.1%, and 91.4% respectively) (Gupta et al. 2002). In another study involving 25 patients, the sensitivity, specificity, and accuracy of detecting malignant effusion were found to be 95%, 67%, and 92% respectively by ^{18}F-FDG PET (Erasmus et al. 2000). PET has a high negative predictive value, which is very useful for confirming and ruling out false-positive results of a CT scan. It also reduces the number of repeat thoracic biopsies. PET is also useful for node imaging when distal metastasis is not seen, and it helps to decide the treatment regime. Compared to CT, PET is highly sensitive for the detection of nodal involvement, however false-negative results in PET can occur for lesions smaller than 1 cm because a critical mass of metabolically active malignant cells is required for PET diagnosis. Nomori et al. conducted a study to evaluate use of ^{18}F-FDG PET for pulmonary nodule scanning wherein, a cohort of 141 patients was subjected to CT scan. CT scan identified 73 malignant and 43 benign nodules that were further subjected to ^{18}F-FDG PET and histopathology (Nomori et al. 2004). Among the 73 malignant nodules, 58 were found to be positive and 15 were negative by ^{18}F-FDG-PET scan. However, those 15 malignant nodules turned out to be false-negative results of ^{18}F-FDG PET scan as IHC determined 12 modules as well differentiated adenocarcinoma and one as moderately differentiated adenocarcinoma. Similarly, out of 43 benign nodules (revealed by CT), 15 were positive in ^{18}F-FDG PET scan, which were later found to be inflammation not lung nodules. The 28 ^{18}F-FDG PET negative nodules included five ground-glass opacity (GGO) benign lung tumors. Though ^{18}F-FDG PET scanning for pulmonary nodules showed an overall sensitivity, specificity, and accuracy of 79%, 65%, and 74%, respectively, for detection of GGO nodules, the sensitivity, specificity, and accuracy dropped to 10%, 20%, and 13%, respectively. A CT scan of 141 patients identified ten malignant nodules with GGO images, comprised of well-differentiated adenocarcinomas, and nine of them (90%) were negative for ^{18}F-FDG-PET scan (Nomori et al. 2004).

This study signifies the ability of ^{18}F-FDG PET to correctly detect both benign and malignant nodules of ≧1 cm in size, but it is not reliable for diagnosing nodules with GGO images or <1 cm in size. Detection of metastases in patients with non-small cell lung cancer (NSCLC) has major implications on management and prognosis where PET imaging plays an important role in identifying extra-thoracic metastasis. Lung cancer often metastasizes to the CNS and to bones. Bone scintigraphy using 99m Technetium methylene diphosphate is widely used to detect bone metastasis and has a good sensitivity (90%) but a low specificity (±60%). When scintigraphy imaging is compared to PET for bone metastasis detection, the latter is reported to have a similar sensitivity (≥90%), but a higher specificity (≥98%) and accuracy (≥96%), and is therefore considered superior to bone scintigraphy in the detection of bone involvement (Bury et al. 1998). ^{18}F-FDG-PET is less suitable for diagnosing brain metastasis and when compared to CT, PET has non-significant difference in sensitivity, specificity, and accuracy for detection of liver metastasis.

Advantages The superior ability to detect nodal metastasis and ability to distinguish benign and malignant pleural effusion by PET imaging has improved noninvasive detection of lung cancer and at an early stage. PET has also improved TNM staging of lung cancer helping in better diagnosis and treatment options. PET is often used to monitor treatment response post chemo or radio therapy. It has been a valuable tool not only for diagnosis but also to follow treatment response and to monitor tumor growth and relapse after treatment.

Disadvantages PET is often performed as complementary imaging for CT in lung cancer since PET imaging is unable to detect anatomical information. Also, failure to differentiate from active inflammation and failure to detect disease in GGO nodules poses a challenge of using PET as a primary detection method in lung cancer diagnosis.

8.3.2.3.3 Magnetic Resonance Imaging in Lung Cancer Management

MRI is not a mainstream imaging modality for lung cancer diagnosis and is only performed, where it is necessary to assess the extent of disease, for patients with superior sulcus tumors. MRI detection of pulmonary nodules is inferior to CT detection, however, MRI yields supplementary morphologic information that is valuable for differential diagnosis, including for sclerosing hemangioma, bronchial carcinoid tumor, tuberculoma, progressive massive fibrosis (Kurihara et al. 2014). There are various dynamic MR techniques and sequences for distinguishing malignant nodules from benign nodules, with reported sensitivities ranging from 94–100%, specificities from 70–96% and accuracies from 80–95%. While comparing whole-body diffusion weighted MRI with ^{18}F-FDG-PET-CT for the M-stage assessment capability, DW-MRI was found to be as effective as integrated ^{18}F-FDG-PET-CT, with a sensitivity of 70%, specificity of 92%, and accuracy of 87.7% (Ohno et al. 2008). Recently, Koyama et al. (2008) compared capabilities of pulmonary nodule detection and differentiation of malignant from benign nodules between non-contrast-enhanced, multi-detector CT and MRI using a 1.5 T system in 161 patients presented with a total of 200 pulmonary nodules. Although the overall detection rate of multi-detector CT was superior to that of respiratory-triggered short tau inversion recovery (STIR) turbo spin echo imaging, there was no significant difference in malignant nodule detection rate between the two methods (Koyama et al. 2008). It is known that pulmonary nodule assessment using MRI has been inferior to multi-detector CT due to motion artifacts from respiratory and cardiac movements. However, development of new sequence algorithms and breath-hold protocols had made whole body MRI feasible in clinical settings with advancement of technology. Turbo spine echo sequence shows many pulmonary nodules, including lung cancers, pulmonary metastases, and low-grade malignancies. Enhancement patterns or blood supply evaluated with dynamic contrast-enhanced (CE) MRI is helpful for diagnosis of pulmonary nodules. It has been suggested that dynamic CE MRI is effective for assessment of tumor angiogenesis (Li, Wang, et al. 2015).

Advantages The lack of ionizing radiation makes MRI a safe tool for repeated dynamic evaluations of tumor perfusion. Dynamic MRI with advanced sequences requires less than 30 seconds of breath-holding for acquisition of all data, a much more comfortable session for a patient.

Disadvantages To date, MRI detection of pulmonary nodules is inferior to CT detection.

8.3.2.3.4 Ultrasound Imaging in Lung Cancer Management

US imaging is not a common practice for lung cancer diagnosis. However, it is a useful imaging modality to guide needle aspiration or biopsy of cervical lymphadenopathy, peripheral tumors in contact with the pleura, distant metastases, and sampling of pleural tissue or fluid. Three minimally invasive methods combined with US imaging used for sampling tissue samples are transthoracic needle aspiration (TTNA), endoscopic ultrasound (EUS) fine needle aspiration (EUS-FNA), endobronchial ultrasound (EBUS) trans-bronchial needle aspiration (EBUS-TBNA), and non-ultrasound-guided TBNA. Mediastinal staging in patients with NSCLC is crucial in dictating surgical vs non-surgical treatment options. Cervical mediastinoscopy is the "gold standard" in mediastinal staging but is an invasive procedure and limited in assessing the posterior subcarinal, lower mediastinal, and hilar lymph nodes. Further, noninvasive mediastinal staging by CT and PET scans generates significant false-negative and false-positive data and requires lymph node tissue confirmation. FNA techniques, with guidance by EBUS and EUS, have thus become more widely acceptable. The combination of EBUS-FNA and EUS-FNA of mediastinal lymph nodes can be a viable alternative to surgical mediastinal staging. (Varela-Lema, Fernandez-Villar, and Ruano-Ravina 2009). Bronchoscopic EBUS is performed as an adjunct procedure following routine bronchoscopy. The EBUS bronchoscope is equipped with a linear probe US at the tip, which allows for visualization of mediastinal structures including lymph nodes, vasculature, and stromal tissue (Varela-Lema, Fernandez-Villar, and Ruano-Ravina 2009). EBUS also allows for real-time direct visualization of mediastinal node puncture and subsequent aspiration. The addition of EBUS after routine bronchoscopy adds approximately 15 to 30 minutes extra to each procedure, depending on the number of nodal stations sampled. In addition to the mediastinal nodal stations (2R, 2L, 4R, 4L, and 7) accessible by cervical mediastinoscopy, the hilar, interlobar, and lobar nodes (levels 10, 11, and 12, respectively) are also accessible by EBUS (Colella et al. 2014). Yasufuku et al. (2011) prospectively evaluated 153 patients who underwent EBUS and cervical mediastinoscopy in back-to-back procedures and demonstrated a specificity and positive predictive value (PPV) of 100% for both techniques. The sensitivity, negative predictive value (NPV), and diagnostic accuracy for EBUS were 81%, 91%, and 93%, respectively, compared with 79%, 90%, and 93% for mediastinoscopy (Yasufuku et al. 2011). EUS-FNA avoids the need for surgical staging procedures in 50% to 70% of patients. Zhang et al. (2013) recently reported the estimated summary measures, a meta-analysis, for quantitative analysis of EBUS-TBNA plus EUS-FNA for mediastinal nodal staging of lung cancer. The combined sensitivity (86%) and specificity (100%) was better than EBUS-TBNA and EUS-FNA alone and with a positive likelihood ratio of 51.8 and negative likelihood ratio of 0.15 (Zhang et al. 2013).

Advantages Both the US mediated endoscopic techniques (EUS and EBUS) show higher sensitivity and specificity when performed in combination than each done alone and are minimally invasive in nature. EUS-FNA plus EBUS-TBNA might replace more invasive methods for evaluating mediastinal node staging of lung cancer in future.

Disadvantages Current barriers to the dissemination of these techniques include initial cost of equipment, lack of access to rapid on-site cytology, and the time required to obtain sufficient skills to perform the technique.

8.3.2.3.5 Multimodality Imaging

PET-CT Application of combined PET-CT imaging in lung cancer is highly dependent on requirement. In contrast to CT being conclusive for T-staging of disease, PET does not have any added advantage for primary tumor imaging in lung cancer. However, PET is important for benign to malignant differentiation and to identify nodal involvement and metastasis. Various studies have reported combined PET-CT imaging to have a better outcome than either of alone in TNM staging of the disease. Co-registered PET-CT imaging decreases false-positive results and improves specificity.

PET-MR Though PET-MR multi-modality imaging for lung cancer is an attractive option considering MRI is able to provide the anatomical information without involvement of radiation exposure, it is yet to be introduced for lung cancer detection.

8.3.2.3.6 Future Directions

Currently, CT and PET remain the standard tools for lung cancer diagnosis. Contrast-enhanced CT and PET has improved TNM staging of disease reflecting better treatment management of the disease. Low-dose CT screening for high-risk individuals holds the key to early detection of disease along with identifying molecular biomarkers, which will improve lung cancer prognosis and reduce mortality. Similarly, PET imaging with new biomarker-based tracer (F18-DOPA, F18-Choline for brain and

CNS metastasis and 18F-fluoromisonidazole for angiogenesis) will result in better patient stratification and treatment.

Recent advancements in fast MR imaging using parallel imaging and compressed sensing technology have demonstrated excellent capability in accelerating MR imaging acquisition. To reduce imaging acquisition to five seconds or lesser, a technique that combines parallel imaging and compressed sensing with optimized imaging parameters and acceleration performance has to be developed. Another promising future technique for imaging pulmonary nodules is an ultrashort echo time (UTE) MR image. UTE MR imaging is also advantageous for lung imaging because it is relatively robust about motion artifacts and therefore high-quality clinical images can be acquired with free-breathing in the limited field-of-view setting despite the regular non-accelerated acquisitions.

8.3.3 Pancreatic Cancer

8.3.3.1 Introduction to Pancreatic Cancer

Among pancreatic malignancies, pancreatic ductal adenocarninoma (PDA) is the most common type and represents 85–90% of all pancreatic neoplasms. Despite a variety of improved diagnostic and therapeutic approaches over the past several decades, it still remains the fourth leading cause of cancer-related deaths in the US and the overall five-year survival rate is only around 5–7% (Shrikhande, Barreto, and Koliopanos 2009, Adamska, Domenichini, and Falasca 2017). Since the ductal epithelial cells are believed to be the cell of origin for PDA, the characteristic features of this disease include mucin production along with the expression of certain cytokeratins (7, 8, 13, 18, and 19), a signature that is easily identifiable through IHC (Hruban and Fukushima 2007). While 60–70% of ductal adenocarcinomas are localized in the head of the gland, the remaining proportion is found to be localized in the body and/or tail. Based on the cell type origin, a set of histologic subtypes have been classified as follows: adenosquamous carcinoma, mucinous noncystic carcinoma, signet-ring cell carcinoma, undifferentiated (anaplastic) carcinoma, and mixed ductal-endocrine carcinoma (Hamilton and Aaltonen 2000). Each subtype has a distinct site of origin and dictates the symptoms, early stage diagnosis, therapy response, and overall survival rate. Therefore, management of PDA requires thorough understanding of the underlined anatomy and biology of each subtype while designing the strategy for therapy. Surgical resection remains the only potentially curative treatment for PDA which is possible only in 15–20% of the cases. About 40% of the patients have the locally advanced non-resectable disease (Zakharova, Karmazanovsky, and Egorov 2012). Earlier, intraoperative assessment was the only way to approximate the resectability of the tumor. However, application of modern imaging techniques has allowed preoperative staging of patients that reveals valuable information about the tumor.

8.3.3.2 Image-Guided Diagnostic Intervention

The complex anatomical and physiological features of the pancreas impose a challenge in the management of pancreatic cancer. This could be considered as one of the major factors responsible for dismal survival rates of the patients since surgical resection of the whole tumor may not be feasible. However, current imaging techniques

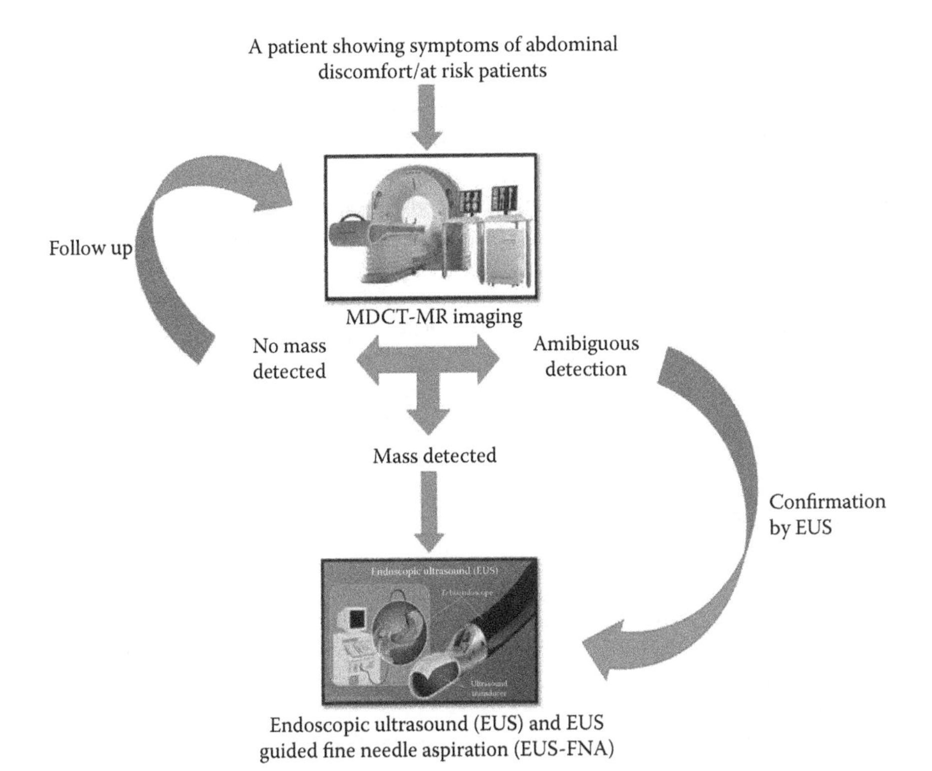

FIGURE 8.1 Imaging-based screening procedure of PDA patients. The screening procedure starts by performing MDCT-MR scan to detect the presence of adominal mass. The patients with a lump in the abdomen are further tested by EUS for confirmation. Whereas the patients showing no mass may be followed up using an MDCT-MR scan. In case of ambiguity, EUS remains the gold standard for diagnosis.

have been proven to be very effective tools in not just the detection of the tumor but also in the overall management of pancreatic cancer. Preoperative imaging is necessary as it yields important information about tumor resectability. Since the pancreas is a highly vascular organ, it is inevitable to take into account the involvement of adjacent blood vessels (superior mesenteric vein, portal vein, and superior mesenteric artery). Imaging can also provide valuable information about nodal involvement and distant metastatic lesions, which can further be utilized to design appropriate treatment regimen for the patients. Figure 8.1 describes the imaging-based screening procedure of patients with PDA. The commonly used imaging techniques are as follows.

8.3.3.2.1 Ultrasound Imaging in Pancreatic Cancer Management

Abdominal US is the first modality of choice as it is cost effective and routinely available. PDA neoplasms occur mostly in the head part of the pancreas and have typical features like hypoechoic mass, dilatation of the pancreatic duct, and dilatation of the bile duct that can be easily observed in an US scan. Moreover, in advanced stage pancreatic carcinoma, ascites accumulation. and liver metastases (>1 cm) may be seen (Miura et al. 2006). However, the presence of gas bubbles, shadowing effect of transverse colon, accumulation of fat etc. can mask the detection of tumors present in the body/tail part of the pancreases. A study of 189 patients with pancreatic head cancer showed that the detection rate of pancreatic cancer with transabdominal US was 82.0%, which was lower than that with CT (93.1%) and endoscopic US (94.7%). The sensitivity of this technique is highly operator dependent and therefore can vary anywhere between 50–90% (Lee and Lee 2014, Conrad and Fernández-del Castillo 2013). Echo-enhanced power doppler sonography is an advanced version of the traditional abdominal US that offers better sensitivity and specificity. In a study performed on 137 patients suffering from PDA, the sensitivity and specificity of echo-enhanced power doppler sonography was found to be 87% and 94% respectively, which is quite promising (Rickes et al. 2002).

Advantages Overall, abdominal US is used for preliminary screening, which requires confirmation through CT. Its noninvasive and cost-effective nature makes it the first line of choice.

Disadvantages The findings may vary depending upon the operator's handling. Physiological conditions like overlying bowel gas, obesity etc. may compromise the sensitivity of identification of the tumor mass. Therefore, further development of contrast agents and standardization of the procedure is necessary.

8.3.3.2.2 Computed Tomography in Pancreatic Cancer Management

After abdominal US, CT is the second imaging modality used to confirm the presence of PDA neoplasms, which provides better resolution and greater sensitivity as compared to US. Moreover, it serves as a reliable diagnostic intervention that is not dependent on the handling of the operator. Owing to the recent technical advances and to its ability to yield reproducible results, CT is considered as the gold standard for diagnosis of pancreatic cancer. The recent CT scan uses triple-phase (non-contrast, arterial, and portal venous) helical multi-detector row CT (MDCT) that enables us to have 1 mm thin sections through the pancreas which are then reconstructed to have a three-dimensional tomogram (Bronstein et al. 2004). Specific algorithms and protocols have also been developed for pancreatic CT that resolves the complexities of the organ by enhancing the contrast. It is possible to achieve almost 100% sensitivity for pancreatic lesions greater than 2 cm through CT imaging. Moreover, involvement of tumor vasculature is assessed and classified into different types through CT upon which the resectability can be judged. Table 8.3 describes the criteria for this classification with interpretation as adapted from Tamm et al. AJR (2003) (Tamm et al. 2003). However, the reliability of the detection may be compromised if other pathophysiological conditions like focal chronic pancreatitis or autoimmune pancreatitis co-exist.

Advantages Overall CT scan is preferred for pancreatic cancer because of its wide availability, noninvasiveness, and easily reproducible images. Most importantly, it offers a preoperative evaluation of the tumor vasculature that is useful in deciding the approach during surgery (Tamm et al. 2003).

TABLE 8.3
Criteria for the Classification of Involvement of Tumor Vasculature on the Basis of CT Scan

Type	Criterion	Interpretation
A	Well separated tumor, no vessels involved	Resectable tumor
B	Tumor separated from distinct parenchyma	Resectable tumor
C	Convex point of contact with adjacent vessels	Ambiguous involvement of the vessels
D	Concave point of contact with adjacent vessels	Tumor can be resected with partial resection of the vessel
E	Tumor surrounding adjacent vessels	No tumor resection possible without negative margin
F	Tumor occlusion with the adjacent vessels	No tumor resection possible without negative margin.

Disadvantages Despite its reproducibility, longitudinal monitoring of the tumor during therapy may yield ambiguous results if the scanning procedure is altered by the technician. Moreover, small metastatic nodules (<2 cm) may not be observed through CT scan.

8.3.3.2.3 Endoscopic Ultrasound in Pancreatic Cancer Management

Endoscopic US uses a high-frequency (7.5- to 12-MHz) sonographic transducer that is introduced into the gastrointestinal tract via a side-viewing endoscope. This modality, therefore, provides real-time cross-sectional images of the gastrointestinal wall with detailed structure of adjacent soft-tissues. It transmits high-frequency sound waves through the upper GI tract to detect abnormalities in the pancreas. EUS is considered to have a similar (or better) sensitivity and specificity for the diagnosis of pancreatic cancer as compared to that of MDCT (Soriano et al. 2004). The important feature of this modality is that subtle pancreatic anomalies like pancreatic duct strictures and small neuroendocrine tumors can be easily picked up and distinguished from other malignant tumors. It is very efficient in detecting tumors smaller than 1 cm, which is difficult for other modalities. It can also distinguish between malignant as well as pre-malignant cystic lesions such as intraductal papillary mucinous neoplasms with its characteristic findings of intramural nodules and its communication with the main pancreatic duct (Santo 2012, Canto et al. 2012). During diagnostic evaluation of the patients, EUS allows the sampling of the tumor mass, regional nodes, liver lesions, ascites, and malignant cyst fluid. Besides, it also assesses tumor resectability. The EUS-FNA technique allows multiple tumor sampling with a minimally invasive procedure and therefore avoids complications due to imaging interventions. In a study carried out on a cohort of 81 patients, the accuracy of spiral CT, EUS, and EUS-FNA was evaluated for diagnosing pancreatic cancer and was found to be 74%, 94%, and 88% respectively. The authors also suggest that EUS with FNA can serve as a valuable extension to newer high-resolution multidetector spiral CT for diagnostic evaluation of patients with suspected pancreatic cancer (Agarwal et al. 2004). In another study, the diagnostic accuracy of EUS-FNA was evaluated in patients with suspected pancreatic cancer along with the assessment of procedure-associated complications. This study demonstrated that it is a highly accurate process for identifying patients with suspected pancreatic cancer, especially when other modalities fail. Major procedure-associated complications are rare and minor complications are similar to those reported for upper endoscopy. In spite of these advantages, the accuracy of EUS still remains dependent upon operator. Further, use of EUS-FNA is still quite limited as it is not yet available at many community hospitals.

Advantages This is a very effective imaging modality as far as the detection of small metastatic tumors is concerned. Liver metastases can also be reliably observed.

Disadvantages However, poor penetration often compromises the diagnosis. Moreover, for staging, CT serves as a better option.

8.3.3.2.4 Endoscopic Retrograde Cholangiopancreatography (ERCP) and Magnetic Resonance Cholangiopancreatography in Pancreatic Cancer Management

ERCP was developed with an intention of visualizing the duodenum and ampulla, as well as delineation of the biliary and pancreatic ductal systems directly. However, its extended application includes collection of brush samples for cytology and intraductal biopsy samples. During brush biopsy, a small brush is passed through the endoscope to rub off cells from the pancreatic duct that can be used for cytological testing. ERCP can be used as a confirmatory test where EUS-FNA results are inconclusive. However, ERCP is an invasive technique and a person undergoing the test may suffer from post ERCP-pancreatitis. Therefore, this technique is restricted to palliating patients with obstructive jaundice from pancreatic cancer to understand the therapeutic indications (van der Gaag et al. 2010).

Magnetic resonance cholangiopancreatography (MRCP) is better suited for outlining the anatomy of the biliary and pancreatic tract as compared to CT. It is also useful in deciding the course of endoscopic therapy. MRI of the pancreas can provide valuable information about solid tumors of the pancreas as well as cystic neoplasms and is generally comparable to MDCT in determining resectability of the tumor (Lee et al. 2010).

Advantages The biggest advantage of this techniques is that it offers multiple biopsies through brush sampling. Treatment response can also be evaluated in terminally ill patients. MRCP offers precise detection of the tumor and adjacent organs.

Disadvantages These modalities are not locally available. ERCP can also cause other clinical complications during the procedure.

8.3.3.2.5 Multimodality Imaging

PET-CT As described previously, CT is also used in conjunction with ^{18}F-FDG-CT for real-time follow up of primary tumor/metastases imaging post treatment (Tamm et al. 2003).

8.3.3.3 Comparison among Various Imaging Modalities

Each of the imaging modalities applied for detection for pancreatic cancer have their respective advantages and disadvantages and hence preference over one another is based upon the intention of the diagnosis as described in detail in Table 8.4. Although the current imaging modalities have improved the overall accuracy, these modalities often fail to detect lesions <1 cm in size. Moreover, some techniques like ERCP have procedure-related risks. Additionally, the high cost of these procedures discourages their use for the follow-up of asymptomatic cases. All of these drawbacks warrant further development in this field that will make imaging safer and cost effective for these patients.

8.3.4 Ovarian Cancer

8.3.4.1 Introduction to Ovarian Cancer

Despite recent advances in the field surgery, chemotherapy, and basic cancer biology, ovarian cancer, the silent lady killer remains as one of the most deadly gynecological malignancies. Ovarian cancer is a complex heterogeneous disease consisting of several subtypes differing in various pathological features. According to the World Health Organization, ovarian cancer is broadly classified into three major subtypes based on the cell of origin, epithelial ovarian cancer (EOC), germ cell ovarian cancer, and sex cord-stromal tumors. EOC is the most prevalent subtype among them (Chen et al. 2003). According to FIGO classification, ovarian cancer is classified into four stages based on spread of the diseases and each stage has varying effect over five-year survival. According to the Surveillance, Epidemiology, and End Results (SEER) database of the National Cancer Institute (NCI), if a case is diagnosed in stage 1 when the tumor is confined to the ovary, the relative five-year survival rate is about 90% while stage 4 patients have relative five-year survival rate of 10–28%. Recently, a more simplistic model of epithelial ovarian cancer based on the aggressiveness and molecular features was proposed by Kurman and Shih Ie (Kurman and Shih Ie 2010). According to this model, EOC can be classified in to two groups, type 1 tumors comprising of low-grade serous, low-grade endometrioid, clear cell, mucinous, and transitional carcinomas. Nearly two–thirds of these tumors carry a mutation in *KRAS*, *BRAF*, and *ERBB2* gene. While type II tumors consisting of high-grade serous, high-grade endometrioid, and undifferentiated carcinomas are more aggressive and represent more advanced stage of the disease. These tumors lack genetic stability and show high percentage of p53 mutation and *CCNE1* gene amplification.

8.3.4.2 Diseases Management

The major challenge remains in field of ovarian cancer is late detection due to its asymptomatic characteristics. The symptoms are generally similar to the symptoms of gastrointestinal abnormalities. Diagnosis mainly relies upon upregulated level of biomarkers like CA-125 and HE 4 (Wilbaux et al. 2014, Li et al. 2009). Advanced ovarian cancer leads to a median CA-125 value of more than 400 U/mL (Van Calster et al. 2011) but CA 125 level alone cannot be used as sole predictor of ovarian cancer

TABLE 8.4
Comparative Look at the Various Imaging Modalities Associated with Pancreatic Cancer Management

Imaging Modality	Abdominal Ultrasound	Computed Tomography	Endoscopic Ultrasound	Endoscopic Retrograde Cholangiopancreatography/ Magnetic Resonance Cholangiopancreatography
Sensitivity	90%	89%–97%	92%	40%–60%
Specificity	95%	95%	96%	91%–100%
Advantages	1. Provides basis for conformation by CT scan 2. Wide availability 3. Noninvasive	1. Wide availability 2. Noninvasive 3. Easily reproducible	1. Tumor masses <3 cm can be reliably detected 2. Small liver metastasis and ascites can be detected	1. Offers repetitive biopsies
Limitations	1. Not reliable for masses <3 cm 2. Fails to distinguish cancer from chronic or autoimmune pancreatitis	1. Operator dependent 2. Detects metastasis with low sensitivity	1. Limited penetration 2. It cannot be used for staging of metastasis	1. It can cause other clinical complications

because other malignancies liked uterine, pancreatic malignancies, and certain benign conditions also lead to increase in CA 125 level. Thus, imaging modalities play a crucial role in detection and diagnosis. The use of invasive techniques is not generally preferred for ovarian cancer though techniques like omentectomy or laparotomy may play a role in biopsies before or after treatment. Once an individual is confirmed for ovarian cancer, surgical removal of tumor mass followed by chemotherapy or neo-adjuvant chemotherapy is prescribed depending on the extent of tumor mass. Combination of cisplatin and paclitaxel treatment is currently the first line chemotherapy for ovarian cancer, which shows a maximum objective response. (Jelovac and Armstrong 2011).

8.3.4.3 Role of Imaging in the Management of Ovarian Cancer

Imaging modalities play a crucial role in every stage of ovarian cancer management, from initial diagnosis to monitoring the surgical and chemotherapy treatment. Each of the standard imaging techniques like US, CT, MRI, and PET has its own advantages and are preferred over one another depending upon the required application in detection, staging, preoperative planning, and surveillance of ovarian cancer.

8.3.4.3.1 Ultrasound Imaging in Ovarian Cancer Management

Trans-vaginal ultra sound (TVUS) imaging is the first line of imaging modality for primary diagnosis of ovarian cancer due to high sensitivity and specificity values. Granberg et al. (1989), for the first time, reported the correlation between US features and the histology of ovarian cancer and concluded that morphologically different tumors have varying risk of malignancy (Granberg, Wikland, and Jansson 1989). With the development of different modes of US imaging, several morphological characteristics like the internal structure of cysts, thickness and extent of septa, echogenicity and acoustic shadow patterns, etc. have been associated with extent of tumor malignancy (Sassone et al. 1991).

Since diagnosis of ovarian cancer is difficult due to its asymptomatic nature, several authors have reported the use of scoring system based on gray-scale (two-dimensional US) images. One of the important scoring system developed British Royal College of Obstetricians and Gynecologists (RCOG) is the Risk of Malignancy Index (RMI). Tumors were divided into benign and malignant on the basis of RMI cut off value of 200, menopausal status, CA125 level, and US image characteristics. This method had a sensitivity of 85% and a specificity of 97% for diagnosing the disease (Jacobs et al. 1990). In the late 1990s, a new method called "pattern recognition" was introduced and ovarian tumors were classified according to their ultrasonographic characteristics such as volume, localization, and associated features like presence of ascites, carcinomatosis, thickness of internal structure (wall, inner contour/papillary projections, septa, solid areas), echogenicity, and the presence of shadow and/or Crescent sign (Klangsin et al. 2013).

With the advancement of technology, US imaging has been multiplexed with color Doppler technique, which leads to an increase in diagnosis accuracy. The Doppler technique is different from gray-scale US imaging. In this technique, sound waves are reflected back by the moving red blood cells, which results in the increase or decrease in the intensity of these waves based on the direction and velocity of the red blood cells (Sofferman 2012). Doppler imaging is used to understand the vascularity of a tumor since malignant tumors are characterized with high level of blood flow and a denser vascular network while benign tumors show little or no vascularity. Several reports suggested the benefit of using color Doppler along with gray-scale US imaging. A study by Alcazar et al. (2003) of imaging 705 adnexal masses in 705 patients showed a superior diagnostic performance of color Doppler compared to other scoring systems simply based on morphological features (Alcazar et al. 2003). In 2013, International Ovarian Tumor Analysis (IOTA) group developed several models based on Doppler and US imaging along with other factors for assessment of malignancy of adnexal tumors (Kaijser et al. 2013).

TVS with colored Doppler is still the first line and best imaging technique used for presurgical assessment of ovarian cancer. This assessment is very important for further steps of ovarian cancer management and primarily aims for the following goals: 1. Confirmation of the benign or malignant nature of the detected tumor/s in the ovaries; 2. Confirmation that the tumor/s primarily originate from ovary and are not metastatic nodule of tumors of gastrointestinal tract or pancreas; 3. Assessment to determine the tumor burden and the metastatic spread of the tumor/s, and; 4. Diagnosis of other complications (Forstner et al. 2010). Trans-abdominal US is preferred when lesions are larger than the field of view of TVS (Forstner, Meissnitzer, and Cunha 2016).

Several multicentric trials had been carried out for the assessment of the importance of US imaging. The UK collaborative trial of ovarian cancer screening aimed to assess the accuracy of US imaging in a large cohort of 202,638 patients and reported the high specificity of US imaging. Another trial by van Nagell et al. (2011) demonstrated that even asymptomatic patients can be diagnosed with US in early stages of ovarian cancer (van Nagell et al. 2011).

Advantages The major advantage of US imaging is that it is an inexpensive, non-invasive, and nonradiative technique. The combination of trans-vaginal and trans-abdominal US with color Doppler imaging allows the detailed diagnosis and staging of ovarian cancer.

Disadvantages US technique cannot be used for proper screening of secondary liver metastasis in presence of large amount of ascites and also this technique is not suitable for bone metastasis resulting from spread of 1–2% ovarian cancer. Tumor infiltration in the retroperitoneal lymph node is often observed in the advanced stage of the disease (Camara and Sehouli 2009). US imaging of retroperitoneal metastasis fails to provide sufficient information in obese patients because of the higher amount of adipose tissue, which causes increased attenuation of sound waves. As a result US beam fails to penetrate the abdomen and retroperitoneal organ thus limiting US information (Modica, Kanal, and Gunn 2011).

8.3.4.3.2 Computed Tomography in Imaging Ovarian Cancer Management

Since the majority of the ovarian cancer patients are diagnosed at stage III or IV, the major way of management is either cytoreductive surgery or neo-adjuvant chemotherapy followed by cyto reduction. The criteria for optimal surgical resectability are versatile and depend on the expertise of the center. As a majority of these advanced patients show abdominal and pelvic metastasis thus screening for metastasis and staging plays a crucial role on effectivity of the cyto-reduction surgery (Bristow et al. 2002). CT plays a major role in staging and decision-making procedure to identify the resectability criteria. Thus, a detailed CT report examining the possible sites of tumor metastasis is deemed important. Ovarian tumors most frequently show pelvic invasion or transcoelmic spread into the peritoneal, subdiaphragmatic, mesenteric, and serosal surfaces (Sahdev 2016). Thus, a detailed examination of these sites is very important. Though US imaging is the most favored imaging modality for initial assessment and detection of ovarian cancer, MDCT is first line of imaging for preoperative staging due to its short examination time, reproducibility, and it provides far superior information on extra abdominal, lymph node, pleural, and pulmonary metastases than obtained through US (Tempany et al. 2000, Forstner 2007). A detailed CT report suggesting size, morphology, uterine endometrial thickening, bladder, bowel invasion or pelvic side-wall invasion, presence and amount of ascites in the pelvis or upper abdomen, and omental metastases is used for understanding the size, shape, and localization of the lesions and helps in determining the optimal resectability R(0) or near complete resectability R(1) during surgery (Forstner et al. 2010).

Several other studies also indicated the importance of CT for predicting surgical outcomes. To investigate the role of CT imaging in prediction of feasibility of optimal cytoreduction, Ferrandina et al. (2009) assessed 195 patients suspected of advanced ovarian cancer. Various parameters like peritoneal thickening, peritoneal implants >2 cm, bowel mesentery involvement, omental cake, pelvic sidewall involvement, and/or hydroureter, suprarenal aortic lymph nodes >1 cm, infrarenal aortic lymph nodes >2 cm, superficial liver metastases >2 cm, and/or intraparenchimal liver metastases of any size, large volume ascites (>500 ml) were used along with age, CA 125 level, and Eastern Cooperative Oncology Group performance status table (ECOG-PS) for prediction of optimal cytoreduction. ECOG-PS describes the overall impact of a disease on the patient's daily life; acting as an indicator of the influence of the disease on an individual's ability to take care of herself, perform daily activities, and their physical ability to work. It grades patients from a scale of 0 to 5 where 0 signifies no significant change in the patient's physical performance before and after the disease while 5 signifies death of the patient (Oken et al. 1982). Using all these criteria, a predictive index (PI) score was calculated for each patient that reliably suggested the usefulness of CT for predicting optimal cytoreduction. It was evident from this study that the predictive value of CT for foretelling optimal cytoreduction improved when CT scan parameters were integrated with ECOG-PS data (Ferrandina et al. 2009). Another retrospective study in cohort of 118 patients aimed to identify features for predicting sub-optimal cytoreduction, showed preoperative CT scan depicting presence of omental extension to stomach or spleen and presence of inguinal or pelvic lymph nodes >2 cm indicating suboptimal cytoreduction procedure (Kim et al. 2014).

Advantages The major advantage of CT is its speed. Thus, a large area of the body can be scanned for abnormalities in a short time. It can generate a whole body scan within a minute (Pelc 2014).

Disadvantages The major limitation of CT is the use of toxic contrast agent, ionizing radiation etc. A multi-institutional retrospective by Axtell et al. (2007) aimed at addressing the specificity and sensitivity of different CT parameters, demonstrated that the CT predictors show high sensitivity and specificity within the original cohort that is used to determine the sensitivity or specificity of the CT parameters. The sensitivity and specificity of these predictors falls drastically when applied to a different cohort (Axtell et al. 2007). Also, the specificity and sensitivity of CT for preoperative evaluation depends upon the anatomical region as shown in the study by Cerci et al. (2016) in a cohort of 114 patients. The sensitivity and specificity of detection of uterine and umbilical spread was 66% and 89% respectively while that of cervical involvement was 100% and 80% respectively (Cerci et al. 2016). Lymph node assessment is another pitfall of CT imaging as it can only differentiates lymph nodes into benign and malignant based on size (Forstner et al. 2010). Another limitation of CT lies in soft tissue contrast and thus is often difficult for differentiating cysts into benign and malignant. Also the sensitivity of CT declines to 25–50% for tumor masses that are less than 1 cm diameter (Coakley et al. 2002).

8.3.4.3.3 Magnetic Resonance Imaging in Ovarian Cancer Management

Characterization of certain adnexal masses remains indeterminate after US and CT imaging. MRI plays a major role in characterization of these adnexal masses. MR imaging has similar sensitivity as US imaging but has higher specificity and accuracy (84% and 89%) compared to US (40% and 64%) (Coakley et al. 2002). Conventional MR imaging produces a sequence of T1-weighted, T2-weighted, and fat suppressed T1-weighted images that provide high resolution anatomic features of the targeted region. MRI imaging features suggesting the presence of papillary projection within cystic masses, presence of ascites, and necrosis are strong predictors of malignancy (Sohaib et al. 2003). The signal intensity of T1, T2, and fat suppressed images varies depending upon the cystic, fibrous, and hemorrhagic content of the tumor. Different characteristics of signal intensity on MRI images is well reviewed by Mohaghegh et al. (2012). Briefly, an adnexal mass is defined as benign when it shows well-defined MR characteristics like high signal intensity on T1-weighted images (indicative of fat or blood) with subsequent loss of signal intensity on fat-suppressed images (indicative of fat), high signal intensity on T1-weighted fat-suppressed images (indicative of blood), and low signal intensity on T2-weighted images (indicative of fibrous tissue or hemosiderin). Generally when an adnexal mass is defined by low signal intensity on T2-weighted image it is classified as noninvasive benign tumor (Mohaghegh and Rockall 2012).

The use of contrast agent like gadolinium in contrast-enhanced MR or dynamic contrast enhanced MR (DCE-MR) imaging further helps in understanding malignant characteristics of adnexal masses by differentiating between solid components and papillae from clots and debris within a cyst through specific enhancement of the signal. Li et al. (2015) showed the importance of dynamic contrast enhanced MRI in a cohort of 48 subjects on the basis of certain criteria like enhancement pattern, time-signal intensity, and signal intensity at initial 60s and time to peak within 200s. There were significant differences for each of these of criteria for malignant and benign tumors. The benign tumors showed a lower signal after initial 60s and prolonged time to peak within 200s compared to malignant ones (Li, Hu et al. 2015).

Diffusion weighted imaging (DWI) distinguishes malignant tumors based on restricted diffusion and the low apparent diffusion coefficient value of the tumor. Diffusion weighted imaging is quite helpful in detecting metastatic spread on the pelvic surface. In a cohort of 32 patients, whole-body MRI and DWI had been shown to have 91% accuracy for peritoneal staging and 87% sensitivity in detection of retroperitoneal lymphadenopathies (Michielsen et al. 2014). The study also showed that MRI and DWI have higher accuracy in characterizing primary tumors and preoperative staging in comparison to PET and CT.

Advantages The major advantage of MRI is the ability to distinguish between benign and malignant tumors with high sensitivity and specificity. With the advancement of technology, DCE MRI, and DIW MRI along with conventional MRI are able to produce improved imaging features that help in characterization of adnexal masses especially for those cases where US and CT diagnosis remain inconclusive. The combination of various MRI techniques together are able to produce a detailed analysis of most of the adnexal mass with accuracy of 95% (Thomassin-Naggara et al. 2011). The main advantage of MRI in comparison to CT images is that MRI has a much higher soft tissue contrast and does not involve use of radiation.

Disadvantages MRI is used as a second line of diagnosis and only after US or CT. Also, MR imaging requires a longer acquisition time and presence of expert for every step of the procedure. MRI also has a low accuracy in staging ovarian cancer in comparison to CT (Sohaib and Reznek 2007).

8.3.4.3.4 Positron Emission Tomography in Imaging Ovarian Cancer Management

The role of PET in early diagnosis of ovarian cancer is not well established though certain studies hint the benefit of PET when combined with CT. A study by Castellucci et al. (2007) aiming to differentiate malignant versus benign tumors on the basis of uptake of ^{18}F-FDG, was reported to have sensitivity of 87% and specificity of 100% (Castellucci et al. 2007). Nam et al. in 2010 compared PET-CT with other imaging modalities like US, CT, and MRI to distinguish between benign, borderline, and malignant lesions in a cohort of 133 patients and reported higher accuracy of PET-CT in comparison to other modalities (Nam et al. 2010). Despite these facts PET-CT is still not considered as a choice for diagnosis in ovarian cancer because of its false-positive and false-negative results. The same study by Castellucci et al. (2007) though reported a specificity of 100%, had three to four false negative cases. Thus, other imaging modalities like US and MRI remains the choice of diagnostic imaging modality. There are certain reports that suggest the benefit of using PET for the diagnosis of infra- and supradiaphragmatic lymph-node metastases (Hynninen et al. 2013). PET plays an important role in treatment planning and assessment of the effectivity of the treatment since metabolic changes may be more indicative of the tumor's response to therapy. ^{18}F-FDG-PET has been found to be very useful in determining the effectiveness of neo-adjuvant chemotherapy in comparison to CA-125 level. Avril et al. (2005) showed that ^{18}F-FDG-PET can predict the outcome and survival better after rounds of chemotherapy by evaluating the metabolic function of the tumor mass. PET-CT imaging has higher sensitivity of 95–97% and specificity of 80–100% for detection of recurrent ovarian cancer (Avril et al. 2005). Several meta-analysis studies showed the accuracy of PET-CT in diagnosis of recurrent ovarian cancer. Thrall et al. (2007) in a cohort of 39 patients retrospectively evaluated the role of PET-CT to detect recurrent diseases and predicted recurrence in three patients out of eight patients having clinical manifestation of recurrence but normal CA 125 levels. All these three patients showed recurrence within six months of follow up (Thrall et al. 2007). Similarly, Gu et al. (2009) reported higher sensitivity and specificity of PET-CT for detection of recurrent disease in comparison to CT or MRI (Gu et al. 2009).

Advantages The major advantage PET combined with anatomic data of CT is increased detection of cancerous lesion. Also, a whole-body scan with PET-CT can identify asymptomatic primary malignant tumors in secondary site/s in the body.

Disadvantages One of the major limitation of PET is inability to detect tumors less than 5 mm (Prakash, Cronin, and Blake 2010) PET-CT has several limitations in detection of abdominal lesion since ^{18}F-FDG-PET is excreted through urinary tract and can accumulate in the bowel thus generating high noise that interferes with the detection of abdominal or pelvic spread of ovarian cancer (Avril 2004). ^{18}F-FDG can also accumulate in the luteal phase of menstrual cycle during ovulation and during inflammation thus hampering the detection of tumors (Kim et al. 2005). The cost of ^{18}F-FDG-PET is also higher than US, CT, and MRI.

8.3.4.4 Comparison among Various Imaging Modalities

Each of imaging modalities have their respective advantages and disadvantages and are preferred over one another based on demand of the diagnostic question as described in detail in Table 8.5.

8.3.4.5 Future Perspectives

US imaging is the first line of imaging in ovarian cancer having high specificity and sensitivity. Its low cost and easy availability makes it a favorable imaging modality of choice. US with color Doppler image provides morphological and perfusion biology of the targeted site (Sofferman 2012). In recent years, much attention has been given to the vascularity or angiogenesis of a tumor. US imaging has very low contrast between blood and tissue thus failing to provide optimal information. Multiplexing with color Doppler helps in improving contrast to some extent but only can detect blood vessels greater than 100–200 μm (Fleischer et al. 2010). To address these issues research has been focused on enhancement of contrast between blood and tissue by use of microbubbles. A microbubble generally contains low-diffusivity gases such as nitrogen or perfluorocarbon with in the coating of biodegradable material (Quaia 2007). These microbubbles are highly echogenic and produce an enhancement in contrast due to acoustic mismatch. Several preclinical studies report the use of microbubbles in cardiovascular and renal disease (Arnold et al. 2010). Along with their imaging advantage microbubbles can be used for therapeutic purpose as carriers of drug or genes. A few studies also indicated a significant difference in contrast enhancement pattern between benign and malignant ovarian tumors. Testa et al. (2009) in a cohort of 72, adnexal masses showed that

TABLE 8.5
Comparative Look at the Various Imaging Modalities Associated with Ovarian Cancer Management

	Ultrasound	Computed Tomography	Magnetic Resonance Imaging	Positron Emission Tomography/ Computed Tomography
Role in Cancer Management	First line imaging modality for detection, staging, follow up and image guided biopsies	Primary role in staging and second line imaging modality for image-guided modality	In characterization of ovarian masses which are remains undetermined in US. Second line of imaging modality for staging after CT	In characterization of lymph node metastasis and follow up of patients undergoing chemotherapy
Sensitivity	96%	87%	91.9%	97.9%
Specificity	90%	84%	88.4%	73%
Cost	Least expensive among other	Much more expensive than US	More expensive than both US and CT	Most expensive among the four
Availability	Easily available	In many hospitals	In advanced hospitals and centers	In few advanced hospitals and centers
Duration of the Procedure	15–20 min	1 min	30–45 min	30 min
Special Preparation before Imaging	None	Fasting for four hours is required	Antiperistaltic agents	Same as CT with one additional hour of rest
Contrast Agent and Radiation Exposure	Generally, none expect in case of contrast enhanced US and no exposure to radiation	Iodine based contrast agent and 10–20 mSv radiation exposure	Gadolinium-based and no exposure	FDG and iodine based and 20–30 mSv radiation exposure.
Associated Hazard or Toxicity	None	Contraindication for iodine-based contrast agent: renal insufficiency, hyperthyroidism, iodine allergy	Claustrophobia and removal of all metal components (Cochlear implants, cardiac pacemaker)	Same as CT

Sources: Fischerova, D., and A. Burgetova, *Best Pract Res Clin Obstet Gynaecol*, 28 (5):697–720, 2014; Timmerman, D. et al., *Ultrasound Obstet Gynecol*, 13 (1):11–6, 1999; Dodge, J. E. et al., *Curr Oncol*, 19 (4):e244–57, 2012; Nam, E. J. et al., *Gynecol Oncol*, 116 (3):389–94, 2010.

malignant tumors had significantly higher peak contrast signal intensity and area under the contrast signal intensity curve than benign and borderline tumors (Testa et al. 2009). Recent studies in ovarian cancer using contrast-enhanced microbubbles have a specificity ranging from 90–96% (Fleischer et al. 2008). Another development in the field of US is development of three-dimensional ultrasound and three-dimensional power Doppler ultrasound that can be used for generating a volumetric image. These images provide similar morphological information to gray-scale US, but they give better information about the dimension and vascularity of adnexal masses. In a study by Alcazar et al. (2013) three-dimensional volumetric US images were used in specific training program for diagnosis of adnexal masses. It was reported that after 170–185 examinations, the observer attained a sensitivity of >95% and a specificity of >90% (Alcazar et al. 2013).

Multiplexing of PET with MR (PET-MR) has been an emerging field of imaging of gynecological cancer. Currently PET-CT is generally used for diagnosis of lymph node metastasis, recurrence, and staging. Multiplexing of MR instead of CT with PET reduces unwanted exposure to radiation without any loss of diagnostic performance. This is especially important for individuals undergoing multiple examinations. PET-MR can also be used for better detection of local and distant metastatic diseases as MR imaging possesses better contrast compared to CT. Different modes of MR imaging along with PET may give better characterization of lesions with high sensitivity and specificity (Ponisio, Fowler, and Dehdashti 2016).

8.4 CONCLUDING REMARKS

Both unimodal (MR, CT, PET-SPECT, and US) and hybrid imaging techniques have been thoroughly integrated in screening and management of glioblastoma, lung, pancreatic, and ovarian cancer. Patients are gaining significant benefits and PFS and OS have been improving. However, we have a long way to go before we have a complete grip on these diseases. Further improvement in instruments with sensitive detectors, comprehensive algorithms, better acquisition sequences, and improvement in probes would lead us to precision medicine. These techniques are yet to detect tumor heterogeneity, a major caveat to provide best therapeutic efficacy, which might be possible in future with diligent efforts in imaging research.

REFERENCES

Adamska, A., A. Domenichini, and M. Falasca. 2017. "Pancreatic ductal adenocarcinoma: Current and evolving therapies." *Int J Mol Sci* 18 (7):E1338.

Agarwal, B., E. Abu-Hamda, K. L. Molke, A. M. Correa, and L. Ho. 2004. "Endoscopic ultrasound-guided fine needle aspiration and multidetector spiral CT in the diagnosis of pancreatic cancer." *Am J Gastroenterol* 99 (5):844–50.

Al-Okaili, R. N., J. Krejza, S. Wang, J. H. Woo, and E. R. Melhem. 2006. "Advanced MR imaging techniques in the diagnosis of intraaxial brain tumors in adults." *Radiographics* 26 Suppl 1:S173–89.

Albright, A. L., R. J. Packer, R. Zimmerman, et al. 1993. "Magnetic resonance scans should replace biopsies for the diagnosis of diffuse brain stem gliomas: a report from the Children's Cancer Group." *Neurosurgery* 33 (6):1026–9; discussion 1029–30.

Alcazar, J. L., L. Diaz, P. Florez, S. Guerriero, and M. Jurado. 2013. "Intensive training program for ultrasound diagnosis of adnexal masses: protocol and preliminary results." *Ultrasound Obstet Gynecol* 42 (2):218–23.

Alcazar, J. L., L. T. Merce, C. Laparte, M. Jurado, and G. Lopez-Garcia. 2003. "A new scoring system to differentiate benign from malignant adnexal masses." *Am J Obstet Gynecol* 188 (3):685–92.

Alexiou, G. A., S. Tsiouris, S. Voulgaris, A. P. Kyritsis, and A. D. Fotopoulos. 2012. "Glioblastoma multiforme imaging: the role of nuclear medicine." *Curr Radiopharm* 5 (4):308–13.

An, S. J., Z. H. Chen, J. Su, et al. 2012. "Identification of enriched driver gene alterations in subgroups of non-small cell lung cancer patients based on histology and smoking status." *PLoS One* 7 (6):e40109.

Araki, T., T. Inouye, H. Suzuki, T. Machida, and M. Iio. 1984. "Magnetic resonance imaging of brain tumors: measurement of T1. Work in progress." *Radiology* 150 (1):95–8.

Arnold, J. R., T. D. Karamitsos, T. J. Pegg, et al. 2010. "Adenosine stress myocardial contrast echocardiography for the detection of coronary artery disease: a comparison with coronary angiography and cardiac magnetic resonance." *JACC Cardiovasc Imaging* 3 (9):934–43.

Avril, N. 2004. "GLUT1 expression in tissue and (18)F-FDG uptake." *J Nucl Med* 45 (6):930–2.

Avril, N., S. Sassen, B. Schmalfeldt, et al. 2005. "Prediction of response to neoadjuvant chemotherapy by sequential F-18-fluorodeoxyglucose positron emission tomography in patients with advanced-stage ovarian cancer." *J Clin Oncol* 23 (30):7445–53.

Axtell, A. E., M. H. Lee, R. E. Bristow, et al. 2007. "Multi-institutional reciprocal validation study of computed tomography predictors of suboptimal primary cytoreduction in patients with advanced ovarian cancer." *J Clin Oncol* 25 (4):384–9.

Becherer, A., G. Karanikas, M. Szabo, et al. 2003. "Brain tumour imaging with PET: a comparison between [18F]fluorodopa and [11C]methionine." *Eur J Nucl Med Mol Imaging* 30 (11):1561–7.

Benvenuti, L., S. Chibbaro, S. Carnesecchi, F. Pulera, and R. Gagliardi. 2005. "Automated three-dimensional volume rendering of helical computed tomographic angiography for aneurysms: an advanced application of neuronavigation technology." *Neurosurgery* 57 (1 Suppl):69–77; discussion 69–77.

Bergstrom, M., H. Lundqvist, K. Ericson, et al. 1987. "Comparison of the accumulation kinetics of L-(methyl-11C)-methionine and D-(methyl-11C)-methionine in brain tumors studied with positron emission tomography." *Acta Radiol* 28 (3):225–9.

Bleichner-Perez, S., F. Le Jeune, F. Dubois, and M. Steinling. 2007. "99mTc-MIBI brain SPECT as an indicator of the chemotherapy response of recurrent, primary brain tumors." *Nucl Med Commun* 28 (12):888–94.

Bristow, R. E., R. S. Tomacruz, D. K. Armstrong, E. L. Trimble, and F. J. Montz. 2002. "Survival effect of maximal cytoreductive surgery for advanced ovarian carcinoma during the platinum era: a meta-analysis." *J Clin Oncol* 20 (5):1248–59.

Bronstein, Y. Lisenko, E. M. Loyer, et al. 2004. "Detection of small pancreatic tumors with multiphasic helical CT." *AJR Am J Roentgenol* 182 (3):619–23.

Burger, P. C., and S. B. Green. 1987. "Patient age, histologic features, and length of survival in patients with glioblastoma multiforme." *Cancer* 59 (9):1617–25.

Bury, T., A. Barreto, F. Daenen, et al. 1998. "Fluorine-18 deoxyglucose positron emission tomography for the detection of bone metastases in patients with non-small cell lung cancer." *Eur J Nucl Med* 25 (9):1244–7.

Camara, O., and J. Sehouli. 2009. "Controversies in the management of ovarian cancer--pros and cons for lymph node dissection in ovarian cancer." *Anticancer Res* 29 (7):2837–43.

Canto, M. I., R. H. Hruban, E. K. Fishman, et al. 2012. "Frequent detection of pancreatic lesions in asymptomatic high-risk individuals." *Gastroenterology* 142 (4):796–804.

Castellucci, P., A. M. Perrone, M. Picchio, et al. 2007. "Diagnostic accuracy of 18F-FDG PET/CT in characterizing ovarian lesions and staging ovarian cancer: correlation with transvaginal ultrasonography, computed tomography, and histology." *Nucl Med Commun* 28 (8):589–95.

Castillo, M., J. K. Smith, L. Kwock, and K. Wilber. 2001. "Apparent diffusion coefficients in the evaluation of high-grade cerebral gliomas." *AJNR Am J Neuroradiol* 22 (1):60–4.

Cerci, Z. C., D. K. Sakarya, M. H. Yetimalar, et al. 2016. "Computed tomography as a predictor of the extent of the disease and surgical outcomes in ovarian cancer." *Ginekol Pol* 87 (5):326–32.

Chen, I. Y., and J. C. Wu. 2011. "Cardiovascular molecular imaging: focus on clinical translation." *Circulation* 123 (4):425–43.

Chen, V. W., B. Ruiz, J. L. Killeen, et al. 2003. "Pathology and classification of ovarian tumors." *Cancer* 97 (10 Suppl):2631–42.

Chen, W., D. H. Silverman, S. Delaloye, et al. 2006. "18F-FDOPA PET imaging of brain tumors: comparison study with 18F-FDG PET and evaluation of diagnostic accuracy." *J Nucl Med* 47 (6):904–11.

Coakley, F. V., P. H. Choi, C. A. Gougoutas, et al. 2002. "Peritoneal metastases: detection with spiral CT in patients with ovarian cancer." *Radiology* 223 (2):495–9.

Cohen, B. A., E. A. Knopp, H. Rusinek, et al. 2005. "Assessing global invasion of newly diagnosed glial tumors with whole-brain proton MR spectroscopy." *AJNR Am J Neuroradiol.* 26 (9):2170.

Colella, S., P. Vilmann, L. Konge, and P. F. Clementsen. 2014. "Endoscopic ultrasound in the diagnosis and staging of lung cancer." *Endosc Ultrasound* no. 3 (4):205–12. doi: 10.4103/2303-9027.144510.

Conrad, C., and C. Fernández-del Castillo. 2013. "Preoperative evaluation and management of the pancreatic head mass." *J Surg Oncol.* 107 (1):23–32.

Dela Cruz, C. S., L. T. Tanoue, and R. A. Matthay. 2011. "Lung cancer: epidemiology, etiology, and prevention." *Clin Chest Med* 32 (4):605–44.

Dodge, J. E., A. L. Covens, C. Lacchetti, et al. 2012. "Management of a suspicious adnexal mass: a clinical practice guideline." *Curr Oncol* 19 (4):e244–57.

Erasmus, J. J., H. P. McAdams, S. E. Rossi, et al. 2000. "FDG PET of pleural effusions in patients with non-small cell lung cancer." *AJR Am J Roentgenol* 175 (1):245–9.

Eskandary, H., M. Sabba, F. Khajehpour, and M. Eskandari. 2005. "Incidental findings in brain computed tomography scans of 3000 head trauma patients." *Surg Neurol* 63 (6):550–3; discussion 553.

Ferda, J., J. Kastner, P. Mukensnabl, et al. 2010. "Diffusion tensor magnetic resonance imaging of glial brain tumors." *Eur J Radiol* 74 (3):428–36.

Ferlay, J., I. Soerjomataram, R. Dikshit, et al. 2015. "Cancer incidence and mortality worldwide: sources, methods and major patterns in GLOBOCAN 2012." *Int J Cancer* 136 (5):E359–86.

Ferrandina, G., G. Sallustio, A. Fagotti, et al. 2009. "Role of CT scan-based and clinical evaluation in the preoperative prediction of optimal cytoreduction in advanced ovarian cancer: a prospective trial." *Br J Cancer* 101 (7):1066–73.

Fischerova, D., and A. Burgetova. 2014. "Imaging techniques for the evaluation of ovarian cancer." *Best Pract Res Clin Obstet Gynaecol* 28 (5):697–720.

Fleischer, A. C., A. Lyshchik, R. F. Andreotti, et al. 2010. "Advances in sonographic detection of ovarian cancer: depiction of tumor neovascularity with microbubbles." *AJR Am J Roentgenol* 194 (2):343–8.

Fleischer, A. C., A. Lyshchik, H. W. Jones, Jr., et al. 2008. "Contrast-enhanced transvaginal sonography of benign versus malignant ovarian masses: preliminary findings." *J Ultrasound Med* 27 (7):1011–8.

Fontana, R. S., D. R. Sanderson, L. B. Woolner, et al. 1986. "Lung cancer screening: the Mayo program." *J Occup Med* 28 (8):746–50.

Forstner, R. 2007. "Radiological staging of ovarian cancer: imaging findings and contribution of CT and MRI." *Eur Radiol* 17 (12):3223–35.

Forstner, R., M. Meissnitzer, and T. M. Cunha. 2016. "Update on imaging of ovarian cancer." *Curr Radiol Rep* 4:31.

Forstner, R., E. Sala, K. Kinkel, et al. 2010. "ESUR guidelines: ovarian cancer staging and follow-up." *Eur Radiol* 20 (12):2773–80.

Fueger, B. J., J. Czernin, T. Cloughesy, et al. 2010. "Correlation of 6-18F-fluoro-L-dopa PET uptake with proliferation and tumor grade in newly diagnosed and recurrent gliomas." *J Nucl Med* 51 (10):1532–8.

Galanis, E., J. C. Buckner, M. J. Maurer, et al. 2006. "Validation of neuroradiologic response assessment in gliomas: measurement by RECIST, two-dimensional, computer-assisted tumor area, and computer-assisted tumor volume methods." *Neuro Oncol* 8 (2):156–65.

Gerard, I. J., M. Kersten-Oertel, K. Petrecca, et al. 2017. "Brain shift in neuronavigation of brain tumors: a review." *Med Image Anal* 35:403–20.

Gerstner, E. R., Z. Zhang, J. R. Fink, et al. 2016. "ACRIN 6684: assessment of tumor hypoxia in newly diagnosed glioblastoma using 18F-FMISO PET and MRI." *Clin Cancer Res* 22 (20):5079–86.

Gould, M. K., C. C. Maclean, W. G. Kuschner, C. E. Rydzak, and D. K. Owens. 2001. "Accuracy of positron emission tomography for diagnosis of pulmonary nodules and mass lesions: a meta-analysis." *JAMA* 285 (7):914–24.

Granberg, S., M. Wikland, and I. Jansson. 1989. "Macroscopic characterization of ovarian tumors and the relation to the histological diagnosis: criteria to be used for ultrasound evaluation." *Gynecol Oncol* 35 (2):139–44.

Gu, P., L. L. Pan, S. Q. Wu, L. Sun, and G. Huang. 2009. "CA 125, PET alone, PET-CT, CT and MRI in diagnosing recurrent ovarian carcinoma: a systematic review and meta-analysis." *Eur J Radiol* 71 (1):164–74.

Gupta, N. C., J. S. Rogers, G. M. Graeber, et al. 2002. "Clinical role of F-18 fluorodeoxyglucose positron emission tomography imaging in patients with lung cancer and suspected malignant pleural effusion." *Chest* 122 (6):1918–24.

Hamilton, S. R., and L. A. Aaltonen. 2000. *Pathology and Genetics of Tumours of the Digestive System*. Vol. 48: IARC Press Lyon.

Hanahan, D., and R. A. Weinberg. 2000. "The hallmarks of cancer." *Cell* 100 (1):57–70.

Hein, P. A., C. J. Eskey, J. F. Dunn, and E. B. Hug. 2004. "Diffusion-weighted imaging in the follow-up of treated high-grade gliomas: tumor recurrence versus radiation injury." *AJNR Am J Neuroradiol* 25 (2):201–9.

Henschke, C. I., D. I. McCauley, D. F. Yankelevitz, et al. 1999. "Early Lung Cancer Action Project: overall design and findings from baseline screening." *The Lancet* 354 (9173): 99–105.

Hruban, R. H., and N. Fukushima. 2007. "Pancreatic adenocarcinoma: update on the surgical pathology of carcinomas of ductal origin and PanINs." *Mod Pathol* 20 Suppl 1:S61–70.

Huang, A. P., J. C. Tsai, L. T. Kuo, et al. 2014. "Clinical application of perfusion computed tomography in neurosurgery." *J Neurosurg* 120 (2):473–88.

Hynninen, J., J. Kemppainen, M. Lavonius, et al. 2013. "A prospective comparison of integrated FDG-PET/contrast-enhanced CT and contrast-enhanced CT for pretreatment imaging of advanced epithelial ovarian cancer." *Gynecol Oncol* 131 (2):389–94.

International Early Lung Cancer Action Program Investigators. 2007. "Computed tomographic screening for lung cancer: individualising the benefit of the screening." *Eur Respir J* 30 (5):843–7.

Jackson, A., J. O'Connor, G. Thompson, and S. Mills. 2008. "Magnetic resonance perfusion imaging in neuro-oncology." *Cancer Imaging* 8:186–99.

Jackson, R. J., G. N. Fuller, D. Abi-Said, et al. 2001. "Limitations of stereotactic biopsy in the initial management of gliomas." *Neuro Oncol* 3 (3):193–200.

Jacobs, I., D. Oram, J. Fairbanks, et al. 1990. "A risk of malignancy index incorporating CA 125, ultrasound and menopausal status for the accurate preoperative diagnosis of ovarian cancer." *Br J Obstet Gynaecol* 97 (10):922–9.

Jelovac, D., and D. K. Armstrong. 2011. "Recent progress in the diagnosis and treatment of ovarian cancer." *CA Cancer J Clin* 61 (3):183–203. 0113.

Jeong, D., C. Malalis, J. A. Arrington, et al. 2015. "Mean apparent diffusion coefficient values in defining radiotherapy planning target volumes in glioblastoma." *Quant Imaging Med Surg* 5 (6):835–45.

Kaijser, J., T. Bourne, L. Valentin, et al. 2013. "Improving strategies for diagnosing ovarian cancer: a summary of the International Ovarian Tumor Analysis (IOTA) studies." *Ultrasound Obstet Gynecol* 41 (1):9–20.

Kapoor, V., B. M. McCook, and F. S. Torok. 2004. "An introduction to PET-CT imaging." *Radiographics* 24 (2):523–43.

Karunanithi, S., P. Sharma, A. Kumar, et al. 2013. "18F-FDOPA PET/CT for detection of recurrence in patients with glioma: prospective comparison with 18F-FDG PET/CT." *Eur J Nucl Med Mol Imaging* 40 (7):1025–35.

Kato, T., J. Shinoda, N. Nakayama, et al. 2008. "Metabolic assessment of gliomas using 11C-methionine, [18F] fluorodeoxyglucose, and 11C-choline positron-emission tomography." *AJNR Am J Neuroradiol* 29 (6):1176–82.

Khan, M. N., A. M. Sharma, M. Pitz, et al. 2016. "High-grade glioma management and response assessment-recent advances and current challenges." *Curr Oncol* 23 (4):e383–91.

Kim, H. J., C. H. Choi, Y. Y. Lee, et al. 2014. "Surgical outcome prediction in patients with advanced ovarian cancer using computed tomography scans and intraoperative findings." *Taiwan J Obstet Gynecol* 53 (3):343–7.

Kim, S. K., K. W. Kang, J. W. Roh, et al. 2005. "Incidental ovarian 18F-FDG accumulation on PET: correlation with the menstrual cycle." *Eur J Nucl Med Mol Imaging* 32 (7):757–63.

Klangsin, S., T. Suntharasaj, C. Suwanrath, D. Kor-Anantakul, and V. Prasartwanakit. 2013. "Comparison of the five sonographic morphology scoring systems for the diagnosis of malignant ovarian tumors." *Gynecol Obstet Invest* no. 76 (4):248–53.

Knoepp, U. W., and J. G. Ravenel. 2006. "CT and PET imaging in non-small cell lung cancer." *Crit Rev Oncol Hematol* 58 (1):15–30.

Kono, K., Y. Inoue, K. Nakayama, et al. 2001. "The role of diffusion-weighted imaging in patients with brain tumors." *AJNR Am J Neuroradiol* 22 (6):1081–8.

Koyama, H., Y. Ohno, A. Kono, et al. 2008. "Quantitative and qualitative assessment of non-contrast-enhanced pulmonary MR imaging for management of pulmonary nodules in 161 subjects." *Eur Radiol* 18 (10):2120–31.

Kurihara, Y., S. Matsuoka, T. Yamashiro, A et al. 2014. "MRI of pulmonary nodules." *AJR Am J Roentgenol* 202 (3):W210–6.

Kurman, R. J., and M. Shih Ie. 2010. "The origin and pathogenesis of epithelial ovarian cancer: a proposed unifying theory." *Am J Surg Pathol* no. 34 (3):433–43.

Lee, E. S., and J. M. Lee. 2014. "Imaging diagnosis of pancreatic cancer: a state-of-the-art." *World J Gastroenterol* 20 (24):7864–77.

Lee, J. K., A. Y. Kim, P. N. Kim, M.-G. Lee, and H. K. Ha. 2010. "Prediction of vascular involvement and resectability by multidetector-row CT versus MR imaging with MR angiography in patients who underwent surgery for resection of pancreatic ductal adenocarcinoma." *Eur J Radiol* 73 (2):310–16.

Lell, M. M., K. Anders, M. Uder, et al. 2006. "New techniques in CT angiography." *Radiographics* 26 Suppl 1:S45–62.

Li, J., S. Dowdy, T. Tipton, et al. 2009. "HE4 as a biomarker for ovarian and endometrial cancer management." *Expert Rev Mol Diagn* 9 (6):555–66.

Li, L., K. Wang, X. Sun, et al. 2015. "Parameters of dynamic contrast-enhanced MRI as imaging markers for angiogenesis and proliferation in human breast cancer." *Med Sci Monit* 21:376–82.

Li, X., J. L. Hu, L. M. Zhu, et al. 2015. "The clinical value of dynamic contrast-enhanced MRI in differential diagnosis of malignant and benign ovarian lesions." *Tumour Biol* 36 (7):5515–22.

Livieratos, L. 2012. "Basic principles of SPECT and PET imaging." In *Radionuclide and Hybrid Bone Imaging*, edited by I. Fogelman, G. Gnanasegaran, and H. van der Wall, 345–59. Berlin, Heidelberg: Springer Berlin Heidelberg.

Mabray, M. C., R. F. Barajas, Jr., and S. Cha. 2015. "Modern brain tumor imaging." *Brain Tumor Res Treat* 3 (1):8–23.

Maravilla, K. R., and W. C. Sory. 1986. "Magnetic resonance imaging of brain tumors." *Semin Neurol* 6 (1):33–42.

Marom, E. M., S. Sarvis, J. E. Herndon, and E. F. Patz. 2002. "T1 lung cancers: sensitivity of diagnosis with fluorodeoxyglucose PET." *Radiology* no. 223 (2):453–9. doi: 10.1148/radiol.2232011131.

Massoud, T. F., and S. S. Gambhir. 2003. "Molecular imaging in living subjects: seeing fundamental biological processes in a new light." *Genes Dev* 17 (5):545–80.

Mattei, L., F. Prada, F. G. Legnani, et al. 2016. "Neurosurgical tools to extend tumor resection in hemispheric low-grade gliomas: conventional and contrast enhanced ultrasonography." *Childs Nerv Syst* 32 (10):1907–14.

McGirt, M. J., A. T. Villavicencio, K. R. Bulsara, and A. H. Friedman. 2003. "MRI-guided stereotactic biopsy in the diagnosis of glioma: comparison of biopsy and surgical resection specimen." *Surg Neurol* 59 (4):279–83.

Michielsen, K., I. Vergote, K. Op de Beeck, et al. 2014. "Whole-body MRI with diffusion-weighted sequence for staging of patients with suspected ovarian cancer: a clinical feasibility study in comparison to CT and FDG-PET/CT." *Eur Radiol* 24 (4):889–901.

Mirsadraee, S., D. Oswal, Y. Alizadeh, A. Caulo, and E. van Beek. 2012. "The 7th lung cancer TNM classification and staging system: review of the changes and implications." *World J Radiol* 4 (4):128–34.

Miura, F., T. Takada, H. Amano, et al. 2006. "Diagnosis of pancreatic cancer." *HPB* 8 (5):337–42.

Modica, M. J., K. M. Kanal, and M. L. Gunn. 2011. "The obese emergency patient: imaging challenges and solutions." *Radiographics* 31 (3):811–23.

Mohaghegh, P., and A. G. Rockall. 2012. "Imaging strategy for early ovarian cancer: characterization of adnexal masses with conventional and advanced imaging techniques." *Radiographics* 32 (6):1751–73.

Morris, Z., W. N. Whiteley, W. T. Longstreth, Jr., et al. 2009. "Incidental findings on brain magnetic resonance imaging: systematic review and meta-analysis." *BMJ* 339:b3016.

Muhm, J. R., W. E. Miller, R. S. Fontana, D. R. Sanderson, and M. A. Uhlenhopp. 1983. "Lung cancer detected during a screening program using four-month chest radiographs." *Radiology* 148 (3):609–15.

Mullins, M. E., G. D. Barest, P. W. Schaefer, et al. 2005. "Radiation necrosis versus glioma recurrence: conventional MR imaging clues to diagnosis." *AJNR Am J Neuroradiol* 26 (8):1967–72.

Nam, E. J., M. J. Yun, Y. T. Oh, et al. 2010. "Diagnosis and staging of primary ovarian cancer: correlation between PET/CT, Doppler US, and CT or MRI." *Gynecol Oncol* 116 (3):389–94.

National Lung Screening Trial Research Team, T. R. Church, W. C. Black, et al. 2013. "Results of initial low-dose computed tomographic screening for lung cancer." *N Engl J Med* 368 (21):1980–91.

Nomori, H., K. Watanabe, T. Ohtsuka, et al. 2004. "Evaluation of F-18 fluorodeoxyglucose (FDG) PET scanning for pulmonary nodules less than 3 cm in diameter, with special reference to the CT images." *Lung Cancer* 45 (1):19–27.

Ohno, Y., H. Koyama, Y. Onishi, et al. 2008. "Non-small cell lung cancer: whole-body MR examination for M-stage assessment—utility for whole-body diffusion-weighted imaging compared with integrated FDG PET/CT." *Radiology* 248 (2):643–54.

Oken, M. M., R. H. Creech, D. C. Tormey, et al. 1982. "Toxicity and response criteria of the Eastern Cooperative Oncology Group." *Am J Clin Oncol* 5 (6):649–55.

Oken, M. M., W. G. Hocking, P. A. Kvale, et al. 2011. "Screening by chest radiograph and lung cancer mortality: the Prostate, Lung, Colorectal, and Ovarian (PLCO) randomized trial." *JAMA* 306 (17):1865–73.

Pelc, N. J. 2014. "Recent and future directions in CT imaging." *Ann Biomed Eng* 42 (2):260–8.

Ponisio, M. R., K. J. Fowler, and F. Dehdashti. 2016. "The emerging role of PET/MR imaging in gynecologic cancers." *PET Clin* no. 11 (4):425–40. doi: 10.1016/j.cpet.2016.05.005.

Prakash, P., C. G. Cronin, and M. A. Blake. 2010. "Role of PET/CT in ovarian cancer." *AJR Am J Roentgenol* 194 (6):W464–70.

Provenzale, J. M., S. Mukundan, and D. P. Barboriak. 2006. "Diffusion-weighted and perfusion MR imaging for brain tumor characterization and assessment of treatment response." *Radiology* 239 (3):632–49.

Quaia, E. 2007. "Microbubble ultrasound contrast agents: an update." *Eur Radiol* 17 (8):1995–2008.

Rees, J. 2003. "Advances in magnetic resonance imaging of brain tumours." *Curr Opin Neurol* 16 (6):643–50.

Rickes, S., K. Unkrodt, H. Neye, K. W. Ocran and W. Wermke. 2002. "Differentiation of pancreatic tumours by conventional ultrasound, unenhanced and echo-enhanced power Doppler sonography." *Scand J Gastroenterol* 37 (11):1313–20.

Ridge, C. A., A. M. McErlean, and M. S. Ginsberg. 2013. "Epidemiology of lung cancer." *Semin Intervent Radiol* 30 (2):93–8.

Sahdev, A. 2016. "CT in ovarian cancer staging: how to review and report with emphasis on abdominal and pelvic disease for surgical planning." *Cancer Imaging* 16 (1):19.

Santo, E. 2012. "Pancreatic cancer imaging: which method?" *Asia-Pacific Cancer Imaging* 1 (1).

Sassone, A. M., I. E. Timor-Tritsch, A. Artner, C. Westhoff, and W. B. Warren. 1991. "Transvaginal sonographic characterization of ovarian disease: evaluation of a new scoring system to predict ovarian malignancy." *Obstet Gynecol* 78 (1):70–6.

Shackelford, D. B., E. Abt, L. Gerken, et al. 2013. "LKB1 inactivation dictates therapeutic response of non-small cell lung cancer to the metabolism drug phenformin." *Cancer Cell* 23 (2):143–58.

Shrikhande, S. V., G. Barreto, and A. Koliopanos. 2009. "Pancreatic carcinogenesis: the impact of chronic pancreatitis and its clinical relevance." *Indian J Cancer* no. 46 (4):288.

Sizoo, E. M., H. R. Pasman, L. Dirven, et al. 2014. "The end-of-life phase of high-grade glioma patients: a systematic review." *Support Care Cancer* 22 (3):847–57.

Smith, J. S., I. Tachibana, S. M. Passe, et al. 2001. "PTEN mutation, EGFR amplification, and outcome in patients with anaplastic astrocytoma and glioblastoma multiforme." *J Natl Cancer Inst Monogr.* 93 (16):1246–56.

Sofferman, R. A. 2012. "Physics and principles of ultrasound." In *Ultrasound of the Thyroid and Parathyroid Glands*, edited by R. A. Sofferman and A. T. Ahuja, 9–19. New York: Springer.

Sohaib, S. A., and R. H. Reznek. 2007. "MR imaging in ovarian cancer." *Cancer Imaging* 7 Spec No A:S119–29.

Sohaib, S. A., A. Sahdev, P. Van Trappen, et al. 2003. "Characterization of adnexal mass lesions on MR imaging." *AJR Am J Roentgenol* 180 (5):1297–304.

Soriano, A., A. Castells, C. Ayuso, et al. 2004. "Preoperative staging and tumor resectability assessment of pancreatic cancer: prospective study comparing endoscopic ultrasonography, helical computed tomography, magnetic resonance imaging, and angiography." *Am J Gastroenterol* 99 (3):492–501.

Stall, B., L. Zach, H. Ning, et al. 2010. "Comparison of T2 and FLAIR imaging for target delineation in high grade gliomas." *Radiat Oncol* 5:5.

Sun, S., J. H. Schiller, and A. F. Gazdar. 2007. "Lung cancer in never smokers—a different disease." *Nat Rev Cancer* 7 (10):778–90. doi: 10.1038/nrc2190.

Tamm, E. P., P. M. Silverman, C. Charnsangavej, and O. B. Evans. 2003. "Diagnosis, staging, and surveillance of pancreatic cancer." *AJR Am J Roentgenol* 180 (5):1311–23.

Tempany, C. M., K. H. Zou, S. G. Silverman, et al. 2000. "Staging of advanced ovarian cancer: comparison of imaging modalities–report from the Radiological Diagnostic Oncology Group." *Radiology* 215 (3):761–7.

Testa, A. C., D. Timmerman, V. Van Belle, et al. 2009. "Intravenous contrast ultrasound examination using contrast-tuned imaging (CnTI™) and the contrast medium SonoVue® for discrimination between benign and malignant adnexal masses with solid components." *Ultrasound Obstet Gynecol* 34 (6):699–710.

Thomassin-Naggara, I., I. Toussaint, N. Perrot, et al. 2011. "Characterization of complex adnexal masses: value of adding perfusion- and diffusion-weighted MR imaging to conventional MR imaging." *Radiology* 258 (3):793–803.

Thrall, M. M., J. A. DeLoia, H. Gallion, and N. Avril. 2007. "Clinical use of combined positron emission tomography and computed tomography (FDG-PET/CT) in recurrent ovarian cancer." *Gynecol Oncol* 105 (1):17–22.

Timmerman, D., P. Schwarzler, W. P. Collins, et al. 1999. "Subjective assessment of adnexal masses with the use of ultrasonography: an analysis of interobserver variability and experience." *Ultrasound Obstet Gynecol* 13 (1):11–6.

Tsui, E. Y., J. H. Chan, R. G. Ramsey, et al. 2001. "Late temporal lobe necrosis in patients with nasopharyngeal carcinoma: evaluation with combined multi-section diffusion weighted and perfusion weighted MR imaging." *Eur J Radiol* 39 (3):133–8.

Ulzheimer, S., and T. Flohr. 2009. "Multislice CT: Current technology and future developments." In *Multislice CT*, edited by M. F. Reiser, C. R. Becker, K. Nikolaou, et al., 3–23. Berlin, Heidelberg: Springer Berlin Heidelberg.

Van Calster, B., L. Valentin, C. Van Holsbeke, et al. 2011. "A novel approach to predict the likelihood of specific ovarian tumor pathology based on serum CA-125: a multicenter observational study." *Cancer Epidemiol Biomarkers Prev* 20 (11):2420–8.

van der Gaag, N. A., E. A. J. Rauws, C. H. J. van Eijck, et al. 2010. "Preoperative biliary drainage for cancer of the head of the pancreas." *N Engl J Med* 362 (2):129–37.

van Nagell, J. R., Jr., R. W. Miller, C. P. DeSimone, et al. 2011. "Long-term survival of women with epithelial ovarian cancer detected by ultrasonographic screening." *Obstet Gynecol* 118 (6):1212–21.

Varela-Lema, L., A. Fernandez-Villar, and A. Ruano-Ravina. 2009. "Effectiveness and safety of endobronchial ultrasound-transbronchial needle aspiration: a systematic review." *Eur Respir J* 33 (5):1156–64.

von Bartheld, C. S., J. Bahney, and S. Herculano-Houzel. 2016. "The search for true numbers of neurons and glial cells in the human brain: A review of 150 years of cell counting." *J Comp Neurol* 524 (18):3865–95.

Vos, M. J., B. N. Tony, O. S. Hoekstra, et al. 2007. "Systematic review of the diagnostic accuracy of 201Tl single photon emission computed tomography in the detection of recurrent glioma." *Nucl Med Commun* 28 (6):431–9.

Wang, Y., and T. Jiang. 2013. "Understanding high grade glioma: molecular mechanism, therapy and comprehensive management." *Cancer Lett* 331 (2):139–46.

Watanabe, M., R. Tanaka, and N. Takeda. 1992. "Magnetic resonance imaging and histopathology of cerebral gliomas." *Neuroradiology* 34 (6):463–9.

Wen, P. Y., and S. Kesari 2008. "Malignant gliomas in adults." *N Engl J Med* 359 (5):492–507.

Whelan, H. T., J. A. Clanton, R. E. Wilson, and N. B. Tulipan. 1988. "Comparison of CT and MRI brain tumor imaging using a canine glioma model." *Pediatr Neurol* 4 (5):279–83.

Wick, W., C. Hartmann, C. Engel, et al. 2009. "NOA-04 randomized phase III trial of sequential radiochemotherapy of anaplastic glioma with procarbazine, lomustine, and vincristine or temozolomide." *J Clin Oncol* 27 (35):5874–80.

Wilbaux, M., E. Henin, A. Oza, et al. 2014. "Prediction of tumour response induced by chemotherapy using modelling of CA-125 kinetics in recurrent ovarian cancer patients." *Br J Cancer* 110 (6):1517–24.

Yasufuku, K., A. Pierre, G. Darling, et al. 2011. "A prospective controlled trial of endobronchial ultrasound-guided transbronchial needle aspiration compared with mediastinoscopy for mediastinal lymph node staging of lung cancer." *J Thorac Cardiovasc Surg* 142 (6):1393–400.

Yitta, S., E. M. Hecht, C. M. Slywotzky, and G. L. Bennett. 2009. "Added value of multiplanar reformation in the multidetector CT evaluation of the female pelvis: a pictorial review." *Radiographics* 29 (7):1987–2003.

Zakharova, O. P., G. G. Karmazanovsky, and V. I. Egorov. 2012. "Pancreatic adenocarcinoma: Outstanding problems." *World J Gastrointest Surg* 4 (5):104–13.

Zanzonico, P. 2012. "Principles of nuclear medicine imaging: planar, SPECT, PET, multimodality, and autoradiography systems." *Radiat Res* 177 (4):349–64.

Zhang, R., K. Ying, L. Shi, L. Zhang, and L. Zhou. 2013. "Combined endobronchial and endoscopic ultrasound-guided fine needle aspiration for mediastinal lymph node staging of lung cancer: a meta-analysis." *Eur J Cancer* 49 (8):1860–7.

Zhuang, H., M. Pourdehnad, E. S. Lambright, et al. 2001. "Dual time point 18F-FDG PET imaging for differentiating malignant from inflammatory processes." *J Nucl Med* 42 (9):1412–7.

9 Multifunctional Nanocarriers as Theranostic Systems for Targeting Cancer

Gaurav Pandey, Rashmi Chaudhari, Vinod Kumar Gupta, and Abhijeet Joshi

CONTENTS

9.1 INTRODUCTION

In accordance with the data from World Health Organization in 2015, a rough estimate of 84 million deaths in the last decade have been caused by cancer (Bukhtoyarov et al. 2015; WHO 2017). Occurrences of cancer and its further development have been classified as a complex and multistep process in which an initiating event (e.g. mutation & aberrant transformation) leads to malignant proliferation. The changes that lead to the transformation of a normal cell to cancerous cells are either inherited mutations and induced mutations by environmental factors like exposure to a certain type of viruses, UV light, X-rays, potentially mutagenic chemicals; or substance abuse like drugs, tobacco, smoking, alcohol. Reported evidence suggests that most of the cancers are not the result of a single contact or an event related to one factor but, usually take more than four to seven such events for the transformation of cells to a premalignant stage and further to a cancerous cell. Also, the changes can take a short time span or several years. It is a well-known phenomenon of cancer cells that they grow rapidly and compete with the other healthy cells present in surrounding tissues for uptake of available nutrients from the bloodstream. This results in an imbalance eliminating normal cells, leading to suppression of primary functions, failure of organs, and finally death of an organism. There is a vast diversity of different types of cancer, but some peculiar features remain common among all the cancerous cells. These abnormal cells have an impaired mechanism of control over cell division with additional functional abnormalities in comparison to normal cells (Furuya et al. 2005; Kawasaki and Player 2005; Bukhtoyarov et al. 2015).

9.1.1 Pathophysiology of Cancer

Several altered molecular events lead to changes in the fundamental properties of cells making them cancerous in nature. Cancerous cells constitute severely impaired/absent preventive checks for cellular overgrowth and invasion. These functional deficits of cells enable them to grow even when signals suppressing cell growth are present. Decrease in cell adhesion properties, upregulation of certain enzymes are a few other steps for propagation and intrusion of cancer cells in normal tissues. Aberrational changes in key proteins responsible for cell division/regulation along with mutation of genes leads to abnormality in functioning of cells. Over time the repair mechanism of the cell gets knocked-down due to increased mutations and due to the higher frequency of mutations with compromised repair mechanism, abnormalities arrive in abundance. Some of these mutations lead to cell death as well, in the premalignant phase before transforming completely into a cancerous cell. Another well-known and interesting fact about cancer is that in a healthy organism some cancer like cells always exists but they do not cause or progress cancer probably due to various built-in molecular balances in cellular systems and vigilant immune systems. It is also very interesting that cancerous cells share a lot of common features with stem cells, which are now being looked upon to be used as therapeutic agent in many severe human disorders. When the transformed cells remain localized and non-intruding they are termed as benign (non-recurrent or non-progressive) otherwise they are considered malignant when they are progressive and unchecked.

Malignancy of cancer tissues leads to formation of new tumor tissues in other sites of the body due to migration of cancer cells in a process called as metastasis. Figure 9.1 summarizes the structural and functional differences between normal and cancer cells. Certain types of genes are more associated with cancers as their expression may be enhanced or inhibited. One such type of gene is called a proto-oncogene, which expresses proteins that normally enhance cell division or inhibits cell death. Mutated forms of these genes that are overexpressed in cancer cells are called oncogenes. Tumor suppressor genes are another type of gene that prevent cell division or lead to cell death. They are inactivated in the cancerous cells. Apart from these, DNA repair genes are responsible for inhibiting mutated cells reaching cancerous stages, however mutation/suppression of them does not allow them to function efficiently.

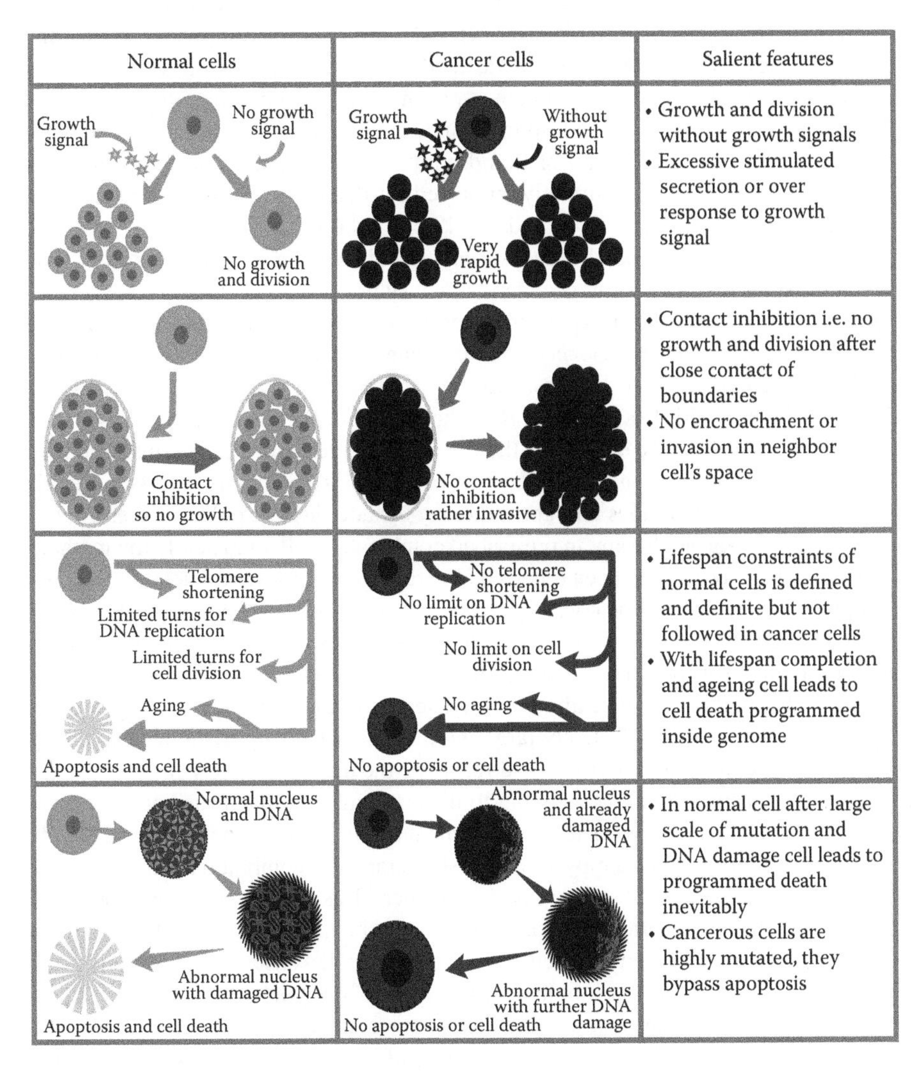

FIGURE 9.1 Differentiating characteristics and cell dynamics of normal and cancer cells.

In the progression of cancer the cellular machinery takes drastic turn from the normal course in the form of overproduced proteins or the inhibition of protein production due to mutations that check cell division and apoptotic process (Weinberg 2001; Miller et al. 1981; Furuya et al. 2005; Bukhtoyarov et al. 2015).

9.1.2 Clinical Methods of Diagnosis and Treatment of Cancer

Several methods are used clinically for the diagnosis of cancer. Due to advancement in technologies, there is an increase in the number of diagnostic tools available for the detection of cancer. Several imaging techniques like computed tomography (CT) scans, magnetic resonance imaging (**MRI**) scans, and X-rays are used for screening of cancer patients. CT scans provide a detailed computerized imaging of organ in question using exposure of X-rays. MRI also produces a computerized imaging of soft tissues and organs by magnetic field. Radiolabeling for positron emission tomography (**PET**) and the use of contrast agents for MRI enhances the accuracy of these imaging techniques. Solid tumors can also be detected using ultrasound technique wherein a tumor of solid or liquid lumps can be characterized accurately. A blood sample analysis is commonly employed for detecting specific tumor markers such as prostate specific antigen and CA-125 in ovarian cancers. Cancerous tissue can also be removed by needle aspiration/surgical excision or in the form of a Pap smear to be observed in a microscope in a biopsy procedure. Endoscopic techniques like laparoscopy have been used for the visualization of abdominal areas using a camera attached to a tube. Genetic mutation can be detected using a chromosomal analysis that will indicate the susceptibility towards cancer (Xie, Lee, and Chen 2010; Muthu et al. 2014; Kelkar and Reineke 2011; Yu, Park, and Jon 2012).

The vast diversity of cancers and their causes have remained impossible to understand in order to develop universal methods for diagnosis and therapy. Most commonly applied treatment modalities for cancer therapy include surgery, radiation therapy, and chemotherapy or gene therapy in prescribed combinations. Paradoxically the methods of therapy themselves have been reported as carcinogenic in several scientific and clinical platforms, which can be considered as a setback in successful treatment. Various reports indicate that damage and detrimental side effects imparted by these treatment strategies affect functioning of other normal cells. The injurious side effects also contribute to the morbidity and death rate associated with cancer. Efficacy of any strategy is directly correlated to selectively targeting tumor tissues, improving bioavailability by overcoming biological barriers, and releasing the drug adequately in a controlled/prolonged or triggered manner. Owing to the challenges in diagnosis and therapy of cancers the most persuasive idea of management of cancers is to combine several techniques of diagnosis and therapy to establish therapeutic compliance where the collateral damage by applied therapy is as detrimental as the disease condition itself. Among various combination approaches improving bioavailability, targeting to tumor tissues, minimizing systemic toxicity of drugs, controlled/sustained/prolonged delivery of drugs, and continuous simultaneous imaging of drug delivery are sought. These will eventually lead to developments in therapeutic compliance, reduced cost and time of therapy (Mironidou-Tzouveleki, Imprialos, and Kintsakis 2011; Anonymous 2016; Bukhtoyarov et al. 2015; Wang 2008; Singh and Nehru 2008; Gold 2011).

9.2 NANOMATERIALS FOR CANCER DIAGNOSIS AND THERAPY

Nanoparticles have served a great deal in reducing the perils of conventional chemotherapy. Small nanoparticles of different materials like polymers, lipids, inorganic materials, and biological materials have been described to be used for management of cancer. The loading of drugs and imaging probes together can serve in developing theranostics for cancer (Figure 9.2). Anticancer nanotherapeutics have come a long way since their first clinical trial in 1980s and the entry of first liposomal doxorubicin in the market in 1995 (Nguyen 2011). Nanoparticles have been in the prime of research due to their advantages, like the capability of encapsulating hydrophobic drugs, increased retention time in the body, reduced toxicity, improved penetration in unreachable organs, and specific cancer targeting. Recent nanotechnological developments have also stirred the old strategies and potential drug molecules, which had to be discontinued due to their poor interaction, adverse effects, and assimilation in biological systems. Several nanomaterial properties like very small size, enhanced permeability and retention in tumors, preferential ligand-receptor mediated

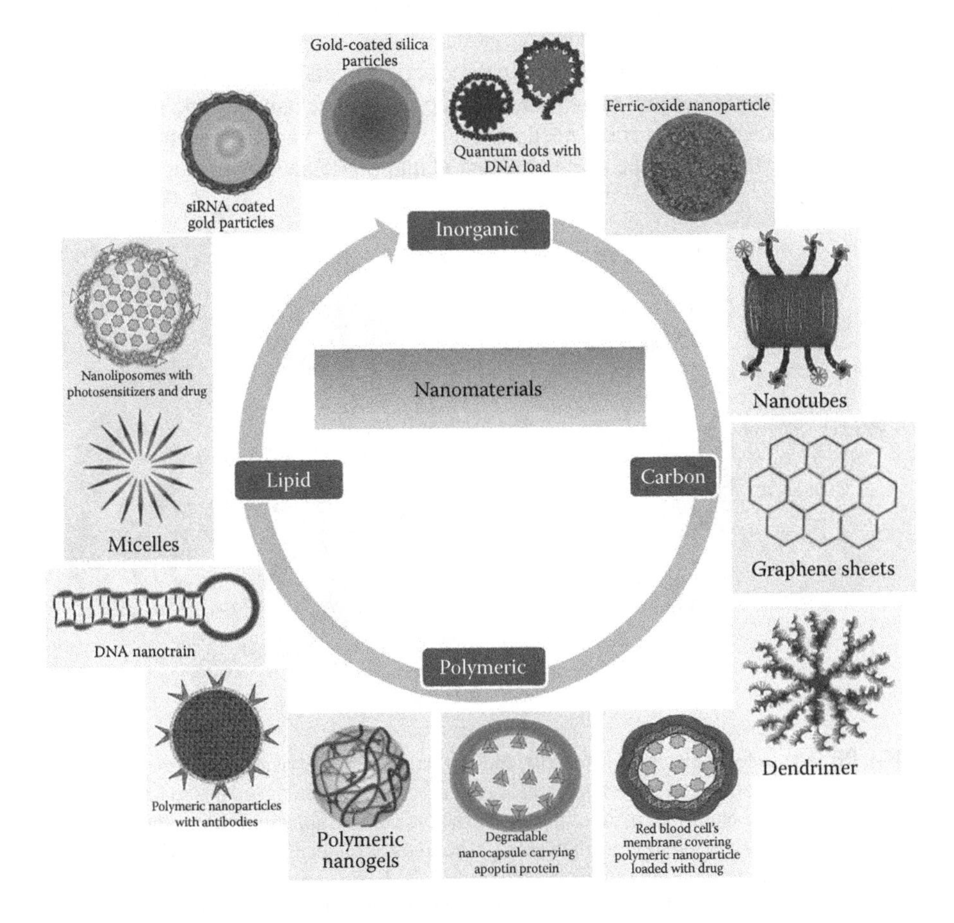

FIGURE 9.2 Nanomaterials used as theranostic carriers.

attachment, and triggered release potential provide them good stead in therapeutic application. Recent advancement in the field of nanomaterials gives firm ground for a hope of better solutions. Several different types of compositions in nanomaterials have been tried to date to fulfill the aims of cancer management (diagnosis and therapy). Drug/imaging agents can be delivered in a localized or a systemic manner. Direct/localized delivery of drugs can be achieved by direct injection of drugs at the tumor sites where as systemic delivery of actives involve administration into blood vessels and transport to the active sites. Localized or direct delivery of drugs and imaging agents although superior is not possible in several cases due to inaccessibility of tumor tissues. In such cases, systemically administered drug remains the only choice where in drug can be transported to tumor tissues mediated passively or actively (Figure 9.3). The success of cancer therapy is in the targeting capability of these nanoformulations. Passive transportation of drugs or drug-loaded nanomaterials can be achieved by controlling and optimizing different parameters like particle size, surface charge, ability to fuse with cells, stability of the complexes, cytotoxicity etc. Attachment of cell-specific ligands to such nanomaterials in addition to optimizing the above parameters can achieve actively targeted nanomaterials. Most investigated nanocarriers include inorganic/organic nanocarriers that function as biocompatible, biodegradable or eliminating nanocarriers (Figure 9.2). Nanocarrier elimination through a renal route is only possible when the nanoparticle size is below 5–8 nm. A larger nanoparticle needs to be degraded *in vivo* to prevent any undue toxicity. This fact led to the use of biodegradable materials for the development of

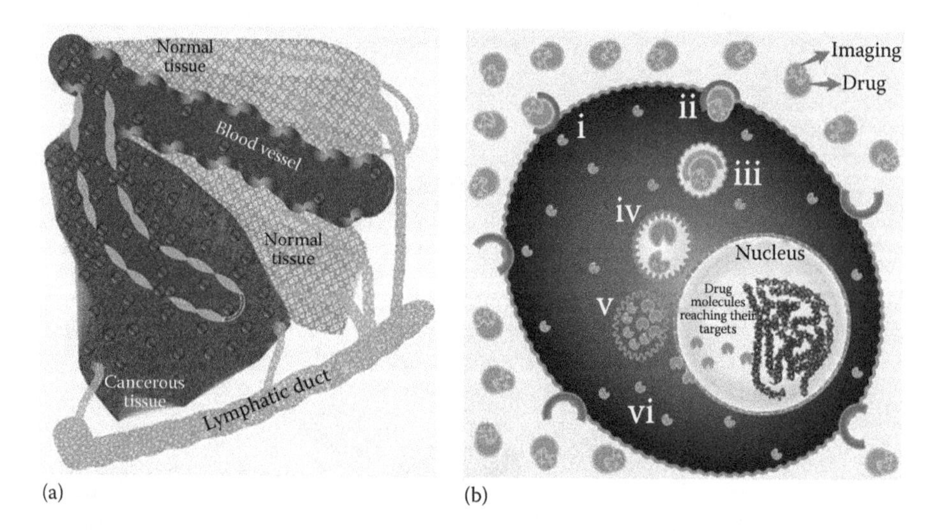

FIGURE 9.3 Mechanisms of transport and uptake of nanocarriers in tumors (a) passive (enhanced permeation and retention effect in cancerous tissues, produced due to leaky vessels thus more deposition of therapeutic in cancer tissues, poor lymphatic drainage of cancerous tissue) and (b) active targeting mechanisms [i) theranostic molecules binding with receptors, ii) uptake of nanoparticles, iii) endosomal ingestion, iv) lysosome degradation and separation of imaging agent and drugs, v) drug release and distribution of drug release in the entire tissue] with specificity and selectivity based on ligands of nanocarriers in tumors.

nanocarriers. Polymer degradation can occur due to physiological conditions in a time-based manner or triggered erosion based on a response to stimuli. A diverse range of materials have been used ranging from inorganic, organic, and biological origin. The current focus of developing novel materials and techniques is towards reduction of toxicity to normal cells by targeting of therapy (Joshi et al. 2018).

9.2.1 Gold Nanoparticles

Inorganic nanoparticles have shown promise in establishing diagnostic and treatment alternatives like magnetic hyperthermia using iron nanoparticles, photo-thermal hyperthermia using gold nanoparticles/nanoshells, quantum dots etc. Quantum dots, which are semiconductor nanoparticles, have been very useful for imaging studies owing to their fluorescence properties. Inorganic nanomaterials used in diagnosis or therapy of cancer are eliminated via a renal route owing to their very small sizes below 5 nm. Among inorganic materials gold nanoparticles have been used to deliver drugs precisely to tumor tissues. Gold colloidal solutions have been modified using tumor necrosis factor alpha (TNF-α), which causes tumor necrosis. Modification using polyethylene glycol (PEG) with thiol derivatives enables them to be transported to tumor tissues without being engulfed by the reticuloendothelial systems. The preferential deposition of these gold nanoparticles can occur due to Enhanced Permeation and Retention (EPR) Effect or due to active targeting using a ligand. Further, multiple drugs can be attached to gold nanoparticles to provide a better efficacy of therapy and minimal side effects. Owing to the properties of surface plasmonic resonance, NIR absorbance, and photo-thermal effect, gold nanoparticles play a significant role as imaging agents in NIR imaging, X-ray, and CT imaging. There these particles generate heat locally to destroy tissue after being activated by near infrared laser of a selected wavelength, which harmlessly crosses through human soft tissues. The nanoparticles are designed to absorb these wavelengths confirming their targeted activation selectively by laser illumination to specific areas, leaving the healthy adjacent tissue. This approach is based on the property of optical tunability (absorbance to one selective wavelength range), converting light energy into heat. Photo-thermal acting nanoparticles also consist of gold nanoshells. Gold is deposited over a template nanoparticle like silver or silica and then the core nanoparticles are removed forming gold nanoshells. Additionally, the core-shell structure formed from gold coated on silica nanoparticles has been effective for photo-thermal therapy when a near infrared laser of a selected wavelength is used. The method achieves light induced targeting and photo-thermal effect of a localized tumor site. The absorbance of a specific light wavelength can be altered and tuned by varying the size of core particles along with the thickness of the outer shell (Figure 9.2). The combination of these two attributes gives specific optical characteristic properties to nanoparticles also allowing them to be customized and tailor made. This photothermal therapy has been found promising in trials under the name of AuroLase or Auroshell in animal models and is supposed to enter human trials soon in near future. The tested animals remained cancer free even after the completion of treatment, and the treatment-associated symptoms were reduced (Anonymous 2016; Xie, Lee, and Chen 2010; Haynes et al. 2016; Lee et al. 2014).

9.2.2 Magnetic Nanoparticles (MNPs)

Another common type of inorganic nanoparticle used are iron nanoparticles (magnetic nanoparticles), which have been very widely used as a theranostic nanocarrier system due to their MRI imaging capability, magnetic hyperthermic effect, magnetic targeting etc. Superparamagnetic coated iron oxide nanoparticles (SPIONs) are often coated with organic materials like polymers, fatty acids, or inorganic shells to improve stability and improve functionality. SPIONs have been widely employed as contrast agents for MRI. The composite structures of magnetic nanoparticles and polymers forming nanoparticles and microparticles have been developed so that a variety of anticancer drugs can be loaded or conjugated to polymers forming multifunctional theranostic carriers having properties of magnetic hyperthermia, multiple drug chemotherapy, and targeting capability. The killing of tumor cells through the heat irradiated from the magnetic nanoparticles attached to them is a direct consequence of the heating properties under an applied alternating magnetic field. Core-shell nanoparticles can generate unique properties that cannot be obtained using other single-component systems. Gold coated iron nanoparticles can bind to "–SH" groups on the particle's surface so that particle size and drug/peptide/protein conjugation can be favorably done. Magnetic nanoparticles have been encapsulated in gels, liposomes, bubbles, carbon nanotubes, and pH and temperature sensitive polymers to generate multifunctional theranostic systems (Anonymous 2016; Xie, Lee, and Chen 2010; Hoare et al. 2009; Murase et al. 2015; Lu et al. 2005; Kumar et al. 2010; Joshi et al. 2011).

9.2.3 Quantum Dots (QD)

Quantum dots are light-emitting semiconductor nanocrystals made of CdSe, CdTe, and PbS, which have unique optical properties such as photo-stability, narrow emission spectrum, and quantum yield (Figure 9.2). The tunability of size and composition can impart a variable emission wavelength in visible to near infrared region (NIR) range. Quantum dots have not been useful for *in vivo* applications due to their issues with biocompatibility and poor penetration of fluorescence emission. QD-based drug delivery has not been investigated mainly due to its toxicity. Limited quantum dots are being researched that are Cd/Pb free such as InAs/ZnSe and InP/ZnSe. An example QD-based theranostic system comprising of QDs, MNPs, and doxorubicin loaded into micelles is developed from PEGylated phospholipids, which has been shown to be successful for tumor targeting in MDA-MB xenograft models (Kim et al. 2012; Yu, Park, and Jon 2012). QDs modified with cysteine have served the purpose of water solubility so that they can be administered systemically. Such nanoparticles have a propensity to be excreted via renal route and not be engulfed in reticuloendothelial system. A QD-Aptamer-DOX conjugate has displayed multifunctional action in cancer imaging, therapy, and therapeutic monitoring. QDs have also been shown to be effective in photodynamic therapy where they can act as photosensitizers or carriers for photosensitizers. Photosensitizers become activated by light and transfer to triplet state energy to nearby oxygen molecules, which causes cell damage. However, the sensitizing effect of QDs is poor in comparison to classic photosensitizers. In spite of these capabilities, the metabolism of QDs and subsequent toxicity profiles need to be addressed for improved biomedical applications.

9.2.4 Carbon Nanotubes (CNTs)

Carbon nanotubes are hydrophobic and have surface with aromatic nature. Due to their hydrophobic nature they can easily anchor using non-covalent interactions to different types of groups. They have shown potential applications in Raman, photoacoustic imaging, and attachment of drugs. Drug delivery can be established by conjugation using covalent links of amide bonds. Several amphiphilic compounds like sodium dodecyl sulfate, triton-X-100, Cetyltrimethylammonium bromide (CTAB), sodium dodecyl benzene sulfonate, and sodium dodecane sulfonic acid are used for dispersing CNTs. CNTs can be internalized in the cells by different mechanisms and the extent of this can be influenced by surface coatings. Dai group applied phospholipid-CNT conjugates in both imaging and therapy. Drugs like paclitaxel can be coupled to nanotubes using cleavable ester bonds, which was tested in a murine 4T1 breast cancer model. The conjugates showed a ten-fold increase in accumulation of medicine in the tumor. NIR optical absorbance of CNTs acts as a promising strategy towards photo-thermal therapy. Once internalized in cells CNTs can be subjected to NIR radiation causing endosomal rupture and a subsequent cell death. Apart from drugs linked to CNTs, gold nanoparticles, and iron nanoparticles have also been linked to CNTs for enrichment of multifunctionality. After systemic administration, CNTs have been found to be entrapped in liver and spleen without degradation and toxicity. It has remained debatable about the safety and long-term usage of CNTs and ability of modifications like surface coating to be effective in surpassing the toxicity issues (Xie, Lee, and Chen 2010; Faraji and Wipf 2009).

9.2.5 Other Inorganic Nanoparticles

Silica nanoparticles are one of the inorganic carriers that are regarded as biocompatible but not biodegradable. Silica nanoparticles, due to their mesoporous nature, provide a platform for the loading of imaging agents or therapeutic molecules making them suitable candidates for theranostic purposes. Silica nanoparticles can also be used for encapsulation of iron oxide nanoparticles, gold nanoparticles, and quantum dots for their respective applications (Figure 9.2). Encapsulation of photosensitizer molecules like 2-devinyl-2-(1-hexyloxyethyl) pyropheophorbide, which are fluorescent in nature, can provide photo-dynamic therapy when irradiated with laser light. Radiation therapy has remained a viable approach to kill the cells, however both normal and cancer cells become affected in this process. This adds to the morbidity and mortality associated with cancers. Several new approaches have been developed to prevent such side effects to normal cells; one of them is the use of nanoparticles of hafnium oxide that are inert in nature until they are activated through an X-ray beam. Exposure to X-rays leads to generation of free radicals that ultimately cause damage to cancer tissues. The energy is dissipated in the surrounding tissues by the means of kinetic energy from emitted electrons. These nanoparticles have been found to be excreted in feces, which reduces renal damage (Lee, Na, and Bae 2003; Song, Wu, and Chang 2015; Uchida et al. 2016; Ihre et al. 2002).

9.2.6 Lipid Nanoparticles

Liposomes are bilayer vesicles enclosing aqueous compartment that are formed using amphiphilic phospholipids and cholesterol. Owing to their structure and capability to encapsulate hydrophobic and hydrophilic components, liposomes act as interesting tools for drug and diagnostic agent delivery. In order to provide stealth properties against opsonization by the immune system and elimination, liposomes are coated with hydrophilic compounds like PEG. Different hydrophilic/hydrophobic drugs have been investigated in liposomal delivery systems. Agents that can be surface coated on liposomes or covalently linked targeting ligands can enable liposomes to be targeted to particular cancer tissues. Certain coatings like TPGS and PEG and targeting ligands like folic acid have been commonly used for improving the circulation time and targeting to specific tumor tissues. Liposomes have also been used to encapsulate different kinds of nanoparticles like polymeric nanoparticles, gold nanoparticles, and iron oxide nanoparticles within the core of liposomes or as a coating over liposomes. Encapsulation of different kinds of nanoparticles in/on liposomes can provide multifunctionality, for example magnetic targeting and magnetic hyperthermia using iron oxide nanoparticles, photo-thermal therapy using gold nanoparticles, imaging of tumors using quantum dots (Wen et al. 2013) or controlled drug release, drug-loaded polymeric nanoparticles. Alternative to the flexible structure of liposomes are solid-lipid nanoparticles that are made of biocompatible lipids that have a rigid structure. Similar to other nanoparticulate structures, solid lipid nanoparticles can provide targeted co-delivery of diagnostic and therapeutic agents. Sub-100 nm solid lipid nanoparticles can cross the blood-brain barrier and can be used for brain-associated cancers. Apart from drugs they can be loaded with NIR quantum dots and conjugated with ligands like cRGD for targeting and live animal imaging applications (Shuhendler et al. 2012). Core-shell structures with solid-lipid nanoparticles have also been formulated using a drug or imaging agent loaded inside the lipoid core along with quantum dots and an siRNA coating acting as a shell over the solid-lipid nanoparticle (Figure 9.2). Such a system acts as an LDL mimetic multimodal optically traceable nanocarrier acting as a theranostic (Bae et al. 2013).

9.2.7 Polymer Nanoparticles

Polymer nanoparticles have been researched a great deal as therapeutic drugs in a developing area of nanomedicine. The most important purpose of nanomedicines is to have controlled release and targeted drug release. Polymeric chains can be used to prolong the circulation time and inhibit uptake by the reticuloendothelial system. Polymeric nanoparticles loaded with drugs accumulate in cancer tissues due to the EPR effect or active targeting using ligands in comparison to the effect on normal tissues (Figure 9.3). Theranostic applications have been implemented using polymeric nanoparticles by loading diagnostic modalities along with therapeutic modalities. Theranostic polymeric carriers can serve the following advantages like biocompatibility, storage stability, protection of loaded drug/diagnostic agent, and controlled drug release (Figure 9.2). Biodegradation is another parameter for the use of polymeric nanoparticles wherein the polymer chains are hydrolyzed into physiological components of the body like alginate, chitosan, PLGA, PLGA-TPGS, and PLGA-PEG are commonly used biodegradable polymers that are metabolized into lactic acid and glycolic acid monomers to be eliminated from metabolic pathways. Some of the polymers can be obtained from natural origins and others can be synthetically prepared. Surface coatings of polymeric nanoparticles using PEG or polyelectrolytes can produce nanoparticles that can avoid uptake from the reticuloendothelial system and are amenable to being conjugated to biologically active compounds for targeting. The use of co-polymers in synthesis of nanoparticles will provide modulated release profiles and degradation patterns for nanocarriers. Similar to lipid nanoparticles, PEGylation improves the circulation time of polymeric nanoparticles as well. Micelles are self-assembling colloidal structure with a hydrophobic core and hydrophilic shell having sizes less than 100 nm. They are the preferred carriers for water-insoluble materials. Therapeutic or diagnostic agents can be put into hydrophobic core of micelles and to the outer hydrophilic layers a targeting agent can be employed. PLA-PEG-based micelles have proven to be effective for delivery of drugs like paclitaxel. TPGS has also been used for preparation of micelles of iron oxide, which can show improved properties of magnetic hyperthermia, uptake, and *in vivo* imaging. Hyaluronic acid along with DOX coated with gold as half shells have been developed for multifunctional theranostic applications in cancers (Kim et al. 2012). Folate or TPGS coatings can provide benefits of targeting to the tumor tissues. Different kinds of inorganic nanoparticles like quantum dots, iron oxide nanoparticles, and gold nanoparticles can be loaded into these polymeric nanoparticles to obtain composite structures having multiple functionalities. These nanoparticles embedded composite structures will produce applications of *in vivo* imaging, magnetic targeting and hyperthermia, photo-thermal therapy, and triggered therapy using external stimuli like light or magnetism. Dendrimers are an example of synthetic polymeric nanomaterials that are highly branched in nature with sizes in the range of 10–100 nm. Converging or diverging structures along with variety of substituents in the form of drugs or imaging agents can be developed for theranostic applications. Several physical properties, degree of polymerization, and chemical group combinations are being investigated for the formation of dendrimers to utilize them as targeted therapeutic carriers to minimize drug toxicity in healthy tissues. The fifth generation G5 type of dendrimers are the most commonly studied as they can impart extended drug encapsulation and stability. Paclitaxel and Cy5.5 dye have been delivered using these nanocarriers and targeting was established by attaching luteinizing hormone releasing hormone (LHRH) peptide targeted to receptors overexpressed on cancer cells. Phthalocyanines-C was encapsulated in the dendrimer and targeted using an LHRH peptide to tumors showed a photodynamic therapeutic potential and fluorescence imaging capability. Owing to the tenability of size and applying a variety of linkages to active compounds dendrimers show huge potential for theranostic applications (Mironidou-Tzouveleki, Imprialos, and Kintsakis 2011; Singh and Nehru 2008; Anonymous 2016).

9.3 TARGETING MECHANISMS: PASSIVE AND ACTIVE

Imaging of anatomical tissues is the most important step in determining the size, shape, and position of an abnormality with high resolution. Computed tomography and MRI have remained the gold standard in imaging of organs. Personalized medicine has not addressed explicitly the targeted imaging in its definition, however it is an important part of diagnosis of diseases. The use of targeted imaging involves the selection of right radiolabeled probe/light emitting molecules for right target for a particular condition. Targeted radiotracers bind to well-defined protein sites like enzymes and receptors. Physiologically receptors are present with a very small density in comparison to enzymes. Distinct imaging advantages can be obtained with low density receptor-ligand interaction, for example the asialoglycoprotein receptor in the liver is present to an extent of 500 nM, which enables MRI to be used for imaging. F-19 can be used for imaging using MRI, as in the case of 2-Fluoro-2 deoxy glucose (FDG), which is commonly used to understand the glucose metabolism. Another successful imaging modality is the use of iron oxide particles linked to visualization of transferrin receptors under MRI. Radionuclides are gamma-emitting molecules that need to be validated for binding to the target specifically. Such targeted imaging methods have been useful to develop radiopharmaceuticals co-administered with pharmaceuticals. Some approaches for using such a combination can help in monitoring disease control points like apoptosis and angiogenesis. They will also aid in monitoring targeting capability of radiopharmaceuticals for a similar target and estimating the downstream biochemical processes affected by drugs. For example, specific neuronal imaging agents have been developed that can monitor different steps in the nor-adrenaline metabolic pathway. [123]-Meta-iodo benzyl guanidine (MIBG) is grabbed by vesicles than [11C]-Meta-hydroxyephedrine (HED) by monoamine transporter.

The fate of nanoparticles is highly dependent on the particle size and surface characteristics. The drug or imaging agent loaded nanoparticulate theranostic should be able to be present in the blood stream for considerable periods without elimination and uptake by reticuloendothelial system. High molecular weight compounds and nanoparticles in the size range of 10–100 nm can become accumulated in the cancer tissues. In order to prolong the circulation time, the surface of nanoparticles should be hydrophilic in nature to escape macrophage capture. This can be achieved by coating hydrophilic polymers or by attachment of PEG. Nanoparticles satisfying the size and surface characteristics have a great probability of reaching the tumor tissue. Passive targeting is mainly dependent on the Enhanced Permeation and Retention Effect produced due to leaky vasculature and unorganized endothelial tumor tissues (Figure 9.3a). This effect is produced due to the enhanced growth of blood vessels supplying the tumor tissues to fulfill their oxygen and nutritional requirements. The imbalance in growth results in disorganized cellular structures and enlarged gap junctions between endothelial cells and poor lymphatic drainage. Nanoparticle formulations like polymeric nanoparticles, micelles, liposomes, etc. can be used to counter passive targeting mechanisms so that toxic effects are reduced. Polymer drug conjugates can be formed to lead high molecular weight like doxorubicin linked to glycol using a spacer of N-cis-aconityl group and carbo-di-imide links. DOX-GC nanoparticles can be formed using polymer conjugates accumulate in tumor tissues due to EPR effect. Similarly, paclitaxel conjugated to chitosan can be cleaved in physiological conditions. Chitosan, due to its cell permeation properties by opening tight junctions between epithelial cells, also helps in the oral administration of paclitaxel. EPR-based targeting, although commonly used, suffers from limitations of tumor types, vascularity, necrosis of tumors, and fibrosis of tissues. Theranostic cationic liposomal carriers loading quantum dots and camptothecin and irinotecan for simultaneous imaging and drug delivery have been described that demonstrated higher efficiency of accumulation in solid tumors exhibiting fluorescence signals. The liposomes were localized up to 24 hours using these cationic nanosystems (Wen et al. 2013). Tumors, owing to their high growth and metabolic rate, lead to the generation of hypoxic conditions and the acidic environment of tumor cells. These tumoral environmental conditions can be used for triggering or targeting the release of nanoparticulate structures in a site-specific manner. For example an albumin-bound doxorubicin incorporating metalloproteinase-2 specific octapeptide sequence was hydrolyzed by the enzymes at tumor sites in an in vitro study. pH sensitive polymeric carriers or liposomes have been utilized for tumor specific release of drugs along with flourescence based imaging (Joshi et al. 2017; Joshi et al. 2018).

Active targeting is the localization of drug therapy on target sites like cell membranes, cytoplasm or nucleus or targeting to particular organs e.g. brain, heart, kidney etc. Targeting to tumors is based on exploring specific antigens and receptors on the surface of cancer tissues (Figure 9.3b). Ligands linked to drugs were one of the primary attempts to target cancer tissues however the chemical linkages have known to reduce the activity of drugs. The process of treatment of cancer using drug-linked antibodies is called chemo-immunotherapy. A similar strategy of targeted radiation therapy in called radio-immunotherapy where in radioactive molecules are attached to antibodies. Different ligands that can be attached to nanoparticles for active targeting can include small molecules like folate and large molecules like RGD peptide, bombesin, A54, luteinizing hormone releasing hormone, proteins like EGFR, VEGF, antibodies, cell-penetrating peptides, myristoylated polyarginine peptides, TAT peptides, herceptin etc. Folic acid targeting has been employed for targeting tumors owing to the high requirement of folic acid in tumor tissues and the overexpression of folate receptors. Folate receptors are found in about 53% of tumor tissues (Saba et al. 2009). In addition to receptor binding, the penetration of nanocarriers inside the cell for effective release of drugs is highly critical. Endocytosis is the most common mechanism of internalization of nanocarriers. Magnetic field-based targeting is another interesting approach that uses magnetic nanoparticles conjugated to anticancer drugs like doxorubicin. An example of a study on such system was done by encapsulating such drug-conjugated structures in silica shells with PEGylation to provide stealth properties (Wu et al. 2011). Coatings over magnetic nanoparticles like carbohydrates, liposomes, hydrogels, lauric/citric acid, and oleic acid make them suitable for attachment of drugs. The use of magnetic nanoparticles provides an additional benefit of magnetic hyperthermia for cancer therapy either alone or to induce magnetically triggered chemotherapy. Therapy for a highly invasive prostate cancer has been attempted using nano-liposomes carrying IPA3 molecules that inhibit PAK1 protein in cancer cells which can slow down the progression of cancer and apoptosis resulting in cell death. Oligonucleotide sequences such as DNA,

RNA are called aptamers, which bear conformations capable of binding to antigens of tumor tissues. Aptamers have been used to enhance the selectivity of drug therapy like docetaxel in PLGA nanoparticles in prostate cancer (Farokhzad et al. 2006). Transferrin and lectin are other ligands that act as transporters to delivery iron and carbohydrates, respectively. Transferrin receptors are overexpressed in tumor tissues compared to normal tissues. Thus, transferrin-conjugated paclitaxel loaded in PLGA nanoparticles showed greater inhibition of MCF-7 cells. On the other hand, lectins are proteins that bind to carbohydrate molecules attached to proteins on the extracellular side of the plasma membrane. Cancer cells express different kinds of glycan that are different from the normal tissues making them suitable for targeting. Two approaches have generally been described using lectins, one of which involves the coating of lectins over nanoparticulate carriers that can interact with cellular carbohydrates and in another coating of carbohydrate molecules against the lectins expressed in cancer cells.

Targeted nanoparticles, through extensive application, have come through as a promising strategy, though several issues remain unexplored. Among the development of drug resistance rendering drugs released from nanoparticles as ineffective. Drug resistance can be addressed by combining multiple drugs, drugs, and gene delivery or developing multifunctional targeted nanoparticulate systems. Biopharmaceutical challenges like stability *in vivo* and pharmacokinetic properties of the payload are least explored. Some polymeric materials like PLGA are not toxic and degrade quickly after administration, however other materials like inorganic materials, quantum dots, carbon nanotubes etc., can persist, not metabolize, and accumulate for years making them potentially toxic for repeated administration. Developing new materials, selecting appropriate materials, optimizing drug loading and release are some other factors that need to be optimized during the development stage. Some of these issues can be solved by developing hybrid structures to incorporate properties of each component and also provide multifunctionality (Yu, Park, and Jon 2012; Shao et al. 2016; Gao 2016; Lammers et al. 2012; Joshi et al. 2017).

9.4 MULTIFUNCTIONAL NANOTHERANOSTICS: STRATEGIES AND ADVANCES

Theranostics is a composite term derived from therapeutics and diagnostics, which have been important pillars of health care. The term was coined around 2002 by John Funkhouser, a medical consultant, in the context of profiling of subjects for diagnosis of cardiovascular diseases.

Theranostics puts the promise of novel and advanced insight into research, diagnosis, and treatment for complex diseases like neurodegenerative disorders as Alzheimer's, metabolic disorders like cystic fibrosis, cellular anomalies like cancer, autoimmune disorders etc. Such diseases have remained difficult to treat due to insufficient explorations. A combined imaging and therapeutic modality can provide a stepping stone for the execution of sufficient explorations in untreatable diseases and disorders (Figure 9.4). The current definition of theranostics encompasses the delivery of therapeutic agents along with imaging of cells and tissues, or detecting pathophysiological biomarkers within the body. Theranostics can also be focused

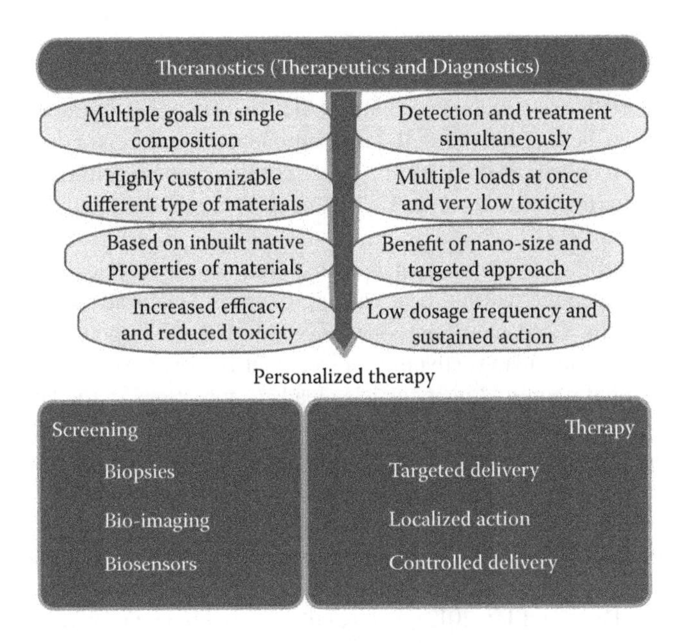

FIGURE 9.4 Theranostics: description and advantages in diagnosis and therapy.

towards developing tailored therapies based on the individual genetic makeup, a concept known as personalized medicine. In the recent developments, the applicability has been extended to multiple utilities like screening, targeting, localization, efficacy enhancement etc. In order to generate such multiple applications in a single package, different materials, and nanoformulations need to be employed. There are several strategies for the application of theranostics in healthcare. The following sections describe individual strategies and some examples for cancer diagnosis and therapy.

9.4.1 Photo-Dynamic Therapy and Photo-Thermal Therapy

Photo-dynamic therapy is another mode of usage of light for therapy especially in cancer using photosensitizer molecules that are administered by systemic or local routes. Photo-dynamic therapy involves the use of light-activated photosensitizers to form reactive oxygen that has the capability of killing cells in response to selective light illumination. Literature cites several photosensitizers' mainly 5-amino levulinic acid (ALA) and its derivatives, Verteporfin (benzoporphyrin derivative) and Photofrin (hematoporphyrin derivatives), which have been shown to be effective in photodynamic therapy (Lou et al. 2006). The main issue with using photosensitizers is the hydrophobicity, which results in uneven pharmacokinetic patterns. Photosensitizers are fluorescent molecules that can be used to locate the disease and performing surgical procedures. According to a report, a phthalocyanine photosensitizer was encapsulated in PEG-coated AuNPs and due to the EPR effect a deeper penetration occurred in the tumor tissues (Kim et al. 2013). The application of photo-dynamic therapy provides dual specificity by localizing the production of reactive oxygen species mediated by light, focusing upon tumor tissues makes it useful as non-cancerous tissues remain unaffected. Tumor selectivity can be increased by the attachment of antibodies to form photo-immuno-conjugates so that they can be targeted to cellular components like receptors, organelles or diseased tissues. Savellano et al. used a benzoporphyrin monoacid ring A (BPD) to PEGylated Cetuximab that binds to EGFR receptor expressed in epithelial cancers (Driel et al. 2016).

Photo-thermal therapy is associated with illumination of optical radiations that are absorbed and converted into heat causing thermal denaturation of proteins and thereby cell death. Gold nanocages have been developed as good candidates for photo-thermal therapy using galvanic replacement between Ag and HAuCl4. The interaction of light with gold nanoparticles leads to vibrations in the lattice causing an increased temperature. The gold nanocages have tunable peaks and can be tuned in NIR region that is an important reason for their suitability for photo-thermal therapy. The popularity of gold nanoparticles for photo-thermal therapy is also governed by the biocompatibility and amenability of functionalization Chen et al. have described functionalization of gold nanocages using anti-HER2 antibodies (trastuzumab) so that they can be targeted to SKBR3 cancer cells (Zhu et al. 2014; Kennedy et al. 2011; Yu, Park, and Jon 2012). Such antibody attachment provides for selective photo-thermal destruction of SKBR3 cells. Irudayaraj et al. used "nano-pearl-necklaces" based on Au nanorods with Fe_3O_4 nanoparticles and with trastuzumab can be used as a theranostic formulation (Wang et al. 2008; Cho et al. 2008).

9.4.2 Image-Guided Drug Delivery

Image-guided drug delivery can employ fluorophore, radionuclides or MRI probes co-immobilized in nanoparticulate carriers. Radionuclides and MRI probes can be described to be providing quantitative information in comparison to fluorophores that provide semi-quantitative data. Image-guided drug delivery can be helpful for the visualization of distribution patterns of active agents entrapped in polymers, monitoring tumor accumulation, visualization of drug release properties, monitoring intratumoral distribution, and assessment of therapeutic efficacy. For example, iodine-labeled hydroxypropyl methacrylamide (HPMA)-based polymers are suitable candidates for scintigraphy, PET, SPECT to provide for analysis of distribution of active ingredients in the body. Similarly, gadolinium-labeled HPMA co-polymers have been used for MR angiography to evaluate the blood capillaries. Polymer functionalization using imaging probes to provide for imaging capability, iodine-tagged hydrazone polymer prodrugs are some of the interesting avenues to be explored for image-guided drug therapy. Doxorubicin and gemcitabine-loaded HPMA carriers were used to study AT1 rat prostate carcinoma with respect to the imaging capability and therapeutic efficiency. Imaging analysis proved that the carriers localized in tumor tissues after being in circulation for a prolonged time and helped in the delivery of drugs. Further it was also described that chemotherapy and radiochemotherapy were synergistically beneficial for therapy of tumors. Targeting of drugs has been achieved by attachment of ligands like amino sugars, hormones, antibodies, and peptides to polymer nanoparticle carriers. Harrington et al.

developed liposomal formulations namely PEGylated doxorubicin-loaded liposomes and PEGylated cisplatin-loaded liposomes that were used as carriers for radio-chemotherapy for animals in single doses along with fractioned radiotherapy together. The results indicate that the animal survived for a longer periods of time than standard radio-chemotherapy (Harrington et al. 2000).

9.4.3 Triggered Delivery of Targeted Theranostic Nanocarriers

Theranostics are believed to be the next breakthrough to solve unmet medical needs. Deployment of theranostics in several disease conditions has been producing fruitful results recently in the fields of imaging, drug delivery, and bio-sensing. An important aspect of drug delivery is bringing about spatial and temporal control of drug delivery, which is the localization of drug delivery at a particular site and controlling the rate of drug release over a period of time, respectively.

Apart from the physical localization of drug delivery at a site, researchers have investigated several triggering systems that can selectively deposit molecules of interest at a particular site.

Noninvasive triggering mechanisms will be a boon to healthcare as they can be formed as point-of-care delivery systems without any manual intervention. Different stimuli used for developing triggered delivery systems include temperature, pH, magnetic or electrical fields, ultrasound, light, or enzymatic action (Figure 9.5). The following sections will describe in depth each of the triggering mechanism for targeting of drug delivery.

9.4.3.1 Light

Light can be used for a number of purposes including targeted imaging, image-guided drug delivery, and photo-triggered drug delivery systems. The use of light can be considered as the simplest form of triggered drug delivery. Optical imaging systems can be considered as excellent tools to obtain spatial resolution and since they are inexpensive they can be converted into portable devices. However, they offer few limitations such as low penetration depth, field, and anatomic resolution. Based on the exposure of ionizing radiations, among all the optical techniques, PET scanning is a highly accurate and noninvasive technique, however it requires control of dosage to avoid toxicity (Corsi et al. 2009). Photo-triggered drug delivery systems are externally activated systems that release the drug at the site and the rate of delivery can be controlled based on the matrix used. Some light-responsive drug delivery systems are for single use (i.e. the light triggers an irreversible structural change that provokes the delivery of the entire dose, for example a photosensitizer system conjugated to a contrasting agent; administered to selectively accumulate in target tissues and later activated by external light trigger for detachment from contrasting agent, leaving it activated for imaging purposes) while others can endure reversible structural changes when cycles of light/dark are applied and they may also act as multi-switchable carriers (releasing the drug in a pulsatile manner, for example photo degradation by laser pulses to a controlled population of liposomes, for releasing the load at a specific site or tissue; carbon nanotubes can be loaded in carriers to produce heat upon laser irradiation over a period of time or in different intervals to curb the growth of

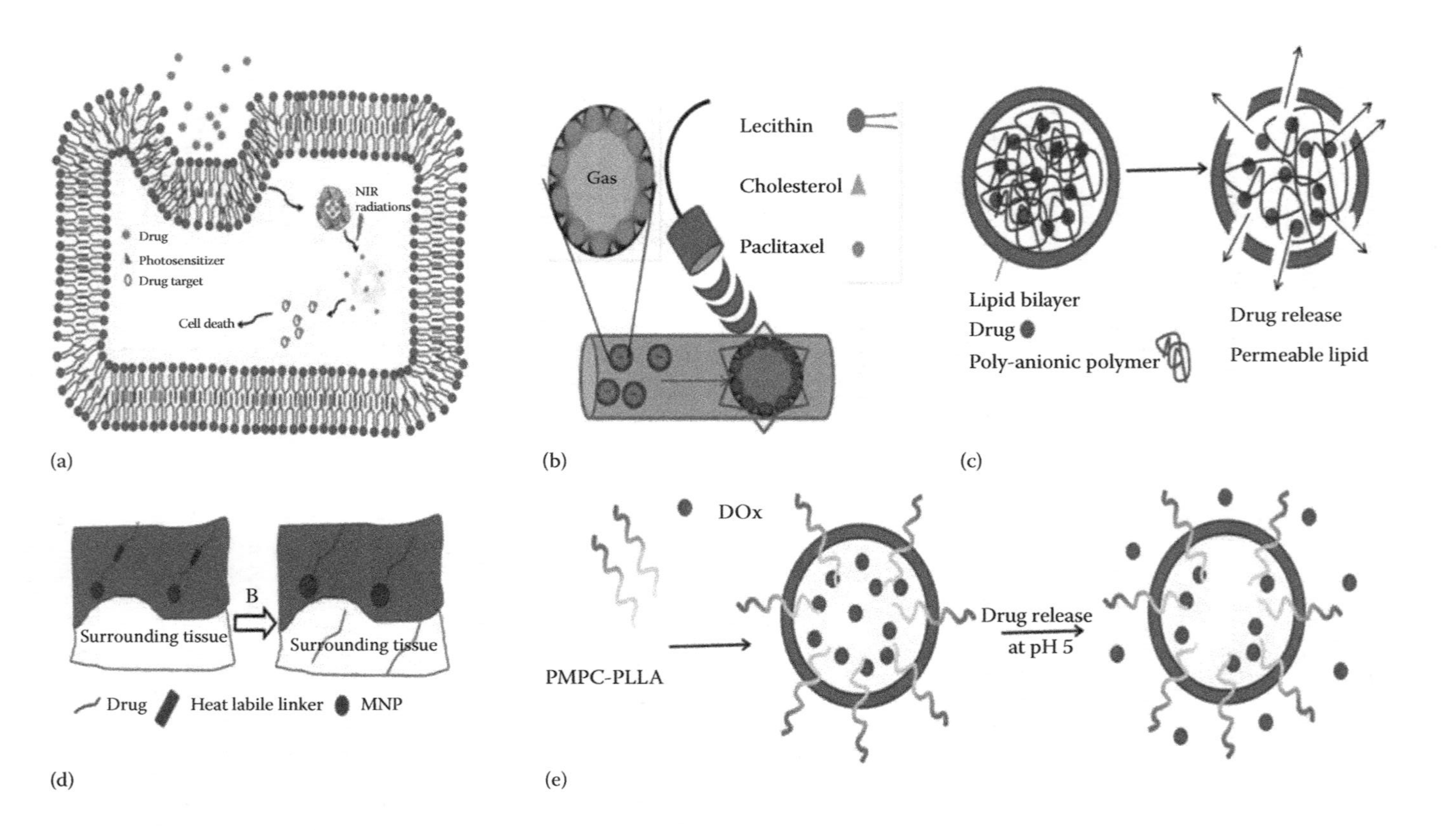

FIGURE 9.5 Different strategies for triggered drug delivery using external and internal stimuli (a) NIR light, (b) ultrasound, (c) enzymes, (d) temperature and (e) pH.

cancerous tissue). Light-controlled drug liberation offers the unique advantage of controlling the location, quantity, and timing of the drug released. Common modes of photon-induced drug delivery include photo-chemical triggering or photo-thermal triggering mechanism. Photo-chemical internalization (PCI) is a technique used for release of drug-engulfed macromolecules into the cytoplasm. As a consequence of light irradiation, cells exposed to photosensitizers result in such type of delivery and are found to be promising for treatment of soft tissue sarcomas and other solid tumors. Norum et al. have described PCI for release of endocytosed bleomycin into the cytoplasm by photo-chemical rupture. Human fibrosarcoma xenograft HT1080 was used in the leg muscles of athymic mice along with di-sulfonated aluminum phthalocyanine. The results indicated that PCI is superior to photodynamic therapy (Norum et al. 2009). Lou et al. have described the application of PCI for release of doxorubicin from endocytic vesicles in MDR cells of cancer cell lines MCF-7 and MCF-7/ADR using a photosensitizer TPPS (di-sulfonated meso-tetraphyenyl-porphine) and light. Exposure of doxorubicin alone gave the IC50 concentrations of 0.1 μM in MCF-7 and 1 μM in MCF-7/ADR, however after photo-chemical therapy followed by doxorubicin the IC50 concentration was 0.1 μM for both cell lines. After PCI and drug treatment it was found doxorubicin was released in the cytoplasm and it entered the cell nuclei in MCF-7 cells without PCI (Lou et al. 2006). Rai et al. (2010). have reviewed studies of Febvay et al. (2010), Pashkovskaya et al. (2010), Lu et al. (2006), Dvir et al. (2010) also presented a novel method of delivery based on size tunable mesoporous silica nanocarriers functionalized and targeted to specific cell surface proteins. Cell-impermeable fluorescent compounds were delivered inside the cytosol as visualized by confocal microscopy. The results obtained revealed that the nanocarriers could be loaded with different kinds of compounds that are impermeable inside cells. Pashkovskaya et al. have demonstrated that damaging the liposome membranes through light can be used for release of drugs using dye release assays based on fluorescence de-quenching of dyes. Similar to liposomes, nanoimpeller-controlled mesostructured silica nanoparticles were developed by Lu et al., which were called light-activated mesoporous silica (LAMS). These nanoparticles were loaded with luminescent dyes and anticancer drugs that were released inside the cancer cells that were illuminated using specific wavelengths, which activated the nanoimpellers. It was observed that for such systems, light intensity, and time of irradiation incident light governs the quantity of molecules released. Pancreatic cancer cell line and colon cancer cell line SW480 were used to study the uptake of particles. Another interesting finding to target nanoparticles was described by Dvir et al. who showed that nanoparticle surface was covalently linked with targeting moiety caged with a photo-removable protecting group. The caging group used in this study was a small peptide YIGSR that was required for adhesion to integrin β1 and could be removed by illumination (Rai et al. 2010).

9.4.3.2 Enzyme

Use of theranostics can have great repercussion on targeting using nanotechnology. Use of enzyme-based triggering is one of the significant biological mechanisms to target a drug to a particular tissue. It is a known fact that enzyme-substrate interaction is highly specific. Enzymes localized in a specific area can provide site-specific imaging and localized drug release. Enzyme-triggered drug delivery systems can be developed using a variety of enzymes belonging to categories like proteases, phospholipases, caspase, matrix metalloproteinase, phosphodiesterase's, amidase, phosphatases, glucose oxidase, urease, hyaluronidase, and trypsin. It is necessary that the nanocarriers used for enzymatic triggering must degrade when exposed. The most widespread enzyme-based triggered drug delivery systems are based on phospholipases and hydrolases. As reported by Zerouga et al. (2002) phospholipases have the ability to hydrolyze lipids present in liposomes and thus deliver the drug. The researchers observed that methotrexate, an anticancer drug attached to docohexanoic acid (DHA) and phosphatidyl choline was hydrolyzed by phospholipase A2 leading to reduction in murine leukemia growth (Davidsen, Vermehren, and Frokjaer 2001; Siddiqui et al. 2005). Matrix metalloproteinase (MMP) enzymes can increase uptake of nanocarriers into solid tumors by assisting the interstitial fluid pressure. The matrix metalloproteinase (MMP) enzymes are mostly tissue specific in their expression. These MMP enzymes also have a behavior of altering the basement membrane and extracellular matrix environment, this increases cancer tissue proliferation and metastasis by the mode of increased inflow of fluids and nutrients along. Apart from previously mentioned characteristics, MMP enzymes also facilitate the expression of several factors uncommon in regular cellular activity for example overexpression of angiogenesis factor (Overall and López-Otín 2002). This property can be effectively used for lipid-based nanocarrier systems like micelles and solid-lipid nanoparticles. It is also observed that phospholipase A2 can be overexpressed in cancers of colon, pancreas, prostate, breast etc. making it worthy of therapeutic intervention for triggered delivery of anticancer drugs (Overall and López-Otín 2002). As reported in the literature, PEGylated lipid (1,2-distearoyl-sn-glycero-phosphoethanol amine-N-methoxy [PEG]-2000) provided rapid release of cisplatin and doxorubicin causing toxicity to Colo 205 colon cancer cells and improved efficacy of treatment in breast cancer xenograft model (MT-3) in mice (Davidsen, Vermehren, and Frokjaer 2001; Anderson 2008). Like phospholipases, proteases also hydrolyse peptide bonds and cleave larger peptides into smaller fragments.

MMP plays a pivotal role in cancer invasion, angiogenesis and metastasis. Through their research findings Wong et al. proposed that MMP-2-responsive nanoparticle aggregates can further be broken down into 10 nm particles. 10 nm particles were found to penetrate deep in tumor tissues even in the presence of interstitial fluid pressure due to enhanced permeation and retention (Wong et al. 2011). Kim et al. have compiled an extensive review on protease-responsive drug delivery as DOX-peptide-coated, magnetic silica nanoparticle conjugates for intracellular delivery *in vitro* (Kim et al. 2013). Chi et al. (Chi et al. 2015) performed surgical removal of orthotopic and metastatic breast tumors by matrix metalloproteinase targeted fluorophores using liposomes developed from lipo-peptides. The lipo-peptides were found to be sensitive to MMP-9 and induced liposome destabilization and thus resulted in the release of the drug. Similar research work was carried out by the Harashima group, they developed PEG masked liposomes with configuration of PEG-peptide-DOPE having a lipoid core that degraded by MMP-2 by destabilization-fusion and aggregation (Hatakeyama et al. 2007). The results also indicated that the liposomes of the said configuration were less stable in circulation as against

PEGylated liposomes. Literature reports also reveal that macromolecular carriers like Glu-Pro-Cit-Gly-Hof-Tyr-Leu cross-linked to DOX showed improved efficiency of prodrugs. Other examples of MNP prodrugs include use of macromolecular carriers like (Pro-Val-Gly-Leu-Ile-Gly), which function as peptide linker to dextran and methotrexate (Chau et al. 2006; Choi et al. 2012). Likewise, Gly-Phe-Leu-Gly structure can be designed to effect drug release from the nanocarriers by the lysosomal enzyme cathepsin B that is found to exist only in the intracellular compartment. After internalization in the cells, the encapsulated drug exerts its effect. This unique feature not only improves the therapeutic efficacy but also tends to reduce any side effects associated with the drug. Investigations have been carried out on this peptide sequence for DOX linked to HPMA nanoparticles through aminoglutethimide (AGM), which acts as an aromatase inhibitor (Vicent et al. 2005).

Another class of enzymes that have potential applications in enzyme-triggered drug delivery systems is glycosidase. Glycosidases are the enzymes that hydrolyze carbohydrates to yield small sugars. An example of an enzyme belonging to this class is α-amylase, which is highly expressed in tumor environment making it suitable to design a drug delivery system containing sugar molecules coated on the nanocarriers. A blooming proof of this concept was shown by linking dextran with phospholipase via succinylation. It was observed that when lactose and starch derivatives coupled to MSNP with fluorophore trapped inside the pores were treated with pancreatin or β-D-galactosidase there was release of dye. Dox-loaded and starch-modified MSNP exerted a high influence on cell survival upon release of drug (Bernardos et al. 2010). Dzamukova showed that halloysite nanotube carriers (50 nm diameters) can be used for the selective delivery of anticancer drug doxorubicin and Brilliant green. In the study, dextrin end stoppers were used to prevent release of Brilliant green; however, when intercellular glycosyl hydrolases degraded the dextrin stoppers the entrapped molecules were released. Since the rate of internalization of nanotubes was dependent on the growth rates of tumor tissues, such delivery systems can prove to be effective and suitable for malignant cancer tissues (Dzamukova et al. 2015).

Another interesting example is that of phosphatase enzyme. In this research, Phosphatase triggered co-assembly can be used to form tumor specific ICG-doped nanofibers for cancer theranostics. NIR absorbance of ICG improves photo-acoustic and photo-thermal properties. In one of the studies, phosphatase instructed delivery was successfully studied ranging from *in vitro*, living cell, tissue mimic, and *in vivo*. The results of the study revealed that ICG (in conjugated form) uptake of tumors was markedly increased 25-fold higher than free ICG (Huang et al. 2015). Davis and Szoka used phosphatase to catalyze liposomes that promoted transfection with plasmid DNA that encoded luciferase, which was involved in a transition of lamellar phase to inverted hexagonal phase due to removal of phosphate group (Davis and Szoka 1998).

9.4.3.3 Temperature

Interesting carriers of choice for stimuli-responsive drug delivery are temperature-responsive polymers. These polymers can be categorized using volume phase transition at a particular temperature. Lower critical solution temperature (LCST) and upper critical solution temperature (UCST) are the parameters that demonstrate their solubility based on temperature. Polymers that precipitate on application of heat have a low LCST and polymers that solubilize upon heating have an UCST. An interesting example is that of hydrogel, which can be characterized using lower gel transition temperature (LGTT) and upper gel transition temperature (UGTT) due to its temperature-responsive nature. Examples of commonly used LCST polymers include N, N-diethylacrylamide (DEAM), N-isopropyl acrylamide (NIPAM), N-vinyl caprolactum, Methyl vinyl ether (MVE) (Schmaljohann 2006). Examples of widely applicable UCST polymeric systems include PEO-b-PPO block copolymers, PEO-b-PPO-b-PEO, PEG-b-PLGA-b-PEG, and combination of acrylamide and acrylic acid. Apart from the LCST other factors affecting the solubilization of polymers include concentration, temperature, molecular weight, and presence of co-solvents or additives. PNIPAM co-polymers have been widely exploited for temperature responsive properties. Ramkisson Ganorkar et al. used PNIPAM-co-BMA-co-AAc in their studies for intestinal delivery of calcitonin (Ramkissoon-Ganorkar et al. 1999). Another example is liposomes that have also been used for thermal triggering using 1, 2-dipalmitoyl-sn-glycero-3-phosphocholine (DPPC), and a lysolipid such as 1-palmitoyl-sn-glycero-3-phosphocholine or 1-stearoylsn-glycero-3-phosphocholine, the researchers concluded that the lysolipid increases the diffusion of drug through gel to liquid phase transition. Thus it was concluded that such liposomes would be useful for mild hyperthermia (Needham, Anyarambhatla, and Kong 2000). Since liposomes are unstable under physiological conditions, thermos-responsive polymers like poly [2-(2-ethoxy) ethoxy ethyl vinyl ether) and poly (N-isopropyl acryl amide) can be used to increase the stability of liposomes and modify their form. However, the above polymers are non-biodegradable and hence are used to a very limited extent. Another interesting example of temperature-triggered delivery system is that of elastin-like polypeptides incorporated in dipalmitoylphosphatidyl choline (DPPC) based liposome. In one of the studies, cyclic arginine-glycine-aspartic acid (cRGD) has been used for targeting angiogenic vasculature and tumor cells. The results for accumulation studies showed that the targeting capability was improved five-fold when compared against non-targeted liposomes. Along with an external heat generating system the system also offered improved performance (Kim et al. 2014).

Dual-responsive behaviors can be achieved with a combination of temperature and pH responsive polymers and this can lead to interesting biomedical applications. An example of such dual-responsive drug delivery was demonstrated by Chilkoti et al. (2002). They designed doxorubicin-polypeptide conjugate for cancer therapy. The LCST behavior was customized to higher temperature so that it underwent a phase transition to become insoluble after reaching the tumor target. The second responsive behavior was generated using an acid labile linker so that DOX was released in a low pH environment (Mo et al. 2015). Liu et al. developed polymerosomes, a novel form of temperature-responsive system. The delivery system comprised of poly (N-vinylcaprolactam)n-poly(dimethylsiloxane)65-poly(N-vinylcaprolactam) co-polymers. These polymersomes offered numerous advantages including properties like being biocompatible, biodegradable, monodisperse, stable at room temperature, tunable to size, and thermally responsive, and they showed a high loading capacity of 40 % (Liu et al. 2015). For delivery of drugs, development of temperature-sensitive gels using polymers is very widely investigated. An example is that of hydrogels, which possess swelling and de-swelling properties and contain insoluble polymer above LCST or UCST. A significant application is the development of ophthalmic formulation using *in-situ* gels that are temperature triggered. One of the studies reported the use of thermosensitive polymer pluronic F-127, HPMC-E50 LV, and anti-glaucoma drug betaxolol HCl were combined together to increase residence time of drug in the eyes for over seven hours. The polymers in the form of an *in-situ* gel showed good ocular tolerance. Such a formulation would offer a great potential in glaucoma therapy (Geethalakshmi et al. 2013). Temperature triggered delivery to eye has also been extensively studied by embedding timolol in HEMA gel, which is a contact lens material. The results revealed that the drug release was prolonged for two to four weeks using cross-linked particles of dimethacrylate, ethylene glycol and propoxylated glyceryl triacrylate. The driving mechanism for drug release was proposed to be due to the change in temperature after lens insertion in the eyes that triggered the drug release (Jung and Chauhan 2012). Multifunctional PEGylated liposomes are also found to be effective as temperature-triggered drug delivery systems and also for magnetic resonance imaging (MRI) developed from poly(2-ethoxy(ethoxyethyl) vinylether) chains with a lower critical solution temperature around 40°C and polyamidoamine G3 dendron-based lipids having Gd3+ chelate residues. In one of the studies, these liposomes were loaded with DOX wherein the drug release was expected to occur above 40°C. Intravenous administration of these formulations resulted in controlled and targeted delivery of DOX and tumor accumulation was observed after eight hours in 26 tumor bearing mice. Additionally, T-1 weighted images of MRI showed accumulation of Gd loaded in tumor tissues (Kono et al. 2011).

9.4.3.4 pH

pH-triggered systems are the most common forms of endogenously acting stimuli or triggered responsive drug delivery systems. This type of system was employed for oral delivery system in order to either increase the absorption of drug from GIT or to protect the drug from the acidic atmosphere by enteric coating. Enteric coating helps to release drug in the alkaline pH of intestine. Recently in a glucose oxidase catalytic reaction, a glucose-responsive drug delivery systems was developed that worked on the pH triggering mechanism since glucose that is converted to gluconic acid leads to a change in pH. This approach can be used to make implantable systems that will control the release of glucose and will be useful in management of glucose in diabetes. The extracellular pH of cancerous cells is slightly more acidic than normal and lies in between 6.5 to 7.2. Similarly, lysosomes are more acidic in nature and this property can be used to develop pH-sensitive carriers. Different acids and bases such as phosphoric acid, carboxylic acids, amines form ionizable polymers. These ionizable polymers can have pH range from 3 and 10. Change in the pH can lead to a conformational change in the polymer, which alters the swelling pattern of polymer. Polymers such as acrylate, methacrylate, and their derivative, maleic anhydride have pH dependent properties. These moieties were employed for the delivery of various drugs including indomethacin, caffeine, and cation protein such as lysosome. Poly L-histidine and amphoteric poly (amido-amine)s have exhibited endo-osmolytic properties and later it has been used for gene delivery (Huh et al. 2012). Bae et al.

examined the weak sulfonamide acid for the extracellular delivery of doxorubicin as triggering agent. Adriamycin with Poly(L-histidine)-b-PEG and PLLA-b-PEG have investigated for targeting extracellular tumor. Adriamycin was released at transition pH that takes place at 6.6. (Lee, Na, and Bae 2003).

9.4.3.5 Ultrasound

Sound waves are better alternative for imaging of soft tissues in ultrasound. To alter the permeability of blood vessels or to increase the localized release of drug, ultrasound acts as a crucial tool. Ultrasound also helps to reduce side effects. Due to change in permeability of blood vessels, nanoparticles, and impermeable drugs can enter the cells easily, e.g. alprazolam. The bioavailability of certain drugs such as cisplatin, doxorubicin, paclitaxel, siRNA, and plasmid DNA can be enhanced by using ultrasound waves. Macro-pinocytosis, caveolin, and clathrin type of mechanism help to extravasate the high molecular weight compounds such as dextran having molecular weight between 4–500 KDa (Meijering et al. 2009). Walton and Shohet (2009) have shown two different mechanisms for permeability: subcavitary oscillation and inertial cavitation through microbubble ultrasound-mediated delivery. The key carriers in the ultrasound-based triggering are the microbubbles that can be used for drug delivery. Due to air or gaseous nature of microbubbles, they show acoustic properties that can be imaged by ultrasound easily. Rupture of these bubbles releases the drug and leads to disappearance of echogenicity, which is another marker for imaging using ultrasound techniques (Deckers and Moonen 2010). Microbubbles containing paclitaxel were used for restenosis in a rabbit iliac balloon injury model along with ultrasound and these were effectively releases drug for visualization (Zhu et al. 2016). Micelles, polymer capsule, and liposomes can also offer an ideal carrier for ultrasound sensitive drug (Ayre et al. 2013). The sono-sensitive and MRI contrast agent was developed from Gd(III) DOTA altered with sono-sensitive liposomes that can be used for the delivery of DOX to a specific site and rate with improved signal intensity at the area of accumulation (Zhu et al. 2016). At different ultrasonic pressure, polymeric PLLA capsules having different shell width can be triggered both *in vivo* and *in vitro* using Evans blue as a model agent.

9.4.3.6 Magnetism

In order to achieve targeting on some specific sites through magnetic nanoparticles, an external magnetic field can be used for the alignment of the magnetic moment in a particular direction. Magnetic nanoparticles (MNP) are perfect contenders for the development of novel theranostic systems due to their attractive properties such as magnetic hyperthermia, negative contrast for MRI, target specific-controlled drug delivery. In MRI, MNP can be used as negative contrast that can be indicated by T2 relaxation phenomena (Murase et al. 2015). Akbarzadeh et al. have reported the synthesis of PNIPAM-MAA grafted magnetic nanoparticles by radical polymerization method for MRI application (Akbarzadeh et al. 2012). Similarly, Wang et al. have developed chitosan and polyethyleneimine (PEI)-coated magnetic micelles for MRI and to deliver nucleic acid. Experimentally it has been shown that superparamagnetic iron oxide nanoparticles (SPIONs) integrated inside the micelle have higher biocompatibility, prolonged circulation time, and improved contrast agent. Moreover, micelles were used effectively for probing in MRI and in drug delivery. SPIONs have various applications in cancer management such as drug load on MNP, target selectivity because of magnetic field or conjunction with antibodies. MRI is considered a better diagnostic tool because of good spatial resolution, outstanding anatomical image, and high soft tissue contrast. The coating of MNP by polymer can modify the release of drug and can offer better stability. Various formulations such as emulsions, contraceptive systems, implants, infusion pumps along with magnetic nanoparticles have shown substantial improvements for targeting and triggering drugs for preferred site. Kumar et al. have demonstrated that Fe_2O_3 nanoparticles coated with chitosan transport localize at a specific site (Kumar et al. 2010; Mishra, Patel, and Tiwari 2010; Kumari, Yadav, and Yadav 2010). Bio-conjugation with protein, ligand or antibodies can offer selectivity to drug delivery system. Iron oxide nanoparticles with PNIPAM composite can be used not only for magnetic hyperthermia but also for drug release because of thermo-sensitive nature of PNIPAM (Purushotham and Ramanujan 2010).

9.5 CONCLUSIONS AND FUTURE PROSPECTS

Nanotheranostic approach of management of diseases appears to be superior to conventional modes of diagnosis and therapies. This fact can be attributed to selective treatment, use of nanosystems that provide deeper penetration in physiological systems and triggered delivery of diagnostic or therapeutic agents. The recent theranostic approaches are still in developmental phase and have shown sufficient promise against cancer. Theranostic systems consist of nanomaterials developed from biomolecules, natural or synthetic organic or inorganic materials or any combinations of them to provide different applications in a single system. Functionalization of nanomaterials in order to localize or control the fate of nanoformulations is another advantage obtained using targeting ligands conjugated to theranostic nanomaterials. Targeting allows for site-specific release and reduced toxicity to other systemic organs improving the therapeutic compliance. Treatment using genetic materials like DNA, siRNA, and oligonucleotides help in interacting with cellular mechanisms and will allow for treatment of diseases with genetic predisposition. Reports of use of theranostics in the clinic and validation of effectiveness of performance in animals and humans need to be visited in great depth before actual practice of theranostics as a healthcare solution. Nanotheranostic approaches are being accepted worldwide across the research community as it projects the expansion of new horizons of adapted and applied therapies for various diseases and disorders with lesser toxicity against any anomaly ranging from gastrointestinal, brain, and cardiovascular ailments. Thus, the use of theranostics allows for efficient healthcare solutions to take care of morbidities in a more precise manner in comparison to conventional diagnostic or therapeutic strategies.

ACKNOWLEDGMENTS

Gaurav Pandey acknowledges fellowship provided by MHRD, Government of India, and IIT Indore. Abhijeet Joshi acknowledges the INSPIRE Faculty award provided by Department of Science and Technology, Government of India.

REFERENCES

Akbarzadeh, A., N. Zarghami, H. Mikaeili, et al. 2012. "Synthesis, Characterization, and in Vitro Evaluation of Novel Polymer-Coated Magnetic Nanoparticles for Controlled Delivery of Doxorubicin." *Nanotechnology, Science and Applications* 5 (1).

Anderson, J. L. 2008. "Lipoprotein-Associated Phospholipase A2: An Independent Predictor of Coronary Artery Disease Events in Primary and Secondary Prevention." *American Journal of Cardiology* 101 (12 SUPPL.).

Anonymous. 2016. "Understanding Nanotechnology in Cancer Treatment." Accessed December 1. www.understandingnano.com/cancer-treatment-nanotechnology.html.

Ayre, A. P., V. J. Kadam, N. M. Dand, and P. B. Patel. 2013. "Polymeric Micelles as a Drug Carrier for Tumor Targeting." *Chronicles of Young Scientists* 4 (2): 94.

Bae, K. H., J. Y. Lee, S. H. Lee, T. G. Park, and Y. S. Nam. 2013. "Optically Traceable Solid Lipid Nanoparticles Loaded with SiRNA and Paclitaxel for Synergistic Chemotherapy with In Situ Imaging." *Advanced Healthcare Materials* 2 (4): 576–84.

Bernardos, A., L. Mondragón, E. Aznar, et al. 2010. "Enzyme-Responsive Intracellular Controlled Release Using Nanometric Silica Mesoporous Supports Capped with 'Saccharides.'" *ACS Nano* 4 (11): 6353–68.

Bukhtoyarov, O. V., D. M. Samarin, O. V. Bukhtoyarov, and D. M. Samarin. 2015. "Pathogenesis of Cancer: Cancer Reparative Trap." *Journal of Cancer Therapy Pathogenesis of Cancer. Cancer Reparative Trap. Jour-Nal of Cancer Therapy* 6 (6): 399–412.

Chau, Y., N. M. Dang, F. E. Tan, and R. Langer. 2006. "Investigation of Targeting Mechanism of New Dextran-Peptide-Methotrexate Conjugates Using Biodistribution Study in Matrix-Metalloproteinase-Overexpressing Tumor Xenograft Model." *Journal of Pharmaceutical Sciences* 95 (3): 542–51.

Chi, C., Q. Zhang, Y. Mao, et al. 2015. "Increased Precision of Orthotopic and Metastatic Breast Cancer Surgery Guided by Matrix Metalloproteinase-Activatable Near-Infrared Fluorescence Probes." *Scientific Reports* 5 (1): 14197.

Chilkoti, A., M. R. Dreher, D. E. Meyer, et al. 2002. "Targeted Drug Delivery by Thermally Responsive Polymers." *Advanced Drug Delivery Reviews* 54 (5): 613–30.

Cho, K., X. Wang, S. Nie, Z. Chen, and D. M. Shin. 2008. "Therapeutic Nanoparticles for Drug Delivery in Cancer." *Clinical Cancer Research* 14 (5): 1310–6.

Choi, K. Y., M. Swierczewska, S. Lee, and X. Chen. 2012. "Protease-Activated Drug Development." *Theranostics* 2 (2): 156–78.

Corsi, F., C. De Palma, M. Colombo, et al. 2009. "Towards Ideal Magnetofluorescent Nanoparticles for Bimodal Detection of Breast-Cancer Cells." *Small* 5 (22): 2555–64.

Davidsen, J., C. Vermehren, and S. Frokjaer. 2001. "Drug Delivery by Phospholipase A 2 Degradable Liposomes." *International Journal of Pharmaceutics* 214 (1–2): 67–9.

Davis, S. C., and F. C. Szoka. 1998. "Cholesterol Phosphate Derivatives: Synthesis and Incorporation into a Phosphatase and Calcium-Sensitive Triggered Release Liposome." *Bioconjugate Chemistry* 9 (6): 783–92.

Deckers, R., and C. T. W. Moonen. 2010. "Ultrasound Triggered, Image Guided, Local Drug Delivery." *Journal of Controlled Release* 148 (1): 25–33.

Driel, P. B. A. A. Van, M. C. Boonstra, M. D. Slooter, et al. 2016. "EGFR Targeted Nanobody-Photosensitizer Conjugates for Photodynamic Therapy in a Pre-Clinical Model of Head and Neck Cancer." *Journal of Controlled Release* 229: 93–105.

Dvir, T., M. R. Banghart, B. P. Timko, et al. 2010. "Photo-Targeted Nanoparticles." *Nano Letters*. 10: 250–4.

Dzamukova, M. R., E. A. Naumenko, Y. M. Lvov, and R. F. Fakhrullin. 2015. "Enzyme-Activated Intracellular Drug Delivery with Tubule Clay Nanoformulation." *Scientific Reports* 5.

Faraji, A. H., and P. Wipf. 2009. "Nanoparticles in Cellular Drug Delivery." *Bioorganic and Medicinal Chemistry* 17 (8): 2950–62.

Farokhzad, O. C., J. Cheng, B. A. Teply, et al. 2006. "Targeted Nanoparticle-Aptamer Bioconjugates for Cancer Chemotherapy in Vivo." *Proceedings of the National Academy of Sciences* 103 (16): 6315–20.

Febvay, S., D. M. Marini, A. M. Belcher, et al. 2010 "Targeted Cytosolic Delivery of Cell-Impermeable Compounds by Nanoparticle-Mediated, Light-Triggered Endosome Disruption." *Nano Letters*, 10: 2211–9.

Furuya, M., M. Nishiyama, Y. Kasuya, S. Kimura, and D. P. Venkatesh. 2005. "Pathophysiology of Tumor Neovascularization." *Vascular Health and Risk Management* 1 (4): 277–90.

Gao, H. 2016. "Progress and Perspectives on Targeting Nanoparticles for Brain Drug Delivery." *Acta Pharmaceutica Sinica B* 6 (4): 268–86.

Geethalakshmi, A., R. Karki, P. Sagi, et al. 2013. "Temperature Triggered in Situ Gelling System for Betaxolol in Glaucoma." *Journal of Applied Pharmaceutical Science* 3 (2): 153–9.

Gold, J. 2011. "What Is Cancer?—The Pathogenesis of Cancer: Cancer as a Normal Protective Device of the Human Body." October 1. http://thepathogenesisofcancer.com/.

Harrington, K. J., G. Rowlinson-Busza, K. N. Syrigos, et al. 2000. "PEGylated liposome-Encapsulated Doxorubicin and Cisplatin Enhance the Effect of Radiotherapy in a Tumor Xenograft Model." *Clinical Cancer Research*, 6 (12): 4939–49.

Hatakeyama, H., H. Akita, K. Kogure, et al. 2007. "Development of a Novel Systemic Gene Delivery System for Cancer Therapy with a Tumor-Specific Cleavable PEG-Lipid." *Gene Therapy*, 14 (1): 68–77.

Haynes, B., Y. Zhang, F. Liu, et al. 2016. "Gold Nanoparticle Conjugated Rad6 Inhibitor Induces Cell Death in Triple Negative Breast Cancer Cells by Inducing Mitochondrial Dysfunction and PARP-1 Hyperactivation: Synthesis and Characterization." *Nanomedicine: Nanotechnology, Biology, and Medicine* 12 (3): 745–57.

Hoare, T., J. Santamaria, G. F. Goya, et al. 2009. "A Magnetically Triggered Composite Membrane for on-Demand Drug Delivery." *Nano Letters* 9 (10): 3651–7.

Huang, P., Y. Gao, J. Lin, et al. 2015. "Tumor-Specific Formation of Enzyme-Instructed Supramolecular Self-Assemblies as Cancer Theranostics." *ACS Nano* 9 (10): 9517–27.

Huh, K. M., H. C. Kang, Y. J. Lee, and Y. H. Bae. 2012. "pH-Sensitive Polymers for Drug Delivery." In *Macromolecular Research*. 20: 224–33.

Ihre, H. R., O. L. P. de Jesus, F. C. Szoka, and J. M. J. Frechet. 2002. "Polyester Dendritic Systems for Drug Delivery Applications: Design, Synthesis, and Characterization." *Bioconjugate Chemistry* 13 (3): 443–52.

Joshi, A., R. Chaudhari, R. Srivastava. 2017. "PH and Urea Estimation in Urine Samples Using Single Fluorophore and Ratiometric Fluorescent Biosensors." *Nature Scientific Reports* 7 (1): 5840.

Joshi, A., J. Kaur, R. Kulkarni, R. Chaudhari. 2018. "In-vitro and Ex-vivo Evaluation of Raloxifene Hydrochloride Delivery Using Nano-transfersome Based Formulations." *Journal of Drug Delivery Science and Technology* 45: 151–8.

Joshi, A., S. Solanki, R. Chaudhari, D. Bahadur, M. Aslam, R. Srivastava. 2011. "Multifunctional Alginate Microspheres for Biosensing, Drug Delivery and Magnetic Resonance Imaging." *Acta Biomaterialia* 7 (11): 3955–63.

Jung, H. J. and A. Chauhan. 2012. "Temperature Sensitive Contact Lenses for Triggered Ophthalmic Drug Delivery." *Biomaterials* 33 (7): 2289–300.

Kawasaki, E. S., and A. Player. 2005. "Nanotechnology, Nanomedicine, and the Development of New, Effective Therapies for Cancer." *Nanomedicine: Nanotechnology, Biology, and Medicine* 1 (2): 101–9.

Kelkar, S. S., and T. M. Reineke. 2011. "Theranostics: Combining Imaging and Therapy." *Bioconjugate Chemistry* 22 (10): 1879–903.

Kennedy, L. C., L. R. Bickford, N. A. Lewinski, et al. 2011. "A New Era for Cancer Treatment: Gold-Nanoparticle-Mediated Thermal Therapies." *Small* 7 (2): 169–83.

Kim, C. S., B. Duncan, B. Creran, and V. M. Rotello. 2013. "Triggered Nanoparticles as Therapeutics." *Nano Today* 8 (4): 439–47.

Kim, K. S., S. J. Park, M. Y. Lee, K. G. Lim, and S. K. Hahn. 2012. "Gold Half-Shell Coated Hyaluronic Acid-Doxorubicin Conjugate Micelles for Theranostic Applications." *Macromolecular Research* 20 (3): 277–82.

Kim, M. S., D. W. Lee, K. Park, et al. 2014. "Temperature-Triggered Tumor-Specific Delivery of Anticancer Agents by cRGD-Conjugated Thermosensitive Liposomes." *Colloids and Surfaces B: Biointerfaces* 116: 17–25.

Kono, K., S. Nakashima, D. Kokuryo, et al. 2011. "Multi-Functional Liposomes Having Temperature-Triggered Release and Magnetic Resonance Imaging for Tumor-Specific Chemotherapy." *Biomaterials* 32 (5): 1387–95.

Kumar, A., P. K. Jena, S. Behera, R. J. Lockey, S. Mohadapatra, and S. Mohapatra. 2010. "Multifunctional Magnetic Nanoparticles for Targeted Delivery." *Nanomedicine: Nanotechnology, Biology, and Medicine* 6 (1): 64–9.

Kumari, A., S. K. Yadav, and S. C. Yadav. 2010. "Biodegradable Polymeric Nanoparticles Based Drug Delivery Systems." *Colloids and Surfaces B: Biointerfaces* 75 (1): 1–18.

Lammers, T., F. Kiessling, W. E. Hennink, and G. Storm. 2012. "Drug Targeting to Tumors: Principles, Pitfalls and (Pre-) Clinical Progress." *Journal of Controlled Release* 161 (2): 175–87.

Lee, E. S., K. Na, and Y. H. Bae. 2003. "Polymeric Micelle for Tumor pH and Folate-Mediated Targeting." *Journal of Controlled Release* 91 (1–2): 103–13.

Lee, S. M., H. J. Kim, S. Y. Kim, et al. 2014. "Drug-Loaded Gold Plasmonic Nanoparticles for Treatment of Multidrug Resistance in Cancer." *Biomaterials* 35 (7): 2272–82.

Liu, F., V. Kozlovskaya, S. Medipelli, et al. 2015. "Temperature-Sensitive Polymersomes for Controlled Delivery of Anticancer Drugs." *Chemistry of Materials* 27 (23): 7945–56.

Lou, P. J., P. S. Lai, M. J. Shich, A. J. MacRobert, K. Berg, and S. G. Brown. 2006. "Reversal of Doxorubicin Resistance in Breast Cancer Cells by Photochemical Internalization." *International Journal of Cancer* 119 (11): 2692–8.

Lu, Z., M. D. Prouty, Z. Quo, V. O. Golub, C. S. S. R Kumar, and Y. M. Lvav. 2005. "Magnetic Switch of Permeability for Polyelectrolyte Microcapsules Embedded with Co@ Aunanoparticles." *Langmuir* 21 (5): 2042–50.

Meijering, B. D. M., L. J. M. Juffermans, A. Van Wamel, et al. 2009. "Ultrasound and Microbubble-Targeted Delivery of Macromolecules Is Regulated by Induction of Endocytosis and Pore Formation." *Circulation Research* 104 (5): 679–87.

Miller, A. B., B. Hoogstraten, M. Staquet, and A. Winkler. 1981. "Reporting Results of Cancer Treatment." *Cancer* 47 (1): 207–14.

Mironidou-Tzouveleki, M., K. Imprialos, and A. Kintsakis. 2011. "Nanotechnology in Cancer Treatment." Proc. SPIE Biosensing and Nanomedicine IV, 809917, 8099.

Mishra, B., B. B. Patel, and S. Tiwari. 2010. "Colloidal Nanocarriers: A Review on Formulation Technology, Types and Applications toward Targeted Drug Delivery." *Nanomedicine: Nanotechnology, Biology, and Medicine* 6 (1): 9–24.

Mo, R., T. Jiang, W. Sun, and R. Gu. 2015. "ATP-Responsive DNA-Graphene Hybrid Nanoaggregates for Anticancer Drug Delivery." *Biomaterials* 50 (1): 67–74.

Murase, K., K. Nishimoto, A. Mimura, M. Aoki, K. Hamanakawa, and N, Banura. 2015. "Application of Magnetic Particle Imaging to Pulmonary Imaging Using Nebulized Magnetic Nanoparticles: Phantom and Small Animal Experiments." *2015 5th International Workshop on Magnetic Particle Imaging (IWMPI).*

Muthu, M. S., D. T. Leong, L. Mei, and S. S. Feng. 2014. "Nanotheranostics—Application and Further Development of Nanomedicine Strategies for Advanced Theranostics." *Theranostics* 4 (6): 660–77.

Needham, D., G. Anyarambhatla, and G. Kong. 2000. "A New Temperature-Sensitive Liposome for Use with Mild Hyperthermia: Characterization and Testing in a Human Tumor Xenograft Model."

Nguyen, K. T. 2011. "Targeted Nanoparticles for Cancer Therapy: Promises and Challenges." *Journal of Nanomedicine & Nanotechnology* 2 (5): 103e.

Norum, O. J., J. V. Gaustad, E. Angell-Petersen, et al. 2009. "Photochemical Internalization of Bleomycin Is Superior to Photodynamic Therapy due to the Therapeutic Effect in the Tumor Periphery." *Photochemistry and Photobiology* 85 (3): 740–9.

Overall, C. M., and C. López-Otín. 2002. "Strategies for Mmp Inhibition in Cancer: Innovations for the Post-Trial Era." *Nature Reviews Cancer* 2 (9): 657–72.

Pashkovskaya, A., E. Kotova, Y. Zorlu, et al. 2010, "Light-Triggered Liposomal Release: Membrane Permeabilization by Photodynamic Action." *Langmuir* 26: 5726–33.

Purushotham, S., and R. V. Ramanujan. 2010. "Thermoresponsive Magnetic Composite Nanomaterials for Multimodal Cancer Therapy." *Acta Biomaterialia* 6 (2): 502–10.

Rai, P., S. Mallidi, X. Zheng, et al. 2010. "Development and Applications of Photo-Triggered Theranostic Agents." *Advanced Drug Delivery Reviews* 62 (11): 1094–124.

Ramkissoon-Ganorkar, C., F. Liu, Mi. Baudyš, and S. W. Kim. 1999. "Modulating Insulin-Release Profile from pH/thermosensitive Polymeric Beads through Polymer Molecular Weight." *Journal of Controlled Release* 59 (3): 287–98.

Saba, N. F., X. Wang, S. Müller, et al. 2009. "Examining Expression of Folate Receptor in Squamous Cell Carcinoma of the Head and Neck as a Target for a Novel Nanotherapeutic Drug." *Head and Neck* 31 (4): 475–81.

Schmaljohann, D. 2006. "Thermo- and pH-Responsive Polymers in Drug Delivery." *Advanced Drug Delivery Reviews* 58 (15): 1655–70.

Shao, D., J. Li, X. Zheng, et al. 2016. "Janus 'nano-Bullets' for Magnetic Targeting Liver Cancer Chemotherapy." *Biomaterials* 100: 118–33.

Shuhendler, A. J., P. Prasad, M. Leung, A. M. Rauth, R. S. Dacosta, and X. Y. Wu. 2012. "A Novel Solid Lipid Nanoparticle Formulation for Active Targeting to Tumor αvβ3 Integrin Receptors Reveals Cyclic RGD as a Double-Edged Sword." *Advanced Healthcare Materials* 1 (5): 600–8.

Siddiqui, R. A., M. Zerouga, M. Wu, et al. 2005. "Anticancer Properties of Propofol-Docosahexaenoate and Propofol-Eicosapentaenoate on Breast Cancer Cells." *Breast Cancer Research: BCR* 7 (5): R645–54.

Singh, O. P., and Nehru, R. M. 2008. "Nanotechnology and Cancer Treatment." *Asian J Exp Sci* 22 (2): 6.

Song, B., C. Wu, and J. Chang. 2015. "Ultrasound-Triggered Dual-Drug Release from Poly(lactic-Co-Glycolic Acid)/mesoporous Silica Nanoparticles Electrospun Composite Fibers." *Regenerative Biomaterials* 2 (4): 229–37.

Uchida, S., H. Kinoh, T. Ishii, et al. 2016. "Systemic Delivery of Messenger RNA for the Treatment of Pancreatic Cancer Using Polyplex Nanomicelles with a Cholesterol Moiety." *Biomaterials* 82: 221–8.

Vicent, M. J., F. Greco, R. I. Nicholson, A. Paul, P. C. Griffiths, and R. Duncan. 2005. "Polymer Therapeutics Designed for a Combination Therapy of Hormone-Dependent Cancer." *Angewandte Chemie—International Edition* 44 (26). WILEY-VCH Verlag: 4061–6.

Walton, C. B., and R. V. Shohet. 2009. "Tiny Bubbles and Endocytosis?" *Circulation Research* 104 (5): 563–5.

Wang, X., L. Yang, Z. Chen, and D. M. Shin. 2008. "Application of Nanotechnology in Cancer Therapy and Imaging." *CA: A Cancer Journal for Clinicians* 58 (2): 97–110.

Wang, X. S.. 2008. "Pathophysiology of Cancer-Related Fatigue." *Clinical Journal of Oncology Nursing* 12 (SUPPL. 5): 11–20.

Weinberg, H., E. Blackburn, B. Druker, M. C. Leland, and R. King. 2001. "Cell Biology and Cancer." *Counseling about Cancer: Strategies for Genetic Counseling*, 1–17.

Wen, C. J., C. T. Sung, I. A. Aljuffali, Y. J. Huang and J. Y. Fang. 2013. "Nanocomposite Liposomes Containing Quantum Dots and Anticancer Drugs for Bioimaging and Therapeutic Delivery: A Comparison of Cationic, PEGylated and Deformable Liposomes." *Nanotechnology* 24 (32): 325101.

WHO. 2017. "Cancer." *February 2017.* www.who.int/mediacentre/factsheets/fs297/en/.

Wong, C., T. Stylianopoulos, J. Cui, W. Chen, C. Chang, and C. Y. Mou. 2011. "Multistage Nanoparticle Delivery System for Deep Penetration into Tumor Tissue." *Proceedings of the National Academy of Sciences* 108 (6): 2426–31.

Wu, S. H., C. Y. Lin, Y. Hung, et al. 2011. "PEGylated Silica Nanoparticles Encapsulating Multiple Magnetite Nanocrystals for High-Performance Microscopic Magnetic Resonance Angiography." *Journal of Biomedical Materials Research—Part B Applied Biomaterials* 99 B (1): 81–8.

Xie, J., S. Lee, and X. Chen. 2010. "Nanoparticle-Based Theranostic Agents." *Advanced Drug Delivery Reviews* 62 (11): 1064–79.

Yu, M. K., J. Park, and S. Jon. 2012. "Targeting Strategies for Multifunctional Nanoparticles in Cancer Imaging and Therapy." *Theranostics* 2 (1): 3–44.

Zerouga, M., W. Stillwell., L. J. Jenski. 2002. "Synthesis of a Novel Phosphatidylcholine Conjugated to Docosahexaenoic Acid and Methotrexate that Inhibits Cell Proliferation." *Anticancer Drugs* 13 (3): 301–11.

Zhu, H., H. Chen, X. Zeng, Z. Wang, X. Zhang and Y. Wu. 2014. "Co-Delivery of Chemotherapeutic Drugs with Vitamin E TPGS by Porous PLGA Nanoparticles for Enhanced Chemotherapy against Multi-Drug Resistance." *Biomaterials* 35 (7): 2391–400.

Zhu, X., J. Guo, C. He, et al. 2016. "Ultrasound Triggered Image-Guided Drug Delivery to Inhibit Vascular Reconstruction via Paclitaxel-Loaded Microbubbles." *Scientific Reports* 6 (November 2015.)

10 Emerging Nanotechnologies for Cancer Immunodiagnosis and Cancer Immunotherapies

Amy M. Wen, Nicole F. Steinmetz, and Sourabh Shukla

CONTENTS

10.1 INTRODUCTION

As described all throughout this book, nanoparticles exhibit novel properties, which largely differ from the bulk materials due to their enhanced surface to volume ratios. These include both physiochemical and biological properties. The nano-dimension not only confers unique electronic, optical, and electrochemical properties to the nanoparticles; the nano size scale also renders these materials comparable to biomolecules, cellular components as well as pathogens. While the former enables unique and highly sensitive interactions with light and matter, the latter enables precise modulation of biomolecular interactions through these tiny machines. This chapter dwells at the interface of these two capabilities of nanomaterials—as tools for diagnosis and biological manipulations. Keeping in mind the immensity of such a topic, we have focused our discussion on the cancer diagnostic and therapeutic aspects through immunological lenses.

Immunology offers a highly specific and sensitive analytical methodology to detect, quantify, and diagnose with great precision. Combining the unique opto-electronic properties of nanomaterials with immune-analytical techniques and tools results in some very sensitive yet simple and portable diagnostic methods and tools for rapid detection of cancer biomarkers. Early detection and characterization of cancer is critical for understanding the underlying disease process, and designing personalized courses of treatment, with the ultimate goal of improving patient survival and treatment prognosis. This gives rise to a need for practical, cheap, an d quick methods for routine screening of early cancer markers in a sensitive and selective manner. There have been great strides made in the development of cancer biosensors in recent years, and the incorporation of nanoparticles has played a critical role in improving the realm of immunodiagnostics. This chapter highlights some of the most recent developments in the field of cancer immunodiagnostics based on nanotechnology.

Nanoparticles also possess unique biological properties—and have been increasingly adapted for biomedical applications, as evident from the rapid growth of the field of nanomedicine. As carriers of insoluble and unstable drug molecules, nanoparticles provide the stability and improve the bioavailability of such therapeutic compounds. With the ability to accumulate in targeted tissues and solid tumors, and rapid clearance, nanoparticles are fast emerging as the carriers of choice to deliver imaging contrast agents for diagnostic imaging. With the emergence of immunotherapies as highly potent adjuvant treatment options for cancer, nanoparticles are uniquely suited to make drastic improvements in treatment strategies based on similar principles of nanomedicine. The second half of this chapter focuses on some of these critical contributions of nanotechnology in cancer immunotherapies.

10.2 CANCER IMMUNODIAGNOSIS

Immunodiagnostics has been an essential technique for the past several decades for cancer detection (Gold et al. 1965). By taking advantage of the strong and highly specific association of antibodies with their respective antigens, accurate detection of certain cancer biomarkers such as prostate-specific antigen (PSA) and carcinoembryonic antigen (CEA) can be achieved (Ludwig et al. 2005, Polascik et al. 1999).

Immunology as a means for diagnostics was first demonstrated in 1960 for detecting serum insulin (Yalow et al. 1960). Not long after, in 1965, Gold and Freedman discovered the presence of CEA, which is normally found in fetal tissues, in the blood of colon cancer patients (Gold et al. 1965). Since then, the field of biomarker-based cancer immunodiagnostics has greatly expanded, and advancements in technology, including new detection techniques and the incorporation of nanoparticles, have significantly improved the sensitivity and specificity of detection (Jayanthi et al. 2017, Malhotra et al. 2016).

In this section, we focus on nanoparticle technologies for detecting protein markers that are correlated with cancer; this includes immunodiagnostics aimed to detect antigens expressed by cancer cells as well as cancer-associated serum autoantibodies. The coupling of nanoparticles to antibody-antigen interactions is a powerful technique that can be further applied to *in vivo* imaging diagnostics, which is examined in more detail in several comprehensive reviews (Bazak et al. 2015, Chen et al. 2016). Our principal aim is to introduce recent developments regarding innovative *in vitro* cancer immunodiagnostics technologies for analysis of blood and other biological samples, which have the advantages of being more convenient, cheaper, and more readily repeated. Specific examples are presented to demonstrate the range of applications, but this is by no means the whole extent of the field, and readers are encouraged to look through the review articles presented here for more detailed information regarding cancer biosensors (Anik et al. 2016, Chikkaveeraiah et al. 2012, Ge et al. 2017, Guo et al. 2015, He et al. 2016, Jayanthi et al. 2017).

Enzyme-linked immunosorbent assays (ELISAs) in microtiter plates have long been the traditional form for immunodiagnostics since its first description in 1971 (Engvall et al. 1971). While conventional ELISAs have been highly useful and robust, steps are being taken toward developing quicker and more automated point-of-care immunosensors (Ge et al. 2017). These immunosensors can implement a variety of transduction methods for signal detection, with electrochemical (Chikkaveeraiah et al. 2012) and optical techniques (He et al. 2016) being among the most popular. The general setup of biosensors involves: 1) a biorecognition molecule for the analyte, typically an antibody in immunodiagnostics, 2) a support on which to immobilize the biorecognition molecule, and 3) a transducer to quantify the signal from binding events, which encompasses electrochemical, optical, and mass-based techniques (Figure 10.1) (Chikkaveeraiah et al. 2012, Jayanthi et al. 2017). Through incorporating nanoparticles and exploiting their high molar absorptivity, lower detection limits and higher sensitivities can be achieved (Kreibig et al. 2013, Zhao et al. 2008). After a brief overview of potential biomarkers for cancer diagnostics, a more detailed account of the range of immunodiagnostic methods involving nanoparticles is provided, followed by discussion on future opportunities for cancer diagnostics.

10.2.1 Tumor Biomarkers

During cancer development, there is a functional gain in oncogenes or loss in tumor suppressor genes that grant a cell undiminished proliferation, and a common consequence is an alteration in the expression profiles of certain protein biomarkers that

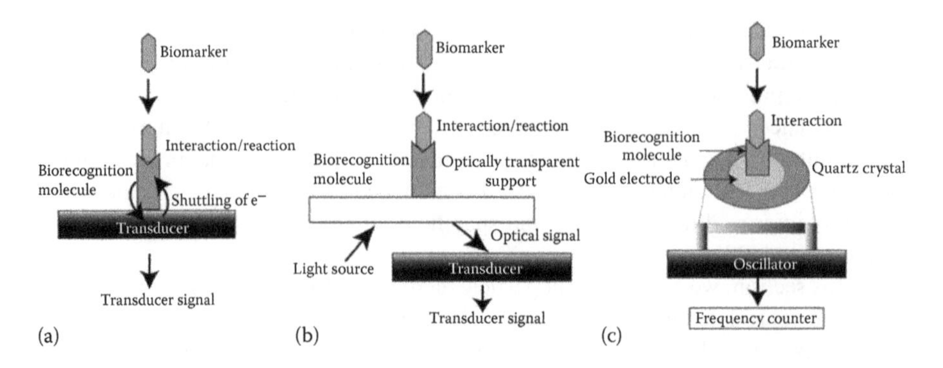

FIGURE 10.1 General schematics of different biosensor transducers. (a) Electrochemical, (b) optical, and (c) mass-based transduction systems. (Reproduced from Jayanthi, V. S. P. K. S. A. et al., *Biosens Bioelectron*, 91:15–23, 2017. With permission.)

can be used to differentiate between healthy and diseased states (Hanahan et al. 2000). There are a wide variety of markers that are regularly applied in the clinic for cancer detection, classification, and surveillance (Table 10.1) (Handy 2009). Some of the markers used in biosensors include PSA, CEA, cancer antigens (CA 125, CA 15-3, CA 19-9, CA 27.29), alpha-fetoprotein (AFP), human epidermal growth factor

TABLE 10.1
Tumor Markers Routinely Utilized in Clinical Laboratories

Cancer Type	Biomarker	Biomarker Type	Clinical Use
Testicular	AFP	Glycoprotein	Staging
Testicular	hCG	Glycoprotein	Staging
Pancreatic	CA 19-9	Carbohydrate	Monitoring
Ovarian	CA 125	Glycoprotein	Monitoring
Colon	CEA	Protein	Monitoring
Colon	EGFR	Protein	Therapy selection
Gastrointestinal	KIT	Protein	Diagnosis and therapy selection
Thyroid	Thyroglobulin	Protein	Monitoring
Prostate	PSA	Protein	Screening, diagnosis, and monitoring
Breast	CA 15-3	Glycoprotein	Monitoring
Breast	CA 27-29	Glycoprotein	Monitoring
Breast	Cytokeratins	Protein	Prognosis
Breast	HER2/neu	Protein	Prognosis and therapy selection
Bladder	NMP22	Protein	Screening and monitoring
Bladder	Fibrin/FDP	Protein	Monitoring
Bladder	BTA	Protein	Monitoring
Bladder	CEA	Protein	Monitoring

Source: Reproduced with permission from Ludwig, J. A., and Weinstein, J. N. *Nat Rev Cancer*, 5:845–56, 2005. With permission.

receptor 2 (HER-2/neu), and human chorionic gonadotropin (hCG). In addition to diagnostics, some of these biomarkers can also be used as tumor-associated antigens (TAAs) for immunotherapy (Buonaguro et al. 2011), as discussed later in this chapter. Of the biomarkers identified, only PSA is routinely used for cancer screening in the clinic, while others are generally applied toward monitoring treatment response and detecting recurrence (Smith et al. 2016).

The primary considerations for the utility of biomarkers are their sensitivity and specificity, in other words their ability to detect the presence of cancer (sensitivity) without any false positives (specificity) (Handy 2009). Ideally, all biomarkers would have 100% sensitivity and specificity for reliable and early detection. However, due to cancer heterogeneity and low abundance of certain biomarkers, no one test is perfect. For example, in the case of PSA testing, high sensitivities of over 90% can be achieved, but the tradeoff is a low specificity of around 25% (Lee et al. 2006). One approach to improve predictions of cancer presence is to combine several individual biomarkers in a panel (Rusling et al. 2010). In such a way, patients not expressing a particular biomarker would still be diagnosed and false positives and negatives would be minimized based on other biomarkers in the panel, consequently improving specificity and sensitivity. To this end, multiplexed assays for detecting multiple protein biomarkers found in cancer are being developed (Rusling et al. 2010).

10.2.2 Electrochemical Techniques

The primary methods for electrochemical immunosensing are electrochemical impedance spectroscopy (EIS), amperometry, and voltammetry (e.g. linear sweep [LSV], differential pulse [DPV], cyclic [CV], and squarewave [SWV]), and they generally employ an applied potential that generates a redox reaction in labeled electroactive species and results in a current that is proportional to the analyte concentration (Anik et al. 2016). There has been a plethora of investigations in the area of immunosensing, and only a small cross-section are described here, which are summarized in Table 10.2.

CEA is a common biomarker for biosensor validation, and a great variety of nanoparticles have been investigated for advancing electrochemical immunosensors. In one approach, core-shell Fe_3O_4@Ag magnetic nanoparticles were used by taking advantage of the Ag shell for adsorption of CEA antibody and the iron oxide core to magnetically incorporate the nanoparticles into a carbon paste on the electrode (Tang et al. 2006). Using potentiometry, detection of CEA at concentrations as low as 0.5 ng/mL was achieved. A similar strategy utilized core-shell Fe_3O_4/SiO_2 detector nanoparticles in a flow-injection device, and CEA in human serum could be detected based on EIS measurements (Pan et al. 2007). There was a linear resistance response between 1.5 and 60 ng/mL and a detection limit of 0.5 ng/mL. A more recent work illustrates the huge improvements in sensitivity that have been achieved since then; the linear detection region for CEA was expanded to a range from 0.1 ng/mL to 1000 ng/mL and the detection limit was reduced down to 0.007 ng/mL (Li et al. 2017). This was made possible by a combination strategy of applying gap-based interdigitated electrodes (IDEs) and polyaniline-Au nanoparticles (PANi-AuNPs). After antibody capture of CEA onto the IDEs, antibody-labeled PANi-AuNPs bound and connected the microgap of IDEs, leading to greatly reduced resistance values.

TABLE 10.2
Examples of Nanoparticle-Based Biosensors for Cancer Immunodiagnostics

Nanoparticle Material	Cancer Biomarker	Detection Method	Dynamic Range	Limit of Detection	Reference
Fe_3O_4@Ag	CEA	Electrochemical	1.5–200 ng/mL	0.5 ng/mL	(Tang et al. 2006)
Fe_3O_4/SiO_2	CEA	Electrochemical	1.5–60 ng/mL	0.5 ng/mL	(Pan & Yang 2007)
AuNPs	CEA	Electrochemical	0.1–1000 ng/mL	7 pg/mL	(Li et al. 2017)
MWCNTs	PSA	Electrochemical	0.2–1 ng/mL; 1–40 ng/mL	20 pg/mL	(Salimi et al. 2013)
IrO	AMACR	Electrochemical	NR; ~0.1–0.3 mg/mL	NR	(Lin et al. 2012)
Fe_3O_4	HER2	Electrochemical	0.01–10 ng/mL; 10–100 ng/mL	0.995 pg/mL	(Emami et al. 2014)
AuNPs	EGFR	Electrochemical	1 pg/mL–1 μg/mL	0.88 pg/mL	(Elshafey et al. 2013)
AuNPs	EpCAM	Electrochemical	45–100,000 cells/mL	28 cells/mL	(Pallela et al. 2016)
AgNPs	EpCAM	Electrochemical	2–2000 pg/mL	0.8 pg/mL (~0.4 cells/mL)	(Bravo et al. 2017)
AgNPs, AuNPs	CEA	Electrochemical	0.01–80 ng/mL	2.8 pg/mL	(Li et al. 2016)
	AFP		0.01–80 ng/mL	3.5 pg/mL	
ZnO	CEA	Electrochemical	0.005–20 ng/mL	1.8 pg/mL	(Ge et al. 2017)
	CA 125		0.01–50 U/mL	3.6 mU/mL	
	CA 15-3		0.01–50 U/mL	3.8 mU/mL	
CdTe/CdS QD	CEA	Fluorescence	250 fM–25 nM	2.5 pM	(Hu et al. 2010)
	AFP		250 fM–25 nM	2.5 pM	
CdTe QDs/CuO	CEA	Fluorescence	5 pg/mL–200 ng/mL	1.4 pg/mL	(Ge et al. 2013)
	AFP		1 pg/mL–100 ng/mL	0.3 pg/mL	
	CA 125		0.2 mU/mL–100 U/mL	0.061 mU/mL	
	CA 15-3		0.1 mU/mL–200 U/mL	0.29 mU/mL	
UCNPs	CEA	Fluorescence	0.1–100 ng/mL	0.89 ng/mL	(Xu et al. 2016)
	AFP		NR	NR	

(Continued)

TABLE 10.2 (CONTINUED)
Examples of Nanoparticle-Based Biosensors for Cancer Immunodiagnostics

Nanoparticle Material	Cancer Biomarker	Detection Method	Dynamic Range	Limit of Detection	Reference
	CA 125		NR	NR	
Latex NPs	PSA	Fluorescence	NR; ~1–20 ng/mL	0.056 ng/mL	(Kupstat et al. 2011)
CdSe/ZnS QDs	AFP	Fluorescence	NR; ~1 ng/mL–1 μg/mL	0.4 ng/mL	(Chen et al. 2012)
CdSe/ZnS QDs	CEA	Fluorescence	NR; ~1–200 ng/mL	0.4 ng/mL	(Chen et al. 2013)
CdSe/ZnS QDs	PSA	Fluorescence	NR; ~1–500 ng/mL	1.6 ng/mL	(Wegner et al. 2013)
CdSe/ZnS QDs	EGFR	Fluorescence	NR; ~50 ng/mL–1 μg/mL	34 ng/mL	(Wegner et al. 2014)
CdSeTe/ZnS QDs	PSA	Fluorescence	2–80 ng/mL	0.8 ng/mL	(Mattera et al. 2016)
CNDs	CEA	Electrochemiluminescence	NR; ~0.05–100 ng/mL	4.0 pg/mL	(Wang et al. 2012)
	AFP		NR; ~0.1–100 ng/mL	0.02 ng/mL	
	CA 19-9		NR; ~0.05–100 U/mL	6.0 mU/mL	
	CA 15-3		NR; ~0.05–100 U/mL	5.0 mU/mL	
AgNPs, Au nanocages	CEA	Electrochemiluminescence	1 pg/mL–50 ng/mL	0.7 pg/mL	(Gao et al. 2015)
CdS QDs	CEA	Electrochemiluminescence	0.57 ng/mL–0.50 μg/mL	NR; ~0.57 ng/mL	(Wu et al. 2012)
	AFP		0.75–75 ng/mL	NR; ~0.75 ng/mL	
	PSA		1.0 ng/mL–1.25 μg/mL	NR; ~1 ng/mL	
CdS:Eu QDs, Au nanorods	CEA	Electrochemiluminescence	0.01 pg/mL–1 ng/mL	0.01 pg/mL	(Liu et al. 2016)
AgNPs	HE4	SPR	10 pM–10 nM	4 pM	(Yuan et al. 2012)
AgNPs	SCCa	SPR	0.1 pM–1 nM	0.125 pM	(Zhao et al. 2014)
AuNPs	AQP1	SPR	NR; ~10 ng/mL–5 μg/mL	10 ng/mL	(Tian et al. 2012)
AuNPs	CEA	SPR	NR; ~10 fM–1 nM	94 fM (~17 pg/mL)	(Lee et al. 2015)

(Continued)

TABLE 10.2 (CONTINUED)
Examples of Nanoparticle-Based Biosensors for Cancer Immunodiagnostics

Nanoparticle Material	Cancer Biomarker	Detection Method	Dynamic Range	Limit of Detection	Reference
	AFP		NR; ~10 fM–1 nM	91 fM (~6 pg/mL)	
	PSA		NR; ~10 fM–1 nM	10 fM (~0.9 pg/mL)	
AuNPs	PSA	SPR	N/A	1 ag/mL	(de la Rica & Stevens 2012)
AuNPs	PSA	SPR	10 fg/mL–100 pg/mL	3.1 fg/mL	(Liu et al. 2014)
AuNPs	PSA	SPR	N/A	1 ag/mL	(Rodríguez-Lorenzo et al. 2012)
AuNPs	VEGF	Light scattering	0.1 pg/mL–10 ng/mL	7 fg/mL	(Li et al. 2013)
AgNPs, AuNPs	hCG	Light scattering	0.05–20 ng/mL	NR	(Wen et al. 2016)
AuNPs	EphA2+ EVs	Light scattering	0.1 ng/μL–10 μg/μL	0.2 ng/μL	(Liang et al. 2017)
AuNPs	AFP	Mass	15.3–600 ng/mL	15.3 ng/mL	(Ding et al. 2007)
AuNPs	CD10	Mass	0.01–10 nM	2.4 pM	(Yan et al. 2015)
AuNPs	Autontibodies	DLS	NR	NR	(Zheng et al. 2015)

Abbreviation: NR: not reported (estimates given if calibration graph was presented); N/A: not applicable.

PSA is another biomarker commonly researched in cancer detection sensor development. One example of an innovative system for PSA detection was established by immobilizing an anti-PSA antibody on a nanocomposite electrode modified with multiwalled carbon nanotubes (MWCNTs) and an ionic liquid (Salimi et al. 2013). DPV detection of PSA gave linear responses in two ranges, between 0.2 and 1 ng/mL and between 1 and 40 ng/mL, with a detection limit of 20 pg/mL. It was further shown that the immunosensor could be used to detect PSA in prostate tissue samples. An alternative and promising prostate cancer biomarker, alpha-methylacyl-CoA racemase (AMACR) was shown to be highly sensitive and specific in differentiating prostate cancer patients (Lin et al. 2012). Electrochemical measurements using iridium oxide-modified electrodes demonstrated that average AMACR levels in the prostate cancer patients was ten-fold higher than either healthy individuals or high grade prostatic intraepithelial neoplasia patients, and a cutoff could be identified to achieve 100% accuracy in separating prostate cancer patients from controls.

Work exploring other cancer biomarkers include the breast cancer marker HER2, and ultrasensitive detection was achieved in plasma samples using antibody-modified Fe_3O_4 nanoparticles overlaid on the electrode surface (Emami et al. 2014). A linear DPV response to HER2 concentrations was observed over the ranges of 0.01 to 10 ng/mL and 10 to 100 ng/mL, with the limit of detection reaching 0.995 pg/mL. A sensitive immunosensor with a wide dynamic range of 1 pg/mL to 1 μg/mL has also been presented for the detection of epidermal growth factor receptor (EGFR), a biomarker for a variety of cancers including lung and ovarian cancer (Elshafey et al. 2013). The impedimetric immunosensor involved the electrodeposition of AuNPs on a gold electrode that increased electrochemical active area by 68% and showed a limit of detection as low as 0.88 pg/mL.

A valuable application for immunodiagnostics is the detection of metastasis, and sensing of epithelial cell adhesion molecule (EpCAM) with an amperometric biosensor has been proposed as a viable method for detecting circulating epithelial carcinoma cells (Pallela et al. 2016). The sensor probe is fabricated by immobilizing a capture antibody on a composite layer formed from a conducting polymer and AuNPs. Following a sandwich-type immunoassay, the binding of EpCAM-positive cancer cells was confirmed using EIS, CV, and chronoamperometry. The dynamic range for cancer cell detection was determined to be in the range of 45 to 100,000 cells/mL with a detection limit of 28 ± 3 cells/mL. The immunosensor could successfully detect EpCAM-expressing metastatic cancer cells in both serum and mixed cell samples, and its capacity to detect of early stage metastatic cells was confirmed by fluorescence-activated cell sorting. In another study, EpCAM was detected using a microfluidic sensor created with silver nanoparticles (AgNPs) attached in the channel that were then covalently bound to EpCAM antibodies (Bravo et al. 2017). Chronoamperometric measurements were able to achieve a linear response range of 2 to 2000 pg/mL, with a limit of detection of 0.8 pg/mL compared to a detection limit of 13.9 pg/mL for a commercial ELISA kit. This corresponded to a capacity to detect as low as four circulating tumor cells measured in 10 mL of blood.

Simultaneous detection of multiple protein biomarkers has also been explored, with good sensitivity and little cross-reactivity, as demonstrated by a recent work for multiplexed electrochemical detection of CEA and AFP (Li et al. 2016).

For detection of AFP, carbon nanospheres were coated with AgNPs and anti-AFP antibodies, while for detection of CEA, carbon nanospheres were coated with a thionine signal tag before coating with AuNPs and anti-CEA antibodies (Figure 10.2a). Antigen entrapment was achieved on the glassy carbon electrode surface using a composite formed from graphene oxide and AuNPs that was then modified with anti-CEA and anti-AFP antibodies (Figure 10.2b). The addition of the decorated carbon nanosphere immunoprobes to the sensing substrate in the presence of the target proteins resulted in two distinguishable peaks when measured with DPV, one at +0.16 V (corresponding to AgNPs and hence AFP) and another at −0.33 V (corresponding to thionine and hence CEA). The linear range of detection was 0.01 to 80 ng/mL, with detection limits for CEA and AFP being 2.8 and 3.5 pg/mL, respectively.

In the world of sensing, microfluidic paper-based analytical devices (μPADs) have been gaining ground in recent years due to their simplicity, portability, low cost, three-dimensional configurability, and unique porous and absorptive surface properties (Ge et al. 2017). As an example, a photoelectrochemical immunoassay using a portable inorganic-organic, light-emitting diode (LED) was integrated in a μPAD for multiplexed detection of CEA, CA 125, and CA 15-3 (Zhang et al. 2014). A six-electrode array was screen printed on the paper, antibodies immobilized into the working electrodes, and LEDs based on ZnO nanorods grown perpendicular to the electrode surface. After LED photoexcitation of the immunosensor array, specific and sensitive detection of the cancer biomarkers was achieved, with detection limits of 1.8 pg/mL for CEA, 3.6 mU/mL for CA 125, and 3.8 mU/mL for CA 15-3.

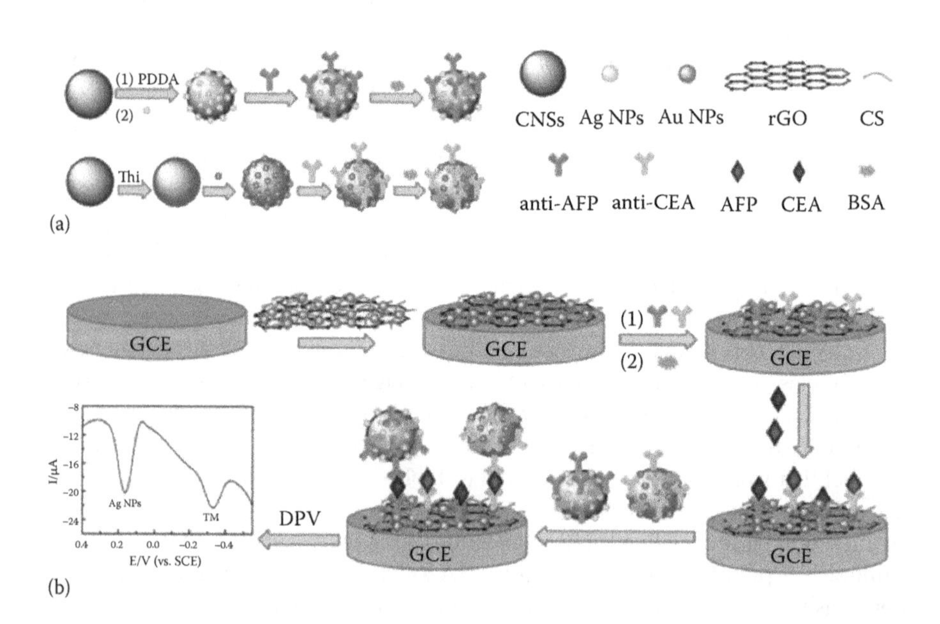

FIGURE 10.2 (a) Preparation process of immunoprobes. (b) Fabrication procedure of the immunosensor. (Reproduced from Li, L. et al., *Anal Biochem*, 505:59–65, 2016. With permission.)

10.2.3 Optical Techniques

10.2.3.1 Fluorescence

Direct measurement using antibodies tagged with fluorescent nanoparticles can be used for sensitive detection of cancer biomarkers, which was demonstrated in a microfluidic protein chip for multiplexed detection of AFP and CEA with femtomolar sensitivity using CdTe/CdS quantum dots (QDs) of different sizes/colors (Hu et al. 2010). QDs are also useful due to their involvement in Förster resonance energy transfer (FRET)-based approaches. For example, a paper-based immunosensor utilized coordination of dithizone energy acceptors at the surface of CdTe QDs to strongly quench its green emission, but release of Cu^{2+} in a sandwich immunoassay involving secondary antibodies conjugated to CuO nanoparticles switched off the FRET pathway in the presence of target biomarkers and resulted in recovered CdTe QD fluorescence (Ge et al. 2013). This strategy gave rise to ultralow detection limits of 3.0×10^{-4} ng/mL, 6.1×10^{-5} U/mL, 2.9×10^{-4} U/mL, and 1.4×10^{-3} ng/mL for AFP, CA 125, CA 15-3, and CEA, respectively. Upconversion nanoparticles (UCNPs) can also be applied for FRET-based sensors and have the added advantage of avoiding autofluorescence and light scattering that may arise from ultraviolet or visible light excitation in paper-based substrates (Xu et al. 2016). Antibody-labeled lanthanide (Ln^{3+})-doped UCNPs were directly printed on filter paper, and introduction of CEA antigen-labeled acceptors resulted in a fluorescent signal at concentrations as low as 0.89 ng/mL. Furthermore, simultaneous detection of multiple biomarkers (CA 125, CEA, and AFP) was also demonstrated.

Time-resolved FRET (TR-FRET) is another strategy that offers increased signal-to-noise ratio due to lack of interference from the excitation light resource and background autofluorescence, and it has been utilized for quantitative detection of PSA (Kupstat et al. 2011), AFP (Chen et al. 2012), and CEA (Chen et al. 2013), among others. Time-resolved fluoroimmunoassays are homogeneous assays that can be performed in solution without need for antigen or antibody immobilization and involve FRET between a donor and acceptor, such as between luminescent terbium chelates (LTCs) and QDs, that are brought together in the presence of the desired protein marker. Variations have been explored to enhance performance and flexibility of this strategy. Time-gated detection, different QD colors, and application of IgG, $F(ab')_2$, and F(ab) antibodies were verified for detection of PSA in 50 μL serum samples with subnanomolar detection limits (Wegner et al. 2013). A similar study demonstrated the feasibility of using nanobodies instead for the detection of EGFR at subnanomolar concentrations in 50 μL buffer or serum samples (Wegner et al. 2014). Another study investigated engineering more compact QDs by surface functionalization with zwitterionic penicillamine and observed an improvement in the detection limit over commercial QDs (Mattera et al. 2016).

10.2.3.2 Electrochemiluminescence

Electrochemiluminescence (ECL) methods are commonly employed in conjunction with paper-based devices and are generally extremely robust with high signal intensity. Simultaneous detection of two analytes with one electrode was accomplished using tris-(bipyridine)-ruthenium(II) ($Ru(bpy)_3^{2+}$) and carbon nanodots (CNDs) as

ECL labels due to their different optimal operational potentials for generating ECL emission (-1.2 V for CNDs and +1.2 V for $Ru(bpy)_3^{2+}$) (Wang et al. 2012). In this study, it was shown that ECL detection could also be triggered using a battery simply by reversing the connection mode for the two measurements, making this system low-cost and disposable. Functional demonstration of AFP, CA 19-9, CA 15-3, and CEA detection using two electrodes was presented, with limits of detection (clinical cutoff values shown in parentheses) of 0.02 ng/mL (25 ng/mL), 6.0 mU/mL (35 U/mL), 5.0 mU/mL (35 U/mL), and 4.0 pg/mL (5 ng/mL), respectively. ECL immunoassays can also be implemented on three-dimensional microfluidic origami devices, as shown in one application for CEA detection that utilized Au nanocages functionalized with $Ru(bpy)_3^{2+}$ for ECL signal amplification on a paper working electrode modified with Ag nanospheres (Ag-PWE) (Gao et al. 2015). The Au nanocages were effective for signal amplification due to presenting a high surface area for $Ru(bpy)_3^{2+}$ adsorption, as both the inner and outer surfaces could be used. Highly sensitive detection of CEA was observed, with a linear range of 0.001-50 ng/mL and a detection limit of 0.7 pg/mL.

Using a FRET-like approach in which energy was transferred from a donor (CdS QDs) to an acceptor ($Ru(bpy)_3^{2+}$), the concept of ECL resonance energy transfer (ECL-RET) was introduced as another method to improve the sensitivity of the assay (Wu et al. 2012). After QD/antigen immobilization on the electrode surface, capture of $Ru(bpy)_3^{2+}$-labeled antibodies resulted in a change in ECL emission from green (QDs) to red ($Ru(bpy)_3^{2+}$), and a competitive immunoassay with antigens of interest reverted the color back to green. Spot arrays with AFP, CEA, and PSA antigens resulted in multiplexed detection and linear ranges of 0.75 ng/mL to 75 ng/mL for AFP, 0.57 ng/mL to 0.50 μg/mL for CEA, and 1.0 ng/mL to 1.25 μg/mL for PSA. A recent study built on the concept of ECL-RET and presented a double-quenching method for sensitive detection of CEA using a combination of a ternary composite of hemin-graphene-Au nanorods with CdS:Eu QDs (Liu et al. 2016). The ECL emission intensity of the QDs in the presence of the ternary composite was quenched first due to their high electrocatalytic activity for the reduction of H_2O_2, a coreactant of QDs-based ECL, and second due to ECL-RET between the Au nanorods and the QDs. The immunoassay had a very low detection limit of 0.01 pg/mL and a linear range from 0.01 pg/mL to 1 ng/mL.

10.2.3.3 Surface Plasmon Resonance

Noble metal nanoparticles such as gold and silver nanoparticles have many unique optical and electromagnetic properties (Eustis et al. 2006). Their surface plasmon oscillation in particular can be taken advantage of for biomarker detection. Localized surface plasmon resonance (LSPR) originates from the collective oscillation of valence electrons at the nanoparticle's surface, which results in unique surface plasmon absorption and strong light-scattering spectra, spectra that are strongly dependent on the size, shape, and composition of the nanoparticle, as well as the refractive index of the surrounding medium (Guo et al. 2015). Different strategies can be applied for detection using plasmonic nanosensors: 1) shift in the plasmon peak due to refractive index change, 2) shift in the plasmon peak due to nanoparticle coupling, and 3) increase in the peak amplitude due to nanoparticle growth (Figure 10.3) (Guo et al. 2015). By taking advantage of the inherent LSPR properties of the nanoparticles, rapid and label-free approaches can be realized.

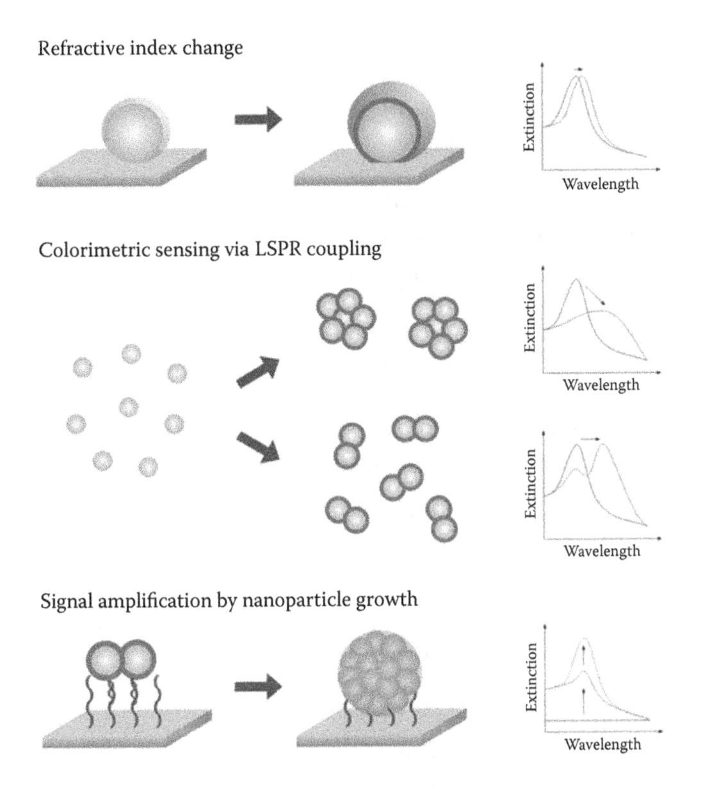

FIGURE 10.3 Illustration of different strategies for plasmonic sensing. Sensing can be accomplished by changing the local dielectric environment around plasmonic nanoparticles. Some strategies include surface modification of the nanoparticles that results in a refractive index change, plasmonic coupling of the nanoparticles, and growth of the nanoparticles. (Reproduced from Guo, L. et al., *Nano Today*, 10:213–39, 2015. With permission.)

Binding of target markers to LSPR-based nanosensors generates a change in the refractive index and a shift in the extinction maximum of the LSPR spectrum that is directly related to the concentration of the analyte. For example, in one study, a rhombic silver nanoparticle array was created on a sensor chip and functionalized with antibodies for the detection of human epididymis secretory protein 4 (HE4), a biomarker for the early diagnosis of ovarian cancer (Yuan et al. 2012). Using this method, HE4 could be rapidly detected in serum from patients with ovarian cancer and with good specificity, effective reproducibility, and long-term stability. Results were comparable to those obtained from ELISAs, and the linear range was between 10 pM and 10 nM, more than sufficient for HE4 assessments where the normal range is considered to be less than 150 pM. A similar study utilized an antibody-modified triangle-shaped silver nanoparticle array to detect squamous cell carcinoma antigen (SCCa) as a tool for cervical cancer detection (Zhao et al. 2014). SCCa was successfully detected in both buffer and human serum at a physiologically relevant range of 0.1 pM to 1 nM, which is compatible with the normal range being below 45 pM. An additional benefit of this biosensor chip was its ability to be regenerated by immersion in 50 mM glycine-HCl (pH 2.0) for three minutes to remove bound SCCa and still generate reproducible results. Another group investigated the application of LSPR for early detection of renal cancer carcinoma (RCC) using a paper substrate adsorbed with gold nanorods (AuNRs) conjugated with aquaporin-1 (AQP1)-specific IgG (Tian et al. 2012). AuNRs were utililized for the high sensitivity of their longitudinal LSPR to refractive index change, and the method allowed for rapid and quantitative detection of AQP1 in urine. The detection limit of this method, 10 ng/mL, matched well with the lower limit of AQP1 in patients with RCC.

Nanoplasmonic biosensors can also be used for multi-analyte detection. This was demonstrated in an investigation where various immuno-gold nanoparticles specific for AFP, CEA, and PSA were site-selectively immobilized on glass slides (Lee et al. 2015). The multiplex biosensor demonstrated excellent sensitivity and selectively when tested using a panel of biomarkers in serum, with very little cross reactivity and a limit of detection of 91 fM, 94 fM, and 10 fM for AFP, CEA, and PSA, respectively.

As an alternative to the above, label free approaches that require appropriate instrumentation to detect the plasmon resonance shift, a colorimetric plasmonic ELISA, have been developed in which the presence of the target results in a shift in solution color from red to blue (de la Rica et al. 2012). The strategy utilizes the plasmon peak shift that occurs due to aggregation of AuNPs. The process is similar to a traditional ELISA, except catalase is the enzyme coupled to the secondary antibody. With increasing analyte concentration and thus increasing catalase concentration, the hydrogen peroxide (H_2O_2) substrate is consumed, which subsequently affects its ability to reduce gold ions into AuNPs and results in growth of aggregated AuNPs of ill-defined morphology. As opposed to a red solution that is generated when non-aggregated AuNPs are formed in the absence of the protein of interest, a blue solution observable by the naked eye results. This strategy was established for the ultrasensitive detection of PSA at concentrations as low as 10^{-18} g/mL. Another ultrasensitive colorimetric immunoassay was demonstrated based on glucose oxidase (GOx)-catalyzed growth of AuNPs for the detection of PSA (Liu et al. 2014). In this instance, biomarker presence resulted in accumulation of GOx, which then catalyzed the generation of H_2O_2 through oxidation of β-D-glucose. The H_2O_2 would then reduce Au ions to AuNPs, resulting in a red solution corresponding to a positive signal. Detection of PSA as low as 4.6 fg/mL was achieved in serum, with a wide linear range from 10 to 10^5 fg/mL.

Although most sensors generate a signal proportional to the analyte concentration, another GOx-based immunoassay approach demonstrated unique inverse sensitivity (Figure 10.4) (Rodríguez-Lorenzo et al. 2012). Using a sandwich immunoassay format, GOx-labeled secondary antibodies were used for detection of PSA bound to Au nanostars conjugated to PSA antibodies. At high GOx concentrations, nucleation of free-standing silver nanocrystals is favored, resulting in only a small shift in the extinction maximum of the Au nanostars. On the other hand, relatively low concentrations of GOx result in the adsorption of Ag ions on the Au nanostars, formation of a homogeneous silver coating, and a significant blue shift in the plasmon resonance

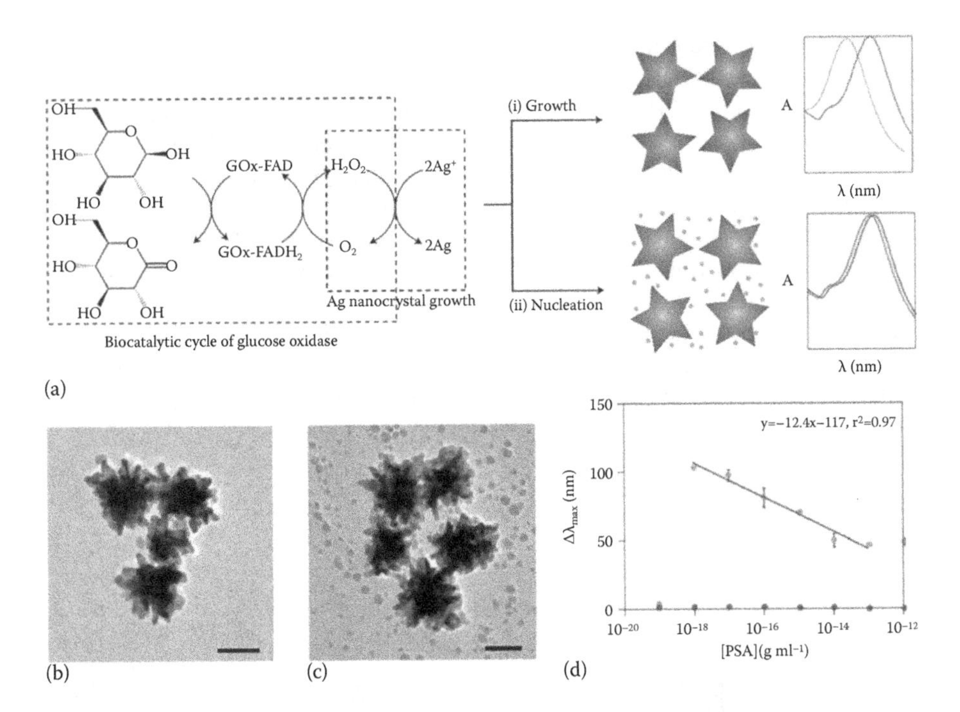

FIGURE 10.4 Plasmonic nanosensor with inverse sensitivity. (a) The detection scheme based on glucose oxidase (GOx)-mediated silver nanocrystal growth at low concentrations and nucleation at high concentrations. (b,c) TEM images of effect of low (10^{-20} g/mL) and high (10^{-14} g/mL) GOx concentrations, respectively. Scale bars = 50 nm. (d) Calibration curve for measuring PSA biomarker concentration. (Reproduced from Rodríguez-Lorenzo, L. et al., *Nat Mater*, 11:604–7, 2012. With permission.)

spectrum. This inverse sensitivity approach thus enables highly sensitive detection of PSA, and concentrations down to 10^{-18} g/mL in whole serum was measured.

10.2.3.4 Light Scattering

In addition to possessing LSPR, gold and silver nanoparticles also exhibit surface-enhanced Raman scattering (SERS), in which these plasmonic nanoparticles can be exploited for an intense enhancement in the Raman scattering signal of reporter molecules (Schlucker 2014). The SERS signature of the nanoparticles can be used in live-cell imaging (Lee et al. 2014, Lee et al. 2012, Narayanan et al. 2015), but we focus here on *in vitro* quantification assays. As an example, a SERS immunosensor has been demonstrated for the detection of vascular endothelial growth factor (VEGF) in clinical blood plasma samples from breast cancer patients at levels similar to those measured using a standard ELISA (Li, Cushing, et al. 2013). Capture antibodies were immobilized on a triangular gold nanoarray, detection antibodies were conjugated to gold nanostar@Raman reporter@silica sandwich nanoparticles, and the resultant proximity of the two sharp gold surfaces led an enhancement in electromagnetic hot spots that strongly amplified the Raman probe's signal. The utility of SERS for immunosensing was also demonstrated for the detection of hCG using a method where the presence of hCG promoted the catalysis of gold shell formation around AgNPs and led to an increase in the SERS intensity of the Raman reporter (Wen et al. 2016). Electrostatic interaction of a positively charged peptide specific for hCG with citrate-capped AgNPs led to aggregation of the nanoparticles, but the presence of hCG released the AgNPs and had a strong catalytic effect on AuNP formation, resulting in a strong SERS signal that increased linearly with hCG concentration between 0.05 and 20 ng/mL.

Enhanced scattering intensity has also been observed when gold nanospheres and nanorods are brought in close proximity to each other, and this principle can be applied for the detection and characterization of extracellular vesicles (EVs) in plasma for tumor diagnostics (Figure 10.5) (Liang et al. 2017). This nanoplasmon-enhanced scattering (nPES) assay used an EV-specific antibody to capture EVs on

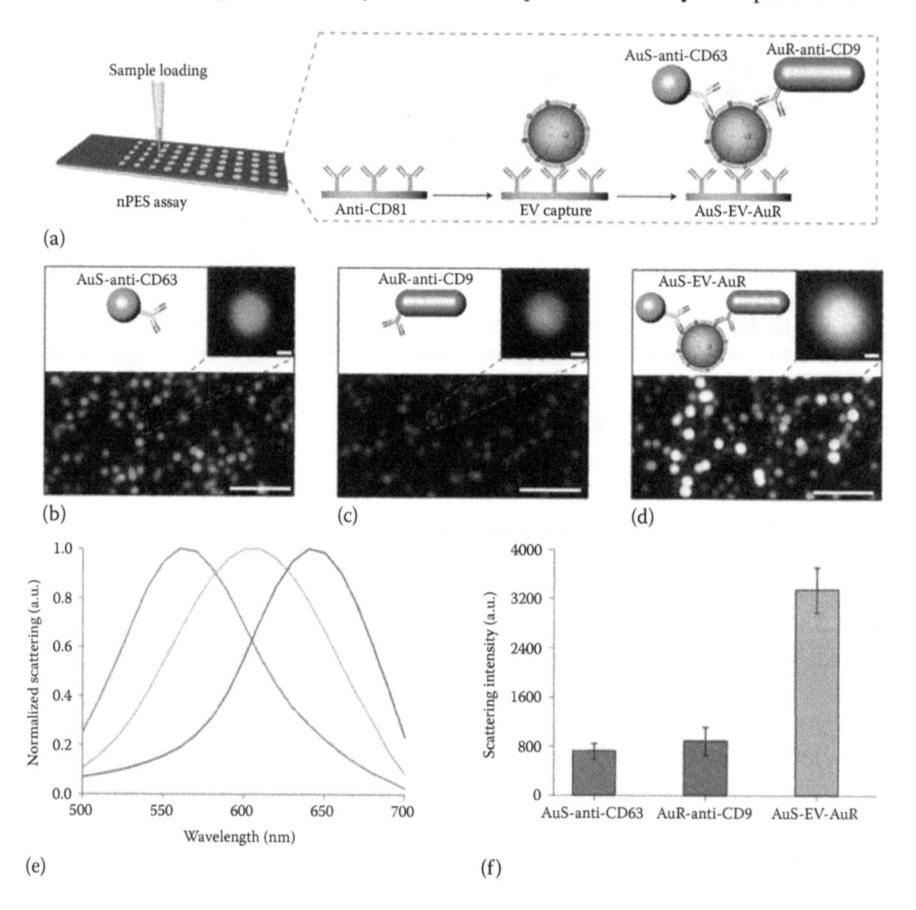

FIGURE 10.5 Nanoplasmon-enhanced scattering assay design for sensing of EVs. (a) Schematic of the assay. (b–d) Dark-field images of gold nanospheres (AuS-anti-CD63; green), gold nanorods (AuR-anti-CD9; red), and nanosphere-EV-nanorod complexes (AuS-EV-AuR; yellow). Scale bars = 2 μm (100 nm for magnified images). (e) Normalized scattering spectra of the spheres, rods, and complex. (f) Scattering intensities of the spheres, rods, and complex. (Reproduced from Liang, K. et al., *Nat Biomed Eng*, 1:0021, 2017. With permission.)

the sensor chip followed by a combination of gold nanorods conjugated to another EV-specific antibody and gold nanospheres conjugated to an antibody to ephrin type-A receptor 2 (EphA2), a pancreatic cancer biomarker, to detect the presence of tumor-derived EVs. Using volumes as low as 1 μL of plasma, the nPES assay was capable of differentiating pancreatic cancer patients from both healthy and pancreatitis patients.

10.2.3.5 Mass-Based Techniques

Mass-based techniques have not been as commonly explored in the context of nanoparticle-enhanced cancer detection, but a few examples can be found in the literature. One method used to improve detection involved coating a piezoelectric immunosensor surface with a hybrid of gold nanoparticles and nanosized hydroxyapatite for a greater surface area for immobilization of AFP antibodies and resulted in a detection range of 15.3–600.0 ng/mL (Ding et al. 2007). An alternative approach involved a sandwich immunoassay on a quartz crystal microbalance, where a capture antibody was used to immobilize CD10, a marker for common acute lymphoblastic leukemia followed by signal amplification with AuNPs decorated with detection antibodies (Yan et al. 2015). The detection range for CD10 was 1.0×10^{-8} M to 1.0×10^{-11} M, and the greater sensitivity could be attributed to the augmented mass of the AuNPs along with the synergistic effect of having multiple antibodies displayed on each particle.

10.2.3.6 Serum Autoantibodies

It has long been known that cancer elicits the production of autoantibodies in patients (Graham et al. 1955), and a recent study exploited this property for cancer detection using only a few drops of blood (Zheng et al. 2015). The relatively simple assay involved incubating citrate-coated AuNPs in blood serum, using a secondary antibody to detect human IgG in the resulting protein corona that formed on the nanoparticles, then quantifying the size of the resultant particles by dynamic light scattering, which correlated with the quantity of IgG present. When blood sera of prostate cancer patients was compared to healthy controls, it was discovered that IgG was more abundant in cancer patients, likely resulting from autoantibodies produced against the cancer, and remarkably this measurement could be used detect early stage prostate cancer with 90–95% specificity and 50% sensitivity. This assay does not specifically identify prostate cancer-specific antibodies, so it likely applies broadly to most cancers and can only be utilized as a quick assay for general cancer screening.

10.2.4 Conclusions and Future Directions in Cancer Immunodiagnostics

By examining the possibilities presented by nanoparticles for cancer immunodiagnostics, it is clear that major headway has been made for rapid, cost effective, and highly sensitive biomarker detection with broad acceptable ranges. Great progress has been made towards early cancer detection, but there is still a need for better specificity in differentiating cancer patients and healthy individuals. Innovations in genomics, proteomics, metabolomics, and imaging in recent years will aid in comprehensive examination of cancer cells for identification of suitable biomarkers (Nair et al. 2014). Such information will eventually allow us to design personalized therapies for patients depending on their specific molecular signatures. Undoubtedly, nanotechnology has the potential to profoundly impact cancer diagnosis and treatment in the near future. While there are still many hurdles remaining to be conquered, the future looks bright.

10.3 NANOTECHNOLOGY AND CANCER IMMUNOTHERAPY

The immune system has evolved to differentiate self from non-self (Jiang et al. 2009, Swann et al. 2007). This capacity enables the immune system to efficiently guard the body against foreign invasions by a coordinated action of innate and adaptive arms that eradicate invading pathogens with high specificity while preventing immune response against self-proteins. However, this very feature hampers the ability of the immune system to deal with transforming or transformed cells in cancer—that are recognized as self, and prevent an immune system-mediated clearance of developing or returning tumors (Zou 2006). Nevertheless, cancer cells accumulate genetic aberrations and express of neo-antigens resulting from breakdown of genetic and translational regulations (Linley et al. 2011, Bright et al. 2014). These neoantigens can differentiate cancer cells from healthy cells and can potentiate an immune response. Presence of weak autonomous cellular or humoral immunity in some cancer patients and its association with reduced severity and aggressiveness of the disease is attributed to the immune system recognizing and responding to cancerous cells (Swann et al. 2007, Vesely et al. 2011). However, under the influence of an intricate milieu of immunomodulators a highly immunosuppressive tumor microenvironment keeps such natural antitumor immunity in check and the tumor prevails (Schreiber et al. 2011). Additionally, under the selection pressure from the immune system, the rapidly evolving cancer cells learn to evade the immune surveillance through intricate immune-editing that includes antigen shedding, negative selection of antigenic cells and downregulation of antigen processing and presenting components (Schreiber et al. 2011). Furthermore, by turning off activated immune cells through negative regulators and promoting regulatory immune cells, cancer cells render antitumor immunity ineffective (Zindl et al. 2010).

Several of these mechanisms by which tumors can evade immune recognition are now being identified. For example, tumor cells can gain survival advantage by downregulating MHC class I expression, which effectively compromises recognition and elimination of cancer cells by tumor antigen specific cytotoxic T-cells (CTLs). Alternatively, through antigen shedding or depletion, cancer cells evolve and escape the surveillance. On the other hand, tumor cells develop resistance to CTL-mediated killing mechanisms by either expressing granzyme-specific serine proteases (Bladergroen et al. 2002), or through expression of decoys for death receptors like Fas and TRAIL, and by overexpression of anti-apoptotic and pro-survival molecules such as BcL-2, STAT3, etc. (Topfer et al. 2011, Ashkenazi 2002).

Another key mechanism is immune suppression by turning off activated immune cells through immune checkpoint regulators such as PD-L1 and CTLA-4, which specifically ligate to inhibitory receptors on T-cells (PD-1) and antigen presenting cells (CD80/86). Inhibitory receptors control immune responses while limiting autoimmunity. However, tumors can take control of these checkpoints for protection from immune attack (Pardoll 2012). Tumor-specific T-cells that exhibit an exhausted, unresponsive phenotype express high levels of inhibitory receptors including CTLA4, PD1, and LAG3 (Woo et al. 2012, Jiang et al. 2015). Intratumoral regulatory T-cells (T_{regs}) promote immunosuppression and express multiple inhibitory receptors (Sakaguchi et al. 2008). Tumor cells can also secret immune suppressor cytokines such as TGF-β that inhibits the activation and differentiation of T-cells and APCs (Filippi et al. 2008). Other mechanisms involve expression of inducible enzymes with immunomodulatory functions such as cyclooxigenase-2 and Indoleamine 2,3-dioxygenase (IDO), whose metabolites suppress T-cell activation and in conjugation with cytokines/chemokines promote immune suppressor cells including Foxp3$^+$ T_{regs}, myeloid-derived suppressor cells (MDSCs) (Gabrilovich et al. 2009) or M2-type macrophages (via interleukin [IL]-4 and IL-13) (Basu et al. 2006). Therefore, it is apparent that the cancer cells and solid tumors can modulate the immune system through a broad range of cellular and molecular effectors to overcome immune-surveillance and persist while still accumulating mutations that would otherwise flag them for elimination.

This growing insight into the interplay between the immune system and tumors has led to the emergence of cancer immunotherapies that are aimed at empowering the immune system to overcome suppressive mechanisms and initiate, rejuvenate, or amplify a self-sustaining cycle of anticancer immune response. The common underlying theme of such therapies is to modulate tumor-immune system interactions to tilt the balance in favor of antitumor immunity by counterbalancing cellular and molecular effectors to facilitate cancer cell recognition and elimination (Blattman et al. 2004). The potential of cancer immunotherapies has been realized by the success of checkpoint blockade therapies and adaptive T-cell therapies. Checkpoint blockade therapies targeting immunosuppressive regulators PD-1 and CTLA-4 are based on antibodies that block these receptors and inhibit subsequent regulatory pathways thereby restoring T-cell function. These therapies have made significant impact on the treatment outcomes in melanoma, renal cell carcinoma and several other malignancies (Pardoll 2012). Similarly, adaptive T-cell therapies are based on chimeric antigen receptors (CARs) and employ patients' own T-cells genetically engineered to recognize tumor antigens and re-introduced in body to fight cancer (Grupp et al. 2013). A wide range of other immunotherapeutic approaches is under pre-clinical development and clinical testing (Table 10.3). These include immunomodulation with cytokines and chemokines, antibodies and small molecular drugs, which can alter the tumor microenvironment and facilitate activation and expansion of effector immune cells (Nicholas et al. 2011). On the other hand, tumor antigen specific therapeutic cancer vaccines are aimed at boosting antitumor immunity via stimulation of an endogenous immune response—cellular and humoral (Winter et al. 2014, Azvolinsky 2013).

TABLE 10.3
Current Approaches to Cancer Immunotherapies

Immunotherapy Strategies/Class	Basic Mechanisms and Advantages	Major Challenges	FDA-Approved Examples/ Cancer Types	References
Immune checkpoint blockade	-Unleashes pre-existing anticancer cellular immune response	-Severe immune-related adverse events common	-Ipilimumab (for advanced melanoma)	(Fellner 2012)
Anti-CTLA-4 monoclonal antibodies and	-Renders other immunotherapies more effective	-Clinical benefits limited to some patients only	-Pembrolizumab (for metastatic melanoma)	(Poole 2014)
Anti-PD1 and anti-PD-L1 antibodies	-Often long lasting immune responses		-Nivolumab (advanced melanoma, advanced lung cancer, bladder cancer)	(Raedler 2015)
Adoptive T-cell therapy (ACT)	-Produces high avidity effector T-cells	-Currently restricted to melanoma		(Rosenberg et al. 2015)
CAR T-cell therapy	-Lymphodepleting conditioning regimen prior to TIL infusion enhances efficacy -Genetic T-cell engineering can broaden TIL scope and affinity	-Lack of long lasting responses -Expensive to produce and time consuming		(Fesnak et al. 2016)
Therapeutic Antibodies	-Specificity to targeted antigens	-Passive therapies require frequent administrations -Development of resistance	-Trastuzumab (targets HER2) -Rituximab (targets CD20 on B cells: B-cell chronic lymphocytic leukemia)	(Hudis 2007) (Smith 2003)
Antibody–drug conjugates (ADCs)	-Can cause apoptosis, antibody-dependent cell cytotoxicity (ADCC) or complement dependent cytotoxicity (CDC)	-Cardiac toxicity	-Ado-trastuzumab emtansine (HER2 breast cancer) -Brentuximab vedotin (Hodgkin lymphoma, non-Hodgkin T-cell lymphoma)	(Lambert et al. 2014) (Berger et al. 2017)

(Continued)

TABLE 10.3 (CONTINUED)
Current Approaches to Cancer Immunotherapies

Immunotherapy Strategies/Class	Basic Mechanisms and Advantages	Major Challenges	FDA-Approved Examples/ Cancer Types	References
Cancer vaccines	-Stimulates the host's immune system—active immunity	-Lack of universal antigens -Often inefficient/weak responses	-Gardasil®, Gardasil 9® (HPV-caused cervical, vulvar, vaginal, and anal cancers)	(Ribeiro-Muller et al. 2014)
Prophylactic vaccines	-Administered in the outpatient clinic -Low toxicities	-Ineffective against immunoediting/antigen shedding	-Cervarix® (cervical cancer caused by HPV)	(Wei et al. 2015) (Rehman et al. 2016)
Therapeutic vaccines	-Patient derived DCs cultured with antigen/GM-CSF -Oncolytic virus therapy		-Sipuleucel-T (metastatic prostate cancer, PAP antigen) -T-VEC (talimogene laherparepvec: for metastatic melanoma)	
Immunomodulators Cytokines and Interferons IL-2, IFN-α	-Stimulates immune system and other cytokines/ chemokines	-Pleotropic effects -Systemic toxicity -Passive, require frequent administration	-Denileukin diftitox (IL-2/diphtheria toxin: for cutaneous T-cell lymphoma)	(Litzinger et al. 2007)
Growth factors GM-CSF Combination therapies	Synergistic mechanism: conventional radiation/ chemotherapy with immunotherapy or combinations of different immunotherapies			(Kang et al. 2016, Khalil et al. 2016, Melero et al. 2015)

10.3.1 Nanoparticles-Mediated Delivery of Immunomodulators to the Tumors

Cytokines and chemokine have broad spectrum of immunomodulatory effects and multiple mechanisms of action. For example, cytokine IL-2 promotes cytotoxic T-lymphocyte proliferation and has been shown to enhance the efficiency of other immunotherapies including vaccines and adoptive immune therapies (Nicholas et al. 2011). Other cytokines such as IL-21 and IL-18 activate a range of immune effector cells including $CD4^+/CD8^+$ cells, natural killer cells and B cells while suppressing regulatory T-cells; these cytokines also enhance production of other stimulatory molecules including IFN-γ, IL-2, TNF-α, GM-CSF, and IL-6 by activated T-cells (Davis et al. 2015, Tian et al. 2014). Type-1 interferons IFN-α and β promote natural killer (NK) cells mediated antitumor activity and suppress allospecific suppressor T-cells and have showed promising therapeutic effects in clinical trials for several malignancies (Zitvogel et al. 2015). Type II interferons (IFN-γ) on the other hand, induce apoptosis and upregulate HLA-I and II, thus restore antigen processing and presentation in cancer cells (Tagawa et al. 2011). Other non-specific immunomodulators include Toll-like Receptor (TLR) agonists that activate TLRs and results in the activation of innate and adaptive responses via dendritic cell (DC) maturation, $CD4^+/CD8^+$ T-cell proliferation and modulation of suppressive immune cell populations (Paulos et al. 2007, Kaczanowska et al. 2013). For example, synthetic oligonucleotides with CpG motifs promotes Th1 polarization and have been used as adjuvants and resulting in enhanced levels of protective cytokines including IL-2, IL-4 and IL-10 (Shirota et al. 2013, Klinman 2004).

Such therapies often involve systemic administration of these potent immunomodulators with broad spectrum of non-specific activity, which poses considerable safety risks and could lead to severe off-target pleotropic effects. These include non-specific lymphocyte activation in circulation resulting in elevated risk of autoimmune and allergic responses and adverse effects on normal tissue metabolism (Vanneman et al. 2012, Kwong et al. 2011, Kwong et al. 2013). For example, IL-1, IL-2, IL-6, TNF-α, and TGF-β could lead to modulation of hepatic metabolisms (Weber et al. 2015). IL-2 can also cause hypothyroidism, thrombocytopenia, anemia, coagulopathy, or impairment of neutrophil chemotaxis, autoimmunity, neurotoxicity, and myocarditis (Atkins et al. 1988, Klempner et al. 1990), whereas systemic administration of IL-2 with adaptively transferred T-cells could also cause multi-organ failures in severe cases. Additionally, the ability of IL-2 to stimulate regulatory T-cells diminishes the beneficial effects of tumor-specific $CD8^+$ cell response (Boyman et al. 2006). At the same time, dose-dependent toxicity including thrombocytopenia, fatigue, and pyrexia has been associated with checkpoint blockade inhibitors (Weber et al. 2015). Likewise, IL-2 administration at high doses cause vascular leak syndrome (VLS; also known as capillary leak syndrome), which is associated with increased vascular permeability, hypotension, pulmonary edema, liver cell damage, and renal failure (Donohue et al. 1983). Similarly, systemic administration of CD40 agonist for DC activation can lead to widespread symptoms of cytokine release syndrome, ocular inflammation, elevated levels of hepatic enzymes, and hematologic toxicities including T-cell depletion. Besides, anti-CD40 therapy has also been linked to long-term immunosuppression mediated by activation-induced apoptosis of $CD4^+$ and $CD8^+$ T-cells (Bartholdy et al. 2007, Berner et al. 2007). Similarly, overexposure to CpG could result in suppression of adaptive T-cell immunity (Heikenwalder et al. 2004, Wingender et al. 2006).

Besides the pleotropic effects and toxicities associated with systemic administration, the underlying challenges of short half-lives, instability, and poor bioavailability of these immunomodulators has also adversely affected their successful clinical translation. It is apparent that these challenges are similar to contemporary drug development processes where drug candidates often fare poorly in terms of molecular stability, circulation longevity and bioavailability in absence of carriers, a role increasingly played by nanoparticles. Nanomedicine is therefore an established field and nanotechnology-based approaches have significantly improved drug delivery and biomolecular imaging and diagnostics (Peer et al. 2007, Mitragotri et al. 2015). Besides improving solubility, stability, and bioavailability of drug and contrast agents, nanocarriers also mitigate toxicities by limiting the off-target accumulations (Lammers et al. 2012). Nearly 50 nanoformulations have been approved by the FDA and ~80 ongoing clinical trials are evaluating nanoparticles-based platforms for drug delivery applications (Bobo et al. 2016). These include polymeric nanoparticles, liposomes, micelles, metal nanoparticles, and protein-based nanoplatforms for diseases ranging from various types of cancer, muscular degeneration, autoimmune disorders, to viral infections and hemophilia. Such nanoformulations of therapeutic molecules outperform conventional formulations in efficiency and specificity of drug delivery and are poised for significant influence on the developing immunotherapies.

10.3.1.1 Delivering Immunomodulators to Tumor Microenvironment

Currently, a wide range of nanoparticle platforms is under preclinical development or clinical evaluation as carriers of immunomodulators, as immunostimulatory vaccine adjuvants, and as depots for sustained release of immune-drugs. Nanoparticles facilitate improved bioavailability of immunomodulatory molecules through prolonged circulation in blood and by providing *in vivo* stability of payload against serum inactivation and enzyme degradation (Christian et al. 2012, Petros et al. 2010). For example, intravenous administrations of liposomal formulations of IFN-γ, IFN-α, IL-2 or TNF-α have been shown to enhance the plasma residence time (van der Veen et al. 1998, Kedar et al. 2000, Christian et al. 2012). On the other hand, intraperitoneal, intramuscular, subcutaneous, or intranasal administration of such liposomal and polymeric formulations have been used to create local cytokine depots and increase residence times of the immunostimulatory payloads at the site (Kedar et al. 1997, Eppstein et al. 1982, Anderson et al. 1992). Furthermore, by engineering nanoformulations that require external or physiological stimuli, another level of control has been incorporated that ensures the release of immunostimulatory cargo only at targeted sites for improved bioavailability and safety (Yuyama et al. 2000, ten Hagen et al. 2002, Heffernan et al. 2009, Haining et al. 2004).

In order to alter the tumor microenvironment and render it favorable for an antitumor response, delivery of immunotherapy to the tumor tissue is a key requirement. Nanoformulations of immunomodulatory or immunostimulatory molecules have been used for preferential accumulation and retention in tumors via the enhanced permeability and retention (EPR) effect resulting from the abnormal, leaky tumor vasculature and impaired lymphatic drainage. Together with prolonged circulation times, such enhancement of tumor accumulation also minimizes off-target systemic toxicities (Davis et al. 2008, Maeda et al. 2013). For example, lipid-coated calcium phosphate nanoparticles (LCP-NPs) have been used for modulation of the tumor microenvironment by delivering TGF-β siRNA to downregulate the levels of this immunosuppressive cytokine within the tumor or to deliver a broad-spectrum anti-inflammatory drug that significantly reduced T_{reg} and MDSC populations within the tumor microenvironment (Xu et al. 2014, Zhao et al. 2015). Similarly, PD-L1 siRNA delivery by polyethylenimine (PEI) liposomes has been employed to knock-down PD-L1 levels in human and mouse ovarian cancer models that resulted in an immunostimulatory phenotype with a subsequent increase in tumor-reactive $CD8^+$ T-cells population and improved mice survival (Cubillos-Ruiz et al. 2009). EPR-mediated accumulation of liposome-encapsulated polymer nanogels has also been utilized for intratumoral delivery of a small molecule drug SB505124 that inhibits cytokine (IL-2) and TGF-β receptor leading to expansion of T-cells and NK cells by blocking key immunosuppressive pathways (Park et al. 2012). Similarly, liposomal delivery of IL-2 has also showed enhanced therapeutic effects with reduced toxicities in a variety of other tumors including liver and lung cancers (Kanaoka et al. 2002, Neville et al. 2001).

While the EPR effect mediated tumor accumulation following systemic delivery is a critical advantage offered by nanocarriers, direct intratumoral administration of such formulations limits the drainage and escape of immunotherapeutic molecules into systemic circulations and therefore minimizes off-target toxicity. For example, immunostimulatory liposomes conjugated with IL-2 and anti-CD137 antibodies targeting activated T-cells lead to improved IL-2 dosing within the tumor via intratumoral administration vs. systemic injections and resulted in a higher ratio of tumor-infiltrating $CD8^+$ T-cells over regulatory T-cells in established melanomas (Kwong et al. 2013). In a similar approach, PEGylated liposome formulation has been used to deliver anti-CD40 antibodies and TLR agonist CpG molecules via intratumoral administration resulting in significant tumor inhibition while sequestering the immunostimulatory payload in cancerous tissue, thus minimizing off-target inflammatory effects (Kwong et al. 2011). Intratumoral administration of CpG payloads on gold nanoparticles has been similarly used to build up high concentration of CpG in the tumor tissue to induce significant macrophage and DC infiltration without requiring administration of a high systemic dose (Lin et al. 2013). Once in the tumor tissues, either through EPR or intratumoral injections, nanoparticles-based immune-drugs can be selectively engulfed by tumor infiltrating immune cells. Such strategies have been used to reprogram the tumor microenvironment. For example, nanocomplexes encapsulating CpG oligonucleotide were efficiently captured by tumor-associated macrophages (TAMs) and resulted in altered macrophage phenotypes, leading to a significant antitumor effect in a hepatoma murine model (Huang et al. 2012). It has also been demonstrated that by using surface shielding strategies such as acid sensitive PEGylation, polymeric micelles carrying immunomodulator payloads can be engineered to evade phagocytic uptake by phagocytic macrophages of the mononuclear phagocyte system (MPS). The long-circulating micelles shed PEG layer in the acidic environment of the tumor and are selectively taken up by the tumor-associated macrophages to achieve similar reprograming of the tumor microenvironment (Zhu et al. 2013). Another approach to deliver therapeutic payloads to the tumor microenvironment is by hitchhiking circulating immune cells that infiltrate tumor tissues. For example, RGD-targeted single-walled carbon nanotubes (SWCNTs) showed enhanced tumor accumulation via hitchhiking Ly6C^{hi} monocytes in the circulation that are recruited to the site of the tumor in response to inflammation (Smith et al. 2014). Similarly, macrophages and monocytes that are home to hypoxic regions of tumors have been loaded with liposome-encapsulated chemotherapeutics to increase tumor delivery of the drug (Choi et al. 2012). In another approach, intracranial injection of cyclyodextrin-based nanoparticles resulted in enhanced accumulation at tumor sites via tumor-associated macrophages (Alizadeh et al. 2010). In a recent study, immunotherapy aimed at stimulating an MHC-1 non-restricted immune response mediated by NK cells has been delivered via YSK05 lipid-containing liposomes. Specifically, Cyclic di-GMP (c-di-GMP), a ligand of the stimulator of interferon genes (STING) signal pathway was delivered using a liposomal carrier to cytosol and resulted in a NK cells mediated antitumor effects (Nakamura et al. 2015).

Another area of great interest in cancer immunotherapy is the requirement for a sustained immunostimulus and recent nanotechnology-based developments are promising. For example, cytokine depots based on polymeric nano- and microparticles have been employed to treat primary tumors with peri- or intratumoral injections. Here, primary tumors serve as the source of antigen while cytokine depot bearing IL-2, IL-12, TNF-α, GM-CSF, IL-18 or combinations thereof activate leukocytes in the tumor microenvironment and promote immunotherapy against primary tumor and metastasized tumor cells (Hanes et al. 2001, Christian et al. 2012). Similarly, treatment of established tumors with IL-12 loaded particles prior to surgical resections promoted systemic antitumor immunity that prevented recurrence and metastasis (Sabel et al. 2001). The concept of immunotherapeutic depot has also been similarly used in therapeutic vaccine DepoVax™ (DPX-0907), which employs a liposome-based platform harboring custom formulated mixtures of CD8$^+$ T-cell peptide epitopes, a tetanus toxoid derived epitope, and an adjuvant of choice (such as a toll-like receptor agonist) to provide signals for improved antigen presentation. The liposomes carry incorporated hydrophilic antigens and adjuvant directly into an oil medium such as Montanide ISA51 VG, entrapping all vaccine ingredients in a form suitable for efficient uptake, processing, and presentation by antigen-presenting cells (APCs). Such DPX-formulated vaccines have been shown to induce effective immune responses after a single-dose administration (Karkada et al. 2014).

10.3.1.2 Delivering Immunomodulators to Immune Cells

While intratumoral administration is an efficient way of concentrating nanocarriers with immune-drug payloads in the tumor tissues, many tumors types are anatomically inaccessible and rely on the passive accumulation of nanoparticles through the EPR effect. However, the overall efficacy of tumor homing based on EPR remains low and only a small fraction of the injected dose accumulates in the tumor—a large fraction is still sequestered in RES organs including liver and spleen (Wilhelm et al. 2016). Delivering immuostimulators or immunomodulators to immune cells could therefore be a more efficient approach. With potential to propagate, even a small population of stimulated immune cells could have a more pronounced immunotherapeutic effect. A growing number of strategies are exploiting the affinity of nanoparticles to circulating or tissue resident immune cells to deliver immunostimulatory payloads and modulate the functioning of DCs, T-cells, and B cells for therapeutic interventions based on a wide range of mechanisms. For example, pluronic-stabilized polypropylene sulfide (PPS) nanoparticles carrying CpG oligonucleotides have been used to activate DCs residing in tumor draining lymph nodes (TDLNs) following subcutaneous injections in melanoma bearing mice. The nanoparticles are taken up by DLN resident APCs, triggering the release of cytokines IL-12 and IL-6 and resulting in increased activation of effector CD4 T-helper cells. Such an approach of administrating activating nanoparticles in tumor proximal sites has been shown to slow the growth of subcutaneous tumors (Reddy et al. 2007). Using a very different approach, magnetic nanoparticles have been used to enhance T-cell activation and expansion for adaptive T-cell therapy. Specifically, dextran-coated iron oxide nanoparticles with surface-coupled MHC-Ig dimers and anti-CD28 antibodies were designed to allow magnetic field-based aggregation of nanoparticles bound to T-cell receptors (TCRs). *Ex vivo* stimulation of T-cells with these particles in the presence of a magnetic field-enhanced TCR clustering reduced the threshold of T-cell activation (Perica et al. 2014). Such innovative approaches can significantly improve the efficacy of adaptive T-cell therapy. In other examples, *in vivo* loading of T-cells with lipid nanoparticle "backpacks" carrying stimulatory cytokines has been developed for *in vivo* priming of adaptively transferred T-cells. The increased T-cell expansion resulting from sustained *in vivo* stimulus resulted in significant enhancements in ACT efficacy while prevented toxicity arising from systemic cytokine activation (Stephan et al. 2010). Similarly, circulating adaptive T-cells targeted *in vivo* by IL-2 loaded liposomes via anti-Thy1 antibodies, have been shown to enhance T-cell proliferation more effectively compared to administration of soluble cytokines (Zheng et al. 2013). These approaches overcome decline in function of transplanted T-cells, particularly in the setting of solid cancers with a highly immunosuppressive microenvironment. Another related development has been the engineering of artificial APCs (aAPCs) based on nano- and micron-sized particles to which proteins required for T-cell activation, such as MHC-epitope complexes, agonist anti-CD3 and agonist anti-CD28, have been conjugated (Eggermont et al. 2014). aAPCs are synthetic mimics of natural antigen-presenting cells that promote T-cell activation and subsequent expansion, both *ex vivo* and *in vivo*. Both spatial and temporal organization of these signals during aAPC/T-cell contact is important for efficient T-cell activation. Nanoparticles based aAPCs have demonstrated a favorable distribution to T-cell-rich regions such as spleens and lymph nodes upon systemic administration (van der Weijden et al. 2014).

10.3.2 Nanoparticles-Based Combination Therapies

Cancer immunotherapies have also been combined with nanoparticles-based conventional chemotherapy or photothermal therapies to generate systemic antitumor immune responses. For example, photothermal ablation of primary tumors with single-walled carbon nanotubes (SWNTs) has been shown to trigger significant adaptive immune response against the cancer cells. The debris of dying tumor cells generates a pool of tumor antigens. When combined with a CTLA4 blockade strategy, such photothermal ablation can induce significant DC maturation in tumor-draining lymph nodes and promote the generation of multiple cellular immunity-related cytokines, and subsequently is able to prevent the development of tumor metastasis (Wang et al. 2014). In another example, PLGA nanoparticles co-encapsulating Indocyanine green (ICG), a photothermal agent, and imiquimod, a TLR-7 agonist was employed to trigger near IR mediated tumor ablation of primary tumors, generating a pool of tumor-associated antigens. In the presence of a TLR-7 agonist, these neo-antigens showed vaccine-like function. In the presence of co-administered CTLA4 blocking antibodies, a subsequent immune response ensued that protected against tumor re-challenge (Chen, Xu, et al. 2016) (Figure 10.6). In yet another example of photoimmunotherapy, ferritin nanocages have also been used to remove carcinoma-associated fibroblasts and suppress extracellular matrix deposition—which in turn leads to increased T-cell infiltration and tumor suppression (Zhen et al. 2017). Based on similar principles, nanoparticle-delivered chemotherapy can lead to an antitumor immune response via dying cancer cells in the presence of immunostimulatory nanoparticles or co-delivered immunomodulators. For example, PLGA nanoparticles containing paclitaxel as the chemotherapy and SP-LPS as the immunostimulatory drug showed both direct cytotoxicity and immunostimulatory activity. Such combined therapies are more effective than either of monotherapies (Roy et al. 2010).

10.3.3 Nanotechnology and Cancer Vaccines

The field of cancer vaccines is another significant avenue that can benefit from advances in nanomaterials and nanotechnology. A wide range of cancer vaccines is under pre-clinical and clinical evaluation for a variety of human malignancies (Irvine et al. 2015, Bolhassani et al. 2011). The basic concept behind cancer vaccine research is to define strategies by which the immune system recognizes tumor-associated antigens that are otherwise not being recognized in the tumor-bearing host. The overarching goal is to deliver tumor-associated antigens to professional APCs to elicit adaptive immune responses mediated by tumor-specific cytotoxic T-cells and antibodies. Such an active immunotherapy approach based on vaccine-triggered endogenous immune response can overcome the shortcomings of passive immunotherapies requiring multiple administrations and offer effective long-term protection against recurring and residual tumors. Cancer vaccines can also mitigate the high cost and compliance issues associated with multiple administrations of passive immunotherapies. However, development of an effective therapeutic vaccine against established disease is challenging, and despite decades of pursuit, establishment of successful vaccination strategies based on proteins, peptides, autologous DCs or tumor cells have largely been unsuccessful (Azvolinsky 2013, Sakaguchi et al. 2008). While vaccines based on autologous cells are costly and technically challenging, peptide-based cancer vaccination suffers from inefficient uptake, processing, and presentation of the delivered epitopes by activated professional APCs (Eggermont et al. 2014, Azvolinsky 2013, Cohen et al. 2009,

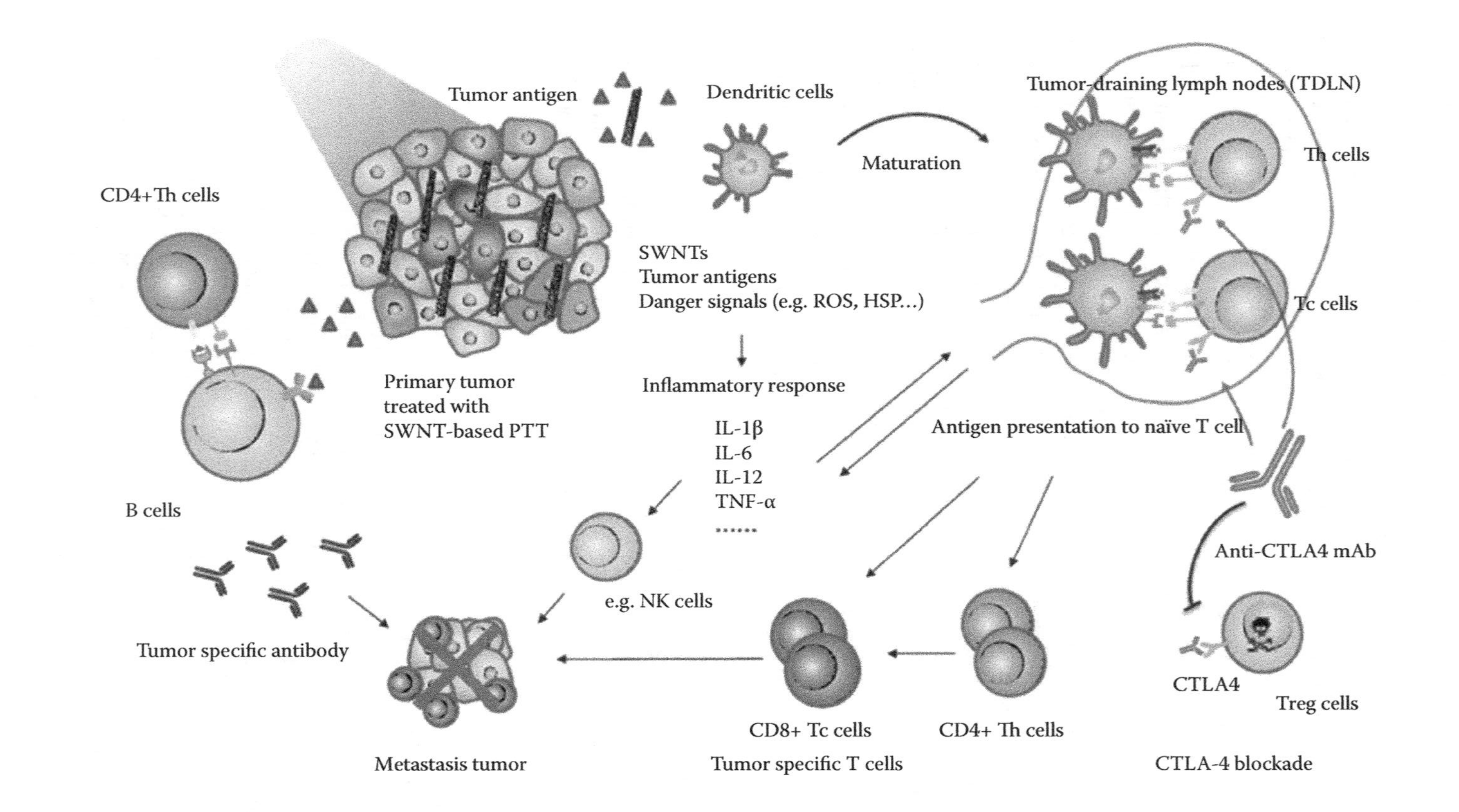

FIGURE 10.6 Combining photothermal therapy with immunotherapy via nanoparticles. The mechanism of antitumor immune responses induced by PLGA-ICG-R837-based photothermal therapy (PTT) in combination with the anti-CTLA-4 mediated checkpoint-blockade. (Reproduced from Chen, Q., Xu, L. et al., *Nat Commun*, 7:13193, 2016. With permission.)

Melero et al. 2014, Rosenberg et al. 2004). Moreover, such vaccination strategies have led to a generation of low-avidity, tumor-specific T-cell responses. Whole protein vaccination with powerful and often poorly tolerated adjuvants; immunostimulatory cytokines such as IL-2 or GM-CSF and/or TLR agonists have failed to induce clinically significant antitumor responses (Altin et al. 2006, Dubensky et al. 2010, Lima et al. 2004). Nanoparticles, with their high payload carrying capacity and affinity towards immune cells are ideally suited for vaccine applications and offer multiple advantages that can help meet the challenges of generating a potent immune response towards tumor antigens. Presenting antigenic epitopes on nanoparticulate carriers not only provides the requisite stability, it also facilitates efficient interactions with key immune cell populations (Wu et al. 2009, Krishnamachari et al. 2011). Nano-vaccine formulations can enhance immunostimulation by promoting multivalent receptor cross-linking and by altering intracellular processing and presentation of antigens. Moreover, by co-delivering antigens and co-stimulatory molecules within the same cellular populations, nanocarriers also provide a more potent immune-stimulus that can generate high-avidity, tumor-reactive T-cells (Fan et al. 2015, Shao et al. 2015). Nanoparticles also reduce the rate of vaccine clearance from the lymphoid tissue and facilitate antigen uptake by antigen-presenting cells. Also, the particulate nature of nanoparticles mimic pathogen-associated molecular patterns (PAMPs) that are perceived as danger signals and drive protective immunity (Irvine et al. 2013). Such patterns are recognized by the pattern recognizing receptors (PPRs) such as TLR on immune cells and facilitate enhanced uptake of nanoparticle-based vaccines by these cells (Gallucci et al. 2001, Matzinger 1994). Activation of PPRs provides immunogenic cues to the immune system instructing it to launch specific response to the antigens carried by the nanoformulation. Nanoparticle-based multivalent antigen displays are also highly efficient in engaging B cell receptors (BCRs) to promote greater signaling, antigen internalization and processing of antigens for presentation to $CD4^+$ T-cells towards an effective antibody response (Bachmann et al. 1993). By engaging TLRs, multivalent nanoparticles-based vaccines can generate a long-lived, high-avidity antibody response mediated co-stimulatory cytokines such as IFN-α and IL-12 (Pasare et al. 2005). Based on such principles, we have recently developed viral nanoparticles-based breast cancer vaccines that could break the tolerance for self-antigen HER2 and effectively generated HER2-specific antibodies (Shukla et al. 2014, Shukla et al. 2017). Similarly, plant viral nanoparticles coupled to a weak idiotypic (Id) tumor antigen have been used as a conjugate vaccine to induce antibody formation against a murine B-cell malignancy (Jobsri et al. 2015). Other virus-like particles have similarly been evaluated as efficient vaccine carriers (Scheerlinck et al. 2008, Riedel et al. 2011, Noad et al. 2003, Plummer et al. 2011, Lee et al. 2016). Professional APCs, particularly DCs, are critical initiators of adaptive immune response, comprising both humoral and cellular responses, and are therefore an important target for cancer vaccines. Nanoparticle-based vaccines are readily taken up by DCs and are associated with enhanced antitumor response as compared to soluble antigens (Fang et al. 2015). Cross-presentation of tumor-associated antigens in a critical step for $CD8^+$ T-cell priming and ensuing CTL response against cancer cells (Joffre et al. 2012). Such cross-presentation of cancer antigens have been demonstrated for a wide range of distinct nanoparticles and new strategies are being developed to improve the efficiency of such cross-presentation (Molino et al. 2013,

Saluja et al. 2014, Sharp et al. 2009). These include strategies employed for cytosolic drug delivery of membrane-impermeable molecules such as via endolysosomal disruption through the proton sponge effect using biodegradable nanogels, (Li, Li, et al. 2013) endolysosomolytic and pH-responsive micelles (Hu et al. 2007), as well as endoplasmic reticulum (ER) targeting approaches where nanoparticles shuttle to cytosol following endosome-ER fusion (Mukai et al. 2011). Nanoparticles-based cancer vaccines can be further improved by targeting DCs and overcoming non-specific uptake of particles by other phagocytic cells. For example, cancer vaccines based on biodegradable poly(lactic-co-glycolic acid) (PLGA) nanoparticles coated with an agonistic αCD40-mAb (NP-CD40), showed highly efficient and selective delivery to DCs *in vivo* and improved priming of $CD8^+$ T-cells against two independent tumor-associated antigens (Rosalia et al. 2015) Additionally, the nature of immune response can be further tuned by targeting a specific subset of DCs (Hardy et al. 2013). For example, TLR7 and TLR9 agonists can convert the tolerogenic plasmacytoid DCs to innate immunostimulatory types, whereas targeting various C-type lectins can modulate variable adaptive response. For instance, DC-SIGN, DEC-205, DNGR-1 and Langerin favors $CD8^+$ T-cell cellular (Th1) responses while $CD4^+$ and B cell humoral (Th2) responses are achieved by targeting of DCIR2 (Cruz et al. 2012). A noteworthy development in nanoparticles-based immunotherapy has been the mRNA vaccines. Nucleotide vaccines, including mRNA vaccines with their intracellular antigen synthesis, have been shown to be potent activators of a cytotoxic immune response. For example, lipid nanoparticles have been used to deliver mRNA coding for tumor antigens gp100 and TRP2 intracellular to the cytosol of APCs including DCs, macrophages and neutrophils to induce a strong CTL response resulting in tumor shrinkage and extended survival (Oberli et al. 2017). In another proof of concept study, liposome mediated delivery of IFN-γ pDNA has been used to enhance antitumor immune response in mice bearing E.G7-OVA tumor, when simultaneously injected at the same subcutaneous site with liposomes carrying OVA antigenic peptide. The liposome combination promoted the infiltration of CTLs to tumors (Yuba et al. 2015). In a recent study, a new strategy to engineer cationic nanoparticles-coated bacterial vectors has been used to efficiently deliver oral DNA vaccine for cancer immunotherapy. By coating live attenuated bacteria with synthetic nanoparticles self-assembled from cationic polymers and plasmid DNA, the protective nanoparticle coating layer is able to facilitate bacteria to effectively escape phagosomes, significantly enhance the acid tolerance of bacteria in stomach and intestines, and greatly promote dissemination of bacteria into blood circulation after oral administration. The oral delivery of DNA vaccines encoding autologous vascular endothelial growth factor receptor 2 (VEGFR2) by this hybrid vector showed significant T-cell activation and cytokine production. Successful inhibition of tumor growth was also achieved by efficient oral delivery of VEGFR2 with nanoparticle-coated bacterial vectors due to angiogenesis suppression in the tumor vasculature and tumor necrosis (Figure 10.7) (Hu et al. 2015).

10.3.3.1 In Situ Vaccination Strategies

A recently approved oncolytic virus therapy talimogene laherparepvec (T-VEC) for metastatic melanoma that cannot be surgically removed has opened new avenues for in situ vaccination-based immunotherapies. T-VEC, a genetically modified live

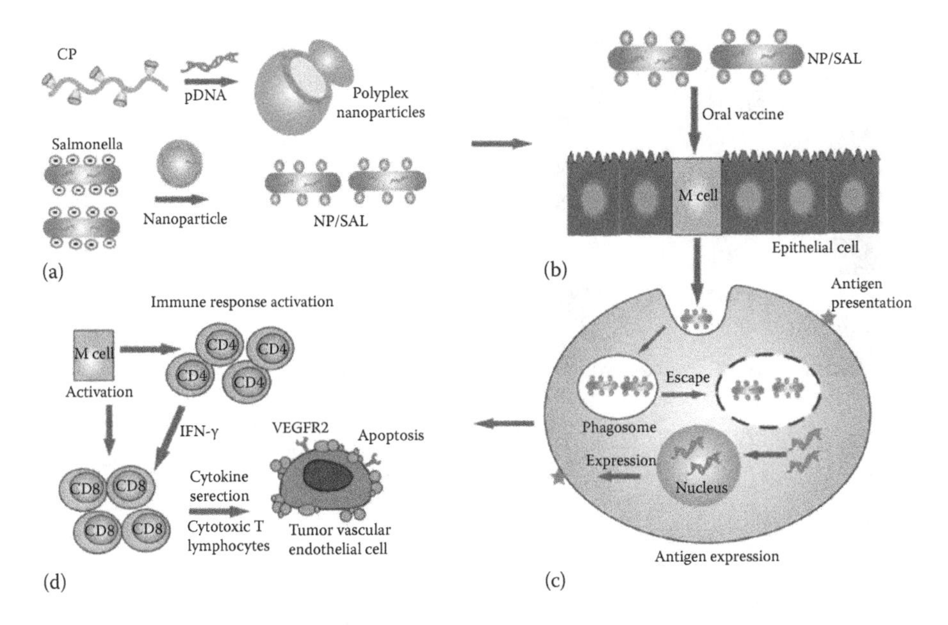

FIGURE 10.7 Schematic illustration of the cationic nanoparticle-coated attenuated Salmonellae for improved antigen expression and antitumor immune response activation. (a) Engineering of polyplex nanoparticle-coated Salmonellae. (b) Oral DNA vaccine delivery mediated by nanoparticle-coated Salmonellae. (c) Intracellular trafficking of nanoparticle-coated Salmonellae and antigen expression. (d) Activation of antitumor immune response. (Reproduced from Hu, Q. et al., *Nano Lett*, 15 (4):2732–9, 2015. With permission.)

oncolytic herpes virus is injected directly into the melanoma lesions, where it preferentially replicates inside dividing cancer cells and exerts antitumor activity through directly mediating cell death and by augmenting local and distant immune responses (Johnson et al. 2015, Rehman et al. 2016).

Based on a similar approach, immunostimulatory nanocarriers based on plant viruses have been recently developed as an immunotherapeutic nanomaterial, where the properties of the nanoparticle itself unlocked a potent antitumor immune response in an *in situ* vaccination approach (Lizotte et al. 2016, Lebel et al. 2016). We have demonstrated that plant-derived virus-like particles cowpea mosaic virus (CPMV) stimulate a potent immune-mediated antitumor response when introduced into the established primary tumors via intratumoral administration (Lizotte et al. 2016). Such immune response to VLPs generated an effective antitumor immune response in mouse models of multiple tumor types, including triple negative breast cancer, disseminated ovarian cancer, and primary and metastatic melanoma (Figure 10.8). Most importantly, the effect is systemic and durable, resulting in immune memory protecting mice from re-challenge. This approach modulates the local microenvironment to relieve immunosuppression and potentiate antitumor immunity against tumor antigens.

Thus, it is evident that by combining nanoengineering principles with immunomodulating payloads or immunostimulatory carriers, improved immunotherapy strategies can be developed. Going forward, tissue- and cell-specificity can be

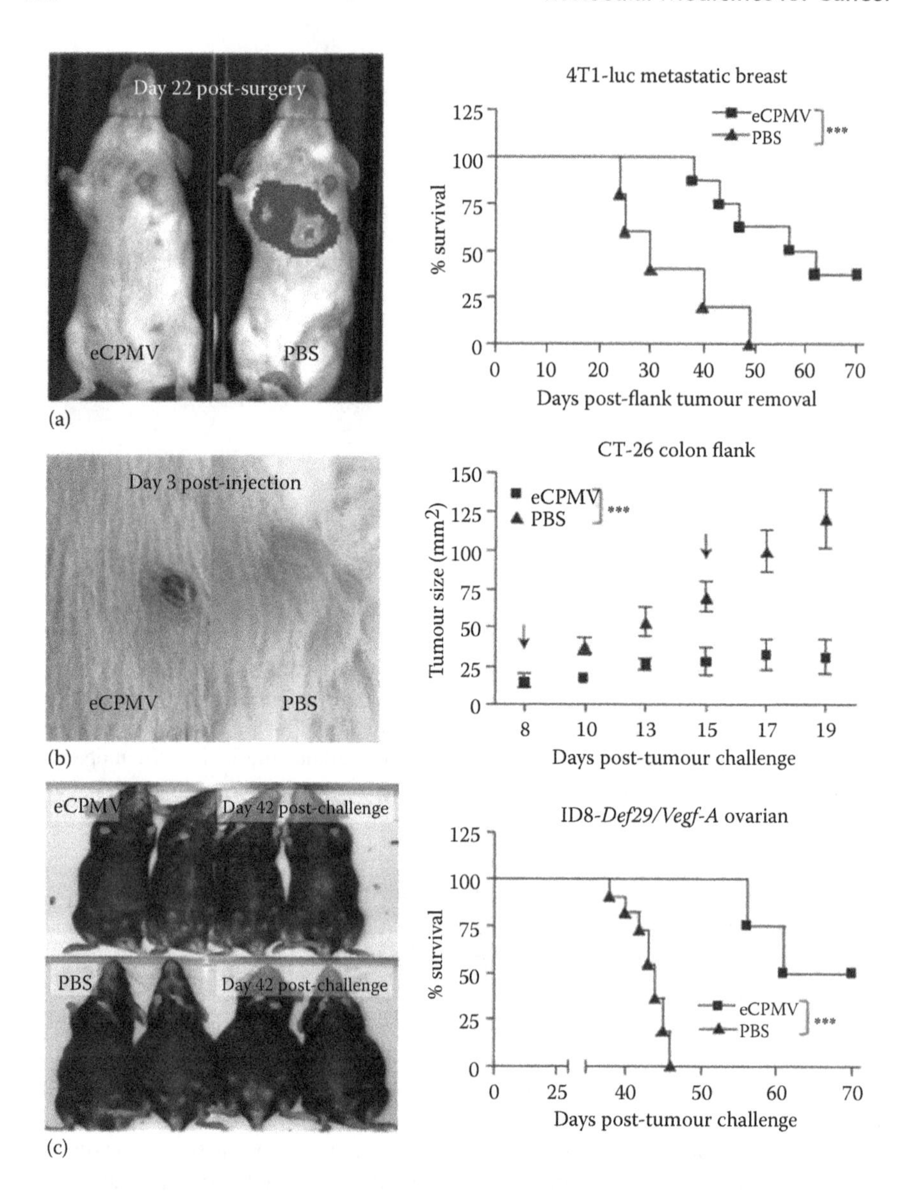

FIGURE 10.8 Immunostimulatory viral nanoparticles eCPMV exhibited successful immunotherapy in metastatic breast, colon, and ovarian carcinoma models. (a) Intratracheally administered eCPMV conferred protection against metastatic lung tumor 22 days following surgical removal of primary 4T1 breast cancer and prolonged survival as compared to mice treated similarly with PBS (IVIS image and Kaplan-Meier curves). (b) Intradermal CT26 colon cancer bearing mice also responded to direct intratumoral eCPMV injections and showed delayed growth as compared to PBS injected tumors. (c) eCPMV nanoparticle immunotherapy also was effective in improving survival in ID8-*Defb29/Vegf*-A ovarian cancer challenged mice. As compared to PBS treated mice, eCPMV treated mice displayed no visible ascites on day 42 post challenge. (Reproduced from Lizotte, P. H. et al., *Nat Nanotechnol*, 11 (3):295–303, 2016. With permission.

tailored by altering nanoparticle shape, size, charge, hydrophobicity, and surface chemistry (Smith et al. 2013, Champion et al. 2009, Kumar et al. 2015).

10.3.4 Challenges and Future Avenues for Cancer Immunotherapies

Even with extensive research efforts and resources only a handful of nanotechnology-based immunotherapies have been successfully translated into the clinic. This translational gap highlights not only the complex relationship of cancer and immunology, but also underlines poor understanding of the biological fate of nanoparticles and safety concerns (Dawidczyk et al. 2014, Stylianopoulos et al. 2015). Undesirable immunotoxicity of nanoparticles, adverse interactions with biomolecules, inflammatory responses, and long-term accumulation at off-target sites remain challenging issues. While moderate success has been achieved in nanoparticles-based delivery of chemotherapies and imaging contrast agents to tumor tissues, rapid clearance and accumulation of a large fraction of nanocarriers in tissues and organs of RES is a reality. Such off-target accumulation of nanocarriers bearing broad-spectrum immunomodulatory cargos could have even severe implications. Since the biological interactions and applicability of nanoparticles are influenced by physiochemical properties and compositions, selection of appropriate nanoparticle platform and its careful evaluation is a critical requirement. Therefore, aspects such as route of administration, biodistribution, pharmacokinetics, range of cellular interactions, and toxicological risk assessments must be thoroughly evaluated in extensive preclinical studies. Similarly, improving nanoparticle stability in circulation, upon serum protein conjugations and under varying pH and redox potentials should also be key to rule out premature release of such broad-spectrum cargo. At the same time, emphasizing the development of safer biodegradable nanoplatforms could facilitate easier translation of such technologies. Lastly, nanomanufacturing and quality control assurance must be considered along the translational pathway. Improved scaled-up production of nanoparticles with excellent physiochemical and compositional reproducibility is a critical bottleneck that must be resolved through innovative technologies and bio-inspired nanomaterials.

10.4 CONCLUDING REMARKS

Nanotechnology has rapidly gained acceptance in our day-to-day life—from consumer goods to paints and coatings, from fertilizers to water purifiers and from cosmetics to food adducts. While advancements in nanotechnology-based product development have been very visible, the field of nanotechnology has been quietly revolutionizing scientific research. As discussed in this chapter, significant improvements have been made in diagnostics methods with greater accuracy and sensitivity based on the optical and electronic properties of nanomaterials. These new methods of biomarkers and metabolite detection are enabling physicians and healthcare workers to diagnose and detect diseases quickly and with greater certainty. On the other hand, amalgamation of nanotechnology and cancer immunotherapy has brought two highly potent and complementary therapeutic modalities together. The multivalency and physiological uniqueness of nanoparticles in conjugation with specificity and potential of immune interactions is enabling manipulation of biological interactions with great precision, which is resulting in the development of newer and safer therapies. Both these fields—immunodiagnostics and immunotherapies—will continue the rapid growth via a highly interdisciplinary research and development effort and will provide huge opportunities for researchers in coming decades.

ACKNOWLEDGMENTS

This work has been supported in part by funding from the Susan G. Komen foundation (CCR14298962 to N. F. Steinmetz).

REFERENCES

Alizadeh, D., Zhang, L., Hwang, J., Schluep. T., and Badie, B. 2010. "Tumor-associated macrophages are predominant carriers of cyclodextrin-based nanoparticles into gliomas." *Nanomedicine* 6 (2):382–90.

Altin, J. G., and Parish, C. R. 2006. "Liposomal vaccines—targeting the delivery of antigen." *Methods* 40 (1):39–52.

Anderson, P. M., Katsanis, E., Sencer, S. F., et al. 1992. "Depot characteristics and biodistribution of interleukin-2 liposomes: importance of route of administration." *J Immunother (1991)* 12 (1):19–31.

Anik, U., and Timur, S. 2016. "Towards the electrochemical diagnosis of cancer: nanomaterial-based immunosensors and cytosensors." *RSC Adv* 6:111831–41.

Ashkenazi, A. 2002. "Targeting death and decoy receptors of the tumour-necrosis factor superfamily." *Nat Rev Cancer* 2 (6):420–30.

Atkins, M. B., Mier, J. W., Parkinson, D. R., et al. 1988. "Hypothyroidism after treatment with interleukin-2 and lymphokine-activated killer cells." *N Engl J Med* 318 (24):1557–63.

Azvolinsky, A. 2013. "Cancer vaccines: always a bridesmaid, never a bride?" *J Natl Cancer Inst* 105 (4):248–9.

Bachmann, M. F., Rohrer, U. H., Kundig, T. M., et al. 1993. "The influence of antigen organization on B cell responsiveness." *Science* 262 (5138):1448–51.

Bartholdy, C., Kauffmann, S. O., Christensen, J. P., and Thomsen, A. R. 2007. "Agonistic anti-CD40 antibody profoundly suppresses the immune response to infection with lymphocytic choriomeningitis virus." *J Immunol* 178 (3):1662–70.

Basu, G. D., Tinder, T. L., Bradley, J. M., et al. 2006. "Cyclooxygenase-2 inhibitor enhances the efficacy of a breast cancer vaccine: role of IDO." *J Immunol* 177 (4):2391–402.

Bazak, R., Houri, M., El Achy, S., Kamel, S., and Refaat, T. 2015. "Cancer active targeting by nanoparticles: a comprehensive review of literature." *J Cancer Res Clin Onco* 141:769–84.

Berger, G. K., McBride, A., Lawson, S., et al. 2017. "Brentuximab vedotin for treatment of non-Hodgkin lymphomas: a systematic review." *Crit Rev Oncol Hematol* 109:42–50.

Berner, V., Liu, H., Zhou, Q., et al. 2007. "IFN-gamma mediates CD4+ T-cell loss and impairs secondary antitumor responses after successful initial immunotherapy." *Nat Med* 13 (3):354–60.

Bladergroen, B. A., Meijer, C. J., ten Berge, R. L., et al. 2002. "Expression of the granzyme B inhibitor, protease inhibitor 9, by tumor cells in patients with non-Hodgkin and Hodgkin lymphoma: a novel protective mechanism for tumor cells to circumvent the immune system?" *Blood* 99 (1):232–7.

Blattman, J. N., and Greenberg, P. D. 2004. "Cancer immunotherapy: a treatment for the masses." *Science* 305 (5681):200–5.

Bobo, D., Robinson, K. J., Islam, J., Thurecht, K. J., and Corric, S. R. 2016. "Nanoparticle-based medicines: a review of FDA-approved materials and clinical trials to date." *Pharm Res* 33 (10):2373–87.

Bolhassani, A., Safaiyan, S., and Rafati, S. 2011. "Improvement of different vaccine delivery systems for cancer therapy." *Mol Cancer* 10:3.

Boyman, O., Surh, C. D., and Sprent, J. 2006. "Potential use of IL-2/anti-IL-2 antibody immune complexes for the treatment of cancer and autoimmune disease." *Expert Opin Biol Ther* 6 (12):1323–31.

Bravo, K., Ortega, F. G., Messina, G. n. A., et al. 2017. "Integrated bio-affinity nano-platform into a microfluidic immunosensor based on monoclonal bispecific trifunctional antibodies for the electrochemical determination of epithelial cancer biomarker." *Clinica Chimica Acta* 464:64–71.

Bright, R. K., Bright, J. D., and Byrne, J. A. 2014. "Overexpressed oncogenic tumor-self antigens." *Hum Vaccin Immunother* 10 (11):3297–305.

Buonaguro, L., Petrizzo, A., Tornesello, M. L., and Buonaguro, F. M. 2011. "Translating tumor antigens into cancer vaccines." *Clin Vaccine Immunol* 18 (1):23–34.

Champion, J. A., and Mitragotri, S. 2009. "Shape induced inhibition of phagocytosis of polymer particles." *Pharm Res* 26 (1):244–9.

Chen, G., Roy, I., Yang, C., and Prasad, P. N. 2016. "Nanochemistry and nanomedicine for nanoparticle-based diagnostics and therapy." *Chem Rev* 116:2826–85.

Chen, M.-J., Wu, Y.-S., Lin, G.-F., et al. 2012. "Quantum-dot-based homogeneous time-resolved fluoroimmunoassay of alpha-fetoprotein." *Anal Chim Acta* 741:100-105.

Chen, Q., Xu, L., Liang, C., et al. 2016. "Photothermal therapy with immune-adjuvant nanoparticles together with checkpoint blockade for effective cancer immunotherapy." *Nat Commun* 7:13193.

Chen, Z. H., Wu, Y. S., Chen, M. J., et al. 2013. "A novel homogeneous time-resolved fluoroimmunoassay for carcinoembryonic antigen based on water-soluble quantum dots." *J Fluores* 23:649–57.

Chikkaveeraiah, B. V., Bhirde, A. A., Morgan, N. Y., Eden, H. S., and Chen, X. 2012. "Electrochemical immunosensors for detection of cancer protein biomarkers." *ACS Nano* 6:6546–61.

Choi, J., Kim, H. Y., Ju, E. J., et al. 2012. "Use of macrophages to deliver therapeutic and imaging contrast agents to tumors." *Biomaterials* 33 (16):4195–203.

Christian, D. A., and Hunter, C. A. 2012. "Particle-mediated delivery of cytokines for immunotherapy." *Immunotherapy* 4 (4):425–41.

Cohen, E. P., Chopra, A., I, O. S., and Kim, T. S. 2009. "Enhancing cellular cancer vaccines." *Immunotherapy* 1 (3):495–504.

Cruz, L. J., Tacken, P. J., Rueda, F., et al. 2012. "Targeting nanoparticles to dendritic cells for immunotherapy." *Methods Enzymol* 509:143–63.

Cubillos-Ruiz, J. R., Engle, X., Scarlett, U. K., et al. 2009. "Polyethylenimine-based siRNA nanocomplexes reprogram tumor-associated dendritic cells via TLR5 to elicit therapeutic antitumor immunity." *J Clin Invest* 119 (8):2231–44.

Davis, M. E., Chen, Z. G., and Shin, D. M. 2008. "Nanoparticle therapeutics: an emerging treatment modality for cancer." *Nat Rev Drug Discov* 7 (9):771–82.

Davis, M. R., Zhu, Z., Hansen, D. M., et al. 2015. "The role of IL-21 in immunity and cancer." *Cancer Lett* 358 (2):107–14.

Dawidczyk, C. M., Kim, C., Park, J. H., et al. 2014. "State-of-the-art in design rules for drug delivery platforms: lessons learned from FDA-approved nanomedicines." *J Control Release* 187:133–44.

de la Rica, R., and Stevens, M. M. 2012. "Plasmonic ELISA for the ultrasensitive detection of disease biomarkers with the naked eye." *Nat Nanotech* 8:1759–64.

Ding, Y., Liu, J., Wang, H., Shen, G., and Yu, R. 2007. "A piezoelectric immunosensor for the detection of α-fetoprotein using an interface of gold/hydroxyapatite hybrid nanomaterial." *Biomaterials* 28:2147–54.

Donohue, J. H., and Rosenberg, S. A. 1983. "The fate of interleukin-2 after in vivo administration." *J Immunol* 130 (5):2203–8.

Dubensky, T. W., Jr., and Reed, S. G. 2010. "Adjuvants for cancer vaccines." *Semin Immunol* 22 (3):155–61.

Eggermont, L. J., Paulis, L. E., Tel, J., and Figdor, C. G. 2014. "Towards efficient cancer immunotherapy: advances in developing artificial antigen-presenting cells." *Trends Biotechnol* 32 (9):456–65.

Elshafey, R., Tavares, A. C., Siaj, M., and Zourob, M. 2013. "Electrochemical impedance immunosensor based on gold nanoparticles-protein G for the detection of cancer marker epidermal growth factor receptor in human plasma and brain tissue." *Biosens Bioelectron* 50:143–9.

Emami, M., Shamsipur, M., Saber, R., and Irajirad, R. 2014. "An electrochemical immunosensor for detection of a breast cancer biomarker based on antiHER2-iron oxide nanoparticle bioconjugates." *Analyst* 139:2858–66.

Engvall, E., Jonsson, K., and Perlmann, P. 1971. "Enzyme-linked immunosorbent assay. II. Quantitative assay of protein antigen, immunoglobulin g, by means of enzyme-labelled antigen and antibody-coated tubes." *Biochim Biophysic Acta* 251:427–34.

Eppstein, D. A., and Stewart, W. E., 2nd. 1982. "Altered pharmacological properties of liposome-associated human interferon-alpha." *J Virol* 41 (2):575–82.

Eustis, S., and El-Sayed, M. A. 2006. "Why gold nanoparticles are more precious than pretty gold: Noble metal surface plasmon resonance and its enhancement of the radiative and nonradiative properties of nanocrystals of different shapes." *Chem Soc Rev* 35:209–17.

Fan, Y., and Moon, J. J. 2015. "Nanoparticle drug delivery systems designed to improve cancer vaccines and immunotherapy." *Vaccines (Basel)* 3 (3):662–85.

Fang, R. H., Kroll, A. V., and Zhang, L. 2015. "Nanoparticle-based manipulation of antigen-presenting cells for cancer immunotherapy." *Small* 11 (41):5483–96.

Fellner, C. 2012. "Ipilimumab (yervoy) prolongs survival in advanced melanoma: serious side effects and a hefty price tag may limit its use." *P T* 37 (9):503–30.

Fesnak, A. D., June, C. H., and Levine, B. L. 2016. "Engineered T-cells: the promise and challenges of cancer immunotherapy." *Nat Rev Cancer* 16 (9):566–81.

Filippi, C. M., Juedes, A. E., Oldham, J. E., et al. 2008. "Transforming growth factor-beta suppresses the activation of CD8+ T-cells when naive but promotes their survival and function once antigen experienced: a two-faced impact on autoimmunity." *Diabetes* 57 (10):2684–92.

Gabrilovich, D. I., and Nagaraj, S. 2009. "Myeloid-derived suppressor cells as regulators of the immune system." *Nat Rev Immunol* 9 (3):162–74.

Gallucci, S., and Matzinger, P. 2001. "Danger signals: SOS to the immune system." *Curr Opin Immunol* 13 (1):114–9.

Gao, C., Su, M., Wang, Y., Ge, S., and Yu. 2015. "A disposable paper-based electrochemiluminescence device for ultrasensitive monitoring of CEA based on Ru(bpy)32+@Au nanocages." *RSC Adv* 5:28324–31.

Ge, S., Ge, L., Yan, M., et al. 2013. "A disposable immunosensor device for point-of-care test of tumor marker based on copper-mediated amplification." *Biosens Bioelectron* 43:425–31.

Ge, S., Zhang, L., Zhang, Y., et al. 2017. "Nanomaterials-modified cellulose paper as a platform for biosensing applications." *Nanoscale* 10.1039/C6NR08846E:-.

Gold, P., and Freedman, S. O. 1965. "Demonstration of Tumor-Specific Antigens in Human Colonic Carcinomata By Immunological Tolerance and Absorption Techniques." *J Expl Med* 121:439–62.

Graham, J. B., and Graham, R. M. 1955. "Antibodies elicited by cancer in patients." *Cancer* 8:409–16.

Grupp, S. A., Kalos, M., Barrett, D., et al. 2013. "Chimeric antigen receptor-modified T-cells for acute lymphoid leukemia." *N Engl J Med* 368 (16):1509–18.

Guo, L., Jackman, J. A., Yang, H.-H., et al. 2015. "Strategies for enhancing the sensitivity of plasmonic nanosensors." *Nano Today* 10:213–39.

Haining, W. N., Anderson, D. G., Little, S. R., et al. 2004. "pH-triggered microparticles for peptide vaccination." *J Immunol* 173 (4):2578–85.

Hanahan, D., and Weinberg, R. A. 2000. "The hallmarks of cancer." *Cell* 100:57–70.

Handy, B. 2009. "The clinical utility of tumor markers." *Laboratory Medicine* 40:99–103.

Hanes, J., Sills, A., Zhao, Z., et al. 2001. "Controlled local delivery of interleukin-2 by biodegradable polymers protects animals from experimental brain tumors and liver tumors." *Pharm Res* 18 (7):899–906.

Hardy, C. L., Lemasurier, J. S., Mohamud, R., et al. 2013. "Differential uptake of nanoparticles and microparticles by pulmonary APC subsets induces discrete immunological imprints." *J Immunol* 191 (10):5278–90.

He, J.-L., Wang, D.-S., and Fan, S.-K. 2016. "Opto-microfluidic immunosensors: from colorimetric to plasmonic." *Micromachines* 7:29.

Heffernan, M. J., Kasturi, S. P., Yang, S. C., Pulendran, B., and Murthy, N. 2009. "The stimulation of CD8+ T-cells by dendritic cells pulsed with polyketal microparticles containing ion-paired protein antigen and poly(inosinic acid)-poly(cytidylic acid)." *Biomaterials* 30 (5):910–8.

Heikenwalder, M., Polymenidou, M., Junt, T., et al. 2004. "Lymphoid follicle destruction and immunosuppression after repeated CpG oligodeoxynucleotide administration." *Nat Med* 10 (2):187–92.

Hu, Q., Wu, M., Fang, C., et al. 2015. "Engineering nanoparticle-coated bacteria as oral DNA vaccines for cancer immunotherapy." *Nano Lett* 15 (4):2732–9.

Hu, Y., Litwin, T., Nagaraja, A. R., et al. 2007. "Cytosolic delivery of membrane-impermeable molecules in dendritic cells using pH-responsive core-shell nanoparticles." *Nano Lett* 7 (10):3056–64.

Huang, Z., Zhang, Z., Jiang, Y., et al. 2012. "Targeted delivery of oligonucleotides into tumor-associated macrophages for cancer immunotherapy." *J Control Release* 158 (2):286–92.

Hudis, C. A. 2007. "Trastuzumab–mechanism of action and use in clinical practice." *N Engl J Med* 357 (1):39–51.

Irvine, D. J., Hanson, M. C., Rakhra, K., and Tokatlian, T. 2015. "Synthetic Nanoparticles for Vaccines and Immunotherapy." *Chem Rev* 115 (19):11109–46.

Irvine, D. J., Swartz, M. A., and Szeto, G. L. 2013. "Engineering synthetic vaccines using cues from natural immunity." *Nat Mater* 12 (11):978–90.

Jayanthi, V. S. P. K. S. A., Das, A. B., and Saxena, U. 2017. "Recent advances in biosensor development for the detection of cancer biomarkers." *Biosens Bioelectron* 91:15–23.

Jiang, H., and Chess, L. 2009. "How the immune system achieves self-nonself discrimination during adaptive immunity." *Adv Immunol* 102:95–133.

Jiang, Y., Li, Y., and Zhu, B. 2015. "T-cell exhaustion in the tumor microenvironment." *Cell Death Dis* 6:e1792.

Jobsri, J., Allen, A., Rajagopal, D., et al. 2015. "Plant virus particles carrying tumour antigen activate TLR7 and Induce high levels of protective antibody." *PLoS One* 10 (2):e0118096.

Joffre, O. P., Segura, E., Savina, A., et al. 2012. "Cross-presentation by dendritic cells." *Nat Rev Immunol* 12 (8):557–69.

Johnson, D. B., Puzanov, I., and Kelley, M. C. 2015. "Talimogene laherparepvec (T-VEC) for the treatment of advanced melanoma." *Immunotherapy* 7 (6):611–9.

Kaczanowska, S., Joseph, A. M., and Davila, E. 2013. "TLR agonists: our best frenemy in cancer immunotherapy." *J Leukoc Biol* 93 (6):847–63.

Kanaoka, E., Takahashi, K., Yoshikawa, T., et al. 2002. "A significant enhancement of therapeutic effect against hepatic metastases of M5076 in mice by a liposomal interleukin-2 (mixture)." *J Control Release* 82 (2–3):183–7.

Kang, J., Demaria, S., and Formenti, S. 2016. "Current clinical trials testing the combination of immunotherapy with radiotherapy." *J Immunother Cancer* 4:51.

Karkada, M., Berinstein, N. L., and Mansour, M. 2014. "Therapeutic vaccines and cancer: focus on DPX-0907." *Biologics* 8:27–38.

Kedar, E., Gur, H., Babai, I., et al. 2000. "Delivery of cytokines by liposomes: hematopoietic and immunomodulatory activity of interleukin-2 encapsulated in conventional liposomes and in long-circulating liposomes." *J Immunother* 23 (1):131–45.

Kedar, E., Palgi, O., Golod, G., Babai, I., and Barenholz, Y. 1997. "Delivery of cytokines by liposomes. III. Liposome-encapsulated GM-CSF and TNF-alpha show improved pharmacokinetics and biological activity and reduced toxicity in mice." *J Immunother* 20 (3):180–93.

Khalil, D. N., Smith, E. L., Brentjens, R. J., and Wolchok, J. D. 2016. "The future of cancer treatment: immunomodulation, CARs and combination immunotherapy." *Nat Rev Clin Oncol* 13 (5):273–90.

Klempner, M. S., Noring, R., Mier, J. W., and Atkins, M. B. 1990. "An acquired chemotactic defect in neutrophils from patients receiving interleukin-2 immunotherapy." *N Engl J Med* 322 (14):959–65.

Klinman, D. M. 2004. "Immunotherapeutic uses of CpG oligodeoxynucleotides." *Nat Rev Immunol* 4 (4):249–58.

Kreibig, U., and Vollmer, M. 2013. "Optical properties of metal clusters." 25.

Krishnamachari, Y., Geary, S. M., Lemke, C. D., and Salem, A. K. 2011. "Nanoparticle delivery systems in cancer vaccines." *Pharm Res* 28 (2):215–36.

Kumar, S., Anselmo, A. C., Banerjee, A., Zakreusky, M., and Mitragotri, S. 2015. "Shape and size-dependent immune response to antigen-carrying nanoparticles." *J Control Release* 220 (Pt A):141–8.

Kupstat, A., Kumke, M. U., and Hildebrandt, N. 2011. "Toward sensitive, quantitative point-of-care testing (POCT) of protein markers: miniaturization of a homogeneous time-resolved fluoroimmunoassay for prostate-specific antigen detection." *Analyst* 136:1029–35.

Kwong, B., Gai, S. A., Elkhader, J., Wittrup, K. D., and Irvine, D. J. 2013. "Localized immunotherapy via liposome-anchored Anti-CD137 + IL-2 prevents lethal toxicity and elicits local and systemic antitumor immunity." *Cancer Res* 73 (5):1547–58.

Kwong, B., Liu, H., and Irvine, D. J. 2011. "Induction of potent anti-tumor responses while eliminating systemic side effects via liposome-anchored combinatorial immunotherapy." *Biomaterials* 32 (22):5134–47.

Lambert, J. M., and Chari, R. V. 2014. "Ado-trastuzumab Emtansine (T-DM1): an antibody-drug conjugate (ADC) for HER2-positive breast cancer." *J Med Chem* 57 (16):6949–64.

Lammers, T., Kiessling, F., Hennink, W. E., and Storm, G. 2012. "Drug targeting to tumors: principles, pitfalls and (pre-) clinical progress." *J Control Release* 161 (2):175–87.

Lebel, M. E., Chartrand, K., Tarrab, E., et al. 2016. "Potentiating cancer immunotherapy using papaya mosaic virus-derived nanoparticles." *Nano Lett* 16 (3):1826–32.

Lee, J. U., Nguyen, A. H., and Sim, S. J. 2015. "A nanoplasmonic biosensor for label-free multiplex detection of cancer biomarkers." *Biosens Bioelectron* 74:341–6.

Lee, K. L., Twyman, R. M., Fiering, S., and Steinmetz, N. F. 2016. "Virus-based nanoparticles as platform technologies for modern vaccines." *Wiley Interdiscip Rev Nanomed Nanobiotechnol* 8 (4):554–78.

Lee, R., Localio, A. R., Armstrong, K., et al. 2006. "A meta-analysis of the performance characteristics of the free prostate-specific antigen test." *Urology* 67 (4):762–8.

Lee, S., Chon, H., Lee, J., et al. 2014. "Rapid and sensitive phenotypic marker detection on breast cancer cells using surface-enhanced Raman scattering (SERS) imaging." *Biosens Bioelectron* 51:238–43.

Lee, S., Chon, H., Yoon, S.-Y., et al. 2012. "Fabrication of SERS-fluorescence dual modal nanoprobes and application to multiplex cancer cell imaging." *Nanoscale* 4:124.

Li, L., Feng, D., and Zhang, Y. 2016. "Simultaneous detection of two tumor markers using silver and gold nanoparticles decorated carbon nanospheres as labels." *Anal Biochem* 505:59–65.

Li, M., Cushing, S. K., Zhang, J., et al. 2013. "Three-dimensional hierarchical plasmonic nano-architecture enhanced surface-enhanced Raman scattering immunosensor for cancer biomarker detection in blood plasma." *ACS Nano* 7:4967–76.

Li, W., Li, L., Li, M., et al. 2013. "Development of a 3D origami multiplex electrochemical immunodevice using a nanoporous silver-paper electrode and metal ion functionalized nanoporous gold–chitosan." *Chemcomm* 49:9540.

Li, X., Yu, M., Chen, Z., Lin, X., and Wu, Q. 2017. "A sensor for detection of carcinoembryonic antigen based on the polyaniline-Au nanoparticles and gap-based interdigitated electrode." *Sens Actuators B Chem* 239:874–82.

Liang, K., Liu, F., Fan, J., et al. 2017. "Nanoplasmonic quantification of tumour-derived extracellular vesicles in plasma microsamples for diagnosis and treatment monitoring." *Nat Biomed Eng* 1:0021.

Lima, K. M., dos Santos, S. A., Rodrigues, J. M., Jr., and Silva, C. L. 2004. "Vaccine adjuvant: it makes the difference." *Vaccine* 22 (19):2374–9.

Lin, A. Y., Almeida, J. P., Bear, A., et al. 2013. "Gold nanoparticle delivery of modified CpG stimulates macrophages and inhibits tumor growth for enhanced immunotherapy." *PLoS One* 8 (5):e63550.

Lin, P. Y., Cheng, K. L., McGuffin-Cawley, J. D., et al. 2012. "Detection of alpha-methylacyl-coa racemase (AMACR), a biomarker of prostate cancer, in patient blood samples using a nanoparticle electrochemical biosensor." *Biosensors (Basel).* 2 (4):377–87.

Linley, A. J., Ahmad, M., and Rees, R. C. 2011. "Tumour-associated antigens: considerations for their use in tumour immunotherapy." *Int J Hematol* 93 (3):263–73.

Litzinger, M. T., Fernando, R., Curiel, T. J., et al. 2007. "IL-2 immunotoxin denileukin diftitox reduces regulatory T-cells and enhances vaccine-mediated T-cell immunity." *Blood* 110 (9):3192–201.

Liu, D., Yang, J., Wang, H.-F., et al. 2014. "Glucose Oxidase-catalyzed growth of gold nanoparticles enables quantitative detection of attomolar cancer biomarkers." *Anal Chem* 86:5800–6.

Liu, J., Cui, M., Zhou, H., and Zhang, S. 2016. "Efficient double-quenching of electrochemiluminescence from CdS:Eu QDs by hemin-graphene-Au nanorods ternary composite for ultrasensitive immunoassay." *Sci Report* 6:1–9.

Lizotte, P. H., Wen, A. M., Sheen, et al. 2016. "In situ vaccination with cowpea mosaic virus nanoparticles suppresses metastatic cancer." *Nat Nanotechnol* 11 (3):295–303.

Ludwig, J. A., and Weinstein, J. N. 2005. "Biomarkers in cancer staging, prognosis and treatment selection." *Nat Rev Cancer* 5:845–56.

Maeda, H., Nakamura, H., and Fang, J. 2013. "The EPR effect for macromolecular drug delivery to solid tumors: Improvement of tumor uptake, lowering of systemic toxicity, and distinct tumor imaging in vivo." *Adv Drug Deliv Rev* 65 (1):71–9.

Malhotra, B. D., Kumar, S., and Mouli Pandey, C. 2016. "Nanomaterials based biosensors for cancer biomarker detection." *J Phys: Conf Ser* 704 012011.

Mattera, L., Bhuckory, S., Wegner, K. D., et al. 2016. "Compact quantum dot-antibody conjugates for FRET immunoassays with subnanomolar detection limits." *Nanoscale* 8 (21):11275–83.

Matzinger, P. 1994. "Tolerance, danger, and the extended family." *Annu Rev Immunol* 12:991–1045.

Melero, I., Berman, D. M., Aznar, M. A., et al. 2015. "Evolving synergistic combinations of targeted immunotherapies to combat cancer." *Nat Rev Cancer* 15 (8):457–72.

Melero, I., Gaudernack, G., Gerritsen, W., et al. 2014. "Therapeutic vaccines for cancer: an overview of clinical trials." *Nat Rev Clin Oncol* 11 (9):509–24.

Mitragotri, S., Anderson, D. G., Chen, X., et al. 2015. "Accelerating the translation of nanomaterials in biomedicine." *ACS Nano* 9 (7):6644–54.

Molino, N. M., Anderson, A. K., Nelson, E. L., and Wang, S. W. 2013. "Biomimetic protein nanoparticles facilitate enhanced dendritic cell activation and cross-presentation." *ACS Nano* 7 (11):9743–52.

Mukai, Y., Yoshinaga, T., Yoshikawa, M., et al. 2011. "Induction of endoplasmic reticulum-endosome fusion for antigen cross-presentation induced by poly (gamma-glutamic acid) nanoparticles." *J Immunol* 187 (12):6249–55.

Nair, M., Singh Sandhu, S., and K Sharma, A. 2014. "Prognostic and predictive biomarkers in cancer." *Curr Cancer Drug Targ* 14:477–504.

Nakamura, T., Miyabe, H., Hyodo, M., et al. 2015. "Liposomes loaded with a STING pathway ligand, cyclic di-GMP, enhance cancer immunotherapy against metastatic melanoma." *J Control Release* 216:149–57.

Narayanan, N., Karunakaran, V., Paul, W., et al. 2015. "Aggregation induced Raman scattering of squaraine dye: implementation in diagnosis of cervical cancer dysplasia by SERS imaging." *Biosens Bioelectron* 70:145–152.

Neville, M. E., Robb, R. J., and Popescu, M. C. 2001. "In situ vaccination against a non-immunogenic tumour using intratumoural injections of liposomal interleukin 2." *Cytokine* 16 (6):239–50.

Nicholas, C., and Lesinski, G. B. 2011. "Immunomodulatory cytokines as therapeutic agents for melanoma." *Immunotherapy* 3 (5):673–90.

Noad, R., and Roy, P. 2003. "Virus-like particles as immunogens." *Trends Microbiol* 11 (9):438–44.

Oberli, M. A., Reichmuth, A. M., Dorkin, J. R., et al. 2017. "Lipid nanoparticle assisted mRNA delivery for potent cancer immunotherapy." *Nano Lett* 17 (3):1326–35.

Pallela, R., Chandra, P., Noh, H. B., and Shim, Y. B. 2016. "An amperometric nanobiosensor using a biocompatible conjugate for early detection of metastatic cancer cells in biological fluid." *Biosens Bioelectron* 85:883–90.

Pan, J., and Yang, Q. 2007. "Antibody-functionalized magnetic nanoparticles for the detection of carcinoembryonic antigen using a flow-injection electrochemical device." *Anal Bioanal Chem* 388:279–86.

Pardoll, D. M. 2012. "The blockade of immune checkpoints in cancer immunotherapy." *Nat Rev Cancer* 12 (4):252–64.

Park, J., Wrzesinski, S. H., Stern, E., et al. 2012. "Combination delivery of TGF-beta inhibitor and IL-2 by nanoscale liposomal polymeric gels enhances tumour immunotherapy." *Nat Mater* 11 (10):895–905.

Pasare, C., and Medzhitov, R. 2005. "Control of B-cell responses by Toll-like receptors." *Nature* 438 (7066):364–8.

Paulos, C. M., Kaiser, A., Wrzesinski, C., et al. 2007. "Toll-like receptors in tumor immunotherapy." *Clin Cancer Res* 13 (18 Pt 1):5280–9.

Peer, D., Karp, J. M., Hong, S., et al. 2007. "Nanocarriers as an emerging platform for cancer therapy." *Nat Nanotechnol* 2 (12):751–60.

Perica, K., Tu, A., Richter, A., et al. 2014. "Magnetic field-induced T-cell receptor clustering by nanoparticles enhances T-cell activation and stimulates antitumor activity." *ACS Nano* 8 (3):2252–60.

Petros, R. A., and DeSimone, J. M. 2010. "Strategies in the design of nanoparticles for therapeutic applications." *Nat Rev Drug Discov* 9 (8):615–27.

Plummer, E. M., and Manchester, M. 2011. "Viral nanoparticles and virus-like particles: platforms for contemporary vaccine design." *Wiley Interdiscip Rev Nanomed Nanobiotechnol* 3 (2):174–96.

Polascik, T. J., Oesterling, J. E., and Partin, A. W. 1999. "Prostate specific antigen: A decade of discovery - What we have learned and where we are going." *J Urology* 162 (2):293–306.

Poole, R. M. 2014. "Pembrolizumab: first global approval." *Drugs* 74 (16):1973-81.

Raedler, L. A. 2015. "Opdivo (Nivolumab): second PD-1 inhibitor receives FDA approval for unresectable or metastatic melanoma." *Am Health Drug Benefits* 8 (Spec Feature):180–3.

Reddy, S. T., van der Vlies, A. J., Simeoni, E., et al. 2007. "Exploiting lymphatic transport and complement activation in nanoparticle vaccines." *Nat Biotechnol* 25 (10):1159–64.

Rehman, H., Silk, A. W., Kane, M. P., and Kaufman, H. L. 2016. "Into the clinic: Talimogene laherparepvec (T-VEC), a first-in-class intratumoral oncolytic viral therapy." *J Immunother Cancer* 4:53.

Ribeiro-Muller, L., and Muller, M. 2014. "Prophylactic papillomavirus vaccines." *Clin Dermatol* 32 (2):235–47.

Riedel, T., Ghasparian, A., Moehle, K., et al. 2011. "Synthetic virus-like particles and conformationally constrained peptidomimetics in vaccine design." *Chembiochem* 12 (18):2829–36.

Rodríguez-Lorenzo, L., de la Rica, R., Álvarez-Puebla, R., et al. 2012. "Plasmonic nanosensors with inverse sensitivity by means of enzyme-guided crystal growth." *Nat Mater* 11:604–7.

Rosalia, R. A., Cruz, L. J., van Duikeren, S., et al. 2015. "CD40-targeted dendritic cell delivery of PLGA-nanoparticle vaccines induce potent anti-tumor responses." *Biomaterials* 40:88–97.

Rosenberg, S. A., and Restifo, N. P. 2015. "Adoptive cell transfer as personalized immunotherapy for human cancer." *Science* 348 (6230):62–8.

Rosenberg, S. A., Yang, J. C., and Restifo, N. P. 2004. "Cancer immunotherapy: moving beyond current vaccines." *Nat Med* 10 (9):909–15.

Roy, A., Singh, M. S., Upadhyay, P., and Bhaskar, S. 2010. "Combined chemo-immunotherapy as a prospective strategy to combat cancer: a nanoparticle based approach." *Mol Pharm* 7 (5):1778–88.

Rusling, J. F., Kumar, C. V., Gutkind, J. S., and Patel, V. 2010. "Measurement of biomarker proteins for point-of-care early detection and monitoring of cancer." *Analyst* 135:2496–511.

Sabel, M. S., Hill, H., Jong, Y. S., et al. 2001. "Neoadjuvant therapy with interleukin-12-loaded polylactic acid microspheres reduces local recurrence and distant metastases." *Surgery* 130 (3):470–8.

Sakaguchi, S., Yamaguchi, T., Nomura, T., and Duo, M. 2008. "Regulatory T-cells and immune tolerance." *Cell* 133 (5):775–87.

Salimi, A., Kavosi, B., Fathi, F., and Hallaj, R. 2013. "Highly sensitive immunosensing of prostate-specific antigen based on ionic liquid-carbon nanotubes modified electrode: Application as cancer biomarker for prostatebiopsies." *Biosens Bioelectron* 42:439–46.

Saluja, S. S., Hanlon, D. J., Sharp, F. A., et al. 2014. "Targeting human dendritic cells via DEC-205 using PLGA nanoparticles leads to enhanced cross-presentation of a melanoma-associated antigen." *Int J Nanomedicine* 9:5231–46.

Scheerlinck, J. P., and Greenwood, D. L. 2008. "Virus-sized vaccine delivery systems." *Drug Discov Today* 13 (19-20):882–7.

Schlucker, S. 2014. "Surface-enhanced raman spectroscopy: concepts and chemical applications." *Angew Chem Int Edit* 53 (19):4756–95.

Schreiber, R. D., Old, L. J., and Smyth, M. J. 2011. "Cancer immunoediting: integrating immunity's roles in cancer suppression and promotion." *Science* 331 (6024):1565–70.

Shao, K., Singha, S., Clemente-Casares, X., et al. 2015. "Nanoparticle-based immunotherapy for cancer." *ACS Nano* 9 (1):16–30.

Sharp, F. A., Ruane, D., Claass, B., et al. 2009. "Uptake of particulate vaccine adjuvants by dendritic cells activates the NALP3 inflammasome." *Proc Natl Acad Sci U S A* 106 (3):870–5.

Shirota, H., and Klinman, D. M. 2013. "Use of CpG oligonucleotides for cancer immunotherapy and their effect on immunity in the tumor microenvironment." *Immunotherapy* 5 (8):787–9.

Shukla, S., Myers, J. T., Woods, S. E., et al. 2017. "Plant viral nanoparticles-based HER2 vaccine: Immune response influenced by differential transport, localization and cellular interactions of particulate carriers." *Biomaterials* 121:15–27.

Shukla, S., Wen, A. M., Commandeur, U., and Steinmetz, N. F. 2014. "Presentation of HER2 epitopes using a filamentous plant virus-based vaccination platform." *J Mater Chem B* 2 (37):6249–58.

Smith, B. R., Ghosn, E. E., Rallapalli, H., et al. 2014. "Selective uptake of single-walled carbon nanotubes by circulating monocytes for enhanced tumour delivery." *Nat Nanotechnol* 9 (6):481–7.

Smith, D. M., Simon, J. K., and Baker, J. R., Jr. 2013. "Applications of nanotechnology for immunology." *Nat Rev Immunol* 13 (8):592–605.

Smith, M. R. 2003. "Rituximab (monoclonal anti-CD20 antibody): mechanisms of action and resistance." *Oncogene* 22 (47):7359–68.

Smith, R. A., Andrews, K., Brooks, D., et al. "Cancer screening in the United States, 2016: A review of current American Cancer Society guidelines and current issues in cancer screening." *CA Cancer J Clin* 66:95–114.

Stephan, M. T., Moon, J. J., Um, S. H., et al. 2010. "Therapeutic cell engineering with surface-conjugated synthetic nanoparticles." *Nat Med* 16 (9):1035–41.

Stylianopoulos, T., and Jain, R. K. 2015. "Design considerations for nanotherapeutics in oncology." *Nanomedicine* 11 (8):1893–907.

Swann, J. B., and Smyth, M. J. 2007. "Immune surveillance of tumors." *J Clin Invest* 117 (5):1137–46.

Tagawa, M., Kawamura, K., Li, Q., et al. 2011. "A possible anticancer agent, type III interferon, activates cell death pathways and produces antitumor effects." *Clin Dev Immunol* 2011:479013.

Tang, D., Yuan, R., and Chai, Y. 2006. "Magnetic core-shell Fe3O4@Ag nanoparticles coated carbon paste interface for studies of carcinoembryonic antigen in clinical immunoassay." *J Phys Chem B* 110:11640–46.

ten Hagen, T. L., Seynhaeve, A. L., van Tiel, S. T., Ruiter, D. J., and Eggermont, A. M. 2002. "Pegylated liposomal tumor necrosis factor-alpha results in reduced toxicity and synergistic antitumor activity after systemic administration in combination with liposomal doxorubicin (Doxil) in soft tissue sarcoma-bearing rats." *Int J Cancer* 97 (1):115–20.

Tian, H., Shi, G., Yang, G., et al. 2014. "Cellular immunotherapy using irradiated lung cancer cell vaccine co-expressing GM-CSF and IL-18 can induce significant antitumor effects." *BMC Cancer* 14:48.

Tian, L., Morrissey, J. J., Kattumenu, R., et al. 2012. "Bioplasmonic paper as a platform for detection of kidney cancer biomarkers." *Anal Chem* 84(22): 9928–34.

Topfer, K., Kempe, S., Muller, N., et al. 2011. "Tumor evasion from T-cell surveillance." *J Biomed Biotechnol* 2011:918471.

van der Veen, A. H., Eggermont, A. M., Seynhaeve, A. L., van, T., and ten Hager, T. L. 1998. "Biodistribution and tumor localization of stealth liposomal tumor necrosis factor-alpha in soft tissue sarcoma bearing rats." *Int J Cancer* 77 (6):901–6.

van der Weijden, J., Paulis, L. E., Verdoes, M., van Hest, J. C. M., and Figdor, C. G. 2014. "The right touch: design of artificial antigen-presenting cells to stimulate the immune system." *Chem Sci* 5 (9):3355–67.

Vanneman, M., and Dranoff, G. 2012. "Combining immunotherapy and targeted therapies in cancer treatment." *Nat Rev Cancer* 12 (4):237–51.

Vesely, M. D., Kershaw, M. H., Schreiber, R. D., and Smyth, M. J. 2011. "Natural innate and adaptive immunity to cancer." *Annu Rev Immunol* 29:235–71.

Wang, C., Xu, L., Liang, C., et al. 2014. "Immunological responses triggered by photothermal therapy with carbon nanotubes in combination with anti-CTLA-4 therapy to inhibit cancer metastasis." *Adv Mater* 26 (48):8154–62.

Wang, S., Ge, L., Zhang, Y., et al. 2012. "Battery-triggered microfluidic paper-based multiplex electrochemiluminescence immunodevice based on potential-resolution strategy." *Lab Chip* 12 (21):4489–98.

Weber, J. S., Yang, J. C., Atkins, M. B., and Disis, M. L. 2015. "Toxicities of immunotherapy for the practitioner." *J Clin Oncol* 33 (18):2092–9.

Wegner, K. D., Jin, Z., Linden, S., Jennings, T. L., and Hildebrandt, N. 2013. "Quantum-dot-based Förster resonance energy transfer immunoassay for sensitive clinical diagnostics of low-volume serum samples." *Acs Nano* 7:7411–9.

Wegner, K. D., Lindén, S., Jin, Z., et al. 2014. "Nanobodies and nanocrystals: highly sensitive quantum dot-based homogeneous FRET immunoassay for serum-based EGFR detection." *Small* 10:734–40.

Wei, X. X., Fong, L., and Small, E. J. 2015. "Prostate cancer immunotherapy with Sipuleucel-T: current standards and future directions." *Expert Rev Vaccines* 14 (12):1529–41.

Wen, G., Liang, X., Liu, Q., Liang, A., and Jiang, Z. 2016. "A novel nanocatalytic SERS detection of trace human chorionic gonadotropin using labeled-free Vitoria blue 4R as molecular probe." *Biosens Bioelectron* 85:450–6.

Wilhelm, S., Tavares, A. J., Dai, Q., et al. 2016. "Analysis of nanoparticle delivery to tumours." *Nat Rev Mater* 1 (5).

Wingender, G., Garbi, N., Schumak, B., et al. 2006. "Systemic application of CpG-rich DNA suppresses adaptive T-cell immunity via induction of IDO." *Eur J Immunol* 36 (1):12–20.

Winter, H., Fox, B. A., and Ruttinger, D. 2014. "Future of cancer vaccines." *Methods Mol Biol* 1139:555–64.

Woo, S. R., Turnis, M. E., Goldberg, M. V., et al. 2012. "Immune inhibitory molecules LAG-3 and PD-1 synergistically regulate T-cell function to promote tumoral immune escape." *Cancer Res* 72 (4):917–27.

Wu, M. S., Shi, H. W., He, L. J., Xu, J. J., and Chen, H. Y. 2012. "Microchip device with 64-site electrode array for multiplexed immunoassay of cell surface antigens based on electrochemiluminescence resonance energy transfer." *Anal Chem* 84:4207–13.

Wu, T. L., and Ertl, H. C. 2009. "Immune barriers to successful gene therapy." *Trends Mol Med* 15 (1):32–9.

Xu, Z., Wang, Y., Zhang, L., and Huang, L. 2014. "Nanoparticle-delivered transforming growth factor-beta siRNA enhances vaccination against advanced melanoma by modifying tumor microenvironment." *ACS Nano* 8 (4):3636–45.

Yalow, R. S., and Berson, S. A. 1960. "Immunoassay of Endogenous Plasma Insulin in Man." *Journal of Clinical Investigation* 39:1157–75.

Yan, Z., Yang, M., Wang, Z., et al. 2015. "A label-free immunosensor for detecting common acute lymphoblastic leukemia antigen (CD10) based on gold nanoparticles by quartz crystal microbalance." *Sens Actuators B Chem* 210:248–53.

Yuan, J., Duan, R., Yang, H., Luo, X., and Xi, M. 2012. "Detection of serum human epididymis secretory protein 4 in patients with ovarian cancer using a label-free biosensor based on localized surface plasmon resonance." *Intl J Nanomed* 7:2921–8.

Yuba, E., Kanda, Y., Yoshizaki, Y., et al. 2015. "pH-sensitive polymer-liposome-based antigen delivery systems potentiated with interferon-gamma gene lipoplex for efficient cancer immunotherapy." *Biomaterials* 67:214–24.

Yuyama, Y., Tsujimoto, M., Fujimoto, Y., and Oku, N. 2000. "Potential usage of thermosensitive liposomes for site-specific delivery of cytokines." *Cancer Lett* 155 (1):71–7.

Zhang, Y., Ge, L., Li, M., et al. 2014. "Flexible paper-based ZnO nanorod light-emitting diodes induced multiplexed photoelectrochemical immunoassay." *ChemComm (Cambridge, England)* 50:1417–9.

Zhao, Q., Duan, R., Yuan, J., et al. 2014. "A reusable localized surface plasmon resonance biosensor for quantitative detection of serum squamous cell carcinoma antigen in cervical cancer patients based on silver nanoparticles array." *Intl J Nanomed* 9:1097–104.

Zhao, W., Brook, M. A., and Li, Y. 2008. "Design of gold nanoparticle-based colorimetric biosensing assays." *Chembiochem* 9:2363–71.

Zhao, Y., Huo, M., Xu, Z., Wang, Y., and Huang, L. 2015. "Nanoparticle delivery of CDDO-Me remodels the tumor microenvironment and enhances vaccine therapy for melanoma." *Biomaterials* 68:54–66.

Zhen, Z., Tang, W., Wang, M., et al. 2017. "Protein nanocage mediated fibroblast-activation protein targeted photoimmunotherapy to enhance cytotoxic T-cell infiltration and tumor control." *Nano Lett* 17 (2):862–9.

Zheng, T. Y., Pierre-Pierre, N., Yan, X., et al. 2015. "Gold nanoparticle-enabled blood test for early stage cancer detection and risk assessment." *ACS Appl Mater Interface* 7 (12):6819–27.

Zheng, Y., Stephan, M. T., Gai, S. A., et al. 2013. "In vivo targeting of adoptively transferred T-cells with antibody- and cytokine-conjugated liposomes." *J Control Release* 172 (2):426–35.

Zhu, S., Niu, M., O'Mary, H., and Cui, Z. 2013. "Targeting of tumor-associated macrophages made possible by PEG-sheddable, mannose-modified nanoparticles." *Mol Pharm* 10 (9):3525–30.

Zindl, C. L., and Chaplin, D. D. 2010. "Immunology. tumor immune evasion." *Science* 328 (5979):697–8.

Zitvogel, L., Galluzzi, L., Kepp, O., Symth, M. J., and Kroemer, G. 2015. "Type I interferons in anticancer immunity." *Nat Rev Immunol* 15 (7):405–14.

Zou, W. 2006. "Regulatory T-cells, tumour immunity and immunotherapy." *Nat Rev Immunol* 6 (4):295–307.

Section III

Oligonucleotides and Gene-Based Therapies

11 Nucleic Acid Nanotherapeutics

Siddharth Jhunjhunwala

CONTENTS

11.1 INTRODUCTION

Nucleic acids are a group of biological macromolecules essential for life. In all living systems (barring viruses), two major types of nucleic acids are present—deoxyribonucleic acid (DNA) and ribonucleic acid (RNA). DNA is the primary carrier of information (in the form of genes) required for the proper functioning of the living system (a cell for example) and RNA functions to convert the information stored in DNA to proteins, although exceptions to these roles exist. Given the functional importance of nucleic acids in a cellular system, it has long been envisaged that they could be used as therapeutics for the treatment of a myriad of diseases including cancer.

Introduction of exogenous nucleic acids into a cell has the potential to alter genetic information in that cell, and this process is often described as gene therapy (Yin et al. 2014). Gene therapy has the potential to be used as a therapeutic for disease treatment, however, its use in the clinic has met with limited success due to challenges associated with delivering nucleic acids into cells. Nucleic acids may be delivered using biological systems (using a virus) or through synthetic systems (non-viral delivery vehicles [Blackburn et al. 2006]). Viral systems are capable of delivering nucleic acids with high efficacy, but are plagued with problems of safety and immunogenicity, and will not be discussed further in this chapter (for more information refer to [Merten and Al-Rubeai 2011]). Non-viral systems while safer and potentially easier to manufacture, generally have low efficacies (Kay 2011; Verma and Weitzman 2005). Advances in nanotechnology based non-viral systems could potentially overcome the aforementioned limitations, and hence improve clinical translational of gene therapy.

In this chapter, first a biological overview of the nucleic acids that are being used as therapeutics is provided. Next nanotechnology-based systems used for delivery of nucleic acids are described. Finally, the challenges associated with *in vivo* and intracellular delivery of nucleic acid nanotherapeutics and a summary of a few new and successful delivery systems that are capable of modulating gene expression are discussed.

11.2 NUCLEIC ACIDS AND THEIR USE AS CANCER THERAPEUTICS

11.2.1 Nucleic Acid Components and Structure

Nucleic acids are naturally occurring polymers that are made up of monomeric nucleotide units linked together by a phosphodiester bond. Nucleotide units in turn comprise three components, a nitrogenous base, a sugar (pentose), and a phosphate group. Five nitrogenous bases are commonly found in nucleotides present in cellular systems, with adenine (A), guanine (G), and cytosine (C) common to both DNA and RNA, and thymine (T) present in DNA while uracil (U) is present in RNA. The pentose sugar, which is a ribose molecule, is common to all nucleotides with one major difference; in DNA a hydroxyl group is absent from the 2' carbon of the ribose, while it is present in RNA. In both DNA and RNA, a phosphate group is attached to the 3' carbon of one nucleotide unit and 5' carbon of another nucleotide unit to form a phosphodiester bond, resulting in a sequence of nucleotide units linked together. Such a sequence of nucleotides is often called a strand.

A variety of bonding interactions may exist between intra-strand (within the same strand) and inter-strand (between two strands) nucleotides. These interactions lead to complex three-dimensional structures of nucleotide strands. Hydrogen bonding between nitrogenous bases is the primary inter-strand interaction, which results in the formation of a double-stranded nucleic acid. DNA molecules primarily exhibit a double-stranded structure. Contrastingly, RNA molecules are generally observed to be present as a single strand, with intra-strand interactions. Exceptions to these structures do exist. Single-stranded DNA (ssDNA) is observed as a temporary state in all cells during replication and transcription. Additionally, some viruses have their genetic material encoded as ssDNA. Similarly, certain class of viruses have their genetic material encoded as a double-stranded RNA (dsRNA). dsRNA may also be found in many cells as regulatory elements that silence the activity of many genes and will be discussed in greater detail later in this chapter.

11.2.2 Deoxyribonucleic Acid (DNA)

The DNA molecule has two major roles inside cellular systems. First, it is the molecule that forms the basis of information inheritance in prokaryotes and eukaryotes. Basically, the structure of DNA enables it to be faithfully replicated and passed on to progeny cells. Altering the sequence of nucleotides in the DNA of a cell results in the altered sequence being inherited by all of its progeny, a feature that may be exploited for therapeutic purposes. Second, the specific nucleotide sequence of DNA forms the code for the production of the workhorses of the cell—proteins. This trait makes DNA an attractive therapeutic in diseased conditions where a protein is either not being produced or is generated abnormally resulting in absence of function.

11.2.3 Ribonucleic Acid (RNA)

RNA molecules have a variety of functions in living systems. In most prokaryotic and eukaryotic systems, RNA molecules may be divided into three distinct types based on their functions: (i) sequences that provide the code for production of a specific poly-peptide chain (termed as messenger RNA or mRNA), (ii) sequences that help translate the code carried by mRNA to a poly-peptide (transfer RNA or tRNA), and (iii) sequences that form an essential part of the ribosomal enzyme machinery that catalyze the translation of the mRNA code to a poly-peptide sequence (termed as ribosomal RNA or rRNA). In addition to these primary functions, recent discoveries have resulted in the recognition of new roles for specific RNA molecules. For example, the short (or small) interfering RNA (siRNA) and microRNA (miRNA) have recently been shown to be involved in regulation of mRNA translation process in the cell. In addition, RNA molecules with other functions such as the long non-coding RNA (lncRNA), piwi interacting RNA (piRNA) and small nucleolar RNA (snoRNA) have also been identified. Of these numerous types of RNA, only siRNA and mRNA will be discussed in this chapter.

11.2.4 DNA and RNA as Cancer Therapeutics

Specifically in cancer, DNA molecules could potentially be used as a therapeutic in many different forms (Ginn et al. 2013; Kozielski, Rui, and Green 2016). In cancerous cells that arise due to the absence or aberrant production of the protein encoded by tumor suppressor genes (proteins that would normally limit the proliferative ability of the cell or induce apoptosis in abnormal cells), delivering the DNA sequence that codes for that specific protein could prevent cancerous cell proliferation or induce apoptosis. Alternately, DNA sequences that code for a protein (antigen) overexpressed in cancerous cells could be delivered to specific immune cells (antigen presenting cells). In turn, these immune cells will activate immune responses against that specific protein, which has the potential to result in immune cell-mediated killing of cancerous cells (DNA vaccines). Further, DNA sequences could be used to generate toxins that have the ability to specifically kill cells they are produced in. Delivering such molecules to cancerous cells has the potential to treat cancer.

Numerous other strategies are being developed for the use of DNA molecules as cancer therapeutics (Ortiz et al. 2012; Yin et al. 2014).

siRNA has the potential to be used in a variety of cancer treatment strategies due to its ability to regulate protein expression (Lytton-Jean et al. 2015; Whitehead, Langer, and Anderson 2009). An example of its potential use is the ability to suppress the expression of proteins derived from oncogenes that are the primary cause

TABLE 11.1
A Summary of Basic Characteristics of DNA, siRNA, mRNA That Would Be Relevant for Their Development as a Cancer Therapeutic

Property	DNA	siRNA	mRNA
Function	Stores genetic information that may be used to produce proteins or alter other cellular activities	Interferes with translational of mRNA, thereby suppressing protein production	Is translated into a specific protein
Location of Activity	Nucleus	Cytoplasm	Cytoplasm
Structure	Double stranded	Double stranded	Single stranded
Size	Large. Generally, hundred thousand to millions of base-pairs long	Small. Generally, 20–25 base-pairs long	Medium. Generally, a few thousand bases long
Stability (Storage)	Stable at room temperature for a few hours in appropriate buffers. Stable under refrigeration and freezing conditions for a longer time	Stable at room temperature for a short time, when stored dry. Stable under refrigeration and freezing conditions for a longer time	Generally unstable but may be stored at room temperature for brief time periods in appropriate buffers. Stable when frozen at or below -80°C
Stability (*in vivo*)	Stability in body fluids dependent on presence of DNase	Generally unstable, but base modification improves stability in body fluids	Unstable in body fluids. Base modification suggested, although remains questionable
In vivo Inflammatory Properties	Extracellular naked DNA can be highly inflammatory. Encapsulated forms are less inflammatory	siRNA made of native nucleotides is highly inflammatory. Base modification reduces immunogenicity	Naked mRNA sequences are highly inflammatory. Effect of base modification is unclear
Bases	Natural bases. Non-natural bases or modifications generally affect activity	Natural bases. Modified bases can be used with little to no effect on activity	Natural bases. Modified bases being tested for effect on activity
Replication Ability	Replicates in the nucleus. Has the potential to be inherited	Cannot replicate	Cannot replicate

of numerous cancers. Alternately, siRNA may also be used to silence the activity of regulatory immune cells, which restrain the activity of effector immune cells acting against aberrant cancerous cells. mRNA molecules also have the potential to be used in cancer treatment due to their ability to produce specific proteins. Similar to the therapeutic uses of DNA, mRNA may be used to synthesize proteins generated by tumor suppressor genes or be used to produce anticancer vaccines (Rausch et al. 2014).

While DNA, siRNA, and mRNA molecules may all be used as cancer therapeutics, they have very different stability, structural, and functional properties. A summary of a few of these basic properties is provided in Table 11.1. These individual characteristics will help us determine the nucleic acid that might be most suitable for a specific type of treatment of a specific type of cancer.

In addition to these nucleic acid molecules being explored as therapeutics, RNA sequences that are a part of the clustered regularly interspaced short palindromic repeats (CRISPR), single guide RNA (sgRNA), have also been suggested as potential cancer therapies. sgRNA coupled with the CRISPR associated protein-9 (Cas9) could be used for the modification of genes that have been aberrantly altered in cancerous cells (Sánchez-Rivera and Jacks 2015). Additionally, the CRISPR-Cas9 system has the potential for expanding research efforts in cancer biology, including the establishment of *in vitro* and *in vivo* models that accurately replicate human disease in the laboratory. While the possible applications of this technology are immense, clinical applications are still in nascent stages of development, and hence will not be discussed further in this chapter.

11.3 OVERVIEW OF NANOPARTICULATE SYSTEMS

A major challenge for most therapeutics has been the delivery of a sufficient dose of the active agent to the cancerous cells or their microenvironment. This challenge is magnified for nucleic acid therapeutics due to their unique physical properties (size and charge), relative lack of stability (when compared to small molecule drugs), and difficulty in providing large doses (cost and immunogenicity issues). Delivery vehicles, primarily nanotechnology-based offer a solution to this problem (Cho et al. 2008; Kozielski, Rui, and Green 2016). Delivery vehicles may broadly be sub-divided into two categories—biological (viral) and synthetic (non-viral). A majority of gene therapy clinical trials have utilized a virus as a delivery system for nucleic acids, and viral vectors have been discussed extensively in the several review articles (Thomas, Ehrhardt, and Kay 2003; Waehler, Russell, and Curiel 2007). Hence, viral nucleic acid delivery systems will not be described in this chapter. Synthetic delivery vehicles, a vast majority of which are nanoparticulate, have been extensively explored for the delivery of nucleic acids as well (Mintzer and Simanek 2009). In the next few subsections, each of these synthetic nano-delivery systems will be discussed.

11.3.1 Polymeric Nanoparticles

Polymeric nanoparticles are amongst the most commonly available delivery systems for a variety of therapeutics including nucleic acids (Pack et al. 2005). Polymers used to fabricate nucleic acid delivery systems can largely be divided based on the technique used to prepare nucleic acid nanoparticles: (i) direct encapsulation of nucleic acids inside the polymer matrix; or (ii) through interactions of cationic polymers with negatively charged nucleic acid that result in the formation of polyplexes; or (iii) through direct covalent conjugation of nucleic acids to the polymers (Figure 11.1).

Encapsulating nucleic acids into degradable polymer matrices has been studied as a potential delivery system. Poly (lactic-co-glycolic) acid (PLGA) has been the most commonly used polymeric matrix for this application (Tinsley-Bown et al. 2000; Woodrow et al. 2009) due to its biological compatibility and status as an FDA approved substance, but other degradable polymers have been employed as well (Little et al. 2004). However, the technique used for encapsulation in these cases employs a strategy that involves the formation of water-in-oil emulsions, which results in nucleic acids coming in contact with a water-oil interface. Such contact may result in loss of structure and function. Hence, alternate delivery strategies with relatively better efficacies have been developed.

The most common polymeric delivery vehicles for nucleic acids are polyplexes that are self-assembled structures formed by the condensation of negatively charged nucleic acids with positively charged polymers. Poly (L-lysine) (PLL) was one of the first cationic polymers to be used for DNA complexation and delivery (Wu and Wu 1987; Wu et al. 1988). However, due to its high cellular toxicity and low efficacy (primarily the absence of endosomal escape), PLL is not used extensively. To overcome some of these challenges, alternate cationic polymers were tested for their efficacy in condensing and delivering DNA. One of the most encouraging successes was polyethylenimine (PEI), which was shown to have the capability to deliver oligonucleotides by Boussif et. al. (Boussif et al. 1995). Since its first development and use, PEI has become the gold standard for nucleic acid delivery to cells both *in vitro* and *in vivo* (Neu, Fischer, and Kissel 2005; Mintzer and Simanek 2009). PEI may be synthesized as either a linear or branched polymer, and both versions are efficient at

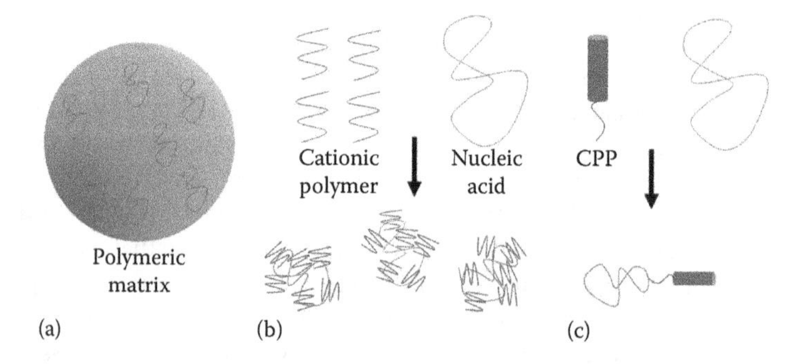

FIGURE 11.1 Schematic representation of nucleic acid—polymeric nanoparticles. (a) Nucleic acids encapsulated in a polymer matrix. (b) Self-assembled particulate formed by the condensation of a cationic polymer with a nucleic acid. (c) Conjugate particulate formed when a cell-penetrating peptide (cpp) is covalently linked to a nucleic acid.

intracellular nucleic acid delivery. PEI's excellent transfection efficacy is the result of high charge density associated with the numerous nitrogen atoms in its structure, which enables endosomal escape inside the cell (Godbey, Wu, and Mikos 1999). The high charge density on PEI is thought to enable osmotic swelling of endosomes that have taken up PEI-nucleic acid complexes, resulting in bursting of the endosomal membrane and escape of the delivery vehicle into the cytoplasm (termed as the proton sponge effect) (Behr 1997; Akinc et al. 2005). The success of PEI as delivery vehicle for nucleic acids resulted in its development into commercial reagents for gene delivery (marketed for laboratory use only) as well as therapeutics that are in clinical trials (Yin et al. 2014). Nevertheless, PEI is not degradable in the body and its accumulation has been shown to result in cellular toxicity. Hence, polymeric substitutes that retain the cationic charge but contain groups that are degradable have been developed. The alternatives include modified PEI with disulfide or ester linkages, cyclodextrin-based polymers, chitosan, polyamidoamine dendrimers, and poly (beta-amino-ester) (PBAE) among others (Mintzer and Simanek 2009; Haensler and Szoka 1993; Lynn and Langer 2000). Of these, PBAE-nucleic acid nanoparticles have been shown to have better efficiency as well as lower toxicities than PEI, and could potentially be designed to target specific cell types (Kozielski, Rui, and Green 2016). Alternately, polymers containing both a hydrophilic and a hydrophobic region that self-assemble into micellar structures (polymeric micelles) in aqueous solutions (Torchilin 2001)are also widely studied for *in vivo* drug delivery applications. Polymeric micelles may be designed to have multiple functionalities that enable effective drug encapsulation, protection from harsh *in vivo* environments, and stimuli-responsiveness. While these systems have primarily been designed for the delivery of hydrophobic small molecules that are trapped in the hydrophobic core of the micelle, they have also been tested as delivery systems for siRNA. The primary strategy employed for entrapment of siRNA in polymeric micelles is the formation of ionic polyplexes, as discussed above (Jhaveri and Torchilin 2014).

Direct covalent conjugation of polymers with oligonucleotides has also been developed as a technique for nucleic acid delivery. The most common polymers that are used for such delivery vehicles are polypeptides. Polypeptides rich in basic amino acid residues have been shown the have the ability to directly cross the cellular plasma membrane, and hence have been termed as cell-penetrating peptides (CPP). Numerous types of CPP have been identified and developed (Gupta, Levchenko, and Torchilin 2005). Most of them contain a peptide sequence rich in lysine and arginine amino acids, and can be naturally occurring or synthetically designed (Karagiannis et al. 2013). For delivery of oligonucleotides, such as siRNA, the CPP are generally covalently conjugated to the nucleic acid sequence. For delivery of larger molecules such as DNA and mRNA, while covalent conjugation might be feasible, most commonly condensed nanocomplexes of CPP-nucleic acid are developed and have been shown to have high efficacy both *in vitro* and *in vivo.* However, a major challenge with currently available CPPs is that the peptide sequences are mostly non-human, which could result in immune responses when used in humans (Karagiannis et al. 2013). Other polymer-oligonucleotide conjugates have also been developed, and one of the clinically advanced candidates is the dynamic polyconjugates (DPC) (Rozema et al. 2007).

11.3.2 Lipid-Based Nanoparticles

Certain lipids can self-assemble in aqueous solvents to form nanoparticulate vesicles that are generally referred to as liposomes. Liposomes are among the earliest and most commonly used nanoparticulate systems for nucleic acid delivery (Mintzer and Simanek 2009; Yin et al. 2014). Liposomes are classified based on either (i) the number of lipid bilayers present- uni-lamellar vesicles (having one bilayer) or multi-lamellar vesicles (having two or more bilayers), or (ii) the ionic characteristics of the lipids used, cationic or neutral. Numerous cationic and neutral lipid reagents (structures in Figure 11.2) are now commercially available for use as nucleic acid delivery systems in the laboratory. A majority of the cationic lipid molecules are similar and are essentially made of three parts: the head group (which is cationic or neutral), the tail group (which is hydrophobic), and linker between these two groups. Modifications to the head group or the tail group results in the generation of different cationic lipid molecules used to prepare liposomes. Mintzer and Simanek describe these modifications in detail (Mintzer and Simanek 2009). Neutral lipid molecules are generally not used in isolation to make liposomes, but are used in combination with other lipids where they help to improve liposome stability and potentially transfection efficacy (Li and Szoka 2007).

While individual lipid molecules have shown efficacy in nucleic acid delivery, most common liposomal formulations are a combination of a variety of lipids. These lipids consist of naturally occurring lipids as well synthetic lipids (Whitehead, Langer, and Anderson 2009). Liposomes made of a mixture of lipids are referred to as stable nucleic acid lipid particles (SNALP) (Zimmermann et al. 2006; Semple et al. 2010). While initially such particles were used for DNA delivery (Wheeler et al. 1999), they were adapted and have been optimized for siRNA delivery (Yin et al. 2014). Although the exact mechanisms of *in vivo* efficacy of these formulations remain to be determined, it is suggested that the serum proteins, especially the apo-lipoproteins, bind to the formulations to affect their *in vivo* uptake. Using these formulations, several clinical trials have been initiated, which have been recently described by Anderson and colleagues (Yin et al. 2014; Lytton-Jean et al. 2015).

Although liposomal nanoparticles are excellent vehicles for nucleic acid delivery, safety issues relating to the use of lipid molecules remain. Toxicity and immunogenicity are the primary safety concerns. Studies have shown that cationic lipid head groups cause cellular toxicity (S. Li et al. 1998), and may result in proinflammatory interferon secretion by immune cells (Ma et al. 2005). Strategies to overcome these concerns are currently being explored, which include PEGylation of the liposomes or modification with other biological polymers that are more biologically compatible. Another approach that has been developed to move past the issue of toxicity is the use of lipid-like materials called lipidoids (Akinc et al. 2008). Many of these modified liposomal formulations have shown remarkable potency in different animal models and are currently being explored as potential therapeutics (Lytton-Jean et al. 2015).

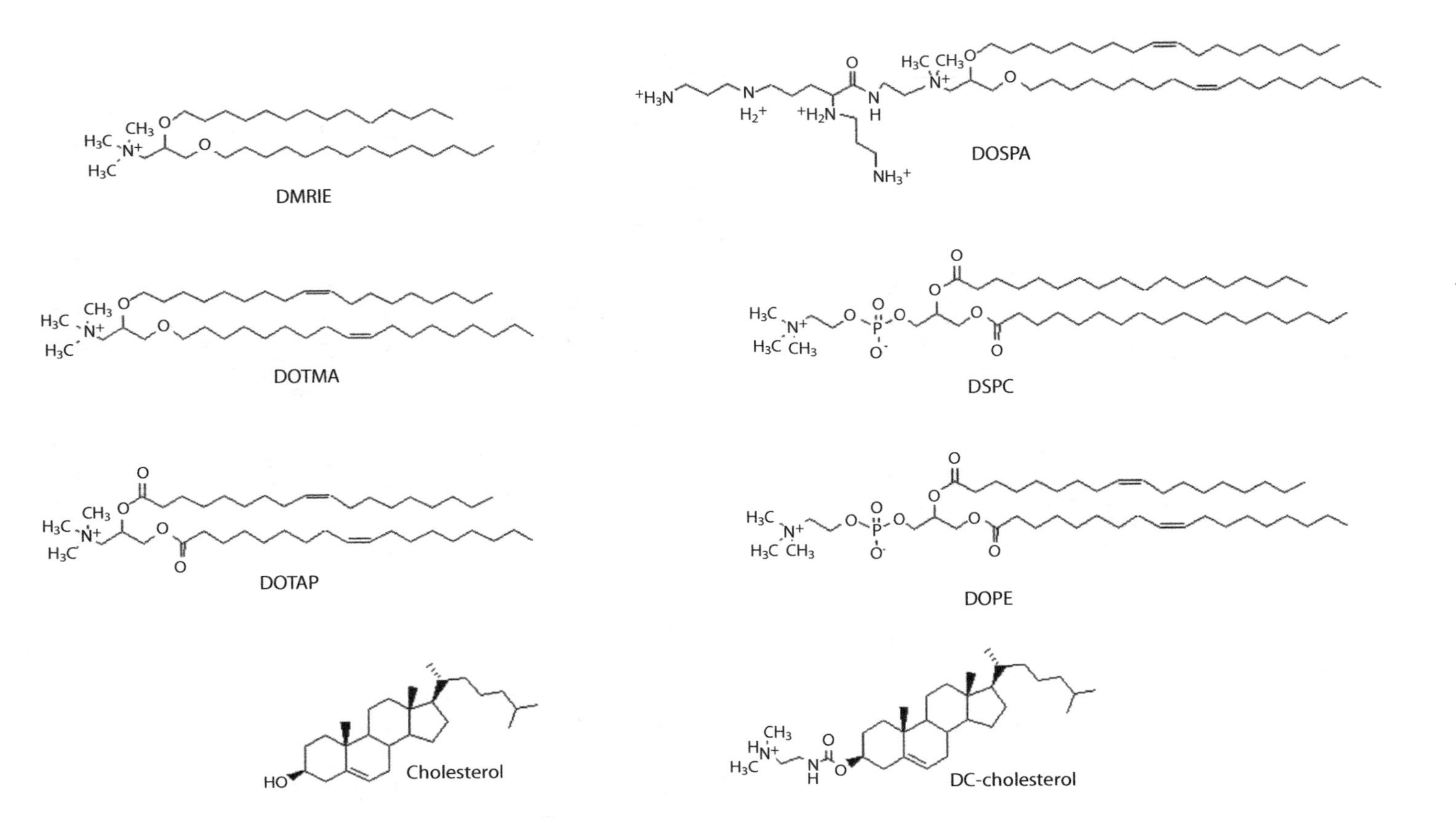

FIGURE 11.2 Chemical structures of some of the most commonly used cationic and neutral lipid molecules for the delivery of nucleic acids. Adapted and reprinted with permission from Macmillan Publishers Ltd.: Nature Reviews Genetics (Yin et al. 2014) Nature Publishing Group.

11.3.3 Nucleic Acid Conjugates

In Section 11.3.1 covalent conjugation of nucleic acids with polymers was discussed. Similar formulations involving direct conjugation of nucleic acids with small molecules have also been developed. Technically these conjugates are not nanoparticles but have been shown to have high efficacy in targeting specific cell types *in vivo*. One of the most clinically advanced conjugates is nucleic acids to N-acetylgalactosamine (GalNAc). GalNAc acts as a ligand that specifically targets hepatocytes, thereby delivering high doses of nucleic acids to these cells. Numerous nucleic acid therapeutics involving conjugation with GalNAc are under clinical trials for a variety of indications (Kanasty et al. 2013). Recent work by Zhang and colleagues shows that carbon particle-based conjugates may also be effective in delivering short nucleic acid sequences like siRNA (L. Zhang et al. 2016). Further, not all nucleic acid conjugates are developed for the purpose of delivering nucleic acids. A classic example is protein-nucleic acid conjugates that are being explored as vaccine candidates, where the protein is an antigen and the nucleic acid is the adjuvant capable of activating immune responses (Bode et al. 2011). Along these lines, Walker and colleagues have fabricated a new oligodeoxynucleotide-antigen conjugate that has the potential to be developed as an antitumor vaccine (Kramer et al. 2017).

11.3.4 Metal Nanoparticles

Metallic nanoparticles have been described as potential nucleic acid delivery vehicles for numerous decades. The earliest reports demonstrate the use of direct particle bombardment (using shock waves) or particle acceleration devices for the delivery of nucleic acids associated with gold nanoparticles (Mintzer and Simanek 2009). However, these techniques were plagued with the problem of causing tissue damage and being unable to deliver to deep tissues in the body. Alternate strategies were developed that involved modification of gold nanoparticle surfaces such that they would readily be taken up by cells through endocytic processes. Rotello and colleagues used this strategy to demonstrate excellent efficacy in delivering covalently conjugated DNA strands to cells (Sandhu et al. 2002; Ding et al. 2014). Other groups have improved upon these techniques using polymeric conjugates to gold nanoparticles that either aid in cellular uptake or improve intracellular detachment of nucleic acids from the gold nanoparticles. Mirkin and colleagues applied similar principles to deliver siRNA to cells using gold nanoparticles. One of the advances made in siRNA delivery was the arrangement of the oligonucleotides in a dense, spherical fashion around the gold particle to improve delivery efficiency. These particulates, termed as spherical nucleic acids have shown tremendous potential for treatment of multiple disorders (Rosi et al. 2006; Cutler, Auyeung, and Mirkin 2012). Similarly, iron oxide nanoparticles have been developed for nucleic acid delivery. Significant advantage of using iron oxide particles is that they have been approved by the US Food and Drug Administration (FDA) for human use as diagnostic agents, they are biologically compatible, they can be visualized using current imaging techniques, and they may be guided using external magnetic fields to specific tissue sites. Nevertheless, these metal nanoparticles are yet to be tested in clinical trials for the delivery of nucleic acids.

In addition to these, metal-based quantum dot nanoparticles have also been used for nucleic acid transfection. The first report on using quantum dot (CdSe/ZnS) for DNA delivery was by Burgess and colleagues (Srinivasan et al. 2006), where they directly conjugated DNA onto lipid-coated quantum dots. Following this approach, self-assembled nanoparticles formed by non-covalent association of DNA with cationic polymer-coated quantum dots has also been attempted (Zhang and Liu 2010). Bhatia and colleagues have shown that similar techniques can be used for siRNA delivery to cancerous cells using quantum dots (Derfus et al. 2007). While quantum dots provide numerous advantages for imaging and tracking of the nucleic acid, questions regarding their efficacy and *in vivo* translatability remain.

11.3.5 Other Nanoparticulate Systems

Inorganic nanoparticles are another class of delivery vehicle that have been developed for nucleic acid delivery, of which silica-based systems are the most widely studied. Silica nanoparticles are attractive due to numerous advantages: (i) nucleic acids can be directly conjugated to them using various chemistries, (ii) porous particles may be prepared in facile manner that allows for encapsulation of nucleic acids, (iii) they have an affinity to bind to lipid molecules on membranes that have the potential to enhance cellular uptake, and (iv) they are biologically compatible and stable both *in vitro* and *in vivo*. Numerous silica-based nucleic acid delivery systems are under development in many laboratories across the world, with potential for clinical translation.

Nucleic acids may also be made to self-assemble into nanoparticles. The ability of nucleic acid strands to form double-stranded structures due to hydrogen bonding between bases allows for the fabrication of nanoparticles made strictly of nucleic acids (Smith et al. 2013). While this technique could be used for both DNA and siRNA molecules (but not for mRNA as it is single stranded), successful, and efficacious intracellular delivery has been shown with siRNA (Shu et al. 2011).

Another system for nucleic acid delivery that has shown promise in laboratory experiments is the use of carbon nanotubes (CNT). Similar to the inorganic and metal nanoparticulate systems, modification of CNT surfaces with polymers enables direct conjugation of nucleic acids to them. Additionally, CNT surfaces may be modified to become cationic, which allows for self-assembly of cationic-CNT and negatively charged nucleic acids. While CNTs *in vitro* transfection ability has never been in doubt, they have also been shown to be capable of delivering siRNA *in vivo* to cancerous cells (Zhang et al. 2006). Nevertheless, questions regarding safety and immunogenicity of CNTs remain, which has affected their clinical translation.

11.4 CHALLENGES ASSOCIATED WITH DELIVERY OF NUCLEIC ACIDS

A majority of drugs taken by humans today are in the form of oral formulations. Unfortunately, most biological drugs (biologics), including nucleic acid therapeutics, cannot be administered to individuals through the oral route. Two main reasons are that nucleic acids are likely to be degraded by the hostile environment of the digestive track (pH and presence of degradative enzymes), and intestinal cells find it challenging to directly absorb (take up into the body) large macromolecules such as nucleic acids. Hence, alternate routes of delivery are being explored. Parenteral delivery, via intravenous, subcutaneous or intra-muscular means, remains one of the most viable options for nucleic acid therapeutics that cannot be taken up through the alimentary canal. However, each of these routes of administration is associated with its own challenges.

Delivery of nucleic acid therapeutics through any of the aforementioned routes results in their delivery to the extracellular space. The human body has evolved to perceive extracellular nucleic acids as a threat from a pathogenic agent or to be associated with a dead cell, both of which result in degradation. Key players in the recognition and degradation process are the body's immune cells. Innate immune cells have numerous cell surface receptors (termed as toll-like receptors or TLRs) that recognize nucleic acids as a threat, resulting in cellular activation. Activated innate immune cells are not only capable of directly taking up and degrading the extracellular nucleic acids, but also result in local or systemic inflammation. Recent progress in the area of chemical modifications to siRNA has helped reduce both degradation and immunogenicity, but the same principles have not worked for mRNA as these modifications make it ineffective. Similar strategies are also not feasible for DNA molecules. To avoid the problems of degradation and immunogenicity, nucleic acid therapeutics need be protected from directly encountering the immune system, which may be achieved using nanoparticle technologies.

In the previous section, numerous nanoparticulate systems for packaging and delivering nucleic acids were described. While these systems, can be effective in surmounting the aforementioned challenges, their have other barriers to overcome. Broadly, nucleic acid nanoparticulates should be capable of the following tasks: (i) delivery to the right cells/tissues (cancer cells or microenvironment) *in vivo*, and (ii) delivery to the right intracellular compartment. Numerous obstacles exist to achieving these tasks (Figure 11.3), as described lucidly by Whitehead et. al. (Whitehead, Langer, and Anderson 2009) for siRNA therapeutics (similar challenges faced by DNA and mRNA therapeutics).

11.4.1 *In Vivo* Targeting

Nucleic acid nanoparticle therapeutics may be directly injected into localized tumors that are easily accessible (for example, localized skin tumors). However, a majority of tumors reside in or have spread to tissues that are not easily accessible. Targeting refers to specifically enriching the therapeutic in these tissues, that may be achieved through many different means.

Often systemic administration (through the bloodstream) is employed to achieve delivery of therapeutic to the tissues deep within the body. Systemic delivery leads to interaction of the nanoparticles with various systems in the body that might result in their degradation. For example, following entry into the bloodstream the particles may be subjected to, (i) coating with serum proteins that might lead to aggregation, (ii) degradation by nucleases present in the serum, (iii) uptake by phagocytic immune cells in circulation, (iv) uptake by phagocytic cells in the liver, and (v) filtration by

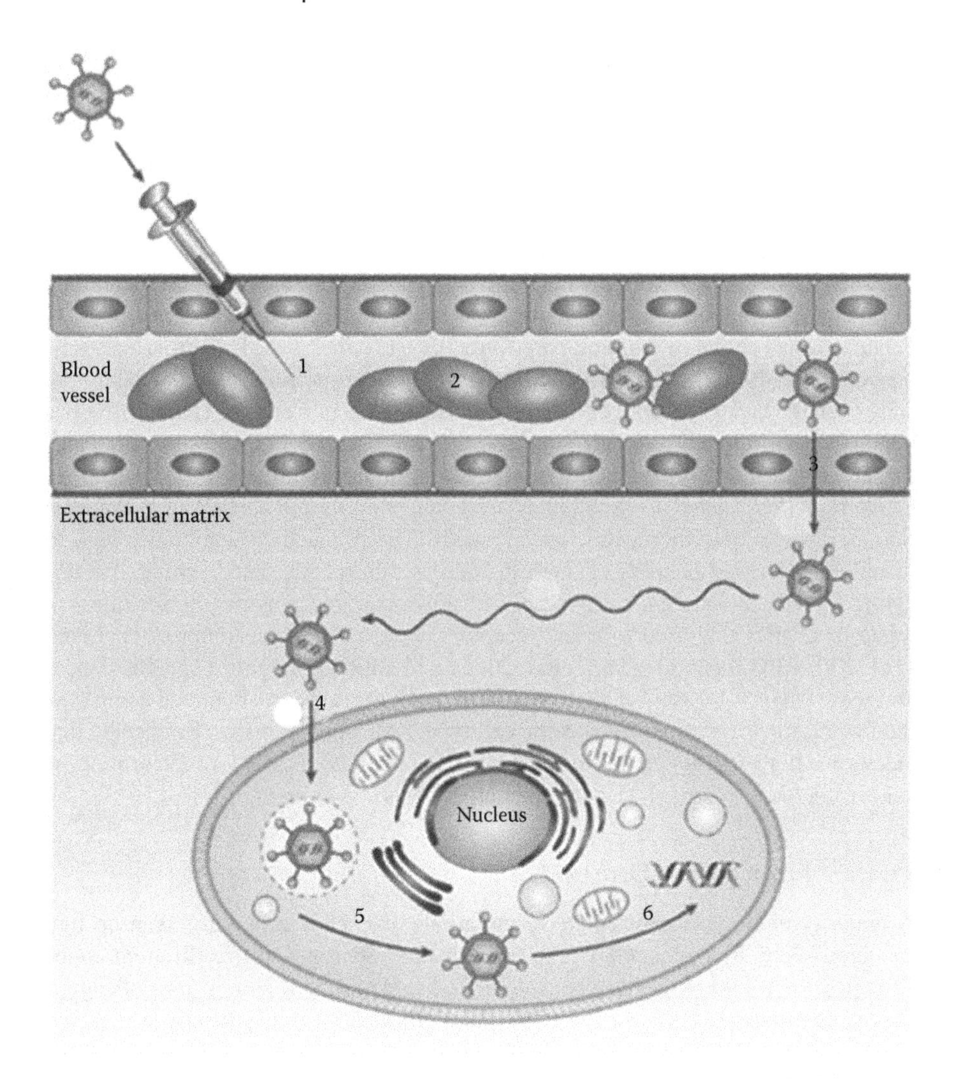

FIGURE 11.3 Barriers to delivery of nucleic acid nanoparticles. the obstacles to delivery are: (1) serum protein coat that promotes particle aggregation, degradation by serum nucleases and filtration by the kidney; (2) phagocytic uptake by immune cells; (3) traverse the endothelium; (4) getting past the cell membrane; (5) escaping the endosome; and (6) unpacking of the nanoparticle. Adapted and reprinted by permission from Macmillan Publishers Ltd.: Nature Reviews Drug Discovery (Whitehead et al. 2009) Nature Publishing Group.

the kidney. A common method to avoid these off-target interactions of nanoparticles has been the use of surface PEGylation (similar to the technique used for certain biologics administered intravenously) (Jokerst et al. 2011). If the nanoparticulate system is able to overcome these challenges, then it remains in circulation and may accumulate in tissues with elevated vascular supply. Accumulation has been shown to occur primarily in organs such as lung, liver, and spleen, but it has been suggested that similar high levels of build-up may be achieved in tumors due to the presence of vasculature that is not fully developed and hence leaky. This phenomenon is termed as the Enhanced Permeation and Retention (EPR) Effect (Matsumura and Maeda 1986; Maeda 2010). The EPR effect is a passive-targeting strategy that utilizes natural *in vivo* processes for increasing the concentration of a nanotherapeutic at the site of interest (tumors in this case).

A recent study by Chan and colleagues, however, questions the validity of the EPR effect. They show that among numerous published studies with sufficient data to quantify the number of nanoparticles in the tumor, the median percentage of nanoparticles present in the tumor was only 0.7% (Wilhelm et al. 2016). This study, coupled with the lack of translation of the EPR effect in clinical studies (and possibly even the lack of its effect in certain mouse models of research) (Park 2016), has resulted in an urgency to develop better targeting modalities. Alternate strategies for targeting nanotherapeutics involve the use of an active-targeting mechanism. In active targeting, ligands or antibodies are conjugated on the surface of a nanoparticle or directly onto therapeutic nucleic acids. These ligands or antibodies bind to receptors expressed on cells of interest, such as tumor cells. Successful active targeting ligands include GalNAc (discussed previously but not specific for tumors), folate molecules that bind to folate receptors overexpressed on tumor cells (Liang et al. 2008), RGD peptide (Park et al. 2012), and antibodies against specific cancer antigens (Palanca-Wessels et al. 2016). Targeting has also been attempted using synthetic molecules that bind to a specific cell type. An example of such a molecule that has shown high efficacy in delivering siRNA to endothelial cells (also present in the cancer microenvironment) (Dahlman et al. 2014) is shown in Figure 11.4.

11.4.2 Intracellular Delivery

A majority of the nanoparticulate formulations of nucleic acids are taken up into cancerous cells by endocytosis. The cells of interest might utilize different endocytic processes for uptake have been described in detail by Kabanov and colleagues (Sahay, Alakhova, and Kabanov 2010). The process of endocytosis results in the therapeutic being entrapped in an endosome. Endosomes undergo a maturation process inside the cell, which is a result of fusion with other intracellular vesicles. The maturation process results in the lowering of intra-endosomal pH and the accumulation of degradative enzymes. These conditions are detrimental for nucleic acid and nanoparticles, leading to degradation of the nucleic acids resulting in lowering of therapeutic efficacy. Additionally, endosomes eventually fuse with the lysosomes, which have an even lower pH and greater number of degradative enzymes. Further, the nanoparticulate cargo in the endosomes may also be recycled and secreted out of the cell through exosomes. Hence, many current-generation nanoparticulates have a built-in capability to lyse endosomes and escape into the cytoplasm (as discussed in Section 11.3.1 on cationic polymeric particles and the proton sponge effect) or are designed to avoid the endosomal uptake process. Avoiding the endosomal uptake process may be achieved through direct fusion with the membrane (for example with use of CPP) or through the creation of temporary pores on the cell membrane that enable nanoparticulate entry through diffusion (Sharei et al. 2013).

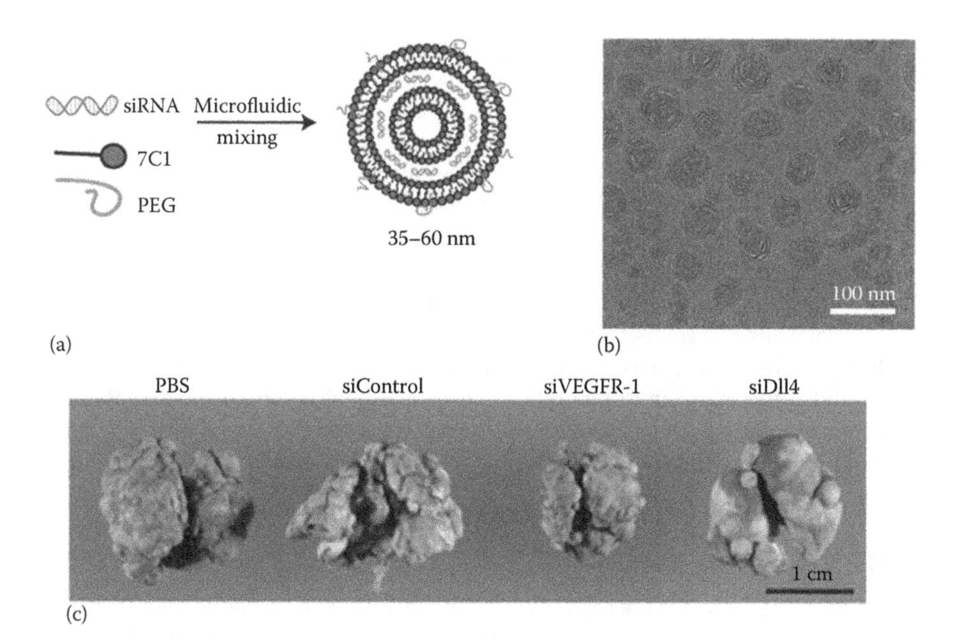

FIGURE 11.4 Nanoparticle system for targeting of endothelial cells. (a) Scheme showing a system for the nanoencapsulation of sirna using lipid-polymer hybrids (7cl being one of the most potent of these molecules). (b) Electron micrograph of the lipid-polymer nanoparticle encapsulating sirna. (c) Images of murine lungs with metastatic tumors (lewis lung carcinoma) following treatment, showing that sirna-based silencing of specific molecules helps reduce the number of metastatic lesions. Reprinted by permission from Macmillan Publishers Ltd.: Nature Nanotechnology (Dahlman et al. 2014) Nature Publishing Group.

siRNA and mRNA therapeutics are able to exert their action in the cytoplasm, so this is their intracellular target. However, the nanoparticulate system used for packaging needs to dissociate itself from the RNA molecules for them to function effectively. While some nanoparticles are designed to fall apart in the cytoplasm or during the endosomal escape process, the exact unpacking mechanisms remain unclear for many nanoparticulate systems. For DNA therapeutics, the intracellular target is the nucleus. In many cells (including cancer cells), the efficacy of the DNA therapeutic is highest when the cell is dividing, as the nuclear envelope has degraded during mitosis. However, this passive mechanism is insufficient for potent DNA delivery to cancer cells. In addition to using such a passive mechanism, strategies such as nuclear targeting sequences and utilization of intra-nuclear protein transport mechanisms have been attempted (Dean, Strong, and Zimmer 2005).

11.5 FUTURE OUTLOOK

The genomics revolution coupled with major advances in tumor biology is providing us with information on new targets for the treatment of a variety of cancers. Nucleic acid therapeutics and their delivery systems will have a prominent role to play in the translation of this data into clinically viable treatment strategies. As discussed in this chapter, numerous nanoparticle-based systems have been designed for nucleic acid delivery, several of which are currently in clinical trials (reviewed in [Yin et al. 2014; Lytton-Jean et al. 2015]). While this fact demonstrates the progress has been made since the development of the first laboratory systems for gene therapy, continued efforts are still required for design of safer systems with higher efficacy.

The ideal delivery system for nucleic acids should be able to package potent doses, overcome the *in vivo* and intracellular barriers for delivery, precisely target the cancerous cells, and avoid the problems of immunogenicity and toxicity. To achieve this goal, we need to better understand the chemical properties of a variety of nanoparticulate systems that would allow us to package increasing amounts of nucleic acid, improve *in vivo* targeting, as well as characterize the biology behind intracellular targeting and comprehend the sources of immunogenicity. Recent innovative work in each of these areas is promising, and the prospect of nanoparticulate-based gene therapy system for the treatment of cancer in humans is within reach.

REFERENCES

Akinc, A., M. Thomas, A. M. Klibanov, et al. 2005. "Exploring Polyethylenimine-Mediated DNA Transfection and the Proton Sponge Hypothesis." *The Journal of Gene Medicine* 7 (5): 657–63.

Akinc, A., A. Zumbuehl, M. Goldberg, et al. 2008. "A Combinatorial Library of Lipid-like Materials for Delivery of RNAi Therapeutics." *Nature Biotechnology* 26 (5): 561–69.

Behr, J.-P. 1997. "The Proton Sponge: A Trick to Enter Cells the Viruses Did Not Exploit." *Chimia International Journal for Chemistry* 51 (1–2): 34–36.

Blackburn, G. M., M. J. Gait, D. Loakes, et al., eds. 2006. *Nucleic Acids in Chemistry and Biology*. 3rd ed. Cambridge: Royal Society of Chemistry. http://ebook.rsc.org/?DOI=10.1039/9781847555380.

Bode, C., G. Zhao, F. Steinhagen, et al. 2011. "CpG DNA as a Vaccine Adjuvant." *Expert Review of Vaccines* 10 (4): 499–511.

Boussif, O., F. Lezoualc'h, M. A. Zanta, et al. 1995. "A Versatile Vector for Gene and Oligonucleotide Transfer into Cells in Culture and in Vivo: Polyethylenimine." *Proceedings of the National Academy of Sciences of the United States of America* 92 (16): 7297–301.

Cho, K., X. Wang, S. Nie, et al. 2008. "Therapeutic Nanoparticles for Drug Delivery in Cancer." *Clinical Cancer Research* 14 (5): 1310–16.

Cutler, J. I., E. Auyeung, and C. A. Mirkin. 2012. "Spherical Nucleic Acids." *Journal of the American Chemical Society* 134 (3): 1376–91.

Dahlman, J. E., C. Barnes, O. F. Khan, et al. 2014. "In Vivo Endothelial siRNA Delivery Using Polymeric Nanoparticles with Low Molecular Weight." *Nature Nanotechnology* 9 (8): 648–55.

Dean, D. A., D. D. Strong, and W. E. Zimmer. 2005. "Nuclear Entry of Nonviral Vectors." *Gene Therapy* 12 (11): 881–90.

Derfus, A. M., A. A. Chen, D.-H. Min, et al. 2007. "Targeted Quantum Dot Conjugates for siRNA Delivery." *Bioconjugate Chemistry* 18 (5): 1391–96.

Ding, Y., Z. Jiang, K. Saha, et al. 2014. "Gold Nanoparticles for Nucleic Acid Delivery." *Molecular Therapy* 22 (6): 1075–83.

Ginn, S. L., I. E. Alexander, M. L. Edelstein, et al. 2013. "Gene Therapy Clinical Trials Worldwide to 2012 - an Update: Gene Therapy Clinical Trials Worldwide to 2012—An Update." *The Journal of Gene Medicine* 15 (2): 65–77.

Godbey, W. T., K. K. Wu, and A. G. Mikos. 1999. "Tracking the Intracellular Path of Poly(ethylenimine)/DNA Complexes for Gene Delivery." *Proceedings of the National Academy of Sciences of the United States of America* 96 (9): 5177–81.

Gupta, B., T. Levchenko, and V. Torchilin. 2005. "Intracellular Delivery of Large Molecules and Small Particles by Cell-Penetrating Proteins and Peptides." *Advanced Drug Delivery Reviews* 57 (4): 637–51.

Haensler, J., and F. C. Szoka. 1993. "Polyamidoamine Cascade Polymers Mediate Efficient Transfection of Cells in Culture." *Bioconjugate Chemistry* 4 (5): 372–9.

Jhaveri, A. M., and V. P. Torchilin. 2014. "Multifunctional Polymeric Micelles for Delivery of Drugs and siRNA." *Frontiers in Pharmacology* 5: 77.

Jokerst, J. V., T. Lobovkina, R. N. Zare, et al. 2011. "Nanoparticle PEGylation for Imaging and Therapy." *Nanomedicine* 6 (4): 715–28.

Kanasty, R., J. R. Dorkin, A. Vegas, et al. 2013. "Delivery Materials for siRNA Therapeutics." *Nature Materials* 12 (11): 967–77.

Karagiannis, E. D., A. M. Urbanska, G. Sahay, et al. 2013. "Rational Design of a Biomimetic Cell Penetrating Peptide Library." *ACS Nano* 7 (10): 8616–26.

Kay, M. A. 2011. "State-of-the-Art Gene-Based Therapies: The Road Ahead." *Nature Reviews Genetics* 12 (5): 316–28.

Kozielski, K. L., Y. Rui, and J. J. Green. 2016. "Non-Viral Nucleic Acid Containing Nanoparticles as Cancer Therapeutics." *Expert Opinion on Drug Delivery* 13 (10): 1475–87.

Kramer, K., N. J. Shields, V. Poppe, et al. 2017. "Intracellular Cleavable CpG Oligodeoxynucleotide-Antigen Conjugate Enhances Anti-Tumor Immunity." *Molecular Therapy* 25 (1): 62–70.

Li, S., M. A. Rizzo, S. Bhattacharya, et al. 1998. "Characterization of Cationic Lipid-protamine–DNA (LPD) Complexes for Intravenous Gene Delivery." *Gene Therapy* 5 (7): 930–37.

Li, W., and F. C. Szoka. 2007. "Lipid-Based Nanoparticles for Nucleic Acid Delivery." *Pharmaceutical Research* 24 (3): 438–49.

Liang, B., M.-L. He, Z.-P. Xiao, et al. 2008. "Synthesis and Characterization of Folate-PEG-Grafted-Hyperbranched-PEI for Tumor-Targeted Gene Delivery." *Biochemical and Biophysical Research Communications* 367 (4): 874–80.

Little, S. R., D. M. Lynn, Q. Ge, et al. 2004. "From the Cover: Poly-Amino Ester-Containing Microparticles Enhance the Activity of Nonviral Genetic Vaccines." *Proceedings of the National Academy of Sciences* 101 (26): 9534–39.

Lynn, D. M., and R. Langer. 2000. "Degradable Poly(β-Amino Esters): Synthesis, Characterization, and Self-Assembly with Plasmid DNA." *Journal of the American Chemical Society* 122 (44): 10761–68.

Lytton-Jean, A. K. R., K. J. Kauffman, J. C. Kaczmarek, et al. 2015. "Cancer Nanotherapeutics in Clinical Trials." In *Nanotechnology-Based Precision Tools for the Detection and Treatment of Cancer*, edited by C. A. Mirkin, T. J. Meade, S. H. Petrosko, et al., 166: 293–322. Cham: Springer International Publishing. 3.

Ma, Z., J. Li, F. He, et al. 2005. "Cationic Lipids Enhance siRNA-Mediated Interferon Response in Mice." *Biochemical and Biophysical Research Communications* 330 (3): 755–59.

Maeda, H. 2010. "Tumor-Selective Delivery of Macromolecular Drugs via the EPR Effect: Background and Future Prospects." *Bioconjugate Chemistry* 21 (5): 797–802.

Matsumura, Y., and H. Maeda. 1986. "A New Concept for Macromolecular Therapeutics in Cancer Chemotherapy: Mechanism of Tumoritropic Accumulation of Proteins and the Antitumor Agent Smancs." *Cancer Research* 46 (12): 6387–92.

Merten, O.-W., and M. Al-Rubeai, eds. 2011. *Viral Vectors for Gene Therapy*. Vol. 737. Methods in Molecular Biology. Totowa, NJ: Humana Press.

Mintzer, M. A., and E. E. Simanek. 2009. "Nonviral Vectors for Gene Delivery." *Chemical Reviews* 109 (2): 259–302.

Neu, M., D. Fischer, and T. Kissel. 2005. "Recent Advances in Rational Gene Transfer Vector Design Based on Poly(ethylene Imine) and Its Derivatives." *The Journal of Gene Medicine* 7 (8): 992–1009.

Ortiz, R., C. Melguizo, J. Prados, et al. 2012. "New Gene Therapy Strategies for Cancer Treatment: A Review of Recent Patents." *Recent Patents on Anti-Cancer Drug Discovery* 7 (3): 297–312.

Pack, D. W., A. S. Hoffman, S. Pun, et al. 2005. "Design and Development of Polymers for Gene Delivery." *Nature Reviews Drug Discovery* 4 (7): 581–93.

Palanca-Wessels, M. C., G. C. Booth, et al. 2016. "Antibody Targeting Facilitates Effective Intratumoral siRNA Nanoparticle Delivery to HER2-Overexpressing Cancer Cells." *Oncotarget* 7(8): 9561–75.

Park, J, K. Singha, S. Son, et al. 2012. "A Review of RGD-Functionalized Nonviral Gene Delivery Vectors for Cancer Therapy." *Cancer Gene Therapy* 19 (11): 741–48.

Park, K. 2016. "Drug Delivery of the Future: Chasing the Invisible Gorilla." *Journal of Controlled Release* 240: 2–8.

Rausch, S., C. Schwentner, A. Stenzl, et al. 2014. "mRNA Vaccine CV9103 and CV9104 for the Treatment of Prostate Cancer." *Human Vaccines & Immunotherapeutics* 10 (11): 3146–52.

Rosi, N. L., D. A. Giljohann, C.S. Thaxton, et al. 2006. "Oligonucleotide-Modified Gold Nanoparticles for Intracellular Gene Regulation." *Science* 312 (5776): 1027–30.

Rozema, D. B., D. L. Lewis, D. H. Wakefield, et al. 2007. "Dynamic PolyConjugates for Targeted in Vivo Delivery of siRNA to Hepatocytes." *Proceedings of the National Academy of Sciences* 104 (32): 12982–7.

Sahay, G., D. Y. Alakhova, and A. V. Kabanov. 2010. "Endocytosis of Nanomedicines." *Journal of Controlled Release* 145 (3): 182–95.

Sánchez-Rivera, F. J., and T. Jacks. 2015. "Applications of the CRISPR–Cas9 System in Cancer Biology." *Nature Reviews Cancer* 15 (7): 387–95.

Sandhu, K. K., C. M. McIntosh, J. M. Simard, et al. 2002. "Gold Nanoparticle-Mediated Transfection of Mammalian Cells." *Bioconjugate Chemistry* 13 (1): 3–6.

Semple, S. C., A. Akinc, J. Chen, et al. 2010. "Rational Design of Cationic Lipids for siRNA Delivery." *Nature Biotechnology* 28 (2): 172–76.

Sharei, A., J. Zoldan, A. Adamo, et al. 2013. "A Vector-Free Microfluidic Platform for Intracellular Delivery." *Proceedings of the National Academy of Sciences* 110 (6): 2082–87.

Shu, D., Y. Shu, F. Haque, et al. 2011. "Thermodynamically Stable RNA Three-Way Junction for Constructing Multifunctional Nanoparticles for Delivery of Therapeutics." *Nature Nanotechnology* 6 (10): 658–67.

Smith, D., V. Schüller, C. Engst, et al.2013. "Nucleic Acid Nanostructures for Biomedical Applications." *Nanomedicine* 8 (1): 105–21.

Srinivasan, C., J. Lee, F. Papadimitrakopoulos, et al. 2006. "Labeling and Intracellular Tracking of Functionally Active Plasmid DNA with Semiconductor Quantum Dots." *Molecular Therapy* 14 (2): 192–201.

Thomas, C. E., A. Ehrhardt, and M. A. Kay. 2003. "Progress and Problems with the Use of Viral Vectors for Gene Therapy." *Nature Reviews Genetics* 4 (5): 346–58.

Tinsley-Bown, A. M., R. Fretwell, A. B. Dowsett, et al. 2000. "Formulation of poly(D,L-Lactic-Co-Glycolic Acid) Microparticles for Rapid Plasmid DNA Delivery." *Journal of Controlled Release* 66 (2–3): 229–41.

Torchilin, V. P. 2001. "Structure and Design of Polymeric Surfactant-Based Drug Delivery Systems." *Journal of Controlled Release* 73 (2–3): 137–72.

Verma, I. M., and M. D. Weitzman. 2005. "Gene Therapy: Twenty-First Century Medicine." *Annual Review of Biochemistry* 74 (1): 711–38.

Waehler, R., S. J. Russell, and D. T. Curiel. 2007. "Engineering Targeted Viral Vectors for Gene Therapy." *Nature Reviews Genetics* 8 (8): 573–87.

Wheeler, J. J., L. Palmer, M. Ossanlou, et al. 1999. "Stabilized Plasmid-Lipid Particles: Construction and Characterization." *Gene Therapy* 6 (2): 271–81.

Whitehead, K. A., R. Langer, and D. G. Anderson. 2009. "Knocking Down Barriers: Advances in siRNA Delivery." *Nature Reviews Drug Discovery* 8 (2): 129–38.

Wilhelm, S., A. J. Tavares, Q. Dai, et al. 2016. "Analysis of Nanoparticle Delivery to Tumours." *Nature Reviews Materials* 1 (5): 16014.

Woodrow, K. A., Y. Cu, C. J. Booth, et al. 2009. "Intravaginal Gene Silencing Using Biodegradable Polymer Nanoparticles Densely Loaded with Small-Interfering RNA." *Nature Materials* 8 (6): 526–33.

Wu, G., J. M. Wilson, F. Shalabyq, et al. 1988. "Receptor-Mediated Gene Delivery and Expression in Vivo." *The Journal of Biological Chemistry* 263 (29): 14621–4.

Wu, G. Y., and C. H. Wu. 1987. "Receptor-Mediated in Vitro Gene Transformation by a Soluble DNA Carrier System." *The Journal of Biological Chemistry* 262 (10): 4429–32.

Yin, H., R. L. Kanasty, A. A. Eltoukhy, et al. 2014. "Non-Viral Vectors for Gene-Based Therapy." *Nature Reviews Genetics* 15 (8): 541–55.

Zhang, L., W. Zheng, R. Tang, et al. 2016. "Gene Regulation with Carbon-Based siRNA Conjugates for Cancer Therapy." *Biomaterials* 104: 269–78.

Zhang, P., and W. Liu. 2010. "ZnO QD@PMAA-Co-PDMAEMA Nonviral Vector for Plasmid DNA Delivery and Bioimaging." *Biomaterials* 31 (11): 3087–94.

Zhang, Z., X. Yang, B. Zeng, et al. 2006. "Delivery of Telomerase Reverse Transcriptase Small Interfering RNA in Complex with Positively Charged Single-Walled Carbon Nanotubes Suppresses Tumor Growth." *Clinical Cancer Research* 12 (16): 4933–9.

Zimmermann, T. S., A. C. H. Lee, A. Akinc, et al. 2006. "RNAi-Mediated Gene Silencing in Non-Human Primates." *Nature* 441 (7089): 111–4.

12 Long Non-Coding RNA and Cancer

Mateja M. Jelen and Damjan Glavač

CONTENTS

12.1 INTRODUCTION

Cancer is predominantly a genetic disease and represents major public health burden worldwide. In 2013, the Global Burden of Disease Cancer Collaboration (GLOBOCAN) reported cancer as the second leading cause of death worldwide, after cardiovascular disease (Fitzmaurice et al. 2015). Among the 14.9 million incident cases reported in 2013, the top five leading types of cancer were: lung cancer (1.8 million), breast cancer (1.8 million in women), colorectal cancer (1.6 million), prostate cancer (1.4 million), and stomach cancer (984,000) (Fitzmaurice et al. 2015). Although the genetic determinants of oncogenesis have been extensively studied, cancer genetics remains complex. Cancer develops as a consequence of dysregulated gene expression and epigenetic changes and is mostly associated with somatic mutations, including single nucleotide polymorphisms (SNPs), deletions, duplications, translocations, copy-number alterations, and rarely reflects inheritable abbreviations in a single gene (Cheetham et al. 2013).

In the last decade a colossal improvement has been made in development and availability of genome-wide deep sequencing technologies, microarrays, and databases. Gene annotations of human genome are being revealed and updated constantly by systematic databases, such as consortium Encyclopedia of DNA Elements (ENCODE) and GENCODE Project (Dunham et al. 2012, Jalali, Gandhi, and Scaria 2016, Harrow et al. 2012). It has become increasingly evident over the last few years that the largest portion of the genome consists of regulatory, non-protein coding RNAs (non-coding RNAs, ncRNAs), previously annotated as "dark matter" or "junk DNA" (Taft et al. 2010, Ponting and Belgard 2010, Cheetham et al. 2013). NcRNAs are heterogeneous, evolutionary conserved molecules that resemble protein-coding genes and have multiple molecular functions across cellular processes, including cancer development. For the purpose of simplicity ncRNAs can be divided in three categories: (1) short ncRNAs (up to 200 bp in length), such as transfer RNAs (tRNAs), micro RNAs (miRNAs) etc., (2) mid-size ncRNAs (up to 300 bp), such as small nucleolar RNAs (snoRNAs), and (3) long ncRNAs (lncRNAs) (more than 200 bp) (Esteller 2011). Interestingly, it has been shown by the majority of genome-wide association studies that cancer risk loci occur outside of protein-coding regions and that locations of most SNPs currently linked to cancer locate in the introns (40 %) of protein-coding genes or intergenic regions (44 %) (Cheetham et al. 2013). Extensive studies and reviews began to arise, associating mutations and dysregulations in lncRNAs with various diseases, including cancer (Wapinski and Chang 2011, Gutschner and Diederichs 2012, Li and Chen 2013, Gupta et al. 2010). Accordingly, lncRNAs represent potential biomarkers for cancer diagnosis, progression, and therapeutic targets.

In this chapter we outline the lncRNA regulatory roles and basis of their genetic association with cancer. Subsequently, we briefly summarize importance of lncRNAs for cancer diagnostics and their therapeutic potential.

12.2 LONG NON-CODING RNAs (LNCRNAs)

The current GENCODE Version 25 (V25), revised in 2016, reported that human genome contains 15,767 lncRNA genes and 27,692 lncRNA transcripts (www.gencodegenes.org/stats/current.html).

LncRNAs are 200–1,000 nucleotides long non-coding transcripts and are involved in numerous essential cellular processes and epigenetic mechanisms (Amaral and Mattick 2008)), such as genomic imprinting (Lee and Bartolomei 2013, Leighton et al. 1995), gene expression stages including transcription, translation, chromatin modification (Mercer and Mattick 2013, Clark and Mattick 2011), protein signaling pathways (Shi et al. 2013), cell cycle control (Wapinski and Chang 2011), cell development and differentiation (Clark and Mattick 2011), and apoptosis control (Wapinski and Chang 2011). In accordance with their functions, lncRNAs are located in the nucleus, while some reside in cytoplasm (Mercer and Mattick 2013). However, some lncRNAs form fragments, which function in different cell locations. For example, lncRNA MALAT1 single gene locus generates a nuclear-retained ncRNA and a small RNA that is transported into cytoplasm (Wilusz, Freier, and Spector 2008). LncRNAs were originally discovered by analyzing the mouse transcriptome through large-scale sequencing of full-length cDNA libraries (Okazaki et al. 2002).

Genetically, lncRNAs can be categorized in five categories: (1) sense, (2) antisense, (3) bidirectional, when the initiation of its expression and a neighboring coding transcript on the opposite strand occurs in the close proximity, (4) intronic, when derived from an intron of a second transcript, and (5) intergenic, when it is located between the two genes (Ponting, Oliver, and Reik 2009).

12.2.1 LncRNA Structure

Similar to protein structure, RNA molecules consist of primary (nucleic acid composition), secondary (duplexes, hairpins, loops, and junctions), and tertiary structures (interactions between distant secondary elements), which provide binding sites for various proteins and play roles throughout the process of gene expression (Wan et al. 2011). Additionally, lncRNAs form duplexes or even triplexes with other DNA or RNA molecules, which also represent binding sites for numerous proteins (Li and Chen 2013). LncRNAs are mostly transcribed by RNA polymerase II, are capped at 5'-end, and undergo 3'-end polyadenylation and splicing (Li and Chen 2013). Four main categories of lncRNAs are showing association with cancer: (1) long intergenic ncRNAs (lincRNAs), (2) natural antisense transcripts (NAT), (3) transcribed ultra-conserved regions (T-UCR), and (4) non-coding pseudogenes (Li and Chen 2013).

12.2.2 LncRNA Regulatory Roles in Cancer

The major role of lncRNAs is probably to effect epigenetic changes in the genome, as many large intergenic human ncRNAs associate with chromatin-modifying complexes that alter gene expression (Khalil et al. 2009). LncRNAs act as tumor suppressors or oncogenes. The mechanisms of their regulatory networks in cancer are diverse. In order to alter gene expressing programs lncRNAs target complexes that modify chromatin and proteins that bind RNA (Tsai, Spitale, and Chang 2011). Since it is known that lncRNAs play critical roles in various biological processes, it is inevitable that lncRNA dysfunctions associate with a wide range of diseases. Several lncRNA online databases exist and are publically available (for details see the review by Gibb et al., 2011 [Gibb, Brown, and Lam 2011]). Recently, a new database for long-non-coding RNA-associated diseases has been launched: The LncRNADisease database, which integrated more than 1,000 lncRNA-disease entries and around 500 lncRNA interaction entries and provided the predicted associated diseases of 1,564 human lncRNAs (Chen et al. 2013). In this review, functions of lncRNAs will be briefly summarized as examples of four groups of molecular mechanisms (Wang and Chang 2011): chromatin-modifying lncRNAs, transcription (de)activators, post-transcriptional regulators, and lncRNAs that act as decoys (Cheetham et al. 2013, Shi et al. 2013, Wang and Chang 2011) (Figures 12.1 and 12.2). As an individual lncRNA may retain functions in other presented groups of mechanism, the presented groups of mechanisms are not mutually exclusive (Wang and Chang 2011).

12.2.2.1 Chromatin Remodeling LncRNAs

HOTAIR is a Hox transcript antisense RNA, located in the HOXC locus and regulating transcription within the HOX locus *in cis* or *trans*. HOTAIR is a lincRNA and interacts with histone modifiers Polycomb Repressive Complex 2 (PRC2) and Lysine-specific histone demethylase 1 (LSD1) and guides them to specific genomic region where it is involved in gene silencing of several metastasis suppressor genes through histone tail methylation (Rinn et al. 2007, Gupta et al. 2010, Tsai et al. 2010). The discovery of an lncRNA that regulates epigenetic silencing has important implications as it may operate in other loci (Rinn et al. 2007). It has been shown that HOTAIR is overexpressed in various cancers, including breast cancer (Chisholm et al. 2012), colorectal cancer (Kogo et al. 2011), hepatocellular carcinoma (HCC) (Yang, Zhou, et al. 2011), pancreatic cancer (Kim et al. 2013), esophageal squamous cell carcinoma (ESCC) (Song and Zou 2016), laryngeal squamous cell carcinoma (Li et al. 2013), nasopharyngeal carcinoma (Nie et al. 2013), and mesenchymal glioma (Zhang et al. 2013). Overexpression of HOTAIR is also prone to lymph node metastasis. The recent meta-analysis has demonstrated higher incidence of lymph node metastasis in patients with HOTAIR overexpression in comparison to patients with lower expression levels of HOTAIR (Cai et al. 2014).

ANRIL is an antisense ncRNA in the INK4 locus that belongs to a class of NAT lncRNAs, which have their transcripts complementary to other RNAs (Li and Chen 2013). ANRIL contains 19 exons, and it is transcribed in the opposite direction from INK4b-ARF-INK4a gene cluster, encoding tumor suppressor proteins $p15^{INK4b}$, $p14^{ARF}$, and $p16^{INK4a}$. ANRIL activates two histone modifiers, PRC1 and PRC2, which in turn triggers re-organization of the chromatin and repression of the INK4b-ARF-INK4a locus (Yu et al. 2008, Pasmant et al. 2007, He et al. 2013). ANRIL was identified as a lncRNA with oncogenic properties, involved in a series of tumors, such as: gastric cancer (Zhang et al. 2014), non-small cell lung cancer (NSCLC) (Nie et al. 2015), HCC (Huang et al. 2015), ESCC (Chen et al. 2014), cervical cancer studied in HeLa and H1299 cells (Naemura et al. 2015), prostate cancer (Yap et al. 2010), melanoma, where ANRIL was originally identified (Pasmant et al. 2007), bladder cancer (Zhu et al. 2015), and serous ovarian cancer (Qiu et al. 2015). It has been shown recently, that ANRIL plays an important role in the DNA damage response. Wan et al. (2013) have demonstrated the upregulation of ANRIL by E2F1 in an ATM-dependent manner and that elevated ANRIL suppresses the expression of INK4B-ARF-INK4A at the late-stage of DNA damage response, which in turn forms a negative feedback loop to the DNA damage response (Wan et al. 2013).

XIST, an X-inactive-specific transcript, is located at the X-chromosome inactivation center and plays a critical role in female development due to its promotion of X-chromosome inactivation. XIST recruits the PRC2 complex to silence the X chromosome (Zhao et al. 2008). TSIX, the antisense ncRNA of XIST, regulates XIST levels during X-inactivation (Lee, Davidow, and Warshawsky 1999). Namely, XIST has been linked to various gynecological diseases and is downregulated in female breast and ovarian cancer cell lines (Kawakami et al. 2004). The breast and ovarian cancer suppressor gene BRCA1 has been shown to interact with XIST and this interaction may imply a novel gender-specific consequence of BRCA1 loss, however, further studies are needed to confirm the relevance of BRCA1/XIST interaction (Ganesan et al. 2004).

12.2.2.2 Transcriptional Co-Activators and Repressors

H19 is located near the insulin-like growth factor 2 (IGF2) gene. H19 and IGF2 are imprinted, thus, the genes are expressed reciprocally: IGF2 is expressed from paternally inherited chromosome and H19 from maternally inherited chromosome (Gabory, Jammes, and Dandolo 2010). It has been shown after screening of murine fetal liver cDNA libraries that H19 is expressed during embryogenesis, inactivated in postnatal period, and re-activated in cancer development (Pachnis, Belayew, and Tilghman 1984). Berteaux et al., have shown that H19 promotes the transition from G1 phase to S phase in breast cancer cells through an active link to the E2F transcription factor 1 (E2F1) (Berteaux et al. 2005). The upregulated expression of H19 has been reported in hypoxic stress and associated with tumorigenesis in broad types of cancers, including breast cancer (Berteaux et al. 2008), lung cancer (Kondo et al. 1995), cervical cancer (Douc-Rasy et al. 1996), ESCC (Gao et al. 2015), bladder cancer (Berteaux et al. 2008), ovarian cancer (Mizrahi et al. 2009), and colorectal cancer (Deng et al. 2014). In breast cancer the H19 promoter is activated by E2F1 and the knock-down of H19 by small interfering RNA (siRNA) blocks wild-type and H19-transfected cells from entering into the S phase (Berteaux et al. 2005). Circulating H19 in plasma has been recently suggested as a novel biomarker for breast cancer (Zhang et al. 2016). Furthermore, it has been verified in gastric cancer tissues that H19 interacts with tumor suppressor p53, which triggers p53 inactivation, and that H19 siRNA treatment participated in apoptosis of AGS cell line, suggesting an application of H19 in therapy of gastric cancer (Yang et al. 2012).

LincRNA-p21 has been characterized as a regulator of cell proliferation, apoptosis, and DNA damage response, and is involved in various diseases (Tang, Zheng, and Xiong 2015). It is located upstream of the p21 and has a key role in regulation of p21 transcriptional levels. This is achieved by lincRNA-p21 interaction with heterogeneous nuclear ribonucleoprotein K (hnRNP-K) and its localization to the p21 promoter, which is essential for binding p53 to the p21 promoter, which in turn initiates the p21 transcription (Dimitrova et al. 2014). LincRNA-p21 was in addition found to be regulated by p53, directly upon DNA damage. Transcriptional regulation of p53-regulated genes by lincRNA-p21 is reached through an interaction with hnRNP-K. LincRNA-p21 localizes the hnRNP-K to promoters of repressed genes and is required for regulation of p53-mediated apoptosis (Huarte et al. 2010). Based on results obtained in their studies, Huarte et al. proposed that lincRNAs may serve as key regulators in transcriptional pathways (Huarte et al. 2010). LincRNA-p21 has been reported to be associated with colorectal cancer (Zhai et al. 2013), skin cancer (Hall et al. 2015), prostate cancer (Isin et al. 2015), and chronic lymphocytic leukemia (Isin et al. 2014). However, there is currently some ambiguity regarding lincRNA-p21 function in cancer development. Recently, Castellano et al., reported that high expression of lincRNA-p21 was associated with poor outcome in patients with NSCLC adenocarcinoma (Castellano et al. 2016). The role of lincRNA-p21 in angiogenesis was studied *in vitro* and researchers observed a global downregulation of angiogenesis-associated genes when lincRNA-p21 was inhibited (Castellano et al. 2016). Conversely, lincRNA-p21 levels were markedly decreased in diffuse large B cell lymphoma (DLBCL) tissues and it has been proposed that lincRNA-p21 predicts

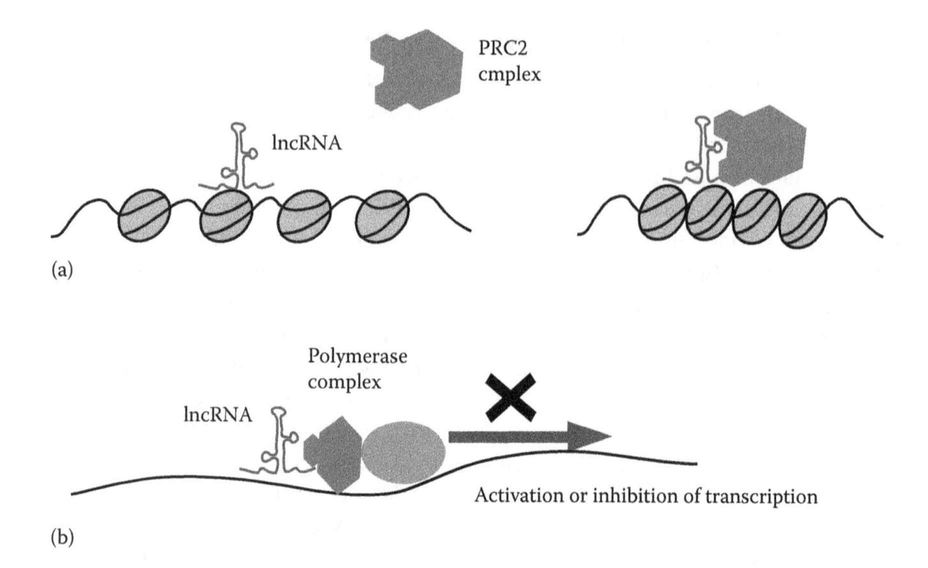

FIGURE 12.1 LncRNA chromatin remodeling and transcription regulation. (a) Chromatin remodeling. (b) Co-activation/repression of transcription.

favorable clinical outcome in DLBCL patients treated with rituximab plus cyclophosphamide/ doxorubicin/ vincristine/ prednisone (R-CHOP) chemotherapy (Peng, Wu, and Feng 2015). Thus, to further elucidate the exact molecular mechanism of action of lincRNA-p21 in different cancer types, additional studies are needed.

Another lncRNA, activated by external stimuli is located in the CDKN1A promoter. The p21-associated ncRNA DNA damage-activated (PANDA) is induced in the p53-manner, in response to DNA damage. PANDA represses apoptosis by interacting with the transcription factor Y subunit alpha (NF-YA) and limits the expression of apoptotic genes (Hung et al. 2011). However, PANDA could be considered as a NF-YA decoy as it sequesters the transcription factor in order to inhibit apoptotic genes. PANDA levels vary in different types of cancers, for example it is selectively upregulated in metastatic ductal carcinomas (Hung et al. 2011).

A steroid receptor RNA activator (SRA) is a transcriptional co-activator of various steroid hormone receptors (Fatima et al. 2015, Cheetham et al. 2013). Due to alternative splicing SRA1 gene exhibits dual roles and encodes a SRAP protein and a lncRNA SRA (Kawashima et al. 2003). SRA impacts tumorigenesis and cancer progression through *trans* activation of genes, interacting with the AF1 domain of nuclear receptors (Lanz et al. 1999), and it has been found to be deregulated and associated with breast cancer (Leygue et al. 1999, Liu et al. 2016), uterine cancer (Lanz et al. 2003), ovarian cancer (Lanz et al. 2003), and prostate cancer (Kawashima et al. 2003).

12.2.2.3 Post-Transcriptional Regulators

MALAT1, a metastasis associated lung adenocarcinoma transcript 1, is a nuclear lincRNA, involved in gene expression regulation as well as post-transcriptional modification. It is evolutionary strongly conserved among mammalian species (Gutschner,

Hammerle, and Diederichs 2013, Ji et al. 2003). MALAT1 interacts with numerous regulators and interaction partners (reviewed by Gutschner et al. [Gutschner, Hammerle, and Diederichs 2013]), thus its function depends on combination of interacting proteins in the respective cell. MALAT1 employs a mechanism of alternative splicing of pre-mRNAs, where it binds with serine/arginine splicing factors and thereby influences the distribution of splicing factors in nuclear speckle domains. Moreover, it has been shown that MALAT1 regulates levels of phosphorylated serine/arginine proteins and that depletion of MALAT1 alters splicing factor localization and activity, which leads to altered pattern of alternative splicing for pre-mRNAs (Tripathi et al. 2010). MALAT1 exhibits a strong association with genes involved in cancer like cellular growth, movement, proliferation, signaling, and immune regulation. Despite years of research, MALAT1 still lacks a molecular mechanism of upregulation in cancer (Gutschner, Hammerle, and Diederichs 2013). MALAT1 is highly expressed in non-small cell lung cancer (NSCLC) (Ji et al. 2003), bladder cancer (Ying et al. 2012), HCC (Lai et al. 2012), colorectal carcinoma (Xu et al. 2011), prostate cancer (Ren et al. 2013), breast cancer (Jadaliha et al. 2016), endometrial stromal sarcoma (ESS) of the uterus (Yamada et al. 2006), osteosarcoma (Gao and Lian 2016), and pancreatic cancer (Pang et al. 2015). Experiments on NSCLC cell lines showed that RNAi-mediated suppression of MALAT1 abolished clonogenic growth, while upon injection of NSCLC xenografts with reduced MALAT1 expression into nude mice, the tumor formation and growth were damaged (Schmidt et al. 2011). It has been demonstrated that MALAT1 overexpression predicts recurrence of HCC in patients with liver transplantation and could serve as a promising therapeutic target (Lai et al. 2012).

12.2.2.4 LncRNA Decoys

Pseudogenes are dysfunctional relatives of normal genes. Since they harbor premature stop codons, indels, and frameshift mutations they cannot be translated into functional proteins (Geisler and Coller 2013). Due to preservation of their nucleotide sequences, their important functional role in cells is not to be excluded. One example is a PTENP1 pseudogene, which is highly homologous to tumor suppressor gene PTEN and contains conserved sites for different miRNAs that target PTEN (Poliseno et al. 2010, Fujii et al. 1999). These conserved elements between the two sequences suggest that PTEN1 and PTEN are subjected to the same post-transcriptional regulation, mediated by miRNAs. Poliseno et al. have studied PTENP1 function in prostate cancer samples and demonstrated that the PTEN1 regulates cellular levels of PTEN and that its 3'UTR region retains the tumor suppressive activity (Poliseno et al. 2010). The research group (Poliseno et al. 2010) examined the ability of PTENP1 functioning as a decoy for miRNAs-targeting PTEN.

GAS5 (growth arrest specific 5) is a growth arrest- and starvation-associated repressor of the glucocorticoid receptor (GR) (Kino et al. 2010). GAS5 binds the DNA-binding domain of the glucocorticoid receptor as it mimics its DNA binding site. The interaction triggers repression of glucocorticoid receptor-mediated transcription, which influences metabolic activities during cell starvation (Kino et al. 2010, Geisler and Coller 2013). GAS5 has been reported to be downregulated in

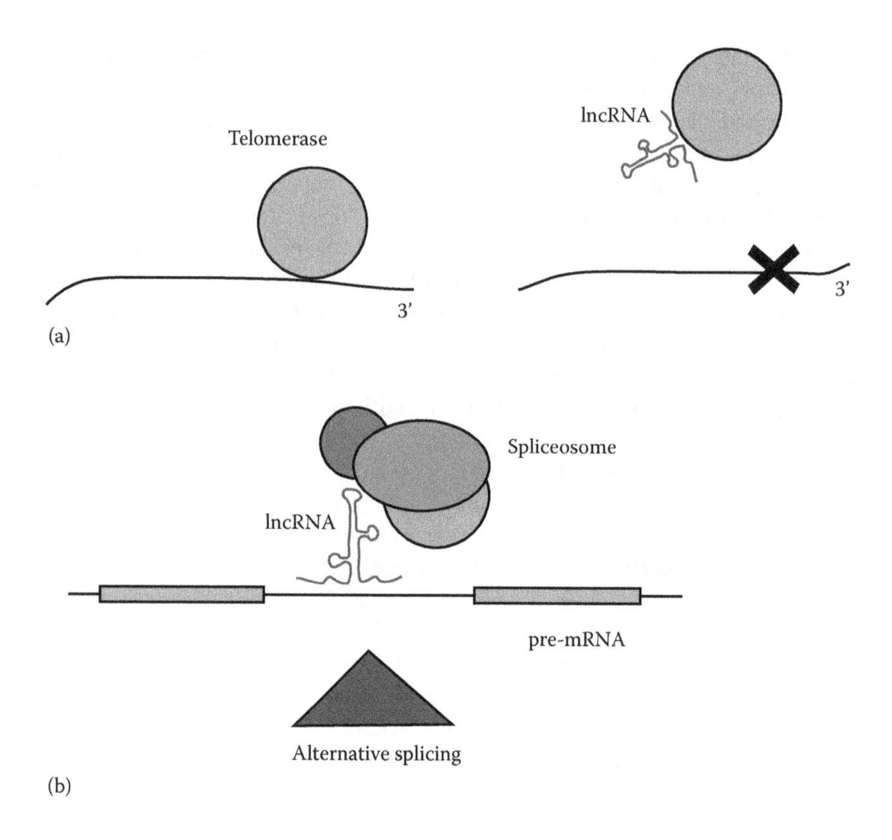

FIGURE 12.2 LncRNA post-transcriptional regulation and decoy mechanism. (a) Decoy mechanism. (b) Post-transcriptional regulation.

breast cancer (Mourtada-Maarabouni et al. 2009) and regulates apoptosis in prostate cancer cell lines (Pickard, Mourtada-Maarabouni, and Williams 2013).

TERRA, a telomeric repeat containing RNA is transcribed from subtelomeric loci towards chromosome ends, telomeres (Redon, Reichenbach, and Lingner 2010, Azzalin et al. 2007). Telomeres are protective ends of chromosomes that shorten progressively during each cell division and finally reach the critical length, which induces cellular senescence. Telomerase enzyme is able to re-extend the 3'end telomeric repeats and it is therefore used by majority of cancerous cells. *In vitro* study by Redon et al. has demonstrated lncRNA TERRA as a telomerase ligand, which directly inhibits human telomerase (Redon, Reichenbach, and Lingner 2010). TERRA is downregulated in cancer cells, which explains the long life of cancer cells that circumvent the cellular senescence and death (Shay and Bacchetti 1997).

12.3 LncRNAs DIAGNOSTIC AND THERAPEUTIC POTENTIAL

As described in previous sections, some lncRNAs act as oncogenic lncRNAs and stimulate cancer growth when upregulated. On the other hand, some lncRNAs express tumor suppressive features and drive cancer growth only when downregulated

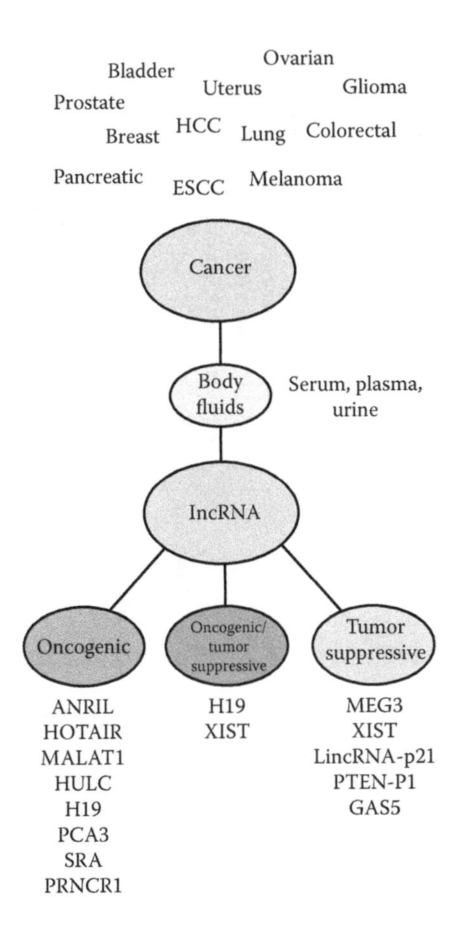

FIGURE 12.3 Oncogenic and tumor-suppressive lncRNAs.

(Figure 12.3). Deregulated expression levels of lncRNAs in numerous cancers indicate their promising role in diagnostics and prognosis of cancer and provide a novel alternative in medical treatment of cancer (Table 12.1).

12.3.1 Biomarkers in Diagnostics

Identification of numerous dysregulated lncRNAs in various types of cancers has raised possibility of using these lncRNAs as potential diagnostic and prognostic biomarkers. Unfortunately, limited data exist on stability of lncRNAs in different body fluids and mechanisms that regulate lncRNA expression remain poorly resolved. However, ncRNAs are often stable in serum and may be measured with reverse transcription quantitative PCR amplification (RT-qPCR) or screened for their entire transcriptome using RNA sequencing (RNAseq). LncRNA profiling tools most widely used are RT-PCR, Northern blotting, lncRNA arrays, RNAseq, and *in situ* hybridization. Nevertheless, lncRNAs in diagnostics have major advantages over

TABLE 12.1
Biological Functions of LncRNAs, Their Mechanisms of Regulation and Associations with Different Cancers

LncRNA	Cancer Type	Biological Function	Mechanism of Regulation	Reference
HOTAIR	Breast, colorectal, HCC, pancreatic, ESCC, laryngeal squamous cell carcinoma, nasopharyngeal, glioma, lymph node metastasis	Repression at the HOXD locus	Chromatin-mediated repression	(Chisholm et al. 2012, Kogo et al. 2011, Yang, Zhou, et al. 2011, Kim et al. 2013, Song and Zou 2016, Li et al. 2013, Nie et al. 2013, Zhang et al. 2013, Cai et al. 2014)
ANRIL	Gastric, ESCC, cervical, prostate, melanoma, bladder, ovarian, lymphoblastic leukemia	Repression at the INK4b-ARF-INK4a locus	Chromatin-mediated repression	(Pasmant et al. 2007, Zhang et al. 2014, Nie et al. 2015, Huang et al. 2015, Chen et al. 2014, Naemura et al. 2015, Yap et al. 2010, Zhu et al. 2015, Qiu et al. 2015)
XIST	Ovarian, breast	X inactivation	Chromatin-mediated repression	(Kawakami et al. 2004)
H19	Breast, lung, cervical, ESCC, bladder, colorectal, ovarian	Promotes cell proliferation	Transcriptional co-activation	(Berteaux et al. 2008, Kondo et al. 1995, Douc-Rasy et al. 1996, Gao et al. 2015, Mizrahi et al. 2009, Deng et al. 2014)
SRA	Breast, ovarian, uterine, prostate	Transcriptional co-activator of steroid receptors	Transcriptional co-activation	(Kawashima et al. 2003, Leygue et al. 1999, Liu et al. 2016, Lanz et al. 2003)
LincRNA-p21	Colorectal, skin, prostate, chronic lymphatic leukemia, NSCLC adenocarcinoma, DLBCL	Regulation of p53 response	Transcriptional repression	(Zhai et al. 2013, Hall et al. 2015, Isin et al. 2015, Isin et al. 2014, Castellano et al. 2016, Peng, Wu, and Feng 2015)

(*Continued*)

TABLE 12.1 (CONTINUED)
Biological Functions of LncRNAs, Their Mechanisms of Regulation and Associations with Different Cancers

LncRNA	Cancer Type	Biological Function	Mechanism of Regulation	Reference
PANDA	Metastasis of ductal carcinoma	Inhibition of apoptosis	Transcriptional repression (decoy mechanism)	(Hung et al. 2011)
TERRA	Cancerous cell lines	Facilitates telomeric heterochromatin formation and inhibits polymerase	Protein inhibition	(Redon, Reichenbach, and Lingner 2010, Shay and Bacchetti 1997)
MALAT1	NSCLC, bladder, colorectal, prostate, breast, ESS of the uterus, osteosarcoma, pancreatic	Regulation of distribution of serine/arginine splicing factors	Scaffolding of sub-nuclear domains	(Ji et al. 2003, Ying et al. 2012, Lai et al. 2012, Xu et al. 2011, Ren et al. 2013, Jadaliha et al. 2016, Yamada et al. 2006, Gao and Lian 2016, Pang et al. 2015)
PTENP1 pseudogene	Prostate	Upregulation of tumor suppressor gene PTEN	Sequestration of miRNAs	(Poliseno et al. 2010)
GAS5	Breast	Repression of glucocorticoid receptor-mediated transcription	Sequestration of glucocorticoid receptor	(Mourtada-Maarabouni et al. 2009)

Abbreviations: NSCLC: non-small cell lung cancer; HCC: hepatocellular carcinoma; ESCC: esophageal squamous cell carcinoma; ESS: endometrial stromal sarcoma; DLBCL: diffuse large B cell lymphoma; HOTAIR: hox transcript antisense RNA; ANRIL: antisense ncRNA in the INK4 locus; XIST: X-inactive-specific transcript; PANDA: P21-associated ncRNA DNA damage-activated; SRA: steroid receptor RNA activator; MALAT1: metastasis associated lung adenocarcinoma transcript 1; GAS5: growth arrest specific 5.

protein-coding RNAs as their measured expression directly indicates levels of active molecules in the cell.

Ideal biomarkers are easily accessible, sampled noninvasively (sampled from body fluids) and therefore need to be highly stable as circulating molecules (Qi and Du 2013). Currently, only potential lncRNA biomarkers have been characterized in body fluids, such as PCA3 in urine samples for prostate cancer diagnosis (Tinzl et al. 2004), lncRNA HULC as a blood biomarker for HCC (Panzitt et al. 2007), HOTAIR as a negative prognostic biomarker in tumor tissues and blood of patients with colorectal carcinoma (Svoboda et al. 2014), circulating H19 in plasma has been recently suggested as a novel biomarker for breast cancer (Zhang et al. 2016), MALAT1 in urine for prostate cancer diagnosis (Wang et al. 2014) etc. Critical reviews summarizing recent studies on different circulating biomarker lncRNAs in different cancers have been published recently (Shi, Gao, and Cao 2016, Fatima et al. 2015, Li and Chen 2013).

12.3.2 Therapeutic Targets

LncRNAs present valuable potential targets in cancer therapeutic opportunities. Two major advantages of lncRNAs need to be mentioned in this context: in comparison to protein-coding genes, lncRNAs are tissue specific and stable in body fluids and tissues, which facilitates the noninvasive targeting. A variety of approaches exist to therapeutically target lncRNAs. Currently, there are two therapeutic RNA targeting methods tested in clinical trials: RNA-mediated interference (Ozcan et al. 2015) and antisense oligonucleotides (Evers, Toonen, and van Roon-Mom 2015). In this section we will briefly outline five promising therapeutic approaches targeting lncRNAs. Therapeutic targeting of lncRNAs was extensively reviewed recently by Fatima et al. and Chery (Chery 2016, Fatima et al. 2015).

12.3.2.1 RNA Interference (RNAi)-Based Therapy

RNAi-mediated inhibition of various targets could be achieved by implementing different RNA molecules (siRNA, shRNA, miRNA). Targeting lncRNAs with siRNAs includes unwinding the siRNA duplex and assembly of RNA-induced silencing complex (RISC) (Li and Chen 2013). It has been shown that downregulation of the lncRNA HOTAIR by siRNA in HCC reduced cell viability and cell invasion, sensitized TNF-α induced apoptosis, and increased the chemotherapeutic sensitivity of cancer cells to cisplatin and doxorubicin (Yang, Zhou, et al. 2011). H19 siRNA treatment has been suggested as an application in therapy of gastric cancer (Yang et al. 2012). Furthermore, Bhan et al. (Bhan et al. 2013) have designed the small interfering sense (siSENSE) DNA oligonucleotide to knock-down the HOTAIR induced apoptosis of transfected MCF7 breast cancer cell line. Knock-down of HOTAIR by siRNA was performed also by Kim et al. (Kim et al. 2013) and revealed the inhibition of tumor growth in mouse xenograft model. SiRNA-mediated downregulation of lncRNA has proven successful also in case of PCA3, which is highly expressed in prostate cancer. In prostate cancer, however, several other lncRNA targets have shown potential for siRNA-mediated cancer inhibition. For example, PRNCR1 and PCGEM1 strongly enhanced proliferation in prostate cancer cells (Yang et al. 2013).

For metastatic prostate cancer, a novel biomarker or potential therapeutic agent lncRNA PCAT18 has been discovered recently (Crea et al. 2014). PCAT18 silencing through siRNA significantly inhibited pancreatic cell proliferation (Crea et al. 2014). In HCC, siRNA/shRNA-mediated inhibition of lncRNAs H19 (Vernucci et al. 2000), HULC (Panzitt et al. 2007), HEIH (Yang, Zhang, et al. 2011), and MVIH (Yuan et al. 2012) have as well resulted in reduced tumor growth, therefore these lncRNAs present potential therapeutic targets. Although, there are currently several RNAi-based therapies in clinical trials, further research in lncRNA-based therapy and further advances in development of safe and effective therapy application are needed. In spite of many challenges, such as rapid degradation, difficulties in siRNA delivery to target sites and unspecific target effects, 26 siRNAs have been already tested in more than 50 clinical trials, involving an extensive list of various diseases (Chery 2016). SiRNA-based therapeutics in clinical trials are presented in a recent review by Ozcan et al. (Ozcan et al. 2015).

12.3.2.2 Antisense Oligonucleotide (ASO)-Mediated Therapy

ASOs are short single stranded oligonucleotides (13–50 bp) that are able to specifically bind to target RNAs in the nucleus or cytoplasm (pre-mRNAs, mRNAs, ncRNAs etc.). ASOs induce cleavage of target transcripts and inhibit mRNA translation. Degradation of lncRNA transcripts includes binding of ASOs to target RNA and form a DNA-RNA heteroduplex that is recognized by RNase H that catalyzes heteroduplex cleavage. ASOs are more stable in comparison to siRNAs, highly soluble in water and their management *in vivo* is less complicated (Li and Chen 2013, Chery 2016). Another advantage of ASOs over siRNAs is that they are single stranded and easier to manufacture. However, both dsRNAs and single-stranded molecules can induce innate immune response, with excessive cytokine release, an undesirable effect in siRNA or ASO-based drugs (Li and Chen 2013, Robbins, Judge, and MacLachlan 2009, Gantier et al. 2008). Their disadvantage is their lack of ability of crossing the blood-brain barrier. ASOs use various mechanisms of action, including inhibition of 5'-cap formation, splicing, steric blocking of translation, and recruitment of the enzyme RNase H. (Chery 2016, Fatima et al. 2015, Parasramka et al. 2016). Knock-down of MALAT1 by ASO has been demonstrated to inhibit the metastasis in lung cancer cells in a mouse xenograft model (Gutschner, Hammerle, and Diederichs 2013). To the best of our knowledge, there are currently three ASO drugs that have been approved by the Federal Drug Administration (Formivirsen [Vitravene], Mipomersen, and Macugen) and that have been used in clinical departments, however they do not influence lncRNA activity (Chery 2016).

12.3.2.3 Plasmid-Based Therapy

Plasmid-based therapy has been innovated recently by Mizrahi et al., 2009 (Mizrahi et al. 2009) who developed the targeted therapy for ovarian cancer mediated by a plasmid BC-819/DTAH19, which harbors the diphtheria toxin subunit A under the control of H19 promoter. The preliminary study showed the applicative role of regulatory sequences of the H19 gene for the development of DNA-based therapy for human ovarian cancer related ascites (Mizrahi et al. 2009). Inventors of this therapeutic approach have published a case report of a progressive ovarian cancer and reported on currently conducting an extensive Phase I study in order to assess safety and efficacy on a larger number of patients (Mizrahi et al. 2010). This method has been explored in several cancers, including bladder cancer (Amit and Hochberg 2010), however, to the best of our knowledge no further steps in clinical trials have been published to date.

12.3.2.4 Gene-Based Therapy

Observations that several lncRNAs become downregulated in certain tumor samples, in comparison to normal tissues, has raised a novel possibility of using lncRNAs to suppress tumor growth. The delivery of tumor suppressor RNAs could be a great potential and help in gene therapy. Although the mentioned approach of targeting ncRNAs brought promising results in cell lines, *in vivo* targeting still requires to overcome a series of challenges, including a point of concern that many lncRNAs are species-specific, therefore human-specific, and cannot be explored in knock-down animal models (Fatima et al. 2015).

12.3.2.5 Small Molecule Inhibitor-Based Therapy

Small molecule inhibitors (SMIs) mask the binding sites on lncRNAs or their various interacting molecules. Typical small molecules that are used to target RNA structure are for example tetracyclines. Inhibition through SMI of viral molecules in HIV and HCV has been also reported (Li and Chen 2013). It has been shown for example in glioblastoma xenograft model that HOTAIR interaction with PRC2 and LSD1 can be blocked with SMI of the PRC complex subunit zeste homolog 2 (EZH2) (Zhang et al. 2015). Researchers found that EZH2 inhibition elicited effects that were consistent with those elicited by HOTAIR-targeted siRNA (Zhang et al. 2015). SMIs have several advantages over RNAi and ASO and siRNAs based methods: they are easier to administer, exhibit better cellular uptake, and are more target specific (Fatima et al. 2015).

Among the above-mentioned therapeutic options that inhibit desired RNA molecule, there are also ribozymes, which are naturally produced molecules and degrade other RNA molecules and aptamers, short DNA or RNA molecules or peptides with stable three-dimensional structure that overcome RNA secondary structures as they rely on fitting shape of their ligands (Li and Chen 2013). LncRNA-targeted therapeutic strategies are presented in Figure 12.4.

Despite the fact that many therapeutic approaches exist in the lncRNA research area, there are four major challenges that remain unresolved in using circulating lncRNAs as cancer biomarkers or therapeutic agents (reviewed by Qi et al.[Qi, Zhou, and Du 2016]): poorly explored complex biological pathways, accurate quantification protocols, delivery and pharmacokinetics, and toxicity issues (Qi, Zhou, and Du 2016).

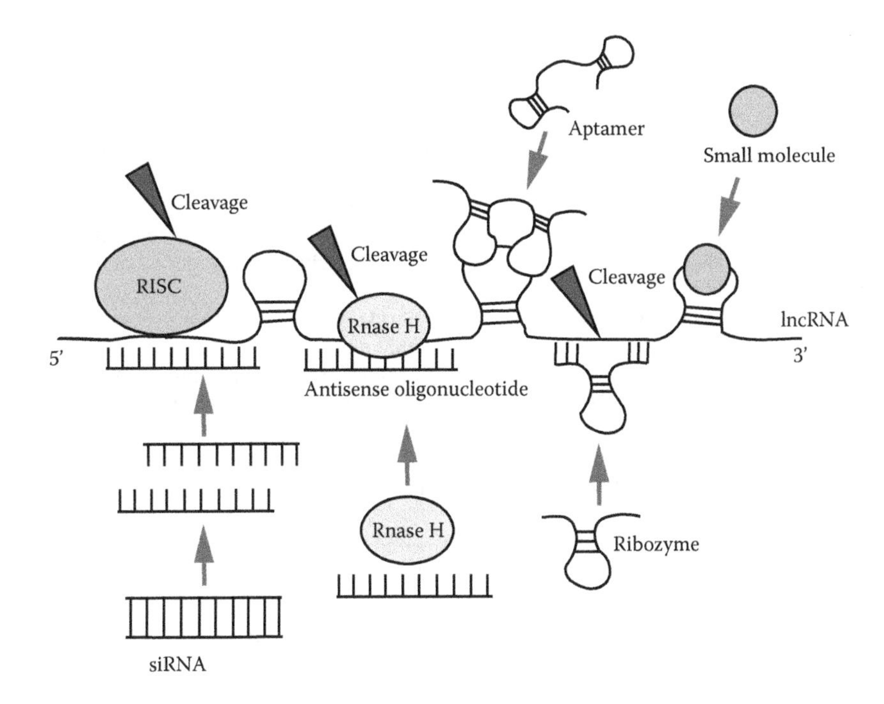

FIGURE 12.4 Potential therapeutic mechanisms of targeting lncRNAs.

12.4 CONCLUSIONS AND PERSPECTIVES

In this chapter we have outlined critical characteristics of lncRNAs, presented their functional roles in different cancers and described their significant potential roles in diagnostics and treatment of cancer. LncRNAs are numerous tissue-specific molecules, dysregulated in different types of cancer and hold evident potential in cancer diagnostics and lncRNA-mediated therapy of cancer. In the past few years the evidence of lncRNA importance has just started to emerge, covering currently only the tip of the "knowledge iceberg." Further studies are required to elucidate reliable lncRNA measurement methods, delivery systems, and to bypass off-target effects. Despite the fact that therapeutic options may be possible in more distant future, currently accomplished knowledge has answered a lot of key questions regarding potential roles of non-coding genome material in human disease. In the process of uncovering novel models of lncRNA regulation, ncRNAs will most probably assist or combat small RNAs and proteins in their clinical applications.

REFERENCES

Amaral, P. P., and J. S. Mattick. 2008. "Noncoding RNA in development." *Mamm Genome* 19 (7–8):454–92.

Amit, D., and A. Hochberg. 2010. "Development of targeted therapy for bladder cancer mediated by a double promoter plasmid expressing diphtheria toxin under the control of H19 and IGF2-P4 regulatory sequences." *J Transl Med* 8:134.

Azzalin, C. M., P. Reichenbach, L. Khoriauli, et al. 2007. "Telomeric repeat containing RNA and RNA surveillance factors at mammalian chromosome ends." *Science* 318 (5851):798–801.

Berteaux, N., N. Aptel, G. Cathala, et al. 2008. "A novel H19 antisense RNA overexpressed in breast cancer contributes to paternal IGF2 expression." *Mol Cell Biol* 28 (22):6731–45.

Berteaux, N., S. Lottin, D. Monte, et al. 2005. "H19 mRNA-like noncoding RNA promotes breast cancer cell proliferation through positive control by E2F1." *J Biol Chem* 280 (33):29625–36.

Bhan, A., I. Hussain, K. I. Ansari, et al. 2013. "Antisense transcript long noncoding RNA (lncRNA) HOTAIR is transcriptionally induced by estradiol." *J Mol Biol* 425 (19):3707–22.

Cai, B., Z. Wu, K. Liao, et al. 2014. "Long noncoding RNA HOTAIR can serve as a common molecular marker for lymph node metastasis: a meta-analysis." *Tumour Biol* 35 (9):8445–50.

Castellano, J. J., A. Navarro, N. Vinolas, et al. 2016. "LincRNA-p21 impacts prognosis in resected non-small cell lung cancer patients through angiogenesis regulation." *J Thorac Oncol.* 11(12):2173–82.

Cheetham, S. W., F. Gruhl, J. S. Mattick, et al. 2013. "Long noncoding RNAs and the genetics of cancer." *Br J Cancer* 108 (12):2419–25.

Chen, D., Z. Zhang, C. Mao, et al. 2014. "ANRIL inhibits p15(INK4b) through the TGFbeta1 signaling pathway in human esophageal squamous cell carcinoma." *Cell Immunol* 289 (1–2):91–6.

Chen, G., Z. Wang, D. Wang, et al. 2013. "LncRNADisease: a database for long-non-coding RNA-associated diseases." *Nucleic Acids Res* 41 (Database issue):D983–6.

Chery, J. 2016. "RNA therapeutics: RNAi and antisense mechanisms and clinical applications." *Postdoc J* 4 (7):35–50.

Chisholm, K. M., Y. Wan, R. Li, et al. 2012. "Detection of long non-coding RNA in archival tissue: correlation with polycomb protein expression in primary and metastatic breast carcinoma." *PLoS One* 7 (10):e47998.

Clark, M. B., and J. S. Mattick. 2011. "Long noncoding RNAs in cell biology." *Semin Cell Dev Biol* 22 (4):366–76.

Crea, F., A. Watahiki, L. Quagliata, et al. 2014. "Identification of a long non-coding RNA as a novel biomarker and potential therapeutic target for metastatic prostate cancer." *Oncotarget* 5 (3):764–74.

Deng, Q., B. He, T. Gao, et al. 2014. "Up-regulation of 91H promotes tumor metastasis and predicts poor prognosis for patients with colorectal cancer." *PLoS One* 9 (7):e103022.

Dimitrova, N., J. R. Zamudio, R. M. Jong, et al. 2014. "LincRNA-p21 activates p21 in cis to promote Polycomb target gene expression and to enforce the G1/S checkpoint." *Mol Cell* 54 (5):777–90.

Douc-Rasy, S., M. Barrois, S. Fogel, et al. 1996. "High incidence of loss of heterozygosity and abnormal imprinting of H19 and IGF2 genes in invasive cervical carcinomas. Uncoupling of H19 and IGF2 expression and biallelic hypomethylation of H19." *Oncogene* 12 (2):423–30.

Dunham, I., Kundaje, A., Aldred, S.F., et al. 2012. "An integrated encyclopedia of DNA elements in the human genome." *Nature* 489 (7414):57–74. doi: 10.1038/nature11247.

Esteller, M. 2011. "Non-coding RNAs in human disease." *Nat Rev Genet* 12 (12):861–74.

Evers, M. M., L. J. Toonen, and W. M. van Roon-Mom. 2015. "Antisense oligonucleotides in therapy for neurodegenerative disorders." *Adv Drug Deliv Rev* 87:90–103.

Fatima, R., V. S. Akhade, D. Pal, et al. 2015. "Long noncoding RNAs in development and cancer: potential biomarkers and therapeutic targets." *Mol Cell Ther* 3:5.

Fitzmaurice, C., D. Dicker, A. Pain, et al. 2015. "The global burden of cancer 2013." *JAMA Oncol* 1 (4):505–27.

Fujii, G. H., A. M. Morimoto, A. E. Berson, et al. 1999. "Transcriptional analysis of the PTEN/MMAC1 pseudogene, psiPTEN." *Oncogene* 18 (9):1765–9.

Gabory, A., H. Jammes, and L. Dandolo. 2010. "The H19 locus: role of an imprinted non-coding RNA in growth and development." *Bioessays* 32 (6):473–80.

Ganesan, S., D. P. Silver, R. Drapkin, et al. 2004. "Association of BRCA1 with the inactive X chromosome and XIST RNA." *Philos Trans R Soc Lond B Biol Sci* 359 (1441):123–8.

Gantier, M. P., S. Tong, M. A. Behlke, et al. 2008. "TLR7 is involved in sequence-specific sensing of single-stranded RNAs in human macrophages." *J Immunol* 180 (4):2117–24.

Gao, K. T., and D. Lian. 2016. "Long non-coding RNA MALAT1 is an independent prognostic factor of osteosarcoma." *Eur Rev Med Pharmacol Sci* 20 (17):3561–5.

Gao, T., B. He, Y. Pan, et al. 2015. "Long non-coding RNA 91H contributes to the occurrence and progression of esophageal squamous cell carcinoma by inhibiting IGF2 expression." *Mol Carcinog* 54 (5):359–67.

Geisler, S., and J. Coller. 2013. "RNA in unexpected places: long non-coding RNA functions in diverse cellular contexts." *Nat Rev Mol Cell Biol* 14 (11):699–712.

Gibb, E. A., C. J. Brown, and W. L. Lam. 2011. "The functional role of long non-coding RNA in human carcinomas." *Mol Cancer* 10:38.

Gupta, R. A., N. Shah, K. C. Wang, et al. 2010. "Long non-coding RNA HOTAIR reprograms chromatin state to promote cancer metastasis." *Nature* 464 (7291):1071–6.

Gutschner, T., and S. Diederichs. 2012. "The hallmarks of cancer: a long non-coding RNA point of view." *RNA Biol* 9 (6):703–19.

Gutschner, T., M. Hammerle, and S. Diederichs. 2013. "MALAT1—a paradigm for long non-coding RNA function in cancer." *J Mol Med (Berl)* 91 (7):791–801.

Hall, J. R., Z. J. Messenger, H. W. Tam, et al. 2015. "Long noncoding RNA lincRNA-p21 is the major mediator of UVB-induced and p53-dependent apoptosis in keratinocytes." *Cell Death Dis* 6:e1700.

Harrow, J., A. Frankish, J. M. Gonzalez, et al. 2012. "GENCODE: the reference human genome annotation for The ENCODE Project." *Genome Res* 22 (9):1760–74.

He, S., W. Gu, Y. Li, and H. Zhu. 2013. "ANRIL/CDKN2B-AS shows two-stage clade-specific evolution and becomes conserved after transposon insertions in simians." *BMC Evol Biol* 13:247.

Huang, M. D., W. M. Chen, F. Z. Qi, et al. 2015. "Long non-coding RNA ANRIL is upregulated in hepatocellular carcinoma and regulates cell proliferation by epigenetic silencing of KLF2." *J Hematol Oncol* 8 (1):57.

Huarte, M., M. Guttman, D. Feldser, et al. 2010. "A large intergenic noncoding RNA induced by p53 mediates global gene repression in the p53 response." *Cell* 142 (3):409–19.

Hung, T., Y. Wang, M. F. Lin, et al. 2011. "Extensive and coordinated transcription of noncoding RNAs within cell-cycle promoters." *Nat Genet* 43 (7):621–9.

Isin, M., E. Ozgur, G. Cetin, et al. 2014. "Investigation of circulating lncRNAs in B-cell neoplasms." *Clin Chim Acta* 431:255–9.

Isin, M., E. Uysaler, E. Ozgur, et al. 2015. "Exosomal lncRNA-p21 levels may help to distinguish prostate cancer from benign disease." *Front Genet* 6:168.

Jadaliha, M., X. Zong, P. Malakar, et al. 2016. "Functional and prognostic significance of long non-coding RNA MALAT1 as a metastasis driver in ER negative lymph node negative breast cancer." *Oncotarget* 7 (26):40418–36.

Jalali, S., S. Gandhi, and V. Scaria. 2016. "Navigating the dynamic landscape of long non-coding RNA and protein-coding gene annotations in GENCODE." *Hum Genomics* 10 (1):35.

Ji, P., S. Diederichs, W. Wang, et al. 2003. "MALAT-1, a novel noncoding RNA, and thymosin beta4 predict metastasis and survival in early-stage non-small cell lung cancer." *Oncogene* 22 (39):8031–41.

Kawakami, T., C. Zhang, T. Taniguchi, et al. 2004. "Characterization of loss-of-inactive X in Klinefelter syndrome and female-derived cancer cells." *Oncogene* 23 (36):6163–9.

Kawashima, H., H. Takano, S. Sugita, et al. 2003. "A novel steroid receptor co-activator protein (SRAP) as an alternative form of steroid receptor RNA-activator gene: expression in prostate cancer cells and enhancement of androgen receptor activity." *Biochem J* 369 (Pt 1):163–71.

Khalil, A. M., M. Guttman, M. Huarte, et al. 2009. "Many human large intergenic noncoding RNAs associate with chromatin-modifying complexes and affect gene expression." *Proc Natl Acad Sci U S A* 106 (28):11667–72.

Kim, K., I. Jutooru, G. Chadalapaka, et al. 2013. "HOTAIR is a negative prognostic factor and exhibits pro-oncogenic activity in pancreatic cancer." *Oncogene* 32 (13):1616–25.

Kino, T., D. E. Hurt, T. Ichijo, N. Nader, and G. P. Chrousos. 2010. "Noncoding RNA gas5 is a growth arrest- and starvation-associated repressor of the glucocorticoid receptor." *Sci Signal* 3 (107):ra8.

Kogo, R., T. Shimamura, K. Mimori, et al. 2011. "Long noncoding RNA HOTAIR regulates polycomb-dependent chromatin modification and is associated with poor prognosis in colorectal cancers." *Cancer Res* 71 (20):6320–6.

Kondo, M., H. Suzuki, R. Ueda, et al. 1995. "Frequent loss of imprinting of the H19 gene is often associated with its overexpression in human lung cancers." *Oncogene* 10 (6):1193–8.

Lai, M. C., Z. Yang, L. Zhou, et al. 2012. "Long non-coding RNA MALAT-1 overexpression predicts tumor recurrence of hepatocellular carcinoma after liver transplantation." *Med Oncol* 29 (3):1810–6.

Lanz, R. B., S. S. Chua, N. Barron, et al. 2003. "Steroid receptor RNA activator stimulates proliferation as well as apoptosis in vivo." *Mol Cell Biol* 23 (20):7163–76.

Lanz, R. B., N. J. McKenna, S. A. Onate, et al. 1999. "A steroid receptor coactivator, SRA, functions as an RNA and is present in an SRC-1 complex." *Cell* 97 (1):17–27.

Lee, J. T., and M. S. Bartolomei. 2013. "X-inactivation, imprinting, and long noncoding RNAs in health and disease." *Cell* 152 (6):1308–23.

Lee, J. T., L. S. Davidow, and D. Warshawsky. 1999. "Tsix, a gene antisense to Xist at the X-inactivation centre." *Nat Genet* 21 (4):400–4.

Leighton, P. A., R. S. Ingram, J. Eggenschwiler, et al. 1995. "Disruption of imprinting caused by deletion of the H19 gene region in mice." *Nature* 375 (6526):34–9.

Leygue, E., H. Dotzlaw, P. H. Watson, et al. 1999. "Expression of the steroid receptor RNA activator in human breast tumors." *Cancer Res* 59 (17):4190–3.

Li, C. H., and Y. Chen. 2013. "Targeting long non-coding RNAs in cancers: progress and prospects." *Int J Biochem Cell Biol* 45 (8):1895–910.

Li, D., J. Feng, T. Wu, et al. 2013. "Long intergenic noncoding RNA HOTAIR is overexpressed and regulates PTEN methylation in laryngeal squamous cell carcinoma." *Am J Pathol* 182 (1):64–70.

Liu, C., H. T. Wu N. Zhu, et al. 2016. "Steroid receptor RNA activator: biologic function and role in disease." *Clin Chim Acta* 459:137–46.

Mercer, T. R., and J. S. Mattick. 2013. "Structure and function of long noncoding RNAs in epigenetic regulation." *Nat Struct Mol Biol* 20 (3):300–7.

Mizrahi, A., A. Czerniak, T. Levy, et al. 2009. "Development of targeted therapy for ovarian cancer mediated by a plasmid expressing diphtheria toxin under the control of H19 regulatory sequences." *J Transl Med* 7:69.

Mizrahi, A., A. Czerniak, P. Ohana, et al. 2010. "Treatment of ovarian cancer ascites by intra-peritoneal injection of diphtheria toxin A chain-H19 vector: a case report." *J Med Case Rep* 4:228.

Mourtada-Maarabouni, M., M. R. Pickard, V. L. Hedge, et al. 2009. "GAS5, a non-protein-coding RNA, controls apoptosis and is downregulated in breast cancer." *Oncogene* 28 (2):195–208.

Naemura, M., C. Murasaki, Y. Inoue, et al. 2015. "Long noncoding RNA ANRIL regulates proliferation of non-small cell lung cancer and cervical cancer cells." *Anticancer Res* 35 (10):5377–82.

Nie, F. Q., M. Sun, J. S. Yang, et al. 2015. "Long noncoding RNA ANRIL promotes non-small cell lung cancer cell proliferation and inhibits apoptosis by silencing KLF2 and P21 expression." *Mol Cancer Ther* 14 (1):268–77.

Nie, Y., X. Liu, S. Qu, et al. 2013. "Long non-coding RNA HOTAIR is an independent prognostic marker for nasopharyngeal carcinoma progression and survival." *Cancer Sci* 104 (4):458–64.

Okazaki, Y., M. Furuno, T. Kasukawa, et al. 2002. "Analysis of the mouse transcriptome based on functional annotation of 60,770 full-length cDNAs." *Nature* 420 (6915):563–73.

Ozcan, G., B. Ozpolat, R. L. Coleman, et al. 2015. "Preclinical and clinical development of siRNA-based therapeutics." *Adv Drug Deliv Rev* 87:108–19.

Pachnis, V., A. Belayew, and S. M. Tilghman. 1984. "Locus unlinked to alpha-fetoprotein under the control of the murine raf and Rif genes." *Proc Natl Acad Sci U S A* 81 (17):5523–7.

Pang, E. J., R. Yang, X. B. Fu, et al. 2015. "Overexpression of long non-coding RNA MALAT1 is correlated with clinical progression and unfavorable prognosis in pancreatic cancer." *Tumour Biol* 36 (4):2403–7.

Panzitt, K., M. M. Tschernatsch, C. Guelly, et al. 2007. "Characterization of HULC, a novel gene with striking up-regulation in hepatocellular carcinoma, as noncoding RNA." *Gastroenterology* 132 (1):330–42.

Parasramka, M. A., S. Maji, A. Matsuda, et al. 2016. "Long non-coding RNAs as novel targets for therapy in hepatocellular carcinoma." *Pharmacol Ther* 161:67–78.

Pasmant, E., I. Laurendeau, D. Heron, et al. 2007. "Characterization of a germ-line deletion, including the entire INK4/ARF locus, in a melanoma-neural system tumor family: identification of ANRIL, an antisense noncoding RNA whose expression coclusters with ARF." *Cancer Res* 67 (8):3963–9.

Peng, W., J. Wu, and J. Feng. 2015. "LincRNA-p21 predicts favorable clinical outcome and impairs tumorigenesis in diffuse large B cell lymphoma patients treated with R-CHOP chemotherapy." *Clin Exp Med.* 17(1):1–8.

Pickard, M. R., M. Mourtada-Maarabouni, and G. T. Williams. 2013. "Long non-coding RNA GAS5 regulates apoptosis in prostate cancer cell lines." *Biochim Biophys Acta* 1832 (10):1613–23.

Poliseno, L., L. Salmena, J. Zhang, et al. 2010. "A coding-independent function of gene and pseudogene mRNAs regulates tumour biology." *Nature* 465 (7301):1033–8.

Ponting, C. P., and T. G. Belgard. 2010. "Transcribed dark matter: meaning or myth?" *Hum Mol Genet* 19 (R2):R162–8.

Ponting, C. P., P. L. Oliver, and W. Reik. 2009. "Evolution and functions of long noncoding RNAs." *Cell* 136 (4):629–41.

Qi, P., and X. Du. 2013. "The long non-coding RNAs, a new cancer diagnostic and therapeutic gold mine." *Mod Pathol* 26 (2):155–65.

Qi, P., X. Y. Zhou, and X. Du. 2016. "Circulating long non-coding RNAs in cancer: current status and future perspectives." *Mol Cancer* 15 (1):39.

Qiu, J. J., Y. Y. Lin, J. X. Ding, et al. 2015. "Long non-coding RNA ANRIL predicts poor prognosis and promotes invasion/metastasis in serous ovarian cancer." *Int J Oncol* 46 (6):2497–505.

Redon, S., P. Reichenbach, and J. Lingner. 2010. "The non-coding RNA TERRA is a natural ligand and direct inhibitor of human telomerase." *Nucleic Acids Res* 38 (17):5797–806.

Ren, S., Y. Liu, W. Xu, et al. 2013. "Long noncoding RNA MALAT-1 is a new potential therapeutic target for castration resistant prostate cancer." *J Urol* 190 (6):2278–87.

Rinn, J. L., M. Kertesz, J. K. Wang, et al. 2007. "Functional demarcation of active and silent chromatin domains in human HOX loci by noncoding RNAs." *Cell* 129 (7):1311–23.

Robbins, M., A. Judge, and I. MacLachlan. 2009. "siRNA and innate immunity." *Oligonucleotides* 19 (2):89–102.

Schmidt, L. H., T. Spieker, S. Koschmieder, et al. 2011. "The long noncoding MALAT-1 RNA indicates a poor prognosis in non-small cell lung cancer and induces migration and tumor growth." *J Thorac Oncol* 6 (12):1984–92.

Shay, J. W., and S. Bacchetti. 1997. "A survey of telomerase activity in human cancer." *Eur J Cancer* 33 (5):787–91.

Shi, T., G. Gao, and Y. Cao. 2016. "Long noncoding RNAs as novel biomarkers have a promising future in cancer diagnostics." *Dis Markers* 2016:9085195.

Shi, X., M. Sun, H. Liu, et al. 2013. "Long non-coding RNAs: a new frontier in the study of human diseases." *Cancer Lett* 339 (2):159–66.

Song, W., and S. B. Zou. 2016. "Prognostic role of lncRNA HOTAIR in esophageal squamous cell carcinoma." *Clin Chim Acta* 463:169–73.

Svoboda, M., J. Slyskova, M. Schneiderova, et al. 2014. "HOTAIR long non-coding RNA is a negative prognostic factor not only in primary tumors, but also in the blood of colorectal cancer patients." *Carcinogenesis* 35 (7):1510–5.

Taft, R. J., K. C. Pang, T. R. Mercer, et al. 2010. "Non-coding RNAs: regulators of disease." *J Pathol* 220 (2):126–39.

Tang, S. S., B. Y. Zheng, and X. D. Xiong. 2015. "LincRNA-p21: implications in Human Diseases." *Int J Mol Sci* 16 (8):18732–40.

Tinzl, M., M. Marberger, S. Horvath, et al. 2004. "DD3PCA3 RNA analysis in urine—a new perspective for detecting prostate cancer." *Eur Urol* 46 (2):182–6.

Tripathi, V., J. D. Ellis, Z. Shen, et al. 2010. "The nuclear-retained noncoding RNA MALAT1 regulates alternative splicing by modulating SR splicing factor phosphorylation." *Mol Cell* 39 (6):925–38.

Tsai, M. C., O. Manor, Y. Wan, et al. 2010. "Long noncoding RNA as modular scaffold of histone modification complexes." *Science* 329 (5992):689–93.

Tsai, M. C., R. C. Spitale, and H. Y. Chang. 2011. "Long intergenic noncoding RNAs: new links in cancer progression." *Cancer Res* 71 (1):3–7.

Vernucci, M., F. Cerrato, N. Besnard, et al. 2000. "The H19 endodermal enhancer is required for Igf2 activation and tumor formation in experimental liver carcinogenesis." *Oncogene* 19 (54):6376–85.

Wan, G., R. Mathur, X. Hu, et al. 2013. "Long non-coding RNA ANRIL (CDKN2B-AS) is induced by the ATM-E2F1 signaling pathway." *Cell Signal* 25 (5):1086–95.

Wan, Y., M. Kertesz, R. C. Spitale, et al. 2011. "Understanding the transcriptome through RNA structure." *Nat Rev Genet* 12 (9):641–55.

Wang, F., S. Ren, R. Chen, et al. 2014. "Development and prospective multicenter evaluation of the long noncoding RNA MALAT-1 as a diagnostic urinary biomarker for prostate cancer." *Oncotarget* 5 (22):11091–102.

Wang, K. C., and H. Y. Chang. 2011. "Molecular mechanisms of long noncoding RNAs." *Mol Cell* 43 (6):904–14.

Wapinski, O., and H. Y. Chang. 2011. "Long noncoding RNAs and human disease." *Trends Cell Biol* 21 (6):354–61.

Wilusz, J. E., S. M. Freier, and D. L. Spector. 2008. "3' end processing of a long nuclear-retained noncoding RNA yields a tRNA-like cytoplasmic RNA." *Cell* 135 (5):919–32.

Xu, C., M. Yang, J. Tian, et al. 2011. "MALAT-1: a long non-coding RNA and its important 3' end functional motif in colorectal cancer metastasis." *Int J Oncol* 39 (1):169–75.

Yamada, K., J. Kano, H. Tsunoda, et al. 2006. "Phenotypic characterization of endometrial stromal sarcoma of the uterus." *Cancer Sci* 97 (2):106–12.

Yang, F., J. Bi, X. Xue, et al. 2012. "Up-regulated long non-coding RNA H19 contributes to proliferation of gastric cancer cells." *Febs J* 279 (17):3159–65.

Yang, F., L. Zhang, X. S. Huo, et al. 2011. "Long noncoding RNA high expression in hepatocellular carcinoma facilitates tumor growth through enhancer of zeste homolog 2 in humans." *Hepatology* 54 (5):1679–89.

Yang, L., C. Lin, C. Jin, et al. 2013. "lncRNA-dependent mechanisms of androgen-receptor-regulated gene activation programs." *Nature* 500 (7464):598–602.

Yang, Z., L. Zhou, L. M. Wu, et al. 2011. "Overexpression of long non-coding RNA HOTAIR predicts tumor recurrence in hepatocellular carcinoma patients following liver transplantation." *Ann Surg Oncol* 18 (5):1243–50.

Yap, K. L., S. Li, A. M. Munoz-Cabello, et al. 2010. "Molecular interplay of the noncoding RNA ANRIL and methylated histone H3 lysine 27 by polycomb CBX7 in transcriptional silencing of INK4a." *Mol Cell* 38 (5):662–74.

Ying, L., Q. Chen, Y. Wang, et al. 2012. "Upregulated MALAT-1 contributes to bladder cancer cell migration by inducing epithelial-to-mesenchymal transition." *Mol Biosyst* 8 (9):2289–94.

Yu, W., D. Gius, P. Onyango, et al. 2008. "Epigenetic silencing of tumour suppressor gene p15 by its antisense RNA." *Nature* 451 (7175):202–6.

Yuan, S. X., F. Yang, Y. Yang, et al. 2012. "Long noncoding RNA associated with microvascular invasion in hepatocellular carcinoma promotes angiogenesis and serves as a predictor for hepatocellular carcinoma patients' poor recurrence-free survival after hepatectomy." *Hepatology* 56 (6):2231–41.

Zhai, H., A. Fesler, K. Schee, et al. 2013. "Clinical significance of long intergenic noncoding RNA-p21 in colorectal cancer." *Clin Colorectal Cancer* 12 (4):261–6.

Zhang, E. B., R. Kong, D. D. Yin, et al. 2014. "Long noncoding RNA ANRIL indicates a poor prognosis of gastric cancer and promotes tumor growth by epigenetically silencing of miR-99a/miR-449a." *Oncotarget* 5 (8):2276–92.

Zhang, J. X., L. Han, Z. S. Bao, et al. 2013. "HOTAIR, a cell cycle-associated long noncoding RNA and a strong predictor of survival, is preferentially expressed in classical and mesenchymal glioma." *Neuro Oncol* 15 (12):1595–603.

Zhang, K., Z. Luo, Y. Zhang, et al. 2016. "Circulating lncRNA H19 in plasma as a novel biomarker for breast cancer." *Cancer Biomark* 17 (2):187–94.

Zhang, K., X. Sun, X. Zhou, et al. 2015. "Long non-coding RNA HOTAIR promotes glioblastoma cell cycle progression in an EZH2 dependent manner." *Oncotarget* 6 (1):537–46.

Zhao, J., B. K. Sun, J. A. Erwin, et al. 2008. "Polycomb proteins targeted by a short repeat RNA to the mouse X chromosome." *Science* 322 (5902):750–6.

Zhu, H., X. Li, Y. Song, et al. 2015. "Long non-coding RNA ANRIL is up-regulated in bladder cancer and regulates bladder cancer cell proliferation and apoptosis through the intrinsic pathway." *Biochem Biophys Res Commun* 467 (2):223–8.

13 miRNA Therapeutics to Target Multiple Molecular Pathways

Current Status, Challenges, and Future Prospects

Lihui Zhu and Guofeng Cheng

CONTENTS

13.1 INTRODUCTION

miRNA, a key component of the small non-coding RNA family, can regulate the expression of target mRNA by binding to 3'-untranslated regions (UTRs) through imperfect and perfect base pairing; as such, they are involved in the regulation of multiple cellular functions. Since their first discovery in *Caenorhabditis elegans* in the early 1990s (Lee et al. 1993; Fire et al. 1998), miRNAs have been recognized as key players in many biological processes. To date, changes in miRNA expression profiles have been noted during the progression of many diseases, particularly caner (Lee et al. 2016; Li and Sarkar 2016; Yonemori et al. 2017). In addition, functional modulation of miRNAs in several animal disease models has indicated that miRNAs are involved in the pathophysiology of different diseases (Guo et al. 2010; van Rooij and Olson 2012). Although few miRNA-based therapeutics have entered human clinical trials, miRNA targeting in some animal models has provided important proof-of-concept for the future development of novel clinical therapies (Li and Sarkar 2016; Yonemori et al. 2017).

The main advantage of miRNA-based therapeutics is that one miRNA can target multiple genes that are potentially involved in several pathways required for cancer initiation, promotion, and progression (Guo et al. 2010; van Rooij and Olson 2012). This provides a biological rationale for the use of a small number of miRNAs to achieve broad silencing of pro-tumorigenic pathways. However, potential side effects and/or off-target effects should be overcome prior to the effective clinical use of miRNA-based therapies. miRNAs and small interfering RNAs (siRNAs) are two main small RNAs designed to affect any gene of interest. Both are short RNA duplexes that target mRNA to produce a gene-silencing effect, yet their mechanisms of action are distinct. siRNAs are derived by processing of long, double-stranded RNAs and are often of exogenous origin and degrade mRNAs; they should be fully complementary to their target sequences (Zeng et al. 2003). In contrast, miRNAs are endogenously encoded, small noncoding RNAs, derived by processing of short RNA hairpins, which can inhibit the translation of mRNAs with partially complementary target sequences; thus, a single miRNA therapy could simultaneously affect multiple genes (Zeng et al. 2003). As a result, the requirements for sequence design and therapeutic applications of siRNAs and miRNAs are different. However, for clinical development, the two types of small RNA molecules to some extent face similar obstacles including instability *in vivo*, delivery limitations, and off-target effects (Lam et al. 2015). In this chapter, we summarize the current understanding of the roles of miRNA in several diseases and their associated molecular pathways that could be targeted for therapeutic proposes. Subsequently, we review miRNAs as potential therapeutic targets for different diseases, specifically for cancer, and the associated challenges that remain.

13.2 ABERRANT DISEASE-ASSOCIATED MECHANISMS OF miRNA BIOGENESIS

miRNA biogenesis and target identification have been described elsewhere (for biogenesis and regulation see Cheng 2015, and for target identification see Bartel 2009). Here, we briefly summarize these processes as follows. The miRNA biogenesis pathway is shown in Figure 13.1. miRNA genes are first derived from precursor transcripts called primary miRNA (pri-miRNA) by RNA polymerase II. The pri-miRNA contains stem-loop structures ranging from hundreds to thousands of nucleotides in length and is further processed in the nucleus by DGCR8 and the ribonuclease Drosha into approximately 70–100 nt long hairpin structures, called precursor miRNA (pre-miRNA) (Han et al. 2004). Next, the exportin 5 exports

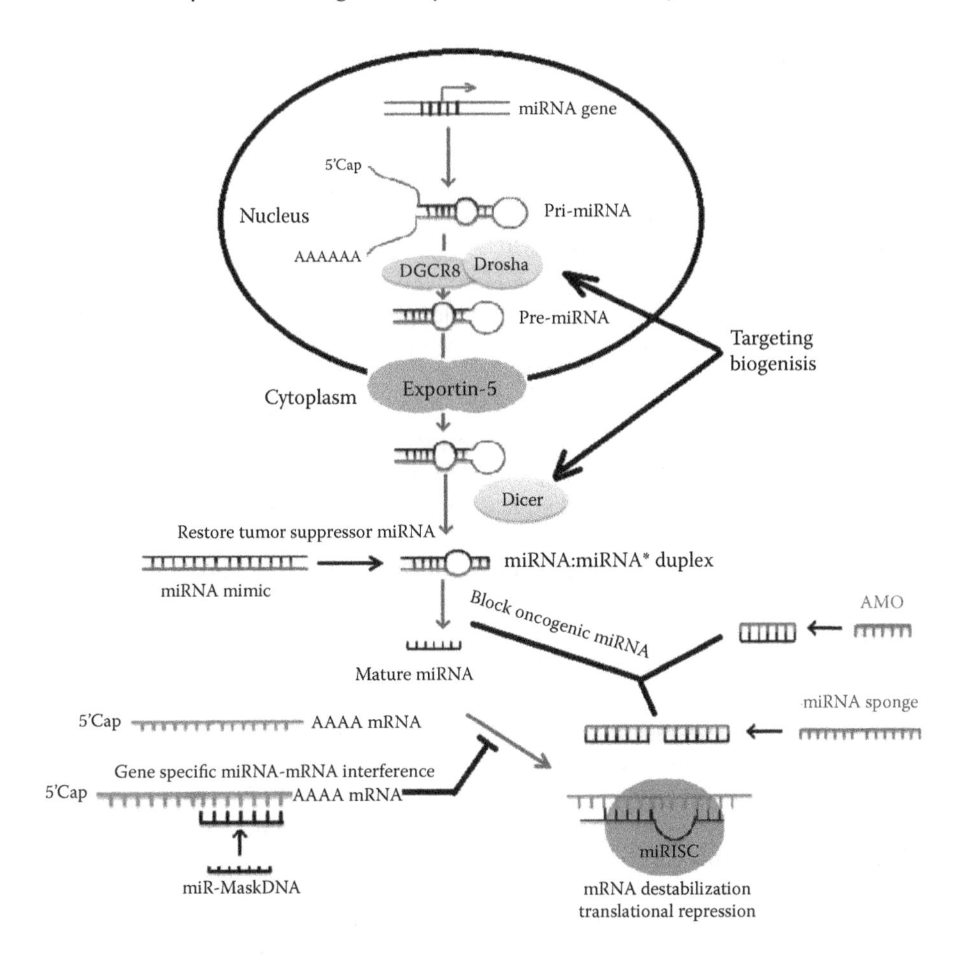

FIGURE 13.1 Schematic diagram of miRNA biogenesis and therapeutic strategies. anti-miRNA oligonucleotides (amo) interact with miRNA to inhibit binding of miRNA to target miRNA; miRNA sponges contain multiple binding sites for a specific miRNA, which in turn block the binding of this miRNA to its targets; miR-mask DNA is complementary to the binding site of a specific miRNA, thereby blocking access of the endogenous miRNA to its binding site, resulting in de-repression of the target gene; miRNA mimics are delivered by nanoparticles or liposomes to enhance the expression of tumor suppressor miRNAs; targeting of miRNA biogenesis components such Dicer and Drosha can be achieved to regulate miRNA maturation.

the pre-miRNA to the cytoplasm where it is further processed by Dicer, an RNase III-like enzyme, into a 19–25 nt double-stranded miRNA duplex. Then, the mature miRNA strand is subsequently incorporated into the RNA-induced silencing complex (RISC), where it directly binds to a member of the AGO protein family. The other strand is referred to as miRNA* and is normally degraded. The RISC can recognize the target mRNA by binding of the seed sequence (position 2 to 8 from the 5'end) of the mature miRNA strand typically to the 3'-UTR of the mRNA, which is the canonical mechanism of miRNA-mediated gene regulation; however, some other "non-canonical" miRNA-mediated mechanisms of mRNA regulation also exist where miRNA interacts with mRNA coding regions, intron-exon junctions, and 5'-UTRs (Orom et al. 2008; Tay et al. 2008). For example, miR-10a has been shown to bind to the 5'-UTR of a target gene and shown to activate rather than to inhibit gene expression (Orom et al. 2008). It has been reported that miR-328 binds to ribonucleoproteins in a seed sequence- and RISC-independent manner, and then interferes with the RNA decoy activity (Eiring et al. 2010). In addition, Calin and co-workers reported that miRNAs (miR-15 and miR-16) also regulate gene expression at the transcriptional level by binding directly to the DNA (Calin et al. 2002). Overall, these studies demonstrate the complexity and diversity of gene expression regulation by miRNAs, which need to be taken into consideration when developing miRNA-based therapies.

Aberrant expression of components involved in miRNA biogenesis is associated with certain diseases (Karube et al. 2005; Merritt et al. 2008). For example, decreased expression of Dicer and Drosha in epithelial ovarian cancers is associated with poor clinical outcome (Merritt et al. 2008) and diminished miRNA expression is associated with poor prognosis in lung cancer (Karube et al. 2005). Downregulation of Dicer, resulting in impaired miRNA processing, has also been shown to enhance tumor cell proliferation and invasion (Han et al. 2010); this was further confirmed by silencing Dicer, Drosha, and exportin-5 in bladder urothelial carcinoma T24 and 5637 cells (Han et al. 2013). Similarly, Dicer overexpression has been reported to be associated with poor prognosis in prostate and esophageal carcinoma (Sugito et al. 2006; Chiosea et al. 2006). In addition, deficiency in exportin-5 might limit miRNA production, as shown by Yi and colleagues (Yi et al. 2003). Moreover, the Ago protein family, which are critical for mRNA degradation via RISC, might also regulate upstream molecules of RNAi machinery, thereby inhibiting miRNA function (Diederichs et al. 2008).

Furthermore, the knock-down of mature oncogenic miRNAs by targeting of Dicer or Drosha or its cofactor DGCR8 has been shown to suppress tumor formation (Leucci et al. 2012; Geng et al. 2014). Additionally, systemic delivery of anti-miR-9 increased Dicer and HuR expression and led to decreased tumor growth in a xenograft model of Hodgkin lymphoma (Leucci et al. 2012). Upregulation of Dicer was shown to suppress miR-103 induced colorectal cancer cell proliferation and migration *in vitro* and HCT-116 xenograft tumor growth *in vivo* (Geng et al. 2014). Moreover, knock-down of Dicer inhibited endothelial cell tumor growth *via* miR-21a-3p targeting of Nox-4 (Gordillo et al. 2014) and downregulation of Drosha-inhibited cell proliferation, colony formation and migration in cervical cancer (Zhou et al. 2013). Furthermore, loss of ΔNp63 (an oncogenic member of the p53 family) resulted in decreased levels of its target, DGCR8, and inhibited the maturation of let-7d and miR-128 in cancer (Napoli et al. 2016). In contrast, aberrant expression of miRNA biogenesis components may also accelerate tumor formation (Dincbas-Renqvist et al. 2009; Han et al. 2010). Global repression of miRNA expression can be induced by inhibition of Dicer and Drosha, and this treatment promotes cellular transformation and tumorigenesis *in vivo* and in tumor cells (Dincbas-Renqvist et al. 2009; Han et al. 2010). The conditional loss of Dicer in mice lung tissues enhanced the development of lung tumors in a K-Ras mouse model (Dincbas-Renqvist et al. 2009).

Aberrant expression of miRNA biogenesis components has also been observed in non-cancerous diseases such as wound healing (Bhattacharya et al. 2015), diabetes (Rahimi et al. 2015), and cardiovascular diseases (Li et al. 2015c; Hartmann et al. 2016). For example, increased expression of Drosha, Dicer, and DGCR8 were observed in pregnant and gestational diabetes mellitus patients (Rahimi et al. 2015). Dicer inhibition was shown to prevent HaCaT cell migration and affect wound closure, suggesting that it is a promising target to address impaired wound healing (Bhattacharya et al. 2015). In cardiovascular disease, Li and colleagues reported that miR-107 could improve angiogenesis through downregulation of Dicer-1, thereby upregulating endogenous VEGF 165 during hypoxia both *in vivo* and *in vitro.* This suggests that miR-107 could serve as a potential therapeutic target for stroke treatment (Li et al. 2015c). In a recent study, downregulation of miR-103 was observed in Dicer-deficient endothelial cells. Notably, blocking the interaction between miR-103 and its target Krüppel-like factor 4 (KLF4) in arteries reduced atherosclerosis and lesional macrophage formation, similar to the effects of Dicer deletion in endothelial cells. This implies that selective inhibition of KLF4 by miR-103 with antisense oligonucleotides could represent a novel approach to treat atherosclerosis (Hartmann et al. 2016). Overall, these studies demonstrate that aberrant expression of components involved in miRNA biogenesis contributes to the progression of certain diseases, indicating a critical role for miRNAs in these disorders.

13.3 MOLECULAR PATHWAYS THAT COULD BE TARGETED BY miRNAS FOR THERAPEUTIC PURPOSES

A deep understanding of cancer biology, especially relating to molecular pathways of tumor metastasis, is a prerequisite for identifying novel and effective targets for therapeutic intervention. To regulate cell proliferation, miRNAs interact with a variety of key signaling axes such as the WNT, TGF-β, p53, and phosphoinositide-3 kinase (PI3K)/serine/threonine kinase (AKT) pathways, among others (Han et al. 2012; Chen et al. 2016c; Peng et al. 2017a). Hence, understanding the biological roles of miRNAs, as they pertain to key molecular pathways that are involved in cancer metastasis, might provide insight into the signaling mechanisms that are essential for cancer development, resulting in novel therapeutic strategies and agents for cancer treatment. Consequently, in Section 13.3, we summarize recent studies on miRNA-associated pathways that could be targeted for cancer treatment.

13.3.1 Targeting of the WNT/β-catenin Signaling Pathway by miRNAs

The WNT/β-catenin pathway is a highly regulated signaling pathway that controls numerous stages of animal development and tissue homeostasis (Clevers 2006). Activation of this pathway begins with WNT proteins binding to Frizzled (FZD) and lipoprotein receptor-related protein (LRP) receptor complex, which halts β-catenin degradation, resulting in accumulation and nuclear translocation of β-catenin (Patil et al. 2009). In the nucleus, β-catenin interacts with the T-cell factor/lymphoid enhancer factor (TCF/LEF) family to increase the expression of a range of downstream targets that include cyclin-D, c-Myc, and CD44 (Patil et al. 2009). Dysregulation of components of the WNT/β-catenin pathway results in developmental disorders and diseases including, but not limited to, cancer (Mahmood et al. 2016). To date, various studies have demonstrated that miRNAs can activate or repress the WNT/β-catenin pathway, at multiple levels, by targeting WNT ligands/receptors and associated proteins including β-catenin and the β-catenin interacting complex as well as other components of this pathway (Figure 13.2). WNT activation increases expression of some miRNAs (such as miR-183, miR-96, and miR-182) or decreases expression of some other miRNAs (such as miR-34) through the interaction between β-catenin and TCF/LEF, and subsequent binding to promoter regions thus activating transcription (Kim et al. 2011a; Leung et al. 2015).

It is well established that WNT activation originates from WNT proteins binding to the Frizzled/LRP5/6 receptor complex; thus, miRNAs that target components of the WNT ligand/receptor complex such as WNT, FZD, and LRP might repress the

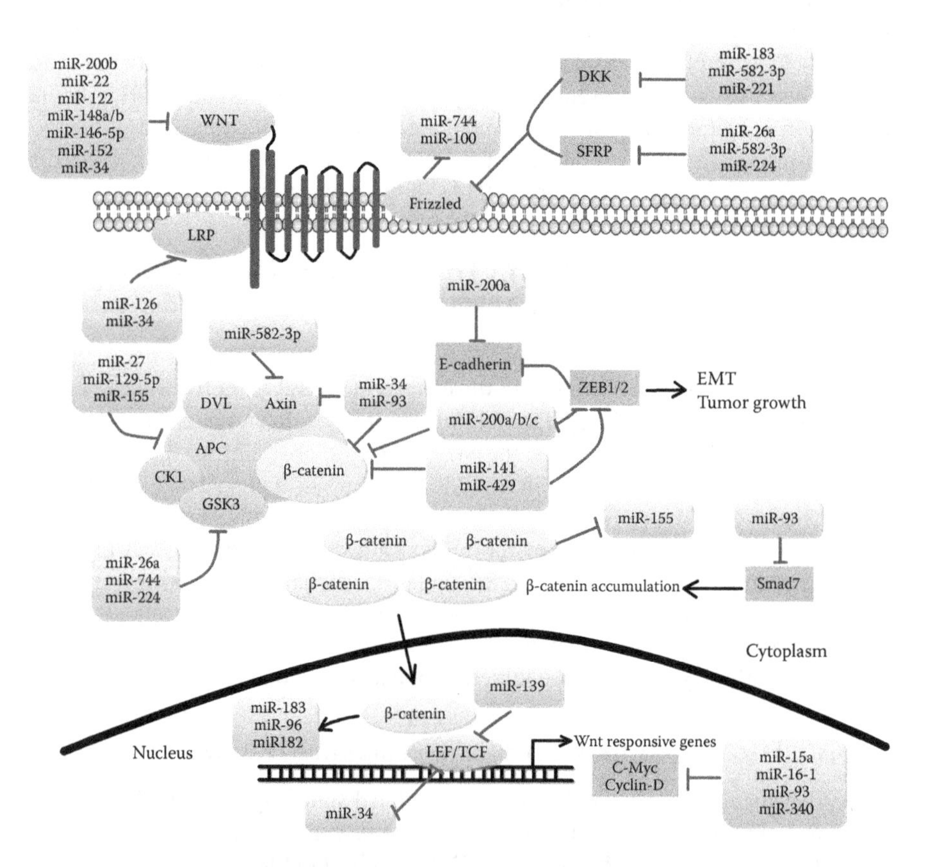

FIGURE 13.2 Major miRNAs involved in the regulation of the wnt/β-catenin signaling pathway. miRNAs activate or repress wnt/β-catenin signaling at multiple levels by targeting wnt ligands/receptors or ligand/receptor associated proteins, including β-catenin, and the β-catenin interacting complex, and multiple wnt signaling pathway components. activating interactions are indicated with arrows, whereas inhibitory interactions are shown as red t-lines.

WNT/β-catenin signaling pathway (Tang et al. 2013; Wen et al. 2015; Zhou et al. 2015; Jiang et al. 2016b; Jiang et al. 2016c). For example, miR-200b and miR-22 have been reported to directly target *WNT-1*-encoding mRNA; in particular, a mutation in the miR-200b and miR-22 "seed region" binding sites was shown to differentially regulate WNT/β-catenin signaling in gastric cancer. Upregulation of miR-22 and miR-200b not only synergistically inhibited gastric cancer growth, but also enhanced the antitumor effect of diallyl disulfide both *in vitro* and *in vivo* (Tang et al. 2013). Similarly, the *WNT-1*-encoding gene was also found to be the direct target of miR-122 and miR-148b in hepatocellular carcinoma; overexpression of these miRNAs inhibited WNT-1 expression, resulting in reduced expression of β-catenin and c-Myc and promotion of HCC cell growth (Zhang et al. 2015a; Ahsani et al. 2016). Other miRNAs such as miR-148a (Jiang et al. 2016c), miR-146a-5p (Du et al. 2015), and miR-152 (Dang et al. 2014) were also shown to target WNT ligand expression to inhibit the WNT/β-catenin signaling pathway. Another group of miRNAs (including miR-744, miR-100, and miR-126) act as tumor suppressors by targeting FZD, glycogen synthase kinase 3β (GSK3β), and LRP (Wen et al. 2015; Zhou et al. 2015; Jiang et al. 2016b). Overexpression of miR-744 promoted WNT/β-catenin signaling by directly targeting negative modulators of this pathway, including secreted FZD1, GSK3β, and transducin-like enhancer of split 3 in pancreatic cancer cells (Zhou et al. 2015). More recently, miR-100 was shown to inhibit the WNT/β-catenin pathway by targeting FZD8 expression, thereby suppressing the migration and invasion of breast cancer cells (Jiang et al. 2016b). In addition, miR-126 was shown to downregulate LRP6 to suppress papillary thyroid carcinoma cell proliferation (Wen et al. 2015).

β-catenin, the central component of the WNT/β-catenin signaling pathway, is involved in signal transduction by directly binding to TCF/LEF transcription factors and activating downstream gene expression to promote the epithelial-mesenchymal transition (EMT), cell proliferation, migration, and metastasis (Valenta et al. 2012). Several miRNAs have been found to suppress the WNT pathway by modulating β-catenin and the β-catenin-interacting complex, which is formed by Axin, the tumor suppressor adenomatous polyposis coli (APC), casein kinase 1 (CK1), and GSK3β. Members of the miR-200 family such as miR-200a/b/c, miR-141, and miR-429 were shown to downregulate β-catenin to inhibit cancer progression (Mongroo and Rustgi 2010; Abedi et al. 2015; Wu et al. 2016). In gastrointestinal cancer, miR-200a was shown to target the E-cadherin repressor, zinc finger ebox binding homeobox (ZEB) (Cong et al. 2013). Additional studies have demonstrated that the miR-200 family can be suppressed by ZEB, since it forms a double-negative feedback loop with miR-200 family (Bracken et al. 2008; Benlhabib et al. 2015). Exogenous injection of a miR-429 mimic dramatically inhibited the migratory ability of urothelial carcinoma cells by reducing ZEB1 and β-catenin expression (Wu et al. 2016). Regarding the β-catenin interacting complex, enhanced miR-582-3p expression was reported to promote cancer stem cell traits in non-small cell lung carcinoma (NSCLC) cells *in vitro* and tumorigenesis and tumor recurrence *in vivo* by simultaneously targeting multiple negative regulators of the WNT/β-catenin pathway, namely dickkopf WNT signaling pathway inhibitor 3 (DKK3), secreted frizzled-related protein 1 (SFRP1), and Axin2 (Fang et al. 2015). Indeed, the *GSK3*β and *SFRP2* genes were demonstrated to be direct targets of miR-224, and knock-down of miR-224 was shown to rescue the expression of GSK3β and SFRP2 and to attenuate WNT/β-catenin-mediated cell metastasis and proliferation *in vitro* and *in vivo* (Li et al. 2016d). In human osteosarcoma, GSK3β was identified as a target of miR-26a, and overexpression of miR-26a induced osteosarcoma cell growth and metastasis, and inhibited GSK3β through direct binding to the 3'-UTR (Qu et al. 2016). Another group of miRNAs, including miR-27 (Park et al. 2014), miR-129-5p (Li et al. 2013a), and miR-155 (Zhang et al. 2013b) was shown to target APC to regulate the WNT/β-catenin pathway.

In addition, miRNAs also modulate other transcriptional factors and co-activators/co-repressors, other than those mentioned above, to affect WNT signaling. For example, miR-15a and miR-16-1 can target *WNT3a*, *Bcl2*, and *cyclin-D1*, resulting in marked inhibition of prostate tumor xenograft growth (Bonci et al. 2008). In another study, miR-320 was shown to suppress the stem cell-like characteristics of prostate cancer cells by downregulating WNT/β-catenin signaling (Hsieh et al. 2013). Another study indicated that miR-183 can interact with *DKK3*, an inhibitor of the WNT signaling pathway, in prostate cancer (Ueno et al. 2013). Furthermore, miR-93 was reported to inhibit the WNT/β-catenin pathway and suppress colorectal cancer development by decreasing the expression of the β-catenin, Axin, c-Myc, and cyclin-D1. Additionally, the MAD-related protein Smad7, required for the nuclear accumulation of β-catenin, was identified as a target gene of miR-93 (Tang et al. 2015). A recent study indicated that miR-340 can target *c-Myc, CTNNB1* (encoding β-catenin), and *Rho/Rho-associated kinase 1* (ROCK 1). Restoration of miR-340 expression in an invasive breast cancer cells suppressed the mRNA and protein expression of these targets and consequently inhibited tumor cell invasion and metastasis (Mohammadi-Yeganeh et al. 2016).

Furthermore, miRNA expressions can be modulated by the WNT signaling pathway through the transcriptional activation of WNT pathway components. It has been reported that the miR-34 family links p53 activity to the WNT pathway through miR-34-specific interactions with the UTRs of multiple genes encoding components of the WNT pathway including *WNT1/3*, β-*catenin*, *LRP6*, *Axin2*, and *LEF1*, thereby inhibiting β-catenin-TCF/LEF-dependent transcriptional activity (Kim et al. 2011a; Liang et al. 2015; Zhu et al. 2015). In hepatocellular carcinoma, hepatitis C virus (HCV)-induced miR-155 expression inhibited hepatocyte apoptosis and promoted cell proliferation upon WNT/β-catenin pathway activation, forced expression of β-catenin, and negative regulation of miR-155 (Zhang et al. 2012). In a subsequent study using a luciferase reporter assay, *TCF-4* was demonstrated to be the direct target of miR-139. Restoration of TCF-4 activity resulted in effects similar to those of miR-139 inhibition in hepatocellular carcinoma (HCC) cells, specifically with regard to enhanced tumor cell growth, migration/invasion, and apoptosis inhibition. Indeed, forced expression of miR-139 suppressed β-*catenin/TCF-4* transcriptional activity by targeting TCF-4 (Gu et al. 2014). Recently, Wang and co-workers found that knock-down of miR-221 in 5-fluorouracil resistant esophageal cancer cells resulted in enhanced antitumor effects and downregulation of the WNT/β-catenin pathway, which was mediated by alterations in the expression of DKK2, the direct target of miR-221(Wang et al. 2016b). Additionally, the negative WNT regulator, SUFU is targeted by the oncogenic miR-194; this interaction was shown to promote gastric cancer cell proliferation and migration by activating WNT/β-catenin signaling (Peng et al. 2017b).

13.3.2 Targeting of the TGF-β/Smad Signaling Pathway by miRNAs

Transforming growth factor beta (TGF-β) acts as a tumor suppressor during cancer initiation but as a tumor promoter during tumor progression; as such, it plays a dual role in cancer during different stages of tumor progression and migration (Chen et al. 2016c). TGF-β binds to its receptors, the type I and type II receptors (TGFRI and TGFRII) leading to the activation of both the Smad family of transcription factors and non-Smad signaling pathways (Heldin et al. 2009). miRNAs have been shown to target both downstream and upstream factors that are involved in TGF-β signaling pathway regulation (Figure 13.3). miR-29 family members such as miR-29a and miR-29b act as downstream inhibitors of TGF-β signaling and collagen production (Qin et al. 2011). In cultured fibroblasts and tubular epithelial cells, Smad3 mediates TGF-β1-induced downregulation of miR-29 by binding to the promoter of miR-29. Overexpression of miR-29b was shown to inhibit TGF-β1-induced expression of collagen I and III in renal tubular cells (Qin et al. 2011). Furthermore, ultrasound-mediated gene delivery of miR-29b *in vivo* either before or after established obstructive nephropathy, was shown to block progressive renal fibrosis (Qin et al. 2011). In primary muscle cells, overexpression of miR-29 and miR-206 resulted in the translational repression of histone deacetylase 4 (HDAC4) in the presence or absence of TGF-β by targeting the 3'-UTR of *HDAC4* (Winbanks et al. 2011). In addition, miR-29a is a downstream target gene of *Smad3* and is negatively modulated by TGF-β/Smad signaling; this has been confirmed by the demonstration of protection against bleomycin or TGF-β1-induced inhibition of miR-29a and fibrosis in Smad3-null mice (Xiao et al. 2012). Recently, TGF-β2 was defined as a novel target of miR-29b. cAMP stimulation enhances, but hypoxia inhibits, the expression of miR-29 in cultured human fetal lung epithelial cells, whereas TGF-β2 expression is coordinately decreased (Guo et al. 2016). Systemic delivery of scAAV8-encoded miR-29a was shown to alleviate fibrosis despite continued exposure to carbon tetrachloride when administered before or after the onset of liver injury. Interestingly, a single injection of 2×10^{11} scAAV8-encoded miR-29a was sufficient to normalize hepatic miR-29a expression for more than four weeks but for less than eight weeks, suggesting that AAV-miR-29 represents a potential therapy for a variety of fibro-proliferative disorders (Knabel et al. 2015).

Members of the miR-200 family have also been shown to target TGF-β2. Studies indicate that during the differentiation of alveolar epithelial type II cell, upregulation of miR-200 is inversely correlated with the expression of its targets including ZEB and TGF-β2. Loss of miR-200 inhibits the expression of thyroid transcription factor-1 (TTF-1) and surfactant proteins and upregulates TGF-β2 and ZEB1 expression, and this effect is reversed by cAMP in type II cells. In addition, overexpression of ZEB1 was shown to suppress TTF-1 and inhibit miR-200 expression, providing evidence of a double-negative feedback loop (miR-200 family and ZEB1) that is regulated by TGF-β (Benlhabib et al. 2015). Subsequently, both miR-200b and miR-200c were

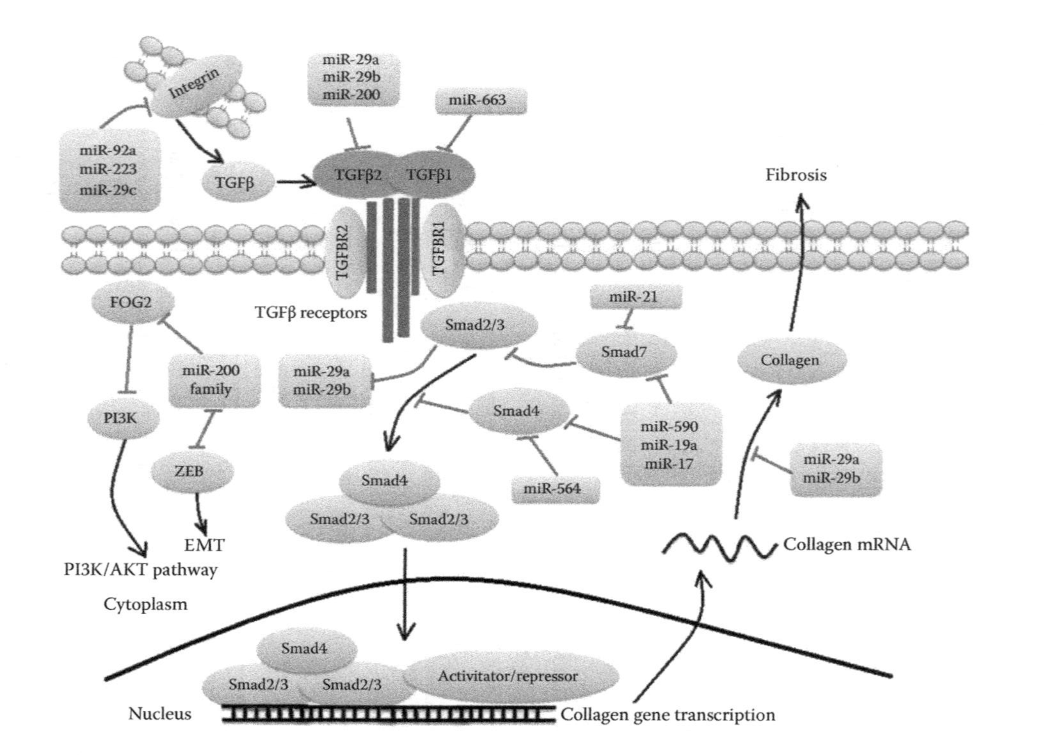

FIGURE 13.3 Major miRNAs involved in the regulation of by the tgf-β/smad signaling pathway. miRNAs can target factors that act both downstream and upstream factors of TGF-β and are involved in this pathway. some miRNAs can directly target TGF-β miR-29a/b and miR-663 or downstream signaling molecules of TGF-β1, namely smads (e.g., miR-21, miR-564, miR-590, miR-19a, and miR-17), to regulate TGF-β signaling. miRNAs such as miR-92a, miR-223, and miR-29c are involved in this signaling pathway by targeting integrins, the main activators of TGF-β. activating interactions are indicated with arrows, whereas inhibitory interactions are shown as red t-lines.

shown to target *FOG2*, an inhibitor of phosphatidylinositol 3-kinase activation, to modulate TGF-β induced Akt activation in glomerular mesangial cells. Suppression of miR-200b/c was shown to attenuate the TGF-β-induced decrease in FOG2 expression and to abrogate the TGF-β-induced increase in protein content per cell (Park et al. 2013). Other targets of miR-200c include Crtap, Fhod1, Smad2, Map3k1, Tob1, 14-3-3γ, 14-3-3β, Smad5, Zfp36, Xbp1, Mapk12, and Snail1, *Crtap, Fhod1, Smad2, Map3k1, Tob1, 14-3-3γ, 14-3-3β, Smad5, Zfp36, Xbp1, Mapk12, and Snail1* which have been experimentally validated by Perdigão-Henriques and colleagues. Among these factors, Smad2 and Smad5 form a complex with ZEB2, whereas 14-3-3β and 14-3-3γ form a complex with Snail1. Furthermore, knock-down of each of the three novel miR-200 target genes (*Smad5, Ywhag* and *Crtap*) results in suppression of cell invasion (Perdigão-Henriques et al. 2016). More recently, the long noncoding RNA, metastasis associated lung adenocarcinoma transcript 1 (MALAT1), was identified as a novel target of miR-200c; TGF-β increases MALAT1 expression by inhibiting miR-200c. In addition, targeting of the miR-200c/MALAT1 axis was shown to inhibit endometrioid endometrial carcinoma growth and EMT-associated protein expression in a xenograft tumor model and *in vitro*. Thus, these results indicate that the miR-200c/MALAT1 axis is a potential target for endometrioid endometrial carcinoma therapy (Li et al. 2016c).

In addition, some miRNAs have been reported to directly target TGF-β1 or downstream signaling molecules, such as Smads. For example, *TGF-β1* was identified as a direct target of miR-663 in glioblastoma; overexpression of TGF-β1 reversed the inhibitory effects of miR-663, promoting proliferation, migration, and invasion in glioblastoma cells (Li et al. 2016b). Wang and co-workers reported that overexpression of miR-21 enhanced TGF-β1-induced EMT by directly downregulating Smad7 (an inhibitory Smad that acts downstream of TGF-β1) and indirectly upregulating Smad3/p-Smad3, which aggravated renal damage during diabetic nephropathy (Wang et al. 2014a). Furthermore, miR-564, a novel tumor-suppressive miRNA, was found to be downregulated in human glioblastoma tissues and cell lines. *TGF-β1* was defined as a direct and functional target of miR-564 and was shown to enhance expression of miR-564 and decrease the expression of p-Smad and Smad4, which act downstream of TGF-β. In contrast, upregulation of miR-564 suppressed TGF-β-mediated proliferation and migration in human U-87 glioblastoma cells, suggesting that miR-564 could serve as a novel therapeutic target for treatment of glioblastoma (Jiang et al. 2016a).

miRNAs also target integrins, which are the main activators of TGF-β. For example, miR-92 targets integrin-β5, a key upstream regulator of TGF-β activation, which leads to inhibition of cancer cell adhesion, invasion, and proliferation. Forced expression of miR-92a in HeyA-8 cells suppressed peritoneal dissemination *in vivo* in ovarian cancer xenograft (Ohyagi-Hara et al. 2013). Overexpression of miR-29c in lung cancer cells led to significant downregulation of matrix metalloproteinase 2 and integrin β1 (Wang et al. 2013a). Additionally, *integrin β1* was also defined as a direct target of miR-29c in gastric cancer cells (Han et al. 2015). Moreover, *integrin β1* was identified as a target of miR-223, and overexpression of miR-223 recapitulated defects in growth factor receptor phosphorylation and AKT signaling via downregulation of integrin β1. Interestingly, reintroduction of integrin β1 into miR-223-overexpressing cells was shown to rescue growth factor signaling and angiogenesis

(Shi et al. 2013). In a recent study, miR-223 was reported to directly interact with both negative (FBXW7 and Acvr2a) and positive (IGF-1R and integrin-β1) regulators of AKT signaling in miR-223-transgenic hearts (a transgenic mouse model of cardiac-specific miR-223 overexpression), and eventually promoted a net increase in the activation of AKT, a key regulator of physiological cardiac hypertrophy. This study suggested that the ultimate phenotypic outcome of miRNA treatment might be decided by secondary net effects of the whole target network rather than effects on several primary direct targets in an organ or tissue (Yang et al. 2016).

13.3.3 Targeting of PI3K/AKT Pathway by miRNAs

miRNAs are also known to regulate tumor metastasis by targeting multiple key components of the PI3K/AKT pathway in cancers (Figure 13.4). Phosphatase and tension homolog (PTEN) is a phosphatase that negatively regulates the PI3K pathway. This protein is involved in tumor angiogenesis, which is mediated mainly through its effects on the PI3K pathway (Li et al. 1997). In colon cancer, PTEN was confirmed to be a target of the miR-17-92 cluster, which is responsible for chemotherapeutic drug resistance and metastasis (Fang et al. 2014). miR-92a was also shown to regulate PTEN expression and induce EMT in colorectal cancer cells (Zhang et al. 2014b). Additionally, PTEN is a functional downstream target of miR-32, which functions by directly targeting the 3'-UTR of *PTEN* in colorectal carcinoma cells (Wu et al. 2013). In liver cancer, *PTEN* was shown to be a direct target of miR-21 in HCC cells. Loss of miR-21 increased the expression of PTEN and decreased HCC proliferation, migration, and invasion, whereas upregulation of miR-21 inhibited PTEN expression and leads to activation of AKT and extracellular signal-regulated kinase pathways (Liu et al. 2011). Moreover, miR-21 and miR-155 are also overexpressed in NSCLC, which promotes the development of the disease, in part by downregulating suppressor of cytokine signaling 1(SOCS1), SOCS6, and PTEN. More importantly, it was shown that combined inhibition of miR-21 and miR-155 was more effective against NSCLC than treatment with a single inhibitor (Xue et al. 2016). miR-221 and miR-222 can also enhance cellular migration and tumorigenicity in breast, gastric, lung, and liver cancer cells by modulating PTEN expression. This occurs through binding to the 3'-UTR of *PTEN* mRNA (Garofalo et al. 2009; Chun-Zhi et al. 2010; Li et al. 2016a). In a recent study, Zhang and co-workers demonstrated the delivery of chemically modified anti-miR-221 to transferrin receptor-overexpressing HepG2 cells using negatively charged liposomes. This led to a 15-fold increase in delivery efficiency compared to that with non-targeting liposome in HepG2 cells. This was also accompanied by the upregulation of miR-221 target genes including *PTEN*, *P27*, and *TIMP metallopeptidase inhibitor 3* (*TIMP3*) (Zhang et al. 2015b). In addition, miR-29a increases HepG2 cell migration by targeting *PTEN* (Kong et al. 2011). Loss of miR-148 was reported to suppress cell proliferation, cell migration, and anchorage independent growth in soft agar and subcutaneous tumor formation in SCID mice through increased PTEN protein and mRNA expression (Yuan et al. 2012). In addition, miR-144 interacts with *PTEN* mRNA and downregulates its

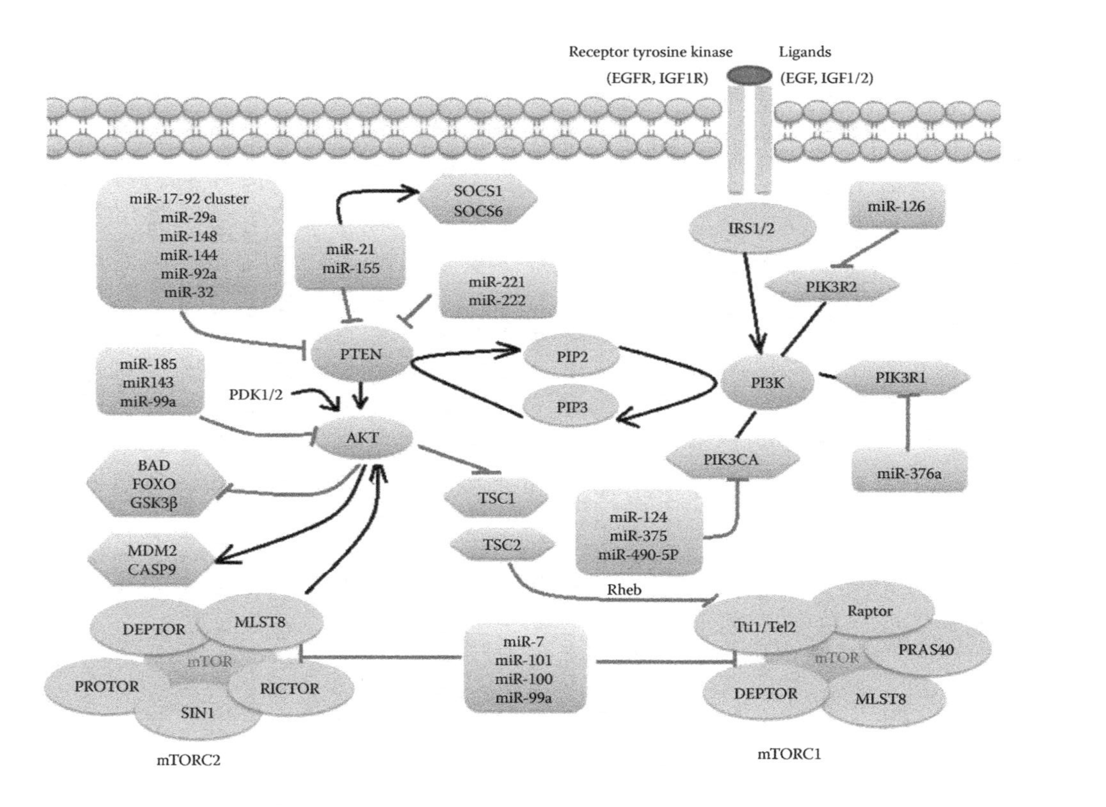

FIGURE 13.4 Major miRNAs interacting with the pi3k/akt pathway. miRNAs are capable of regulating tumor metastasis by targeting multiple key components including the negative regulator pten, a key downstream target of akt, and other downstream targets of akt, such as mtor, and pi3k, all of which are involved in the pi3k/ akt pathway in cancer. arrows and t-shaped lines indicate positive and negative interactions, respectively. pten: phosphatase and tension homolog; pi3k: phosphoinositide-3 kinase; mtor: mammalian target of rapamycin.

expression to promote cell proliferation, migration and invasion in nasopharyngeal carcinoma cells (Zhang et al. 2013a).

Additionally, miRNAs have also been shown to regulate PI3K/AKT pathway by targeting PTEN. For example, repression of *PTEN* by miR-144 leads to increased p-AKT activity (Zhang et al. 2013a). Indeed, *AKT* is a direct target of several miRNAs such as miR-143, miR-99a, and miR-185. Upregulation of miR-143 was also shown to impair insulin-stimulated AKT activation and glucose homeostasis (Jordan et al. 2011) and enforced expression of miR-185 or miR-99a exhibited antitumor effects in NSCLC (Li et al. 2015b; Yu et al. 2015b).

Phosphorylated AKT activates a multitude of downstream targets, including the mammalian target of rapamycin (mTOR), BCL2-associated agonist of cell death, Caspase9, Forkhead box O3 proteins, Tuberous sclerosis 1 (TSC1), Mouse double minute 2 (MDM2), and GSK3β (Danielsen et al. 2015). Among these targets, mTOR is a critical regulator of cell growth, proliferation, migration, invasion, and functions through two complexes, mTORC1 and mTORC2 (Alqurashi et al. 2013). Some miRNAs regulate mTOR in tumor cells, including miR-99a (Cui et al. 2012), miR-7 (Wang et al. 2013c), miR-100 (Xu et al. 2013), and miR-101 (Riquelme et al. 2016). In human renal cell carcinoma, restoration of miR-99a levels was shown to greatly attenuate tumor cell growth *in vitro* and *in vivo* by directly targeting the 3'-UTR of *mTOR* (Cui et al. 2012). Subsequently, inhibition of miR-7 was reported to activate mTOR signaling and promote adult β-cell proliferation in both mouse and human primary islets, implicating miR-7 as a potential therapeutic for diabetes (Wang et al. 2013c). Knock-down of mTOR was shown to enhance the antitumor effect of miR-100 in bladder cancer (Xu et al. 2013). Furthermore, exogenous expression of miR-101-2, miR-125b-2, and miR-451a inhibits the PI3K/AKT/mTOR pathway in gastric cancer through repression of their putative targets mTOR, PIK3CB, and TSC1, respectively (Riquelme et al. 2016).

miRNAs also target phosphoinositide-3 kinase (PI3K) in cancers. For example, miR-126 binds to the 3'-UTR of *phosphatidylinositol-3-kinase* regulatory subunit beta (*p85*β, also called as *PI3K*β), which is involved in the stabilization and propagation of PI3K signaling in colon cancers (Guo et al. 2008). PI3KR2 (PI3K regulatory subunit 2), a negative regulator of angiogenesis that acts through vascular endothelial growth factor (VEGF) pathway inhibition, can be targeted by miR-126 in breast cancer (Zhu et al. 2011). It was shown that overexpression of miR-126 suppresses, whereas inhibition of miR-126 promotes, SGC-7901 cell proliferation by balancing the expression of its target genes, specifically *PI3KR2*, *polo like kinase 2* (*PLK2*), and *Crk* (Zhu et al. 2011). In liver cancer, *PIK3CA* is a target of miR-124, demonstrated by the data that increased levels of miR-124 markedly suppressed cell proliferation by inducing G1-phase cell-cycle arrest and by downregulating PIK3CA *in vitro* and in xenograft animals (Lang and Ling 2012). *In vivo* and *in vitro* evidence supports that overexpression of miR-375 results in inhibition of tumor growth and the PI3K/AKT signaling pathway via direct binding to the 3'-UTR of *PIK3CA* in colorectal cancer (Wang et al. 2014b). Recently, miR-490-5p was confirmed to directly bind the 3'-UTR of *PIK3CA* mRNA and reduce the expression of PIK3CA. Knockdown of PIK3CA blocked whereas overexpression of PIK3CA reversed the inhibitory effect of miR-490-5p on renal cancer cell tumorigenicity (Chen et al. 2016a).

Furthermore, overexpression of miR-376a inhibits proliferation and induces apoptosis *in vitro* and *in vivo* by suppressing p85α (PIK3R1) (Zheng et al. 2012). In addition, a recent study in hepatocellular carcinoma suggested that *PI3KR1* is a direct target of miR-486-5p (Huang et al. 2015).

The role of the PI3K/AKT signaling in cancer metastasis has been deeply investigated in recent years, and several drugs targeting this pathway are under development (Danielsen et al. 2015). The miRNA targets discussed above, such as PIK3CA, PTEN, and AKT, are already considered attractive pharmacological targets, and this has been summarized by Danielsen et al. (2015). However, the development of secondary resistance, unanticipated feedback effects, and pathway cross-talk has complicated the efforts to design therapeutically effective compounds (Yu and Grady 2012). Because these targets can be regulated by miRNAs, miRNA-based therapies might be a good option for cancer treatment. Despite great advances in the understanding of cancer biopathology, miRNA-mediated regulation of multiple signaling pathways involved in cancer metastasis remains largely uncharacterized. Much work is required to decipher the functions of miRNAs with regard to these signaling molecules in cancers, in addition to delineating the cross-talk between miRNAs and other signaling networks implicated in cancer development.

13.3.4 Targeting of p53 Pathway by miRNAs

As a transcription factor, the tumor suppressor p53 is the most frequently altered gene in human cancers. It acts as a multifunctional transcription factor that controls DNA replication and repair and cell cycle progression in addition to playing important roles in tumor suppression, mostly through its downstream target genes (Liao et al. 2014). On the one hand, p53 regulates the transcriptional expression and the biogenesis of its downstream miRNAs, which makes these miRNAs function as effectors in cell cycle progression, proliferation, apoptosis, and EMT. These regulatory mechanisms contribute to the tumor suppressive function of p53 (Tarasov et al. 2007; Suzuki et al. 2009). On the other hand, miRNAs can function as mediators to modulate p53 expression and activity through direct targeting of its 3'-UTR or indirectly by inhibiting p53-modulator proteins (Pollutri et al. 2016). Thus, p53 might regulate miRNA transcription and processing, and miRNAs can conversely regulate p53 activity and expression (Table 13.1).

13.3.4.1 miRNAs as Effectors of the p53 Pathway

p53 can directly bind to the promoters of miRNA-encoding genes and control transcription. For example, miR-34a/b/c, which belongs to the highly conserved miR-34 family, was first defined as a direct transcriptional target of p53 (Tarasov et al. 2007). The promoters of each of these miRNA transcription units harbor functional p53 binding sites (Tarasov et al. 2007). Inhibition of miR-34 family members significantly reduces p53-dependent apoptosis in cell lines, whereas enhanced expression of these miRNAs induces cell-cycle arrest, apoptosis, or cellular senescence by suppressing the expression of various cell cycle-related genes including *cyclin-D*, *E2F transcription factors* (*E2Fs*), and *c-Myc*, thereby promoting tumorigenesis (Bommer et al. 2007; Chang et al. 2007; Tazawa et al. 2007). Restoration of miR-34 levels can

TABLE 13.1
List of Selected miRNAs Related to p53 Pathway

miRNAs	Expression	miRNA/p53 Relationship	Targets	miRNA Functions
miR-16	↓	Effector	*Weel, Chkl*	Tumor suppressor miRNA inducing cell cycle arrest, apoptosis (Lezina et al. 2013)
miR-26a	↓	Effector	*Weel, Chkl, cyclin D2, cyclin E2*	Tumor suppressor miRNA inducing cell cycle arrest, apoptosis (Zhou et al. 2016)
miR-630	↓	Effector	*CDC7*	Tumor suppressor miRNA inducing cell cycle arrest, apoptosis (Cao et al. 2014)
miR-490-3p	↓	Effector	*CDK1*	Tumor suppressor miRNA inducing cell cycle arrest, apoptosis (Chen et al. 2015)
miR-23b	↓	Effector	*cyclin G1*	Tumor suppressor miRNA inducing cell cycle arrest, apoptosis (Yan et al. 2016)
miR-34a	↓	Effector	*cyclin-D, E2F, c-Myc, Bcl2, MDM4, NR4A2*	Tumor suppressor miRNA decreasing tumor invasion and metastasis (Bommer et al. 2007; Chang et al. 2007; Tazawa et al. 2007; Okada et al. 2014; Beard et al. 2016)
miR-24	↑	Regulator	*p53*	Oncogenic miRNA increasing cell proliferation and tumor growth (Chen et al. 2016b)
miR-125b	↑	Regulator	*p53*	Oncogenic miRNA increasing cell proliferation and tumor growth (Le et al. 2009)
miR-33	↑	Regulator	*p53*	Oncogenic miRNA increasing cell proliferation and tumor growth (Herrera-Merchan et al. 2010)
miR-504	↑	Regulator	*p53*	Oncogenic miRNA increasing cell proliferation and tumor growth (Hu et al. 2010)
miR-18b	↑	Regulator	*MDM2*	Oncogenic miRNA increasing cell proliferation and tumor growth (Dar et al. 2013)
miR-339-5p	↓	Regulator	*MDM2*	Tumor suppressor miRNA decreasing tumor invasion and metastasis (Zhang et al. 2014a)
miR-660	↓	Regulator	*MDM2*	Tumor suppressor miRNA decreasing tumor invasion and metastasis (Fortunato et al. 2014)

(*Continued*)

TABLE 13.1 (CONTINUED)
List of Selected miRNAs Related to p53 Pathway

miRNAs	Expression	miRNA/p53 Relationship	Targets	miRNA Functions
miR-340	↓	Regulator	*MDM2*	Tumor suppressor miRNA decreasing tumor invasion and metastasis (Huang et al. 2016)
miR-1827	↓	Regulator	*MDM2*	Tumor suppressor miRNA decreasing tumor invasion and metastasis (Zhang et al. 2016a)
miR-17-5p/20a	↓	Regulator	*p21,TP53INP1*	Tumor suppressor miRNA decreasing tumor invasion and metastasis (Wang et al. 2013b)
miR-214	↓	Regulator	*COP1*	Tumor suppressor miRNA decreasing tumor invasion and metastasis (Heishima et al. 2015)

promote apoptosis in response to p53 activation through inhibition of the expression of anti-apoptotic proteins such as Bcl2, which prevents tumorigenesis (Ji et al. 2008). However, miR-34 deletion alone does not result in p53 dependent cell cycle arrest, apoptosis, or p53-dependent spontaneous tumorigenesis in mice. This is possibly because redundant pathways downstream of p53 compensate for miR-34 loss *in vivo* (Concepcion et al. 2012). A positive feedback loop between p53 and miR-34a during tumor suppression was demonstrated by Okada and co-workers in Kras-induced lung adenocarcinoma in mice. Here, the authors found that p53-miR-34a positive feedback was at least in part due to miR-34a-mediated repression of MDM4 (a negative p53 regulator). When this feedback loop was disrupted, oncogenic consequences were observed *in vivo* (Okada et al. 2014). More recently, an orphan nuclear receptor, nuclear receptor subfamily 4 group A member 2 (NR4A2), which is overexpressed in cancer and promotes cell proliferation, migration, and chemoresistance, was defined as the direct target of miR-34a and was shown to play a crucial role in the p53-miR-34 network. In this study, the authors found that exogenous miR-34 or activation of the p53 pathway decreased NR4A2 expression, whereas overexpression of NR4A2 blocked the induction of p53 target genes, including miR-34a, providing new evidence for cancer-related miRNA regulation of NR4A2 and a possible feedback mechanism involving p53, miR-34, and NR4A2 (Beard et al. 2016). Because miR-34 has been considered for clinical therapeutics, restoring miR-34a expression by epigenetic therapy or through delivery of miR-34a mimics is a promising therapeutic strategy to treat human cancer.

In addition to the miR-34 family, other tumor suppressor miRNAs can also act as p53 effectors; these include miR-16 (Lezina et al. 2013), miR-26a (Zhou et al. 2016), miR-630 (Cao et al. 2014), miR-490-3p (Chen et al. 2015), and miR-23b (Yan et al. 2016). Most of these miRNAs function by targeting several G1/S and G2 checkpoint proteins, including cyclin dependent kinase 1 (CDK1), CDC7, checkpoint kinase 1 (Chk1), cyclin G1, and cyclin E2 in response to p53 activation. For example, miR-16 and miR-26a were reported to target the G1/S checkpoint kinases Weel and Chk1 in response to p53 activation by genotoxic stress (Lezina et al. 2013). Indeed, miR-26a regulates mouse hepatocyte proliferation by directly targeting the 3'-UTR of *cyclin D2/cyclin E2* (Zhou et al. 2016). Similarly, in lung cancer, miR-630 was reported to inhibit proliferation by targeting CDC7 (Cao et al. 2014). Overexpression of miR-490-3p promotes G1/S or G2/M arrest and apoptosis by directly targeting CDK1; this is accompanied by reduced *Bcl-xL* and *cyclin D1* mRNA and protein expression, which was shown to inhibit ovarian epithelial carcinoma tumorigenesis and progression (Chen et al. 2015). More recently, cyclin G1, the key component of p53-dependent G1-S and G2 checkpoints, was defined as the direct target of miR-23b. Upregulation of miR-23b significantly decreases the expression of cyclin G1, urokinase, survivin, Bcl-xL, and matrix metallopeptidase-9 (Yan et al. 2016). Furthermore, miR-23b was shown to inhibit tumor growth and to suppress cyclin G1 expression *in vitro*, suggesting that miR-23b is a potentially novel target for regulating ovarian carcinoma progression (Yan et al. 2016).

In addition, p53 can regulate the maturation of miRNAs by modulating RISC complex function (Suzuki et al. 2009). The first evidence of the direct involvement of p53 in miRNA biogenesis was described by Suzuki and colleagues in 2009 (Suzuki et al. 2009). They found that the pri-miRNA processing activity of Drosha for pri-miR-16-1, miR-143, and miR-145 increased in a p53-dependent manner under DNA damage-inducing conditions in the HCT116 colon carcinoma cell line. During this process, p53 interacts with the Drosha complex through an association with the DEAD-box RNA helicase p68 to facilitate the processing of primary miRNAs to precursor miRNAs. Moreover, the authors demonstrated that overexpression of three tumor-derived transcriptionally inactive p53 mutants in the DNA binding domain (R273H, R175H, C135Y) in the p53-null HCT116 cell line decreased the precursor and mature miRNA levels (miR-16-1, miR-143 and miR-206) and decreased the interaction between Drosha and p68 (Suzuki et al. 2009). These data suggest that transcription-independent regulation of miRNA biogenesis is embedded in an anti-oncogenic program governed by p53. Recently, Garibaldi and colleagues discovered a novel mechanism underlying the mutant p53-dependent modulation of miRNA processing. They found that endogenous mutant p53 proteins (R273H and R175H) directly bind and sequester RNA helicases p72/82 through their N-terminal domains, interfering with Drosha-pri-miRNA association and inhibiting the biogenesis of a subset of miRNAs that is positively regulated by p72 in cancer cells (Garibaldi et al. 2016). It was also reported that loss of miRNAs biogenesis resulted in increased DNA damage and p53 activity, revealing a reciprocal connection between p53 and miRNA pathways (Mudhasani et al. 2008).

13.3.4.2 miRNAs as Mediators of the p53 Pathway

A subset of miRNAs, including miR-24 (Chen et al. 2016b), miR-125b (Le et al. 2009), miR-33 (Herrera-Merchan et al. 2010), and miR-504 (Hu et al. 2010), has been shown to modulate the activity of p53 through direct targeting of p53 (Table 13.1), whereas some other miRNAs target regulators of p53, such as MDM2 (Zhang et al. 2016a), MDM4 (Hoffman et al. 2014), and COP1 (Heishima et al. 2015), to indirectly regulate the activity and function of p53. This highlights the pivotal role of miRNAs in the p53 network and cancer. For example, miR-125a/b acts as negative regulators of p53, transcriptionally repressing *p53* mRNA (Le et al. 2009). Similarly, miR-504 decreases p53 expression and inhibits p53-mediated apoptosis and cell cycle arrest in response to stress, which strongly highlights the contribution of miR-504 to tumorigenesis (Hu et al. 2010). miR-33 was also shown to downregulate p53 by binding to two conserved motifs in the 3'-UTR of *p53* to control hematopoietic stem cell self-renewal (Herrera-Merchan et al. 2010). More recently, Chen and co-workers found that miR-24 can function as an oncogene to promote cell invasion in hepatocellular carcinoma by downregulation of p53, suggesting that miR-24 might be a potential therapeutic target for treatment of HCC (Chen et al. 2016b).

The E3 ubiquitin ligase MDM2 is the most important negative regulator of p53, and it binds directly to mediate ubiquitination-dependent degradation of p53. MDM4 shares structural homology with MDM2 and represses *p53* transcriptional activity by directly binding to the trans-activation domain or by promoting MDM2-mediated degradation (Wade et al. 2013). Recent studies indicated that miRNAs can regulate the p53 pathway by targeting MDM2 and MDM4. For example, miR-18b (Dar et al. 2013), miR-339-5p (Zhang et al. 2014a), miR-660 (Fortunato et al. 2014), miR-340 (Huang et al. 2016), and miR-1827 (Zhang et al. 2016a) target MDM2 to increase p53 levels and function, thereby suppressing tumorigenesis. Similarly, miR-34a also decreases p53 protein levels through seed-matching sequences in the 3'-UTR of *MDM4* (Okada et al. 2014). Indeed, miR-34a is a direct target of *p53* and therefore forms a positive feedback loop with p53. Furthermore, miR-17-5p/20a indirectly downregulates MDM2 by targeting p21 and tumor protein p53-induced nuclear protein 1, promoting gastric cancer cell proliferation (Wang et al. 2013b). In addition, the COP1 E3 ubiquitin ligase, a negative regulator of p53, was shown to be a direct target of miR-214, and it functions as a tumor suppressor in many cancers. Restoration of miR-214 expression was shown to reduce cell growth and to induce apoptosis in canine hemangiosarcoma cells (Heishima et al. 2015).

It is necessary to also point out that p53-regulated miRNAs regulate not only the p53 pathway, but also other pathways such as the Notch (Hsu et al. 2016), TGF-β/Smad (Li et al. 2016e), PI3K/AKT (Jiang et al. 2014), and Wnt/β-catenin signaling pathways (Kim et al. 2011a). Therefore, the cross-talk between miRNAs and the p53 network requires further investigation to expand our understanding of the roles of p53 in tumor suppression and to promote the utilization of miRNAs in diagnostic and therapeutic strategies for cancers.

13.4 CURRENT STRATEGIES FOR miRNA-BASED THERAPIES IN SEVERAL DISEASES

One advantage of miRNA-based therapies over targeting of a single oncogene is the ability of a single miRNA to target many genes and signaling pathways at the same time. Therefore, targeting by a single miRNA could lead to dramatic results due to the ability to affect numerous biological processes. Importantly, individual miRNAs, combinations of miRNAs, and even artificial miRNAs can be designed specifically to target entire biological pathways, making them good candidates for therapeutic intervention when compared to classical single target approaches (Van Rooij et al. 2012). Generally, the most frequently proposed modality of miRNA-targeted therapy is to silence overexpressed miRNAs through antisense miRNA oligonucleotides (anti-miRs) (Krutzfeldt et al. 2005) or to restore levels of downregulated miRNAs by applying miRNA mimics. A schematic diagram of miRNA therapeutic strategies and a summary of current progress are provided in Figure 13.1 and Table 13.2, respectively.

13.4.1 Anti-miRs

Anti-miRs are chemically modified oligoribonucleotides that are widely used to inhibit miRNAs expression *in vitro* and *in vivo*. These chemically modified oligonucleotides contain sequences with reverse complementarity to mature miRNAs, and can be delivered by subcutaneous or intravenous injection, possibly by packaging them into lipid-based and/or antibody-conjugated nanoparticles. These molecules effectively block the association of miRNAs with their targets, and might therefore be used for disease treatment (Krutzfeldt et al. 2005). To date, some chemical modifications that enhance the cellular uptake and stability of anti-miRs have been reported, including 2-O-methyl (2'-OMe)-modification or locked nucleic acid (LNA) modification, in which the 2'-oxygen and the 4' carbon of the ribose moiety are covalently linked. This results in the formation of a thermodynamically strong duplex with the complementary RNA, thereby enhancing cellular oligonucleotide cellular uptake and *in vivo* stability, thus optimizing effectiveness (Weiler et al. 2006; Petersen and Wengel 2003). However, the action of anti-miRNA oligonucleotides (AMO) is sequence-specific but not gene-specific, which might elicit off-target side effects and unwanted toxicity. Alternatively, a novel strategy called "miR-mask" was established by Xiao and colleagues (Xiao et al. 2007). The miR-mask is a single-stranded 22-nt antisense molecule with perfect complementarity to the endogenous miRNA binding site in the target gene.

13.4.2 miR-Mask

Rather than binding to the target miRNA similarly to an AMO, the miR-mask directly binds to the endogenous miRNA-binding site in the 3'-UTR of the target mRNA via its complementarity. This blocks the access of endogenous miRNA to its binding site, thereby de-repressing the target gene (Wang 2011). The miR-mask approach is a valuable supplement to the AMO technique, which might be more appropriate for studying the specific outcome of target gene regulation by endogenous miRNA. For example, in a recent study, the DENN/MADD domain containing 2D (DENND2D) was confirmed to be a direct target of miR-522, which is upregulated in NSCLC samples and cells. A miR-Mask designed to be fully complementary to the target DENND2D sequence of miR-522 reversed the effects of miR-522 on tumor growth and metastasis, indicating that miR-522 represents a potential novel therapy for NSCLC (Zhang et al. 2016b).

TABLE 13.2
Selected Strategies of Reinstatement or Inhibition of miRNA Expression for Therapeutic Proposes

Classification	Purposes	Advantages	Limitations	Clinical Status
miRNA mimics	miRNA restoration	Chemically modified double stranded miRNA molecules	Must be recognized by Argonaute protein and incorporated into RISCs to exert their action Do not tolerate extensive modification, susceptible to nuclease degradation	Formulated miR-34a mimics with liposome (MRX34) in the Phase I clinical (NCT01829971) miR-16 mimic evaluated for malignant pleural mesothelioma treatment is in the Phase I clinical (NCT02369198)
Anti-miRs oligonucleotides (AMOs)	miRNA inhibition	Chemically modified oligoribonucleotides designed to be complementary to the target miRNA Delivered by means of subcutaneous or intravenous injection nanoparticles	Elicit off-target side effects and unwanted toxicity	Preclinical
locked nucleic acid (LNA) modification of oligonucleotides	miRNA inhibition	Display higher aqueous solubility and increase metabolic stability for *in vivo* delivery	Only moderate efficiency for miRNA inhibition	Anti-miR-122 (SPC3649, NCT00979927) for HCV treatment in the Phase II
miR-Mask	miRNA inhibition	Perfect complementarity to the binding site for an endogenous miRNA in the 3'-UTR of target gene, blocking the access of endogenous miRNA to its binding site so as to de-repress the target gene Reduces the off-target effects and is highly target specific	N/A	Preclinical
miRNA sponges	miRNA inhibition	Block a whole family of related miRNAs that have the same target sites	Degraded by Argonaute 2 in the RISC and hold weaker inhibitory activity	Preclinical

13.4.3 miRNA Sponges

Artificial miRNA sponges can also be used to achieve miRNA loss of function in diseases, particularly in cancer cells (Brown et al. 2007; Ebert et al. 2007; Tay et al. 2015). miRNA sponges, which contain binding site complementary to the miRNAs of interest, are competitive inhibitors that are expressed from strong promoters and are designed based on base pairing in the seed region (nucleotides 2–8 from the 5' end of the miRNA). The molecules are either perfectly complimentary or contain mismatches in the middle position, which most likely results in a more stable interaction with miRNAs (Ebert and Sharp 2010). These binding sites are usually tandem repeats of identical sites that are designed to target either single specific miRNAs or miRNA family members sharing the same seed region; this allows these molecules to block a whole family of related miRNAs that have the same target sites. To achieve stable miRNA sponge activity, miRNA sponge expression cassettes usually consist of an RNA polymerase II promoter to drive transcription, a reporter gene to monitor efficacy, and multiple miRNA binding sites downstream of the reporter gene in the 3'-UTR. To deliver the miRNA sponge expression cassette into cells, viral and non-viral gene transfer vectors have been used, resulting in inhibitory effects on miRNA functions. For *in vivo* studies, viral vectors are used to deliver sponge constructs to mouse tissues (Du et al. 2009), whereas non-viral delivery systems consisting of plasmids and synthetic and usually positively charged materials such as liposomes, polysaccharides, polypeptides, and artificial polymers have been used in cells (Ando et al. 2013; Celec and Gardlik 2017). The advantage of this technique compared to anti-miRs is that it can better inhibit functional classes of miRNAs, whereas the antisense inhibitors appear to be specific for one miRNA because they depend on extensive sequence complementarity beyond the seed region (Esau 2008). In addition, because many cells are resistant to uptake of oligonucleotides both *in vitro* and *in vivo*, delivery of the sponge transgene by a viral vector might warrant consideration.

13.4.4 miRNA Mimics

A therapeutic response can be achieved by overexpressing some miRNAs. One experimentally proven way to enhance miRNA effects is using an miRNA mimic. These molecules are double-stranded RNAs in which one strand (guide strand) is identical to the sequence of a mature miRNA and the other (passenger strand) is deliberately designed to have partial complementary, thus functionally mimicking the mature endogenous miRNA. Because the miRNA mimics possess the same structure as the miRNA duplex, they can be directly loaded into RISC complex, and the passenger strand is cleaved by Ago2. The guide strand remains and matches the complementary sequences of the target mRNA, leading to mRNA cleavage or translational repression (Chen et al. 2015). However, miRNA mimics have limitations with regard to therapeutic applications. Similar to endogenous miRNAs, miRNA mimics need to be recognized by Argonaute and incorporated into RISCs to exert their effect. In contrast to anti-miRs, which can be chemically modified to enhance stability and binding, miRNA mimics do not tolerate extensive modification. Additionally, miRNA mimics are still susceptible to nuclease degradation, can be targeted by the innate immune system, and affect non-target tissues (Rachagani et al. 2015). In light of these aspects, to further enhance therapeutic delivery, recent advances have been achieved in lipid-based delivery systems for enhancing the uptake of miRNA mimics. Liposomes of smaller diameter (<100 nm) enable loading with a high drug-to-lipid ratio. The first miRNA mimics to be tested in humans was delivered intravenously to treat liver cancer using a liposome formulation (Ling et al. 2013). Liposomes have also been successfully used to deliver miR-34a mimic (MRX34) intravenously in a mouse model of NSCLC, resulting in a reduction in tumor growth, without significant renal or hepatic toxicity and without triggering an immune response (Trang et al. 2011). The delivery of miRNA mimics in disease settings via adenoviral methods also warrants consideration for enhancing miRNA construct (Kota et al. 2009). To date, various serotypes of adenoviruses with different affinities for various organs to aid in the targeting of miRNAs have proven effective in murine disease models (Kota et al. 2009).

miRNA-based therapeutic approaches have shown promise in ongoing preclinical trials for the treatment of certain diseases; however, there are only a few miRNA drugs in clinical trials, including one antagomiR of miR-122 (SPC3649, NCT00979927), which is currently in Phase II, and one liposomal-formulation of double-stranded mimic of the tumor suppressor miR-34 (MRX34, NCT01829971), which is in Phase I for treatment of liver cancer (Lanford et al. 2010; Mirna Therapeutics 2013; Bouchie 2013). In addition, a miR-16 mimic is also in Phase I clinical trials for malignant pleural mesothelioma (MPM) treatment (NCT02369198; Kao et al. 2015). After MRX34 systemic injection, significant tumor growth inhibition and regression were observed in mice with orthotopic Hep3B and HuH7 liver cancer xenografts (Daige et al. 2014). In particular, systemic injection of MRX34 into immunocompetent mice did not elicit apparent changes in cytokine profiles or result in any toxicity. Additionally, significant tumor growth repression was observed in mouse xenograft models of different cancers, such as melanoma, lymphoma, multiple myeloma, and NSCLC, and breast, prostate, and pancreatic cancer (reviewed by Bader 2012). Systemic miR-34a delivery in a mouse model was not associated with any severe toxicity, based on phenotypic and pathologic observations (Li et al. 2015a; Craig et al. 2012), suggesting that miR-34a could be used for other diseases in future clinical trials. Second, coupling of a miR-16 mimic with an anti-EGFR antibody (EnGeneIC Delivery Vehicle, EnGeneIC, New York, USA), restored miR-16 expression in MPM cells, *in vitro* and *in vivo,* leading to significant tumor growth inhibition (Reid et al. 2013), providing the basis for the ongoing MesomiR 1 Phase I trial for MPM treatment using TargomiR, an experimental therapy consisting of synthetic miR-16-based microRNA mimics packaged in EDV nanocells and targeted with anti-EGFR antibodies (MacDiarmid and Brahmbhatt 2011). Kao and colleagues subsequently published results from a cohort of six patients with MPM, in which TargomiR treatment was shown to be well-tolerated and safe (Kao et al. 2015). The third, alternative miRNA-based therapeutic to reach the clinic is an LNA-modified anti-miR (SPC3649) directed against the liver-specific miRNA miR-122. This miRNA is known to positively regulate HCV replication through direct interaction with the 5'-UTR of the HCV genome (Li et al. 2013b). In preclinical studies, it was reported that treatment with SPC3649 provided long-lasting suppression of HCV replication without side effects in chimpanzees chronically infected with HCV (Lanford et al. 2010). Furthermore, safety studies on SPC3649 showed no apparent toxic effects and no changes in liver morphology or function (Elmen et al. 2008; Hildebrandt-Eriksen et al. 2012). Two Phase I studies (NCT00688012 for a single ascending dose and NCT00979927 for multiple ascending doses) were performed to evaluate the safety of SPC3649 (Lindow and Kauppinen 2012). In a Phase II study, patients with chronic HCV genotype 1 infection receiving five weekly subcutaneous injections of anti-miR-122 showed a dramatic dose-dependent reduction in HCV RNA levels together with a decrease in plasma cholesterol levels, without evidence of viral resistance. In addition, dose-limiting toxicity and desensitizing mutations in the HCV genome were not observed (Janssen et al. 2013). Although proof-of-concept studies have demonstrated the tumor suppressor function of miR-122, it is necessary to point out that strategies for miR-122 silencing might not be suitable for the treatment of patients with dual HBV and HCV infection because endogenous miR-122 depletion resulted in increased production of HBV in transfected cells (Hao et al. 2013).

Overall, these miRNA-based therapeutic strategies have been well tolerated, without apparent inflammatory reactions in human clinical trials. Furthermore, recent studies have reported that co-delivery of chemotherapeutics and cancer-related miRNAs such as miR-34a, miR-21, and miR-145, or the combination of miRNAs with other anticancer therapies such as siRNA, results in potential synergistic effects on tumor growth inhibition (Kopczynska 2015; Devulapally et al. 2015; Kim et al. 2011b). Nanocarrier-mediated co-delivery systems have been the most commonly used systems for the delivery of chemotherapeutic drugs in recent years. For instance, combination of siRNA and miR-34a delivery by GC4 (phage identified internalizing)-targeted nanoparticles additively enhanced the anticancer effect when compared with treatment with siRNAs and miR-34a alone (Chen et al. 2010). Formulation of anti-miR-21 and 4-Hydroxytamoxifen co-loaded biodegradable polymer nanoparticles had dose-dependent anti-proliferative effects that were significantly higher than the effect of treatment with 4-hydroxytamoxifen alone (Devulapally et al. 2015). More recently, a core-shell nanocarrier coated with cationic albumin was developed to simultaneously deliver docetaxel and miRNA-34a into breast cancer cells. The results indicated that this new nanoplatform effectively improved anticancer efficacy by inhibiting tumor growth and metastasis, and provides a promising strategy for treatment of metastatic breast cancer (Zhang et al. 2017). Although various nanocarriers have been developed, there is much attention needed to develop the rational ratio of miRNA, gene, and drug, or the interaction between therapeutic agents and carriers. Consequently, it is also necessary to design new strategies combining miRNA-based therapies with other chemotherapeutic agents or anticancer drugs to evaluate the effect of these combinations. Further, long-term and large-scale studies are warranted. Notably, in addition to miRNA drugs that are currently in clinical trials, other miRNAs are being tested in preclinical settings, such as miR-143 and miR-145 for human colorectal tumors (Akao et al. 2010), let-7 for hepatocellular carcinoma (Liu et al. 2014), miR-16 for Alzheimer's disease (Parsi et al. 2015), and miR-221 for multiple myeloma (Gulla et al. 2016). Several challenges such as delivery, stability, and avoiding the activation of immune responses, need to be overcome; these issues will be discussed as follows.

13.5 CHALLENGES FOR THE CLINICAL DEVELOPMENT OF miRNA-BASED THERAPIES

Despite recent improvements in the understanding of miRNAs, the translation of bench-based findings to clinical practice remains a considerable task (Gavrilov and Saltzman 2012). Several challenges such as off-target effects, cytotoxicity, immunogenicity, and biological instability need to be overcome to meet the demands of clinical therapy (Iorio and Croce 2012).

Since a single miRNAs can target multiple genes and each gene can be targeted by multiple miRNAs, it is not surprising that there are some off-target effects observed with miRNA-based therapies. Factors contributing to these effects include miRNA-like repression, competition for limited resources of the endogenous miRNA pathway, incorporating passenger strands into Ago during loading, and immune responses (Bofill-De Ros and Gu 2016). In addition, systemic administration of anti-miRs will also inhibit the targeted miRNA in other tissues or cell types, which could lead to adverse effects in non-diseased tissues (Van Rooij et al. 2012). To overcome these drawbacks, it is important to understand the functional interactions between these targets as well as the contribution of each to the observed phenotype. Additionally, the design of novel carrier platforms as well as small changes in miRNA sequences (specific base modifications within the siRNA duplex) can dramatically reduce off-target effects while maintaining desirable on-target effects (Jackson et al. 2006). In addition, bioinformatics-based algorithms can also be used to predict off-target effects on a genome-wide scale as suggested by Sigoillot et al. (2012).

miRNA mimics and anti-miRs accumulate predominantly in the liver and kidneys, necessitating substantially higher doses to achieve efficacy in the cardiovascular system and other organs. This raises challenges with respect to achieving intracellular concentrations in tissues that are sufficient to evoke a therapeutic effect without causing hepatic and renal toxicity (Van Rooij et al. 2012). Since miRNAs act on numerous targets, an individual miRNA might have beneficial activity in one tissue but adverse activity in another. In addition, miRNAs might produce different effects in different locations due to regional differences in the cellular environment. In this regard, appropriate modification and conjugation methods for target delivery can reduce such effects.

Due to the physiological and immunological features of human body, miRNA mimics and anti-miRs are subject to renal clearance and phagocytic activity by monocytes and macrophages. Consequently, the initial action of miRNA interference appears to be delayed and to occur with low efficiency (Grueter et al. 2012; Guzman-Villanueva et al. 2012; Van Rooij et al. 2012). In addition, many of the miRNAs involved in tissue pathology have also been implicated as tumor suppressors *in vivo*. Although the potential long-term use of anti-miRs could enable effective treatment of chronic diseases, periodic dosing could lead to anti-miR accumulation in different tissues and the inability to rapidly reverse the activities of these molecules (Grueter et al. 2012). Moreover, many disease conditions involve the aberrant expression of multiple miRNAs, and thus targeting one single miRNA might not be sufficient to correct the phenotype. The effects of any individual miRNA on a single target might be subtle; however, the combinatorial effects of one miRNA on multiple mRNA targets within a regulatory network could profoundly change the output of a pathway.

The biological instability of miRNA mimics or anti-miRs is one of other weaknesses toward clinical translation of these. miRNA mimics are susceptible to nuclease degradation, and they cannot be extensively modified. In general, non-modified "naked" oligonucleotides are rapidly degraded by the abundant nucleases such as serum RNase A in the bloodstream (Raemdonck et al. 2008). Furthermore, the naked miRNAs are cleared rapidly from the body *via* renal clearance, leading to a short half-life in the systemic circulation (Yu et al. 2009). The reticuloendothelial system is another major barrier, in which Kupffer cells of the liver and spleen macrophages eliminate these oligonucleotides from the circulation system, which results in non-specific uptake by innate immune cells such as monocytes and macrophages (Garzon et al. 2010). One strategy to overcome this obstacle and increase the bioavailability of miRNA therapeutics is site-specific delivery by direct injection or other techniques into the target tissue, which has been performed by intratumor injection into subcutaneous tumors. Second, the degradation and elimination of miRNAs by nucleases, renal clearance, or phagocytic immune cells can be minimized by optimizing particle size, surface charge, and chemical modification of the miRNAs. Finally, various targeting ligands and cell-penetrating moieties can be employed to increase tissue permeation of the miRNAs.

13.6 FUTURE PROSPECTS

Owing to the complexity and challenges associated with miRNA research, the development of miRNA-based therapeutics is in its infancy. The use of this class of therapeutics represents a new method to treat some of the more challenging diseases including chronic infection, cancer, and other non-cancer diseases; however, further research is needed to determine the optimal formulations and to achieve precise delivery, avoiding unwanted effects that could result from targeting important genes in healthy tissues.

Before these miRNA-based therapies can be subjected to clinical trials, some strategies should be considered. First, further large-animal studies are needed to prove the utility of miRNA therapeutics. Second, different available chemistries, used to modulate (overexpress/silence) miRNAs, should be compared in various models and tissues, and toxicology testing must be performed. Furthermore, the identification of individual miRNA targets, especially in disease conditions, needs to be identified, as alteration of the targets during disease could affect the efficacy of miRNA-based therapeutics. Finally, apart from these clinical delivery platforms, several novel siRNA or miRNA delivery systems are under development. One such effort includes conjugation of spherical nucleic acid nanoparticles with gold nanoparticle cores, adenoviruses, and sponges, which has shown promise in several preclinical studies (Wang et al. 2016a; Kim et al. 2011b; Tay et al. 2015). If the above approaches continue to show success and enter the clinic, the use of miRNAs as therapeutics could emerge as a key technology for the treatment of cancer and other diseases.

In addition, given the ever-expanding number of miRNAs, understanding their functional roles is important. However, much remains to be learned regarding the precise mechanisms of miRNA activity within complex disease pathways and whether such modest regulators can, indeed, be modulated in cancer. To fully reveal the mechanisms of miRNA function and to clinically apply miRNAs to tumor therapy, a substantial amount of further work is required. Technologies that integrate RNA sequencing, proteomics, and systems biology of gene networks will enable a more comprehensive assessment and understanding of miRNAs. This will provide exciting opportunities to develop new treatment options for cancer management. Future work on miRNA pathways as therapeutic targets might also enable the development of chemotherapy regimens that are more specifically tailored to treat the disease process.

ACKNOWLEDGMENTS

This study was, in part or in whole, supported by National Natural Science Foundation of China (Grants No. 31472187, 31672550 and 31502056) and the Sci & Tech Innovation Program of the Chinese Academy of Agricultural Sciences.

REFERENCES

Abedi, N., S. Mohammadi-Yeganeh, A. Koochaki, et al. 2015. miR-141 as potential suppressor of beta-catenin in breast cancer. *Tumour Biol* 36 (12):9895–9901.

Ahsani, Z., S. Mohammadi-Yeganeh, V. Kia, et al. 2016. WNT1 gene from WNT signaling pathway is a direct target of miR-122 in hepatocellular carcinoma. *Appl Biochem Biotechnol* 181 (3):884–897.

Akao, Y., Y. Nakagawa, I. Hirata, et al. 2010. Role of anti-oncomirs miR-143 and -145 in human colorectal tumors. *Cancer Gene Ther* 17 (6):398–408.

Alqurashi, N., S. M. Hashimi, and M. Q. Wei. 2013. Chemical inhibitors and microRNAs (miRNA) targeting the mammalian target of Rapamycin (mTOR) pathway: Potential for novel anticancer therapeutics. *Int J Mol Sci* 14 (2):3874–3900.

Ando, H., A. Okamoto, M. Yokota, et al. 2013. Polycation liposomes as a vector for potential intracellular delivery of microRNA. *J Gene Med* 15 (10):375–383.

Bader, A. G. 2012. miR-34-a microRNA replacement therapy is headed to the clinic. *Front Genet* 3:120.

Bartel, D. P. 2009. MicroRNAs: target recognition and regulatory functions. *Cell* 136 (2):215–233.

Beard, J. A., A. Tenga, J. Hills, et al. 2016. The orphan nuclear receptor NR4A2 is part of a p53-microRNA-34 network. *Sci Rep* 6:25108.

Benlhabib, H., W. Guo, B. M. Pierce, et al. 2015. The miR-200 family and its targets regulate type II cell differentiation in human fetal lung. *J Biol Chem* 290 (37): 22409–22422.

Bhattacharya, S., R. Aggarwal, V. P. Singh, et al. 2015. Down-regulation of miRNAs during delayed wound healing in diabetes: Role of Dicer. *Mol Med* 21 (1): 847–860.

Bofill-De Ros, X., and S. Gu. 2016. Guidelines for the optimal design of miRNA-based shRNAs. *Methods* 103:157–166.

Bommer, G. T., I. Gerin, Y. Feng, et al. 2007. p53-mediated activation of miRNA34 candidate tumor-suppressor genes. *Curr Biol* 17 (15):1298–1307.

Bonci, D., V. Coppola, M. Musumeci, A., et al. 2008. The miR-15a-miR-16-1 cluster controls prostate cancer by targeting multiple oncogenic activities. *Nat Med* 14 (11):1271–1277.

Bouchie, A. 2013. First microRNA mimic enters clinic. *Nat Biotechnol* 31 (7):577.

Bracken, C. P., P. A. Gregory, N. Kolesnikoff, et al. 2008. A double-negative feedback loop between ZEB1-SIP1 and the microRNA-200 family regulates epithelial-mesenchymal transition. *Cancer Res* 68 (19):7846–7854.

Brown, B. D., B. Gentner, A. Cantore, S., et al. 2007. Endogenous microRNA can be broadly exploited to regulate transgene expression according to tissue, lineage and differentiation state. *Nat Biotechnol* 25 (12):1457–1467.

Calin, G. A., C. D. Dumitru, M. Shimizu, et al. 2002. Frequent deletions and down-regulation of micro- RNA genes miR15 and miR16 at 13q14 in chronic lymphocytic leukemia. *Proc Natl Acad Sci U S A* 99 (24):15524–15529.

Cao, J. X., Y. Lu, J. J. Qi, et al. 2014. MiR-630 inhibits proliferation by targeting CDC7 kinase, but maintains the apoptotic balance by targeting multiple modulators in human lung cancer A549 cells. *Cell Death Dis* 5:e1426.

Celec, P., and R. Gardlik. 2017. Gene therapy using bacterial vectors. *Front Biosci (Landmark Ed)* 22:81–95.

Chang, T. C., E. A. Wentzel, O. A. Kent, K., et al. 2007. Transactivation of miR-34a by p53 broadly influences gene expression and promotes apoptosis. *Mol Cell* 26 (5):745–752.

Chen, K., J. Zeng, K. Tang, et al. 2016a. miR-490-5p suppresses tumour growth in renal cell carcinoma through targeting PIK3CA. *Biol Cell* 108 (2):41–50.

Chen, L., L. Luo, W. Chen, et al. 2016b. MicroRNA-24 increases hepatocellular carcinoma cell metastasis and invasion by targeting p53: miR-24 targeted p53. *Biomed Pharmacother* 84:1113–1118.

Chen, S., X. Chen, Y. L. Xiu, et al. 2015. MicroRNA-490-3P targets CDK1 and inhibits ovarian epithelial carcinoma tumorigenesis and progression. *Cancer Lett* 362 (1): 122–130.

Chen, W., S. Zhou, L. Mao, et al. 2016c. Crosstalk between TGF-beta signaling and miRNAs in breast cancer metastasis. *Tumour Biol* 37 (8):10011–10019.

Chen, Y., D. Y. Gao, and L. Huang. 2015. *In vivo* delivery of miRNAs for cancer therapy: challenges and strategies. *Adv Drug Deliv Rev* 81:128–141.

Chen, Y., X. Zhu, X. Zhang, et al. 2010. Nanoparticles modified with tumor-targeting scFv deliver siRNA and miRNA for cancer therapy. *Mol Ther* 18 (9):1650–1656.

Cheng, G. 2015. Circulating miRNAs: roles in cancer diagnosis, prognosis and therapy. *Adv Drug Deliv Rev* 81:75–93.

Chiosea, S., E. Jelezcova, U. Chandran, et al. 2006. Up-regulation of Dicer, a component of the MicroRNA machinery, in prostate adenocarcinoma. *Am J Pathol* 169 (5):1812–1820.

Chun-Zhi, Z., H. Lei, Z. An-Ling, et al. 2010. MicroRNA-221 and microRNA-222 regulate gastric carcinoma cell proliferation and radioresistance by targeting PTEN. *BMC Cancer* 10:367.

Clevers H. 2006. Wnt/beta-catenin signaling in development and disease. *Cell* 127 (3):469–480.

Concepcion, C. P., Y. C. Han, P. Mu, et al. 2012. Intact p53-dependent responses in miR-34-deficient mice. *PLoS Genet* 8 (7):e1002797.

Cong, N., P. Du, A. Zhang, et al. 2013. Downregulated microRNA-200a promotes EMT and tumor growth through the wnt/beta-catenin pathway by targeting the E-cadherin repressors ZEB1/ZEB2 in gastric adenocarcinoma. *Oncol Rep* 29 (4):1579–1587.

Craig, V. J., A. Tzankov, M. Flori, et al. 2012. Systemic microRNA-34a delivery induces apoptosis and abrogates growth of diffuse large B-cell lymphoma in vivo. *Leukemia* 26 (11):2421–2424.

Cui, L., H. Zhou, H. Zhao, et al. 2012. MicroRNA-99a induces G1-phase cell cycle arrest and suppresses tumorigenicity in renal cell carcinoma. *BMC Cancer* 12:546.

Daige, C. L., J. F. Wiggins, L. Priddy, et al. 2014. Systemic delivery of a miR34a mimic as a potential therapeutic for liver cancer. *Mol Cancer Ther* 13 (10):2352–2360.

Dang, Y. W., J. Zeng, R. Q. He, et al. 2014. Effects of miR-152 on cell growth inhibition, motility suppression and apoptosis induction in hepatocellular carcinoma cells. *Asian Pac J Cancer Prev* 15 (12):4969–4976.

Danielsen, S. A., P. W. Eide, A. Nesbakken, et al. 2015. Portrait of the PI3K/AKT pathway in colorectal cancer. *Biochim Biophys Acta* 1855 (1):104–121.

Dar, A. A., S. Majid, C. Rittsteuer, et al. 2013. The role of miR-18b in MDM2-p53 pathway signaling and melanoma progression. *J Natl Cancer Inst* 105 (6):433–442.

Devulapally, R., T. V. Sekar, and R. Paulmurugan. 2015. Formulation of anti-miR-21 and 4-Hydroxytamoxifen co-loaded biodegradable polymer nanoparticles and their antiproliferative effect on breast cancer cells. *Mol Pharm* 12 (6):2080–2092.

Diederichs, S., S. Jung, S. M. Rothenberg, et al. 2008. Coexpression of Argonaute-2 enhances RNA interference toward perfect match binding sites. *Proc Natl Acad Sci U S A* 105 (27):9284–9289.

Dincbas-Renqvist, V., G. Pepin, M. Rakonjac, et al. 2009. Human Dicer C-terminus functions as a 5-lipoxygenase binding domain. *Biochim Biophys Acta* 1789 (2):99–108.

Du, C., C. Liu, J. Kang, et al. 2009. MicroRNA miR-326 regulates TH-17 differentiation and is associated with the pathogenesis of multiple sclerosis. *Nat Immunol* 10 (12):1252–1259.

Du, J., X. Niu, Y. Wang, et al. 2015. MiR-146a-5p suppresses activation and proliferation of hepatic stellate cells in nonalcoholic fibrosing steatohepatitis through directly targeting Wnt1 and Wnt5a. *Sci Rep* 5:16163.

Ebert, M. S., J. R. Neilson, and P. A. Sharp. 2007. MicroRNA sponges: competitive inhibitors of small RNAs in mammalian cells. *Nat Methods* 4 (9):721–726.

Ebert, M. S., and P. A. Sharp. 2010. MicroRNA sponges: progress and possibilities. *RNA* 16 (11):2043–2050.

Eiring, A. M., J. G. Harb, P. Neviani, et al. 2010. miR-328 functions as an RNA decoy to modulate hnRNP E2 regulation of mRNA translation in leukemic blasts. *Cell* 140 (5):652–665.

Elmen, J., M. Lindow, S. Schutz, et al. 2008. LNA-mediated microRNA silencing in non-human primates. *Nature* 452 (7189):896–899.

Esau, C. C. 2008. Inhibition of microRNA with antisense oligonucleotides. *Methods* 44 (1):55–60.

Fang, L., J. Cai, B. Chen, et al. 2015. Aberrantly expressed miR-582-3p maintains lung cancer stem cell-like traits by activating Wnt/beta-catenin signalling. *Nat Commun* 6:8640.

Fang, L., H. Li, L. Wang, et al. 2014. MicroRNA-17-5p promotes chemotherapeutic drug resistance and tumour metastasis of colorectal cancer by repressing PTEN expression. *Oncotarget* 5 (10):2974–2987.

Fire, A., S. Xu, M.K. Montgomery, et al. 1998. Potent and specific genetic interference by double-stranded RNA in *Caenorhabditis elegans*. *Nature* 391(6669):806–811.

Fortunato, O., M. Boeri, M. Moro, et al. 2014. Mir-660 is downregulated in lung cancer patients and its replacement inhibits lung tumorigenesis by targeting MDM2-p53 interaction. *Cell Death Dis* 5:e1564.

Garibaldi, F., E. Falcone, D. Trisciuoglio, et al. 2016. Mutant p53 inhibits miRNA biogenesis by interfering with the microprocessor complex. *Oncogene* 35 (29):3760–3770.

Garofalo, M., G. Di Leva, G. Romano, et al. 2009. miR-221&222 regulate TRAIL resistance and enhance tumorigenicity through PTEN and TIMP3 downregulation. *Cancer Cell* 16 (6):498–509.

Garzon, R., G. Marcucci, and C. M. Croce. 2010. Targeting microRNAs in cancer: rationale, strategies and challenges. *Nat Rev Drug Discov* 9 (10):775–789.

Gavrilov, K., and W. M. Saltzman. 2012. Therapeutic siRNA: principles, challenges, and strategies. *Yale J Biol Med* 85 (2):187–200.

Geng, L., B. Sun, B. Gao, et al. 2014. MicroRNA-103 promotes colorectal cancer by targeting tumor suppressor DICER and PTEN. *Int J Mol Sci* 15 (5):8458–8472.

Gordillo, G. M., A. Biswas, S. Khanna, et al. 2014. Dicer knockdown inhibits endothelial cell tumor growth via microRNA 21a-3p targeting of Nox-4. *J Biol Chem* 289 (13):9027–9038.

Grueter, C. E., E. van Rooij, B. A. Johnson, et al. 2012. A cardiac microRNA governs systemic energy homeostasis by regulation of MED13. *Cell* 149 (3):671–683.

Gu, W., X. Li, and J. Wang. 2014. miR-139 regulates the proliferation and invasion of hepatocellular carcinoma through the WNT/TCF-4 pathway. *Oncol Rep* 31 (1):397–404.

Gulla, A., M. T. Di Martino, M. E. Gallo Cantafio, et al. 2016. A 13 mer LNA-i-miR-221 inhibitor restores drug sensitivity in Melphalan–refractory multiple myeloma cells. *Clin Cancer Res* 22 (5):1222–1233.

Guo, C., J. F. Sah, L. Beard, et al. 2008. The noncoding RNA, miR-126, suppresses the growth of neoplastic cells by targeting phosphatidylinositol 3-kinase signaling and is frequently lost in colon cancers. *Genes Chromosomes Cancer* 47 (11):939–946.

Guo, H., N. T. Ingolia, J. S. Weissman, et al. 2010. Mammalian microRNAs predominantly act to decrease target mRNA levels. *Nature* 466 (7308):835–840.

Guo, W., H. Benlhabib, and C. R. Mendelson. 2016. The MicroRNA 29 family promotes type II cell differentiation in developing lung. *Mol Cell Biol* 36 (16):2141.

Guzman-Villanueva, D., I. M. El-Sherbiny, D. Herrera-Ruiz, et al. 2012. Formulation approaches to short interfering RNA and MicroRNA: challenges and implications. *J Pharm Sci* 101 (11):4046–4066.

Han, C., G. Wan, R. R. Langley, et al. 2012. Crosstalk between the DNA damage response pathway and microRNAs. *Cell Mol Life Sci* 69 (17):2895–2906.

Han, J., Y. Lee, K. H. Yeom, et al. 2004. The Drosha-DGCR8 complex in primary microRNA processing. *Genes Dev* 18 (24):3016–3027.

Han, L., A. Zhang, X. Zhou, et al. 2010. Downregulation of Dicer enhances tumor cell proliferation and invasion. *Int J Oncol* 37 (2):299–305.

Han, T. S., K. Hur, G. Xu, et al. 2015. MicroRNA-29c mediates initiation of gastric carcinogenesis by directly targeting ITGB1. *Gut* 64 (2):203–214.

Han, Y., Y. Liu, Y. Gui, and Z. Cai. 2013. Inducing cell proliferation inhibition and apoptosis via silencing Dicer, Drosha, and Exportin 5 in urothelial carcinoma of the bladder. *J Surg Oncol* 107 (2):201–205.

Hao, J., W. Jin, X. Li, et al. 2013. Inhibition of alpha interferon (IFN-alpha)-induced microRNA-122 negatively affects the anti-hepatitis B virus efficiency of IFN-alpha. *J Virol* 87 (1):137–147.

Hartmann, P., Z. Zhou, L. Natarelli, et al. 2016. Endothelial Dicer promotes atherosclerosis and vascular inflammation by miRNA-103-mediated suppression of KLF4. *Nat Commun* 7:10521.

Heishima, K., T. Mori, H. Sakai, et al. 2015. MicroRNA-214 Promotes Apoptosis in Canine Hemangiosarcoma by Targeting the COP1-p53 Axis. *PLoS One* 10 (9):e0137361.

Heldin, C. H., M. Landstrom, and A. Moustakas. 2009. Mechanism of TGF-beta signaling to growth arrest, apoptosis, and epithelial-mesenchymal transition. *Curr Opin Cell Biol* 21 (2):166–176.

Herrera-Merchan, A., C. Cerrato, G. Luengo, et al. 2010. miR-33-mediated downregulation of p53 controls hematopoietic stem cell self-renewal. *Cell Cycle* 9 (16):3277–3285.

Hildebrandt-Eriksen, E. S., V. Aarup, R. Persson, et al. 2012. A locked nucleic acid oligonucleotide targeting microRNA 122 is well-tolerated in cynomolgus monkeys. *Nucleic Acid Ther* 22 (3):152–161.

Hoffman, Y., D. R. Bublik, Y. Pilpel, et al. 2014. miR-661 downregulates both MDM2 and MDM4 to activate p53. *Cell Death Differ* 21 (2):302–309.

Hsieh, I. S., K. C. Chang, Y. T. Tsai, et al. 2013. MicroRNA-320 suppresses the stem cell-like characteristics of prostate cancer cells by downregulating the Wnt/beta-catenin signaling pathway. *Carcinogenesis* 34 (3):530–538.

Hsu, K. W., W. L. Fang, K. H. Huang, et al. 2016. Notch1 pathway-mediated microRNA-151-5p promotes gastric cancer progression. *Oncotarget* 7 (25):38036–38051.

Hu, W., C. S. Chan, R. Wu, et al. 2010. Negative regulation of tumor suppressor p53 by microRNA miR-504. *Mol Cell* 38 (5):689–699.

Huang, K., Y. Tang, L. He, et al. 2016. MicroRNA-340 inhibits prostate cancer cell proliferation and metastasis by targeting the MDM2-p53 pathway. *Oncol Rep* 35 (2):887–895.

Huang, X. P., J. Hou, X. Y. Shen, et al. 2015. MicroRNA-486-5p, which is downregulated in hepatocellular carcinoma, suppresses tumor growth by targeting PIK3R1. *FEBS J* 282 (3):579–594.

Iorio, M. V., and C. M. Croce. 2012. microRNA involvement in human cancer. *Carcinogenesis* 33 (6):1126–1133.

Jackson, A. L., J. Burchard, D. Leake, et al. 2006. Position-specific chemical modification of siRNAs reduces "off-target" transcript silencing. RNA 12 (7):1197–1205.

Janssen, H. L., H. W. Reesink, E. J. Lawitz, et al. 2013. Treatment of HCV infection by targeting microRNA. *N Engl J Med* 368 (18):1685–1694.

Ji, Q., X. Hao, Y. Meng, et al. 2008. Restoration of tumor suppressor miR-34 inhibits human p53-mutant gastric cancer tumorspheres. *BMC Cancer* 8:266.

Jiang, C., F. Shen, J. Du, et al. 2016a. MicroRNA-564 is downregulated in glioblastoma and inhibited proliferation and invasion of glioblastoma cells by targeting TGF-beta1. *Oncotarget*:56200–56208.

Jiang, H., W. Wu, M. Zhang, et al. 2014. Aberrant upregulation of miR-21 in placental tissues of macrosomia. *J Perinatol* 34 (9):658–663.

Jiang, Q., M. He, S. Guan, et al. 2016b. MicroRNA-100 suppresses the migration and invasion of breast cancer cells by targeting FZD-8 and inhibiting Wnt/beta-catenin signaling pathway. *Tumour Biol* 37 (4):5001–5011.

Jiang, Q., M. He, M. T. Ma, et al. 2016c. MicroRNA-148a inhibits breast cancer migration and invasion by directly targeting WNT-1. *Oncol Rep* 35 (3):1425–1432.

Jordan, S. D., M. Kruger, D. M. Willmes, et al. 2011. Obesity-induced overexpression of miRNA-143 inhibits insulin-stimulated AKT activation and impairs glucose metabolism. *Nat Cell Biol* 13 (4):434–446.

Kao, S. C., M. Fulham, K. Wong, et al. 2015. A significant metabolic and radiological response after a novel targeted microRNA-based treatment approach in malignant pleural mesothelioma. *Am J Respir Crit Care Med* 191 (12):1467–1469.

Karube, Y., H. Tanaka, H. Osada, et al. 2005. Reduced expression of Dicer associated with poor prognosis in lung cancer patients. *Cancer Sci* 96 (2):111–115.

Kim, N. H., H. S. Kim, N. G. Kim, et al. 2011a. p53 and microRNA-34 are suppressors of canonical Wnt signaling. *Sci Signal* 4 (197):ra71.

Kim, S. J., J. S. Oh, J. Y. Shin, et al. 2011b. Development of microRNA-145 for therapeutic application in breast cancer. *J Control Release* 155 (3):427–434.

Knabel, M. K., K. Ramachandran, S. Karhadkar, et al. 2015. Systemic delivery of scAAV8-encoded MiR-29a ameliorates hepatic fibrosis in carbon tetrachloride-treated mice. *PLoS One* 10 (4):e0124411.

Kong, G., J. Zhang, S. Zhang, et al. 2011. Upregulated microRNA-29a by hepatitis B virus X protein enhances hepatoma cell migration by targeting PTEN in cell culture model. *PLoS One* 6 (5):e19518.

Kopczynska, E. 2015. Role of microRNAs in the resistance of prostate cancer to docetaxel and paclitaxel. *Contemp Oncol (Pozn)* 19 (6):423–427.

Kota, J., R. R. Chivukula, K. A. O'Donnell, E. A. et al. 2009. Therapeutic microRNA delivery suppresses tumorigenesis in a murine liver cancer model. *Cell* 137 (6):1005–1017.

Krutzfeldt, J., N. Rajewsky, R. Braich, et al. 2005. Silencing of microRNAs in vivo with 'antagomirs'. *Nature* 438 (7068):685–689.

Lam, J. K., M. Y. Chow, Y. Zhang, et al. 2015. siRNA versus miRNA as therapeutics for gene silencing. *Mol Ther Nucleic Acids* 4:e252.

Lanford, R. E., E. S. Hildebrandt-Eriksen, A. Petri, et al. 2010. Therapeutic silencing of microRNA-122 in primates with chronic hepatitis C virus infection. *Science* 327 (5962):198–201.

Lang, Q., and C. Ling. 2012. MiR-124 suppresses cell proliferation in hepatocellular carcinoma by targeting PIK3CA. *Biochem Biophys Res Commun* 426 (2):247–252.

Le, M. T., C. Teh, N. Shyh-Chang, et al. 2009. MicroRNA-125b is a novel negative regulator of p53. *Genes Dev* 23 (7):862–876.

Lee, J. Y., D. S. Ryu, W. J. Kim, et al. 2016. Aberrantly expressed microRNAs in the context of bladder tumorigenesis. *Investig Clin Urol* 57 Suppl 1:S52–59.

Lee, R. C., R. L. Feinbaum, and V. Ambros. 1993. The C. elegans heterochronic gene lin-4 encodes small RNAs with antisense complementarity to lin-14. *Cell* 75 (5):843–854.

Leucci, E., A. Zriwil, L. H. Gregersen, et al. 2012. Inhibition of miR-9 de-represses HuR and DICER1 and impairs Hodgkin lymphoma tumour outgrowth in vivo. *Oncogene* 31 (49):5081–5089.

Leung W. K., M. He, A. W. Chan, et al. 2015. Wnt/β-Catenin activates MiR-183/96/182 expression in hepatocellular carcinoma that promotes cell invasion. *Cancer Letters* 362 (1):97–105.

Lezina, L., N. Purmessur, A. V. Antonov, et al. 2013. miR-16 and miR-26a target checkpoint kinases Weel and Chk1 in response to p53 activation by genotoxic stress. *Cell Death Dis* 4:e953.

Li, B., Y. Lu, H. Wang, et al. 2016a. miR-221/222 enhance the tumorigenicity of human breast cancer stem cells via modulation of PTEN/Akt pathway. *Biomed Pharmacother* 79:93–101.

Li, J., M. Lam, and B. Reproducibility Project: Cancer. 2015a. Registered report: The microRNA miR-34a inhibits prostate cancer stem cells and metastasis by directly repressing CD44. *Elife* 4:e06434.

Li, J., C. Yen, D. Liaw, et al. 1997. PTEN, a putative protein tyrosine phosphatase gene mutated in human brain, breast and prostate cancer. *Science* 275:1943–1947.

Li, M., L. Tian, L. Wang, et al. 2013a. Down-regulation of miR-129-5p inhibits growth and induces apoptosis in laryngeal squamous cell carcinoma by targeting APC. *PLoS One* 8 (10):e77829.

Li, Q., Q. Cheng, Z. Chen, et al. 2016b. MicroRNA-663 inhibits the proliferation, migration and invasion of glioblastoma cells via targeting TGF-beta1. *Oncol Rep* 35 (2):1125–1134.

Li, Q., C. Zhang, R. Chen, et al. 2016c. Disrupting MALAT1/miR-200c sponge decreases invasion and migration in endometrioid endometrial carcinoma. *Cancer Lett* 383 (1):28–40.

Li, S., Y. Ma, X. Hou, et al. 2015b. MiR-185 acts as a tumor suppressor by targeting AKT1 in non-small cell lung cancer cells. *Int J Clin Exp Pathol* 8 (9):11854–11862.

Li, T., Q. Lai, S. Wang, et al. 2016d. MicroRNA-224 sustains Wnt/beta-catenin signaling and promotes aggressive phenotype of colorectal cancer. *J Exp Clin Cancer Res* 35:21.

Li, Y., L. Mao, Y. Gao, et al. 2015c. MicroRNA-107 contributes to post-stroke angiogenesis by targeting Dicer-1. *Sci Rep* 5:13316.

Li, Y., T. Masaki, D. Yamane, et al. 2013b. Competing and noncompeting activities of miR-122 and the 5' exonuclease Xrn1 in regulation of hepatitis C virus replication. *Proc Natl Acad Sci U S A* 110 (5):1881–1886.

Li, Y., and F. H. Sarkar. 2016. MicroRNA targeted therapeutic approach for pancreatic cancer. *Int J Biol Sci* 12 (3):326–337.

Li, Y., X. Zhang, D. Chen, et al. 2016e. Let-7a suppresses glioma cell proliferation and invasion through TGF-beta/Smad3 signaling pathway by targeting HMGA2. *Tumour Biol* 37 (6):8107–8119.

Liao, J. M., B. Cao, X. Zhou, et al. 2014. New insights into p53 functions through its target microRNAs. *J Mol Cell Biol* 6 (3):206–213.

Liang, J., Y. Li, G. Daniels, et al. 2015. LEF1 targeting EMT in prostate cancer invasion is regulated by miR-34a. *Mol Cancer Res* 13 (4):681–688.

Lindow, M., and S. Kauppinen. 2012. Discovering the first microRNA-targeted drug. *J Cell Biol* 199 (3):407–412.

Ling, H., M. Fabbri, and G. A. Calin. 2013. MicroRNAs and other non-coding RNAs as targets for anticancer drug development. *Nat Rev Drug Discov* 12 (11):847–865.

Liu, L. Z., C. Li, Q. Chen, et al. 2011. MiR-21 induced angiogenesis through AKT and ERK activation and HIF-1alpha expression. *PLoS One* 6 (4):e19139.

Liu, Y. M., Y. Xia, W. Dai, et al. 2014. Cholesterol-conjugated let-7a mimics: antitumor efficacy on hepatocellular carcinoma in vitro and in a preclinical orthotopic xenograft model of systemic therapy. *BMC Cancer* 14:889.

MacDiarmid, J. A., and H. Brahmbhatt. 2011. Minicells: versatile vectors for targeted drug or si/shRNA cancer therapy. *Curr Opin Biotechnol* 22 (6):909–916.

Mahmood, S., A. Bhatti, N. A. Syed, et al. 2016. The microRNA regulatory network: a far-reaching approach to the regulate the Wnt signaling pathway in number of diseases. *J Recept Signal Transduct Res* 36 (3):310–318.

Merritt, W. M., Y. G. Lin, et al. 2008. Dicer, Drosha, and outcomes in patients with ovarian cancer. *N Engl J Med* 359 (25):2641–2650.

Mirna Therapeutics. 2013. A multicenter phase I study of MRX34, microRNA miR-RX34 liposomal injection NCT01829971. *ClinicalTrials. gov Identifier.*

Mohammadi-Yeganeh, S., M. Paryan, E. Arefian, et al. 2016. MicroRNA-340 inhibits the migration, invasion, and metastasis of breast cancer cells by targeting Wnt pathway. *Tumour Biol* 37 (7):8993–9000.

Mongroo, P. S., and A. K. Rustgi. 2010. The role of the miR-200 family in epithelial-mesenchymal transition. *Cancer Biol Ther* 10 (3):219–222.

Mudhasani, R., Z. Zhu, G. Hutvagner, et al. 2008. Loss of miRNA biogenesis induces p19Arf-p53 signaling and senescence in primary cells. *J Cell Biol* 181 (7):1055–1063.

Napoli, M., A. Venkatanarayan, P. Raulji, et al. 2016. DeltaNp63/DGCR8-dependent microRNAs mediate therapeutic efficacy of HDAC inhibitors in cancer. *Cancer Cell* 29 (6):874–888.

Ohyagi-Hara, C., K. Sawada, S. Kamiura, et al. 2013. miR-92a inhibits peritoneal dissemination of ovarian cancer cells by inhibiting integrin alpha5 expression. *Am J Pathol* 182 (5):1876–1889.

Okada, N., C. P. Lin, M. C. Ribeiro, et al. 2014. A positive feedback between p53 and miR-34 miRNAs mediates tumor suppression. *Genes Dev* 28 (5):438–450.

Orom, U. A., F. C. Nielsen, and A. H. Lund. 2008. MicroRNA-10a binds the 5'UTR of ribosomal protein mRNAs and enhances their translation. *Mol Cell* 30 (4):460–471.

Park, J. T., M. Kato, H. Yuan, et al. 2013. FOG2 protein down-regulation by transforming growth factor-beta1-induced microRNA-200b/c leads to Akt kinase activation and glomerular mesangial hypertrophy related to diabetic nephropathy. *J Biol Chem* 288 (31):22469–22480.

Park, M. G., J. S. Kim, S. Y. Park, et al. 2014. MicroRNA-27 promotes the differentiation of odontoblastic cell by targeting APC and activating Wnt/beta-catenin signaling. *Gene* 538 (2):266–272.

Parsi, S., P. Y. Smith, C. Goupil, et al. 2015. Preclinical evaluation of mir-15/107 family members as multifactorial drug targets for Alzheimer's disease. *Mol Ther Nucleic Acids* 4:e256.

Patil, M. A., S. A. Lee, E. Macias, et al. 2009. Role of cyclin D1 as a mediator of c-Met- and beta-catenin-induced hepatocarcinogenesis. *Cancer Res* 69 (1):253–261.

Peng, Y., X. Zhang, X. Feng, et al. 2017a. The crosstalk between microRNAs and the Wnt/beta-catenin signaling pathway in cancer. *Oncotarget* 8(8):14089–14106

Peng, Y., X. Zhang, Q. Ma, et al. 2017b. MiRNA-194 activates the Wnt/beta-catenin signaling pathway in gastric cancer by targeting the negative Wnt regulator, SUFU. *Cancer Lett* 385:117–127.

Perdigão-Henriques, R., F. Petrocca, G. Altschuler, et al. 2016. miR-200 promotes the mesenchymal to epithelial transition by suppressing multiple members of the Zeb2 and Snail1 transcriptional repressor complexes. *Oncogene* 35 (2):158–172.

Petersen, M., and J. Wengel. 2003. LNA: a versatile tool for therapeutics and genomics. *Trends Biotechnol* 21 (2):74–81.

Pollutri, D., L. Gramantieri, L. Bolondi, et al. 2016. TP53/MicroRNA Interplay in Hepatocellular Carcinoma. *Int J Mol Sci* 17 (12).

Qin, W., A. C. Chung, X. R. Huang, et al. 2011. TGF-beta/Smad3 signaling promotes renal fibrosis by inhibiting miR-29. *J Am Soc Nephrol* 22 (8):1462–1474.

Qu, F., C. B. Li, B. T. Yuan, et al. 2016. MicroRNA-26a induces osteosarcoma cell growth and metastasis via the Wnt/beta-catenin pathway. *Oncol Lett* 11 (2):1592–1596.

Rachagani, S., M. A. Macha, N. Heimann, et al. 2015. Clinical implications of miRNAs in the pathogenesis, diagnosis and therapy of pancreatic cancer. *Adv Drug Deliv Rev* 81:16–33.

Raemdonck, K., R. E. Vandenbroucke, J. Demeester, et al. 2008. Maintaining the silence: reflections on long-term RNAi. *Drug Discov Today* 13 (21-22):917–931.

Rahimi, G., N. Jafari, M. Khodabakhsh, et al. 2015. Upregulation of microRNA processing enzymes Drosha and Dicer in gestational diabetes mellitus. *Gynecol Endocrinol* 31 (2):156–159.

Reid, G., M. E. Pel, M. B. Kirschner, et al. 2013. Restoring expression of miR-16: a novel approach to therapy for malignant pleural mesothelioma. *Ann Oncol* 24 (12):3128–3135.

Riquelme, I., O. Tapia, P. Leal, et al. 2016. miR-101-2, miR-125b-2 and miR-451a act as potential tumor suppressors in gastric cancer through regulation of the PI3K/AKT/mTOR pathway. *Cell Oncol (Dordr)* 39 (1):23–33.

Shi, L., B. Fisslthaler, N. Zippel, et al. 2013. MicroRNA-223 antagonizes angiogenesis by targeting beta1 integrin and preventing growth factor signaling in endothelial cells. *Circ Res* 113 (12):1320–1330.

Sigoillot, F. D., S. Lyman, J. F. Huckins, et al. 2012. A bioinformatics method identifies prominent off-targeted transcripts in RNAi screens. *Nat Methods* 9 (4):363–366.

Sugito, N., H. Ishiguro, Y. Kuwabara, et al. 2006. RNASEN regulates cell proliferation and affects survival in esophageal cancer patients. *Clin Cancer Res* 12 (24):7322–7328.

Suzuki, H. I., K. Yamagata, K. Sugimoto, et al. 2009. Modulation of microRNA processing by p53. *Nature* 460 (7254):529–533.

Tang, H., Y. Kong, J. Guo, et al. 2013. Diallyl disulfide suppresses proliferation and induces apoptosis in human gastric cancer through Wnt-1 signaling pathway by up-regulation of miR-200b and miR-22. *Cancer Lett* 340 (1):72–81.

Tang, Q., Z. Zou, C. Zou, et al. 2015. MicroRNA-93 suppress colorectal cancer development via Wnt/beta-catenin pathway downregulating. *Tumour Biol* 36 (3):1701–1710.

Tarasov, V., P. Jung, B. Verdoodt, et al. 2007. Differential regulation of microRNAs by p53 revealed by massively parallel sequencing: miR-34a is a p53 target that induces apoptosis and G1-arrest. *Cell Cycle* 6 (13):1586–1593.

Tay, F. C., J. K. Lim, H. Zhu, et al. 2015. Using artificial microRNA sponges to achieve microRNA loss-of-function in cancer cells. *Adv Drug Deliv Rev* 81:117–127.

Tay, Y., J. Zhang, A. M. Thomson, et al. 2008. MicroRNAs to Nanog, Oct4 and Sox2 coding regions modulate embryonic stem cell differentiation. *Nature* 455 (7216):1124–1128.

Tazawa, H., N. Tsuchiya, M. Izumiya, et al. 2007. Tumor-suppressive miR-34a induces senescence-like growth arrest through modulation of the E2F pathway in human colon cancer cells. *Proc Natl Acad Sci U S A* 104 (39):15472–15477.

Trang, P., J. F. Wiggins, C. L. Daige, et al. 2011. Systemic delivery of tumor suppressor microRNA mimics using a neutral lipid emulsion inhibits lung tumors in mice. *Mol Ther* 19 (6):1116–1122.

Ueno, K., H. Hirata, V. Shahryari, et al. 2013. microRNA-183 is an oncogene targeting Dkk-3 and SMAD4 in prostate cancer. *Br J Cancer* 108 (8):1659–1667.

Valenta, T., G. Hausmann, and K. Basler. 2012. The many faces and functions of beta-catenin. *EMBO J* 31 (12):2714–2736.

Van Rooij, E., and E. N. Olson. 2012. MicroRNA therapeutics for cardiovascular disease: opportunities and obstacles. *Nat Rev Drug Discov* 11 (11):860–872.

Wade, M., Y. C. Li, and G. M. Wahl. 2013. MDM2, MDMX and p53 in oncogenesis and cancer therapy. *Nat Rev Cancer* 13 (2):83–96.

Wang, H., Y. Zhu, M. Zhao, et al. 2013a. miRNA-29c suppresses lung cancer cell adhesion to extracellular matrix and metastasis by targeting integrin beta1 and matrix metalloproteinase2 (MMP2). *PLoS One* 8 (8):e70192.

Wang, J. Y., Y. B. Gao, N. Zhang, et al. 2014a. miR-21 overexpression enhances TGF-beta1-induced epithelial-to-mesenchymal transition by target smad7 and aggravates renal damage in diabetic nephropathy. *Mol Cell Endocrinol* 392 (1–2):163–172.

Wang, M., H. Gu, H. Qian, et al. 2013b. miR-17-5p/20a are important markers for gastric cancer and murine double minute 2 participates in their functional regulation. *Eur J Cancer* 49 (8):2010–2021.

Wang, X., L. Hao, H. F. Bu, et al. 2016a. Spherical nucleic acid targeting microRNA-99b enhances intestinal MFG-E8 gene expression and restores enterocyte migration in lipopolysaccharide-induced septic mice. *Sci Rep* 6:31687.

Wang, Y., J. Liu, C. Liu, et al. 2013c. MicroRNA-7 regulates the mTOR pathway and proliferation in adult pancreatic beta-cells. *Diabetes* 62 (3):887–895.

Wang, Y., Q. Tang, M. Li, et al. 2014b. MicroRNA-375 inhibits colorectal cancer growth by targeting PIK3CA. *Biochem Biophys Res Commun* 444 (2):199–204.

Wang, Y., Y. Zhao, A. Herbst, et al. 2016b. miR-221 Mediates chemoresistance of esophageal adenocarcinoma by direct targeting of DKK2 expression. *Ann Surg* 264 (5):804–814.

Wang, Z. 2011. The principles of MiRNA-masking antisense oligonucleotides technology. *Methods Mol Biol* 676:43–49.

Weiler, J., J. Hunziker, and J. Hall. 2006. Anti-miRNA oligonucleotides (AMOs): ammunition to target miRNAs implicated in human disease? *Gene Ther* 13 (6):496–502.

Wen, Q., J. Zhao, L. Bai, et al. 2015. miR-126 inhibits papillary thyroid carcinoma growth by targeting LRP6. *Oncol Rep* 34 (4):2202–2210.

Winbanks, C. E., B. Wang, C. Beyer, et al. 2011. TGF-beta regulates miR-206 and miR-29 to control myogenic differentiation through regulation of HDAC4. *J Biol Chem* 286 (16):13805–13814.

Wu, C. L., J. Y. Ho, S. C. Chou, et al. 2016. MiR-429 reverses epithelial-mesenchymal transition by restoring E-cadherin expression in bladder cancer. *Oncotarget* 7 (18):26593–26603.

Wu, W., J. Yang, X. Feng, et al. 2013. MicroRNA-32 (miR-32) regulates phosphatase and tensin homologue (PTEN) expression and promotes growth, migration, and invasion in colorectal carcinoma cells. *Mol Cancer* 12:30.

Xiao, J., X. M. Meng, X. R. Huang, et al. 2012. miR-29 inhibits bleomycin-induced pulmonary fibrosis in mice. *Mol Ther* 20 (6):1251–1260.

Xiao, J., B. Yang, H. Lin, et al. 2007. Novel approaches for gene-specific interference via manipulating actions of microRNAs: examination on the pacemaker channel genes HCN2 and HCN4. *J Cell Physiol* 212 (2):285–292.

Xu, C., Q. Zeng, W. Xu, et al. 2013. miRNA-100 inhibits human bladder urothelial carcinogenesis by directly targeting mTOR. *Mol Cancer Ther* 12 (2):207–219.

Xue, X., Y. Liu, Y. Wang, et al. 2016. MiR-21 and MiR-155 promote non-small cell lung cancer progression by downregulating SOCS1, SOCS6, and PTEN. *Oncotarget* 7(51):84508–84519.

Yan, J., J. Y. Jiang, X. N. Meng, et al. 2016. MiR-23b targets cyclin G1 and suppresses ovarian cancer tumorigenesis and progression. *J Exp Clin Cancer Res* 35:31.

Yang, L., Y. Li, X. Wang, et al. 2016. Overexpression of miR-223 Tips the Balance of Pro- and Anti-hypertrophic Signaling Cascades toward Physiologic Cardiac Hypertrophy. *J Biol Chem* 291 (30):15700–15713.

Yi, R., Y. Qin, I. G. Macara, et al. 2003. Exportin-5 mediates the nuclear export of pre-microRNAs and short hairpin RNAs. *Genes Dev* 17 (24):3011–3016.

Yonemori, K., H. Kurahara, K. Maemura, et al. 2017. MicroRNA in pancreatic cancer. *J Hum Genet* 62(1):33–40.

Yu, B., X. Zhao, L. J. Lee, et al. 2009. Targeted delivery systems for oligonucleotide therapeutics. *AAPS J* 11 (1):195–203.

Yu, M., and W. M. Grady. 2012. Therapeutic targeting of the phosphatidylinositol 3-kinase signaling pathway: novel targeted therapies and advances in the treatment of colorectal cancer. *Therap Adv Gastroenterol* 5 (5):319–337.

Yu, S. H., C. L. Zhang, F. S. Dong, et al. 2015. miR-99a suppresses the metastasis of human non-small cell lung cancer cells by targeting AKT1 signaling pathway. *J Cell Biochem* 116 (2):268–276.

Yuan, K., Z. Lian, B. Sun, et al. 2012. Role of miR-148a in hepatitis B associated hepatocellular carcinoma. *PLoS One* 7 (4):e35331.

Zeng, Y., R. Yi, and B. R. Cullen. 2003. MicroRNAs and small interfering RNAs can inhibit mRNA expression by similar mechanisms. *Proc Natl Acad Sci U S A* 100 (17):9779–9784.

Zhang, C., J. Liu, C. Tan, et al. 2016a. microRNA-1827 represses MDM2 to positively regulate tumor suppressor p53 and suppress tumorigenesis. *Oncotarget* 7 (8):8783–8796.

Zhang, C., J. Liu, X. Wang, et al. 2014a. MicroRNA-339-5p inhibits colorectal tumorigenesis through regulation of the MDM2/p53 signaling. *Oncotarget* 5 (19):9106–9117.

Zhang, G., H. Zhou, H. Xiao, et al. 2014b. MicroRNA-92a functions as an oncogene in colorectal cancer by targeting PTEN. *Dig Dis Sci* 59 (1):98–107.

Zhang, J. G., Y. Shi, D. F. Hong, et al. 2015a. MiR-148b suppresses cell proliferation and invasion in hepatocellular carcinoma by targeting WNT1/beta-catenin pathway. *Sci Rep* 5:8087.

Zhang, L. Y., V. Ho-Fun Lee, A. M. Wong, et al. 2013a. MicroRNA-144 promotes cell proliferation, migration and invasion in nasopharyngeal carcinoma through repression of PTEN. *Carcinogenesis* 34 (2):454–463.

Zhang, L., X. Yang, Y. Lv, et al. 2017. Cytosolic co-delivery of miRNA-34a and docetaxel with core-shell nanocarriers via caveolae-mediated pathway for the treatment of metastatic breast cancer. *Sci Rep* 7:46186.

Zhang, T., Y. Hu, J. Ju, et al. 2016b. Downregulation of miR-522 suppresses proliferation and metastasis of non-small cell lung cancer cells by directly targeting DENN/MADD domain containing 2D. *Sci Rep* 6:19346.

Zhang, W., F. Peng, T. Zhou, et al. 2015b. Targeted delivery of chemically modified anti-miR-221 to hepatocellular carcinoma with negatively charged liposomes. *Int J Nanomedicine* 10:4825–4836.

Zhang, X., M. Li, K. Zuo, et al. 2013b. Upregulated miR-155 in papillary thyroid carcinoma promotes tumor growth by targeting APC and activating Wnt/beta-catenin signaling. *J Clin Endocrinol Metab* 98 (8):E1305–1313.

Zhang, Y., W. Wei, N. Cheng, et al. 2012. Hepatitis C virus-induced up-regulation of microRNA-155 promotes hepatocarcinogenesis by activating Wnt signaling. *Hepatology* 56 (5):1631–1640.

Zheng, Y., L. Yin, H. Chen, et al. 2012. miR-376a suppresses proliferation and induces apoptosis in hepatocellular carcinoma. *FEBS Lett* 586 (16):2396–2403.

Zhou, J., J. Cai, Z. Huang, et al. 2013. Proteomic identification of target proteins following Drosha knockdown in cervical cancer. *Oncol Rep* 30 (5):2229–2237.

Zhou, J., W. Q. Ju, X. P. Yuan, et al. 2016. miR-26a regulates mouse hepatocyte proliferation via directly targeting the 3' untranslated region of CCND2 and CCNE2. *Hepatobiliary Pancreat Dis Int* 15 (1):65–72.

Zhou, W., Y. Li, S. Gou, et al. 2015. MiR-744 increases tumorigenicity of pancreatic cancer by activating Wnt/beta-catenin pathway. *Oncotarget* 6 (35):37557–37569.

Zhu L., J. Gao, K. Huang, et al. 2015. miR-34a screened by miRNA profiling negatively regulates Wnt/beta-catenin signaling pathway in Aflatoxin B1 induced hepatotoxicity. *Sci Rep* 5:16732

Zhu, N., D. Zhang, H. Xie, et al. 2011. Endothelial-specific intron-derived miR-126 is down-regulated in human breast cancer and targets both VEGFA and PIK3R2. *Mol Cell Biochem* 351 (1-2):157–164.

14 Alliance of Lipids with siRNA

Opportunities and Challenges for RNAi Therapy

Anu Puri, Mathias Viard, Paul Zakrevsky, Lorena Parlea, Krishna Pal Singh, and Bruce A. Shapiro

CONTENTS

14.1 INTRODUCTION

Modulation of gene expression by RNA interference (RNAi) in mammalian cells is an important event that was first realized by Fire and Mello two decades ago (Fire et al. 1998). The process of RNAi is a post-transcriptional gene silencing event that can be triggered by a range of RNAi inducers using multiple pathways. Widely studied RNAi inducers include small non-coding RNA molecules such as short interfering RNAs (siRNAs), microRNAs (miRNAs), short hairpin RNAs (shRNAs) and piwi-interacting RNAs (piRNAs) (Rana 2007). Among these, small interfering RNA (siRNA) remains at the forefront of ongoing efforts towards the development of RNA-based therapeutics (Figure 14.1a, siRNA structures) (Ozcan et al. 2015, Bobbin and Rossi 2016)

It is encouraging that several clinical trials are currently being conducted for RNAi therapeutics within this relatively short duration since the initiation of such efforts (Ozcan et al. 2015) and is reviewed in (Wittrup and Lieberman 2015, Pecot et al. 2011, Wu et al. 2014). RNAi therapeutics under clinical development have targeted a multitude of diseases such as viral infections, cancer, cardiovascular diseases, and eye-related disorders. The delivery route of naked and modified siRNA molecules (without a carrier) is geared towards topical applications (such as eye or skin diseases). First clinical trials for systemic delivery of non-modified siRNA to treat solid tumors utilized cyclodextrins as carriers, however, these studies were later terminated (Davis et al. 2010). Concurrently, siRNA-conjugates bearing covalently linked multivalent targeting ligands such as N-acetylgalactosamine (GalNAc) were developed and are currently in clinical trials to treat liver diseases (Nair et al. 2014, Zimmerman et al. 2017); outcome awaited. In parallel, a variety of lipid-based siRNA delivery systems (LNPs) were developed; mainstream clinical trials for systemic delivery of siRNA utilized LNPs in their own accord (reviewed in Rossi, Lieberman, Sood) (Ozcan et al. 2015, Bobbin and Rossi 2016, Wittrup and Lieberman 2015).

14.1.1 Intracellular Processing of siRNA

siRNAs are relatively small RNA moieties, generally constructed as a ~21 base pair (bp) duplex with two nucleotide 3'-overhangs. Cellular processing pathways for the biological function of siRNAs have been studied and provide strategies for their therapeutic exploitation. In the cell, siRNA duplexes enter the RNA interference (RNAi) pathway where they are incorporated into the RNA-induced silencing complex (RISC) (Figure 14.1b). At this point, the "passenger" strand (sense strand) of the siRNA duplex is degraded, while the "guide" strand (anti-sense strand) is retained and used to target a complementary endogenous mRNA sequence for cleavage by the Ago2 component of RISC (Matranga et al. 2005, Liu et al. 2004). The decision of which of the two strands ultimately becomes the guide strand is largely influenced by the thermodynamic stability of each strand's 5' end (Schwarz et al. 2003, Khvorova et al. 2003). Slightly longer RNA duplexes (25–30 bp) termed Dicer substrate interfering RNAs (DsiRNAs) can be used in place of conventional 21-mer siRNAs. DsiRNAs require cellular processing by the ribonuclease-III enzyme Dicer before loading into the RISC complex. However, in some cases DsiRNAs display superior silencing efficacy compared to siRNAs and

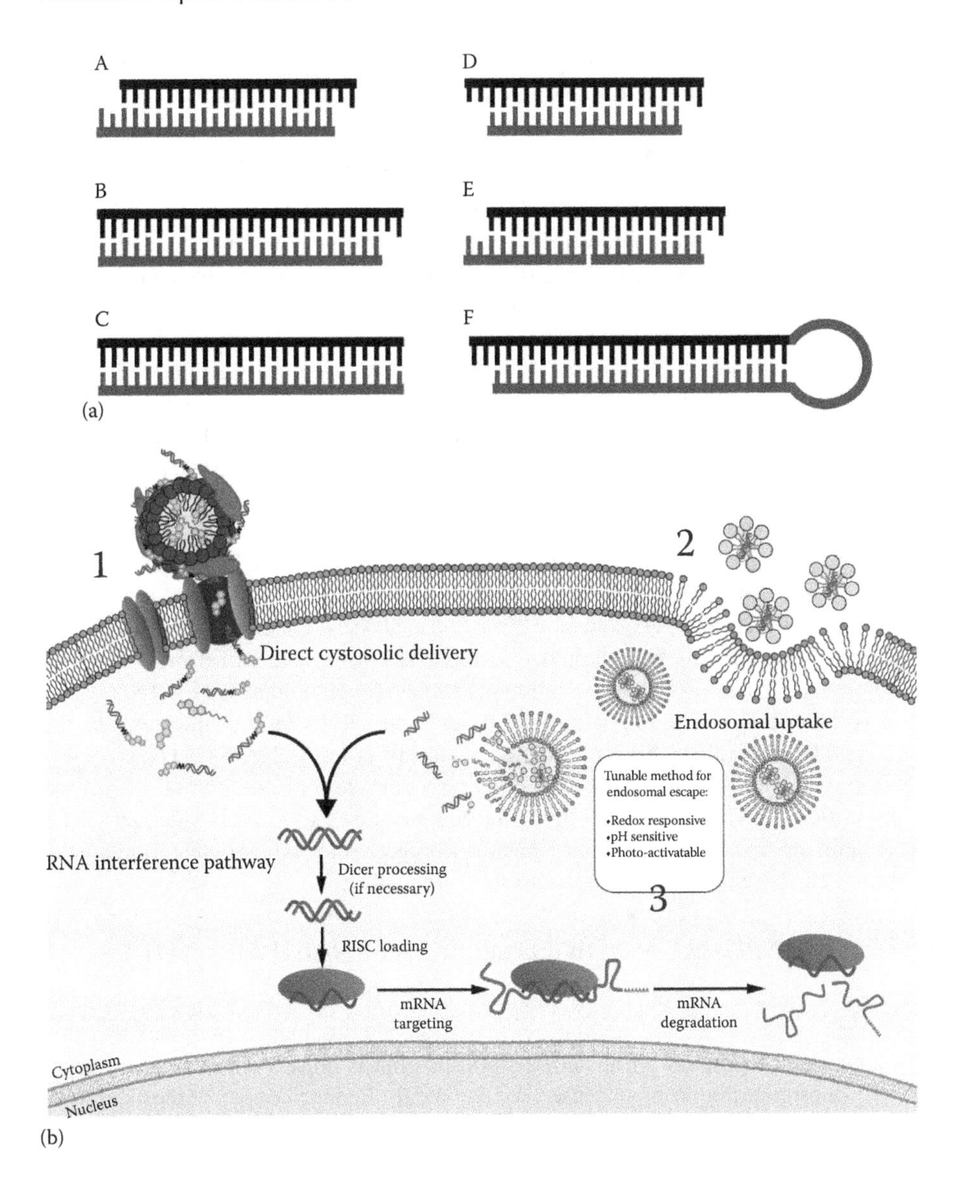

FIGURE 14.1 (a) Basic siRNA structures. (A) Conventional siRNA (21-mer) (B) Dicer substrate RNA with 3' overhang (DsiRNA, 25/27-mer) (C) blunt end DsiRNA (27-mer) (D) Asymmetric siRNA; (E) Internally segmented siRNA; (F) short hairpin RNA. (b) Cellular entry pathways of siRNA, RNA-NPs and siRNA-LNPs. Intracellular uptake of various siRNA assemblies can either occur via (1) direct interactions the with the plasma membrane resulting in cytoplasmic delivery of the siRNA or via the (2) endocytic route. Endosomal release of siRNA can be achieved by indicated strategies (3) for further processing, RISC loading and mRNA degradation.

can be designed to induce polarity in the duplex that helps determine which strand ultimately becomes the guide strand (Kim et al. 2005, Rose et al. 2005).

14.1.2 siRNA-Based Therapeutics, General Considerations

RNAi-based therapies are currently under intensive research for treatment of diseases such as cancer and pathogen infections, either on their own merit or in combination with small molecule drugs (Kanasty et al. 2013, Ozpolat et al. 2010, Musacchio and Torchilin 2013). The effectiveness of naked, unmodified RNA for its biological activity (such as RNAi and gene modulation) is faced with challenges such as degradation by exo- and endo-nucleases, short half-life, and clearance by kidneys and off-target effects (Kanasty et al. 2013, Akhtar and Benter 2007). As expected, the high negative charge of the siRNA creates an obstacle for direct interaction with cellular membranes and subsequently intracellular uptake (Pecot et al. 2011, Musacchio and Torchilin 2013). To evade these limitations, investigators have developed various approaches for siRNA delivery. These include (i) chemical modifications of siRNAs, (ii) the design of self-assembled RNA nanostructures with multiple functionalities (Afonin et al. 2011, Afonin et al. 2014a, Afonin et al. 2014b, Guo 2010), and (iii) utilization of delivery agents (carriers) such as viral vectors and non-viral vectors (Haque and Guo 2015, Shu et al. 2014, Parlea et al. 2016a).

In this review, we have attempted to summarize current research efforts in siRNA delivery. Here, we will cover the following topics: (i) development of siRNA-based RNAi therapeutics including "modified bases" and siRNA-lipid conjugates (chemically modified siRNAs), (ii) self-assembled nanostructures (RNA-NPs) bearing multifunctional siRNAs, and (iii) siRNA-lipid nanoparticles (siRNA-LNPs) that utilize unmodified siRNA and tunable lipids for enhanced cytosolic siRNA delivery. The next sections provide background, basics of design strategies, rationale, fundamentals and current biological applications for each platform.

14.2 CHEMICALLY MODIFIED siRNAs AS RNAI THERAPEUTICS

Chemical modification of siRNA can impart several beneficial characteristics such as increased resistance to enzymatic degradation and modulation of duplex thermodynamics. Incorporation of nucleotide modification is most often done at the level of the phosphoramidite moiety used for chemical oligonucleotide synthesis, allowing for site-specific incorporation of the desired modification. However, nucleotides containing sugar modifications at the 2' position can be introduced during enzymatic RNA synthesis using specific concentrations of divalent metal ions for transcription (Afonin et al. 2012), or by exploiting a mutant RNA polymerase (Padilla and Sousa 2002),, and nucleotides containing substitutions at non-bridging backbone oxygens can be incorporated by wildtype RNA polymerase (Hall et al. 2004).

14.2.1 Chemical Modifications of siRNA

Various chemical modifications of siRNA have been explored that confer advantageous characteristics over unmodified siRNA (Figure 14.2) (Shukla et al. 2010, Deleavey and Damha 2012). Modifications to the ribose sugar including, but not limited to, 2'-O-methyl, 2'-fluoro and locked nucleic acids, have been shown increase resistance

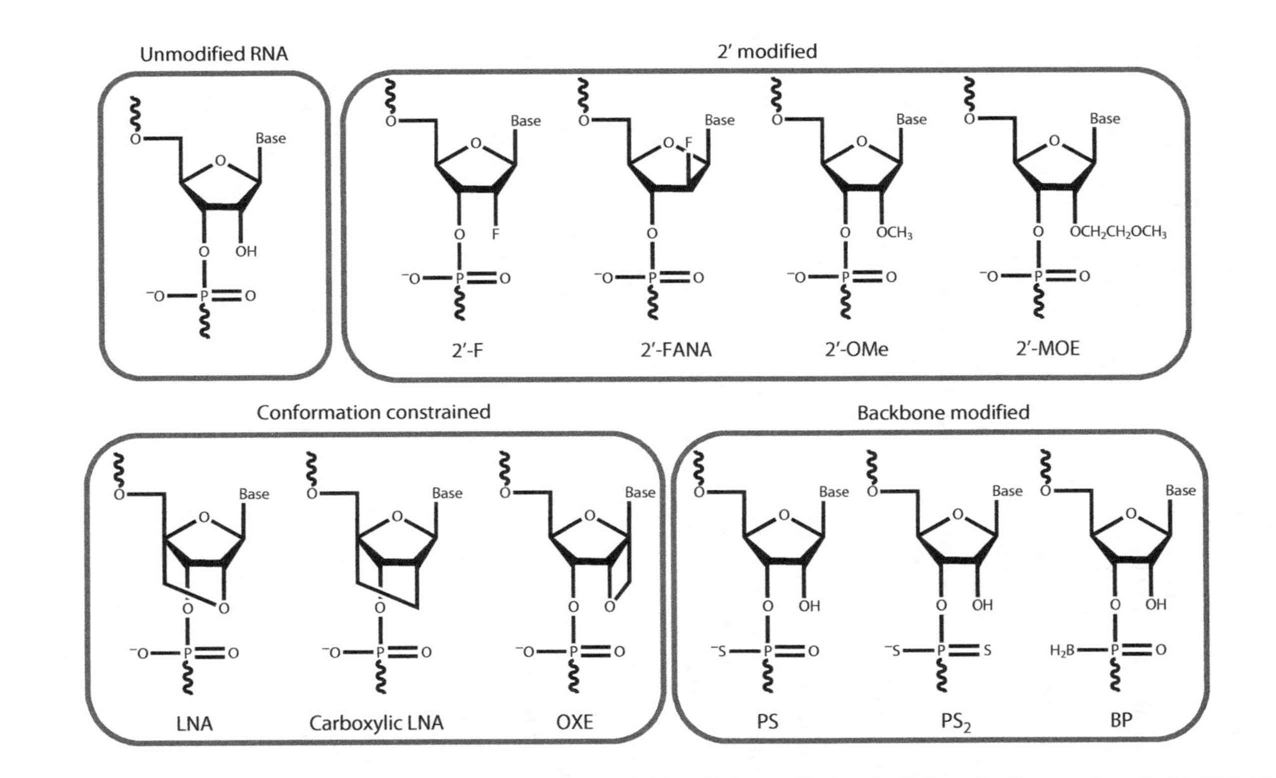

FIGURE 14.2 Common modifications applied to RNA. Unmodified bases, 2'-F: 2'-fluoro; 2'-FANA: 2'-fluoro arabinonucleotide; 2'-OMe: 2'-O methyl; 2'-MOE: 2'- O-methoxyethyl; LNA: Locked nucleic acid containing a 2'-O, 4'-C methylene bridge; carboxylic LNA: 2'C, 4'C carbocyclic locked nucleic acid; OXE: Oxetane constrained nucleic acid; PS: phosphothioate; PS2: phosphodithioate; BP: Boranophosphate.

to ribonuclease digestion (Choung et al. 2006, Elmen et al. 2005, Layzer et al. 2004) (modification to the phosphate backbone, such as phosphorothioate or boranophosphate inclusion, can also increase the ability of siRNAs to resist nuclease degradation (Hall et al. 2004, Choung et al. 2006). These substitutions of non-bridging backbone oxygens are likely to increase nuclease resistance by altering the charge distribution on the electrophilic phosphorus atom and the surrounding environment (Shaw et al. 2003). Chemical modifications can be useful to avoid activation of an innate immune response. Several different ribose modifications, particularly 2'-O-methyl, have been used to reduce the extent of cytokine production in response to siRNA both *in vitro* and *in vivo* (Souid et al. 2007, Deleavey et al. 2010, Robbins et al. 2007, Morrissey et al. 2005). Importantly, commonly used siRNA modifications do not significantly reduce efficacy, and some have been shown to increase siRNA potency and reduce the required dose for effective gene silencing (Elmen et al. 2005, Allerson et al. 2005). The favorable properties attributed to many of these modifications, such as increased stability in serum and reduction of the innate immune response, can also be implemented in the slightly longer DsiRNAs (Collingwood et al. 2008).

14.2.2 siRNA-Lipid Conjugates

Chemical modifications of naked siRNA yield nuclease resistant and serum stable siRNA, which are suitable for *in vitro* assays and cell culture experiments. However, their utility *in vivo* remains a challenge due to fast clearance from circulation (within minutes). The development of large RNA-based nanoparticles has been shown to overcome these clearance limitations (Shu et al. 2014, Haque et al. 2012). Development of siRNA conjugates exhibiting enhanced half-life in circulation has been demonstrated to be a promising avenue of research. siRNA conjugates linked to biomolecules such as peptides (Moschos et al. 2007, Turner et al. 2007, Gandioso et al. 2017, Dong et al. 2014), polymers (Jung et al. 2010), aromatic molecules (Kubo et al. 2012c, Nishina et al. 2008a), lipid-polymer hybrid systems (Shi et al. 2014, Yang et al. 2012), and lipids have been developed (Ku et al. 2016, Lee et al. 2016) for improved pharmacokinetic properties, enhanced cellular/tissue uptake and enhanced RNAi activity *in vitro* and *in vivo* (De Paula et al. 2007, Raouane et al. 2012).

Historically, covalent coupling of lipids to small molecule drugs (Zaro 2015, Trevaskis et al. 2015, Adhikari et al. 2017, Irby et al. 2017), and other pharmaceutical agents such as therapeutic antibodies, antibody domains etc. (Tomita et al. 2012, Manjappa et al. 2011, Ying et al. 2014) has proven to be a successful strategy to improve bioavailability and the therapeutic potential of these drugs. The choice and selection of lipids for a given class of small molecule drugs is dictated by factors such as feasible chemical reaction methods for lipid-drug conjugation, desired solubility, and the target tissues being investigated.

14.2.2.1 Structural Determinants of siRNA for Lipid Conjugation

Both siRNA and DsiRNA have been examined for conjugation with lipids and lipid-like molecules. It appears that modification of the 3' end of the sense strand of siRNA or the 5' end of the antisense strand of siRNA is the preferred choice for the conjugation (Jeong et al. 2009) (Figure 14.3). Most studies available to date for

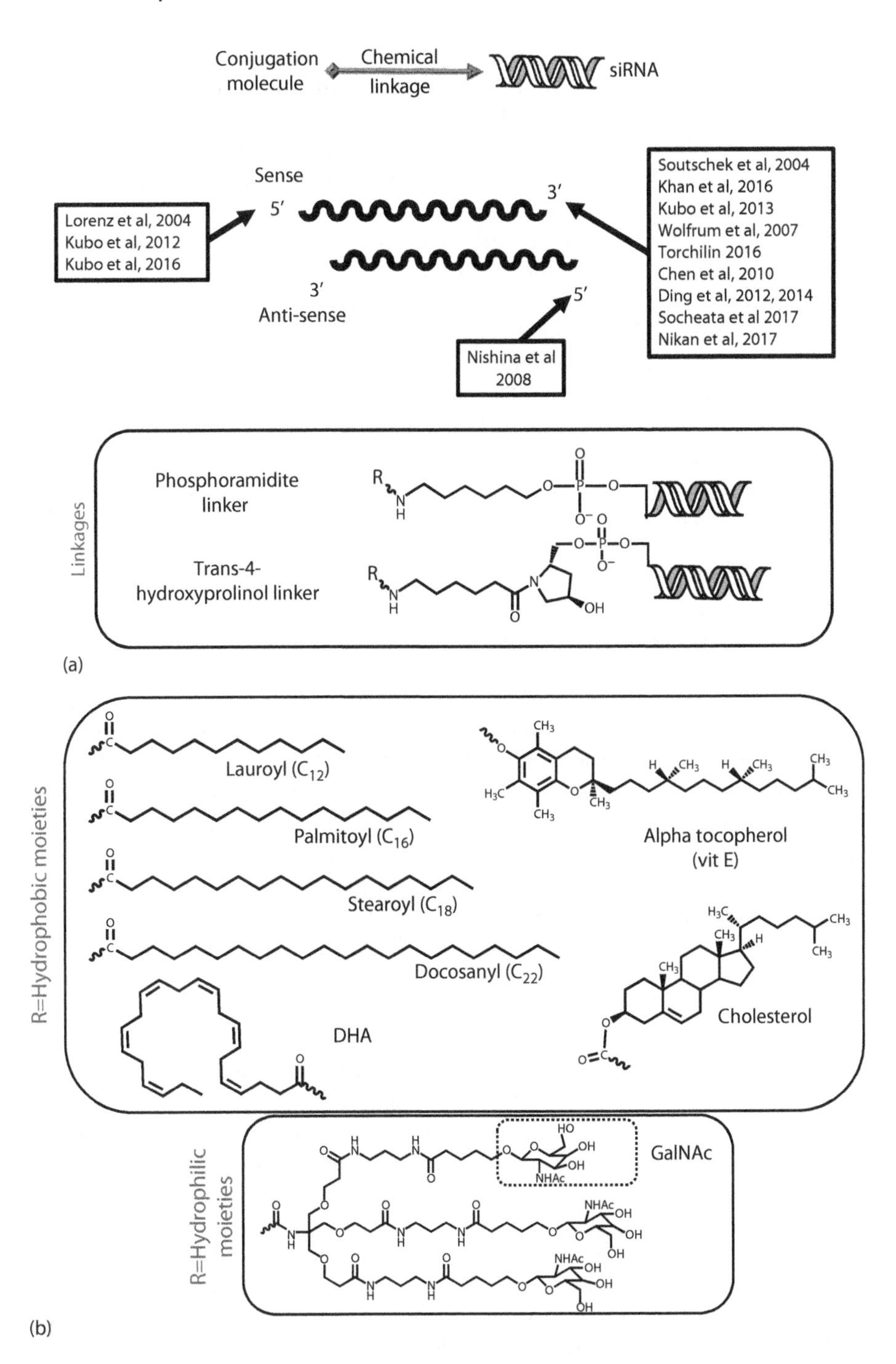

FIGURE 14.3 Various siRNA conjugates. (a) illustrates the preferred site(s) of siRNA modifications examined to date. (b) Linker chemistries, as well as the hydrophobic and hydrophilic molecules commonly conjugated to siRNA.

lipid-siRNA conjugates are designed with an objective to enhance the retention and stability of siRNA in the blood circulation as well as improved intracellular localization of the conjugates (Table 14.1) (De Paula et al. 2007).

Primary considerations for the development of siRNA-lipid conjugates include (i) selection of which strands for linking to the lipid molecules (see Figure 14.3a), (ii) linkage chemistry with no or minimal effects on intracellular processing of siRNA, (iii) enhanced stability of siRNA in blood circulation (iv) choice of lipids with built-in cleavable bonds for site-specific dissociation of siRNA, and (v) preferred uptake of these conjugates by target tissues.

To date, several siRNA-lipid conjugates have been developed and the structure-function activity of the lipids for optimal siRNA activity has been elucidated (Raouane et al. 2012, Wolfrum et al. 2007). The classical lipid molecules examined are sterols (Ding et al. 2012), fatty acids (with various chain lengths and degrees of unsaturation), and phospholipids (Musacchio et al. 2010). In addition, two other classes of molecules, polyethylene glycol (PEG) and α-tocopherol (Nishina et al. 2008a, Nishina et al. 2008b) linked to siRNA have also been developed.

14.2.2.2 Cholesterol, a Preferred Choice for Conjugation

siRNA-cholesterol conjugates are one of the most studied partners among lipid/siRNA conjugate systems. The choice of cholesterol is based on some of its unique features including the chemical structure and biological function, such as its ability to complex with plasma proteins thereby enhancing the half-life of cholesterol-conjugated siRNA in circulation. Selective association of siRNA-cholesterol conjugates with blood lipoproteins (such as HDL and LDL) also bears an advantage for receptor-mediated siRNA delivery to tissues expressing corresponding receptors. Lipoprotein-associated cholesterol has been demonstrated to be critical for their transport to desired tissues (primarily liver). Intracellular uptake of these particles is dependent on the expression of lipoprotein receptors (Wolfrum et al. 2007). Hence, siRNA-cholesterol conjugates are likely to be suitable candidates as RNAi therapeutics for the diseases of the liver. The choice of type of siRNA (ds siRNA or DsiRNA) as well as site/strand selection (3' vs 5') for conjugation of cholesterol has been diverse in studies conducted thus far (Figure 14.3). siRNA-cholesterol conjugates platforms for treatment of various diseased tissues/organs are described below.

Lorenz et al. (2004) and Soutschek et al. (2004) demonstrated the utility of lipid-conjugated siRNA for *in vitro* and *in vivo* gene silencing respectively. Conjugation of a double-stranded siRNA with cholesterol as well as other lipids was achieved via a phosphorothioate linkage at the 5' end of the sense strand (Lorenz et al. 2004) (Figure 14.3). Interestingly, these derivatives performed well for gene silencing in liver cells without any additional transfection agent (Table 14.1, Figure 14.3). In a similar study, Soutschek et al., selected the 3' end of the sense strand of the siRNA molecule for conjugation of cholesterol (Soutschek et al. 2004). In either case, cholesterol modifications did not adversely impact siRNA gene silencing activity. These studies successfully demonstrated gene silencing of endogenously expressed Apo B in mice upon systemic administration of siRNA-cholesterol conjugates (Figure 14.4a, reproduced with permission). There was significantly enhanced half-life of siRNA-cholesterol conjugates ($t_{1/2}$ about 95 minutes) as compared to that of naked siRNA

TABLE 14.1
Lipid-Conjugated siRNAs

RNA System	Strand Modification	Lipid Moiety[a]	Linkage[b]	Delivery Agent[c]	Target Gene[d]	Ref. No.
siRNA	5' (sense)	Chol, FA	phosphoramidite	None	β-gal In vitro	Lorenz at al 2004
siRNA	3' (sense)	Chol	phosphoramidite	none	ApoB (In vitro & in vivo)	Soutschek et al. 2004
siRNA	3' (sense)	Chol, FA	Trans-4-hydroxyprolinol	None, HDL/LDL	ApoB (In vitro & in vivo)	Wolfrum et al. 2007
siRNA, DsiRNA	5' (sense) 3' (sense)	Chol	Amine NHS coupling	L2K	Luciferase, VEGF (In vitro)	Kubo et al. 2012b
DsiRNA	5' (sense)	FA	Amine NHS coupling	L2k	eGFP, VEGF *(In vitro & in vivo)*	Kubo et al. 2016, Kubo et al. 2011
siRNA	3' (sense)	PE	thiol modification	PEG-PE micelles	eGFP (In vitro)	Musacchio et al. 2010
siRNA	3' (sense)	Chol	Not reported	L2k	Mstn (In vitro & in vivo)	Khan et al. 2016
DsiRNA	5' (antisense)	Toco	(possibly direct conjugation?)	none	ApoB (in vitro, in vivo)	Nishina et al. 2008a
siRNA	3' (sense)	Chol	TEG	None, Exosomes	Htt-mRNA *(in vitro, in vivo)*	Byrne et al. 2013, Alterman et al. 2015, Didiot et al. 2016
siRNA	5' (sense)	Chol	TEG	Exosomes	HuR *(in vitro)*	O'Loughlin et al. 2017

[a] Chol, cholesterol., FA, Fatty acid., PE, phosphatidylethanolamine., Toco, Tocopherol.
[b] TEG, tetra-ethylene glycol.
[c] LDL, low density lipoprotein., HDL, high density lipoprotein., L2k, lipofectamine2000.
[d] ApoB, apolipoprotein B., eGFP, green fluorescent protein, VEGF, vascular endothelial growth factor., Mstn, muscle-specific gene myostatin., Htt-mRNA, huntingtin mRNA., HuR, human antigen R.

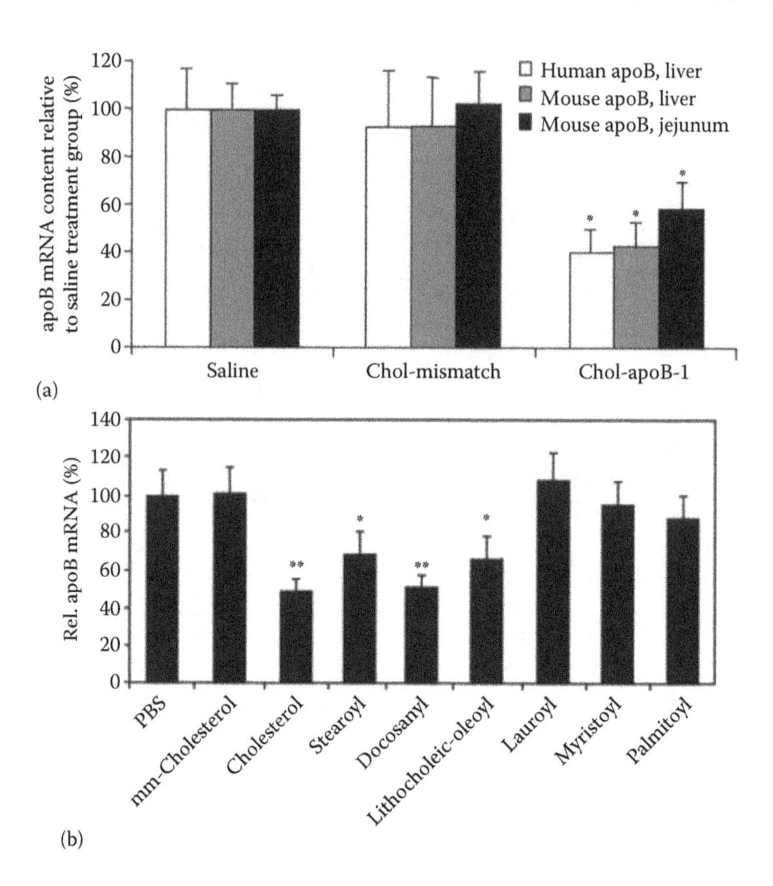

FIGURE 14.4 (a) *In vivo* silencing of murine and human apoB mRNA in mice transgenic for human apoB. Reduction of human and mouse apoB mRNA levels in mice transgenic for human apoB that received saline (n = 8), chol-mismatch-siRNA (n = 8) and chol-apoB-1-siRNA (n = 8). Statistical analysis was by ANOVA with Bonferroni post-hoc t-test, one-tailed. Asterisk, $P < 0.0001$ compared with saline and chol-mismatch-siRNA control animals. Error bars illustrate s.d. of the mean. Reprinted by permission from Macmillan Publishers Ltd: [Nature Biotechnology] (Soutschek et al. 2004), copyright (2004). (b) Lipophilic siRNA onjugates have different *in vivo* activities. *In vivo* silencing of apoB mRNA by lipid-conjugated siRNAs. Liver apoB mRNA levels were normalized to GAPDH mRNA 24 h after three daily intravenous injections of saline or 50 mg/kg stearoyl-siRNA-apoB1, dodecyl-siRNA-apoB1, lithocholic-oleyl-siRNA-apoB1- or docosanyl-siRNA-apoB1 (n = 5 per group). Data show apoB mRNA levels as a percentage of the saline treatment group and is expressed as the mean ± the s.d. Data marked with asterisks are statistically significant relative to the saline treatment group as calculated by ANOVA without replication, alpha value 0.05 (*, $P < 0.05$; **, $P < 0.005$). (Reprinted by permission from Macmillan Publishers Ltd. *Nat Biotechnol*, Wolfrum, C., et al., *Nat Biotechnol*, **25**(10): p. 1149–57, 2007, copyright 2007.)

($t_{1/2}$ about six minutes) in circulation after intravenous injections in mice (Figure 14.4a, reprinted with permission). Recent studies by Khan et al., demonstrated that a highly modified cholesterol-siRNA conjugate significantly reduced the muscle-specific gene myostatin (Mstn) upon systemic administration of this conjugate (Khan et al. 2016).

The Mstn protein down-regulation was observed in the muscles as well as in blood circulation. These studies are likely to have therapeutic potential for muscle-related diseases. In another study, a novel class of hydrophobically modified asymmetric siRNAs (hsiRNAs) were developed by Khvorova and colleagues (Byrne et al. 2013) to improve stability and promote cellular internalization. hsiRNA contain a number of nucleotide modifications and are linked to cholesterol at the 3' end of the sense strand of siRNA. The hsiRNAs functioned without a carrier molecule when delivered locally into the brain and silences Huntington mRNA indicating their implications in neurogenerative disorders (Alterman et al. 2015). The hsiRNA loaded into exosomes, when administered into mouse brain exhibited bilateral distribution and silencing of Huntington mRNA in a recent study by the same group (Didiot et al. 2016).

Synthesis and biological activity of siRNA-cholesterol conjugates bearing a tunable linker that contained cleavable and reducible sulfhydryl linkages was reported by Chen et al. They demonstrated an advantage over non-cleavable alkyl linker for direct CNS delivery to oligodendrocytes and had an advantage *in vivo* in rat studies (Chen et al. 2010) (Figure 14.3).

Kubo et al. (led by Toshio Seyama's group) developed a cholesterol conjugation technology using 27 nucleotide double-stranded RNAs with Dicer substrate potency (Kubo et al. 2013, Kubo et al. 2012a). Here, the 5' end of the sense strand of the dsRNA was modified with an amine and then linked to either cholesterol or fatty acids. The same group also studied the effect of asymmetric lipid conjugation with siRNA on gene silencing activity (described below).

14.2.2.3 siRNA-Fatty Acid Conjugates

Fatty acids, being part of phospholipids, typically linear in structure and bearing a select conjugation site, are often considered a preferred choice to enhance the hydrophobicity of the water-soluble molecules. Pioneering research by Stoffel and colleagues demonstrated the utility of fatty acids conjugated siRNA for RNAi therapy (Wolfrum et al. 2007). In this study, conjugation of siRNA was done to fatty acids of various chain lengths without or with a variable degree of unsaturation. siRNA-linked to longer fatty acids exhibited superior apoB mRNA knock-down *in vivo*. However, cholesterol-siRNA conjugates excelled among all lipid conjugates tested and the presence of unsaturation in the fatty acids did not appear to have a significant effect. In this study, the authors also reported that pre-assembly of a cholesterol-siRNA conjugate with a high-density lipoprotein (HDL) had a remarkable effect (eight- to 15-fold enhancement) on the silencing of apoB protein expression *in vivo* (Figure 14.4b, reprinted with permission).

With an aim of delineating the structure-function activity correlates of siRNA-lipid conjugates with the Dicer recognition (see Figure 14.1b), a series of symmetric and asymmetric C16-fatty acid-siRNA conjugates were synthesized (Kubo et al. 2013). siRNAs of varying lengths of sense and antisense strands (21, 23, and 25-nt) with a 2-nt overhang at each 3'-end (C16-siRNA) or 2 nt DNA overhangs at the 3'-blunt end (C16-LoRNA) were conjugated to the C16-fatty acid. Kubo and colleagues continued their efforts towards conjugation of fatty acids using various chain lengths (Kubo et al. 2016). Their studies included fatty acid conjugation via an amine/NHS coupling reaction to 21-nt or 27-nt-DsiRNA, using the 5' sense strands of siRNA (Figure 14.3). They demonstrated both in cell culture systems and *in vivo* that the efficiency of C16-DsiRNA was superior as compared to short-chain fatty acid-linked DsiRNAs. The lipid-modified siRNAs still needed assistance from a carrier molecule (such as Lipofectamine 2000) for their optimal activity (Table 14.1). In contrast, a previous report by Wolfrum et al., demonstrated that siRNA-cholesterol conjugates performed superior than palmitic-acid conjugated siRNA. These discrepancies may be explained due to different site selection of siRNA and the type of chemical linkages used for generating these conjugates.

14.2.2.4 siRNA-Tocopherol Conjugates

Utilization of bioactive molecules with specific biological functions, discrete transport mechanisms *in vivo*, and abilities to complex with blood components such as lipoproteins is a rational approach for delivery of pharmaceuticals. α-Tocopherol (Vitamin E), a non-lipidic molecule, possessing an affinity for fats, meets the criteria of a suitable agent for exploitation in the development of nanomedicine (Duhem et al. 2014). This molecule is nontoxic *in vivo* and does not have endogenous cellular genesis. Therefore, Nishina et al., developed siRNA-tocopherol conjugates with an objective to target the liver *in vivo*. They used covalent linkages to the antisense strand of 27/29mer siRNA at the 5' end, designated as TOC-siRNA (apoB) (Nishina et al. 2008a, Murakami et al. 2015). To our knowledge, TOC-siRNA is the only available example where ligand conjugation is done using an antisense strand of the siRNA (Figure 14.3). Intravenous injections of TOC-siRNA (apoB) resulted in significant reduction of apoB messenger RNA and no obvious side effects were observed (Nishina et al. 2008a). The same group further used a postprandial enteral delivery route to enhance the specific uptake of this conjugate in the liver. Toc-siRNA was incorporated into linoleic acid and PEG-60 hydrogenated castor oil mixed micelles prior to *in vivo* administration in mice. Authors suggested that TOC-siRNA conjugates were delivered to the liver via the lymphatic pathway. Efficient gene silencing and reduction in LDL-cholesterol was also demonstrated (Murakami et al. 2015). Taken together, these and similar conjugates may find suitability for oral delivery of siRNA.

14.3 RNA-BASED NANOSTRUCTURES AS RNAi THERAPEUTICS (RNA-NPs)

The application of RNA-based nanostructures is growing in popularity as a platform for delivery of RNA-based therapeutics. Use of RNA nanostructures affords the possibility of combining several functional RNA moieties into a single deliverable entity. This style of delivery platform has several advantages over the delivery of multiple individual RNAs. The use of RNA nanostructures allows the user to increase the siRNA payload delivered in a single uptake event and ensure the stoichiometry and co-delivery of multiple distinct siRNAs. Their multivalent nature enables the siRNA harboring particle to also be coupled with additional therapeutic or diagnostic entities. Additionally, RNA-based nanostructures are larger in size than monomeric siRNA duplexes, making the nanostructures more likely to avoid renal clearance.

Numerous approaches to the assembly of RNA nanostructures have been used and vary in their degree of complexity. The use of single-stranded, complementary "sticky" ends is one relatively straightforward approach used to link multiple distinct RNA moieties and create a single amorphous RNA assembly. This approach can be applied to assemble multiple siRNAs, creating a linear gene-like structure (Liu et al. 2012; Yingling and Shapiro 2007), or to tether siRNAs to domains with a different function, such as a targeting aptamer (Zhou et al. 2013).

Greater control over the dimensions, programmable assembly, and functional capacity of deliverable RNA nanostructures can be achieved through the functionalization of a structurally defined, "rigid" core scaffold. The resulting nanostructures are highly reproducible, an important characteristic for therapeutic development. Typically, one of two methods are used to design RNA scaffolds. One approach is to create completely *de novo* sequences that can assemble into two-dimensional or three-dimensional geometric objects through intermolecular helix formation. This technique has been used to create triangles, squares, pentagons, hexagons, and cubes that can be further functionalized with siRNA containing sequences (Bui et al. 2017, Afonin et al. 2010, Bindewald et al. 2011) (Figure 14.5a, i). The second approach utilizes structural motifs that are found in natural RNA molecules as building blocks to construct defined structures, with supramolecular structures often assembled from monomeric units by use of extended sticky regions or kissing loops. This technique has been used to design various two-dimensionals geometries (Parlea et al. 2016b), create fully programmable hexameric RNA nanorings (Grabow et al. 2011), and engineer multiple nanostructures based on the phi29 packaging RNA (pRNA) (Shu et al. 2014) (Figure 14.5a, ii–iii).

RNA nanorings and nanocubes are both hexavalent scaffolds, able to harbor six distinct functional RNAs. These scaffolds are each assembled from six (or ten) individual strands and are fully programmable in their assembly, allowing for customization of the identity, position, and copy number of each functional moiety added to the scaffold. As such, the scaffold can be designed to harbor six copies of the same siRNA, one copy of six different siRNAs, or with other functional RNAs, or any combination in between. Importantly, appending siRNAs to the core nanostructure scaffold does not hinder its cellular uptake or silencing efficacy (Figure 14.5b). RNA nanorings complexed with Lipofectamine 2000 performed as well or better than equimolar concentrations of free siRNA in the knock-down of stably expressed eGFP in cultured breast cancer cells. Intratumoral injections of an siRNA harboring nanoring complexed with bolaamphiphile lipids also displayed slightly improved silencing efficacy over free siRNA complexes in xenograft breast cancer tumor models. RNA nanorings have also been shown to be effective against HIV infection, as targeting six distinct viral proteins with six siRNAs packaged together in the nanoring significantly reduced protein expression levels and viral reverse transcriptase activity in infected HeLa cells (Afonin et al. 2014). RNA nanocubes have also been engineered to be a potent gene silencing platform. Nanocubes complexed with Lipofectamine 2000 have demonstrated potent silencing against stably expressed eGFP in cultured breast cancer cells, as well reduction of HIV infection among a population of 293T cells (Afonin et al. 2015). Both nanoring and nanocube scaffolds have shown the ability to successfully incorporate other functional entities including

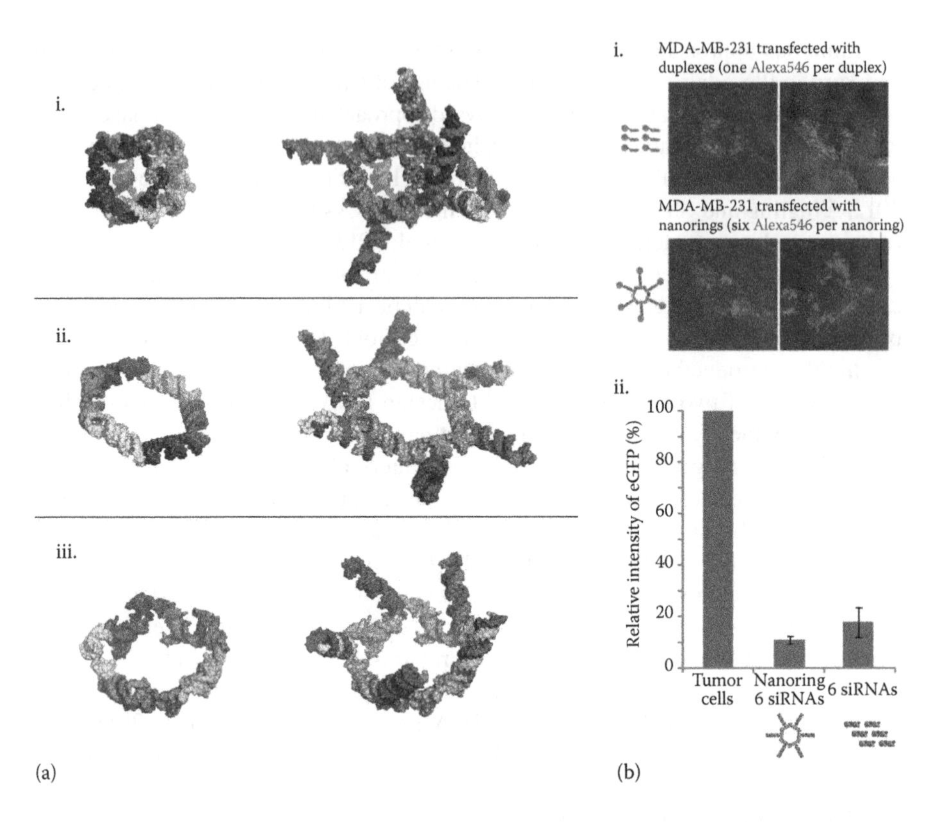

FIGURE 14.5 (a) RNA-based nanostructures for siRNA delivery. (Left) RNA-based nanostructures can be generated through the assembly of multiple RNA strands. Such structures include (i) RNA nanorings, (ii) RNA nanocubes and (iii) novel structures engineered from the phi29 DNA-packaging motor RNA (pRNA). Each RNA strand is indicated by a different color. (Right) RNA nanostructures can be used as a scaffolding for multiple DsiRNAs, or other functional RNAs. DsiRNA helices are highlighted. (b) RNA-based nanostructures for siRNA delivery. Result for RNA nanorings harboring siRNAs that illustrate its efficiency as a RNAi-therapeutic platform. (i) Uptake of Alexa546 labeled siRNA by MDA-MB-231 cells, as either free siRNA duplexes or incorporated within the nanorings scaffold, as indicated (adapted from Figure 3, reference [Afonin et al. 2014]). (ii) *In vivo* eGFP gene silencing in MDA-MB-231 tumor-bearing mice using siRNA duplexes or siRNA-nanorings as indicated (adapted from Figure 5, reference [Afonin et al. 2015]).

fluorescent aptamers, proteins and dyes, targeting aptamers, and the incorporation of RNA-DNA hybrid duplexes that allow for an inducible RNAi response (Afonin et al. 2010, Afonin et al. 2014, Afonin et al. 2015, Afonin et al. 2013).

pRNA-derived nanostructures are multivalent, and range in size from simple divalent dimer structures, to multivalent junctions, rings, and higher order dendrimers (Sharma et al. 2016, Shu et al. 2013). pRNA-based scaffolds have proven successful in siRNA delivery in many different contexts. Experiments performed in cell culture using pRNA-X scaffolds (an "X"-shaped RNA scaffold consisting of four helical domains emanating from two central, adjacent pRNA three-way junctions) harboring four siRNAs complexed with Lipofectamine 2000 have demonstrated significant downregulation of both plasmid-encoded luciferase and endogenous survivin transcripts in HeLa cells (Haque et al. 2012), while Lipofectamine 2000 delivery of chimeric pRNA dimers harboring an siRNA against a viral protease were successful in increasing the viability of cells infected with cox-sackie virus B3 (Zhang et al. 2009). Most development towards *in vivo* therapeutic application of pRNA nanostructures has focused on aptamer or small-molecule targeted delivery in the absence of any lipid carrier. This is largely outside the scope of this review. However, systemic delivery of such siRNA harboring pRNA nanostructures have been successful in treating multiple mouse models including gastric (Cui et al. 2015) and intracranial glioblastoma (Lee et al. 2015) xenograft tumors.

14.4 LIPID-BASED siRNA DELIVERY SYSTEMS: GENERAL CONCEPTS

14.4.1 Background

A popular alternate approach that obviates the need of direct chemical modifications of siRNA relies on the development and utilization of suitable molecules and/or molecular scaffolds (often referred to as delivery vehicles, agents and carriers) to deliver siRNA to its targeted site. The reader is referred to excellent reviews published in recent years, summarizing the status of siRNA-based therapies including various platforms, design strategies, potential utility, pitfalls etc., (Bobbin and Rossi 2016, Wittrup and Lieberman 2015, Kim et al. 2005, Musacchio and Torchilin 2013, Sarisozen et al. 2015, Tseng et al. 2009, Cheng and Mahato 2009, Sarisozen et al. 2016, Ganta et al. 2008, Jabr-Milane et al. 2008, Zhang et al. 2008, Alexis et al. 2008, Ferrari 2005, Nguyen and Szoka 2012, Li and Szoka 2007, Mahato 2000).

Essentials of an ideal delivery agent for siRNA-based therapies, have mainly stemmed from our existing knowledge of long-studied nano drug delivery systems (Zhang et al. 2008, Puri et al. 2009, Allen and Cullis 2013, Torchilin 2007, Connor et al. 1986). Desired characteristics of a nanocarrier, such as its low toxicity, biodegradability, acquiescent surface properties for preferred accumulation in the disease organs (reduced non-specific biodistribution), stability, and timed-clearance rates from the circulation are some important properties commonly shared by both drug and siRNA delivery systems.

14.4.2 siRNA-LNP Systems

Lipids, an integral part of cell membranes, have been designed by nature to perform important biological functions including intracellular compartmentalization of organelles, cellular transport, cell signaling, control of membrane potential mechanisms, and maintenance of cellular architectures etc. Assembly and organization of naturally occurring lipid molecules (including phospholipids, sphingolipids, and cholesterol) is relatively well-understood at the molecular level by examination of model membrane systems (such as supported bilayers and liposomes). Phospholipids have the propensity to organize into vesicular structures of desired size, surface, and physical characteristics with their ability to accommodate payloads of drugs and bear molecular beacons at their surface (Haran et al. 1993). Therefore, lipids have found their niche as suitable delivery agents for drugs (Puri et al. 2009, Kohli et al. 2014, Allen and Cullis 2004, Torchilin 2014) and nucleic acids including DNA and RNA (Santel et al. 2006, Lee et al. 2013). It may be noted that various lipid molecules, in lieu of phospholipids (which are glycerol backbone based), have been developed for delivery of siRNA. These lipid molecules bear positively charged groups of choice in the polar region, include a linker and hydrophobic chains of various lengths and degrees of unsaturation. Design and synthesis of these cationic lipids is based on geometric considerations to favor efficient transfection to accomplish optimal siRNA function.

The complexes of cationic lipids and siRNA are created by electrostatic interactions between the positively charged head groups of the lipids and the negatively charged phosphates of siRNA. The resulting siRNA-lipid assemblies are commonly termed "lipoplexes." Additional lipid molecules, categorized as "helper lipids" are often included in the siRNA-lipid assemblies. The purpose of helper lipids is to facilitate the disassembly of siRNA from the lipid environment and their intracellular disposition. This function helps in localized destabilization of the plasma membrane or endosomal membrane (see Figure 14.1b). Endosomal escape of the lipoplexes into the cytosol can be achieved by multiple mechanisms. We discuss siRNA release mechanisms that are relatively well-studied later in this section.

Due to the discrete features of siRNA (small size and negative charge), additional constraints dictate design considerations for the creation of optimal siRNA-nanocarrier assemblies. These include development of suitable protocols for siRNA loading within the hydrophilic core of a chosen carrier relying on non-covalent or hydrophobic interactions for its encapsulation. However, mainstream siRNA-LNP designs at present are primarily based on direct electrostatic associations of the siRNA with a positively charged delivery agent of choice. Chemical structures of some of the classical and recently developed positively charged lipids are shown in Figure 14.6. (Kanasty et al. 2013, Ozpolat et al. 2010, Sparks et al. 2012, Lee et al. 2013, Santel et al. 2006). Among these, 1,2-dioleoyl-3-trimethylammonium-propane chloride, (DOTAP) (Stamatatos et al. 1988) and 1,2-di-O-octadecenyl-3-trimethylammonium propane chloride (DOTMA) (Felgner et al. 1987) are popular molecules for cell culture studies (Simberg et al. 2004) and have been tested in small animal studies (Fan et al. 2015, Singh et al. 2000). Studies using DOTAP and DOTMA have been extensively reviewed previously (Tseng et al. 2009, Tam et al. 2013). Recently, two distinct classes of lipids, namely oxime ether (Gupta et al. 2015a), and bola lipids (Gupta et al. 2015b) have emerged as viable lipids for improved siRNA delivery. *In vitro* and *in vivo* studies from this laboratory and by Heldman and colleagues allude to the potential utility of bolaamphiphiles for RNAi therapy (Stern et al. 2014, Kim et al. 2013, Dakwar et al. 2012). Clinical success of bolas as siRNA carriers, however, remains to be seen in the near future. Studies related to these lipids were recently discussed in detail elsewhere (Gupta et al. 2017).

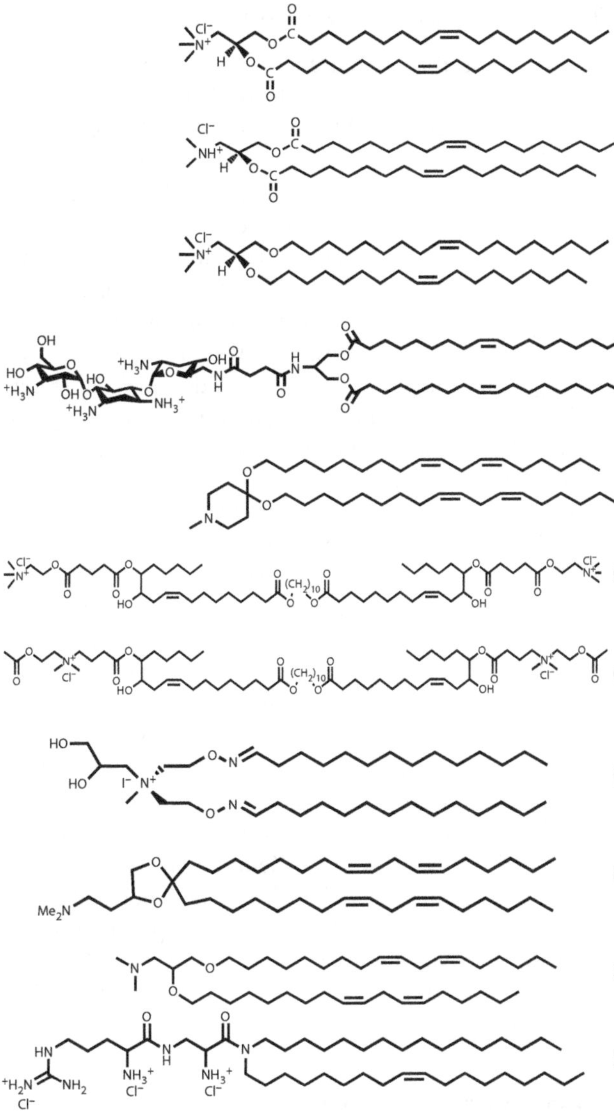

(i) DOTAP

(ii) DODAP

(iii) DOTMA

(iv) DOSST

(v) YSK05 (MEND)

(VI) Bola (GLH-19)

(VII) Bola (GLH-20)

(VIII) OEL

(IX) Dlin-KC2-DMA

(X) DLinDMA (SNALP lipid)

(XI) AtuFECT01

FIGURE 14.6 Chemical structures of representative lipids used for siRNA delivery. DOTAP, 1,2-dioleoyl-3-trimethylammonium-propane (Stamatatos et al. 1988); DODAP, 1,2-dioleoyl-3-dimethylammonium-propane (Leventis and Silvius 1990); DOTMA, 1,2-di-O-octadecenyl-3-trimethylammonium propane (Felgner et al. 1987); DOSST, Di-oleyl-succinyl-serinyl-tobramycin (Habrant et al. 2016); YSK05, a cationic pH sensitive lipid with tertiary amine (Sato et al. 2012); GLH 19 and GLH 20, Bola lipids with non-hydrolyzable and hydrolyzable acetylcholine head groups respectively (Grinberg et al. 2008); OEL, Oxime ether lipid (Biswas et al. 2011); Dlin-KC2-DMA, Cationic lipid (Semple et al. 2010); DlinDMA, lipid of SNALP (Heyes et al. 2005); AtuFECT01, cationic lipid (Santel et al. 2006).

14.4.3 siRNA-LNPs, Intracellular Delivery Pathways

As mentioned earlier, overcoming the limited endosomal escape of endocytosed siRNA-LNPs to the cytosol remains a challenge in the field among other limitations faced by RNAi therapeutics (Tseng et al. 2009, Sarisozen et al. 2016). Intracellular delivery of siRNA can be achieved by two main pathways (Figure 14.1b).

The first mechanism relies on direct siRNA expulsion from the nanoparticles into the cytosol primarily by fusion of the lipid membrane of the nanoparticle with the plasma membrane of the target cells. Here, membrane interacting, conformation-specific molecules such as cell-penetrating peptides (Deshayes et al. 2005, Endoh and Ohtsuki 2009, Torchilin 2008), fusogenic lipids (such as cardiolipin), or positively charged lipids play a role to promote cytosolic delivery (Cullis and K.B. 1979, Viard and Puri 2015). Triggers for direct cytosolic siRNA delivery can result from specific interactions of LNPs with the cellular receptors or subtle changes in the LNPs by the slightly acidic pH of the tumor microenvironment (Bertrand et al. 2014, Danhier 2016, Hobbs et al. 1998).

Recent studies by Ding et al. described a nanosystem for direct cytosolic delivery of HDL-reconstituted siRNA-cholesterol for target specific delivery to treat tumor angiogenesis, bypassing the endocytic route for entry (Ding et al. 2014). Direct transfer of siRNA-cholesterol occurred by the scavenger receptor B1 and was facilitated by HDL. The second pathway for cellular entry of siRNA-LNPs relies on preferential intracellular uptake of the LNPs via endocytic mechanisms. In the latter case, dislodging of the siRNA from the endocytosed nanoparticles for its placement into the cytosol becomes a vital consideration for successful RNAi therapy (Figure 14.1b).

In general, LNPs contain structurally defined lipids with an element of built-in positive charged moieties to enable electrostatic interactions with negatively charged siRNA molecules. However, the positively charged formulations may exhibit non-specific toxic effects *in vivo* (Knudsen et al. 2015, Kedmi et al. 2010, Senior et al. 1991), presumably by undesired immune activation. Therefore, Sood and colleagues developed neutral liposomes (consisting of DOPC) with encapsulation of siRNA against the EphA2 gene (EPHARNA); entering Phase I clinical trials (Wagner et al. 2017). Various tunable LNPs to facilitate escape of siRNA from the endosomes examined to date are presented in the next section.

14.5 TUNABLE LIPID-BASED RNAi THERAPEUTICS FOR siRNA DELIVERY

Lipid-based nanoparticles for on-demand release of loaded drugs *in vitro*, *in vivo*, and in clinical settings have paved the way to partial success (Puri et al. 2009, Puri 2013). Design principles and mechanisms underlying triggerable LNPs for drug delivery can *in part* be translated to triggerable LNPs for siRNA delivery. At present, the tunable siRNA-LNP field can be considered to be only in its juvenile stage while exploring multi-pronged strategies. Efforts to this end can be categorized into the following main platforms (i) development of tunable chemical linkers to generate siRNA-lipid conjugates, and (ii) utilization of stimuli-sensitive lipids to formulate LNPs for efficient siRNA release in the cytosol from the endosomes. These strategies offer distinct molecular mechanisms for endosomal escape, often termed "switching mechanisms." It may be noted that the "switching mechanisms" may lead to reversible or irreversible modifications in the designed molecule or the nanosystem.

An ideal design of carrier lipid-based RNAi nanotherapeutics aims at (i) reasonable siRNA protection, (ii) efficient and preferential intracellular uptake by tumor cells, (iii) enhanced tumor accumulation, (iv) a built-in (tunable) mechanism for siRNA release from the endosomes, and (v) decomplexation from the carrier nanoparticle in the cytosol.

As mentioned earlier, poor endosomal release of the siRNA (plus nanoparticle) is one of the major challenges posing limitations for successful RNAi therapy. Ongoing research efforts to solve this seemingly very difficult problem are aiming at multifaceted platforms to develop "Tunable RNAi Therapeutics." The common principles that rely on the design of tunable lipid-based nanotherapeutics are primarily based on currently used strategies for tunable nanoparticles for on-demand release of small molecule drugs in the field of cancer nanomedicine (Ganta et al. 2008, Puri 2013, McCoy et al. 2010, Zhu and Torchilin 2013, Kale and Torchilin 2007, Torchilin 2009). Following intracellular uptake of siRNA-lipoplexes by the cells, factors that influence RNAi therapy efficiency include precise cellular location of the released siRNA within the cell, as well as the kinetics, and the extent of the siRNA release. It is also important that the modalities/treatments that are applied for siRNA release exert minimal effect on the biological activity of siRNA itself, and do not compromise the host cell machinery.

Currently available research reports that primarily focus on development of nanoparticles that will facilitate siRNA/nanoparticle release from the endosome can be broadly classified into two categories, namely (i) internal and (ii) external stimulus. The internal trigger depends on the unique biology of the disease (such as altered pH, glutathione levels, and overexpressed enzymes). An external trigger, as the name suggests, relies on an outside source to destabilize the endosome and/or the tunable nanoparticle (such as heat, light, magnetic field). Both approaches with their pros and cons have advanced in the cancer nanomedicine field for small molecule drugs (Viard and Puri 2015, Puri 2013). However, at present, the field for on-demand delivery of siRNA-LNP is at a very early stage of development (Salzano et al. 2015). Various platforms including pH-triggered molecular switches (Walsh et al. 2013), redox switches (Vulugundam et al. 2015, Aytar et al. 2013), light switches (Oliveira et al. 2007), polymerizable switches based on surfactants (Wang et al. 2007), and cleavable PEG molecules (Jung et al. 2010) are under development. We describe tunable lipid systems that contain reducible and ionizable groups to facilitate endosomal escape below.

14.5.1 siRNA-LNPs with Tunable Sulfhydryl Linkages

Glutathione (GSH), a tripeptide, is a vital molecule in biological systems, known to maintain the intracellular redox state and modulate oxidative stress among other functions etc. (Jones et al. 2002, Sies 1999). Intracellular GSH levels are reported to be significantly elevated in cancer cells; leveraging for exploitation in favor of GSH to develop tunable RNAi therapeutics containing reducible bonds for site-specific delivery of small molecule drugs (Musacchio et al. 2010, Kim et al. 2010, Vader et al. 2011). To take advantage of the differential of GSH levels in cancer cells, Musacchio et al., developed a novel siRNA-thio phosphatidyl ethanol amine (siRNA-s-s-PE) to accomplish efficient siRNA release from nanoparticles in cell culture assays and in circulation (Musacchio et al. 2010) (Figure 14.7a). They used a functionalized GFP-siRNA (2-pyridyl disulfide activated siRNA, siRNA-SPDP). The functionalization was done at the 3' end of the sense strand and reaction with PE-SH resulted in siRNA-s-s-PE. The molecule was assembled in micelles by addition of a PEG lipid. siRNA-s-s-PE was significantly superior when packaged with the PEG-lipid as

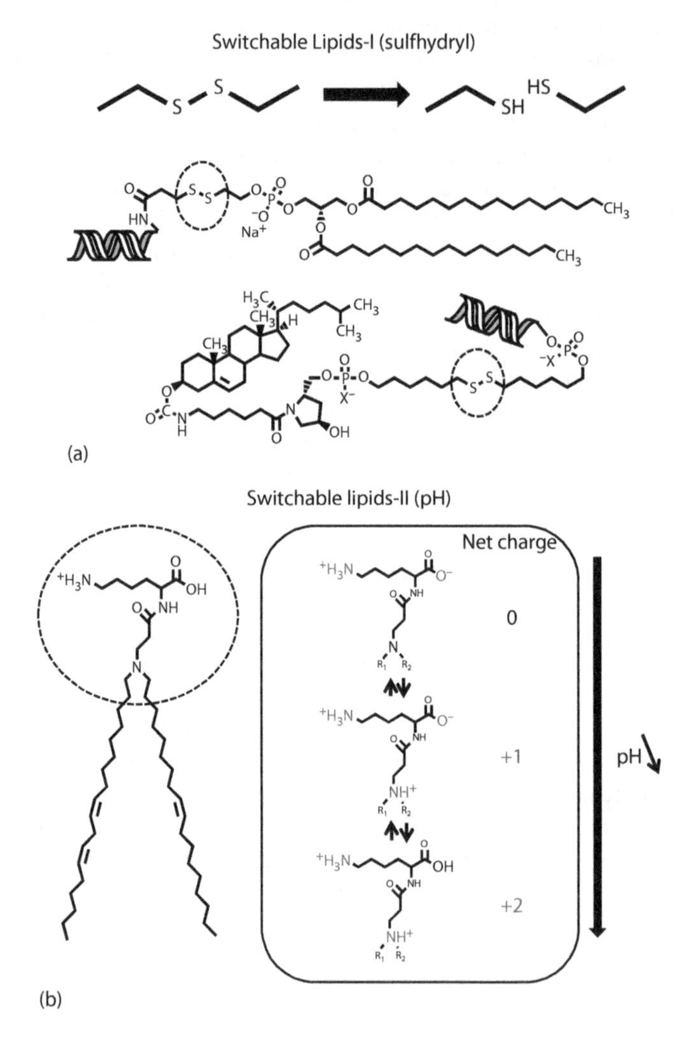

FIGURE 14.7 (a) Tunable lipids used for enhanced siRNA delivery. Glutathione responsive lipids, siRNA-S-S-PE (top) (Musacchio et al. 2010)., siRNA-cholesterol conjugate with S-S tether (Chen et al. 2010). (b) Tunable lipids used for enhanced siRNA delivery. Ionizable lysine-based pH responsive lipid. (From Walsh, C.L., et al., *Bioconjug Chem*, **24**(1): p. 36–43, 2013.)

compared to Lipofectamine 2000. In another study, siRNA-cholesterol conjugates were synthesized via a cleavable disulfide at the 3' end of the sense strand of siRNA (Chen et al. 2010). It was demonstrated that the siRNA-conjugate containing the biocleavable S-S linker was significantly efficient in gene silencing of a phosphodiesterase expressed in oligodendrocytes upon direct CNS-delivery to rats. These studies demonstrate the utility of cleavable sulfhydryl linkers in the field of RNAi therapeutics, and future research will substantiate the translation aspects of this strategy.

14.5.2 pH-Responsive Ionizable Lipids

The design, synthesis, and utility of pH responsive lipids to deliver siRNA is a widely studied area in the realm of tunable RNAi therapeutics. The very existence of slightly acidic pH in the tumor environment (Danhier 2016, Tannock and Rotin 1989) and pH gradients in the endosomes (Liu and Huang 2013, Akinc and Battaglia 2013) bring in attractive modalities for escape of the endocytosed nanoparticles. An ideal ionizable lipid molecule should contain a head group with pKa values responsive to the chosen pH environment for protonation and deprotonation. In addition, the ability of the protonated lipid to assume a preferred conformation (such as hexagonal H_{II} phase) to interact with the inner monolayer of endosomal lipids for membrane destabilization is desirable (Culls and K.B. 1979, Viard and Puri 2015). Various lipid-based systems for endosomal escape of siRNA-LNPs are briefly described below.

Previous studies have shown that incorporation of DOPE, that goes into the hexagonal (H_{II}) phase at low pH, mediates the fusion of the endosomal membrane with the siRNA-lipoplexes, resulting in release of the siRNA into the cytosol (Litzinger and Huang 1992, Farhood et al. 1995). Heyes et al. were the first to synthesize a series of cationic lipid analogs with varying chain lengths and degrees of unsaturation and formulated SNALPs. Using Neuro2A cells stably expressing the luciferase gene, it was demonstrated that the degree of saturation of cationic lipids plays a role in lipid, fusogenicity, cellular uptake, and gene silencing. Efficient endosomal release was suggested for the observed high gene silencing of luciferase in Neuro2A cells (Heyes et al. 2005).

In a follow-up study, Semple et al. utilized a medicinal chemistry approach to rationally design lipids for siRNA delivery taking into consideration the putative *in vivo* mechanism of action of ionizable cationic lipids. Using DLinDMA (a component of SNALPs) as the bench mark, a series of cationic lipids were synthesized to screen gene silencing activity *in vivo*. The lipid DLin-KC2-DMA (Figure 14.7b) formulated in liposomes outperformed other synthesized lipids presumably due to its pKa (6.7 ± 0.08) showing *in vivo* siRNA activity at doses as low as 0.01 mg/kg in rodents and 0.1 mg/kg in non-human primates (Semple et al. 2010). Therefore, these ionizable pH-sensitive lipids are likely to prove to be useful as components of RNAi therapeutics *in vivo*.

In another study, Walsh et al. took a novel approach to synthesize ionizable lipids that contain a lysine head group linked to a long-chain dialkyl amine via an amide coupling reaction (Walsh et al. 2013). Using a simple synthesis scheme, variations in the ionizable head groups, hydrophobic tails, and linker regions were introduced. The introduction of a second ionizable amine group in the core of the lipid had an effect on the pKa, membrane destabilization, endosomal escape, and siRNA delivery into cytoplasm. Interestingly, these lipids performed well in cell culture assays as demonstrated by knock-down of the luciferase gene in HeLa-luc cells. In contrast, these lipids did not show a significant silencing of Factor VII when tested *in vivo* in mouse studies. Under similar conditions, the SNALP lipid, DLinDMA showed significant knock-down of the Factor VII *in vivo*. This study brings an interesting point for further investigation.

14.6 siRNA-LNPs IN CLINICAL TRIALS

A prerequisite for successful RNAi therapeutics requires delivery of the siRNA-loaded nanocarrier into the cell interior, as the site of action of siRNA is in the cytosol. As described in preceding sections, GalNac-siRNA conjugates and cholesterol-siRNA conjugates have paved the way for initial clinical trials for diseases of the liver. Here, GalNac conjugates were delivered subcutaneously to evade the requirement of a carrier to facilitate siRNA delivery. Cholesterol-siRNA conjugates also exploited their association with lipoproteins in blood plasma for transport to the liver, and hence could be delivered via the intravenous route. For diseases not involving the liver, carrier-assisted siRNA delivery becomes important. In recent years, a number of cationic lipids have entered clinical trials for safety and/or efficacy studies (Figure 14.6, DLinDMA & AtuFECT01) (Xu and Wang 2015). A liposomal formulation Atu027, (prepared from cationic AtuFECT01, a helper lipid [DPhyPE] and the PEGylated lipid [DSPE-PEG2000]), assembled with blunt ended negatively charged 23mer RNA ODN was tested for advanced solid tumors (Strumberg et al. 2012, Schultheis et al. 2014). Another promising lipid-based platform called stable nucleic acid-lipid particles (SNALPs) (Santel et al. 2006) was originally described by Heyes et al. (2005) and Morrissey et al. (2005). SNALPs contain an ionizable cationic lipid, 1,2-dilinoleyloxy-3-dimethylaminopropane (DLinDMA, Figure 14.6-x), a helper lipid and a PEG-lipid. The design principle of SNALPs is based on the ability of DLinDMA to undergo protonation at low pH, and assembly of siRNA in the core of the nanoparticle. Using SNALPs, siRNAs for two cell cycle regulating proteins, polo like kinase (PLK1) and kinesin spindle protein (KSP) were tested in a hepatic cancer model (Hep3B, Neuro2a hepatic tumor) (Judge et al. 2009), with induction of apoptosis, significant inhibition of the PLK1-mRNA and KSP-mRNA expression levels resulting in enhanced survival rates of tumor bearing mice. Recently, Alnylam and Sanofi reported a successful Phase III clinical trial for the rare disease RNAi candidate patisiran, showing highly significant improvement in hereditary ATTR (hATTR) amyloidosis with polyneuropathy in patients. Their report embodies a landmark in the development of RNAi therapeutics and marks the first Phase III study for an RNAi drug (clinical trial number NCT02510261). Their formulation was based on SNALP lipids and was administered by intravenous injections.

14.7 SUMMARY AND PERSPECTIVES

RNAi-based nanotherapeutics are currently of huge interest as gene modulation tools to treat cancer, infections, and other related diseases. At present, diseases of the liver have been studied in detail using siRNA-cholesterol conjugates and have met with success. Naked siRNA molecules with suitable modifications may find their applications for diseases of skin and eye, when the therapeutic can be applied topically. Design of various molecules to carry siRNA to desired sites is an area under constant development for intended applications as injectable siRNAs. In this regard, extensive research has been done exploiting versatile lipid molecules as carriers for siRNA delivery. siRNA-lipid partnerships can be multifold as presented in this review. Hence, it's not surprising that several lipid-based siRNA nanoparticles are being examined in clinical trials. A recent successful Phase III clinical trial study included siRNA-LNPs. However, availability of tunable siRNA-LNPs for enhanced and localized delivery is subject to future research efforts. Construction of RNA-based scaffolds (RNA nanostructures) is a relatively new field of research and has begun to show promise using *in vitro* and *in vivo* model systems. The very advantage of RNA nanostructures to accommodate multiple siRNA, molecular beacons and targeting ligands within a single architecture can be considered a leap forward in the realm of RNAi therapeutics.

ACKNOWLEDGMENTS

This research has been funded in whole or in part with federal funds from the Frederick National Laboratory for Cancer Research, National Institutes of Health, under contract HHSN261200800001E. The content of this publication does not necessarily reflect the views or policies of the Department of Health and Human Services, nor does mention of trade names, commercial products or organizations imply endorsement by the US government. This research was supported [in part] by the Intramural Research Program of NIH, Frederick National Lab, and Center for Cancer Research.

REFERENCES

Adhikari, P., et al., *Nano lipid-drug conjugate: an integrated review.* Int J Pharm, 2017. **529**(1–2): p. 629–41.

Afonin, K.A., et al., *In vitro assembly of cubic RNA-based scaffolds designed in silico.* Nat Nanotechnol, 2010. **5**(9): p. 676–82.

Afonin, K.A., et al., *Design and self-assembly of siRNA-functionalized RNA nanoparticles for use in automated nanomedicine.* Nat Protoc, 2011. **6**(12): p. 2022–34.

Afonin, K.A., et al., *Co-transcriptional assembly of chemically modified RNA nanoparticles functionalized with siRNAs.* Nano Lett, 2012. **12**(10): p. 5192–5.

Afonin, K.A., et al., *Activation of different split functionalities on re-association of RNA-DNA hybrids.* Nat Nanotechnol, 2013. **8**(4): p. 296–304.

Afonin, K.A., et al., *Multifunctional RNA nanoparticles.* Nano Lett, 2014. **14**(10): p. 5662–71.

Afonin, K.A., et al., *Computational and experimental characterization of RNA cubic nanoscaffolds.* Methods, 2014a. **67**(2): p. 256–65.

Afonin, K.A., et al., *In silico design and enzymatic synthesis of functional RNA nanoparticles.* Acc Chem Res, 2014b. **47**(6): p. 1731–41.

Afonin, K.A., et al., *Triggering of RNA interference with RNA-RNA, RNA-DNA, and DNA-RNA nanoparticles.* ACS Nano, 2015. **9**(1): p. 251–9.

Akhtar, S. and I.F. Benter, *Nonviral delivery of synthetic siRNAs in vivo.* J Clin Invest, 2007. **117**(12): p. 3623–32.

Akinc, A. and G. Battaglia, *Exploiting endocytosis for nanomedicines.* Cold Spring Harb Perspect Biol, 2013. **5**(11): p. a016980.

Alexis, F., et al., *New frontiers in nanotechnology for cancer treatment.* Urol Oncol, 2008. **26**(1): p. 74–85.

Allen, T.M. and P.R. Cullis, *Drug delivery systems: entering the mainstream.* Science, 2004. **303**(5665): p. 1818–22.

Allen, T.M. and P.R. Cullis, *Liposomal drug delivery systems: from concept to clinical applications.* Adv. Drug Deliv. Rev, 2013. **65**(1): p. 36–48.

Allerson, C.R., et al., *Fully 2'-modified oligonucleotide duplexes with improved in vitro potency and stability compared to unmodified small interfering RNA.* J Med Chem, 2005. **48**(4): p. 901–4.

Alterman, J.F., et al., *Hydrophobically modified siRNAs silence huntingtin mRNA in primary neurons and mouse brain.* Mol Ther Nucleic Acids, 2015. **4**: p. e266.

Aytar, B.S., et al., *Redox-based control of the transformation and activation of siRNA complexes in extracellular environments using ferrocenyl lipids.* J Am Chem Soc, 2013. **135**(24): p. 9111–20.

Bertrand, N., et al., *Cancer nanotechnology: the impact of passive and active targeting in the era of modern cancer biology.* Adv Drug Deliv Rev, 2014. **66**: p. 2–25.

Biswas, S., et al., *Hydrophobic oxime ethers: a versatile class of pDNA and siRNA transfection lipids.* Chem Med Chem, 2011. **6**(11): p. 2063–9.

Bindewald, E., et al., *Multistrand RNA secondary structure prediction and nanostructure design including pseudoknots.* ACS Nano, 2011. **5**(12): p. 9542–51.

Bobbin, M.L. and J.J. Rossi, *RNA interference (RNAi)-based therapeutics: delivering on the promise?* Annu Rev Pharmacol Toxicol, 2016. **56**: p. 103–22.

Bui, M.N., et al., *Versatile RNA tetra-U helix linking motif as a toolkit for nucleic acid nanotechnology.* Nanomedicine, 2017. **13**(3): p. 1137–46.

Byrne, M., et al., *Novel hydrophobically modified asymmetric RNAi compounds (sd-rxRNA) demonstrate robust efficacy in the eye.* J Ocul Pharmacol Ther, 2013. **29**(10): p. 855–64.

Chen, Q., et al., *Lipophilic siRNAs mediate efficient gene silencing in oligodendrocytes with direct CNS delivery.* J Control Release, 2010. **144**(2): p. 227–32.

Cheng, K. and R.I. Mahato, *siRNA delivery and targeting.* Mol Pharm, 2009. **6**(3): p. 649–50.

Choung, S., et al., *Chemical modification of siRNAs to improve serum stability without loss of efficacy.* Biochem Biophys Res Commun, 2006. **342**(3): p. 919–27.

Collingwood, M.A., et al., *Chemical modification patterns compatible with high potency Dicer-substrate small interfering RNAs.* Oligonucleotides, 2008. **18**(2): p. 187–200.

Connor, J., N. Norley, and L. Huang, *Biodistribution of pH-sensitive immunoliposomes.* Biochim Biophys Acta, 1986. **884**(3): p. 474–81.

Cui, D., et al., *Regression of gastric cancer by systemic injection of RNA nanoparticles carrying both ligand and siRNA.* Sci Rep, 2015. **5**: p. 10726.

Cullis, P.R. and K.B. de, *Lipid polymorphism and the functional roles of lipids in biological membranes.* Biochim Biophys Acta 1979 Dec, 1979. **559**(4): p. 399–420.

Dakwar, G.R., et al., *Delivery of proteins to the brain by bolaamphiphilic nano-sized vesicles.* J Control Release, 2012. **160**(2): p. 315–21.

Danhier, F., *To exploit the tumor microenvironment: Since the EPR effect fails in the clinic, what is the future of nanomedicine?* J Control Release, 2016. **244**(Pt A): p. 108–121.

Davis, M.E., et al., *Evidence of RNAi in humans from systemically administered siRNA via targeted nanoparticles.* Nature, 2010. **464**(7291): p. 1067–70.

Deleavey, G.F., et al., *Synergistic effects between analogs of DNA and RNA improve the potency of siRNA-mediated gene silencing.* Nucleic Acids Res, 2010. **38**(13): p. 4547–57.

Deleavey, G.F. and M.J. Damha, *Designing chemically modified oligonucleotides for targeted gene silencing.* Chem Biol, 2012. **19**(8): p. 937–54.

De Paula, D., M.V. Bentley, and R.I. Mahato, *Hydrophobization and bioconjugation for enhanced siRNA delivery and targeting.* RNA, 2007. **13**(4): p. 431–56.

Deshayes, S., et al., *Cell-penetrating peptides: tools for intracellular delivery of therapeutics.* Cell Mol Life Sci, 2005. **62**(16): p. 1839–49.

Didiot, M.C., et al., *Exosome-mediated delivery of hydrophobically modified siRNA for huntingtin mRNA silencing.* Mol Ther, 2016. **24**(10): p. 1836–47.

Ding, Y., et al., *A biomimetic nanovector-mediated targeted cholesterol-conjugated siRNA delivery for tumor gene therapy.* Biomaterials, 2012. **33**(34): p. 8893–905.

Ding, Y., et al., *Direct cytosolic siRNA delivery by reconstituted high density lipoprotein for target-specific therapy of tumor angiogenesis.* Biomaterials, 2014. **35**(25): p. 7214–27.

Dong, Y., et al., *Lipopeptide nanoparticles for potent and selective siRNA delivery in rodents and nonhuman primates.* Proc Natl Acad Sci U S A, 2014. **111**(11): p. 3955–60.

Duhem, N., F. Danhier, and V. Preat, *Vitamin E-based nanomedicines for anti-cancer drug delivery.* J Control Release, 2014. **182**: p. 33–44.

Elmen, J., et al., *Locked nucleic acid (LNA) mediated improvements in siRNA stability and functionality.* Nucleic Acids Res, 2005. **33**(1): p. 439–47.

Endoh, T. and T. Ohtsuki, *Cellular siRNA delivery using cell-penetrating peptides modified for endosomal escape.* Adv Drug Deliv Rev, 2009. **61**(9): p. 704–9.

Fan, Y., et al., *Cationic liposome-hyaluronic acid hybrid nanoparticles for intranasal vaccination with subunit antigens.* J Control Release, 2015. **208**: p. 121–9.

Farhood, H., N. Serbina, and L. Huang, *The role of dioleoyl phosphatidylethanolamine in cationic liposome mediated gene transfer.* Biochim Biophys Acta, 1995. **1235**(2): p. 289–95.

Felgner, P.L., et al., *Lipofection: a highly efficient, lipid-mediated DNA-transfection procedure.* Proc Natl Acad Sci U S A, 1987. **84**(21): p. 7413–7.

Ferrari, M., *Nanovector therapeutics.* Curr. Opin. Chem. Biol., 2005. **9**(4): p. 343–6.

Fire, A., et al., *Potent and specific genetic interference by double-stranded RNA in Caenorhabditis elegans.* Nature, 1998. **391**(6669): p. 806–11.

Gabizon, A.A., *Pegylated liposomal doxorubicin: metamorphosis of an old drug into a new form of chemotherapy.* Cancer Invest., 2001. **19**(4): p. 424–36.

Gandioso, A., et al., *Efficient siRNA-peptide conjugation for specific targeted delivery into tumor cells.* Chem Commun (Camb), 2017. **53**(19): p. 2870–3.

Ganta, S., et al., *A review of stimuli-responsive nanocarriers for drug and gene delivery.* Journal Controlled Release, 2008. **126**(3): p. 187–204.

Grabow, W.W., et al., *Self-assembling RNA nanorings based on RNAI/II inverse kissing complexes.* Nano Lett, 2011. **11**(2): p. 878–87.

Grinberg, S., et al., *Synthesis of novel cationic bolaamphiphiles from vernonia oil and their aggregated structures.* Chem Phys Lipids, 2008. **153**(2): p. 85–97.

Guo, P., *The emerging field of RNA nanotechnology.* Nat Nanotechnol, 2010. **5**(12): p. 833–42.

Gupta, K., A. Puri, and A. Shapiro Bruce, *Functionalized non-viral cationic vectors for effective siRNA induced cancer therapy*, in *DNA and RNA nanotechnology.* 2017. p. 1.

Gupta, K., et al., *Oxime ether lipids containing hydroxylated head groups are more superior siRNA delivery agents than their nonhydroxylated counterparts.* Nanomedicine (Lond), 2015a. **10**(18): p. 2805–18.

Gupta, K., et al., *Bolaamphiphiles as carriers for siRNA delivery: From chemical syntheses to practical applications.* J Control Release, 2015b. **213**: p. 142–51.

Habrant, D., et al., *Design of ionizable lipids to overcome the limiting step of endosomal escape: application in the intracellular delivery of mRNA, DNA, and siRNA.* J Med Chem, 2016. **59**(7): p. 3046–62.

Hall, A.H., et al., *RNA interference using boranophosphate siRNAs: structure-activity relationships.* Nucleic Acids Res, 2004. **32**(20): p. 5991–6000.

Haque, F., et al., *Ultrastable synergistic tetravalent RNA nanoparticles for targeting to cancers.* Nano Today, 2012. **7**(4): p. 245–257.

Haque, F. and P. Guo, *Overview of methods in RNA nanotechnology: synthesis, purification, and characterization of RNA nanoparticles.* Methods Mol Biol, 2015. **1297**: p. 1–19.

Haran, G., et al., *Transmembrane ammonium sulfate gradients in liposomes produce efficient and stable entrapment of amphipathic weak bases.* Biochim.Biophys. Acta, 1993. **1151**(2): p. 201–15.

Heyes, J., et al., *Cationic lipid saturation influences intracellular delivery of encapsulated nucleic acids.* J Control Release, 2005. **107**(2): p. 276–87.

Hobbs, S.K., et al., *Regulation of transport pathways in tumor vessels: role of tumor type and microenvironment.* Proc Natl Acad Sci U S A, 1998. **95**(8): p. 4607–12.

Irby, D., C. Du, and F. Li, *Lipid-drug conjugate for enhancing drug delivery.* Mol Pharm, 2017. **14**(5): p. 1325–38.

Jabr-Milane, L., et al., *Multi-functional nanocarriers for targeted delivery of drugs and genes.* J Controlled Release, 2008. **130**(2): p. 121–8.

Jeong, J.H., et al., *siRNA conjugate delivery systems.* Bioconjug Chem, 2009. **20**(1): p. 5–14.

Jones, D.P., et al., *Redox analysis of human plasma allows separation of pro-oxidant events of aging from decline in antioxidant defenses.* Free Radic Biol Med, 2002. **33**(9): p. 1290–300.

Judge, A.D., et al., *Confirming the RNAi-mediated mechanism of action of siRNA-based cancer therapeutics in mice.* J Clin Invest, 2009. **119**(3): p. 661–73.

Jung, S., et al., *Gene silencing efficiency of siRNA-PEG conjugates: effect of PEGylation site and PEG molecular weight.* J Control Release, 2010. **144**(3): p. 306–13.

Kale, A.A. and V.P. Torchilin, *Design, synthesis, and characterization of pH-sensitive PEG-PE conjugates for stimuli-sensitive pharmaceutical nanocarriers: The effect of substitutes at the hydrazone linkage on the pH stability of PEG-PE conjugates.* Bioconjugate Chem., 2007. **18**(2): p. 363–70.

Kanasty, R., et al., *Delivery materials for siRNA therapeutics.* Nat Mater, 2013. **12**(11): p. 967–77.

Kedmi, R., N. Ben-Arie, and D. Peer, *The systemic toxicity of positively charged lipid nanoparticles and the role of Toll-like receptor 4 in immune activation.* Biomaterials, 2010. **31**(26): p. 6867–75.

Khan, T., et al., *Silencing myostatin using cholesterol-conjugated siRNAs induces muscle growth.* Mol Ther Nucleic Acids, 2016. **5**(8): p. e342.

Khvorova, A., A. Reynolds, and S.D. Jayasena, *Functional siRNAs and miRNAs exhibit strand bias.* Cell, 2003. **115**(2): p. 209–16.

Kim, D.H., et al., *Synthetic dsRNA Dicer substrates enhance RNAi potency and efficacy.* Nat Biotechnol, 2005. **23**(2): p. 222–6.

Kim, H.-K., et al., *Enhanced siRNA delivery using cationic liposomes with new polyarginine-conjugated PEG-lipid.* Int. J. Pharm., 2010. **392**(1): p. 141–7.

Kim, T., et al., *In silico, in vitro, and in vivo studies indicate the potential use of bolaamphiphiles for therapeutic siRNAs delivery.* Mol Ther Nucleic Acids, 2013. **2**: p. e80.

Knudsen, K.B., et al., *In vivo toxicity of cationic micelles and liposomes.* Nanomedicine, 2015. **11**(2): p. 467–77.

Kohli, A.G., et al., *Designer lipids for drug delivery: from heads to tails.* J Controlled Release, 2014. **190**: p. 274–87.

Ku, S.H., et al., *Chemical and structural modifications of RNAi therapeutics.* Adv. Drug Deliv. Rev., 2016. **104**: p. 16–28.

Kubo, T., et al., *Palmitic acid-conjugated 21-nucleotide siRNA enhances gene-silencing activity.* Mol Pharm, 2011. **8**(6): p. 2193–203.

Kubo, T., et al., *Lipid-conjugated 27-nucleotide double-stranded RNAs with Dicer-substrate potency enhance RNAi-mediated gene silencing.* Mol Pharm, 2012a. **9**(5): p. 1374–83.

Kubo, T., et al., *Amino-modified and lipid-conjugated Dicer-substrate siRNA enhances RNAi efficacy.* Bioconjug Chem, 2012b. **23**(2): p. 164–73.

Kubo, T., et al., *SiRNAs conjugated with aromatic compounds induce RISC-mediated antisense strand selection and strong gene-silencing activity.* Biochem Biophys Res Commun, 2012c. **426**(4): p. 571–7.

Kubo, T., et al., *Gene-silencing potency of symmetric and asymmetric lipid-conjugated siRNAs and its correlation with Dicer recognition.* Bioconjug Chem, 2013. **24**(12): p. 2045–57.

Kubo, T., K. Yanagihara, and T. Seyama, *In vivo RNAi Efficacy of palmitic acid-conjugated Dicer-substrate siRNA in a subcutaneous tumor mouse model.* Chem Biol Drug Des, 2016. **87**(6): p. 811–23.

Layzer, J.M., et al., *In vivo activity of nuclease-resistant siRNAs.* RNA, 2004. **10**(5): p. 766–71.

Lee, J.M., T.J. Yoon, and Y.S. Cho, *Recent developments in nanoparticle-based siRNA delivery for cancer therapy.* BioMed Res. Int., 2013. **2013**: p. 782041.

Lee, S.H., et al., *Current preclinical small interfering RNA (siRNA)-based conjugate systems for RNA therapeutics.* Adv. Drug Deliv. Rev., 2016. **104**: p. 78–92.

Lee, T.J., et al., *RNA nanoparticle as a vector for targeted siRNA delivery into glioblastoma mouse model.* Oncotarget, 2015. **6**(17): p. 14766–76.

Leventis, R. and J.R. Silvius, *Interactions of mammalian cells with lipid dispersions containing novel metabolizable cationic amphiphiles.* Biochim Biophys Acta, 1990. **1023**(1): p. 124–32.

Li, W. and F.C. Szoka, Jr., *Lipid-based nanoparticles for nucleic acid delivery.* Pharm Res, 2007. **24**(3): p. 438–49.

Liu, J., et al., *Argonaute2 is the catalytic engine of mammalian RNAi.* Science, 2004. **305**(5689): p. 1437–41.

Liu, X. and G. Huang, *Formation strategies, mechanism of intracellular delivery and potential clinical applications of pH-sensitive liposomes.* Asian Journal of Pharmaceutical Sciences, 2013. **8**(6): p. 319–28.

Liu, X., et al., *Efficient delivery of sticky siRNA and potent gene silencing in a prostate cancer model using a generation 5 triethanolamine-core PAMAM dendrimer.* Mol Pharm, 2012. **9**(3): p. 470–81.

Litzinger, D.C. and L. Huang, *Phosphatidylethanolamine liposomes: drug delivery, gene transfer and immunodiagnostic applications.* Biochim Biophys Acta, 1992. **1113**(2): p. 201–27.

Lorenz, C., et al., *Steroid and lipid conjugates of siRNAs to enhance cellular uptake and gene silencing in liver cells.* Bioorg. Med. Chem. Lett., 2004. **14**(19): p. 4975–77.

Mahato, R.I., *Challenges of turning nucleic acids into therapeutics.* Adv Drug Deliv Rev, 2000. **44**(2-3): p. 79–80.

Manjappa, A.S., et al., *Antibody derivatization and conjugation strategies: application in preparation of stealth immunoliposome to target chemotherapeutics to tumor.* J Control Release, 2011. **150**(1): p. 2–22.

Matranga, C., et al., *Passenger-strand cleavage facilitates assembly of siRNA into Ago2-containing RNAi enzyme complexes.* Cell, 2005. **123**(4): p. 607–20.

McCoy, C.P., et al., *Triggered drug delivery from biomaterials.* Expert Opin. Drug Deliv, 2010. **7**(5): p. 605–16.

Morrissey, D.V., et al., *Potent and persistent in vivo anti-HBV activity of chemically modified siRNAs.* Nat Biotechnol, 2005. **23**(8): p. 1002–7.

Moschos, S.A., et al., *Lung delivery studies using siRNA conjugated to TAT(48-60) and penetratin reveal peptide induced reduction in gene expression and induction of innate immunity.* Bioconjug Chem, 2007. **18**(5): p. 1450–9.

Murakami, M., et al., *Enteral siRNA delivery technique for therapeutic gene silencing in the liver via the lymphatic route.* Sci Rep, 2015. **5**: p. 17035.

Musacchio, T. and V.P. Torchilin, *siRNA delivery: from basics to therapeutic applications.* Front Biosci (Landmark Ed), 2013. **18**: p. 58–79.

Musacchio, T., et al., *Effective stabilization and delivery of siRNA: reversible siRNA-phospholipid conjugate in nanosized mixed polymeric micelles.* Bioconjug Chem, 2010. **21**(8): p. 1530–6.

Nair, J.K., et al., *Multivalent N-Acetylgalactosamine-conjugated siRNA localizes in hepatocytes and elicits robust RNAi-mediated gene silencing.* J.Am. Chem.Soc., 2014. **136**(49): p. 16958–61.

Nguyen, J. and F.C. Szoka, *Nucleic acid delivery: the missing pieces of the puzzle?* Acc Chem Res, 2012. **45**(7): p. 1153–62.

Nishina, K., et al., *Efficient in vivo delivery of siRNA to the liver by conjugation of alpha-tocopherol.* Mol Ther, 2008. **16**(4): p. 734–40.

Oliveira, S., et al., *Photochemical internalization enhances silencing of epidermal growth factor receptor through improved endosomal escape of siRNA.* Biochim Biophys Acta, 2007. **1768**(5): p. 1211–7.

O'Loughlin, A.J., et al., *Functional delivery of lipid-conjugated siRNA by extracellular vesicles.* Mol Ther, 2017. **25**(7): p. 1580–7.

Ozcan, G., et al., *Preclinical and clinical development of siRNA-based therapeutics.* Adv Drug Deliv Rev, 2015. **87**: p. 108–19.

Ozpolat, B., A.K. Sood, and G. Lopez-Berestein, *Nanomedicine based approaches for the delivery of siRNA in cancer.* J Intern Med, 2010. **267**(1): p. 44–53.

Padilla, R. and R. Sousa, *A Y639F/H784A T7 RNA polymerase double mutant displays superior properties for synthesizing RNAs with non-canonical NTPs.* Nucleic Acids Res, 2002. **30**(24): p. e138.

Parlea, L., et al., *Cellular delivery of RNA nanoparticles.* ACS Comb Sci, 2016a. **18**(9): p. 527–47.

Parlea, L., et al., *Ring catalog: A resource for designing self-assembling RNA nanostructures.* Methods, 2016b. **103**: p. 128–37.

Pecot, C.V., et al., *RNA interference in the clinic: challenges and future directions.* Nat Rev Cancer, 2011. **11**(1): p. 59–67.

Puri, A., et al., *Lipid-based nanoparticles as pharmaceutical drug carriers: from concepts to clinic.* Crit Rev. Ther. Drug Carrier Syst, 2009. **26**(6): p. 523–80.

Puri, A., *Phototriggerable liposomes: current research and future perspectives.* Pharmaceutics, 2013. **6**(1): p. 1–25.

Rana, T.M., *Illuminating the silence: understanding the structure and function of small RNAs.* Nat Rev Mol Cell Biol, 2007. **8**(1): p. 23–36.

Raouane, M., et al., *Lipid conjugated oligonucleotides: a useful strategy for delivery.* Bioconjug Chem, 2012. **23**(6): p. 1091–104.

Robbins, M., et al., *2'-O-methyl-modified RNAs act as TLR7 antagonists.* Mol Ther, 2007. **15**(9): p. 1663–9.

Rose, S.D., et al., *Functional polarity is introduced by Dicer processing of short substrate RNAs.* Nucleic Acids Res, 2005. **33**(13): p. 4140–56.

Salzano, G., D.F. Costa, and V.P. Torchilin, *siRNA delivery by stimuli-sensitive nanocarriers.* Curr Pharm Des, 2015. **21**(31): p. 4566–73.

Santel, A., et al., *A novel siRNA-lipoplex technology for RNA interference in the mouse vascular endothelium.* Gene Ther., 2006. **13**(16): p. 1222–34.

Sarisozen, C., G. Salzano, and V.P. Torchilin, *Recent advances in siRNA delivery.* Biomol Concepts, 2015. **6**(5-6): p. 321–41.

Sarisozen, C., G. Salzano, and V.P. Torchilin, *Lipid-based siRNA delivery systems: challenges, promises and solutions along the long journey.* Curr Pharm Biotechnol, 2016. **17**(8): p. 728–40.

Sato, Y., et al., *A pH-sensitive cationic lipid facilitates the delivery of liposomal siRNA and gene silencing activity in vitro and in vivo.* J Control Release, 2012. **163**(3): p. 267–76.

Schultheis, B., et al., *First-in-human phase I study of the liposomal RNA interference therapeutic Atu027 in patients with advanced solid tumors.* J Clin Oncol, 2014. **32**(36): p. 4141–8.

Schwarz, D.S., et al., *Asymmetry in the assembly of the RNAi enzyme complex.* Cell, 2003. **115**(2): p. 199–208.

Semple, S.C., et al., *Rational design of cationic lipids for siRNA delivery.* Nat Biotechnol, 2010. **28**(2): p. 172–6.

Senior, J.H., K.R. Trimble, and R. Maskiewicz, *Interaction of positively-charged liposomes with blood: implications for their application in vivo.* Biochim Biophys Acta, 1991. **1070**(1): p. 173–9.

Sharma, A., et al., *Controllable self-assembly of RNA dendrimers.* Nanomedicine, 2016. **12**(3): p. 835–844.

Shaw, B.R., et al., *Reading, writing, and modulating genetic information with boranophosphate mimics of nucleotides, DNA, and RNA.* Ann N Y Acad Sci, 2003. **1002**: p. 12–29.

Shi, J., et al., *Hybrid lipid-polymer nanoparticles for sustained siRNA delivery and gene silencing.* Nanomedicine, 2014. **10**(5): p. 897–900.

Shu, Y., et al., *Fabrication of 14 different RNA nanoparticles for specific tumor targeting without accumulation in normal organs.* RNA, 2013. **19**(6): p. 767–77.

Shu, Y., et al., *Stable RNA nanoparticles as potential new generation drugs for cancer therapy.* Adv Drug Deliv Rev, 2014. **66**: p. 74–89.

Shukla, S., C.S. Sumaria, and P.I. Pradeepkumar, *Exploring chemical modifications for siRNA therapeutics: a structural and functional outlook.* Chem Med Chem, 2010. **5**(3): p. 328–49.

Sies, H., *Glutathione and its role in cellular functions.* Free Radic Biol Med, 1999. **27**(9–10): p. 916–21.

Sioud, M., G. Furset, and L. Cekaite, *Suppression of immunostimulatory siRNA-driven innate immune activation by 2'-modified RNAs.* Biochem Biophys Res Commun, 2007. **361**(1): p. 122–6.

Soutschek, J., et al., *Therapeutic silencing of an endogenous gene by systemic administration of modified siRNAs.* Nature, 2004. **432**(7014): p. 173–8.

Sparks, J., et al., *Versatile cationic lipids for siRNA delivery.* J Control Release, 2012. **158**(2): p. 269–76.

Stamatatos, L., et al., *Interactions of cationic lipid vesicles with negatively charged phospholipid vesicles and biological membranes.* Biochemistry, 1988. **27**(11): p. 3917–25.

Stern, A., et al., *Steric environment around acetylcholine head groups of bolaamphiphilic nanovesicles influences the release rate of encapsulated compounds.* Int J Nanomedicine, 2014. **9**: p. 561–74.

Strumberg, D., et al., *Phase I clinical development of Atu027, a siRNA formulation targeting PKN3 in patients with advanced solid tumors.* Int J Clin Pharmacol Ther, 2012. **50**(1): p. 76–8.

Tam, Y.Y., S. Chen, and P.R. Cullis, *Advances in lipid nanoparticles for siRNA Delivery.* Pharmaceutics, 2013. **5**(3): p. 498–507.

Tannock, I.F. and D. Rotin, *Acid pH in tumors and its potential for therapeutic exploitation.* Cancer Res, 1989. **49**(16): p. 4373–84.

Tomita, U., et al., *Poly(ethylene glycol)-lipid-conjugated antibodies enhance dendritic cell phagocytosis of apoptotic cancer cells.* Pharmaceuticals (Basel), 2012. **5**(5): p. 405–16.

Torchilin, V.P., *Nanocarriers.* Pharmaceutical Research, 2007. **24**(12): p. 2333–4.

Torchilin, V.P., *Tat peptide-mediated intracellular delivery of pharmaceutical nanocarriers.* Adv. Drug Deliv. Rev., 2008. **60**(4-5): p. 548–58.

Torchilin, V.P., *Multifunctional and stimuli-sensitive pharmaceutical nanocarriers.* Eur. J. Pharm. Biopharm., 2009. **71**(3): p. 431–44.
Torchilin, V.P., *Multifunctional, stimuli-sensitive nanoparticulate systems for drug delivery.* Nat. Rev.. Drug discovery, 2014. **13**(11): p. 813–27.
Trevaskis, N.L., L.M. Kaminskas, and C.J. Porter, *From sewer to saviour - targeting the lymphatic system to promote drug exposure and activity.* Nat Rev Drug Discov, 2015. **14**(11): p. 781–803.
Tseng, Y.C., S. Mozumdar, and L. Huang, *Lipid-based systemic delivery of siRNA.* Adv Drug Deliv Rev, 2009. **61**(9): p. 721–31.
Turner, J.J., et al., *RNA targeting with peptide conjugates of oligonucleotides, siRNA and PNA.* Blood Cells Mol Dis, 2007. **38**(1): p. 1–7.
Vader, P., et al., *Disulfide-based poly(amido amine)s for siRNA delivery: effects of structure on siRNA complexation, cellular uptake, gene silencing and toxicity.* Pharm Res, 2011. **28**(5): p. 1013–22.
Viard, M. and A. Puri, *Stimuli-Sensitive liposomes.* Adv Planar Lipid Bilayers Liposomes, 2015. **22**: p. 1–41.
Vulugundam, G., et al., *Efficacious redox-responsive gene delivery in serum by ferrocenylated monomeric and dimeric cationic cholesterols.* Org Biomol Chem, 2015. **13**(14): p. 4310–20.
Wagner, M.J., et al., *Preclinical mammalian safety studies of EPHARNA (DOPC nanoliposomal EphA2-targeted siRNA).* Mol Cancer Ther, 2017. **16**(6): p. 1114–23.
Walsh, C.L., et al., *Synthesis, characterization, and evaluation of ionizable lysine-based lipids for siRNA delivery.* Bioconjug Chem, 2013. **24**(1): p. 36–43.
Wang, X.L., et al., *Novel polymerizable surfactants with pH-sensitive amphiphilicity and cell membrane disruption for efficient siRNA delivery.* Bioconjug Chem, 2007. **18**(6): p. 2169–77.
Wittrup, A. and J. Lieberman, *Knocking down disease: a progress report on siRNA therapeutics.* Nat Rev Genet, 2015. **16**(9): p. 543–52.
Wolfrum, C., et al., *Mechanisms and optimization of in vivo delivery of lipophilic siRNAs.* Nat Biotechnol, 2007. **25**(10): p. 1149–57.
Wu, S.Y., et al., *RNAi therapies: drugging the undruggable.* Sci Transl Med, 2014. **6**(240): p. 240 ps7.
Xu, C.-f. and J. Wang, *Delivery systems for siRNA drug development in cancer therapy.* Asian J Pharm Sci, 2015. **10**(1): p. 1–12.
Yang, X.Z., et al., *Single-step assembly of cationic lipid-polymer hybrid nanoparticles for systemic delivery of siRNA.* ACS Nano, 2012. **6**(6): p. 4955–65.
Ying, T., et al., *Engineered Fc based antibody domains and fragments as novel scaffolds.* Biochim Biophys Acta, 2014. **1844**(11): p. 1977–82.
Yingling, Y.G. and A.B. Shapiro. *Computational Design of an RNA Hexagonal Nanoring and an RNA Nanotube.* Nano let, 2007. **7**: p. 2328–2334. https://pubs.acs.org/doi/abs/10.1021/nl070984r
Zaro, J.L., *Lipid-based drug carriers for prodrugs to enhance drug delivery.* AAPS J, 2015. **17**(1): p. 83–92.
Zhang, H.M., et al., *Targeted delivery of anti-coxsackievirus siRNAs using ligand-conjugated packaging RNAs.* Antiviral Res, 2009. **83**(3): p. 307–16.
Zhang, L., et al., *Nanoparticles in medicine: therapeutic applications and developments.* Clin. Pharmacol. Ther, 2008. **83**(5): p. 761–9.
Zhou, J., et al., *Functional in vivo delivery of multiplexed anti-HIV-1 siRNAs via a chemically synthesized aptamer with a sticky bridge.* Mol Ther, 2013. **21**(1): p. 192–200.
Zhu, L. and V.P. Torchilin, *Stimulus-responsive nanopreparations for tumor targeting.* Integr. Biol. (Camb.), 2013. **5**(1): p. 96–107.
Zimmermann, T.S., et al., Clinical proof of concept for a novel hepatocyte-targeting GalNAc-siRNA conjugate. Mol Ther, 2017. **25**(1): p. 71–78.

15 Lipid Nanocarriers for RNAi-Based Cancer Therapy

Ranganayaki Muralidharan, Anish Babu, Narsireddy Amreddy, Janani Panneerselvam, Meghna Mehta, Anupama Munshi, and Rajagopal Ramesh

CONTENTS

15.1 INTRODUCTION

For many decades, small molecule inhibitors have been developed. To a degree, these inhibitors have been successful in treating cancer, although they are nonspecific and toxic against normal tissues. RNA interference (RNAi) can circumvent these problems when introduced as a novel therapeutic strategy in the clinic (Haussecker 2014). Mechanistically, RNAi inhibits expression of target mRNA, which can be achieved by introducing chemically synthesized small interfering RNA (siRNA; Siomi and Siomi 2009; Hannon and Rossi 2004). siRNAs have been synthesized to target oncogenes that are involved in cellular processes, including cell proliferation, metastasis, invasion, and angiogenesis. siRNA-based therapeutics can modulate the expression of a target protein with high specificity, providing a promising strategy to treat cancer. The clinical application of siRNAs has been limited due to its biological instability, degradation by endogenous enzymes, and aggregation with proteins in the serum. Furthermore, these negatively charged siRNAs have minimal cellular internalization due to electrostatic repulsion against the negatively charged cell membrane. Therefore, translation into the clinic depends on the development of appropriate delivery vehicles. To overcome these difficulties, the ideal siRNA delivery carrier should have a long circulation time, be able to pass through the vascular endothelium to reach the tumor site, and possess efficient endosomal escape to reach the cytoplasm (Wang et al. 2010). Many nanocarriers, polymers, liposomes, and inorganic nanomaterials have been developed to address these challenges to siRNA delivery. Among them, lipid-based nanocarriers are the most developed, due to biocompatibility and the ability to biodegrade.

15.1.1 Lipid Nanocarriers for siRNA Delivery

Since viral carriers trigger undesirable immunogenic responses, non-viral nanocarriers are found to be the best choice for safe siRNA delivery. Lipid and phospholipid nanocarriers have a natural tendency to interact with the cellular membrane to facilitate siRNA uptake. Liposomes are generally formed by the self-assembly of lipids that contain hydrophilic head groups and hydrophobic tails. The presence of one or more amines on the polar head group makes the lipid more cationic, which facilitates the binding of negatively charged DNA/RNA. Different classes of lipid-based nanocarriers have been studied for RNA/DNA delivery. Figure 15.1 shows the chemical structure of commonly used lipids for liposome preparation for siRNA delivery. Neutral lipids, like cholesterol (Muralidharan et al. 2016) DOPE (Dioleoylphosphatidylethanolamine) (Mochizuki et al. 2013) or (Dioleoylphosphatidylcholine) (Nakamura, Kuroi, and Harashima 2015), are added to stabilize the siRNA delivery system. DOPE is not only neutral but also fusogenic in nature and acts as helper lipid that stabilizes the liposome and enhances the penetration of the liposome and endosomal escape of the cargo. However, the proper mechanism of endosomal escape with DOPE is not fully understood yet. In a typical study, when combined with a cationic head group ethylene diamine and DOPE

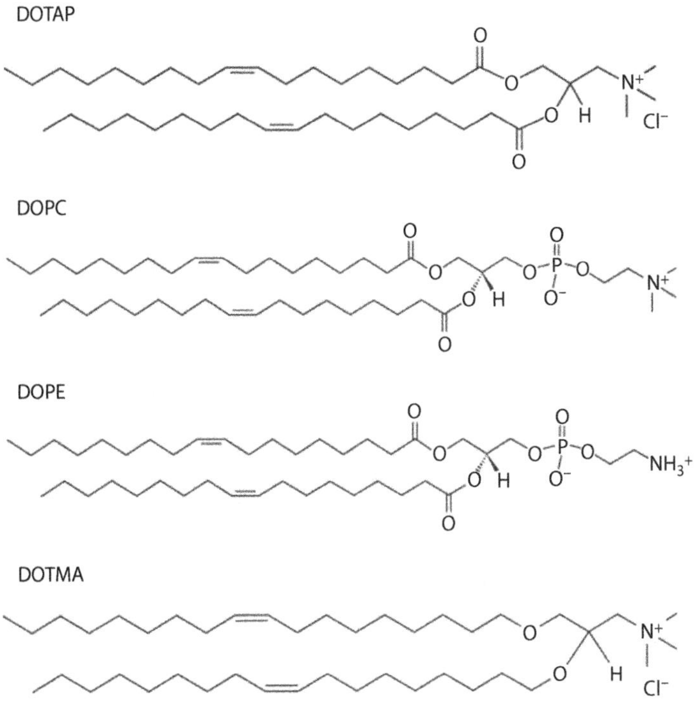

DOTAP

DOPC

DOPE

DOTMA

FIGURE 15.1 Examples of commonly used lipids for liposome preparation for siRNA delivery and their chemical structures. These include DOTAP (dioleoyl trimethylammonium-propane), DOPC (Dioleoylphosphatidylcholine), DOPE (Dioleoylphosphatidylethanolamine) and DOTMA (dioleyloxypropyl trimethylammonium chloride).

in 1:1 ratio, a suitable structural transition in the lipoplex occurred that helped its improved transfection efficiency (Mochizuki et al. 2013).

Dioleoylphosphatidylcholine (DOPC) also, as a helper lipid, yields higher transfection efficiencies with cationic lipids, considered as a result of conformation shift in their structure at low pH (Balazs and Godbey 2011; Hafez and Cullis 2001). To improve the siRNA loading efficiency, helper cationic polymer protamine is added to pre-condense the siRNA in the core of the liposomes. Differences in the type of the lipid used in the formulation yield different types of lipid nanoparticles. Many factors, including the structure, surface charge, degree of PEGylation, and conjugation of targeting ligand(s), affect the delivery of siRNA *via* liposomal delivery system (Figure 15.2). In this chapter we describe the contributions of liposomes and lipid-based nanocarriers in the delivery of siRNA. Further, our research efforts in liposome-based siRNA are also described with the emphasis on cancer therapy.

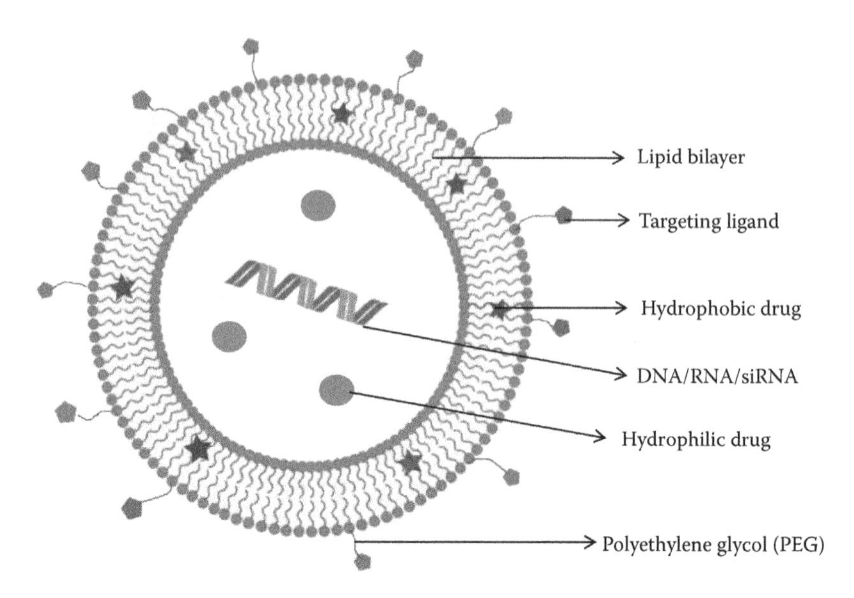

FIGURE 15.2 Structure and design of liposomes for DNA/siRNA/drug delivery. Liposomes can be surface functionalized through PEGylation to promote receptor-mediated endocytosis via targeting ligands. Chemotherapeutic drug can be encapsulated into the aqueous phase, incorporated into the lipid bilayer, or conjugated to the liposome surface.

15.2 METHODS FOR LIPOSOME PREPARATION

Liposomes are prepared using many procedures for the delivery of nucleic acid therapeutics. Nucleic acid molecules such as siRNA, plasmid DNA, or oligonucleotides are complexed with liposomes via either passive loading or by active loading. In passive loading, nucleic acid materials are basically encapsulated inside the liposomes, whereas in active loading the nucleic acid materials are allowed to interact with the lipids and are added to fully formed and intact liposomes (MacLachlan 2007). Commonly used methods of liposome preparation for nucleic acid therapeutics are: (a) thin-film hydration, (b) sonication, (c) ethanol injection and, (d) reverse phase evaporation. Detailed protocols for each methods for liposome synthesis can be found in a recent article by Akbarzadeh et al. 2013.

Briefly, in these methods, the lipids are generally dissolved in a mixture of organic solvents (chloroform, methanol, or ethanol) and are coated inside a round-bottom flask under vacuum and ambient temperature. The thin film formed in this process can be hydrated by the addition of buffers or water, or during the removal of organic solvents or detergents, allowing rapid self-assembly of lipids to form vesicles. At this stage the lipid vesicles are in general of multilamellar and heterogeneous. Then in the next stage to control the particle size and to make multilamellar vesicles to uni- or bi-lamellar liposomes, post-processing procedures such as extrusion, sonication, controlled freeze-thaw cycles or high-pressure homogenization are required. In the extrusion method the liposomal solution is passed through polycarbonate membranes of various pore sizes (nm) under mechanical pressure using specialized devices many times until the desired size of the particles are achieved. Since this method is time-consuming, simple methods such as sonication are extensively used in liposome preparation. In sonication, the heterogeneous lipid vesicles are subjected to ultrasound waves using a bath sonicator or probe sonicator under controlled conditions to reduce the particle size and to obtain small unilamellar vesicles (SUV) from multilamellar (MLV) vesicles. The harsh environment of sonication may not be appropriate for the nucleic acid molecules passively loaded to the liposomes because of the harshness of the procedure.

Another simple method of liposome preparation is the ethanol injection technique, where lipids dissolved in ethanol are forcefully injected into a buffer. This will lead to spontaneous formation of MLVs, but not SUVs, and generates a heterogeneous population of lipid vesicles of various sizes. However for siRNA/DNA encapsulation this method is not preferred as it can precipitate the nucleic acids causing inactivation of their biological activity.

Recent progress in liposome technology emphasizes the high loading capacity of payload and reproducible monodispersed structures for rapid clinical translation purposes. The reverse phase evaporation method is a recent technique, allowing for encapsulation of large amount of water soluble material in the liposome core. In this technique an organic phase containing lipids in which hydrophobic materials are dissolved is mixed with an aqueous phase in which hydrophilic materials are present to form inverted micelles, then organic phase is slowly eliminated to form a viscous gel phase. The gel phase is then hydrated to form liposomes. The liposomes synthesized by the reverse phase evaporation method have high aqueous volume-to-lipid ratios, almost four time higher than MLVs synthesized by other methods (MacLachlan 2007).

15.3 PARAMETERS INFLUENCING CELLULAR ENTRY AND GENE DELIVERY EFFICIENCY OF LIPOSOMES

The cellular entry and gene delivery efficiency of liposomal nanocarriers is influenced by several factors that include: particle size, shape, surface charge, ligand modification for targeted siRNAdelivery, PEGylation degree etc., (Xia, Tian, and Chen 2016). It is important to note that the particle size determines the biological fate and activity of the liposomes *in vivo*. Taking into account the high vascular permeability around tumor tissues particle sizes under 200 nm are preferable for *in vivo* administration. As suggested by Li and Szoka, a particle size <100 nm helps in overcoming the *in vivo* biological barriers for liposome siRNA complexes (Li and Szoka 2007). These barriers include uptake by reticuloendothelial systems (RES), entry to tumor mileu, extracellular matrix, cell membrane, and intra cellular barriers. In a different study, a cationic liposome of ~150 nm size has shown its lung accumulation upon intra-tracheal administration after a period of 72 hours, indicating a slightly higher permeability in lung tissues. However, lung accumulation was time dependent, as at a later time point the liposomes were mostly accumulated in liver (Ishiwata et al. 2000). Surface charge of liposomes is another important factor that influences the gene silencing efficiency of liposome-siRNA complexes (Campbell et al. 2002). Cationic liposomes are extensively used in transfection reagents because of their high transfection efficiency for siRNA or DNA. However, they are toxic too, because of their ability to fuse with negatively charged membranes non-specifically. Anionic liposomes and neutral liposomes are highly regarded as less toxic, but their transfection efficiencies are less compared to cationic liposomes. These systems are further elaborated in following section(s). Polyethylene glycol modification (PEGylation) and the degree of PEGylation are important factors that determine the stability of liposomes in the bloodstream. PEG chain length and density of coating are important in effective transport of liposomes in the circulation. In general, PEGylation reduces the charge of liposomes, to make it to near neutral. This will prolong the circulation time of liposomes (stealth property) by delaying clearance by RES. Liposomes are also made by neutral lipids. However, neutral lipids are supportive of cationic lipids to enhance the transfection efficiency. More details about these systems are presented in Section 15.4. In a nutshell, the charge of the liposome is an important factor in gene delivery efficiency. Finally, to improve the cell-type based transfection efficiency, receptors overexpressed in cell surfaces are explored, by modifying liposomes with ligands specific for those receptors. Details describing suitable examples are presented in Section 15.6.

15.3.1 Cellular Entry Mechanism of Nanocarriers for siRNA Delivery

Endocytosis is the prominent mechanism of cellular entry by nanocarriers. Based on the above-mentioned parameters, an siRNA-loaded nanocarrier may enter cells through macropinocytosis, clathrin-mediated endocytosis (CME), or caveolae-mediated endocytosis. In macropinocytosis, siRNA carriers are internalized through large vesicles called macropinosomes, and undergo endosomal escape through leaky macropinosomes or fuse with lysosomes to release siRNA. They are growth factor-induced, actin-driven endocytosis and largely depend upon cell type. In CME, an siRNA nanocarrier enters the cell through the formation of clathrin-coated vesicles, which fuse with early endosomes and the buffering capacity of nanocarrier material should allow siRNA to escape from endosomes before its maturity and fusion with lysosomes. The initial stage of this cellular entry pathway depends on nanoparticle-cell surface receptor interaction. However, a fraction of the nanoparticles may enter lysosomes resulting in inactivation and degradation of the siRNA carried. This is because CME or micropinocytosis pathways have been shown to deliver some ligands and their receptors ultimately to lysosomes (Pereira et al. 2015). Caveolae mediated endocytosis begins with the formation of caveolae using the caveolin-1 protein present in the plasma membrane. The resulting vesicles, caveosomes, are non-acidic and will not undergo any change in pH. Most caveosomes are directly transported to the cytosol, bypassing normal lysosomal degradation. However, the exact molecular mechanism by which the caveosomes reach the cytosol is unclear (Khalil et al. 2006; Medina-Kauwe, Xie, and Hamm-Alvarez 2005).

The type of nanocarrier material and the physicochemical properties also determine the cellular entry pathway for siRNA delivery. For instance, superparamagnetic iron-oxide nanoparticles and silica-coated, iron-oxide nanoparticles, which are negatively charged, preferentially enter the cells through caveolin-1 or cdc42 based endocytosis pathways (Bohmer and Jordan 2015). It was also reported that gold nanoparticles with different sizes, coated with PEG or other polymers, polystyrene nanoparticles etc., enter cells through caveolin-mediated pathways. The hydrophobicity of a nanoparticle is also an important factor that determines the cellular entry of nanoparticles. For example cholesterol-modified pullulan nanoparticles preferentially enter cells through clathrin-mediated endocytosis and micropinocytosis (Jiang et al. 2013). However, hydrophobically modified glycol chitosan nanoparticles enter cells through several distinct pathways involving clathrin-mediated endocytosis, caveolae-mediated endocytosis, and micropinocytosis (Nam et al. 2009). A direct membrane fusion strategy that bypasses the conventional cell entry pathways has also been used, depending on the nanocarrier material and its interaction with cell membrane (Yang et al. 2016). They have used complementary coiled-coiled lipopeptides inserted into the lipid bilayer of liposomes which resulted in targeted membrane fusion accompanied by drug release into the cell cytoplasm (Yang et al. 2016). Therefore, physicochemical properties of nanoparticles play a role, though it is not fully elucidated for many nanoparticle systems, in determining the mechanism of cell entry pathways and drug or gene delivery. Though the above section is not specific to liposomes, the specific examples serve to give an insight to the general mechanisms of cell uptake by nanocarriers for gene delivery.

Since endosomal escape is a challenge for siRNA/oligonucleotides stability, nanoparticle material that chooses a specific cellular entry pathway other than the endosome pathway or the ability of nanoparticles to help siRNA to escape from endosomal acidic environment influence their successful cytoplasmic delivery. Examples of some important strategies for endosomal escape of siRNA using liposome delivery vehicles are described in Section 15.4.

Different categories of liposomes have been extensively studied for siRNA delivery in mammalian cells. Overall cationic liposomes show high transfection efficiency both *in vitro* and *in vivo* compared to non-cationic liposomes. Cationic liposomes interact with anionic nucleic acids to form positively charged lipoplexes that can easily interact with the negatively charged cell membrane to facilitate the cellular entry. Besides the charge of liposomes, lipid composition, particle size distribution, and charge ratio may also influence the uptake of the particles. Cellular uptake can take place either through direct fusion with the plasma membrane or through endocytosis. Cell surface binding of cationic liposomes alone is not sufficient to enter into the cells without the endocytosis process. The proposed flip-flop mechanism explains the dissociation of nucleic acids from the lipoplexes and escape from the endosomes into cytosol (Zelphati and Szoka 1996). After the cellular uptake, lipoplexes fuse with endosomes and form endosomal vesicles. The close proximity of the endosomal membrane and lipoplexes facilitates the electrostatic interaction between the cationic lipoplexes and the anionic lipids of the endosomal membrane. It destabilizes the lipid bilayer due to the formation of a charge-neutralized ion pair. As a result, the nucleic acids will be released into the cytoplasm.

15.4 TYPES OF LIPOSOMES

15.4.1 Cationic Liposomes

15.4.1.1 Cationic Liposomes/Lipoplexes

Cationic liposomes are positively charged; the hydrophilic head group is linked to a hydrophobic tail that contains alkyl chains of various lengths. The number of amine groups present on the head group determines the charge of the liposomes. These liposomes are synthesized by simply mixing the cationic liposomes with neutral helper lipid and siRNA in the desired ratio. The complexes, thus formed by aggregation of the siRNAs with the liposomes, eventually rupture the lipid bilayer to wrap around the siRNA complex and form multilamellar structures. These complexes protect the siRNA from degradation and show enhanced cellular uptake. Studies have shown that the slightly positively charged lipoplexes interact more efficiently with the negatively charged cell membrane, and demonstrate higher transfection efficiency (Zhang, McIntosh, and Grinstaff 2012). The protection of DNA from nucleases and improved transfection efficiency of the lipoplex depend on the lipid/nucleic acid ratio. The lipid/nucleic acid ratio determines the structure and controls the size of cationic lipoplexes (Almofti et al. 2003), the biophysical characteristics of which are evaluated using transmission electron microscopy. Relatively small complexes are formed at high positive charge ratios with a particle size around 200 nm. However, when the net charge of the liposome-siRNA complexes nears neutrality, large aggregates are formed, mostly in micrometer sizes (Almofti et al. 2003).

DOTAP (dioleoyl trimethylammoniumpropane) and DOTMA (dioleyloxypropyl trimethylammonium chloride) are the two monovalent cationic liposomes that are commonly used for siRNA delivery. The strong cationic quaternary ammonium group present in DOTAP and DOTMA aids in the formation of cationic lipoplexes and lipopolyplexes. These lipoplexes showed high cellular uptake, resulting in enhanced gene silencing. DNA/lipoplexes bind to the cell membrane through non-specific ionic interactions and enter the cells through membrane-associated proteoglycans. These lipoplexes demonstrated efficient endosomal escape due to the ion-pair mechanism. The ion-pair mechanism is the strong electrostatic interaction between the cationic lipoplexes and the anionic lipid phosphatidylserine that is present in the cytoplasmic side of the endosomal membrane. Anionic lipids on the endosomal membrane diffuse into lipoplexes and destabilize the ion pairs, thus disassembling the lipoplexes to release the siRNA into the cytoplasm. This ion-pair mechanism leads to successful escape of siRNA from endosomes. The formulation using 100% DOTAP produces inefficient delivery of DNA/RNA due to a greater positive charge on the liposome, which prevents the counter ion exchange. Protonation of DOTAP at a neutral pH requires more energy to remove the DNA/RNA from the liposomes. Due to its biocompatibility, cholesterol was chosen to stabilize the DOTAP. This formulation showed a two-to-four-fold increase in transfection efficiency and a four-fold decrease in cytotoxicity (Balazs and Godbey 2011).

Khatri et al. investigated the effect of different lipid compositions that influence stability and toxicity of lipoplexes (Khatri et al. 2014). The lipoplex (D_2CH) system had the following components: dioleoyl-trimethylammoniumpropane (DOTAP), dioleoyl-*sn*-glycero-3-phosphoethanolamine (DOPE), hydrogenated soya phosphocholine (HSPC), cholesterol, and methoxy (polyethyleneglycol)$_{2000}$–1,2-distearoyl-*sn*-glycero-3-phosphoethanolamine (mPEG$_{2000}$–DSPE). Compared to the conventional gene delivery system Lipofectamine-2000, the D_2CH system showed significantly low toxicity up to an N/P ratio of 7.5 with siRNA. The lipoplexes generated showed three times less hemolytic potential compared to Lipofectamine 2000. Further the lipoplexes were well tolerated (up to 300 mg lipid/kg) by the mice models studied. Desmet et al. recently introduced novel lipoplexes, which are elastic liposomes; flexible and highly deformable to allow easy penetration through pores much smaller than their size (Desmet et al. 2016). These elastic lipoplexes with a composition of DOTAP, DOPE, cholesterol, and EtOH (named as DDC642), were used to deliver siRNA for topical application, especially psoriasis. Their DDC642 liposomes formed stable complexes with both siRNA and miRNA molecules and demonstrated high penetration in skin cells.

15.4.1.2 Lipopolyplexes

Lipopolyplexes were designed to circumvent the large size and cytotoxicity problems of lipoplexes. These new carriers show better nucleic acid condensation for better siRNA delivery. Lipopolyplexes combines the advantages of both polyplexes and lipoplexes, as they are ternary nanocomplex composed of cationic liposome, polycation, and nucleic acid. They are categorized based on the types of polycationic materials used in the creation of lipopolyplex (i) cationic polymer based (eg: PEI and its derivatives) lipopolyplexes and (ii) cationic polypetide (eg: poly-l-lysine)-based lipopolyplexes (Rezaee et al. 2016).

Lipopolyplexes were prepared by condensing nucleic acids with the help of polycation into a polyplex and entrapped within the liposomes. Various polymers, such as polyethylenimine, poly-L-lysine, protamine, chitosan, and polyallylamine, have been used in lipopolyplex formulation (Chen et al. 2009). The recent review by Rezaee et al. (2016) described different kinds of lipopolyplexes based on the above-mentioned categories in detail. They showed promising transfection efficiencies and safety features for these lipopolyplexes and are superior to conventional cationic liposomes and polymeric gene delivery systems. PEIs of different lengths, either branched or linear are available and can be functionalized by simple methods. These are able to condense the nucleic acids by electrostatic interaction between the positively charged PEIs and negatively charged nucleic acid into homogeneous particles. The positively charged PEI/DNA complexes could bind to the negatively charged cell membrane and result in high transfection efficiency. Garcia et al. developed a lipopolyplex using DOTAP:chol with various PEIs (linear and branched PEI) and DNA (Garcia et al. 2007). They demonstrated that a higher lipid/DNA ratio resulted in better transfection efficiency. De Wolf et al. synthesized cationic liposomes using DOTAP/DOPE with various modified PEIs (PEG-PEI) for siRNA delivery (de Wolf et al. 2007). They showed the influence of various cationic vectors in tumor accumulation and biodistribution. The overall tumor accumulation and circulation kinetics were not significantly different from free non-complexed siRNA. However, intratumoral distribution of siRNA was influenced by the type of cationic carriers used.

Among the studied nanocarriers only PEG-PEI improved the circulation kinetics of pDNA. Their study concluded that the intratumoral, carrier-induced changes are more significant in predicting the fate of siRNA or pDNA, than the advantage provided by the carrier system during transport in circulation until it reaches the tumor (de Wolf et al. 2007).

Lipopolyplexes escape from the endosomes through both the ion-pair formation and proton sponge effect. After the endocytosis, the cationic lipopolyplex and the anionic lipids in the endosomal membrane would form an ion pair and destabilize the membrane. The electrostatic interaction could further facilitate the formation of the inverted hexagonal phase and disrupt the endosomal membrane. The cationic polymers in the core of the lipopolyplexes behave as a sponge and become protonated to withstand the acidification of the endosomes. As a result, more protons will be pumped into the endosomes to lower the pH and the passive entry of counter chloride ions increases the osmotic pressure that results in the rupture of endosomal membrane.

The multifunctional envelope-type nanodevice (MEND) is another example of a lipopolyplex. MEND lipopolyplexes have nucleic acids condensed with a polycation to form a polycation/nucleic acid core, which is then protected by the lipid bilayer decorated with various functional moieties, like targeting ligands. The condensed nucleic acids in the core minimize the size and increase the loading efficiency, protecting the nucleic acids from nucleases (Hatakeyama et al. 2011). In another approach, lipopolyplexes were prepared by mixing DOTAP with a chitosan/DNA polyplex, resulting in a 70-fold increase in the transfection efficiency (Wang 2012).

15.4.1.3 SNALP

Stable nucleic acid lipid particles (SNALP) are formulated using an ionizable cationic lipid and a neutral helper lipid that is further coated by a PEGylated lipid, which enhances the interaction with the plasma membrane, thus promoting cellular uptake and endosomal escape (Semple et al. 2010). The siRNA is encapsulated in the shell of a lipid bilayer that is a mixture of cationic lipids and fusogenic lipids. The PEG coating in the cationic bilayer protects the nucleic acid core from degradation by nucleases. The chemical modification of siRNA at 3' and 5' with 2'-OH and 2'-O-methyl was reported to prevent the immune response and enhanced the stability of siRNA (Burnett and Rossi 2012; Siomi and Siomi 2009). Morrissey et al. (2005) developed a SNALP system using a chemically modified siRNA-targeting hepatitis B virus, in which the liposome consisted of DSPC:chol:PEG-C-DMA:DLinDMA at 20:48:2:30 mol percent. They observed that the intravenous injection of SNALP at 3 mg/kg/day for three days significantly reduced the HBV DNA in the serum (Jayaraman et al. 2012).

Alnylam Pharmaceuticals developed a first-generation SNALP comprised of two different siRNAs to target two genes in liver cancer. One of the siRNAs targeted the mRNA of vascular endothelial growth factor (VEGF), while the other siRNA inhibited the mRNA of kinesin spindle protein (KSP). The SNALP was tested in 41 patients with treatment-resistant disease, in a study that lasted 26 months. One patient had complete remission, but the mRNA knock-down was not observed (Kanasty et al. 2013). Phase I/II studies of modified DLinDMA LNP to target polo-like kinase 1 (PLK1) gene showed improved circulation time and enhanced tumor accumulation (Zatsepin, Kotelevtsev, and Koteliansky 2016). Di Martino et al. 2014, used SNALP to improve the delivery efficiency of miR-34a mimics towards multiple myeloma cells. They prepared SNALP having the composition, DSPC/CHOL/DODAP/PEG2000-Cer16 (molar ratio 25/45/20/10). To prepare SNALP carrying miR-34a, the preheated (65°C, two to three minutes) lipid mixture in solution was rapidly added into the aqueous solution containing miR34a under continuous stirring. The resulting SNALP-miR-34a complex was extruded several times through a polycarbonate membrane of 100 nm pore size and purified by dialysis and column chromatography. The SNALP-miR-34a complex was shown to efficiently deliver the miRNA mimics into tumor cells both *in vitro* and *in vivo*. SNALP-based delivery of miRNA mimics increased survival rate of mice bearing tumors. Another advantage was that the SNALP-based miR-34a mimic did not cause any significant non-specific toxicity in treated animals, indicating its safety as a gene delivery system. While the use of SNALP for miRNA delivery is exciting, additional studies however are warranted prior to advancing SNALP-based miRNA therapy for cancer.

15.4.1.4 pH-Sensitive Liposomes

pH-sensitive liposomes are considered promising nanocarriers for intracellular gene delivery. Like the cationic liposomes, pH-sensitive liposomes are unstable in the circulation and are sequestered by phagocytes in the reticuloendothelial system. To increase the circulation time, cationic liposomes are modified with polyethylene glycol (PEG), a process called PEGylation. Although PEGylation has various advantages, it prevents the interaction between liposomes and the cell membrane. To overcome this problem, a pH-sensitive bond is introduced between the hydrophilic PEG and the hydrophobic lipid moiety. When exposed to lysosomal acidic conditions, the pH-sensitive bond will be degraded, releasing the lipid moiety. The pH in early endosomes is about 6.0 and reduces to 5.0 in late endosomes. Even in weak acids pH-sensitive bonds can be cleaved, which promotes the endosomal escape. DLinDMA, 1, 2-dilinoleyloxy-3-dimethylaminopropane: DLin-KC2-DMA, 2,2-dilinoleyl-4-(2-dimethylaminoethyl)-[1,3]-dioxolane, and YSK05 are common cationic pH-sensitive lipids used to prepare pH sensitive liposomes (Sato et al. 2012)

15.4.1.5 Lipidoids

Lipidoids are new class of chemically synthesized, lipid-like materials used for siRNA delivery (Akinc et al. 2008). More than 1,200 structurally different lipidoids have been developed and tested in animal models. Lipidoids are pH-sensitive and can be synthesized with customized head and tail groups. PEGylated lipidoids C12-200 with PEG 2K at 1.5 mol% showed the downregulation of multiple genes (Frank-Kamenetsky et al. 2008). Akinc et al. developed a novel lipidoid, LNP01, with cholesterol and PEGylated lipid; this lipidoid silenced multiple genes in various animal models (Akinc et al. 2009).

15.4.2 Anionic Liposomes

Anionic liposomes are suggested as an alternative to cationic liposomes due to their low toxicity. However conventional anionic liposomes have limitations such as short circulation time due to complement activation and high macrophage uptake; and poor tumor penetration, and poor cellular uptake in cancer cells due to repulsive force against anionic membranes. The negatively charged head group of anionic lipids prevents the electrostatic interaction between the phosphate backbone of DNA and the anionic head group. However, strategies have been developed to improve the cellular uptake by combining anionic and neutral lipids in the formulation. For instance, liposomes formulated using a 17:83 ratio of anionic lipid DOPG, 1,2-di-(9Z-octadecenoyl)-sn-glycero-3-phospho-(1'-rac-glycerol), and the neutral lipid DOPE showed efficient cellular uptake. Divalent cations can be incorporated to condense the nucleic acids during anionic lipid complexation. A stable anionic lipoplex of DOPG can be synthesized using Ca^{2+} ions. Application of this lipoplex produced 70% gene knock-down. A high concentration of Ca^{2+} ions turns the lipoplexes into a cationic complex and facilitates the cellular uptake through proteoglycan-mediated endocytosis (Patil, Rhodes, and Burgess 2004). The association of siRNA within the lipoplexes depends on the anionic lipid/siRNA molar charge ratios (Srinivasan and Burgess 2009). Based on these ratios siRNA is either complexed or encapsulated within the anionic lipoplexes (Kapoor and Burgess 2012). Reports suggests that at low anionic lipid/siRNA molar charge ratios the siRNA/lipoplexes are in the 'complexed' form whereas at higher ratios they exist in both the 'complexed' as well as 'encapsulated' forms. The same group further optimized the anionic liposome formulation in separate study based on 1 μg/mL lipid (40:60 (DOPG/DOPE m/m), 2.4 mM calcium and 10 nM siRNA, which showed ~70% gene knock-down without being cytotoxic (Kapoor and Burgess 2012). Overall, anionic liposomes using newer formulation strategies like those discussed above may offer a safer alternative to cationic liposomes for *in vitro* as well as *in vivo* siRNA delivery.

15.4.3 Neutral Liposomes

The most widely used neutral liposomes are DOPC (Dioleoylphosphatidylcholine) and DOPE (Dioleoylphosphatidylethanolamine), in combination with cholesterol. Neutral liposomes encapsulate siRNA in its aqueous core. Dried thin films of lipid mixture formed in a rotary evaporator are dispersed in a buffer containing siRNA or other nucleic acid materials to form siRNA/nucleic acid material-loaded neutral liposomes. Nogueira et al. (2017) utilized this strategy to encapsulate locked nucleic acids (LNA, synthetically modified siRNA) in folate modified PEGylated, DOPE-derived neutral liposomes to target Mcl-1 protein in macrophages. They reported that this neutral liposome loaded with Mcl-1 LNAs efficiently silenced Mcl-1 gene causing downregulation of respective protein and apoptosis induction in macrophages. Neutral liposomes composed of DOPE have longer blood circulation and lower toxicity and can effectively deliver siRNA 10–30-fold higher than cationic liposomes. Intravenous or intraperitoneal injection of DOPE-based nanoliposomes encapsulated with siRNAs specific for Bcl-2, IL-8, EphA2, or FAK showed a significant reduction in tumor volume and the target gene in animal models (Wyrozumska et al. 2015; Landen et al. 2005; Merritt et al. 2008). These nanoliposomes showed no toxicity towards normal cells, including fibroblasts and bone marrow cells, making neutral liposomes attractive for further development in siRNA delivery.

15.5 PEGylation OF LIPOSOMES FOR siRNA DELIVERY

An ideal siRNA nanocarrier should provide protection from blood nucleases and extend the circulation time, which would enhance the accumulation of siRNA in tumors. Regardless of the size and charge of the liposomes, upon intravenous administration, these liposomes interact with negatively charged serum proteins and form an aggregate that accumulates in the lungs, liver, and spleen. PEGylation increases the stability of nanocarriers and extend their circulation time. The hydrophilic PEG polymer provides a shielding effect around the nanocarrier and decreases the opsonization effect. This will minimize the recognition by macrophages, which will lead to increased blood circulation time. PEGylation has been shown to increase stability and extend blood circulation time. Both the length of the PEG chain and the density of PEGylation on liposomes affect the stability and circulation time. PEGylated liposomes with shorter PEG chains are negatively charged, and thus will be cleared by macrophages, reducing their circulation time. However, very long PEG chains will reduce the cellular uptake of siRNA and the endosomal escape. Hence, medium-sized PEGs are commonly used (Pozzi et al. 2014). A higher molar ratio of PEG to total lipid gives higher surface coverage, providing more steric shielding and creating a stealth property. Liposomes with higher PEG density will create a better steric barrier between the macrophage and liposomes and have a longer circulation time. It has been shown that ~10.6 mol % of PEG2K modification of liposome-polycation-DNA (LPD) nanoparticles have significantly enhanced the tumor accumulation compared to naked LPD; the tumors showed the uptake of 70–80% of ID/g. Additionally, the liver and lungs took only up to 10–20% of ID/g which shows the selective uptake of the nanoparticle system (Li and Huang 2009; Li and Huang 2010). Dos Santos et al. (2007) demonstrated that incubation of various liposomal formulations (0.0 to 3.5 μmol lipid) with mouse serum for ten minutes at 37°C and liposomes with adsorbed proteins were separated by size exclusion chromatography. They observed that 2% PEGylated liposomes in serum showed the least interaction with serum proteins compared with unmodified liposomes and liposome with various PEG mol% (0.5, 1, 2, and 5 mol%) investigated, aiding in lower aggregation tendency (Dos Santos et al. 2007).

In order to improve the therapeutic outcome, repeated administration of therapeutic doses is recommended. Single doses of PEGylated liposomes showed a longer circulation time, whereas subsequent doses of PEGylated liposomes rapidly cleared from circulation, due to accelerated blood circulation (ABC phenomenon). The first injection of PEGylated liposomes produces IgM antibodies secreted by activated B cells; repeated injections of PEGylated liposomes interact with residual IgM antibodies in the serum, which activate the complement system, and are cleared by macrophages (Wang et al. 2015). Dams et al. reported that the first systemic administration of empty PEGylated liposomes showed improved therapeutic effect (Dams et al. 2000).

PEG-derivatized lipids have offered the ability to control the release of the therapeutic drug at a rate determined by the length of the lipid anchor. The shorter the length of the acyl chain, the faster the dissociation of PEG from the lipid bilayer will be, resulting in high transfection efficiency. This was demonstrated by the stabilized plasmid lipid nanoparticle (SPLP) synthesized using ceramide-anchored CerC8-PEG2000, which showed efficient transfection compared with CerC20-PEG2000. However, systemic administration of CerC8-PEG2000 was unsatisfactory, due to the loss of positive charge (Mok, Lam, and Cullis 1999).

Semple et al. (2010) demonstrated that PEGylated pH-sensitive liposomes containing DLin-KC2-DMA lipid showed efficient tumor accumulation in animal models. PEGylated SNALP cationic pH-sensitive liposomes have very high siRNA encapsulation efficiency. They possess this property because the membrane permeability is not affected by the PEGylation during synthesis, which allows the penetration of siRNA into the liposomal core. These particles have a neutral surface charge, increased circulation time, and enhanced tumor accumulation (Semple et al. 2010). Real-time intracellular trafficking analysis showed that these particles entered the cell through both clathrin-mediated endocytosis and macropinocytosis. However, only 1–2% of the cationic liposomes entered the cytosol from endosomes; this process could be further enhanced by improving the endosomal escape mechanism.

15.6 TUMOR TARGETED DELIVERY OF LIPOSOMES

Targeted delivery of therapeutics into tumor cells, tumor vasculature, and/or the endothelium is a successful approach to cancer treatment. Recent advances in the development of targeted siRNA delivery systems using cationic lipids and polymers increased the therapeutic dosage of drug that was delivered and reduced the off-target effects. The targeting moiety can be directly conjugated to siRNA or incorporated into nanoparticles. Ligands, including antibodies, aptamers, and peptides, are traditionally used for targeted delivery of siRNA to specific tissues. The covalent attachment of a targeting ligand for receptors at the end of the PEG chain that will interact with the receptors overexpressed on the cancer cells is a common strategy. Targeting receptors should be expressed only on malignant cells, with few-to-none expressed on normal cells. Frequently targeted receptors include Her2, folate receptor alpha, transferrin, and epidermal growth factor receptors.

The folic acid receptor is a potential molecular target for human cancers. It is upregulated in a range of human tumors, including lung, ovarian, nasopharyngeal, and colorectal cancers.

Folate receptors are glycoproteins and exist in two isoforms: FR-alpha and FR-beta. Studies have shown that higher expression of FR-alpha is associated with poor patient survival. As a targeting ligand, folic acid has several advantages: high stability, low molecular weight, and high affinity for folate receptors. These advantages make FR-alpha a promising target for cancer therapy. He et al. (2013) synthesized FR-alpha targeting liposomes encapsulated with shRNA specific for the CLDN3 gene. The use of folate-modified liposomes F-P-LP/CLDN3 using DOTAP, Chol, mPEG-suc-Chol and F-PEG-suc-Chol produced a significant reduction of tumor growth in an ovarian tumor model. Folate receptor α (FRα)-targeted nano-liposomes (FLP) was designed to target FR-alpha to deliver exogenous pigment epithelium-derived factor (PEDF) genes, which showed enhanced antitumor effect in a model of cervical cancer. Phase I\II trials of the farletuzumab monoclonal antibody that targets FR-alpha in lung cancer showed that it was well tolerated as a single agent and in combination, without causing additional toxicity. These observations suggest that higher levels of FR-alpha expression in lung cancer could be exploited with FR-targeted nanocarriers.

The transferrin receptor is one of the most-studied tumor-targeting receptors in lung cancer therapy. The transferrin receptor (TfR), CD71, is a transmembrane glycoprotein. Its increased expression in many cancer types is associated with poor prognosis. CD71 is responsible for iron transport and is present in normal cells at minimal levels. It can be internalized efficiently through clathrin-coated vesicles, which makes it a potential target for nucleic acid and drug delivery (Tortorella and Karagiannis 2014). Radiation enhances the delivery of transferrin-conjugated lipoplexes in lung cancer cells. This effect is inhibited by the treatment of cells with free transferrin, indicating specificity for transferrin-conjugated lipoplexes. Studies have shown that the electrostatic association of transferrin with cationic liposomes enhanced the transfection of siRNA. Mendonca et al. (2010) demonstrated that by using a TfR-targeted liposome, BCR-ABL siRNA effectively knocked-down the target gene in K562 and LAMA 84 myeloid leukemia cells. They also demonstrated that the TfR-targeted liposome system can be used to co-encapsulate an anticancer drug and the siRNA for the treatment of chronic myeloid leukemia (Mendonca et al. 2010).

Recent advances in the development of different delivery platforms for siRNA led the researchers to adopt a different strategy to use naturally produced lipid vesicles, namely exosomes, as siRNA carriers (Srivastava et al. 2015). Exosomes and liposomes have some common features including a phospholipid bilayer composed of biocompatible materials that has high encapsulation efficiency for siRNAs (Antimisiaris, Mourtas, and Papadia 2017). The phospholipid bilayer acts as a key mediator and facilitates the transfer of genetic and biochemical information between cells through various surface adhesion molecules and ligands. Due to its unique composition, the uptake of these nanocarriers are very efficient in host cells and it has been shown that the exosome-mediated delivery systems are very well tolerated both *in vitro* and *in vivo* (Antimisiaris, Mourtas, and Papadia 2017; Srivastava et al. 2015; Srivastava et al. 2016). These nanosized vesicles are capable of permeabilizing the biological membranes compared to other delivery systems. Different strategies have been adopted to load siRNAs into exosomes including electroporation, incubation at elevated temperature, freeze-thaw cycles, sonication, and extrusion. Further investigations are required to obtain large quantities of highly purified exosomes. While exosomes as natural gene and drug carriers hold promise, exosome-based gene delivery faces challenges such as lack of efficient and scalable method to load exosomes or extracellular vesicles with RNA molecules of interest. A recent study demonstrated that by conjugating cholesterol molecules to siRNA the loading efficiency in extracellular vesicles can be significantly increased (O'Loughlin et al. 2017). This study optimized that incubating around 15 cholesterol-conjugated siRNA molecules loaded per extracellular vesicle at ambient temperature for one hour resulted in efficient concentration dependent silencing of the target gene *HuR* in cancer cells. While such systems based on exosomes are promising for gene delivery, it seems too early to determine the potential challenges and their feasibility of clinical translation.

Previous studies from our laboratory showed the efficacy of DOTAP:chol cationic liposomes in delivering the therapeutic tumor suppressor gene p53 and FHIT (Fragile Histidine Triad) in an animal model of metastatic lung cancer. We demonstrated that the intravenous injection of a DOTAP:chol/DNA complex enhanced the transgene expression in lung tumor cells compared with normal cells and explained the effective uptake of liposomes complex by the tumor cells. In another study, systemic delivery of a DOTAP:chol/FUS1 DNA complex in an animal model of metastatic lung cancer suppressed tumor growth and prolonged animal survival (Ito et al. 2004). A Phase I clinical trial was also conducted using a DOTAP:chol/FUS1 nanoparticle to treat human patients with lung cancer (Lu et al. 2012). To this DOTAP:chol nanoplatform we carried out modification using ligands such as transferrin (Tf) and folic acid (FA), to target the respective receptors overexpressed in lung cancer cells. Section 15.7 describes our recent efforts in development and efficiency evaluation of folate- or transferrin-modified liposomes for delivery of siRNA towards the molecular target HuR.

15.7 TARGETED LIPID NANOPARTICLES FOR siRNA DELIVERY IN LUNG CANCER: A CASE STUDY

HuR is an RNA-binding protein and a member of the embryonic lethal abnormal vision (ELAV)-like protein gene family. HuR is a nucleocytoplasmic shuttling protein that is predominantly localized in the nucleus. When HuR is activated by various stimuli, it binds to the AU-rich elements (AREs) at the 3' and 5' untranslated region (UTR) of several mRNAs, and is then transported from the nucleus to the cytoplasm, where it stabilizes the mRNA from degradation and enhances protein translation (Brennan and Steitz 2001). HuR overexpression was shown to be a poor prognostic marker, and its overexpression was correlated with poor survival of patients with lung cancer. Many cancer-related oncogenes, cytokines, growth factors, and invasion factors have been characterized as HuR targets (Abdelmohsen et al. 2007; Galban et al. 2008; Wang et al. 2013). These reports suggest that HuR might be a novel and promising therapeutic target and that HuR inhibition may provide a better prognosis for lung cancer (Mehta et al. 2016; Huang et al. 2016; Romeo et al. 2016). Hence, we hypothesized that siRNA-mediated HuR knock-down will modulate a multitude of oncoproteins and may serve as a druggable target in lung cancer.

To test our hypothesis, we designed an FR-alpha-targeted nanoparticle (FNP) to deliver HuR-siRNA and tested its efficacy in lung cancer cells *in vitro* and *in vivo*. FNP was synthesized using a DOTAP:chol lipid nanoparticle and DSPE-PEG-folate was conjugated using the post insertion technique. Biophysical characterization studies showed that the FNPs were negative to slightly positive in surface charge, with hydrodynamic diameters in the range of 200 nm to 300 nm (Figure 15.3). Transfection efficiency studies demonstrated that uptake of FNPs was high in H1299 cells that overexpress the folate receptor compared with A549 cells that have low

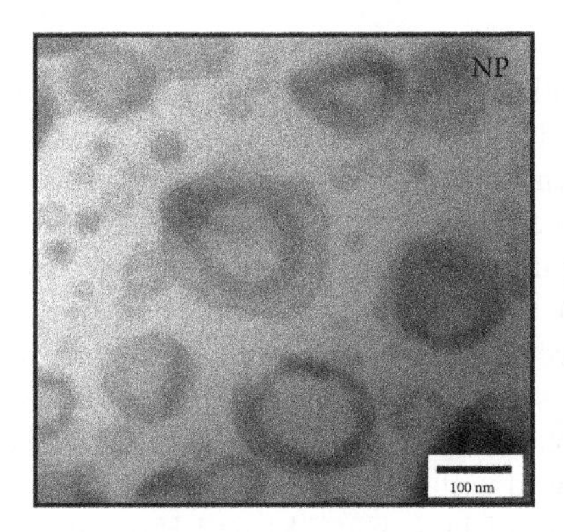

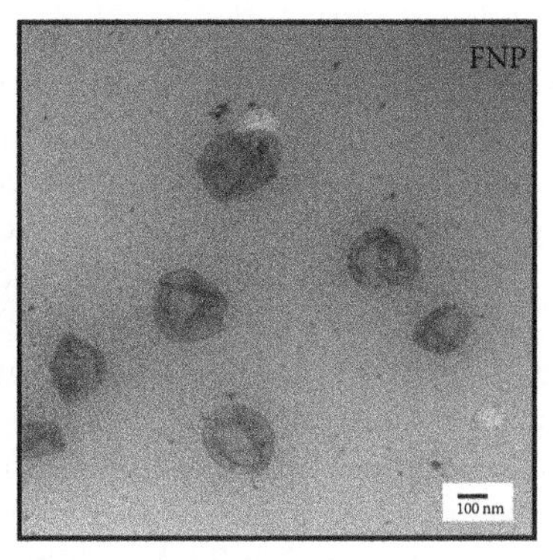

FIGURE 15.3 Transmission electron microscope (TEM) images of NP (magnification, 70,000X) and HuR-FNP (magnification, 25,000X). *Scale bar* denotes 100 nm.

expression. Competitive inhibition studies conducted in H1299 cells suggested that the intracellular delivery of HuR-FNP was efficiently taken up *via* folate receptor-mediated endocytosis. Cell viability studies showed that HuR-FNP significantly inhibited H1299 cell growth by inducing a G1 phase cell-cycle arrest, compared with C-FNP. Additionally, HuR-FNP exhibited significantly higher cytotoxicity against H1299 cells than A549 cells. The reduction in cell viability was correlated with a marked decrease in HuR mRNA and protein expression in H1299 cells. Further, a reduction in the expression of HuR-regulated oncoproteins (cyclin D1, cyclin E, Bcl-2) and an increase in p27 tumor suppressor protein were observed in HuR-FNP-treated cells. Finally, cell migration was significantly inhibited in HuR-FNP-treated H1299 cells (Muralidharan et al. 2016). Since our *in vitro* studies demonstrated the selective uptake of HuR-FNP in folate-receptor positive lung tumor cells, we next conducted pilot *in vivo* studies in subcutaneous xenografts. Intravenous injection of DY647HuR-FNP demonstrated the selective accumulation of FNPs in the tumor site at 24 hours after injection.

We adopted the second strategy to develop lipid nanoparticles to target transferrin receptors (TfRs) that are overexpressed in lung cancer. Thiol-modified transferrin was chemically conjugated into DSPE-PEG, which was post inserted into DOTAP: chol lipid nanoparticles to form Tf-NP. We used Tf-NP to deliver the siRNA specific for HuR (HuR-TfNP). Biophysical analysis showed that Tf-NPs were uniform in size (about 200–300 nm) and slightly positive in charge. Our *in vitro* studies showed the specific delivery of siRNA to the TfR-overexpressing A549 lung tumor cells, which in turn displayed reduced expression of HuR at the protein and mRNA levels. To evaluate the effectiveness of Tf-NP, the Tf-NPs were loaded with fluorescently (DY647)-labeled HuR siRNA and were administered intravenously to an animal model of metastatic lung cancer. The circulation of Tf-NPs was monitored by spectral imaging, which revealed the accumulation of siRNA at the tumor site 24 hours post-injection. To test the functionality of HuR siRNA, HuR-TfNP was injected intravenously into A549 tumor-bearing nude mice. We harvested the tumors at 24 hours after injection and compared the HuR protein levels in the C-TfNP and HuR-TfNP treated groups. The results demonstrated a significant reduction in HuR mRNA and protein levels, as well as a reduction in other HuR-regulated oncogenes (Bcl-2, cyclin D1, COX2). To investigate the therapeutic efficacy of systemic administration of HuR-TfNP in an A549-luc animal model of metastatic lung cancer, we monitored the tumor growth through bioluminescent imaging. A significant reduction in the average radiance was observed six weeks after treatment. We also observed a reduction in the metastatic tumor nodules, clearly demonstrating the therapeutic efficacy of HuR-TfNP (Muralidharan et al. 2017).

Next, we attempted to codeliver two different siRNAs to target two different genes. For that the FNP delivery system was tested. One of the siRNAs targeted the mRNA of High Mobility Group A1 protein (HMGA1), which is highly expressed in lung cancer tissues compared with normal lung tissues. The second siRNA targeted HuR mRNA. Inhibition of both HMGA1 and HuR suppresses several other oncogenes that are involved in different stages of cancer progression (Figure 15.4). The codelivery of both siRNAs is expected to produce an enhanced therapeutic effect in lung cancer therapy. Our study is important in terms of liposomes efficiency, since liposomes are known for co-delivery of siRNA, pDNA or other anticancer agents, however delivery of multiple siRNAs are not common in the literature. A typical study used lipid-coated polymer nanoparticles for multiple siRNA delivery recently (Hasan et al. 2012). They prepared the lipid-coated nanoparticles carrying siRNAs using PRINT® technology and obtained uniform and mono-disperse siRNA encapsulated nanoparticles. They tested the nanoparticles carrying luciferase siRNA for evaluating the transfection potential of the lipid-coated nanoparticles. Further study in prostate cancer cell lines with KF11 targeting siRNA-loaded nanoparticles showed good delivery and targeted gene

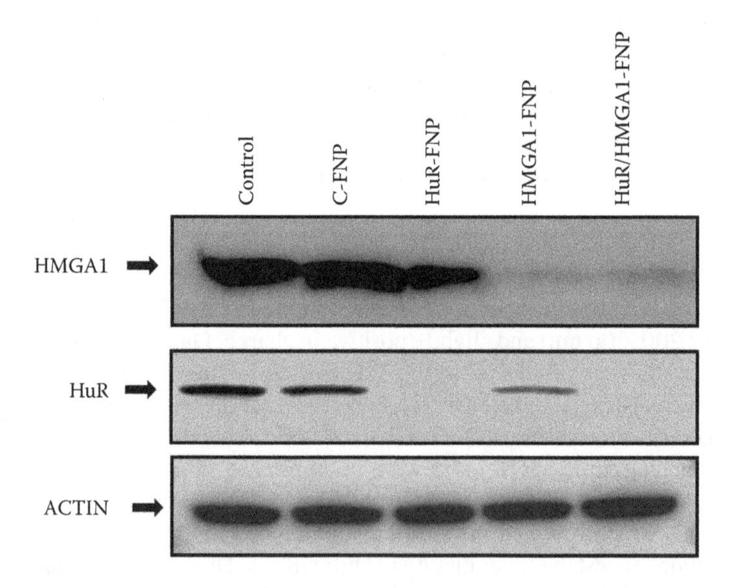

FIGURE 15.4 Western blot analysis showing reduced protein expression of both HuR and HMGA1 in H1299 cells treated with folate-targeted nanoparticle (FNP) containing HuR and HMGA1siRNAs.

silencing efficiencies compared to control siRNA loaded lipid-coated nanoparticles. Our next attempt was to target the epidermal growth factor receptor (EGFR), which is overexpressed in many solid tumors, including non-small cell lung cancer (NSCLC). High EGFR expression has been associated with poor prognosis and resistance to therapy. Gefitinib, a tyrosine kinase inhibitor, targeted towards EGFR has demonstrated antitumor efficacy as a monotherapy for patients with NSCLC (Nurwidya, Takahashi, and Takahashi 2016). Gefitinib is a weak base and has a high affinity for human plasma proteins, which limits the accumulation of drug at the tumor site. Since we demonstrated the therapeutic effect of transferrin-modified liposomes, we attempted to encapsulate gefitinib in the lipid bilayer (Zhou et al. 2012). Gefitinib was encapsulated in cationic liposome (DOTAP:chol) by the thin-film hydration method (GEF-NP), and the cytotoxicity was measured in HCC827 lung tumor cells. Free gefitinib led to higher toxicity than GEF-NP, indicating stable sequestration of gefitinib by the cationic liposome. GEP-NP was then complexed with HuRsiRNA to form GEF/HuR-TfNP, which showed selective and enhanced uptake in A549 and HCC827 lung cancer cells. Our preliminary *in vitro* studies demonstrated that targeted delivery of HuRsiRNA and Gefitinib using TfNP suppressed cell proliferation and inhibited EGFR signaling and knock-down of HuR-regulated proteins (unpublished data). Additional *in vitro* and *in vivo* studies to demonstrate the efficacy of GEF/HuR-TfNP are currently under way.

15.8 CONCLUSION

Significant developments have been made in RNAi technology for clinical applications. However, the targeted delivery of therapeutic siRNA against specific genes remains a challenge. Lipid-based nanocarriers are the most widely used and have shown the most efficient siRNA delivery *in vitro.* SNALP, pH-sensitive liposomes are the most successful *in vivo* siRNA delivery system, with an extended circulation time and efficient tumor penetration. Table 15.1 summarizes recent examples of liposome nanocarriers used in gene delivery applications. PEGylated cationic liposomes showed enhanced cellular uptake and can be used for siRNA delivery in animal models. PEGylation reduces the endosomal escape and increases the circulation time. Further incorporation of targeting ligands would significantly increase the delivery of siRNA. Although a liposomal-based siRNA delivery system showed promise for cancer therapy, several challenges need to be resolved. For rapid clinical translation of the liposomes as gene carriers they need to overcome challenges such as toxicity of the lipid materials, off-target effects, immune-stimulatory effects and stability issues *in vivo.* The mechanisms by which the cationic liposomes enter the cells are not fully understood. More research is needed to determine the amount of siRNA reaching the tumor site and to locate the siRNA molecules that escape during circulation. Using natural liposome-like vehicles such as extracellular vesicles or exosomes offers new strategies for developing rationally designed nanocarriers for gene delivery. These studies would guide us to develop and design nanocarriers that would substantially improve the therapeutic value of the gene therapeutics.

TABLE 15.1
Recent Examples of Liposome Nanocarriers Used in Gene Delivery Applications

Nanoparticles	Lipid Structure	Payload	Source
Cationic liposomes/ Lipoplexes	DOTAP DOTMA DOPE HSPC	DNA/RNA	Balazs and Godbey, 2011 Khatri et al. 2014 Desmet et al. 2016
Lipopolyplexes	DOTAP:chol DOTAP/DOPE MEND	DNA siRNA	Chen et al. 2009 de Wolf et al. 2007 Rezaee et al. 2016 Garcia et al. 2007 Hatakeyama et al. 2011
SNALP	DSPC:chol:PEG-C-DMA:DLinDMA DSPC/CHOL/DODAP	siRNA	Jayaraman et al. 2012 Martino et al. 2014
pH-sensitive liposomes	DLinDMA		Sato et al. 2012
Lipidoids	C12-200 LNP01		Frank-Kamenetsky et al. 2008 Akinc et al. 2009
Anionic liposomes	DOPG DOPG/DOPE	DNA	Patil et al. 2004 Kapoor and Burgess, 2012
Neutral liposomes	DOPC DOPE	siRNA	Nogueira et al. 2017 Wyrozumska et al. 2015, Landen et al. 2005, Merritt et al. 2008
Exosomes	–	siRNA	O'Loughlin et al. 2017

CONFLICT OF INTEREST

The authors declare no competing financial interests.

ACKNOWLEDGMENTS

The study was supported in part by a grant received from the National Institutes of Health R01 CA167516, an Institutional Development Award (IDeA) from the National Institute of General Medical Sciences (P20 GM103639) of the National Institutes of Health, and by funds received from the Presbyterian Health Foundation Seed Grant, Presbyterian Health Foundation Bridge Grant, Stephenson Cancer Center Seed Grant, and Jim and Christy Everest Endowed Chair in Cancer Developmental Therapeutics, The University of Oklahoma Health Sciences Center. The authors thank Ms. Kathy Kyler at the office of Vice President of Research, OUHSC, for editorial assistance. Rajagopal Ramesh is an Oklahoma TSET Research Scholar and holds the Jim and Christy Everest Endowed Chair in Cancer Developmental Therapeutics.

REFERENCES

Abdelmohsen, K., A. Lal, H. H. Kim, et al. 2007. "Posttranscriptional orchestration of an anti-apoptotic program by HuR." *Cell Cycle* 6 (11):1288–92.

Akbarzadeh, A., R. Rezaei-Sadabady, S. Davaran, et al. 2013. "Liposome: classification, preparation, and applications." *Nanoscale Res Lett* 8 (1):102.

Akinc, A., M. Goldberg, J. Qin, et al. 2009. "Development of lipidoid-siRNA formulations for systemic delivery to the liver." *Mol Ther* 17 (5):872–9.

Akinc, A., A. Zumbuehl, M. Goldberg, et al. 2008. "A combinatorial library of lipid-like materials for delivery of RNAi therapeutics." *Nat Biotechnol* 26 (5):561–9.

Almofti, M. R., H. Harashima, Y. Shinohara, et al. 2003. "Cationic liposome-mediated gene delivery: biophysical study and mechanism of internalization." *Arch Biochem Biophys* 410 (2):246–53.

Antimisiaris, S., S. Mourtas, and K. Papadia. 2017. "Targeted si-RNA with liposomes and exosomes (extracellular vesicles): how to unlock the potential." *Int J Pharm* 525 (2):293–312.

Balazs, D. A., and W. Godbey. 2011. "Liposomes for use in gene delivery." *J Drug Deliv* 2011:326497.

Bohmer, N., and A. Jordan. 2015. "Caveolin-1 and CDC42 mediated endocytosis of silica-coated iron oxide nanoparticles in HeLa cells." *Beilstein J Nanotechnol* 6:167–76.

Brennan, C. M., and J. A. Steitz. 2001. "HuR and mRNA stability." *Cell Mol Life Sci* 58 (2):266–77.

Burnett, J. C., and J. J. Rossi. 2012. "RNA-based therapeutics: current progress and future prospects." *Chem Biol* 19 (1):60–71.

Campbell, R. B., D. Fukumura, E. B. Brown, et al. 2002. "Cationic charge determines the distribution of liposomes between the vascular and extravascular compartments of tumors." *Cancer Res* 62 (23):6831–6.

Chen, Y., J. Sen, S. R. Bathula, et al. 2009. "Novel cationic lipid that delivers siRNA and enhances therapeutic effect in lung cancer cells." *Mol Pharm* 6 (3):696–705.

Dams, E. T., P. Laverman, W. J. Oyen, et al. 2000. "Accelerated blood clearance and altered biodistribution of repeated injections of sterically stabilized liposomes." *J Pharmacol Exp Ther* 292 (3):1071–9.

de Wolf, H. K., C. J. Snel, F. J. Verbaan, et al. 2007. "Effect of cationic carriers on the pharmacokinetics and tumor localization of nucleic acids after intravenous administration." *Int J Pharm* 331 (2):167–75.

Desmet, E., S. Bracke, K. Forier, et al. 2016. "An elastic liposomal formulation for RNAi-based topical treatment of skin disorders: Proof-of-concept in the treatment of psoriasis." *Int J Pharm* 500 (1–2):268–74.

Di Martino, M. T., V. Campani, G. Misso, et al. 2014. "In vivo activity of miR-34a mimics delivered by stable nucleic acid lipid particles (SNALPs) against multiple myeloma." *PLoS One* 9 (2):e90005.

Dos Santos, N., C. Allen, A.M. Doppen, et al. 2007. Influence of poly(ethylene glycol) grafting density and polymer length on liposomes: relating plasma circulation lifetimes to protein binding. *Biochim Biophys Acta* 1768 (6):1367–77

Frank-Kamenetsky, M., A. Grefhorst, N. N. Anderson, et al. 2008. "Therapeutic RNAi targeting PCSK9 acutely lowers plasma cholesterol in rodents and LDL cholesterol in nonhuman primates." *Proc Natl Acad Sci U S A* 105 (33):11915–20.

Galban, S., Y. Kuwano, R. Pullmann, Jr., et al. 2008. "RNA-binding proteins HuR and PTB promote the translation of hypoxia-inducible factor 1alpha." *Mol Cell Biol* 28 (1):93–107.

Garcia, L., M. Bunuales, N. Duzgunes, et al. 2007. "Serum-resistant lipopolyplexes for gene delivery to liver tumour cells." *Eur J Pharm Biopharm* 67 (1):58–66.

Hafez, I. M., and P. R. Cullis. 2001. "Roles of lipid polymorphism in intracellular delivery." *Adv Drug Deliv Rev* 47 (2–3):139–48.

Hannon, G. J., and J. J. Rossi. 2004. "Unlocking the potential of the human genome with RNA interference." *Nature* 431 (7006):371–8.

Hasan, W., K. Chu, A. Gullapalli, et al. 2012. "Delivery of multiple siRNAs using lipid-coated PLGA nanoparticles for treatment of prostate cancer." *Nano Lett* 12 (1):287–92.

Hatakeyama, H., H. Akita, E. Ito, et al. 2011. "Systemic delivery of siRNA to tumors using a lipid nanoparticle containing a tumor-specific cleavable PEG-lipid." *Biomaterials* 32 (18):4306–16.

Haussecker, D. 2014. "Current issues of RNAi therapeutics delivery and development." *J Control Release* 195:49–54.

He, Z. Y., X. W. Wei, M. Luo, et al. 2013. "Folate-linked lipoplexes for short hairpin RNA targeting claudin-3 delivery in ovarian cancer xenografts." *J Control Release* 172 (3):679–89.

Huang, Y. H., W. Peng, N. Furuuchi, et al. 2016. "Delivery of therapeutics targeting the mRNA-binding protein HuR using 3DNA nanocarriers suppresses ovarian tumor growth." *Cancer Res* 76 (6):1549–59.

Ishiwata, H., N. Suzuki, S. Ando, et al. 2000. "Characteristics and biodistribution of cationic liposomes and their DNA complexes." *J Control Release* 69 (1):139–48.

Ito, I., L. Ji, F. Tanaka, et al. 2004. "Liposomal vector mediated delivery of the 3p FUS1 gene demonstrates potent antitumor activity against human lung cancer in vivo." *Cancer Gene Ther* 11 (11):733–9.

Jayaraman, M., S. M. Ansell, B. L. Mui, et al. 2012. "Maximizing the potency of siRNA lipid nanoparticles for hepatic gene silencing in vivo." *Angew Chem Int Ed Engl* 51 (34):8529–33.

Jiang, L., X. Li, L. Liu, et al. 2013. "Cellular uptake mechanism and intracellular fate of hydrophobically modified pullulan nanoparticles." *Int J Nanomedicine* 8:1825–34.

Kanasty, R., J. R. Dorkin, A. Vegas, et al. 2013. "Delivery materials for siRNA therapeutics." *Nat Mater* 12 (11):967–77.

Kapoor, M., and D. J. Burgess. 2012. "Physicochemical characterization of anionic lipid-based ternary siRNA complexes." *Biochim Biophys Acta* 1818 (7):1603–12.

Khalil, I. A., K. Kogure, H. Akita, et al. 2006. "Uptake pathways and subsequent intracellular trafficking in nonviral gene delivery." *Pharmacol Rev* 58 (1):32–45.

Khatri, N., D. Baradia, I. Vhora, et al. 2014. "Development and characterization of siRNA lipoplexes: Effect of different lipids, in vitro evaluation in cancerous cell lines and in vivo toxicity study." *AAPS PharmSciTech* 15 (6):1630–43.

Landen, C. N., Jr., A. Chavez-Reyes, C. Bucana, et al. 2005. "Therapeutic EphA2 gene targeting in vivo using neutral liposomal small interfering RNA delivery." *Cancer Res* 65 (15):6910–8.

Li, S., L. Huang. 2009 Nanoparticles evading the reticuloendothelial system: role of the supported bilayer. *Biochim Biophys Acta* 1788:2259–66.

Li, S. D., L. Huang. 2010. "Stealth nanoparticles: high density but sheddable PEG is a key for tumor targeting" *J Control Release* 145 (3):178–81.

Li, W., and F. C. Szoka, Jr. 2007. "Lipid-based nanoparticles for nucleic acid delivery." *Pharm Res* 24 (3):438–49.

Lu, C., D. J. Stewart, J. J. Lee, et al. 2012. "Phase I clinical trial of systemically administered TUSC2(FUS1)-nanoparticles mediating functional gene transfer in humans." *PLoS One* 7 (4):e34833.

MacLachlan, I. 2007. "Liposomal formulations for nucleic acid delivery." In *Antisense Drug Technology Principles, Strategies, and Applications, Second Edition*, Edited by S. T. Crooke, CRC Press 2007, pp. 237–270.

Medina-Kauwe, L. K., J. Xie, and S. Hamm-Alvarez. 2005. "Intracellular trafficking of non-viral vectors." *Gene Ther* 12 (24):1734–51.

Mehta, M., K. Basalingappa, J. N. Griffith, et al. 2016. "HuR silencing elicits oxidative stress and DNA damage and sensitizes human triple-negative breast cancer cells to radiotherapy." *Oncotarget*. doi: 10.18632/oncotarget.11706.

Mendonça, L. S., J. N. Moreira, M. C. de Lima, et al. 2010. "Co-encapsulation of anti-BCR-ABL siRNA and imatinib mesylate in transferrin receptor-targeted sterically stabilized liposomes for chronic myeloid leukemia treatment." *Biotechnol Bioeng* 107 (5):884–93.

Merritt, W. M., Y. G. Lin, W. A. Spannuth, et al. 2008. "Effect of interleukin-8 gene silencing with liposome-encapsulated small interfering RNA on ovarian cancer cell growth." *J Natl Cancer Inst* 100 (5):359–72.

Mochizuki, S., N. Kanegae, K. Nishina, et al. 2013. "The role of the helper lipid dioleoylphosphatidylethanolamine (DOPE) for DNA transfection cooperating with a cationic lipid bearing ethylenediamine." *Biochim Biophys Acta* 1828 (2):412–8.

Mok, K. W., A. M. Lam, and P. R. Cullis. 1999. "Stabilized plasmid-lipid particles: factors influencing plasmid entrapment and transfection properties." *Biochim Biophys Acta* 1419 (2):137–50.

Morrissey, D. V., K. Blanchard, L. Shaw, et al. 2005. "Activity of stabilized short interfering RNA in a mouse model of hepatitis B virus replication." *Hepatology* 41(6):1349–56.

Muralidharan, R., A. Babu, N. Amreddy, et al. 2017. "Tumor-targeted nanoparticle delivery of HuR siRNA inhibits lung tumor growth in vitro and in vivo by disrupting the oncogenic activity of the RNA-binding protein HuR." *Mol Cancer Ther* doi: 10.1158/1535-7163.MCT-17-0134.

Muralidharan, R., A. Babu, N. Amreddy, et al. 2016. "Folate receptor-targeted nanoparticle delivery of HuR-RNAi suppresses lung cancer cell proliferation and migration." *J Nanobiotechnology* 14 (1):47.

Nakamura, T., M. Kuroi, and H. Harashima. 2015. "Influence of endosomal escape and degradation of alpha-galactosylceramide loaded liposomes on CD1d antigen presentation." *Mol Pharm* 12 (8):2791–9.

Nam, H. Y., S. M. Kwon, H. Chung, et al. 2009. "Cellular uptake mechanism and intracellular fate of hydrophobically modified glycol chitosan nanoparticles." *J Control Release* 135 (3):259–67.

Nogueira, E., J. Freitas, A. Loureiro, et al. 2017. "Neutral PEGylated liposomal formulation for efficient folate-mediated delivery of MCL1 siRNA to activated macrophages." *Colloids Surf B Biointerfaces* 155:459–465.

Nurwidya, F., F. Takahashi, and K. Takahashi. 2016. "Gefitinib in the treatment of nonsmall cell lung cancer with activating epidermal growth factor receptor mutation." *J Nat Sci Biol Med* 7 (2):119–23.

O'Loughlin, A. J., I. Mager, O. G. de Jong, et al. 2017. "Functional delivery of lipid-conjugated siRNA by extracellular vesicles." *Mol Ther.* doi: 10.1016/j.ymthe.2017.03.021.

Patil, S. D., D. G. Rhodes, and D. J. Burgess. 2004. "Anionic liposomal delivery system for DNA transfection." *AAPS J* 6 (4):e29.

Pereira, P., S. S. Pedrosa, J. M. Wymant, et al. 2015. "siRNA inhibition of endocytic pathways to characterize the cellular uptake mechanisms of folate-functionalized glycol chitosan nanogels." *Mol Pharm* 12 (6):1970–9.

Pozzi, D., V. Colapicchioni, G. Caracciolo, et al. 2014. "Effect of polyethyleneglycol (PEG) chain length on the bio-nano-interactions between PEGylated lipid nanoparticles and biological fluids: from nanostructure to uptake in cancer cells." *Nanoscale* 6 (5):2782–92.

Rezaee, M., R. K. Oskuee, H. Nassirli, et al. 2016. "Progress in the development of lipopolyplexes as efficient non-viral gene delivery systems." *J Control Release* 236:1–14.

Romeo, C., M. C. Weber, M. Zarei, et al. 2016. "HuR contributes to TRAIL resistance by restricting death receptor 4 expression in pancreatic cancer cells." *Mol Cancer Res* 14 (7):599–611.

Sato, Y., H. Hatakeyama, Y. Sakurai, et al. 2012. "A pH-sensitive cationic lipid facilitates the delivery of liposomal siRNA and gene silencing activity in vitro and in vivo." *J Control Release* 163 (3):267–76.

Semple, S. C., A. Akinc, J. Chen, et al. 2010. "Rational design of cationic lipids for siRNA delivery." *Nat Biotechnol* 28 (2):172–6.

Siomi, H., and M. C. Siomi. 2009. "On the road to reading the RNA-interference code." *Nature* 457 (7228):396–404.

Srinivasan, C., and D. J. Burgess. 2009. "Optimization and characterization of anionic lipoplexes for gene delivery." *J Control Release* 136 (1):62–70.

Srivastava, A., N. Amreddy, A. Babu, J Pannerselvam, et al. 2016. "Nanosomes carrying doxorubicin exhibit potent anticancer activity against human lung cancer cells." *Sci Rep* 6:38541.

Srivastava, A., J. Filant, K. M. Moxley, et al. 2015. "Exosomes: a role for naturally occurring nanovesicles in cancer growth, diagnosis and treatment." *Curr Gene Ther* 15 (2):182–92.

Tortorella, S., and T. C. Karagiannis. 2014. "Transferrin receptor-mediated endocytosis: a useful target for cancer therapy." *J Membr Biol* 247 (4):291–307.

Wang, B., S. Zhang, S. Cui, et al. 2012. "Chitosan enhanced gene delivery of cationic liposome via non-covalent conjugation." *Biotechnol Lett* 34 (1):19–28.

Wang, C., X. Cheng, Y. Su, et al. 2015. "Accelerated blood clearance phenomenon upon cross-administration of PEGylated nanocarriers in beagle dogs." *Int J Nanomedicine* 10:3533–45.

Wang, J., Y. Guo, H. Chu, et al. 2013. "Multiple functions of the RNA-binding protein HuR in cancer progression, treatment responses and prognosis." *Int J Mol Sci* 14 (5):10015–41.

Wang, J., Z. Lu, M. G. Wientjes, and J. L. Au. 2010. "Delivery of siRNA therapeutics: barriers and carriers." *AAPS J* 12 (4):492–503.

Wyrozumska, P., J. Meissner, M. Toporkiewicz, et al. 2015. "Liposome-coated lipoplex-based carrier for antisense oligonucleotides." *Cancer Biol Ther* 16 (1):66–76.

Xia, Y., J. Tian, and X. Chen. 2016. "Effect of surface properties on liposomal siRNA delivery." *Biomaterials* 79:56–68.

Yang, J., A. Bahreman, G. Daudey, et al. 2016. "Drug delivery via cell membrane fusion using lipopeptide modified liposomes." *ACS Cent Sci* 2 (9):621–30.

Zatsepin, T. S., Y. V. Kotelevtsev, and V. Koteliansky. 2016. "Lipid nanoparticles for targeted siRNA delivery - going from bench to bedside." *Int J Nanomedicine* 11:3077–86.

Zelphati, O., and F. C. Szoka, Jr. 1996. "Mechanism of oligonucleotide release from cationic liposomes." *Proc Natl Acad Sci U S A* 93 (21):11493–8.

Zhang, X. X., T. J. McIntosh, and M. W. Grinstaff. 2012. "Functional lipids and lipoplexes for improved gene delivery." *Biochimie* 94 (1):42–58.

Zhou, X., B. Yung, Y. Huang, et al. 2012. "Novel liposomal gefitinib (L-GEF) formulations." *Anticancer Res* 32 (7):2919–23.

16 Advancements in Polymeric Systems for Nucleic Acid Delivery

Vinayak Sadashiv Mharugde, Sudeep Pukale, Saurabh Sharma, Anupama Mittal, and Deepak Chitkara

CONTENTS

16.1 INTRODUCTION

RNA interference (RNAi) has been suggested as a potential treatment method to improve current chemotherapeutic regimens. It is a sequence-specific, post-transcriptional gene silencing mechanism in animals and plants that targets mRNA encoded by the mutant gene. RNA-based strategies are useful in targeting the mutations that results in a gain of function wherein RNA levels are modified and includes the use of antisense oligonucleotide, triplex-forming oligonucleotides, aptamers, trans-splicing, segmental trans-splicing, ribozymes, DNAzymes, siRNA, and miRNA (Chitkara, Singh, & Mittal, 2016). Among these, siRNA and miRNA have generated a lot of interest as they could be easily synthesized, do not require genome integration, and thus could curtail potential problems of insertional mutagenesis. These are 20–25 base pair-long RNA oligonucleotides that are incorporated into the pre-RISC (RNA-induced silencing complex) followed by the cleavage-dependent or independent release of the passenger strand forming the guide strand containing RISC. The guide strand guides RISC to the complementary or near complementary region of the target mRNA. siRNA with a perfect match to its target cleaves the target mRNA via the endonuclease Ago2 whereas miRNA, with an imperfect match to its target, induces mRNA degradation and translational inhibition (Rao et al., 2013) as shown in Figure 16.1.

Both miRNA and siRNA have shown enormous potential as cancer therapeutics, however their delivery, using traditional methods, may not achieve the expected therapeutic response due to biological barriers. On systemic administration, these therapeutics have to overcome several barriers including the enzymes (RNAase), tissue interstitial environment, vascular wall, intercellular tissue junction, and cytoplasmic membrane of the target cells, followed by escape from the endosome, and then incorporation into the RISC. Owing to their high molecular weight, high hydrophilicity, and negative charge, naked siRNA/miRNA are impermeable through biomembranes resulting in inefficient transfection (Chitkara, Singh, & Mittal, 2016). Physical approaches such as needle injection, coated microneedles, electroporation, gene guns, ultrasound, and hydrodynamic delivery have been explored however have limitations in their clinical translation. For example, needle injection is the simplest physical method for gene delivery by direct injection of DNA through a needle into the tissue. The major application of this strategy is DNA vaccination, however, this strategy may result in localized pain, edema, and bleeding at the injection site. Electroporation is another method that uses short pulses with high voltage to carry DNA across the cell membrane. This shock is thought to cause temporary formation of pores in the cell membrane, allowing DNA molecules to pass through. However, inaccessibility of the electrodes to the internal organs limits *in vivo* use for gene transfer to solid tissues. Particle bombardment through a gene gun is an effective and rapid tool to deliver exogenous materials into living tissue. DNA is deposited on the surface of gold particles, which are then accelerated by pressurized helium gas and expelled onto cells or tissue. Gas pressure, particle size, and dosing frequency are critical factors in gene guns (Zhu & Mahato, 2010). This method is easy, fast, and versatile however, it could cause cell damage. Another approach is to use the vectors that could carry the payload to the desired site. Vectors (both viral and non-viral) could deliver the gene of interest at the target site after intravenous administration. Several types of viruses, including retrovirus, adenovirus, and adeno-associated virus (AAV), have been modified for use as vectors. Although viral vectors show efficient transfection and integrate into the host genome (retroviral vectors) leading to long-term gene expression and could efficiently transduce both dividing and non-dividing cell, they may cause immune stimulation that often limits their *in vivo* application.

Non-viral vectors including lipidic (lipoplexes), polymeric (polyplexes, micelleplexes), and bioconjugates have also been reported to efficiently deliver nucleic acid therapeutics for cancer treatment (Zhu & Mahato, 2010). Among these, polymeric systems have gained interest due to their tailor-made properties, biocompatibility, biodegradability, and non-immunogenicity. Cationic polymers including polyethylenimine (PEI) and poly L-lysine (PLL) form compact polyplexes with nucleic acids by means of electrostatic interaction and protect their degradation as well as mask their negative charges. They also facilitate cell attachment, subsequent internalization by endocytosis or membrane fusion, and endosomal escape by proton sponge

(Zhu & Mahato, 2010). Polymeric nanoparticles prepared using biodegradable polymers such as PLGA have also been used for nucleic acid delivery. Another polymeric carrier system, micelleplexes, have also gained acceptance for delivery of siRNA/miRNA with therapeutic purpose. These are nanometric core-shell structures formed by amphiphilic block copolymers, wherein the inner core is formed by hydrophobic block that encapsulate the poorly water-soluble drugs while the hydrophilic blocks of the outer portion generally form a dense hydrophilic shell. Cationic charge on the polymer enables complexation with the oligonucleotides. In this chapter we will focus on the developments that have taken place in polymeric systems for siRNA/miRNA.

16.2 BARRIERS TO NUCLEIC ACID THERAPEUTICS

In order to understand the design elements of polymeric carriers, it is necessary to know of biological barriers that hinder the nucleic acid delivery. Delivery of siRNA/miRNA is a multi-step process, which includes cellular uptake, escape from

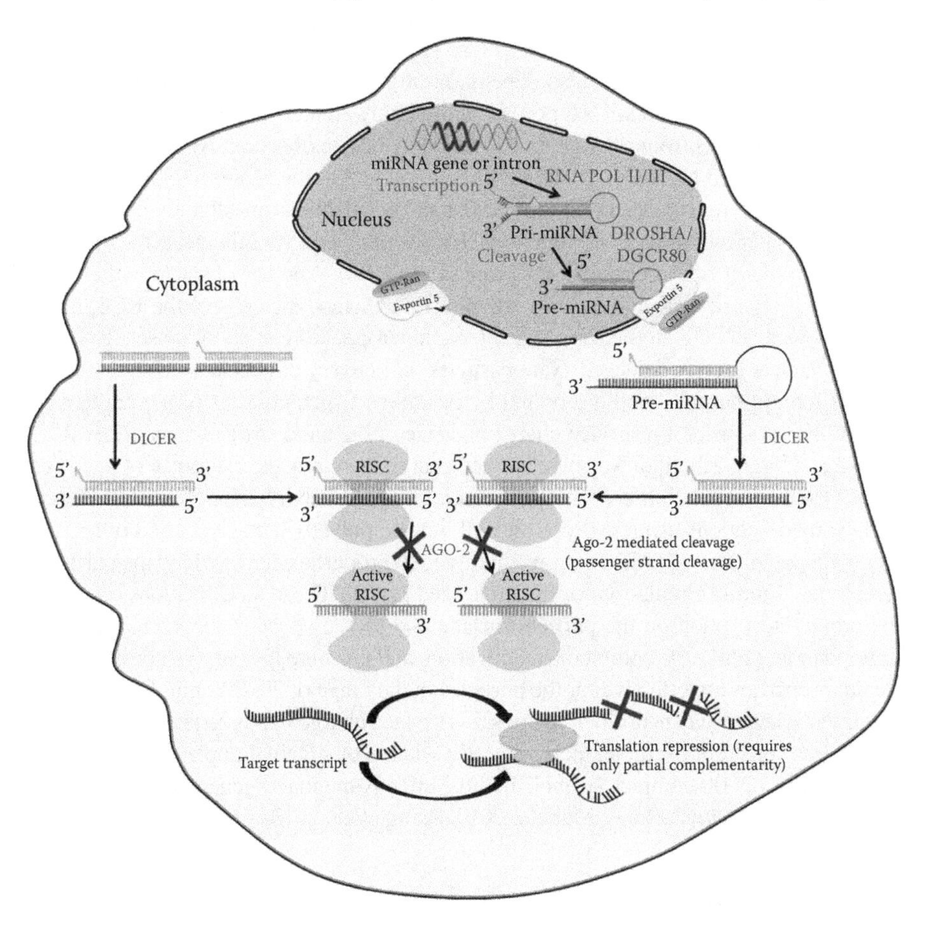

FIGURE 16.1 Mechanism of miRNA expression, gene regulation and function. (From Chitkara, D. et al., *Advanced Drug Delivery Reviews*, 81:34–52, 2015.)

degradation vesicles, and intracellular movement. During the delivery process, these therapeutics face several extracellular and intracellular barriers that may compromise their efficacy.

16.2.1 Extracellular Barriers

Aggregation of complexes in the extracellular environment is the first major obstacle. Vector-siRNA complexes having a size range between 50–200 nm mostly enter cells by endocytosis or pinocytosis. The size and charge of particles depend upon various factors such as type of polymer, concentration of polymer, buffer, and pH. These may aggregate in the extracellular environment resulting in increased size and hence reduced uptake. To avoid aggregation, the surface charge of nanoparticles is modified by incorporating PEG or sugar molecules (e.g., cyclodextrin and hyaluronic acid) (Nimesh, Gupta, & Chandra, 2011). In a study by Merkel et al. increasing the PEG molecular weight incorporated into polymer based PEGylated micelles (from 2 to 20 kDa) was shown to prevent aggregation and adsorption to blood components, leading to increased circulation time *in vivo* (Merkel et al., 2009).

Barriers to siRNA delivery also depend on targeted organs and the administration route. Systemic delivery of siRNA poses a significantly greater challenge, as these therapeutics have to travel from site of administration to the site of action. After i.v. injection, they are distributed to organs via blood circulation and at the same time also undergo elimination. Several studies have reported that naked siRNA is unstable in the presence of a high concentration of serum due to RNAases thus requiring the need for a carrier that can protect it from degradation (Wang et al., 2010). Another obstacle is the rapid clearance by the reticuloendothelial system (RES). Phagocytic cells of the RES, more specially Kupffer cells in the liver and splenic macrophages can endocytose siRNA as well as carriers used to deliver it. Nanoparticles for delivery of nucleic acids are considered as foreign particles and are bound by opsonins, which consists of a complement system. Factors including surface charge and size of the nanoparticles may affect RES uptake and biodistribution. Negative surface charge increases the clearance of particles from systemic circulation compared to neutral or positively charged particles. Hence surface modifications using hydrophilic and flexible polyethylene glycol and other surfactant copolymers, e.g. poloxamer, result in stealth properties due to which the particles remain in systemic circulation for the prolonged period. These modifications can limit the protein adsorption on the particle surface and thereby protect the vectors against opsonization, reduce the complement activation, and promote the cargo stability. These stealth properties are effective for the particles in the range of 70–200 nm. On the other hand, PEGylation may neutralize the positive surface charge that is required for siRNA uptake into cells. For example, increasing PEGylation of siRNA-lipoplexes from 1–2 to 5 mol % PEG 2000 completely abolished the siRNA-mediated gene silencing against PTEN protein *in vitro* (Zhu & Mahato, 2010; Wang et al., 2010).

16.2.2 Intracellular Barriers

Intracellular barriers exist in several processes including vesicle escape, intracellular trafficking, and vector unpacking that is required for siRNA/miRNA delivery as

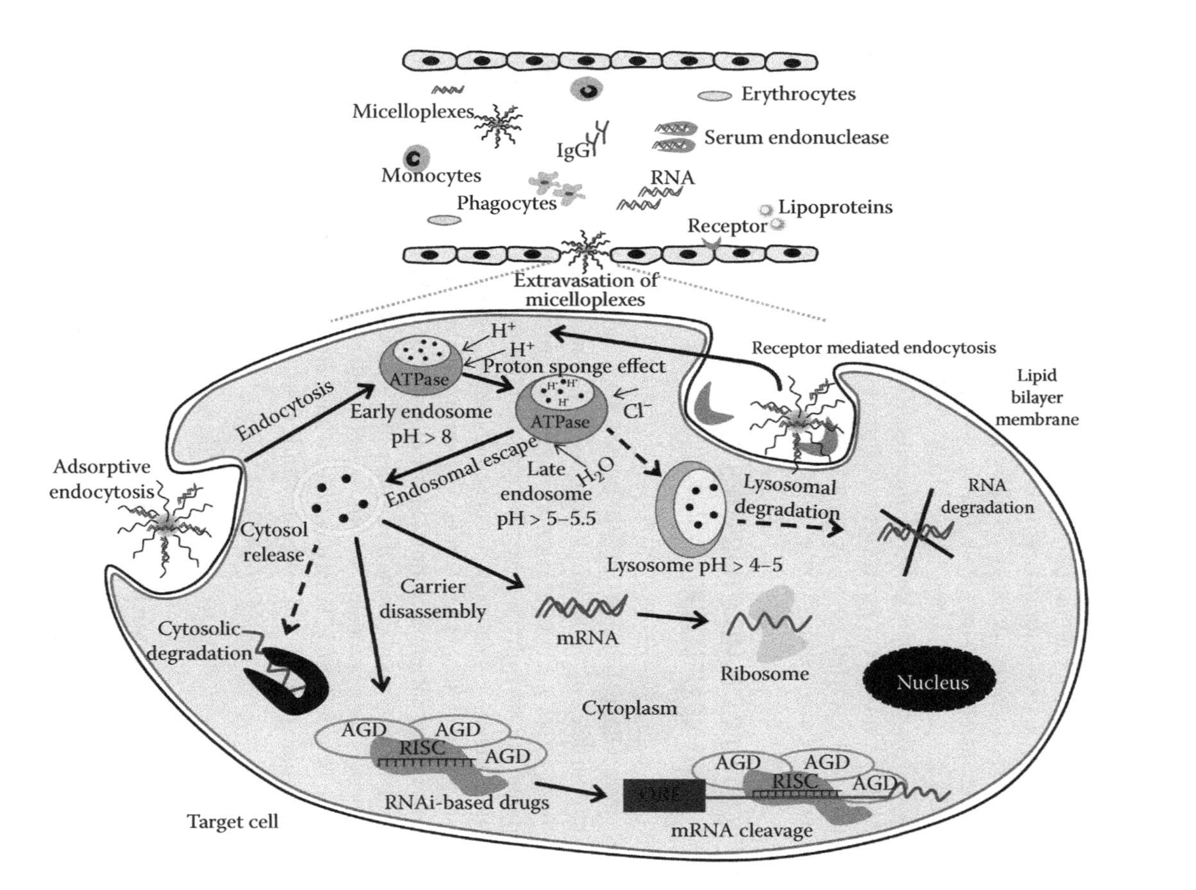

FIGURE 16.2 Major barriers to *in vivo* delivery of nucleic acids. (From Pereira, P. et al., *Expert Opinion on Drug Delivery*, 14:353–371, 2017.)

shown in Figure 16.2. siRNA is transported across the interstitial space to the target cells. After reaching the target cell, siRNA undergoes internalization by endocytosis, a process that involves siRNA being encapsulated in endocytic vesicles that later fuse with endosomes. siRNA must escape from the endosomes and should be released from its carrier to cytosol in order to be loaded onto RISC. Various mechanisms have been proposed for the endosomal escape of vectors, one of the strategies suggests of physical disruption of the negatively charged endosomal membrane by direct interaction with a cationic polymer. Polyamidoamine, poly L-lysine, and dendrimers have been shown to undergo endosomal escape through this mechanism. Positively charged nanoparticles could also bind to anionic microtubules and move to the nuclear membrane along with the cytoskeletal network that enhanced cytoplasmic trafficking. Further unpacking of the polyplexes should occur to release the cargo that will then incorporate itself into the RISC complex. In addition to the release of siRNA/miRNA from the vectors, it is necessary that the polymer should be degraded and excreted out to prevent accumulation of high molecular weight polycations inside cells, which improves their cell viability (Wang et al., 2010).

16.3 POLYMER CHARACTERISTICS REQUIRED FOR GENE DELIVERY

An ideal cationic carrier has to fulfill a series of delivery functions in the extracellular and intracellular transport of siRNA/miRNA. The polymer has to compact siRNA into particles of virus-like dimensions that can migrate through blood circulation to the target site, it has to protect nucleic acid from degradation and against undesired interactions with the biological environment, to facilitate target cell binding and internalization, endosomal escape, and trafficking the cytoplasmic environment. Also, carriers should be non-toxic, non-immunogenic, and biodegradable. Factors influencing nucleic acid binding affinity that are inherent in the chemical structure of polymers are mentioned in Table 16.1. The use of polymers as synthetic non-viral carriers for introducing siRNA/miRNA into cells appears very appealing. Polymers can be generated in large quantities with chemically defined structures and non-immunogenic properties. Different chemical structures and molecular weights

TABLE 16.1
Factors Influencing Nucleic Acid Binding Affinity

S. No.	Factors
i	Number of charge groups per single polymer
ii	The type of charge groups such as primary, secondary and tertiary amino groups
iii	The spacing of charge groups within the polymer
iv	Hydrophobicity of the cationic carrier
v	The degree of branching in the polymer backbone
vi	Ionic strength of polyplex solution
vii	Technical process of polyplex formation

may be applied to tailor-made polymers with optimized characteristics for extracellular delivery of nucleic acid to target tissues and subsequent intracellular delivery into target cells. Further, advantages of a polymeric system over viral vectors are the low cost and consistent standard of production, higher biosafety (less immunogenic as compared to viruses such as adenoviruses), and high flexibility.

Polymeric gene delivery vectors must maintain their physical stability in the presence of serum proteins and high ionic strength, and also protect siRNA from nucleases that are present in the extracellular spaces. The nanoparticles that are formed either by condensation, encapsulation or complexation of nucleic acid have distinct characteristics and varying transfection efficiencies, making them suitable for different gene delivery applications. Plank et al. have demonstrated that a minimum length of six to eight cationic amino acids (lysine, arginine) is required to compact the nucleic acid (DNA) into polyplex structures active in gene delivery utilizing polylysine polymers. Depending on the size and affinity of polycation this may also result in an increased positive charge on the polyplex that promotes cellular uptake and transfection efficiency. However, an increase in charge also results in toxicity due to destabilization and loss of integrity of cellular membranes and the presence of excess free polycation (Kloeckner, Wagner, & Ogris, 2006). The influence of charge group type and spacing was evaluated by Davis and colleagues using carbohydrate-containing polycations. They demonstrated that the distance between the carbohydrate unit and the charge groups in the backbone, and also the types of amino group (quaternary amines vs. amidine groups) were the factors that influence the carrier's transfection efficiency *in vitro*. The study demonstrated that as the amidine charge center is moved further from the carbohydrate unit within the polycation structure an increase in toxicity results. Further inclusion of larger carbohydrate species within polycation backbone reduces toxicity. A series of quaternary ammonium polycations containing N,N,N',N'-tetramethyl-1,6-hexanediamine, d-trehalose, and beta-cyclodextrins were synthesized in order to elucidate the effects of charge center type on gene delivery. In all cases it was found that the quaternary ammonium analogs exhibited lower gene expression values and similar toxicities as those of their amidine analogs. Transfection experiments conducted in the presence of chloroquine revealed an increased gene expression from quaternary ammonium containing polycations and not from amidine analogs (Reineke & Davis, 2003).

The degree of polymer branching has significant effect on flexibility of these macromolecules and their ability to form complexes and transport of nucleic acid. For example, in evaluating histidine/lysine copolymers (HK) in combination with liposomal carriers, Chen et al. showed that the degree of branching was a major factor in determining the transfection efficiency. In the transformed cell line (MDA-MB-231), branched HK polymers were more effective than the linear HK polymers, however linear HK polymers enhanced gene expression in primary cell lines more effectively. Further, the differences in the linear and branched polymers were not due to initial cellular uptake or size of complexes. Also, there was a strong association between the optimal type of HK polymer and the pH of the endocytic vesicles. In the cell cultures, linear polymers showed the best effects wherein the endocytic vesicles were strongly acidic with pH below 5. Conversely, in the cell lines in which the branched polymers were optimal transfection agents, the pH of endocytic vesicles was above 6 (Vasir & Labhasetwar, 2008; Kloeckner, Wagner, & Ogris, 2006).

16.4 POLYMERS FOR NUCLEIC ACID DELIVERY

16.4.1 Cationic Polymers

Cationic polymers readily bind and condense polyanionic nucleic acids and have thus been widely used as transfection reagents for genes, oligonucleotides, and siRNA/miRNA. Several structural modifications have been reported in the cationic polymers varying from linear (chitosan and linear polyethylenimine [PEI]) to branched (branched PEI) and cross-linked-type polymers (cross-linked poly [amino acid]) in order to improve upon their properties particularly their transfection efficiency and toxicity profiles (Figure 16.3).

16.4.1.1 Polyethylenimine (PEI)

Polyethylenimine (PEI) consists of a repeating amine group with a two-carbon (CH_2CH_2) aliphatic spacer. Polyethylenimine is a highly positively charged polymer and is most frequently used because of an excellent transfection efficiency *in vitro* and significant transfection *in vivo*. Structurally, PEI exists in either linear or branched forms, with linear PEI containing all secondary amines in its backbone except the terminal groups, while branched PEI contains primary, secondary, and tertiary amino groups at the ratio of 1:1:1 (Zhu & Mahato, 2010; Kloeckner, Wagner, & Ogris, 2006). Owing to the presence of different types of amino groups with different pKa values, these are protonated at different levels at a given pH. This confers PEI with a superior buffering capacity over a wide range of pH resulting in the well-known "Proton Sponge Effect." Unprotonated amines after endocytosis will be protonated at acidic pH in the endosome, which increases the influx of protons, chloride ions, and water into endosome. The increased osmotic pressure causes the endosome to swell and rupture, thereby releasing the endosomal content into the cytoplasm.

PEI polymers with different molecular weights and degrees of branching have been synthesized and evaluated *in vivo* and *in vitro*. It has been reported that the transfection efficiency increases with an increase in molecular mass, however, this also results in increased cytotoxicity. A suitable molecular mass of PEI for complexation with nucleic acid is between 5 and 25 kDa (Zhu & Mahato, 2010; Nimesh, Gupta, & Chandra, 2011). A highly branched polymer (BPEI; 25 kDa) has been widely used since it effectively forms complexes with even large DNA molecules leading to homogeneous spherical particles with a size of ~100 nm or less that are capable of transfecting cells efficiently *in vitro* as well as *in vivo*. It also provides significantly higher protection against nuclease degradation in comparison to other polycations, such as poly(L-lysine), possibly due to its high charge density and more efficient complexation ability. The efficiency of BPEI-derived vectors and their cytotoxicity effects depend on material characteristics like the molecular weight, the cationic charge density, the degree of branching, buffer capacity, and polyplex properties, such as the particle size and zeta potential. A large amount of positive charge results in high toxicity and is one of the major limiting factors for its *in vivo*

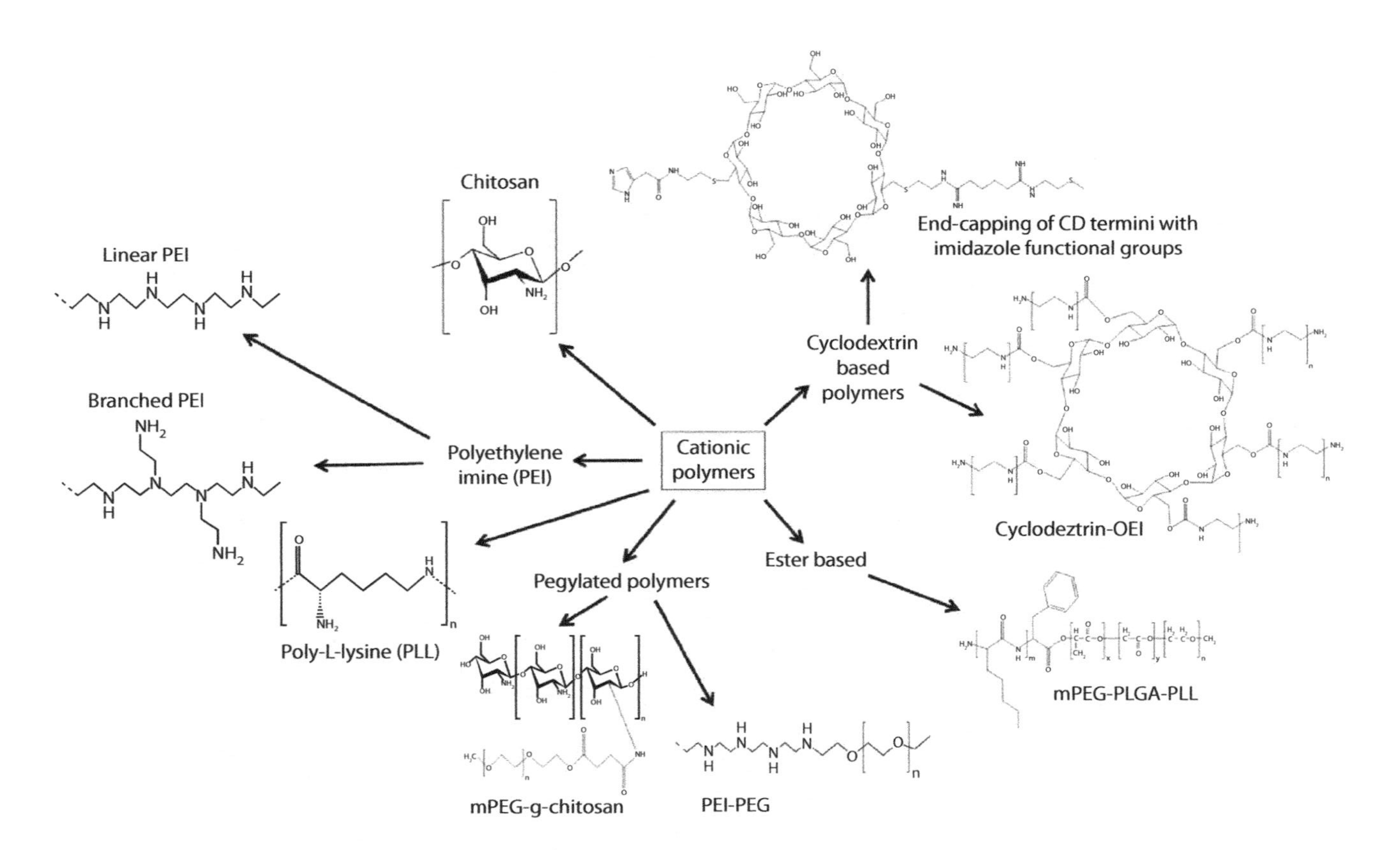

FIGURE 16.3 Cationic polymers for nucleic acid delivery.

application (Nimesh, Gupta, & Chandra, 2011). Grayson et al. evaluated the efficiency of different PEI structures for siRNA delivery in a model system, and determined the biophysical and structural characteristics of PEI at different N/P ratios, i.e. the ratio of concentration of nitrogen atom (N) of polycation to the phosphate groups (P) of DNA, as the characteristic of complex composition. siRNA delivery with PEI was observed within a narrow window of conditions, and only with the BPEI 25 KDa at an N/P ratio of six and eight with 200 nM siRNA. Also, it was suggested that the complex stability might govern transfection efficiency. Further, zeta potential and the size of the PEI/siRNA complexes were correlated to transfection efficiency (Grayson, Doody, & Putnam, 2006). Biodegradable disulfides containing PEI derivatives showed great potential as siRNA vectors for the treatment of cancer. Liu et al. reported a hyperbranched disulfide cross-linked PEI (LPEI-SS) based on linear PEI (LPEI) and ring opening reaction of propylene sulfide. Further, its efficiency was investigated as a siRNA vector, *in vitro* with a luciferase reporter gene system, and *in vivo* as antitumor efficacy of survivin-targeted siRNA [siRNA(sur)] on murine breast cancer model. Results from the hemolysis assay and cytotoxicity proved that lPEI-SS/siRNA polyplexes showed favorable cell and blood compatibility even with high positive zeta potential of 42 mV and an average size of around 230 nm. Further, cellular uptake of lPEI-SS/siRNA polyplexes was significantly improved as compared to the parent LPEI due to its higher branching degree (Liu et al., 2016).

The most widely used approaches to PEI functionalization are PEGylation or conjugation of other biomacromolecules such as polysaccharides, and functionalization with hydrophobic moieties. The degree of PEGylation and molecular weight of PEG strongly influence the properties of the conjugate. PEGylation provides a reduction of toxicity caused by PEI along with providing the stealth property to the polyplex formed after complexation with the nucleic acid. It has been shown that a large amount of PEG chains impedes electrostatic interaction between polycation and polyanion and thus hampers condensation. Mao et al. showed that the stability and size of siRNA-PEI polyplexes were clearly influenced by PEG structure. A high degree of substitution was obtained by using a lower molecular weight PEG (550 Da), resulted in large complexes (300–400 nm) that showed condensation behavior only at N/P ratio of 15. Other polyplexes with PEG molecular weights ranging from 5 to 25 KDa and a PEI control showed similar sizes (150 nm) and complete condensation was reached at N/P ratio at 3. Further protection against RNase digestion was reported to be insufficient in the case of bare PEI/siRNA polyplexes in comparison with their PEGylated counterpart (Merkel et al., 2009).

Dong et al. have reported nanocomplexes consisting of polyethyenimine-3-maleimidopropionic acid hydrazide doxorubicin and polyethylenimine-polyethylene-folate for codelivery of doxorubicin and anti-VEGF siRNA. These components were mixed with the siRNA to obtain actively targeted codelivery nanocomplexes that showed antitumor efficiency after intravenous injections in a subcutaneous MCF-7 tumor xenograft model (Dong et al., 2015).

16.4.1.2 Poly L-lysine

Poly-L-lysine (PLL) is a cationic polypeptide with amino acid L-lysine, as a repeating unit. It is a biocompatible and biodegradable polymer consisting of only primary amines that are positively charged at physiological pH and interact ionically with the negatively charged siRNA (Wang et al., 2017). PLL/siRNA polyplexes have a high positive zeta potential, interact electrostatically with a negatively charged cell surface and are taken up by absorptive endocytosis process (Buyens et al., 2010). The uptake efficiency can be enhanced by covalently coupling the PLL with a ligand that can specifically target cells and promote receptor-mediated uptake. Further, PLL is not biologically inert; for example, polyplexes have been shown to affect cellular processes during transfection by enhancing pinocytosis, phagocytosis and cell division. PLL comes in a variety of sizes and is usually specified as the average number of polylysine molecules within defined solution rather than a specifically defined number of lysine molecules per polylysine molecule (Kircheis et al., 2001). Although PLL can initiate siRNA uptake into cells, often gene expression is low owing to a number of factors such as degradation in the cytoplasm, ineffective endosomal release, limited migration to the nucleus, and insufficient nuclear uptake. All these factors can decrease transfection efficiency; however, a major disadvantage with PLL is that it does not have an intrinsic endosomal release activity. Hence, complexes are often not released or escaped from endosomes/lysosomes into the cytoplasm (Kircheis et al., 2001). Therefore, different strategies have been developed to increase endosomal release by disrupting the endosomal lipid bilayer such as the addition of lysosomotropic agents (e.g., chloroquine) into the cell medium that has increased gene expression in a number of cell lines. Chloroquine accumulates in the vesicles and leads to swelling of the endosome by raising pH and destabilizing it. Furthermore, it may help in the release of siRNA from PLL and increase the gene expression. Similarly, addition of glycerol results in enhanced gene expression in cell lines, probably by weakening of the endosomal lipid bilayer and allowing the polylysine/siRNA complex release into the cytoplasm. In a study by Klink et al., 5% glycerol and 100 μM chloroquine were used as enhancing agents to improve the efficiency of lactosylated poly-l-lysine wherein, approximately 89% transfection efficiency was obtained with 7.5 μg plasmid complexed to 22.5 μg of lactosylated poly-l-lysine (Klink et al., 2003).

16.4.1.3 Chitosan

Chitosan is a non-toxic, biocompatible, and biodegradable polycationic polymer produced from naturally occurring chitin by partial deacetylation. It is a linear polysaccharide, which is composed of deacetylated unit, D-glucosamine, and acetylated unit, glucosamine, linked with randomly distributed β-(1,4) glycosidic bonds. The size and zeta potential, stability, morphology, and biological effect of its polyplexes strongly depend on chitosan's molecular mass, degree of deacetylation, pH, and N/P ratio. The positive charge of chitosan is imparted by deacetylation process, and thus degree of deacetylation is important for nucleic acid delivery. It has been shown that a lower degree of deacetylation caused low transfection efficiency. Further size of polyplexes increases with increasing molecular mass of chitosan resulting in increase in transfection efficiency (Zhu & Mahato, 2010). Bowman and Leong studied chitosans of different molecular masses ranging from 7 to 540 kDa and found that particle size of chitosan/DNA complexes prepared at an N/P ratio 6:1 with plasmid concentration of 100 μg/ml increased from 100 to 500 nm as the molar mass of chitosan was increased (Bowman & Leong, 2006).

Katas et al. prepared chitosan nanoparticles by two methods of ionic cross-linking *i.e.* simple complexation and ionic gelation using sodium triphosphate (TPP). Both methods produced nanosized particles, less than 500 nm depending on the type, molecular weight as well as the concentration of chitosan. It was observed that the method of association of chitosan with siRNA played an important role in gene silencing effect in CHO K1 and HEK 293 cells wherein nanoparticles of chitosan-tripolyphosphate (chitosan-TPP) entrapping siRNA were observed to be better vectors for siRNA delivery compared to chitosan-siRNA complexes, due to their high binding capacity and loading efficiency (Katas & Alpar, 2006). In another study, Lee et al. prepared siRNA encapsulated chitosan nanoparticles using a coacervation method in presence of polyguluronate (PG). The mean diameter of siRNA-loaded, chitosan-based nanoparticles ranged from 110 to 430 nm, depending on the weight ratio between chitosan and siRNA. These nanoparticles were not only efficient in delivering siRNA to HEK 290FT and HeLa cells but also showed low toxicity (Lee et al., 2009).

In order to further improve upon the properties of chitosan, PEG grafting have been employed. Reduced protein adsorption and successful transfection efficiency depend on PEG chain lengths and surface coverage. According to Zheng et al., a PEG density of greater than 7 mol% was necessary for higher shielding against non-specific interactions of chitosan-based complexes with proteins and cells. Various PEGylated chitosans with increasing degree of substitution were described as being successful, ranging from 0.5 to 10% or <28%. High PEG density is also known to replace a certain percentage of primary amino group necessary for nanoplex formation, as well as decrease the transfection efficiency due to reduced cellular uptake (Zheng et al., 2013).

16.4.1.4 Cyclodextrin-Based Cationic Polymers

Cyclodextrins (CDs) are naturally occurring cyclic oligosaccharides composed of six (α-CD), 7(β-CD) or eight (γ-CD) D (+)- glucose units, linked by α-1,4-linkages. Cyclodextrins are topologically represented as torus-like macro-rings with a hydrophilic outer surface and hydrophobic inner cavity that could form inclusion complexes with water insoluble molecules. Therefore, CDs and their derivatives are used in pharmaceutical formulations to enhance stability, solubility, and absorption of small molecule drugs (Zhu & Mahato, 2010). Different strategies have been reported to form cyclodextrin-based polymers for a variety of applications including gene delivery. In one such strategy, polycationic oligomers were synthesized by step-growth polymerization between diamine-bearing cyclodextrin monomers and dimethyl suberimidate, yielding oligomers with amidine functional groups. Strong basicity of these amidine groups mediates efficient condensation of nucleic acid with the cyclodextrin polymers (CDPs) at an N/P ratio as low as three. End-capping of the polymer termini with imidazole functional groups can cause endosomal escape, resulting in improved delivery efficacy of both siRNA and plasmid DNA (Li, Xue, & Mao, 2016).

Davis and colleagues have developed a novel system, based on (i) condensation of siRNA with cyclodextrin containing polycations (CDPs) and (ii) modification of CDP-siRNA by forming inclusion complexes with adamantine that in turn was utilized for the attachment of the functional groups, PEG-transferrin. This formulation has indicated efficacy in the delivery of siRNA against oncogenes to inactivate the tumor growth in a murine model of metastatic Ewing's sarcoma. Continuous three daily doses of the CDP-siRNA-adamantine-PEG-transferrin formulation carrying two types of siRNAs specific for the gene encoding ribonucleotide reductase subunit M2 (RRM2) slowed tumor development. Several factors contributed to the translation of this delivery system includes low toxicity of the cationic polymer, condensation with nucleic acids, the steric stabilization by PEGylation of CDP-siRNA nanoparticles in a stable and non-covalent fashion, and the inclusion of ligand to improve *in vivo* uptake and efficacy (Guo et al., 2010).

A new class of cationic supramolecules, CD-based polyrotaxanes has also been designed for gene delivery. The new cationic polyrotaxane was composed of multiple OEI-grafted β-CDs that were threaded on PPO blocks and blocked on the two ends of a pluronic (PEO-PPO-PEO) triblock copolymer. Cationic polyrotaxanes effectively condensed plasmid DNA to nanoparticles and showed low cytotoxicity and high transfection efficiency in BHK-21 and MES-SA cells. Transfection efficiencies mediated by the cationic polyrotaxanes increased with an increase in the N/P ratio. Hence, higher transfection efficiencies along with lower cytotoxicity make these new cationic polyrotaxanes promising carriers for gene delivery (Yang et al., 2009).

Another recent approach is oligocationic cyclodextrins (CDs) that were modified with alkylimidazole, pyridylamino, methoxyethylamino, or primary amine groups at 6-position of the glucose units. The oligocationic cyclodextrins neutralize nucleic acid to form stable nanoparticulate polyplexes. The transfection efficiency of the cyclodextrin was dependent on the substituents present, with the most efficient having either an amino, butylimidazole group or pyridylamino at the 6-position (Wang et al., 2010).

16.4.2 Polyester-Based Carriers

Polyesters are synthetic polymers containing ester as a functional group in a polymeric backbone. They are hydrophobic, insoluble in water and soluble in organic solvents like ethyl acetate, dichloromethane, and chloroform. PLGA is the most extensively investigated copolymer for controlled drug release. PLGA nanocapsules have been used to encapsulate DNA, antisense oligodeoxynucleotides, and siRNA in order to improve stability and provide controlled release. Nanoparticles of PLGA are solid; they are stable and able to protect nucleic acids from degradation during circulation in the bloodstream. Further, the solid phase allows long-term storage and convenient use in the clinic. However, major drawbacks of PLGA nanoparticles is low siRNA delivery efficiency, inactivation of macromolecules, and a slow release rate that may hinder its application in the delivery of nucleic acids. The hydrophobicity and molecular mass of PLGA also have profound influence on the encapsulation efficiency of nucleic acids (Mittal et al., 2007). Two different approaches have been used in order to load nucleic acids into PLGA nanoparticles, (i) encapsulation into the core of nanoparticles, and (ii) adsorption onto the surface-modified, cationic PLGA NPs via electrostatic interactions. By optimizing the encapsulation procedures, delivery systems with sufficient loading, affording protection as well as controlled release profile of the nucleic acid could be prepared (Cun et al., 2010).

With regards to formulation development, it is very challenging to efficiently encapsulate high amount of hydrophilic macromolecules like siRNA into nano-sized PLGA particles mainly due to hydrophobic nature of the PLGA and the absence of electrostatic interactions between siRNA and PLGA. The addition of cationic excipients into a PLGA matrix may improve encapsulation of nucleic acid. Katas et al., incorporated PEI into PLGA particles by a spontaneous modified emulsification-diffusion method. Incorporation of PEI into PLGA particles with the PLGA to PEI weight ratio 29:1 was found to produce spherical and positively charged nanoparticles where the type of polymer and the type and concentration of the surfactant affected their physical properties. Particle size around 100 nm was obtained when 5% PVA was used as a stabilizer. PLGA-PEI nanoparticles were able to completely bind siRNA at an N/P ratio 20:1 and provided protection against nuclease degradation. *In vitro* cell culture studies revealed that PLGA-PEI nanoparticles with adsorbed siRNA could efficiently silence the targeted gene in mammalian cells, better than PEI alone, with acceptable cell viability. In another study, Su. et al. reported PEI-modified nanoparticles for co-delivery of paclitaxel (TAX) and siRNA for cancer therapy. Paclitaxel, the hydrophobic anticancer drug, was encapsulated in PLGA-PEI nanoparticles. Fluorescence measurements were used to confirm that stat3 siRNA (S3SI) and paclitaxel were delivered simultaneously to lung cancer cells (A549 and A549/T12) and siRNA-based silencing of the stat3 gene using PLGA-PEI-TAX-S3SI nanoparticles rendered cancer cells more sensitive to paclitaxel and produced more cellular apoptosis than did PLGA-PEI-TAX. Murata et al. also reported sustained release PLGA microspheres encapsulating anti-VEGF siRNA/cationic polymer complexes and observed the inhibition of tumor growth (Katas, Cevher, & Alpar, 2009; Amreddy et al., 2017).

Apart from PEI, chitosan-modified PLGA nanoparticles have also been successfully used to deliver antisense oligonucleotides to lung cancer in recent studies. Chitosan-modified PLGA nanoparticles were 130 nm in size with an adjustable positive surface charge. Antisense oligonucleotide and a 2-O-methyl-RNA (OMR) for human telomerase gene was electrostatically bound to chitosan-PLGA nanoparticles, and efficient cellular uptake was observed. The chitosan content, binding efficiency, stability, and cell uptake efficiency of the chitosan-PLGA nanoparticles were evaluated for OMR delivery in a follow-up study by the same group. The researchers observed that the cellular uptake and transfection efficiency of OMR was dependent on the chitosan content in the nanoparticle. Further, these nanoparticles were non-toxic and efficiently inhibited telomerase activity (Amreddy et al., 2017). Du et al. reported efficient delivery of siRNA via the use of biodegradable nanoparticles made from monomethoxypoly(ethylene glycol)-poly(lactic-co-glycolic acid)-poly-L-lysine (mPEG-PLGA-PLL) triblock copolymers. SEM confirmed that the mPEG-PLGA-PLL nanoparticles had a spherical structure. Further mean diameter of blank nanoparticles was 140 ± 62 nm, with a zeta potential of +11.4 mV. The size of nanoparticles loaded with siRNA increased to 151 ± 74 nm, while the zeta potential of nanoparticles decreased to -0.13 mV. The encapsulation efficiency of siRNA in mPEG-PLGA-PLL nanoparticles was 86.06%. Cultured human lung cancer cells,

SPC-A1-GFP, showed that nanoparticles loaded with Cy3-labeled siRNA had much higher intracellular siRNA delivery efficiency than siRNA alone and lipofectamine-siRNA complexes. Further, the gene silencing efficiency of mPEG-PLGA-PLL nanoparticles was also higher than that of lipofectamine while showing no cytotoxicity (Du et al., 2012).

Misra et al. prepared DOX/siPGP loaded nanoparticles of PLGA using cationic stabilizers such as dimethyldidodecylammonium bromide (DMAB) by an O/W emulsion/solvent evaporation method. PLGA and DOX containing organic phase was emulsified into DMAB (0.5%) polyvinyl alcohol (1%) and d-α-tocopheryl polyethylene glycol 1000 succinate (TPGS; 0.5%) containing aqueous phase. Nanoparticles were observed to be 243 nm in size and possessed a surface potential of 36.33 mV. Confocal microscopy results showed higher siRNA internalization through siRNA-loaded nanoparticles than native siRNA in MCF-7/ADR MDR breast cancer cells. The qRT-PCR studies and Western blot results demonstrated a significant decrease in p-gp expression after treatment with siPGP-loaded nanoparticles (Malathi et al., 2015).

16.4.3 Micelleplexes

Micelleplexes have gained acceptance as potential nanocarriers for delivery of nucleic acids because of their excellent tissue penetrating ability and ability to provide controlled release. These are nanometric core-shell structures formed by amphiphilic cationic block copolymers with the core consisting of hydrophobic block, whereas the outer shell containing hydrophilic block and can interact with different types of bioactive molecules, such as targeting moieties and siRNA/miRNA. The electrostatic interaction between cationic charge of micelles with anionic siRNA/miRNA molecules result in the formation of a micelle-gene complex. Further, hydrophobic small molecules could also be loaded in the hydrophobic core of the micelleplexes for simultaneous drug-gene delivery.

It is possible to design polymeric micelles with different functionalities such as stimuli sensitivity, desirable release properties, targetability etc. depending on the amphiphilic copolymer architecture, the core-shell forming material, molecular weight, chain length, chain density, and the proportion of hydrophilic/hydrophobic segment (Pereira et al., 2017). Modification of their surface bears a significant impact on their stability and biologic interaction, in drug absorption, bioavailability, and internalization into the cells via receptor-mediated endocytosis. The copolymers with tumor-targeted, stimuli-sensitive release properties, like acid- or glutathione-sensitive cleavage bonds could be chosen for site-specific controlled delivery. Sun et al. engineered an acid-sensitive Dlinkm group copolymer micelleplex based on self-assembly of PEG-Dlinkm-R9-PCL polymers that interacted with siRNA to form Dm-NP/siRNA through electrostatic bonds. This complex protected siRNA in serum, increased its circulation time and enhanced uptake of siCKD4 in A549 lung tumor cells. Further, these nanoparticles improved gene silencing efficiency and antitumor activity *in vivo* through pH-controlled delivery of therapeutic siRNA (Amreddy et al., 2017; Sun et al., 2015). Zhu et al. delivered VEGF siRNA and paclitaxel simultaneously into the prostate cancer cells (PC-3) using cationic poly(2-(N,N-dimethylaminoethyl) methacrylate)-polycaprolactone-poly(2-(N,N-dimethylaminoethyl methacrylate) (PDMAEMA-PCL-PDMAEMA) triblock copolymers. These triblock copolymers formed nano-sized micelles in water with a positive charge ranging from +29.3 to 35.5 mV. A gel retardation assay showed that micelles could effectively complex siRNA at and above N/P ratios of 4:1 and 2:1. GFP siRNA complexed with micelle exhibited significantly enhanced gene silencing efficiency as compared to that formulated with 20 kDa PDMAEMA or 25 kDa branched PEI in GFP expressed MDA-MB-435-GFP cells. Further, micelles loaded with paclitaxel displayed higher drug efficacy than free paclitaxel in PC3 cells, due to improved cellular uptake (Zhu et al., 2010). In another study, Zheng et al. used poly(ethylene glycol)-b-poly(L-lysine)-b-poly(L-leucine) peptide micelles of mean diameter 121.3 nm for co-delivery of docetaxel and siRNA-Bcl-2. Hydrophobic poly(L-leucine) enabled entrapment of docetaxel while the poly(L-lysine) cationic backbone complex with negatively charged siRNA at N/P ratio of 10:1 (Zheng et al., 2013).

Cationic polymeric micelles could also be prepared from copolymers containing cationic pendant groups onto the hydrophobic blocks. Xiong and Levasanifar integrated multiple functionalities onto PEG-PCL copolymer for co-delivery of siRNA and doxorubicin. Short polyamines were attached onto PCL blocks for siRNA complexation and doxorubicin was conjugated via pH-sensitive hydrazine linkage. Further, an integrin αvβ3-specific ligand (RGD4C) and a cell-penetrating peptide (TAT) were attached to the PEG block to confer the virus mimetic shell on these micelleplexes. These polymeric micelles simultaneously delivered DOX and MDR-1 siRNA to MDA-MB-435 cells and also showed *in vivo* targeting of αvβ3-positive tumors (Xiong & Lavasanifar, 2011). The Mahato group has synthesized polycarbonate-based copolymers with cationic pendant groups for co-delivery of miRNAs with small molecules. A cationic chain including tetraethylenepentamine, spermine or dimethyldipropylenetriamine was conjugated onto the hydrophobic polycarbonate block containing free carboxyl groups. A lipid moiety was also conjugated to the polymer backbone to enable micelleplex formation. This system efficiently delivered miR-205/gemcitabine combination to heterotopic xenograft model of pancreatic cancer (Mittal et al., 2014).

16.5 CONCLUSIONS AND FUTURE PERSPECTIVES

It is encouraging to recognize the tremendous progress that has been made towards developing multifaceted polymeric systems for gene delivery. Delivery systems that show efficacy *in vivo* exhibit great diversity in structure, size, chemistry, and overall approach to delivery problems. Although a variety of delivery systems have been developed in the laboratory, challenges remain in translating the full potential of RNAi to the clinic. Nanoparticles are rapidly emerging as systems of choice for *in vivo* delivery of siRNA/miRNA. These polymeric systems can efficiently protect nucleic acid from degradation by nucleases in an in vivo environment. It is difficult to design new delivery vehicles until a clear structure-activity relationship between transfection agents and transfection efficiency is established. The ideal nano-carrier system should be able to achieve long circulation time, biocompatibility, low immunogenicity, selective targeting and efficient penetration of barriers such as the vascular endothelium and blood-brain barrier, self-regulated release without serious side effects. Cationic polymers, both naturally occurring and synthetic ones with versatile functionalities have immense potential to evolve as nucleic acid delivery systems for clinical applications. The incorporation of surface PEGylation and cell-specific targeting ligands in the carriers may improve the pharmacokinetics, biodistribution, and selectivity of miRNA/siRNA therapeutics. Primary focus on screening the existing derivatives and formulation optimization is needed. Also, there is need to carry out more clinical studies as most of the studies to date have been conducted *in vitro* and in animal models. The effect of strategies for incorporation of nucleic acids into different vectors and their routes of administration on transfection efficiency of different vectors need to be investigated. Overall, an enormous amount of effort has to be done both in the synthesis of better polymers and the design of efficient ways to complex nucleic acids. A collaborative attempt by academicians and industry groups is further needed in order to develop industrially viable methods for preparation of efficient gene delivery systems.

REFERENCES

Amreddy, N., A. Babu, R. Muralidharan, et al. 2017. Polymeric nanoparticle-mediated gene delivery for lung cancer treatment. *Topics in Current Chemistry (Cham)*. 2: 35-017-0128-5.

Bowman, K. and K. W. Leong. 2006. Chitosan nanoparticles for oral drug and gene delivery. *International Journal of Nanomedicine*. 2:117–128.

Buyens, K., M. Meyer, E. Wagner, et al. 2010. Monitoring the disassembly of siRNA polyplexes in serum is crucial for predicting their biological efficacy. *Journal of Controlled Release*. 1:38–41.

Chitkara, D., A. Mittal, R.I. Mahato. 2015. miRNAs in pancreatic cancer: therapeutic potential, delivery challenges and strategies. *Advanced Drug Delivery Reviews*. 81:34–52.

Chitkara, D., S. Singh, and A. Mittal. 2016. Nanocarrier-based co-delivery of small molecules and siRNA/miRNA for treatment of cancer. *Therapeutic Delivery*. 4:245–255.

Cun, D., C. Foged, M. Yang, et al. 2010. Preparation and characterization of poly(dl-lactide -co-glycolide) nanoparticles for siRNA delivery. *International Journal of Pharmaceutics*. 1:70–75.

Dong, D., W. Gao, Y. Liu, et al. 2015. Therapeutic potential of targeted multifunctional nanocomplex co-delivery of siRNA and low-dose doxorubicin in breast cancer. *Cancer Letters*. 2:178–186.

Du, J., Y. Sun, Q. S. Shi, et al. 2012. Biodegradable nanoparticles of mPEG-PLGA-PLL triblock copolymers as novel non-viral vectors for improving siRNA delivery and gene silencing. *International Journal of Molecular Sciences*. 1:516–533.

Grayson, A. C., A. M. Doody, and D. Putnam. 2006. Biophysical and structural characterization of polyethylenimine-mediated siRNA delivery *in vitro*. *Pharmaceutical Research*. 8:1868–1876.

Guo, J., K. A. Fisher, R. Darcy, et al. 2010. Therapeutic targeting in the silent era: advances in non-viral siRNA delivery. *Molecular Biosystems*. 6(7):1143–1161.

Katas, H. and H. O. Alpar. 2006. Development and characterisation of chitosan nanoparticles for siRNA delivery. *Journal of Controlled Release*. (2):216–225.

Katas, H., E. Cevher, and H. O. Alpar. 2009. Preparation of polyethyleneimine incorporated poly(d,l-lactide-co-glycolide) nanoparticles by spontaneous emulsion diffusion method for small interfering RNA delivery. *International Journal of Pharmaceutics.* 2:144–154.

Kircheis, R., T. Blessing, S. Brunner, et al. 2001. Tumor targeting with surface-shielded ligand--polycation DNA complexes. *Journal of Controlled Release.* 72:165–170.

Klink, D., Q. C. Yu, M. C. Glick, et al. 2003. Lactosylated poly-l-lysine targets a potential lactose receptor in cystic fibrosis and non-cystic fibrosis airway epithelial cells. *Molecular Therapy.* 7:73–80.

Kloeckner, J., E. Wagner, and M. Ogris. 2006. Degradable gene carriers based on oligomerized polyamines. *European Journal of Pharmaceutical Sciences.* 29:414–425.

Lee, D. W., K. S. Yun, H. S. Ban, et al. 2009. Preparation and characterization of chitosan/polyguluronate nanoparticles for siRNA delivery. *Journal of Controlled Release: Official Journal of the Controlled Release Society.* 139:146–152.

Li, J., S. Xue, and Z.-W. Mao. 2016. Nanoparticle delivery systems for siRNA-based therapeutics. *Journal of Materials Chemistry B.* 4:6620–6639.

Liu, S., W. Huang, M. J. Jin, et al. 2016. Inhibition of murine breast cancer growth and metastasis by survivin-targeted siRNA using disulfide cross-linked linear PEI. *European Journal of Pharmaceutical Sciences.* 82:171–182.

Malathi, S., P. Nandhakumar, V. Pandiyan, et al. 2015. Novel PLGA-based nanoparticles for the oral delivery of insulin. *International Journal of Nanomedicine.* 10:2207–2218.

Merkel, O. M., D. Librizzi, A. Pfestroff, et al. 2009. Stability of siRNA polyplexes from poly(ethylenimine) and poly(ethylenimine)-g-poly(ethylene glycol) under *in vivo* conditions: effects on pharmacokinetics and biodistribution measured by fluorescence fluctuation spectroscopy and single photon emission computed tomography (SPECT) imaging. *Journal of Controlled Release.* 138:148–159.

Mittal, A., D. Chitkara, S. W. Behrman, et al. 2014. Efficacy of gemcitabine conjugated and miRNA-205 complexed micelles for treatment of advanced pancreatic cancer. *Biomaterials.* 35:7077–7087.

Mittal, A., D. Chitkara, N. Kumar, et al. 2007. Polymeric carriers for regional drug therapy. Smart Polymers: *Applications in Biotechnology and Biomedicine.* 2:359.

Nimesh, S., N. Gupta, and R. Chandra. 2011. Cationic polymer based nanocarriers for delivery of therapeutic nucleic acids. *Journal of Biomedical Nanotechnology.* 7:504–520.

Pereira, P., M. Barreira, J. A. Queiroz, et al. 2017. Smart micelleplexes as a new therapeutic approach for RNA delivery. *Expert Opinion on Drug Delivery.* 14:353–371.

Rao, D. D., Z. Wang, N. Senzer, et al. 2013. RNA interference and personalized cancer therapy. *Discovery Medicine.* 15:101–110.

Reineke, T. M. and M. E. Davis. 2003. Structural effects of carbohydrate-containing polycations on gene delivery. 2. Charge center type. *Bioconjugate Chemistry.* 14:255–261.

Sun, C. Y., S. Shen, C. F. Xu, et al. 2015. Tumor acidity-sensitive polymeric vector for active targeted siRNA delivery. *Journal of the American Chemical Society.* 137:15217–15224.

Vasir, J. K. and V. Labhasetwar. 2008. Preparation of biodegradable nanoparticles and their use in transfection. *CSH Protocols* 2008: pdb.prot4888.

Wang, G., X. Gao, G. Gu, et al. 2017. Polyethylene glycol-poly(epsilon-benzyloxycarbonyl-l-lysine)-conjugated VEGF siRNA for antiangiogenic gene therapy in hepatocellular carcinoma. *International Journal of Nanomedicine.* 12:3591–3603.

Wang, J., Z. Lu, M. G. Wientjes, et al. 2010. Delivery of siRNA therapeutics: barriers and carriers. *The AAPS Journal.* 12:492–503.

Xiong, X. B. and A. Lavasanifar. 2011. Traceable multifunctional micellar nanocarriers for cancer-targeted co-delivery of mdr-1 siRNA and doxorubicin. *ACS Nano.* 5(6): 5202–5213.

Yang, C., H. Li, X. Wang, et al. 2009. Cationic supramolecules consisting of oligoethylenimine-grafted alpha-cyclodextrins threaded on poly(ethylene oxide) for gene delivery. *Journal of Biomedical Materials Research*. Part A. 89:13–23.

Zheng, C., M. Zheng, P. Gong, et al. 2013. Polypeptide cationic micelles mediated co-delivery of docetaxel and siRNA for synergistic tumor therapy. *Biomaterials*. 34:3431–3438.

Zhu, C., S. Jung, S. Luo, et al. 2010. Co-delivery of siRNA and paclitaxel into cancer cells by biodegradable cationic micelles based on pdmaema-pcl-pdmaema triblock copolymers. *Biomaterials*. 31:2408–2416.

Zhu, L. and R. I. Mahato. 2010. Lipid and polymeric carrier-mediated nucleic acid delivery. *Expert Opinion on Drug Delivery*. 7:1209–1226.

17 Self-Assembling Programmable RNA Nanoparticles

From Design and Characterization to Use as an siRNA Delivery Platform

Brandon Roark, Morgan Chandler, Faye Walker, Lucia Milanova, Viktor Viglasky, Martin Panigaj, and Kirill A. Afonin

CONTENTS

17.1 MOTIVATION FOR NUCLEIC ACID NANOTECHNOLOGY

Nucleic acid nanotechnology benefits from the ability of RNA (or DNA) to form both canonical Watson-Crick (WC) *e.g.*, G-C and A-U (T for DNA) and non-canonical base pairings. Non- canonical base pairs, however, are mostly characteristic for RNAs and include 12 basic geometric families thus leading to a diverse set of structural motifs (Leontis and Westhof 1998, 2001, Leontis *et al.* 2002, Leontis and Westhof 2003). Currently, the existing repertoire of available RNA structural motifs allows for the rational design and construction of various programmable nucleic acid nanoparticles (NPs) that can be further used for a broad range of biomedical applications. Recent research shows that both DNA and RNA NPs have been used as nanoscaffolds and delivery vehicles for diverse functionalities used for the regulation of cellular function and gene expression, most notably gene silencing, utilizing small-interfering RNAs (siRNAs), Dicer substrates (DS), and micro RNAs (miRNAs), as well as functional aptamers, ribozymes, fluorophores, and small molecules (Rose *et al.* 2005, Guo 2010, Gross 2013, Afonin *et al.* 2013a, Mohri *et al.* 2014, Kumar *et al.* 2016, Jasinski *et al.* 2017). NPs comprised of RNA and RNA-DNA hybrids offer an advantage over DNA-only NPs in that they are able to utilize a wider array of interacting motifs and functional parts that are, in most cases, RNA-based. Also, the ability of RNA to adopt a more thermodynamically stable and compact A-form helix promotes the thermal stability of RNA NPs. The geometry of an RNA helix is characterized by a shallow minor and deep major groove and has its unique properties due to the C3'-endo orientation of the ribose sugar. The 2'-OH of the ribose also allows RNA to act as an efficient hydrogen bond donor or acceptor in tertiary interactions, thus expanding the possibilities of hydrogen bonding to the 12 geometric families mentioned before (Strobel and Doudna 1997, Tamura and Holbrook 2002, Guo *et al.* 2012).

The overall structures of RNA NPs are promoted by locally formed tertiary interactions and structural motifs such as internal and terminal loops, bulges, three- and four-way junctions, turns, and pseudoknots, just to name a few, that are often stabilized by divalent cations such as magnesium (Mg^{2+}). Reorganizing RNA motifs and testing the encoded sequences that exhibit specific assemblies with unique properties led to the birth of the field known as RNA nanotechnology (Guo *et al.* 1998, Chworos *et al.* 2004, Jaeger *et al.* 2001, Jaeger and Chworos 2006, Guo 2010, Geary *et al.* 2011, Grabow *et al.* 2011, Dabkowska *et al.* 2015, Jedrzejczyk *et al.* 2017). The design principles and *in vitro* synthesis of RNA NPs have evolved immensely through the years since the discovery of manipulating bacteriophage packaging RNA (pRNA) derived from the phi29 DNA packaging motor (Guo *et al.* 1998, Guo *et al.* 2012) and the pioneering concept of RNA architectonics (Jaeger and Leontis 2000, Jaeger *et al.* 2001, Chworos *et al.* 2004, Jaeger and Chworos 2006, Afonin and Leontis 2006, Afonin *et al.* 2012b).

Lately, functionalized RNA NPs have accrued great importance in the field of RNA nanotechnology and nanomedicine (Afonin *et al.* 2013a, Afonin *et al.* 2014a, Dao *et al.* 2015, El Tannir *et al.* 2015, Parlea *et al.* 2016, Jasinski *et al.* 2017). The design principles of RNA NPs rely on X-ray crystallography and nuclear magnetic resonance databases to align and connect tertiary motifs using helical linkages.

These motifs are then extracted and reassembled either manually or computationally with the goal of making a desired shape and eliminating any undesirable secondary folding and tertiary interactions (Jaeger *et al.* 2001, Jaeger and Chworos 2006, Afonin *et al.* 2013a). While the chemical synthesis only allows production of RNA sequences below 80 nucleotides (nts), enzymatic *in vitro* transcription can be used for the synthesis of much longer RNA strands. Subsequent characterization eventually leads to the design of primary optimized sequences for assembly of programmable NPs (Jaeger *et al.* 2001, Chworos *et al.* 2004, Nasalean *et al.* 2006). Novel RNA NPs have been designed to exhibit different functionalities such as ribozymes, therapeutic siRNAs, and RNA and DNA aptamers, which can be encoded directly into the optimized sequence without the interruption of the assembled RNA NP (Jaeger *et al.* 2001, Khaled *et al.* 2005, Liu *et al.* 2005, Jaeger and Chworos 2006, Saito and Inoue 2007, 2009, Shukla *et al.* 2011, Grabow *et al.* 2011, Afonin *et al.* 2011, Ohno *et al.* 2011, Shu *et al.* 2011, Abdelmawla *et al.* 2011, Haque *et al.* 2012, Guo *et al.* 2012, Shu *et al.* 2013, Qiu *et al.* 2013, Afonin *et al.* 2014b, Afonin *et al.* 2014c, Afonin *et al.* 2014d, Osada *et al.* 2014, Shu *et al.* 2014, Khisamutdinov *et al.* 2014, Feng *et al.* 2014, El Tannir *et al.* 2015, Dao *et al.* 2015, Afonin *et al.* 2014c, Zhang *et al.* 2015, Shu *et al.* 2015, Rychahou *et al.* 2015, Li *et al.* 2015, Lee *et al.* 2015, Binzel *et al.* 2016, Stewart *et al.* 2016, Parlea *et al.* 2016, Afonin *et al.* 2016, Bui *et al.* 2017, Halman *et al.* 2017). Three main topics are covered in this chapter. We will first describe the design principles of different NPs and common characterization techniques used in RNA nanotechnology. Then, we will briefly explore the molecular mechanism of RNA interference (RNAi) and discuss biological barriers that impede the use of RNAi nucleic acid nanotherapeutics. Finally, we will consider possible solutions to overcome these obstacles and the use of exogenously derived vehicular transport of nucleic acid NPs.

17.2 DESIGN AND PRODUCTION OF RNA NPs

RNAs that are used as scaffolds to enter the composition of RNA NPs are usually designed *de novo* using canonical WC and non-canonical interactions or by directly incorporating well-defined RNA motifs that may employ any type of interactions (Guo *et al.* 1998, Bindewald *et al.* 2008, Shapiro *et al.* 2008, Severcan *et al.* 2009, Guo 2010, Bindewald *et al.* 2011, Shu *et al.* 2013, Bindewald *et al.* 2016). The overall engineering principles of RNA NPs can be described by two main designing strategies, as exemplified by the nanorings and nanocubes shown in Figure 17.1 (Afonin *et al.* 2010, Grabow *et al.* 2011, Afonin *et al.* 2011, Afonin *et al.* 2014e). The first approach, used for nanoring design, relies on principles formulated for RNA architectonics, also called tectoRNAs, (Jaeger and Leontis 2000, Jaeger *et al.* 2001, Chworos *et al.* 2004, Jaeger and Chworos 2006, Nasalean *et al.* 2006). It requires the folding of individual rationally designed monomers by forming intramolecular hydrogen bonds and the formation of tertiary motifs (called "kissing loops" in the case of nanorings) with further stabilization of intermolecular interactions, that promote the assembly of nanoparticles in the presence of Mg^{2+}. A second strategy, represented by nanocubes, uses *in silico*-designed, single-stranded RNAs (ssRNAs)

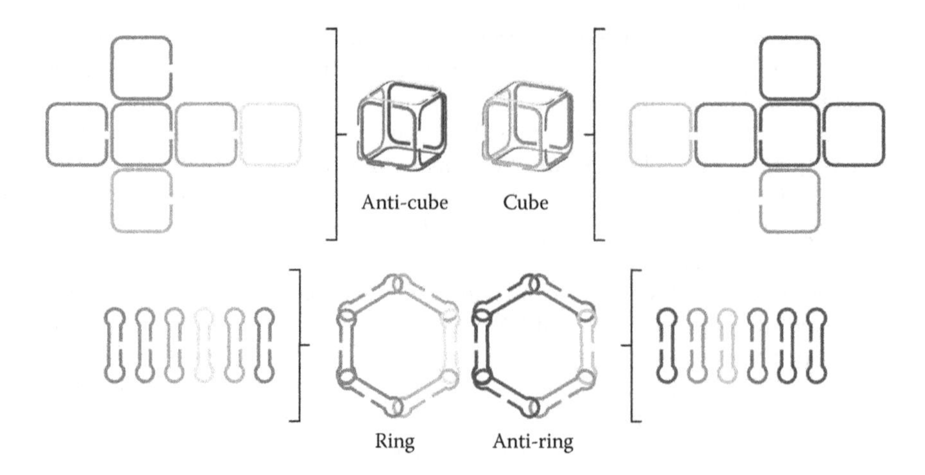

FIGURE 17.1 Designing strategies in RNA nanotechnology. RNA nanocubes and nanorings along with their reverse complement (anti) design principles. The nanorings were designed utilizing the RNA architectonics approach, where monomer pre-folding reveals the RNAI/IIi kissing loops that in the presence of Mg^{2+} further interact with angles of 120° between each monomer, forming the nanorings. Nanocubes were assembled computationally or manually using WC base pairing, while avoiding tertiary interactions. Both RNA NPs can be assembled in high yields *via* one-pot or co-transcriptional self-assembly. (From Grabow, W. W. et al., *Nano Lett.*, 11 (2):878–87, 2011; Afonin, K. A. et al., *Nat. Nanotechnol.*, 5 (9):676–82, 2010; Afonin, K. A. et al., *Nano Lett.*, 12 (10):5192–5, 2012a.)

that lack any intramolecular secondary structure and can only have intermolecular interactions with their cognate partner strands. These interactions can be stabilized by metal ions but do not necessarily require the presence of Mg^{2+}. Once designed, the RNA sequences can be tested in a variety of modeling programs (*i.e.*, RNA2D3D (Martinez *et al.* 2008), NanoTiler (Bindewald *et al.* 2008), and NUPACK (Zadeh *et al.* 2010)) to avoid any problems associated with RNA misfolding (Afonin *et al.* 2014a). More details on available design strategies can be found in our previously published work (Afonin *et al.* 2011, Afonin *et al.* 2014e, Afonin *et al.* 2014a). The reverse complements for the cube and ring strands can be assembled into anti-cube and anti-ring NPs allowing design of dynamic nanostructures. This concept was recently introduced by Halman *et al.* 2017. RNA NPs can be easily produced *in vitro* by following several simple techniques. Those steps can be briefly summarized as the following: PCR-amplified template DNAs are transcribed with T7 RNA polymerase and after gel purification and precipitation, recovered RNAs undergo melting and annealing steps in the presence of monovalent (Na^+/K^+) and divalent (Mg^{2+}) cations. Metal ions stabilize the negative charges of the RNA secondary and tertiary structures (Afonin *et al.* 2011, Afonin *et al.* 2013a). Some RNA tertiary structures, for example the three-way junction motif (3WJ) widely used in RNA nanotechnology, were reported to be resistant to denaturation even in the presence of 8 M urea and exhibit high melting temperatures (Guo 2010, Binzel *et al.* 2014). The 3WJs have been used to assemble a variety of RNA nanoparticle polygons with different geometries, diverse branched motifs, and various sizes using the assembly of loop-loop interactions between pRNA monomers that form tri-heptamers, palindromic sequences complementary to the "foot" region of the pRNA, and branched 3WJs (Shu *et al.* 2013). Reports on further expansion of pRNA-based polygons have changed the number of external and internal strands in the nano-assemblies by fine-tuning the angles of the pRNA 3WJ motif to 60°, 90°, and 108°, resulting in the assembly of triangle, square, pentagon, and tetrahedron-shaped NPs as shown in Figure 17.2 (Khisamutdinov *et al.* 2014, Jasinski *et al.* 2014, Li *et al.* 2016). Additionally, the hydrodynamic diameters of square-shaped RNA NPs can be fine-tuned simply by adjusting the RNA duplexes in the "foot" region of phi29 pRNA (Jasinski *et al.* 2014). Recently, the development of poly-uracils as helical linkers introduced a simple way to design RNA, DNA, and RNA/DNA polygons with fine-tunable chemical, thermodynamic, and immunological properties (Bui *et al.* 2017).

It is important to mention that most of the RNA NPs can also be produced co-transcriptionally by introducing the specifically designed DNA templates in the transcription mixture and incubating it for several hours at 37°C (Afonin *et al.* 2010). Additionally, we demonstrated that by simply changing the composition of the transcription buffer, the introduction of chemically modified nucleotides into the RNA nanoparticle composition becomes possible (Afonin *et al.* 2012a).

RNA nanoparticle functionalities have been effective in confirming assemblies, determining proximity using fluorescence-based assays, and observing regulation of intracellular functions. However, ensuring the conditional activation of embedded functionalities only in the presence of a specific stimulus increases the level of NP regulation. Recently, different RNA, DNA, and RNA/DNA hybrid NPs (Afonin *et al.* 2013b, Afonin *et al.* 2014b, Afonin *et al.* 2014d, El Tannir *et al.* 2015, Dao *et al.* 2015, Halman *et al.* 2017) with split functionalities have been adopted from the

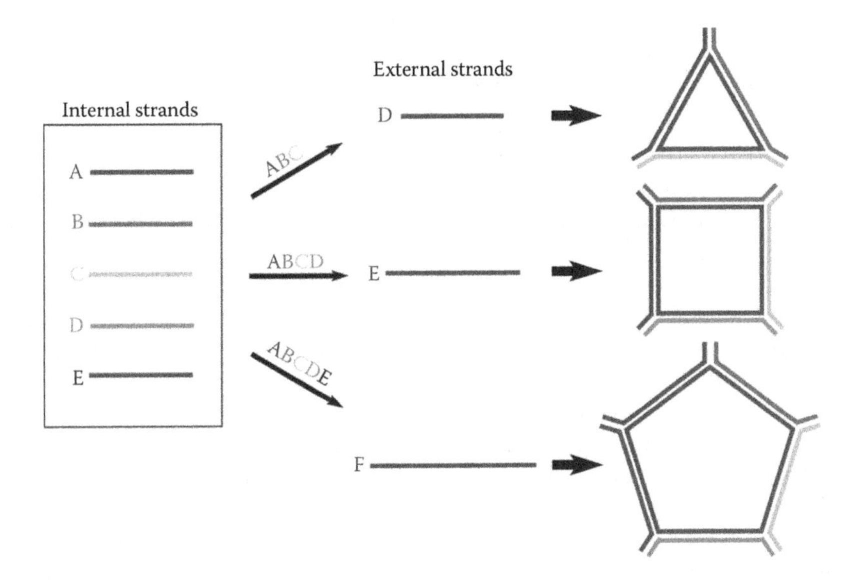

FIGURE 17.2 pRNA polygonal RNA NPs designed by extending the three-way junction angle from 60° to 108°. (Adapted from Jasinski, D. L. et al., *ACS Nano*, 8 (8):7620–9, 2014.)

concept of split-proteins (Shekhawat and Ghosh 2011). In this concept, each individual subunit is split and therefore non-functional, but can be conditionally reassembled with its cognate partner and become activated. Likewise, a response is activated when cognate nucleic acid strands recognize their complements. As a proof of concept, we initially studied RNA/DNA hybrids that can carry and activate different split functionalities. For that, we designed the individual inactive hybrids to interact with each other and re-associate through the interactions of short single-stranded DNA toeholds. The toehold interactions are driven thermodynamically by a difference in free energies between the hybrids and the resulting RNA and DNA duplexes (Afonin *et al.* 2013b). A novel expansion of these split non-functional subunits used the re-association of RNA and DNA NPs and their reverse complements anti-particles as recalled in Figure 17.1. This re-association allowed for the activation of transcription, conditional gene silencing, FRET, and assembly of functional aptamers (Halman *et al.* 2017). The development of the concept, design principles, and experimental verifications of dynamic RNA NPs have opened a whole new field of dynamic RNA nanotechnology (Afonin *et al.* 2013b, Afonin *et al.* 2014a, Afonin *et al.* 2014c, Afonin *et al.* 2014d, Dao *et al.* 2015, Afonin *et al.* 2015, Afonin *et al.* 2014c, Rogers *et al.* 2015, El Tannir *et al.* 2015, Roark *et al.* 2016, Bindewald *et al.* 2016, Groves *et al.* 2016, Afonin *et al.* 2016, Halman *et al.* 2017). This is only one piece of the puzzle involved in establishing nucleic acid NPs for the exogenous delivery and conditional activation of multiple split functionalities in cells. Next, we briefly explain the principles and utilization of common characterization techniques found in nucleic acid nanotechnology.

17.3 COMMON CHARACTERIZATION TECHNIQUES IN DNA/RNA NANOTECHNOLOGY

17.3.1 Electrophoretic Gel Mobility Shift Assays

One of the main characterization techniques used by research groups in the field of RNA nanotechnology is the electrophoretic mobility shift assay (EMSA). These assays are especially important for direct visualization and comparative analysis of various RNA and DNA NPs assembled under the same conditions. Electrophoresis is the movement and separation of macromolecules through a polymer (either polyacrylamide or agarose) under the influence of an electric field. Two main forces exerted on the macromolecule, electrostatic and frictional, are in equilibrium and take into account parameters such as viscosity and ionic radii, resulting in the electrophoretic velocity (Buszewski *et al.* 2013). The electrophoretic velocity is determined by the charge-to-size ratio of molecules, which influences the migration of the NPs as shown in Figure 17.3. The visualization of NPs is achieved by staining the gels with ethidium bromide or SYBR Green I/II (Afonin *et al.* 2010, Guo *et al.* 2012).

In experimental set up, the prepared polymer gel is placed into a buffer solution either in the presence of Mg^{2+} (for native-PAGE) or, if needed, with the addition of ethylenediaminetetraacetic acid (EDTA), which chelates Mg^{2+} in agarose gels.

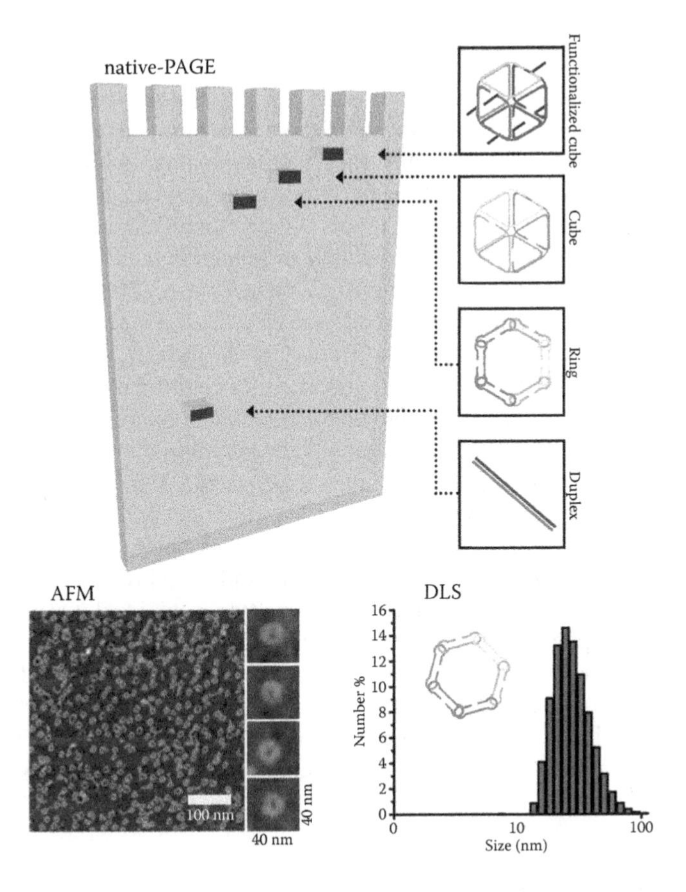

FIGURE 17.3 Characterization techniques including EMSA with corresponding sample bands, AFM, and DLS of RNA nanorings. The size-to-charge ratio of RNA/DNA NPs with and without functionalities is governed by electrostatic and frictional forces, which determine the extent of migration through a polymer matrix. This image depicts native-polyacrylamide gel electrophoresis, or native-PAGE. On native-PAGE, all RNA and DNA NPs remain assembled and appear as sharp bands. Hexameric RNA nanorings were imaged with tapping-mode AFM and DLS is shown of RNA nanorings with hydrodynamic radii histogram.

Typically, polyacrylamide with pore sizes between 5–100 nm is used to separate smaller nucleic acids. The size is determined by both concentration of acrylamide and acrylamide:bis-acrylamide ratio. Bis-acrylamide will change the cross-linking between the acrylamide subunits, optimizing pore size and resolution. The ratio usually range from 19:1 to 37.5:1 with the lower ratio more suitable for fibrous molecules and small RNA assemblies and the higher ratio applied for bulkier three-dimensional RNA NPs. In addition, the electrophoretic mobility through the native gel matrix can be influenced by a vast array of RNA secondary and tertiary conformations as shown in Figure 17.3. Thus, matrix composition, applied voltage, and running time all have strong influences on the resolution of RNA NP assemblies, making this technique widely modifiable.

17.3.2 Temperature Gradient Gel Electrophoresis and UV Melt

Temperature gradient gel electrophoresis (TGGE) is another type of electrophoretic mobility assay that utilizes a temperature gradient to denature the assembled or folded nucleic acids and to further determine their thermodynamic stabilities and measure their melting temperatures (T_m). T_m is the temperature at which half of the hydrogen bonding interactions within a nucleic acid NP are disrupted and is dependent on the size, charge, primary sequence, stability of secondary and tertiary motifs, and buffer conditions (Strobel and Doudna 1997, Víglaský *et al.* 2000, Makovets 2013). When TGGE is run in perpendicular mode (the electric field is aligned perpendicular to the temperature gradient), the sample migrates through the polymer matrix and non-linear/sigmoidal band patterns are indicative of conformational changes due to thermal unfolding of base pairs stabilizing the folded structure. The method allows detection of not only two-state but also multi-state conformational transitions.

This melting behavior can also be confirmed through UV-melting of nucleic acids by measuring the absorbance at ~260 nm of a nucleic acid NP as a function of temperature. As the temperature is ramped, bases unstack and are exposed to the solvent, causing an increase in the absorbance due to hyperchromacity (Darby *et al.* 2002). However, for complex assemblies, UV-melting is not ideal (Owczarzy 2005). For example, despite the experimental UV-melting profiles that produced similar T_ms compared to computationally determined T_ms of RNA-DNA hybrid nanocubes (Halman *et al.* 2017), complex 2D and 3D structures can result in overlapping and unresolved transitions. Therefore, the combination of gel assays (TGGE), UV-melt, and fluorescence assays can be the most accurate method of T_m assessments (Shu *et al.* 2011, Darby *et al.* 2002, Li *et al.* 2016).

17.3.3 Atomic Force Microscopy (AFM)

AFM has become a standard characterization instrument in the field of RNA nanotechnology. This technique allows for the visualization and determination of the relative sizes and shapes of RNA nanoparticle assemblies. For example, 3D RNA nanoprisms, RNA polygons, RNA/DNA nanocubes, and RNA dendrimers have all been characterized by AFM (Shapiro *et al.* Patent No. 2010148085, Grabow *et al.* 2011, Shu *et al.* 2013, Khisamutdinov *et al.* 2016, Bui *et al.* 2017, Halman *et al.* 2017, Sajja *et al.* 2018). AFM is used to image sample topography by measuring intermolecular forces at atomic resolution. This is completed by using a micro-machined cantilever probe with a sharp tip that is mounted to a piezoelectric actuator, which scans across a sample at either constant force or height and changes due to the contours of the sample (Basso *et al.* 1998, Shlyakhtenko *et al.* 2003, Jalili and Laxminarayana 2004, Lyubchenko *et al.* 2011, Shlyakhtenko *et al.* 2013). A sensitive photodetector is positioned to detect reflections off the end-point of the cantilever, providing differences in light intensity to either an upper or lower photodiode.

There are three modes—non-contact, contact, and intermittent contact (tapping)—of AFM operations, each with its own advantages and disadvantages. In non-contact mode, the cantilever is oscillated at its natural resonance frequency to detect attractive van der Waals forces between the tip and sample due to changes in amplitude, phase, or frequency; this allows topographical information to be extrapolated. This mode is suitable for studying elastic and soft biological materials and has improved resolution compared to contact AFM, which degrades the sample (Basso *et al.* 1998, Sebastian *et al.* 1999, Jalili and Laxminarayana 2004). A drawback is that the oscillating probe could become trapped in the fluid layer, decreasing the resolution because it can no longer measure the distance between tip and sample. In contact mode, repulsive forces between the sample and tip dominate due to the increasing interatomic distances. Surface topography is acquired in either constant height or constant force and since constant force is the most common, it will be discussed in detail. The forces between the sample and tip are kept constant by comparing the deflection on the cantilever to a set value. The difference in the signals is used to actuate the piezoelectric element by applying a voltage. This voltage is used as a measure of the surface topography as a function of the lateral position. This mode is not suitable for measuring biological materials because the sheer force due to contact between the sample and tip can cause surface alterations of the sample during scans (Lyubchenko *et al.* 2011). Thus, the most utilized operation for AFM is the tapping mode (Zhong *et al.* 1993). This mode uses characteristics from both non-contact and contact to determine surface topography. The piezoelectric actuator applies a force that causes the cantilever to vibrate/oscillate at a particular amplitude, usually in the range of 20–100 nm when the tip is not in contact with the sample. As the tip is allowed to scan over the sample, peaks and troughs on the surface cause a decrease or increase in the amplitude that is detected by the optical signal. The differences between measured and set references are applied to the piezoelectric actuator to measure the surface anomalies in the vertical direction as a function of the tip's lateral position, thus interpreting the surface topography (Jalili and Laxminarayana 2004).

17.3.4 Dynamic Light Scattering (DLS)

DLS, also referred to as photon correlation spectroscopy, is a hydrodynamic technique that measures the constant random Brownian motion exhibited by all atoms and molecules as a function of the intensity of scattered light from a laser placed at either 90° or 173° (backscattering) (Morrison *et al.* 1985, Pecora 2000, Malvern Instruments 2012). The signal's time correlation function is computed using an auto correlator. By relating the time correlation decay function to the scattering vector (q) as a function of time, the translational diffusion coefficient (D) is used in the Stokes-Einstein relation as shown in Table 17.1 to calculate the hydrodynamic diameter (D_H). The basis of the technique is through the understanding that small particles move more rapidly and cause faster decaying fluctuations, compared to larger particles that move more slowly. These differences are calculated by the auto correlator using different algorithms (*e.g.*, method of cumulants and inverse Laplace transform provided from the CONTIN software) (Provencher 1982, Pecora 2000). Sizes of non-spherical NPs are somewhat difficult to determine, but common workarounds involve calculating the size of a spherical particle that has the same translational diffusion coefficient as the non-spherical particle (ISO13321 1996). The hydrodynamic radii or diameter of several polygonal RNA NPs with and without functionalization have been determined that agree with many computational models (Afonin *et al.* 2010, Afonin *et al.* 2012a, Khisamutdinov *et al.* 2014, Afonin *et al.* 2014e, Jasinski *et al.* 2014).

TABLE 17.1
Brief Overview of the Chemical and Physical Properties and Relevant Calculated Parameters Measured by the Described Characterization Techniques

Common Characterization Techniques in Nucleic Acid Nanotechnology

Technique	Chemical and Physical Properties	Relevant Calculated Parameters
Electrophoretic mobility shift assays	Size: charge ratio influenced by electrostatic and frictional forces	Size, charge, and retention factor (Rf)
TGGE and UV-melt	Thermodynamic stability based on size, charge, primary sequence, and stability of 2°/3° motifs	Melting temperature T_m, standard enthalpy/entropy, and Gibb's free energy from vant' Hoff plots[1]
Atomic force microscopy	Attractive or repulsive intermolecular forces between sample and tip	Relative topography (size and shape)
Dynamic light scattering	Size related to light scattering events from Brownian motion of nanoparticles	Hydrodynamic diameter from the Stokes-Einstein equation[2]
Förester Resonance Energy transfer	Molecular dynamic interactions determined from dipole coupled donor and acceptor fluorophores	FRET efficiency and Förester radius (R_0)[3]

Sources: Marky, L. A., and K. J. Breslauer, *Biopolymers*, 26 (9):1601–20, 1987; Sahoo, H., *J. Photochem. Photobiol.* C, 12 (1):20–30, 2011 Sep.

[1] $\frac{1}{T_m} = \frac{2R}{\Delta H^\circ}\ln[C]_{total} + \frac{\Delta S^\circ - 2R\ln(6)}{\Delta H^\circ}$ where T_m is the melting temperature, ΔH° is the standard enthalpy, R is the gas constant, $[C]_{total}$ is the total concentration, and ΔS° is the standard entropy. Note that a plot of $\frac{1}{T_m}$ vs. $\ln[C]_{total}$ gives a linear relationship that can be used to determine the thermodynamic properties which can be used to determine the standard Gibb's free energy $\Delta G^\circ = \Delta H^\circ - T\Delta S^\circ$.

[2] $D_H = \frac{k_B T}{3\kappa\eta D}$ where D_H is the hydrodynamic diameter in nm, k_B is Boltzmann's constant, T is the absolute temperature, η is the viscosity of medium, and D is the translational diffusion coefficient.

[3] $R_0 = \left[8.79\times10^{-5}\left(\kappa^2\eta^{-4}Q_D J(\lambda)\right)\right]^{1/6}$ Å) where R_0 is the Förester radius which is determined from the input parameters. Experimentally $E_{FRET} = 1 - \left(\frac{I_{DA}}{I_D}\right)$ can be found from the intensity of the donor in the presence of acceptor (I_{DA}) and the intensity of just the donor.

17.3.5 Förster Resonance Energy Transfer (FRET)

FRET is used in RNA nanotechnology to measure molecular dynamic changes between NPs in solution and in cells as a function of time (Afonin *et al.* 2013b, Afonin *et al.* 2014d, Afonin *et al.* 2014c, Halman *et al.* 2017). FRET involves the radiationless energy transfer of an energetically excited fluorophore (donor) to another molecule (acceptor) through dipole coupling. The two-step process is dependent on the Förster radius (R_0) which is determined as 50% of FRET efficiency between a FRET pair. Particular considerations when performing FRET involve donor emission and acceptor absorbance overlap (overlap integral J(λ)), distance between FRET pairs (should fall between 1–10 nm), and the parallel alignment of dipole moments (Sahoo 2011).

An illustration of electrophoresis, AFM, and DLS can be found in Figure 17.3 and a summary of all the techniques can be found in Table 17.1. It is important to note that these are not the only techniques that researchers use to characterize nucleic acid NPs, but these are the most common. In-depth descriptions of the theories, mathematical derivations, and instrumentation can be found elsewhere (Cann 1996, Pecora 2000, Jalili and Laxminarayana 2004, Owczarzy 2005, Sahoo 2011, Binzel *et al.* 2014).

Next, we will explore the natural gene silencing pathway of RNA interference (RNAi), describe potential barriers for the application of exogenously introduced siRNA therapeutics into the clinic, and finally, combat these barriers with innovative approaches for delivery vehicles transporting nucleic acid NPs.

17.4 RNA NPs FOR DELIVERY AND ACTIVATION OF RNA INTERFERENCE

The machinery of RNAi has a huge therapeutic potential. Unlike typical small-molecule drugs, small-interfering RNAs (siRNAs) can be specifically designed to silence nearly any gene in the body by utilizing a pre-existing cellular pathway. Andrew Fire, Craig Mello, and their colleagues showed that double-stranded RNA (dsRNA) induced the RNAi silencing phenomenon more than either the sense or antisense RNAs alone (Fire *et al.* 1998, Liu and Paroo 2010). Briefly, the two main commonly used inducers of RNAi are micro RNAs (miRNAs) and siRNAs (Figure 17.4). The main difference is their individualized processing, but once loaded onto the RNA-induced silencing complex (RISC), the different pathways converge. Genome-encoded primary miRNA transcripts (pri-miRNA) contain a large number of hairpins that are processed by RNase III family enzymes (*e.g.*, Drosha). This produces 65–70 nt precursor miRNA (pre-miRNA) that is transported out of the nucleus into the cytoplasm by Exportin-5 (Lund and Dahlberg 2006, Saini *et al.* 2007, Kim and Kim 2007, Wilson and Doudna 2013). Once in the cytoplasm, pre-miRNAs are truncated by Dicer into dsRNAs 21–25 nts long (while exogenously introduced siRNAs are already of this size) with 2 nt-long 3'-side overhangs. The mature miRNA is then presented to one of the four Argonaute (Ago) family proteins found in humans to generate RNA-induced silencing complex, or RISC. The complex selects the strand with less thermodynamically stable base-pairing at its 5'-end as the guide strand (Tomari *et al.* 2004, Wilson and Doudna 2013, Wittrup and Lieberman 2015). Ago 2 is the only protein of the family that exhibits enzymatic slicer activity using complementarity through the guide (or antisense) strand to the messenger RNA (mRNA) target site. Otherwise, Ago 1, 3, and 4 induce translational repression leading to deadenylation of the messenger RNA poly A-tail and degradation; the former is not required for gene silencing (Jackson and Standart 2007, Wilson and Doudna 2013). RNAi has been shown to be a specific and robust mechanism of gene regulation, with vast implications for the future treatment of genetic maladies.

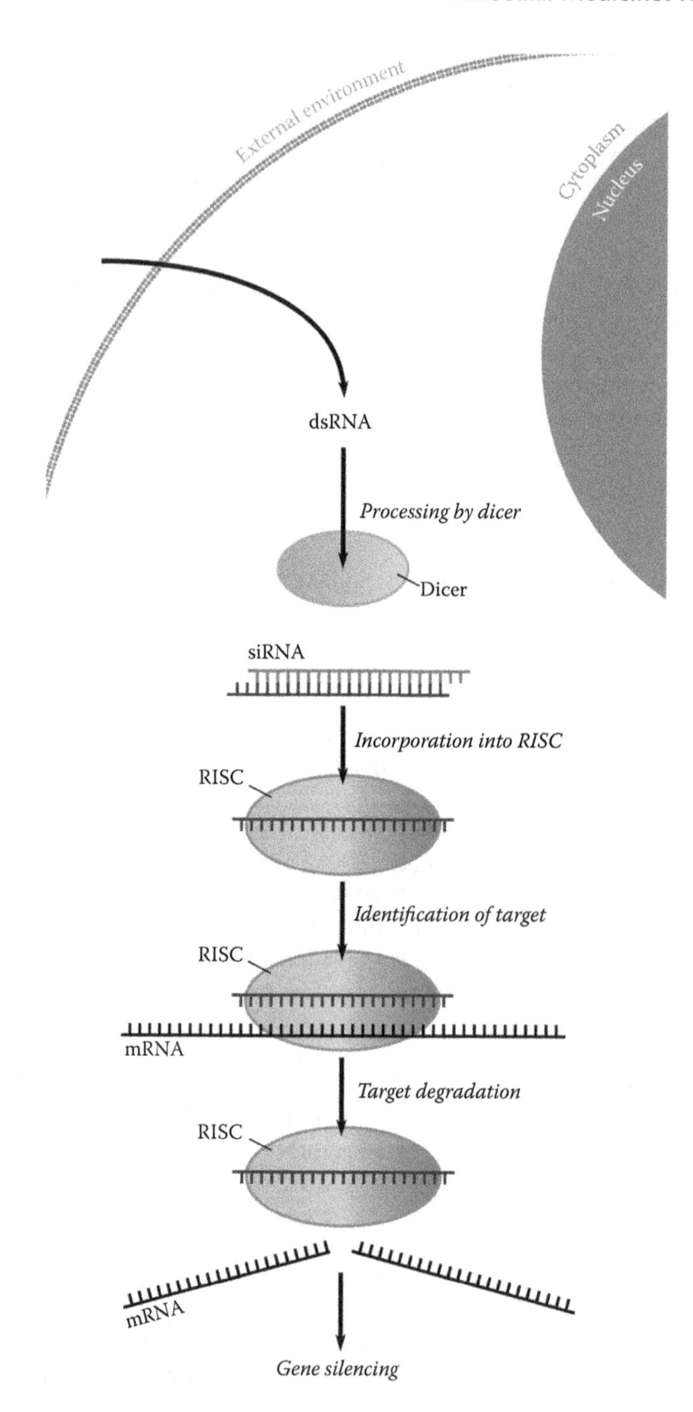

FIGURE 17.4 RNAi pathway. (Adapted from Wilson, R. C., and Je. A. Doudna, "Molecular Mechanisms of RNA Interference," in *Annual Review of Biophysics, Vol 42*, edited by K. A. Dill, 217–39, 2013; Wittrup, A., and J. Lieberman, *Nat. Rev. Genet.*, 16 (9):543–52, 2015.)

17.4.1 Obstacles Encountered in Nucleic Acid Therapeutics

Ideally, drug treatments are optimally safe and effective when activity of medication is restricted to diseased cells, and functional RNA NPs are no exception. Specific siRNA delivery to the affected tissue helps to minimize deregulation of the targeted gene in healthy cells, as well as to decrease the probability of off-target effects. The proper level of delivered siRNA must be high enough to silence expression of the target protein in respect to protein turnover, yet low enough to avoid saturation of the RNAi pathway. However, before taking effect in targeted cells, exogenous siRNAs have to overcome many biological barriers within a patient's anatomy and physiology. The large anionic and hydrophilic character of RNA molecules prevents them from diffusing across the negatively charged hydrophobic cell membranes, entering the cytoplasm, and activating RNAi machinery. The bloodstream is another challenging environment for nucleic acid NPs that may lead to disassembly, elimination from circulation by trans-vascular transport (extravasation), renal clearance, uptake by macrophages, or aggregation by serum proteins (Malek *et al.* 2009, Kettler *et al.* 2014, Xu and Wang 2015). Also, serum nucleases are an imminent threat for non-modified RNA after systemic administration since they rapidly degrade RNAs, thus decreasing their therapeutic potential.

There are two main mechanisms used to circumvent the difficulties of siRNA transport: chemical modification of the RNA backbone and incorporation of RNA into synthetic carriers. The former can greatly extend the half-life of siRNA in plasma, which is a brief ten minutes or less (El Tannir *et al.* 2015, Ku *et al.* 2016). Though sizeable resources in discovery, translational, and clinical research have been invested in applying gene silencing to humans, the development of clinically suitable delivery methods for siRNA still remains a bottleneck. This section will address the technological progress in siRNA technology to tackle the issue of effective and nontoxic siRNA delivery for therapeutic applications.

17.4.2 Chemical Modifications of RNA

Incorporation of chemically-modified nucleotides (Figure 17.5), such as 2'-fluoropyrimidines or 2'-methoxy nucleotides, into RNA building blocks significantly extends the lifetime of RNA NPs in the human blood serum (Afonin *et al.* 2012a). Another example is the 2'-F modified A9g aptamer that is stable in 100% human serum up to eight hours until slow degradation starts, and still, 50% of RNA survives after a week (Dassie *et al.* 2014). Modifications can also include locked nucleic acids (LNAs) with a methylene bridge from the 2' oxygen to the 4' carbon (Elmén *et al.* 2005, Mathe and Perigaud 2008, Stein *et al.* 2010); unlocked nucleic acids (UNAs), which lack a bond between the 2' and 3' carbons (Langkjær *et al.* 2009, Vaish *et al.* 2011); and 2'-aminoethyl (2'-AE), 2'-H (2'-deoxy), 2'-O-methyl (2'-OMe), 2'-O-methoxyethyl (2'-MOE), or 2'-O-guanidinopropyl (2'-GP) substitutions (Morrissey *et al.* 2005, Deleavey *et al.* 2010, Engels 2013). The small size of fluorine accommodates the folding and assembling of natural RNA while reducing immunogenicity (Liu *et al.* 2010, Afonin *et al.* 2012a, Afonin *et al.* 2014a). Larger, more complex modifications (*e.g.*, 2'-O-MOE and 2'-O-allyl)

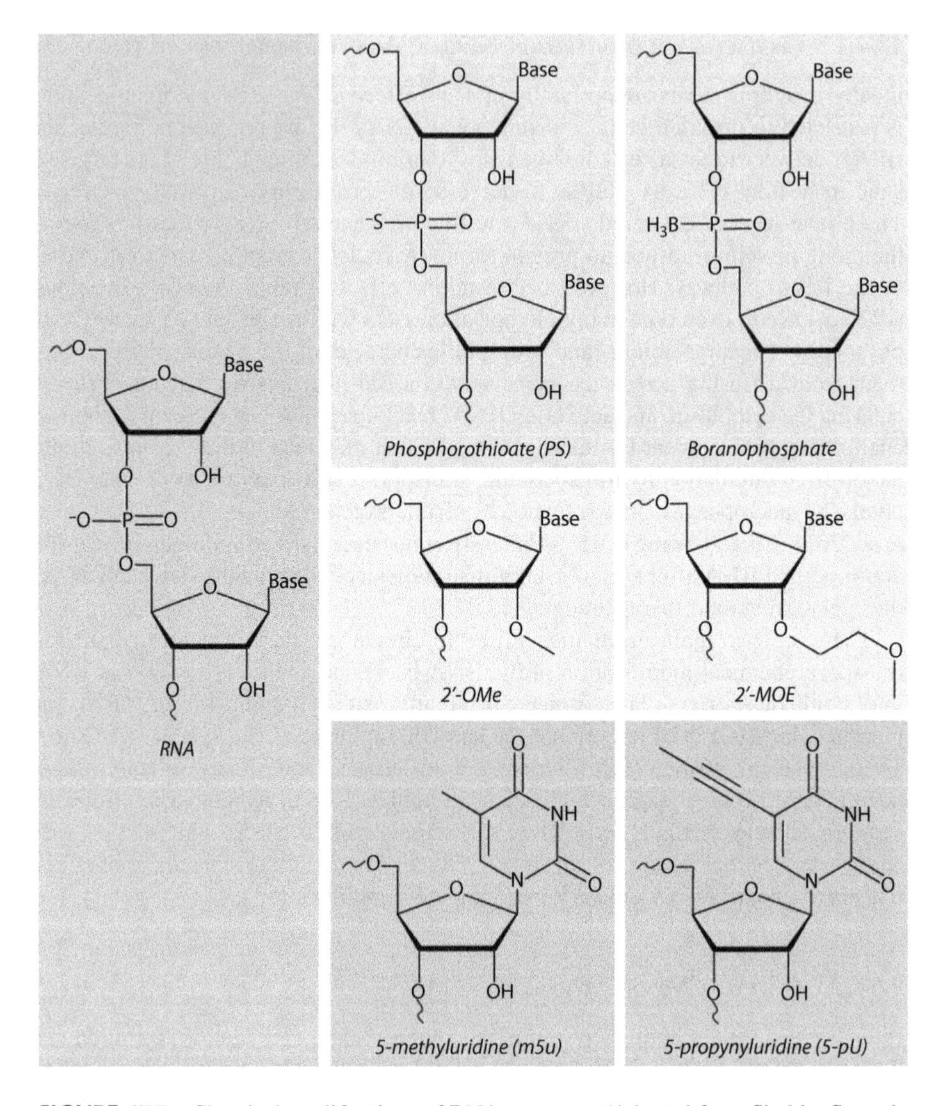

FIGURE 17.5 Chemical modifications of RNA structure. (Adapted from Shukla, S. et al., *Chem. Med. Chem.*, 5 (3):328–49, 2010.)

likewise reduce immune effects, but concomitantly decrease the thermodynamic stability of RNA and may alter the structure of RNA NPs. These sterically bulky groups distort the RNA helix, negatively affecting interactions with RISC and the target mRNA (Bramsen *et al.* 2009). While the outcome of RNAi is sensitive to artificial chemical groups in siRNAs, it seems that the replacement of ribose residues in siRNA molecules may stabilize and improve their *in vivo* activity in comparison to unmodified siRNAs (Morrissey *et al.* 2005). However, the increased potency is not observed in all cases (Layzer *et al.* 2004).

The most common chemical modifications may be at various positions of the ribose ring but approaches that modify the nucleobase itself are also used in gene silencing.

Using 5-methyluridine (m5U) and 5-propynyluridine (5-pU) in the guide strand led to increased serum stability, as well as increased RNAi activity in the case of the former (Terrazas and Kool 2009). Related substitutions of 2-thiouridine and pseudouridine enhanced thermostability and increased siRNA activity when incorporated at the 3' end of the guide strand (Kumar and Davis 1997). Extending this idea of increased stability, modifications with 2,4-difluorotoluene or 5-nitroindole at central regions of the passenger strand also augment silencing (Rand *et al.* 2005, Peacock *et al.* 2011). On the other side, replacement of standard nucleotides for their analogs and/or linking to heterogeneous molecules may interfere with the functional moieties of self-assembled nanostructures. For example, the sequence and associated structure of aptamers is crucial for their high-affinity binding to target molecules. Therefore, nucleotide substitution in aptamer sequences is likely to interfere with folding and may reduce target binding. This problem could be resolved by using a library of chemically modified oligonucleotides during the aptamer selection. Not only does this solution provide RNAs with enhanced nuclease resistance, it also permits the engineering of aptamers with novel conformations and target-binding surfaces otherwise unavailable in aptamers with natural nucleotides (Keefe and Cload 2008).

Chemical modifications in the phosphate linkage can also improve RNA stability while retaining silencing activity: these include boranophosphates, in which the phosphodiester oxygen is replaced with a borane moiety (Hall *et al.* 2006); phosphorothioates, wherein the bridging molecule is a sulfur rather than an oxygen (Braasch *et al.* 2003, Jahns *et al.* 2015); and phosphotriester groups in short interfering ribonucleic neutrals (siRNNs) (Meade *et al.* 2014). Additionally, morpholino nucleoside analogues (Zhang *et al.* 2009), peptide nucleic acids (PNAs) (Boffa *et al.* 2005, Stein *et al.* 2010), and polycarbamate nucleic acids (PCNAs) (Madhuri and Kumar 2010) have been applied to siRNAs for therapeutic purposes. Caution remains a critical factor when introducing modifications into either strand of an siRNA duplex. Selective insertion of modified bases into RNA duplexes can abolish immunostimulatory activity of the delivery vehicles (Robbins *et al.* 2009), but it could come at the cost of reduced RNAi behavior (Hanning *et al.* 2013). A substitution at one position of the guide strand may reduce the negative effects of off-target silencing (Jackson *et al.* 2006), whereas another may interfere entirely with the RNAi process (Leuschner *et al.* 2006).

17.4.3 Internalization

The size, shape, composition, and connectivity of self-assembled RNA and DNA NPs substantially differ from naturally occurring nucleic acids and influence the stability of NPs. It has been shown that while triangles tend to be fragile, tetrahedra resist deformation (Keum and Bermudez 2009) and remain substantially intact within cultured human embryonic kidney cells for at least 48 hours post-transfection (Walsh *et al.* 2011). Furthermore, it seems that survival of 2D and 3D DNA structures does not depend on the cell type (healthy vs. cancerous) (Mei *et al.* 2011). Also, it was reported that one structure (DNA 18, 24, or 35 helix bundles) is selectively sensitive to degradation by different endo- and exonucleases. However, authors caution that their observation holds true just for the structures they used. Interestingly, 2 ng of helix bundles were resistant to degradation for approximately 60 minutes in comparison with 65 ng of plasmid DNA that was destroyed within five minutes (Castro *et al.* 2011). Most NPs delivered to cells localize after 12 hours in the lysosome, where they get digested after 60 hours (Shen *et al.* 2012).

RNA NPs need more than nuclease-resistance to reach their target tissue. Depending on the size of the NP, they can be cleared through the kidneys (<5 nm) (Van de Water *et al.* 2006), or accumulated in the liver, spleen, and bone marrow (Ilium *et al.* 1982). Trans-vascular transport (extravasation) properties of capillary endothelia differ in various organs and tissues representing thus another barrier for efficient delivery of nanotherapeutics (Choi *et al.* 2007). The hydrodynamic properties dictated by size and shape determine the ability of NPs to marginate towards the wall of the blood vessels and subsequently escape from capillaries as well as promote the uptake by macrophages in the reticuloendothelial organs and endocytosis by target cells (Toy *et al.* 2014). The simplicity of nucleic acids' primary structure allows for the precise synthetic control over their size, geometry, and composition. However, despite this advantage, to the best of our knowledge, there is no study showing hemodynamic properties of DNA and RNA NPs.

After surmounting all other barriers, RNA and DNA therapeutic NPs must finally traverse the endosomal membrane to reach their site of action in the cytosol or nucleus. Endocytosis is the most common route of internalization for siRNA delivery systems (Kanasty *et al.* 2012). Even so, the actual mechanism for siRNA particle uptake can vary, and may take place through pathways separate from endocytosis (Au *et al.* 2016). To date, it is unclear how naked nucleic acid assemblies can pass through the membrane of the endosome. The trafficking of engulfed NPs, either bound to the target cell receptor or free, can be complex. Endocytosed cargo can end up degraded in lysosomes, or exported out of the cell, or, ideally, can escape from the endosome and successfully carry on its pre-programmed function. The nature of the receptor, cell type, and the physiological status of the cell determine intracellular transporting and processing of delivered DNA/RNA assemblies (Juliano *et al.* 2008, Roepstorff *et al.* 2009).

Many different cellular factors influence internalization of RNA and DNA NPs: *e.g.*, nanoparticle size, shape, charge, and distribution of ligands. From a medicinal point of view, the most critical factor for internalized nucleic acid-based nanotherapeutics is to avoid triggering innate cellular defense mechanisms. Every endocytosed nucleic acid therapeutic is subject to the cellular innate immune surveillance system, a line of defense that spreads from the cell membrane through the endosome to the cytoplasm. Innate defensive mechanisms have evolved to recognize viral or bacterial pathogens including their RNA and DNA (Whitehead et al. 2011). The structure and patterns of RNA sequences are recognized as foreign pathogens by two general classes of receptors: toll-like receptors (TLRs) and cytoplasmic receptors (Takeuchi and Akira 2010). Once these receptors detect and identify RNA as a pathogen, they induce a strong interferon response. The family of nucleic acid-sensing toll-like receptors, such as TLR3, TLR7, TLR8, and TLR9 can sample the endolysosomal lumen for nucleic acids. It seems that TLR3 recognizes dsRNA in a sequence-independent fashion, while TLR7 and TLR8 are assumed to sense ssRNAs with particularly U- and G- rich regions. TLR9 detects unmethylated cytosine-guanosine (CpG) DNA motifs of bacterial origin (Brencicova and Diebold 2013). Many non-immune cells (epithelial cells and fibroblasts) that do not express TLRs are still able to trigger an innate immune response that proves the existence of an additional surveillance system. This second line of defense contains RNA sensors that are localized in the cytosol and includes the RIG-I-like receptor (RLR) family, 2'-5'-oligoadenylate synthetase (OAS), and dsRNA-dependent protein kinase R (PKR).

When analyzing data from different sources, it should be taken into consideration that laboratory cell lines commonly used in RNAi studies and nonimmune cells do not seem to express TLR7 and TLR8 and do not respond to their ligands. Furthermore, TLR7/8 and 9 expressions in immune cells differ between mice and humans (Wu and Chen 2014). Interestingly, 2′-O-methyl modified RNAs suppress the immunostimulatory activity of siRNAs (Judge *et al.* 2006, Robbins *et al.* 2007). In a detailed safety evaluation of the anti-PSMA A9g RNA aptamer, Dassie *et al.* have not found either any apparent changes in general appearance and behavior of treated mice nor effects on blood cells or abnormalities in major organs (Dassie *et al.* 2014). While the cell specific A9g aptamer has not elicited any expression of inflammatory cytokines, interferons, or viral RNA recognition genes, control non-binding A9g.6 induced a slight increase in both IFN-β/γ levels. The difference between the A9g.6 and A9g is only one nucleotide that most probably led to the structural difference in an A9g.6 aptamer, and as an outcome, caused the increase of IFN-β/γ expression. Therefore, it is important to understand that any switching of modular parts in nucleic acid NPs may have immunostimulatory potential. In addition, if *in silico*-designed RNA and DNA scaffolds are used for the siRNA delivery, it is hard to extrapolate to what extent artificial nucleic acid-based motifs can activate the innate defense system that has primarily evolved to recognize invading viral and microbial agents.

Our own lab recently assessed how simply varying the ratio of RNA to DNA strands in cube NPs affects the immune response of primary human peripheral blood mononuclear cells. In tests with the pro-inflammatory cytokines and chemokines IL-1β, TNFα, IL-8, and MIP-1α, all cube constructs triggered the expression of IFNα, IL-8, and MIP-1α. However, cubes composed entirely of RNA strands were more potent immunostimulants than other tested particles. Overall, for this particular type of nanoparticle, we observed a correlation between the increasing number of RNA strands in cube particles and higher immunogenicity properties. The ability to change immunostimulatory potential based on the DNA vs. RNA strand ratio allows for different usages of cubic assemblies. DNA cubes (negligible induction of proinflammatory cytokines) become suitable for drug delivery, whereas RNA cubes with optimal immunomodulatory properties could be potentially used for vaccines and immunotherapy (Halman *et al.* 2017). All in all, we can exploit immunostimulatory properties of certain nucleic acid structures/sequence motifs to trigger innate immunity or boost defense mechanisms (Bourquin *et al.* 2007, Khisamutdinov *et al.* 2014, Radovic-Moreno *et al.* 2015, Johnson *et al.* 2017, Hong *et al.* 2018). On the other side, specific RNA motifs (chemically unmodified), *e.g.*, derived from viruses, suppress cellular immunity (Manokaran *et al.* 2015) and can be potentially used to block the innate immune system.

Interestingly, applications of modified nucleotides such as 5'-BrU, 5'-IU, or 5'-O-methyl modified RNA not only offer protection against serum nucleases, but also avoid triggering the signaling cascade leading to immune response (Chen *et al.* 2008, Watts *et al.* 2008). Amongst these non-natural nucleotides, 2'-methoxy modifications have shown great success in inhibiting TLR-mediated signaling without diminishing RNAi potency. The same effect can be observed after conjugation of free 5' and 3' ends with streptavidin-biotin, inverted thymidine, amine, polyethylene glycol (PEG), cholesterol, fatty acids, and related delivery materials. Similarly, capping of the 5' end of single- or double-stranded RNA may help RNA to pass through the cellular surveillance system. If synthetic individual strands of siRNA are transcribed *in vitro* by phage polymerases, they lack 5' capping with a methyl-guanosine nucleotide (Kato *et al.* 2006). As the result, human proteins generate an immune response under the assumption that this RNA is from a virus (Hornung *et al.* 2006, Pichlmair *et al.* 2006). Efforts to produce 5'-triphosphate siRNA (Zlatev *et al.* 2013, Thillier *et al.* 2015) or similar bioisosteres (Kenski *et al.* 2012, Zlatev *et al.* 2016) have demonstrated the therapeutic potential of synthetic siRNA molecules.

17.5 CARRIERS USED FOR DELIVERY OF FUNCTIONAL RNA NPs

Nano-formulations consist of functional RNA cargo that is either covalently linked or physically adsorbed onto a synthetic nanoparticle through non-covalent interactions. This association aims to improve biodistribution of RNA cargo and allow its transport from the blood circulation to the cytosol (and later to RISC machinery) (Geng *et al.* 2007, Petros and DeSimone 2010). Ways to improve internalization initially focused on conjugation with cholesterols (Juliano and Carver 2015), but have expanded in recent years to include materials such as small molecules, aptamers, lipids, peptides, proteins, and polymers (Xu and Wang 2015). These elements disguise and direct the NPs to target sites so that they are less prone to clearance at the system and organ levels (Hu *et al.* 2014a). Figure 17.6 outlines the overall strategy of incorporating siRNA into a general delivery vehicle introduced into a cell (Table 17.2).

17.5.1 Lipids and Polymers

Lipids (*i.e.*, fatty acids, cholesterol, bile acids, phospholipids, bolaamphiphiles and glycolipids) are the molecules commonly used for delivery of siRNAs and functional RNA NPs (Zimmermann *et al.* 2006, Sato *et al.* 2008, Afonin *et al.* 2008, Kim *et al.* 2013, Gupta *et al.* 2015, Sun *et al.* 2015, Kim *et al.* 2016). Lipid complexes tended to filter into the liver rather than the site of interest. This accumulation of delivery particles at non-target tissues and organs is typically seen as one of the hazards of the bloodstream but could actually help siRNAs reach the liver while avoiding elimination from circulation, uptake by macrophages, and aggregation by serum proteins (Malek *et al.* 2009, Kettler *et al.* 2014, Xu and Wang 2015). Lipoprotein particles in blood serum will bind to cholesterol-siRNA conjugates and deliver them to the liver for target knock-down *in vivo* (Lorenz *et al.* 2004, Soutschek *et al.* 2004, Wolfrum *et al.* 2007, Nishina *et al.* 2008). Apart from liver targeting, cholesterol has been chemically coupled to siRNAs to create delivery particles that reach the brain (Kuwahara *et al.* 2011) and thyroid tumors (Raouane *et al.* 2011). There are many choices of carbohydrates for conjugation, galactose being the most widely investigated. Several groups have targeted hepatocytes by conjugating siRNAs to one or more galactose or galactose-derived molecules, facilitating delivery to hepatic vessels (Aviñó *et al.* 2011, Nair *et al.* 2014). The N-acetylgalactosamine derivative

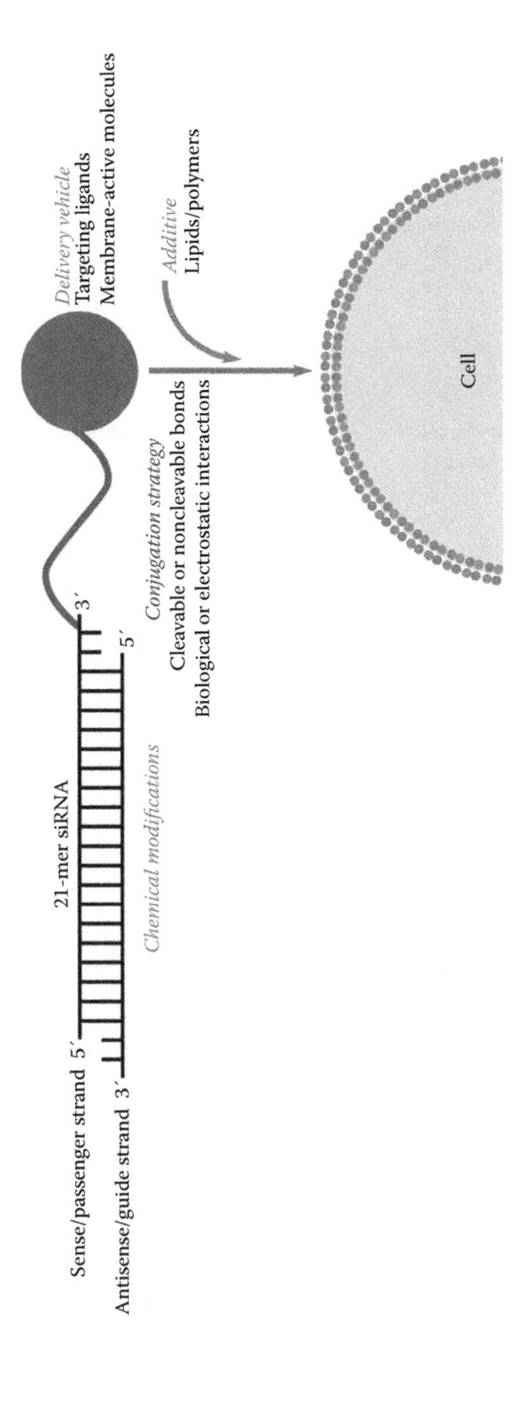

FIGURE 17.6 Overall conjugation strategy of incorporating siRNA for the delivery using a generalized delivery vehicle with additives that can assist the vehicle into the cell membrane. (Adapted from Lee, S. H. et al., *Adv. Drug Del. Rev.*, 104:78–92, 2016; Engels, J. W., *New Biotechnology*, 30 (3):302–7, 2013.)

TABLE 17.2
Brief Overview of the Different Types of Carriers for the Delivery of RNA NPs

	Chemical Components	RNA NP Delivery Approaches
Lipids	Fatty acids, cholesterol, bile acids, phospholipids, bolaamphiphiles, and glycolipids	Filtered to the liver. Avoid uptake of macrophages and aggregation of serum protein. Delivery through the blood brain barrier.
Polymers	Polyethyleneimine poly(lactic-co-glycolic acid)	Diversity, degrees of conjugation, and cleavability. Extends half-life of siRNAs. Combined with targeting ligands and incorporation of other functionalities.
Antibodies	Protein complexes for specific antigens	Cell specific internalization. Can trigger the immune response and expensive to manufacture.
Aptamers	Short single-stranded DNA or RNA oligonucleotides	Easily scalable and less expensive. Easily linked through sequence design. Target many different targets.
Peptides	Cell penetrating, targeting, and lytic peptides	Chemical linkages through disulfide, amide, and thioether/maleimide bonds.

of galactosamine (Baenziger and Maynard 1980) has been featured prominently in multivalent systems (Lee *et al.* 1983, Stokmaier *et al.* 2009, Severgnini *et al.* 2012), going so far as to produce positive results in clinical trials under Alnylam Pharmaceuticals (Nair *et al.* 2014).

In terms of variety, biocompatible polymers like PEG offer a huge diversity of structures, degrees of conjugation, cleavability, and molecular weights to extend the actions of siRNA. PEGylation greatly extends the half-life of siRNA in serum, but the steric bulk interferes with translocation (Lee *et al.* 2016). PEG can be combined with targeting ligands to overcome this problem, while bringing the benefit of incorporating different functional moieties into a single molecule. These include fusing PEG and PLGA with siRNA-phospholipids (Liu *et al.* 2014), linking PEG to siRNA to membrane-active cationic polymers (Rozema *et al.* 2007) in Dynamic PolyConjugates (DPCs) (Wong *et al.* 2012, Wooddell *et al.* 2013, Rozema *et al.* 2015), conjugating mannose-6-phosphate to PEG and siRNA (Zhu and Mahato 2010), or lactolysating PEG-siRNA conjugates (Oishi *et al.* 2005).

17.5.2 Targeted Delivery by Antibodies and Aptamers

Therapeutic delivery molecules are able to counter the obstacles of bodily clearance at the system and organ levels by incorporating antibodies, aptamers, or other ligands to guide them to their intended location *in vivo* (Davis *et al.* 2010). Conjugation to cell-selective ligands helps the nanoparticle undergo cellular internalization. Specific ligands include antibodies to the prostate-specific membrane antigen (PSMA) on prostate cancer cells (Patri *et al.* 2004) used in one of the first antibody-aptamer conjugates and a protamine-antibody fusion protein to target

HIV-1 (Song *et al.* 2005). Since these early examples, further developments have led to fusion proteins against leukocytes (Peer and Shimaoka 2009) and T-cells to suppress HIV infection. Antibodies do suffer from potential immunogenicity and manufacturing difficulties, which is why aptamers have arisen as alternative affinity reagents.

Aptamers are short, single-stranded DNA/RNA oligonucleotides that bind to selected targets with an affinity comparable to monoclonal antibodies. Aptamers are selected from libraries of up to ~10^{16} randomly synthesized sequences through a method termed Systematic Evolution of Ligands by EXponential enrichment (SELEX) (Ellington and Szostak 1990, Tuerk and Gold 1990). This *in vitro* concept selects nucleic acid aptamers with high affinities to the chosen target under desired conditions. Potential target-binding sequences are amplified in each cycle until one or few sequences form an entire population. Negative selection can be performed in parallel to eliminate nonspecifically binding. Moreover, the ability to control the preparation methods, oligonucleotide libraries, and post-process modifications helps to streamline and customize the production of aptamers. In comparison with antibodies, the synthesis of aptamers is an easily scalable and time-saving process. The advantage of aptamers is that any nucleic acid with therapeutic potential can be simply linked to an aptamer sequence, resulting in a bivalent molecule endowed with a targeting aptamer moiety and a functional RNA or DNA moiety. These targets could be transferrin receptors, integrins (Hussain *et al.* 2013), chemokine receptors (Wheeler *et al.* 2011), glycoproteins (Neff *et al.* 2011), antigen receptors (Herrmann *et al.* 2014, Rajagopalan *et al.* 2017), growth factor receptors (Zhang *et al.* 2014b), cyclic peptides (Alam *et al.* 2011), nucleolins (Lai *et al.* 2014), and many other medically relevant targets.

Since the first description of an aptamer-siRNA delivery approach in 2006 (McNamara *et al.* 2006), DNA- and RNA-based aptamers have been shown *in vitro* and *in vivo* to function as delivery agents for therapeutic cargo involving various gene-regulating oligonucleotides (siRNAs, miRNAs, decoys, antimiRs, deoxyribozymes, *etc.*) (Panigaj and Reiser 2015). For instance, one potential target is transferrin receptors (Mathew *et al.* 2015, Zhou and Rossi 2017). In addition to specific target binding, many aptamers elicit antagonistic or agonistic responses upon receptor recognition that, in combination with transported therapeutic payload, have a potential synergistic effect (Esposito *et al.* 2014).

While many different chimeric aptamers have been tested so far, only several reports have been published describing the aptamer-mediated delivery of 3D DNA or RNA assemblies. Nucleic acid NPs are very well-suited to serve as scaffolds equipped with multiple functional modules like aptamers and antisense oligonucleotides. They offer programmability and control over the size, geometry, and composition of the final nucleic acid system. Furthermore, other chemical components (such as fluorophores, chemotherapeutics, *etc.*) can be associated with NPs in addition to intrinsic nucleic acid-based functional moieties (Haque *et al.* 2012, Afonin *et al.* 2014b). Besides bearing multiple functionalities, the programmable 3D structure of nucleic acid scaffolds allows positioning of individual moieties in a rational manner to prevent steric hindrance, trigger FRET, and predefine their stoichiometry.

17.5.3 Peptides

Three types of peptides—cell-penetrating peptides (CPPs), targeting peptides, and lytic peptides—have been applied toward peptide-siRNA conjugates (Torchilin 2008). The conjugation linkage itself could be a disulfide, amide, thioether, or thiol-maleimide bond between a functionalized end of the siRNA and the carboxy terminus or cysteine residue of the peptide. The well-characterized HIV-1 Tat transactivator protein (Frankel and Pabo 1988) has shown promise for the delivery of RNAi (Chiu *et al.* 2004, Moschos *et al.* 2007, Meade and Dowdy 2007, Stewart *et al.* 2008, Nam *et al.* 2012, Meade *et al.* 2014), though control of the attachments and dosage are necessary to avoid inciting an immune response with CPPs. Penetratin (Nam *et al.* 2012) transportan (Muratovska and Eccles 2004), skin penetrating and cell entering (SPACE) (Hsu and Mitragotri 2011, Chen *et al.* 2014) are several other penetrating peptides that improve siRNA delivery. Targeting peptides bind to cell-surface receptors to similarly increase siRNA uptake, as in the case of conjugation with growth factors (Cesarone *et al.* 2007), hormones (Detzer *et al.* 2009), and folic acids (Zhang *et al.* 2014a). A third avenue for translocation into the cytosol is through lytic peptides. One major issue with siRNA-lytic peptide conjugations is toxicity, but masking systems can help the biologically-relevant cargo within the delivery molecule remain inactive until reaching its intended location *in vivo* (Davis et al. 2010). Masking peptides with dimethyl maleic anhydride (Meyer *et al.* 2009) or shielding NPs with PEG (Dohmen *et al.* 2012) offer a few options to disguise the conjugate molecules.

17.6 SELF-ASSEMBLED NUCLEIC ACID NPs AS POTENTIAL THERAPEUTICS

From its start, RNA nanotechnology has represented a rapidly developing and promising part of nanomedicine. However, despite the advancement in computational design and structure prediction that let us create variably shaped RNA scaffolds, relatively few works have attempted to target self-assembled RNA NPs into recipient cells. In one of the pioneering works on therapeutic RNA nanotechnology, Guo *et al.* (Guo *et al.* 2005) targeted CD4-overexpressing T-cells with the pRNA derived from the DNA-packaging motor of bacteriophage phi29. The subsequently forming dimer contained one monomer RNA that harbored the CD4-specific aptamer and a second monomer siRNA that was complementary to survivin mRNA. Cells treated with dimer RNA NPs had decreased viability that was attributed to targeted delivery, although neither survivin mRNA nor protein levels were investigated. The formation of dimers through loop-loop interactions enabled the use of double-stranded helical domains at 5′/3′ ends of pRNA in both interacting pRNAs independently. Exchanging end helical regions for an aptamer or siRNA does not affect pRNA structure. As such, Hu *et al.* created dimeric pRNAs composed of one monomer displaying the FB4 aptamer specific to the mouse transferrin receptor and a second monomer for siRNA targeting ICAM-1 mRNA (FRS-NPs).

ICAM-1 is overexpressed on brain endothelium in pathological conditions, where it facilitates the migration of detrimental leukocytes into the brain. However, blocking of leukocyte adhesion to ICAM-1 reduces the intensity of brain damage.

Treatment of TNF-α stimulated bEND5 cells, an *in vitro* inflammatory model, with FRS-NPs reversed the increase of ICAM-1 expression as well as lowered ICAM-1 levels under oxygen-glucose deprivation/reoxygenation conditions. Furthermore, FRS-NPs blocked adhesion of monocytes under both treatments (Hu *et al.* 2014b). Similarly, the loop-loop interlocking interaction of chimeric pRNA-gp120 with pRNA bearing anti-tat/rev siRNA has been shown to deliver fluorescently-labeled siRNA in a cell-type specific manner *in vitro* (Zhou et al. 2011).

Alternatively, aptamer targeting of nucleic acid NPs to cancer cells can be mediated through folic acid (FA). Cancer cells overexpress the folate receptor to maintain increased demands for FA. Recently, Zhou *et al.* chemically linked FA and anti-miRNA-21 LNA to the 3-way junction (3WJ) core from pRNA (FA-3WJ-LNA-miR21). MicroRNA-21 is upregulated in most types of cancers, where it silences tumor suppressor genes (Chan *et al.* 2005, Papagiannakopoulos and Kosik 2008). Therefore, it is an appealing target for treatment with antisense oligonucleotides. Lee *et al.* were able to systemically administer FA-3WJ-LNA-miR21, specifically and effectively targeting human glioblastoma cells in a murine xenograft model. No accumulation of NPs was observed in normal brain tissue. Although treatment with the FA-3WJ-LNA-miR21 improved survival of mice, the growth of the tumor was not suppressed even after repeated application of NPs (Lee *et al.* 2017).

While Guo *et al.* extensively explored therapeutic possibilities of RNA NPs inspired by natural structures of pRNAs, Afonin *et al.* took advantage of *in silico*-designed and experimentally verified hexameric RNA nanorings (Yingling and Shapiro 2007, Afonin *et al.* 2011, Afonin *et al.* 2014a). To demonstrate that nanorings can be targeted to specific cells *in vitro*, nanorings containing up to five copies of the J18 RNA aptamer binding to the human epidermal growth factor receptor (EGFR) were assembled (Figure 17.7). Nanorings with higher numbers of aptamers per nanoparticle provided a higher binding affinity to target cells. Unfortunately, any subsequent internalization and functional effect of siRNAs delivered by aptamer-nanorings were not studied. *In vivo* experiments were carried out in athymic nude mice bearing xenograft tumors expressing eGFP. Five days after administration of functionalized but non-targeted nanorings associated with bolaamphiphilic cationic carriers, *ex vivo* analysis of silencing efficiencies showed a significant decrease in eGFP fluorescence. In comparison to monomeric and dimeric aptamer-siRNA conjugates, multiple cell-specific aptamers allow for increasing rates of intracellular nanoparticle transport. Building upon this concept, Yoo *et al.* chained a DNA aptamer against Mucin 1 and anti-GFP or Bcl-2 mRNA siRNA conjugates in a multivalent, comb-like shape (Yoo *et al.* 2014). To date, several self-assembled DNA NPs have been also used as simple scaffolds to display antisense DNA oligonucleotides (Keum *et al.* 2011) or deliver siRNAs by binding to folate receptors (Lee *et al.* 2012).

17.7 CONCLUSIONS AND FUTURE PROSPECTS

In conclusion, RNA nanotechnology is an emerging field with much room for creativity, study, and innovation. Nanomedicine has already benefited from the *in-silico* design principles, cost effective synthesis methods, and controllability that underlie

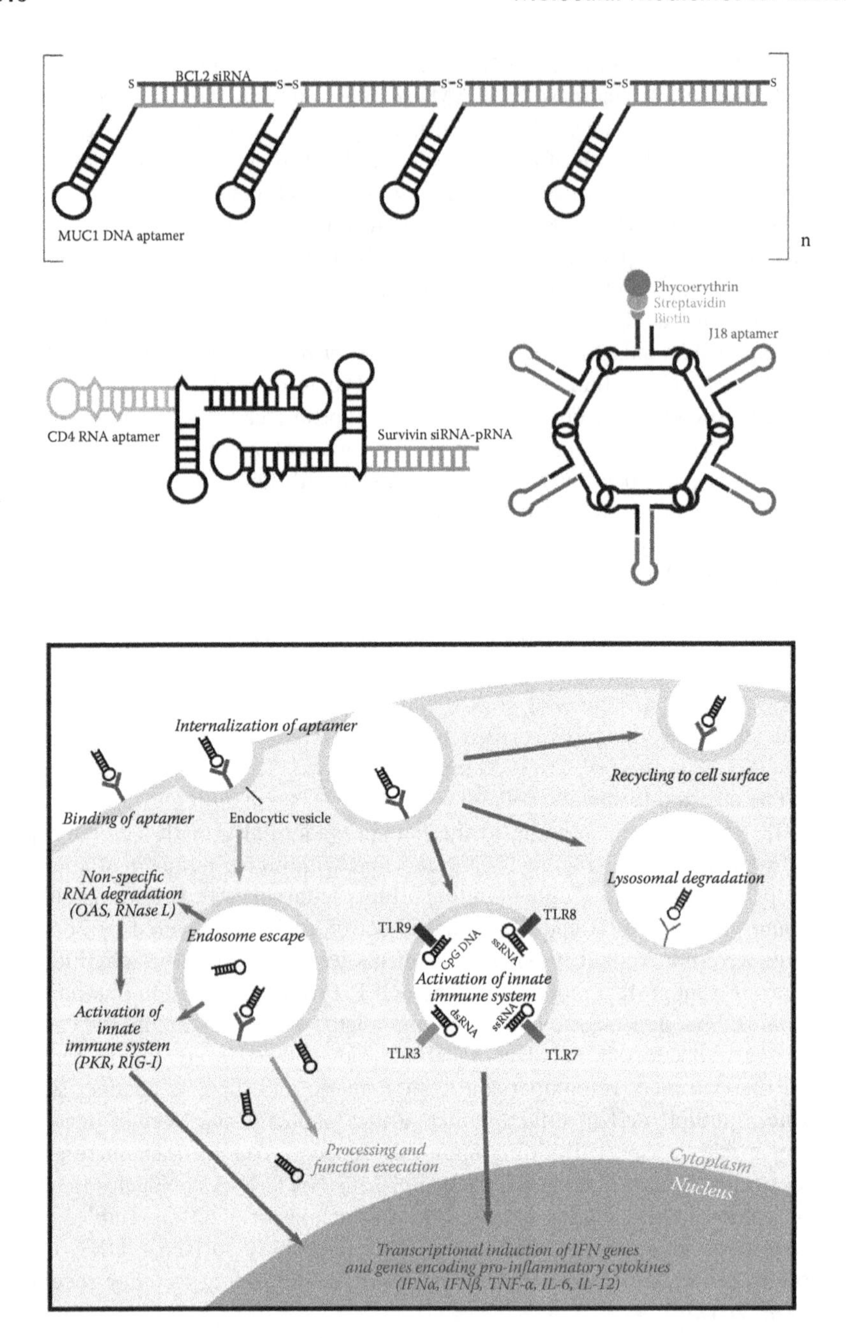

FIGURE 17.7 Aptamers and trafficking A: DNA aptamer against Mucin 1 conjugated to anti GFP or anti Bcl-2 siRNAs in a multivalent comb-like shape; antisense siRNA strands coupled through disulfide bonds are base paired to the DNA aptamer-siRNA sense strand conjugate. B: Dimeric pRNA aptamer formed by interaction of the left- and right-hand loops of the interlocking domain. C: Trafficking of internalized nucleic-acid therapeutics. (Adapted from Panigaj, M., and J. Reiser, *DNA and RNA Nanotechnology*, 2 (1):41–51, 2015.)

RNA delivery systems. A vast array of characterization techniques have contributed to our understanding of RNA nanoparticle assembly and the application of NP carriers for exogenous delivery of siRNAs. There have been immense developments of the RNA/DNA NP library in formulating precise structures, increasing nuclease stability, and reducing off-target effects. Still, the use of nucleic acid nanotherapeutics in modern medicine has many hurdles to overcome: systemic delivery remains a challenge due to biological barriers and elicitation of innate immune responses. Innovations from multiple laboratories worldwide have synergistically expanded the library of DNA/RNA nanoscaffolds with multiple functionalities. Advances in the clinical implementation of nucleic acid nanotherapeutics concentrate on effective delivery vehicles for encapsulating siRNAs. The current approach to use DNA/RNA nanoscaffolds, which offers precise control over the stoichiometry of cargo molecules and activation of multiple functionalities for the trafficking and delivery of siRNAs could be the future of personalized medicine. Lastly, we can trust that researchers in the interdisciplinary field of RNA nanotechnology will continue making strides to combat malevolent diseases through the use of functional RNAs.

REFERENCES

Abdelmawla, S., S. Guo, L. Zhang, et al. 2011. "Pharmacological characterization of chemically synthesized monomeric phi29 pRNA nanoparticles for systemic delivery." *Mol. Ther.* 19 (7):1312–22.

Afonin, K. A., E. Bindewald, A. J. Yaghoubian, et al. 2010. "In vitro assembly of cubic RNA-based scaffolds designed in silico." *Nat. Nanotechnol.* 5 (9):676–82.

Afonin, K. A., E. Bindewald, M. Kireeva, et al. 2015. "Computational and experimental studies of reassociating RNA/DNA hybrids containing split functionalities." *Methods Enzymol.* 553:313–34.

Afonin, K. A., D. J. Cieply, and N. B. Leontis. 2008. "Specific RNA self-assembly with minimal paranemic motifs." *J. Am. Chem. Soc.* 130 (1):93–102.

Afonin, K. A., R. Desai, M. Viard, et al. 2014d. "Co-transcriptional production of RNA-DNA hybrids for simultaneous release of multiple split functionalities." *Nucleic Acids Res.* 42 (3):2085–97.

Afonin, K. A., W. W. Grabow, F. M. Walker, et al. 2011. "Design and self-assembly of siRNA-functionalized RNA nanoparticles for use in automated nanomedicine." *Nat. Protoc.* 6 (12):2022–34.

Afonin, K. A., W. K. Kasprzak, E. Bindewald, et al. 2014a. "In silico design and enzymatic synthesis of functional RNA nanoparticles." *Acc. Chem. Res.* 47 (6):1731–41.

Afonin, K. A., W. Kasprzak, E. Bindewald, et al. 2014e. "Computational and experimental characterization of RNA cubic nanoscaffolds." *Methods (San Diego, Calif.)* 67 (2):256–65.

Afonin, K. A., M. Kireeva, W. W. Grabow, et al. 2012a. "Co-transcriptional assembly of chemically modified RNA nanoparticles functionalized with siRNAs." *Nano Lett.* 12 (10):5192–5.

Afonin, K. A., and N. B. Leontis. 2006. "Generating new specific RNA interaction interfaces using C-loops." *J. Am. Chem. Soc.* 128 (50):16131–7.

Afonin, K. A., Y. P. Lin, E. R. Calkins, et al. 2012b. "Attenuation of loop-receptor interactions with pseudoknot formation." *Nucleic Acids Res.* 40 (5):2168–80.

Afonin, K. A., B. Lindsay, and B. A. Shapiro. 2013a. "Engineered RNA nanodesigns for applications in RNA nanotechnology." *DNA and RNA Nanotechnology* 1 (1).

Afonin, K. A. M. Viard, I. Kagiampakis, et al. 2014c. "Triggering of RNA interference with RNA–RNA, RNA–DNA, and DNA–RNA nanoparticles." *ACS Nano* 9 (1):251–9.

Afonin, K. A., M. Viard, A. Y. Koyfman, et al. 2014b. "Multifunctional RNA nanoparticles." *Nano Lett.* 14 (10):5662–71.

Afonin, K. A., M. Viard, A. N. Martins, et al. 2013b. "Activation of different split functionalities on re-association of RNA-DNA hybrids." *Nat. Nanotechnol.* 8 (4):296–304.

Afonin, K. A., M. Viard, P. Tedbury, et al. 2016. "The use of minimal RNA toeholds to trigger the activation of multiple functionalities." *Nano Lett.* 16 (3):1746–53.

Alam, M. R., X. Ming, M. Fisher, et al. 2011. "Multivalent cyclic RGD conjugates for targeted delivery of small interfering RNA." *Bioconjugate Chem.* 22 (8):1673–81.

Au, J. L.-S., B. Z. Yeung, M. G. Wientjes, et al. 2016. "Delivery of cancer therapeutics to extracellular and intracellular targets: determinants, barriers, challenges and opportunities." *Adv. Drug Del. Rev.* 97:280–301.

Aviñó, A., S. M. Ocampo, R. Lucas, et al. 2011. "Synthesis and in vitro inhibition properties of siRNA conjugates carrying glucose and galactose with different presentations." *Mol. Divers.* 15 (3):751–7.

Baenziger, Ja. U., and Y. Maynard. 1980. "Human hepatic lectin. Physiochemical properties and specificity." *J. Biol. Chem.* 255 (10):4607–13.

Basso, M., L. Giarre, M. Dahleh, and I. Mezic. 1998. "Numerical analysis of complex dynamics in atomic force microscopes." Control Applications, 1998. Proceedings of the 1998 IEEE International Conference on.

Bindewald, E., K. Afonin, L. Jaeger, et al. 2011. "Multistrand RNA secondary structure prediction and nanostructure design including pseudoknots." *ACS Nano* 5 (12):9542–51.

Bindewald, E., K. A. Afonin, M. Viard, et al. 2016. "Multistrand structure prediction of nucleic acid assemblies and design of RNA switches." *Nano Lett.* 16 (3):1726–35.

Bindewald, E., C. Grunewald, B. Boyle, et al. 2008. "Computational strategies for the automated design of RNA nanoscale structures from building blocks using NanoTiler." *J. Mol. Graphics Modell.* 27 (3):299–308.

Binzel, D. W., E. F. Khisamutdinov, and P. Guo. 2014. "Entropy-driven one-step formation of Phi29 pRNA 3WJ from three RNA fragments." *Biochemistry* 53 (14):2221–31.

Binzel, D. W., Y. Shu, H. Li, et al. 2016. "Specific delivery of MiRNA for high efficient inhibition of prostate cancer by RNA nanotechnology." *Mol. Ther.* 24 (7):1267–77.

Boffa, L. C., G. Cutrona, M. Cilli, et al. 2005. "Therapeutically promising PNA complementary to a regulatory sequence for c-myc: pharmacokinetics in an animal model of human Burkitt's lymphoma." *Oligonucleotides* 15 (2):85–93.

Bourquin, C., L. Schmidt, V. Hornung, et al. 2007. "Immunostimulatory RNA oligonucleotides trigger an antigen-specific cytotoxic T-cell and IgG2a response." *Blood* 109 (7):2953–60.

Braasch, D. A., S. Jensen, Y. Liu, et al. 2003. "RNA interference in mammalian cells by chemically-modified RNA." *Biochemistry* 42 (26):7967–75.

Bramsen, J. B., M. B. Laursen, A. F. Nielsen, et al. 2009. "A large-scale chemical modification screen identifies design rules to generate siRNAs with high activity, high stability and low toxicity." *Nucleic Acids Res.* 37 (9):2867–81.

Brencicova, E., and S. S. Diebold. 2013. "Nucleic acids and endosomal pattern recognition: how to tell friend from foe?" *Front Cell Infect. Microbiol.* 3:37.

Bui, M. N., M. Brittany Johnson, M. Viard, et al. 2017. "Versatile RNA tetra-U helix linking motif as a toolkit for nucleic acid nanotechnology." *Nanomedicine.* 10.1016/j.nano.2016.12.018.

Buszewski, B., E. Dziubakiewicz, and M. Szumski, Eds. 2013. *Electromigration techniques: theory and practice.* Berlin and Heidelberg: Springer-Verlag.

Cann, J. R. 1996. "Theory and practice of gel electrophoresis of interacting macromolecules." *Anal. Biochem.* 237 (1):1–16.

Castro, C. E., F. Kilchherr, D.-N. Kim, et al. 2011. "A primer to scaffolded DNA origami." *Nat. Methods* 8 (3):221–9.

Cesarone, G., O. P. Edupuganti, C.-P. Chen, et al. 2007. "Insulin receptor substrate 1 knockdown in human MCF7 ER+ breast cancer cells by nuclease-resistant IRS1 siRNA conjugated to a disulfide-bridged D-peptide analogue of insulin-like growth factor 1." *Bioconjugate Chem.* 18 (6):1831–40.

Chan, J. A., A. M. Krichevsky, and K. S. Kosik. 2005. "MicroRNA-21 is an antiapoptotic factor in human glioblastoma cells." *Cancer Res.* 65 (14):6029–33.

Chen, M., M. Zakrewsky, V. Gupta, et al. 2014. "Topical delivery of siRNA into skin using SPACE-peptide carriers." *J. Controlled Release* 179:33–41.

Chen, P. Y., L. Weinmann, D. Gaidatzis, et al. 2008. "Strand-specific 5′-O-methylation of siRNA duplexes controls guide strand selection and targeting specificity." *RNA* 14 (2):263–74.

Chiu, Y.-L., A. Ali, C. Chu, et al. 2004. "Visualizing a correlation between siRNA localization, cellular uptake, and RNAi in living cells." *Chem. Biol.* 11 (8):1165–75.

Choi, H. S., W. Liu, P. Misra, et al. 2007. "Renal clearance of quantum dots." *Nat. Biotechnol.* 25 (10):1165–70.

Chworos, A., I. Severcan, A. Y. Koyfman, et al. 2004. "Building programmable jigsaw puzzles with RNA." *Science* 306 (5704):2068–72.

Dabkowska, A. P., A. Michanek, L. Jaeger, et al. 2015. "Assembly of RNA nanostructures on supported lipid bilayers." *Nanoscale* 7 (2):583–96.

Dao, B. N., M. Viard, A. N. Martins, et al. 2015. "Triggering RNAi with multifunctional RNA nanoparticles and their delivery." *DNA and RNA Nanotechnology* 1 (1):27–38.

Darby, R. A. J., M. Sollogoub, C. McKeen, et al. 2002. "High throughput measurement of duplex, triplex and quadruplex melting curves using molecular beacons and a LightCycler." *Nucleic Acids Res.* 30 (9):e39–e.

Dassie, J. P., L. I. Hernandez, G. S. Thomas, et al. 2014. "Targeted inhibition of prostate cancer metastases with an RNA aptamer to prostate-specific membrane antigen." *Mol. Ther.* 22 (11):1910–22.

Davis, M. E., J. E. Zuckerman, C. H. Choi, et al. 2010. "Evidence of RNAi in humans from systemically administered siRNA via targeted nanoparticles." *Nature* 464 (7291):1067–70.

Deleavey, G. F., J. K. Watts, T. Alain, et al. 2010. "Synergistic effects between analogs of DNA and RNA improve the potency of siRNA-mediated gene silencing." *Nucleic Acids Res.* 38 (13):4547–57.

Detzer, A., M. Overhoff, W. Wünsche, et al. 2009. "Increased RNAi is related to intracellular release of siRNA via a covalently attached signal peptide." *RNA* 15 (4):627–36.

Dohmen, C., D. Edinger, T. Fro'hlich, et al. 2012. "Nanosized multifunctional polyplexes for receptor-mediated siRNA delivery." *ACS Nano* 6 (6):5198–208.

El Tannir, Z., K. A. Afonin, and B. A. Shapiro. 2015. "RNA and DNA nanoparticles for triggering RNA interference." *RNA Dis.* 2 (3).

Ellington, A. D., and J. W. Szostak. 1990. "In vitro selection of RNA molecules that bind specific ligands." *Nature* 346 (6287):818.

Elmén, J., H. Thonberg, K. Ljungberg, et al. 2005. "Locked nucleic acid (LNA) mediated improvements in siRNA stability and functionality." *Nucleic Acids Res.* 33 (1):439–47.

Engels, J. W. 2013. "Gene silencing by chemically modified siRNAs." *New Biotechnology* 30 (3):302–7.

Esposito, C. L., S. Catuogno, and V. De Franciscis. 2014. "Aptamer-mediated selective delivery of short RNA therapeutics in cancer cells." *J. RNAi Gene Silencing* 10:500.

Feng, L., S. K. Li, H. Liu, et al. 2014. "Ocular delivery of pRNA nanoparticles: distribution and clearance after subconjunctival injection." *Pharm. Res.* 31 (4):1046–58.

Fire, An., S. Xu, M. K. Montgomery, et al. 1998. "Potent and specific genetic interference by double-stranded RNA in Caenorhabditis elegans." *Nature* 391 (6669):806–11.

Frankel, A. D., and C. O. Pabo. 1988. "Cellular uptake of the tat protein from human immunodeficiency virus." *Cell* 55 (6):1189–93.
Geary, C., A. Chworos, and L. Jaeger. 2011. "Promoting RNA helical stacking via A-minor junctions." *Nucleic Acids Res.* 39 (3):1066–80.
Geng, Y., P. Dalhaimer, S. Cai, et al. 2007. "Shape effects of filaments versus spherical particles in flow and drug delivery." *Nat. Nanotechnol.* 2 (4):249–55.
Grabow, W. W., P. Zakrevsky, K. A. Afonin, et al. 2011. "Self-assembling RNA nanorings based on RNAI/II inverse kissing complexes." *Nano Lett.* 11 (2):878–87.
Gross, M. 2013. "DNA nanotechnology gets real." *Current Biology: CB* 23 (3):R95–8.
Groves, B., Y. J. Chen, C. Zurla, et al. 2016. "Computing in mammalian cells with nucleic acid strand exchange." *Nat. Nanotechnol.* 11 (3):287–94.
Guo, P. 2010. "The emerging field of RNA nanotechnology." *Nat. Nanotechnol.* 5 (12):833–42.
Guo, P., F. Haque, B. Hallahan, et al. 2012. "Uniqueness, advantages, challenges, solutions, and perspectives in therapeutics applying RNA nanotechnology." *Nucleic Acid Ther.* 22 (4):226–45.
Guo, P., C. Zhang, C. Chen, et al. 1998. "Inter-RNA interaction of phage phi29 pRNA to form a hexameric complex for viral DNA transportation." *Mol. Cell* 2 (1):149–55.
Guo, S., N. Tschammer, S. Mohammed, et al. 2005. "Specific delivery of therapeutic RNAs to cancer cells via the dimerization mechanism of phi29 motor pRNA." *Hum. Gene Ther.* 16 (9):1097–109.
Gupta, K., K. A. Afonin, M. Viard, et al. 2015. "Bolaamphiphiles as carriers for siRNA delivery: from chemical syntheses to practical applications." *J. Control. Release* 213:142–51.
Hall, A. H. S., J. Wan, A. Spesock, et al. 2006. "High potency silencing by single-stranded boranophosphate siRNA." *Nucleic Acids Res.* 34 (9):2773–81.
Halman, J. R., E. Satterwhite, B. Roark, et al. 2017. "Functionally-interdependent shape-switching nanoparticles with controllable properties." *Nucleic Acids Res.* 45 (4):2210–20.
Hanning, J. E., H. K. Saini, M. J. Murray, et al. 2013. "Lack of correlation between predicted and actual off-target effects of short-interfering RNAs targeting the human papillomavirus type 16 E7 oncogene." *Br. J. Cancer* 108 (2):450–60.
Haque, F., D. Shu, Y. Shu, et al. 2012. "Ultrastable synergistic tetravalent RNA nanoparticles for targeting to cancers." *Nano Today* 7 (4):245–57.
Herrmann, A., S. J. Priceman, M. Kujawski, et al. 2014. "CTLA4 aptamer delivers STAT3 siRNA to tumor-associated and malignant T cells." *J. Clin. Invest.* 124 (7):2977–87.
Hong, E., J. R. Halman, A. B. Shah, et al. 2018. "Structure and composition define immunorecognition of nucleic acid nanoparticles." *Nano Lett.* 18 (7):4309–4321.
Hornung, V., J. Ellegast, S. Kim, et al. 2006. "5'-Triphosphate RNA is the ligand for RIG-I." *Science* 314 (5801):994–7.
Hsu, T., and S. Mitragotri. 2011. "Delivery of siRNA and other macromolecules into skin and cells using a peptide enhancer." *Proc. Natl. Acad. Sci. U S A* 108 (38):15816–21.
Hu, C.-M. J., R. H. Fang, B. T. Luk, et al. 2014a. "Polymeric nanotherapeutics: clinical development and advances in stealth functionalization strategies." *Nanoscale* 6 (1):65–75.
Hu, J., F. Xiao, X. Hao, et al. 2014b. "Inhibition of monocyte adhesion to brain-derived endothelial cells by dual functional RNA chimeras." *Mol Ther. Nucleic Acids* 3:e209.
Hussain, A. F., M. K. Tur, and S. Barth. 2013. "An aptamer–siRNA chimera silences the eukaryotic elongation factor 2 gene and induces apoptosis in cancers expressing αvβ3 integrin." *Nucleic Acid Ther.* 23 (3):203–12.
Ilium, L., S. S. Davis, C. G. Wilson, et al. 1982. "Blood clearance and organ deposition of intravenously administered colloidal particles. The effects of particle size, nature and shape." *Int. J. Pharm.* 12 (2-3):135–46.
ISO13321, IS. 1996. "Methods for determination of particle size distribution part 8: Photon correlation spectroscopy." *International Organisation for Standardisation (ISO).*

Jackson, A. L., J. Burchard, D. Leake, et al. 2006. "Position-specific chemical modification of siRNAs reduces "off-target" transcript silencing." *RNA* 12 (7):1197–205.

Jackson, R. J., and N. Standart. 2007. "How do microRNAs regulate gene expression." *Sci. STKE* 367 (re1).

Jaeger, L., and A. Chworos. 2006. "The architectonics of programmable RNA and DNA nanostructures." *Curr. Opin. Struct. Biol.* 16 (4):531–43.

Jaeger, L., and N. B. Leontis. 2000. "Tecto-RNA: one-dimensional self-assembly through tertiary interactions." *Angew. Chem. Int. Ed. Engl.* 39 (14):2521–4.

Jaeger, L., E. Westhof, and N. B. Leontis. 2001. "TectoRNA: modular assembly units for the construction of RNA nano-objects." *Nucleic Acids Res.* 29 (2):455–63.

Jahns, H., M. Roos, J. Imig, et al. 2015. "Stereochemical bias introduced during RNA synthesis modulates the activity of phosphorothioate siRNAs." *Nat. Commun.* 6.

Jalili, N., and K. Laxminarayana. 2004. "A review of atomic force microscopy imaging systems: application to molecular metrology and biological sciences." *Mechatronics* 14 (8):907–45.

Jasinski, D., F. Haque, D. W. Binzel, et al. 2017. "The advancement of the emerging field of RNA nanotechnology." *ACS Nano* 10.1021/acsnano.6b05737.

Jasinski, D. L., E. F. Khisamutdinov, Y. L. Lyubchenko, et al. 2014. "Physicochemically tunable polyfunctionalized RNA square architecture with fluorogenic and ribozymatic properties." *ACS Nano* 8 (8):7620–9.

Jedrzejczyk, D., E. Gendaszewska-Darmach, R. Pawlowska, et al. 2017. "Designing synthetic RNA for delivery by nanoparticles." *J. Phys. Condens. Matter* 29 (12):123001.

Johnson, M. B., J. R. Halman, E. Satterwhite, et al. 2017. "Programmable nucleic acid based polygons with controlled neuroimmunomodulatory properties for predictive QSAR modeling." *Small* 13 (42):1701255.

Judge, A. D., G. Bola, A. C. H. Lee, et al. 2006. "Design of noninflammatory synthetic siRNA mediating potent gene silencing in vivo." *Mol. Ther.* 13 (3):494–505.

Juliano, R. L., and K. Carver. 2015. "Cellular uptake and intracellular trafficking of oligonucleotides." *Adv. Drug Del. Rev.* 87:35–45.

Juliano, R., M. R. Alam, V. Dixit, et al. 2008. "Mechanisms and strategies for effective delivery of antisense and siRNA oligonucleotides." *Nucleic Acids Res.* 36 (12):4158–71.

Kanasty, R. L., K. A. Whitehead, A. J. Vegas, et al. 2012. "Action and reaction: the biological response to siRNA and its delivery vehicles." *Mol. Ther.* 20 (3):513–24.

Kato, H., O. Takeuchi, S. Sato, et al. 2006. "Differential roles of MDA5 and RIG-I helicases in the recognition of RNA viruses." *Nature* 441 (7089):101–5.

Keefe, A. D., and S. T. Cload. 2008. "SELEX with modified nucleotides." *Curr. Opin. Chem. Biol.* 12 (4):448–56.

Kenski, D. M., A. T. Willingham, H. J. Haringsma, et al. 2012. "In vivo activity and duration of short interfering RNAs containing a synthetic 5′-phosphate." *Nucleic Acid Ther.* 22 (2):90–5.

Kettler, K., K. Veltman, D. van de Meent, et al. 2014. "Cellular uptake of nanoparticles as determined by particle properties, experimental conditions, and cell type." *Environ. Toxicol. Chem.* 33 (3):481–92.

Keum, J.-W., J.-H. Ahn, and H. Bermudez. 2011. "Design, assembly, and activity of antisense DNA nanostructures." *Small (Weinheim an Der Bergstrasse, Germany)* 7 (24):3529–35.

Keum, J.-W., and H. Bermudez. 2009. "Enhanced resistance of DNA nanostructures to enzymatic digestion." *Chem. Commun.* (45):7036–8.

Khaled, A., S. Guo, F. Li, et al. 2005. "Controllable self-assembly of nanoparticles for specific delivery of multiple therapeutic molecules to cancer cells using RNA nanotechnology." *Nano Lett.* 5 (9):1797–808.

Khisamutdinov, E. F., D. L. Jasinski, H. Li, et al. 2016. "Fabrication of RNA 3D nanoprisms for loading and protection of small RNAs and model drugs." *Advanced Materials (Deerfield Beach, Fla.)* 28 (45):10079–87.

Khisamutdinov, E. F., H. Li, D. L. Jasinski, et al. 2014. "Enhancing immunomodulation on innate immunity by shape transition among RNA triangle, square and pentagon nanovehicles." *Nucleic Acids Res.*:gku516.

Kim, H. J., A. Kim, K. Miyata, et al. 2016. "Recent progress in development of siRNA delivery vehicles for cancer therapy." *Adv. Drug Deliv. Rev.* 104:61–77.

Kim, J. E., S. Yoon, B. R. Choi, et al. 2013. "Cleavage of BCR-ABL transcripts at the T315I point mutation by DNAzyme promotes apoptotic cell death in imatinib-resistant BCR-ABL leukemic cells." *Leukemia* 27 (8):1650–8.

Kim, Y.-K., and V. Narry Kim. 2007. "Processing of intronic microRNAs." *EMBO J.* 26 (3).

Ku, S. H., S. D. Jo, Y. K. Lee, et al. 2016. "Chemical and structural modifications of RNAi therapeutics." *Adv. Drug Del. Rev.* 104:16–28.

Kumar, R. K., and D. R. Davis. 1997. "Synthesis and studies on the effect of 2-thiouridine and 4-thiouridine on sugar conformation and RNA duplex stability." *Nucleic Acids Res.* 25 (6):1272–80.

Kumar, V., S. Palazzolo, S. Bayda, et al. 2016. "DNA nanotechnology for cancer therapy." *Theranostics* 6 (5):710–25.

Kuwahara, H., K. Nishina, K. Yoshida, et al. 2011. "Efficient in vivo delivery of siRNA into brain capillary endothelial cells along with endogenous lipoprotein." *Mol. Ther.* 19 (12):2213–21.

Lai, W.-Y., W.-Y. Wang, Y.-C. Chang, et al. 2014. "Synergistic inhibition of lung cancer cell invasion, tumor growth and angiogenesis using aptamer-siRNA chimeras." *Biomaterials* 35 (9):2905–14.

Langkjær, N., A. Pasternak, and J. Wengel. 2009. "UNA (unlocked nucleic acid): a flexible RNA mimic that allows engineering of nucleic acid duplex stability." *Biorg. Med. Chem.* 17 (15):5420–5.

Layzer, J. M., A. P. McCaffrey, A. K. Tanner, et al. 2004. "In vivo activity of nuclease-resistant siRNAs." *RNA* 10 (5):766–71.

Lee, H., A. K. Lytton-Jean, Y. Chen, et al. 2012. "Molecularly self-assembled nucleic acid nanoparticles for targeted in vivo siRNA delivery." *Nat. Nanotechnol.* 7 (6):389–93.

Lee, S. H., Y. Y. Kang, H.-E. Jang, et al. 2016. "Current preclinical small interfering RNA (siRNA)-based conjugate systems for RNA therapeutics." *Adv. Drug Del. Rev.* 104:78–92.

Lee, T. J., F. Haque, M. Vieweger, et al. 2015. "Functional assays for specific targeting and delivery of RNA nanoparticles to brain tumor." *Methods in Molecular Biology (Clifton, N.J.)* 1297:137–52.

Lee, T. J., J. Y. Yoo, D. Shu, et al. 2017. "RNA nanoparticle-based targeted therapy for glioblastoma through inhibition of oncogenic miR-21." *Mol. Ther.* 25 (7):1544–55.

Lee, Y. C., R. R. Townsend, M. R. Hardy, et al. 1983. "Binding of synthetic oligosaccharides to the hepatic Gal/GalNAc lectin. Dependence on fine structural features." *J. Biol. Chem.* 258 (1):199–202.

Leontis, N. B., and E. Westhof. 1998. "Conserved geometrical base-pairing patterns in RNA." *Q. Rev. Biophys.* 31 (4):399–455.

Leontis, N. B., and E. Westhof. 2001. "Geometric nomenclature and classification of RNA base pairs." *RNA* 7 (4):499–512.

Leontis, N. B., and E. Westhof. 2003. "Analysis of RNA motifs." *Curr. Opin. Struct. Biol.* 13 (3):300–8.

Leontis, N. B., J. Stombaugh, and E. Westhof. 2002. "The non-Watson–Crick base pairs and their associated isostericity matrices." *Nucleic Acids Res.* 30 (16):3497–531.

Leuschner, P. J. F., S. L. Ameres, S. Kueng, et al. 2006. "Cleavage of the siRNA passenger strand during RISC assembly in human cells." *EMBO Reports* 7 (3):314–20.

Li, H., T. Lee, T. Dziubla, et al. 2015. "RNA as a stable polymer to build controllable and defined nanostructures for material and biomedical applications." *Nano Today* 10 (5):631–55.

Li, H., K. Zhang, F. Pi, et al. 2016. "Controllable self-assembly of RNA tetrahedrons with precise shape and size for cancer targeting." *Adv. Mater.* 28 (34):7501–7.

Liu, H., Y. Li, A. Mozhi, et al. 2014. "SiRNA-phospholipid conjugates for gene and drug delivery in cancer treatment." *Biomaterials* 35 (24):6519–33.

Liu, J., S. Guo, M. Cinier, et al. 2010. "Fabrication of stable and RNase-resistant RNA nanoparticles active in gearing the nanomotors for viral DNA packaging." *ACS Nano* 5 (1):237–46.

Liu, Q., and Z. Paroo. 2010. "Biochemical principles of small RNA pathways." *Annu. Rev. Biochem.* 79:295–319.

Liu, Y., C. Lin, H. Li, et al. 2005. "Aptamer-directed self-assembly of protein arrays on a DNA nanostructure." *ANIE Angewandte Chemie International Edition* 44 (28):4333–8.

Lorenz, C., P. Hadwiger, M. John, et al. 2004. "Steroid and lipid conjugates of siRNAs to enhance cellular uptake and gene silencing in liver cells." *Biorg. Med. Chem. Lett.* 14 (19):4975–7.

Lund, E., and J. E. Dahlberg. 2006. "Substrate selectivity of exportin 5 and Dicer in the biogenesis of microRNAs." *Cold Spring Harbor Symp. Quant. Biol.* 71:59–66.

Lyubchenko, Y. L., L. S. Shlyakhtenko, and T. Ando. 2011. "Imaging of nucleic acids with atomic force microscopy." *Methods* 54 (2):274–83.

Madhuri, V., and V. A. Kumar. 2010. "Design, synthesis and DNA/RNA binding studies of nucleic acids comprising stereoregular and acyclic polycarbamate backbone: polycarbamate nucleic acids (PCNA)." *Org. Biomol. Chem.* 8 (16):3734–41.

Makovets, S. 2013. "DNA electrophoresis methods and protocols." 2013. *Methods Mol. Biol.* 1054:11–43.

Malek, A., O. Merkel, L. Fink, et al. 2009. "In vivo pharmacokinetics, tissue distribution and underlying mechanisms of various PEI (–PEG)/siRNA complexes." *Toxicol. Appl. Pharmacol.* 236 (1):97–108.

Malvern Instruments. 2012. "Dynamic light scattering: an introduction in 30 minutes." *Technical Note Malvern, MRK656-01*:1–8.

Manokaran, G., E. Finol, C. Wang, et al. 2015. "Dengue subgenomic RNA binds TRIM25 to inhibit interferon expression for epidemiological fitness." *Science (New York, N.Y.)* 350 (6257):217–21.

Marky, L. A., and K. J. Breslauer. 1987. "Calculating thermodynamic data for transitions of any molecularity from equilibrium melting curves." *Biopolymers* 26 (9):1601–20.

Martinez, H. M., J. V. Maizel Jr, and B. A. Shapiro. 2008. "RNA2D3D: a program for generating, viewing, and comparing 3-dimensional models of RNA." *J. Biomol. Struct. Dyn.* 25 (6):669–83.

Mathe, C., and C. Perigaud. 2008. "Recent approaches in the synthesis of conformationally restricted nucleoside analogues." *Eur. J. Org. Chem.* 2008 (9):1489–505.

Mathew, A., T. Maekawa, and D. Sakthikumar. 2015. "Aptamers in targeted nanotherapy." *Curr. Top Med. Chem.* 15 (12):1102–14.

McNamara, J. O., 2nd, E. R. Andrechek, Y. Wang, et al. 2006. "Cell type-specific delivery of siRNAs with aptamer-siRNA chimeras." *Nat. Biotechnol.* 24 (8):1005–15.

Meade, B. R., and S. F. Dowdy. 2007. "Exogenous siRNA delivery using peptide transduction domains/cell penetrating peptides." *Adv. Drug Del. Rev.* 59 (2):134–40.

Meade, B. R., K. Gogoi, A. S. Hamil, et al. 2014. "Efficient delivery of RNAi prodrugs containing reversible charge-neutralizing phosphotriester backbone modifications." *Nat. Biotechnol.* 32 (12):1256–61.

Mei, Q., X. Wei, F. Su, et al. 2011. "Stability of DNA origami nanoarrays in cell lysate." *Nano Lett.* 11 (4):1477–82.

Meyer, M., C. Dohmen, A. Philipp, et al. 2009. "Synthesis and biological evaluation of a bioresponsive and endosomolytic siRNA—polymer conjugate." *Mol. Pharm.* 6 (3):752–62.

Mohri, K., M. Nishikawa, Y. Takahashi, et al. 2014. "DNA nanotechnology-based development of delivery systems for bioactive compounds." *Eur. J. Pharm. Sci.* 58:26–33.

Morrison, I. D., E. F. Grabowski, and C. A. Herb. 1985. "Improved techniques for particle size determination by quasi-elastic light scattering." *Langmuir Langmuir* 1 (4):496–501.

Morrissey, D. V., K. Blanchard, L. Shaw, et al. 2005. "Activity of stabilized short interfering RNA in a mouse model of hepatitis B virus replication." *Hepatology (Baltimore, Md.)* 41 (6):1349–56.

Moschos, S. A., S. W. Jones, M. M. Perry, et al. 2007. "Lung delivery studies using siRNA conjugated to TAT (48–60) and penetratin reveal peptide induced reduction in gene expression and induction of innate immunity." *Bioconjugate Chem.* 18 (5):1450–9.

Muratovska, A., and M. R. Eccles. 2004. "Conjugate for efficient delivery of short interfering RNA (siRNA) into mammalian cells." *FEBS Lett.* 558 (1–3):63–8.

Nair, J. K., J. L. S. Willoughby, A. Chan, et al. 2014. "Multivalent N-acetylgalactosamine-conjugated siRNA localizes in hepatocytes and elicits robust RNAi-mediated gene silencing." *J. Am. Chem. Soc.* 136 (49):16958–61.

Nam, H. Y., J. Kim, S. W. Kim, et al. 2012. "Cell targeting peptide conjugation to siRNA polyplexes for effective gene silencing in cardiomyocytes." *Mol. Pharm.* 9 (5):1302–9.

Nasalean, L., S. Baudrey, N. B. Leontis, et al. 2006. "Controlling RNA self-assembly to form filaments." *Nucleic Acids Res.* 34 (5):1381–92.

Neff, C. P., J. Zhou, L. Remling, et al. 2011. "An aptamer-siRNA chimera suppresses HIV-1 viral loads and protects from helper CD4+ T cell decline in humanized mice." *Sci. Transl. Med.* 3 (66):66ra6–ra6.

Nishina, K., T. Unno, Y. Uno, et al. 2008. "Efficient in vivo delivery of siRNA to the liver by conjugation of α-tocopherol." *Mol. Ther.* 16 (4):734–40.

Ohno, H., T. Kobayashi, R. Kabata, et al. 2011. "Synthetic RNA-protein complex shaped like an equilateral triangle." *Nat. Nanotechnol.* 6 (2):116–20.

Oishi, M., Y. Nagasaki, K. Itaka, et al. 2005. "Lactosylated poly (ethylene glycol)-siRNA conjugate through acid-labile β-thiopropionate linkage to construct pH-sensitive polyion complex micelles achieving enhanced gene silencing in hepatoma cells." *J. Am. Chem. Soc.* 127 (6):1624–5.

Osada, E., Y. Suzuki, K. Hidaka, et al. 2014. "Engineering RNA-protein complexes with different shapes for imaging and therapeutic applications." *ACS Nano* 8 (8):8130–40

Owczarzy, R.. 2005. "Melting temperatures of nucleic acids: discrepancies in analysis." *Biophys. Chem.* 117 (3):207–15.

Panigaj, M., and J. Reiser. 2015. "Aptamer guided delivery of nucleic acid-based nanoparticles." *DNA and RNA Nanotechnology* 2 (1):41–51.

Papagiannakopoulos, T., and K. S. Kosik. 2008. "MicroRNAs: regulators of oncogenesis and stemness." *BMC Med.* 6 (1):15.

Parlea, L., A. Puri, W. Kasprzak, et al. 2016. "Cellular delivery of RNA nanoparticles." *ACS Comb. Sci.* 18 (9):527–47.

Patri, A. K., A. Myc, J. Beals, et al. 2004. "Synthesis and in vitro testing of J591 antibody–dendrimer conjugates for targeted prostate cancer therapy." *Bioconjugate Chem.* 15 (6):1174–81.

Peacock, H., A. Kannan, P. A. Beal, et al. 2011. "Chemical modification of siRNA bases to probe and enhance RNA interference." *J. Org. Chem.* 76 (18):7295–300.

Pecora, R. 2000. "Dynamic light scattering measurement of nanometer particles in liquids." *J. Nanopart. Res.* 2 (2):123–31.

Peer, D., and M. Shimaoka. 2009. "Systemic siRNA delivery to leukocyte-implicated diseases." *Cell Cycle* 8 (6):853–9.

Petros, R. A., and J. M. DeSimone. 2010. "Strategies in the design of nanoparticles for therapeutic applications." *Nat. Rev. Drug Discov.* 9 (8):615–27.

Pichlmair, A., O. Schulz, C. P. Tan, et al. 2006. "RIG-I-mediated antiviral responses to single-stranded RNA bearing 5'-phosphates." *Science* 314 (5801):997–1001.

Provencher, S. W. 1982. "CONTIN: a general purpose constrained regularization program for inverting noisy linear algebraic and integral equations." *Comput. Phys. Commun.* 27 (3):229–42.

Qiu, M., E. Khisamutdinov, Z. Zhao, et al. 2013. "RNA nanotechnology for computer design and in vivo computation." *Philos. Trans. A Math Phys. Eng. Sci.* 371 (2000): 20120310.

Radovic-Moreno, A. F., N. Chernyak, C. C. Mader, et al. 2015. "Immunomodulatory spherical nucleic acids." *Proc. Natl. Acad. Sci. U S A* 112 (13):3892–7.

Rajagopalan, A., A. Berezhnoy, B. Schrand, et al. 2017. "Aptamer-targeted attenuation of IL-2 signaling in CD8+ T cells enhances antitumor immunity." *Mol. Ther.* 25 (1):54–61.

Rand, T. A., S. Petersen, F. Du, et al. 2005. "Argonaute2 cleaves the anti-guide strand of siRNA during RISC activation." *Cell* 123 (4):621–9.

Raouane, M., D. Desmaele, M. Gilbert-Sirieix, et al. 2011. "Synthesis, characterization, and in vivo delivery of siRNA-squalene nanoparticles targeting fusion oncogene in papillary thyroid carcinoma." *J. Med. Chem.* 54 (12):4067–76.

Roark, B. K., L. A. Tan, A. Ivanina, et al. 2016. "Fluorescence blinking as an output signal for biosensing." *ACS Sensors* 1 (11):1295–300.

Robbins, M., A. Judge, L. Liang, et al. 2007. "2'-O-methyl-modified RNAs act as TLR7 antagonists." *Mol. Ther.* 15 (9):1663–9.

Robbins, M., A. Judge, and I. MacLachlan. 2009. "siRNA and innate immunity." *Oligonucleotides* 19 (2):89–102.

Roepstorff, K., M. V. Grandal, L. Henriksen, et al. 2009. "Differential effects of EGFR ligands on endocytic sorting of the receptor." *Traffic* 10 (8):1115–27.

Rogers, T. A., G. E. Andrews, L. Jaeger, et al. 2015. "Fluorescent monitoring of RNA assembly and processing using the split-spinach aptamer." *ACS Synth. Biol.* 4 (2):162–6.

Rose, S. D., D.-H. Kim, M. Amarzguioui, et al. 2005. "Functional polarity is introduced by Dicer processing of short substrate RNAs." *Nucleic Acids Res.* 33 (13):4140–56.

Rozema, D. B., A. V. Blokhin, et al. 2015. "Protease-triggered siRNA delivery vehicles." *J. Controlled Release* 209:57–66.

Rozema, D. B., D. L. Lewis, D. H. Wakefield, et al. 2007. "Dynamic polyconjugates for targeted in vivo delivery of siRNA to hepatocytes." *Proc. Natl. Acad. Sci. U S A* 104 (32):12982–7.

Rychahou, P., F. Haque, Y. Shu, et al. 2015. "Delivery of RNA nanoparticles into colorectal cancer metastases following systemic administration." *ACS Nano* 9 (2):1108–16.

Sahoo, H. 2011. "Förster resonance energy transfer–A spectroscopic nanoruler: Principle and applications." *J. Photochem. Photobiol. C*, 12 (1):20–30.

Saini, H. K., S. Griffiths-Jones, and A. J. Enright. 2007. "Genomic analysis of human microRNA transcripts." *Proc. Natl. Acad. Sci. U S A* 104 (45):17719–24.

Saito, H., and T. Inoue. 2007. "RNA and RNP as new molecular parts in synthetic biology." *J. Biotechnol.* 132 (1):1–7.

Saito, H., and T. Inoue. 2009. "Synthetic biology with RNA motifs." *Int. J. Biochem. Cell Biol.* 41 (2):398–404.

Sajja, S., M. Chandler, D. Fedorov, et al. 2018. "Dynamic behavior of RNA nanoparticles analyzed by AFM on a mica/air interface." *Langmuir.* doi: 10.1021/acs.langmuir.8b00105.

Sato, Y., K. Murase, J. Kato, et al. 2008. "Resolution of liver cirrhosis using vitamin A–coupled liposomes to deliver siRNA against a collagen-specific chaperone." *Nat. Biotechnol.* 26 (4):431–42.

Sebastian, A., M. V. Salapaka, D. J. Chen, et al. 1999. "Harmonic analysis based modeling of tapping-mode AFM." American Control Conference, 1999. Proceedings of the 1999.

Severcan, I., C. Geary, L. Jaeger, et al. 2009. "Computational and experimental RNA nanoparticle design." In *Automation in Genomics and Proteomics: An Engineering Case-Based Approach*, edited by G. Alterovitz, M. Ramoni and R. Benson, 193–220. Hoboken, NJ: Wiley Publishing.

Severgnini, M., J. Sherman, A. Sehgal, et al. 2012. "A rapid two-step method for isolation of functional primary mouse hepatocytes: cell characterization and asialoglycoprotein receptor based assay development." *Cytotechnology* 64 (2):187–95.

Shapiro, B. A., E. Bindewald, W. Kasprzak, et al. 2008. "Protocols for the In silico Design of RNA Nanostructures." In *Nanostructure Design Methods and Protocols*, edited by E. Gazit and R. Nussinov, 93–115. Totowa, NJ: Humana Press.

Shapiro, B. A., Y. Yingling, E. Bindewald, et al. Patent No. 2010148085. "RNA Nanoparticles and Methods of Use."

Shekhawat, S. S., and I. Ghosh. 2011. "Split-protein systems: beyond binary protein–protein interactions." *Curr. Opin. Chem. Biol.* 15 (6):789–97.

Shen, X., Q. Jiang, J. Wang, et al. 2012. "Visualization of the intracellular location and stability of DNA origami with a label-free fluorescent probe." *Chem. Commun. (Camb.)* 48 (92):11301–3.

Shlyakhtenko, L. S., A. A. Gall, A. Filonov, et al. 2003. "Silatrane-based surface chemistry for immobilization of DNA, protein-DNA complexes and other biological materials." *Ultramicroscopy* 97 (1–4):279–87.

Shlyakhtenko, L. S., A. A. Gall, and Y. L. Lyubchenko. 2013. "Mica functionalization for imaging of DNA and protein-DNA complexes with atomic force microscopy." *Methods Mol. Biol.* 931:295–312.

Shu, D., E. F. Khisamutdinov, L. Zhang, et al. 2014. "Programmable folding of fusion RNA in vivo and in vitro driven by pRNA 3WJ motif of phi29 DNA packaging motor." *Nucleic Acids Res.* 42 (2):e10.

Shu, D., Y. Shu, F. Haque, et al. 2011. "Thermodynamically stable RNA three-way junction for constructing multifunctional nanoparticles for delivery of therapeutics." *Nat. Nanotechnol.* 6 (10):658–67.

Shu, D., H. Li, Y. Shu, et al. 2015. "Systemic delivery of anti-miRNA for suppression of triple negative breast cancer utilizing RNA nanotechnology." *ACS Nano* 9 (10):9731–40.

Shu, Y., F. Haque, D. Shu, et al. 2013. "Fabrication of 14 different RNA nanoparticles for specific tumor targeting without accumulation in normal organs." *RNA* 19 (6):767–77.

Shukla, G. C., F. Haque, Y. Tor, et al. 2011. "A boost for the emerging field of RNA nanotechnology." *ACS Nano* 5 (5):3405–18.

Shukla, S., C. S. Sumaria, and P. I. Pradeepkumar. 2010. "Exploring chemical modifications for siRNA therapeutics: a structural and functional outlook." *Chem. Med. Chem.* 5 (3):328–49.

Song, E., P. Zhu, S.-K. Lee, et al. 2005. "Antibody mediated in vivo delivery of small interfering RNAs via cell-surface receptors." *Nat. Biotechnol.* 23 (6):709–17.

Soutschek, J., A. Akinc, B. Bramlage, et al. 2004. "Therapeutic silencing of an endogenous gene by systemic administration of modified siRNAs." *Nature* 432 (7014):173–8.

Stein, C. A., J. B. Hansen, J. Lai, et al. 2010. "Efficient gene silencing by delivery of locked nucleic acid antisense oligonucleotides, unassisted by transfection reagents." *Nucleic Acids Res.* 38 (1):e3–e.

Stewart, J. M., M. Viard, H. K. Subramanian, et al. 2016. "Programmable RNA microstructures for coordinated delivery of siRNAs." *Nanoscale* 8 (40):17542–50.

Stewart, K. M., K. L. Horton, and S. O. Kelley. 2008. "Cell-penetrating peptides as delivery vehicles for biology and medicine." *Org. Biomol. Chem.* 6 (13):2242–55.

Stokmaier, D., O. Khorev, B. Cutting, et al. 2009. "Design, synthesis and evaluation of monovalent ligands for the asialoglycoprotein receptor (ASGP-R)." *Biorg. Med. Chem.* 17 (20):7254–64.

Strobel, S. A., and J. A. Doudna. 1997. "RNA seeing double: close-packing of helices in RNA tertiary structure." *Trends Biochem. Sci.* 22 (7):262–6.

Sun, Q., Z. Kang, L. Xue, et al. 2015. "A collaborative assembly strategy for tumor-targeted siRNA delivery." *J. Am. Chem. Soc.* 137 (18):6000–10.

Takeuchi, O., and S. Akira. 2010. "Pattern recognition receptors and inflammation." *Cell* 140 (6):805–20.

Tamura, M., and S. R. Holbrook. 2002. "Sequence and structural conservation in RNA ribose zippers." *J. Mol. Biol.* 320 (3):455–74.

Terrazas, M., and E. T. Kool. 2009. "RNA major groove modifications improve siRNA stability and biological activity." *Nucleic Acids Res.* 37 (2):346–53.

Thillier, Y., C. Sallamand, C. Baraguey, et al. 2015. "Solid-phase synthesis of oligonucleotide 5′-(α-P-Thio) triphosphates and 5′-(α-P-Thio)(β, γ-methylene) triphosphates." *Eur. J. Org. Chem.* 2015 (2):302–8.

Tomari, Y., C. Matranga, B. Haley, et al. 2004. "A protein sensor for siRNA asymmetry." *Science (New York, N.Y.)* 306 (5700):1377–80.

Torchilin, V. P. 2008. "Cell penetrating peptide-modified pharmaceutical nanocarriers for intracellular drug and gene delivery." *Peptide Science* 90 (5):604–10.

Toy, R., P. M. Peiris, K. B. Ghaghada, et al. 2014. "Shaping cancer nanomedicine: the effect of particle shape on the in vivo journey of nanoparticles." *Nanomedicine (London, England)* 9 (1):121–34.

Tuerk, C., and L. Gold. 1990. "Systematic evolution of ligands by exponential enrichment: RNA ligands to bacteriophage T4 DNA polymerase." *Science* 249 (4968):505–10.

Vaish, N., F. Chen, S. Seth, et al. 2011. "Improved specificity of gene silencing by siRNAs containing unlocked nucleobase analogs." *Nucleic Acids Res.* 39 (5):1823–32.

Van de Water, F. M., O. C. Boerman, et al. 2006. "Intravenously administered short interfering RNA accumulates in the kidney and selectively suppresses gene function in renal proximal tubules." *Drug Metab. Disposition* 34 (8):1393–7.

Víglaský, V., M. Antalík, J. Bagel'ová, et al. 2000. "Heat-induced conformational transition of cytochrome c observed by temperature gradient gel electrophoresis at acidic pH." *Electrophoresis* 21 (5):850–8.

Walsh, A. S., H. Yin, C. M. Erben, et al. 2011. "DNA cage delivery to mammalian cells." *ACS Nano.* 5 (7):5427–32.

Watts, J. K., G. F. Deleavey, and M. J. Damha. 2008. "Chemically modified siRNA: tools and applications." *Drug Discovery Today* 13 (19):842–55.

Wheeler, L. A., R. Trifonova, V. Vrbanac, et al. 2011. "Inhibition of HIV transmission in human cervicovaginal explants and humanized mice using CD4 aptamer-siRNA chimeras." *J Clin. Invest.* 121 (6):2401–12.

Whitehead, K. A., J. E. Dahlman, R. S. Langer, et al. 2011. "Silencing or stimulation? siRNA delivery and the immune system." *Annu. Rev. Chem. Biomol. Eng.* 2:77–96.

Wilson, R. C., and J. A. Doudna. 2013. "Molecular Mechanisms of RNA Interference." In *Annual Review of Biophysics, Vol 42*, edited by K. A. Dill, 217–39.

Wittrup, A., and J. Lieberman. 2015. "Knocking down disease: a progress report on siRNA therapeutics." *Nat. Rev. Genet.* 16 (9):543–52.

Wolfrum, C., S. Shi, K. Narayanannair Jayaprakash, et al. 2007. "Mechanisms and optimization of in vivo delivery of lipophilic siRNAs." *Nat. Biotechnol.* 25 (10):1149–57.

Wong, S. C., J. J. Klein, H. L. Hamilton, et al. 2012. "Co-injection of a targeted, reversibly masked endosomolytic polymer dramatically improves the efficacy of cholesterol-conjugated small interfering RNAs in vivo." *Nucleic Acid Ther.* 22 (6):380–90.

Wooddell, C. I., D. B. Rozema, M. Hossbach, et al. 2013. "Hepatocyte-targeted RNAi therapeutics for the treatment of chronic hepatitis B virus infection." *Mol. Ther.* 21 (5):973–85.

Wu, J., and Z. J. Chen. 2014. "Innate immune sensing and signaling of cytosolic nucleic acids." *Annu. Rev. Immunol.* 32:461–88.

Xu, C.-F., and J. Wang. 2015. "Delivery systems for siRNA drug development in cancer therapy." *Asian Journal of Pharmaceutical Sciences* 10 (1):1–12.

Yingling, Y. G., and B. A. Shapiro. 2007. "Computational design of an RNA hexagonal nanoring and an RNA nanotube." *Nano Lett.* 7 (8):2328–34.

Yoo, H., H. Jung, S. A. Kim, et al. 2014. "Multivalent comb-type aptamer-siRNA conjugates for efficient and selective intracellular delivery." *Chem. Commun. (Camb.)* 50 (51):6765–7.

Zadeh, J. N., C. D. Steenberg, J. S. Bois, et al. 2010. "NUPACK: Analysis and design of nucleic acid systems." *J. Comput. Chem.* 32 (1):170–3.

Zhang, C. Y., P. Kos, K. Müller, et al. 2014a. "Native chemical ligation for conversion of sequence-defined oligomers into targeted pDNA and siRNA carriers." *J. Controlled Release* 180:42–50.

Zhang, H., F. Pi, D. Shu, et al. 2015. "Using RNA nanoparticles with thermostable motifs and fluorogenic modules for real-time detection of RNA folding and turnover in prokaryotic and eukaryotic cells." *Methods Mol. Biol.* 1297:95–111.

Zhang, N., C. Tan, P. Cai, et al. 2009. "RNA interference in mammalian cells by siRNAs modified with morpholino nucleoside analogues." *Biorg. Med. Chem.* 17 (6):2441–6.

Zhang, X., H. Liang, Y. Tan, et al. 2014b. "A U87-EGFRvIII cell-specific aptamer mediates small interfering RNA delivery." *Biomed. Rep.* 2 (4):495–9.

Zhong, Q., D. Inniss, K. Kjoller, et al. 1993. "Fractured polymer/silica fiber surface studied by tapping mode atomic force microscopy." *Surf. Sci.* 290 (1–2):L688–L92.

Zhou, J., and J. Rossi. 2017. "Aptamers as targeted therapeutics: current potential and challenges." *Nat. Rev. Drug Discov.* 16 (3):181–202.

Zhou, J., Y. Shu, P. Guo, et al. 2011. "Dual functional RNA nanoparticles containing phi29 motor pRNA and anti-gp120 aptamer for cell-type specific delivery and HIV-1 inhibition." *Methods (San Diego, Calif.)* 54 (2):284–94.

Zhu, L., and R. I. Mahato. 2010. "Targeted delivery of siRNA to hepatocytes and hepatic stellate cells by bioconjugation." *Bioconjugate Chem.* 21 (11):2119–27.

Zimmermann, T. S., A. C. H. Lee, A. Akinc, et al. 2006. "RNAi-mediated gene silencing in non-human primates." *Nature* 441 (7089):111–4.

Zlatev, I., D. J. Foster, J. Liu, et al. 2016. "5′-C-Malonyl RNA: small interfering RNAs modified with 5′-monophosphate bioisostere demonstrate gene silencing activity." *ACS Chem. Biol.* 11 (4):953–60.

Zlatev, I., J. G. Lackey, L. Zhang, et al. 2013. "Automated parallel synthesis of 5′-triphosphate oligonucleotides and preparation of chemically modified 5′-triphosphate small interfering RNA." *Biorg. Med. Chem.* 21 (3):722–32.

18 DNA Repair and Epigenetics in Cancer

Loredana Zocchi and Claudia A. Benavente

CONTENTS

18.1 DNA REPAIR IN CANCER

Cells are under constant genotoxic pressure from both endogenous and exogenous sources. It has been estimated that every day a single human cell has to endure tens of thousands of DNA lesions (Jackson and Bartek 2009). This damage needs to be repaired to avoid detrimental mutations, blockage of replication and transcription, and chromosomal breakage. DNA repair is the collection of the multiple and diverse ways through which living cells identify alterations in the chemistry of their DNA molecules and correct the damage to restore the integrity of their genome. In cancer, DNA repair serves as a significant barrier that can prevent preneoplastic cells from progressing through malignant transformation. The importance of DNA repair in preventing cancer was first demonstrated in the study of patients with xeroderma pigmentosum (XP), a rare autosomal recessive genetic disorder characterized by extreme sensitivity to ultraviolet (UV) rays caused by a deficiency in the ability to repair damage caused by sunlight (Cleaver 1968, Setlow et al. 1969). Individuals with XP exhibit skin malignancies and cancer at a young age. Further support for the critical role of DNA repair in preventing cancer in humans came from the discovery of other DNA repair mechanisms, summarized in this chapter (Figure 18.1). The mechanism through which DNA is repaired depends on the type and extent of the DNA damage. In mammalian cells, there are six major DNA repair pathways with unique—but sometimes overlapping—functions, to mend the damage caused by exogenous DNA-damaging agents (including chemotherapy and radiotherapy) and damage caused by normal endogenous cellular processes (Kelley and Fishel 2008).

18.1.1 Direct Reversal (DR)

In humans, there is only one type of DNA damage that can be repaired by direct chemical reversal. This mechanism can only repair one type of lesion and does not involve breakage of the phosphodiester backbone; thus, not requiring a template for the repair. The DR pathway removes alkyl groups (CH_3-) at the O^6 position of guanine by direct transfer to O^6-methylguanine-DNA methyltransferase (MGMT) (Tano et al. 1990, Natarajan et al. 1992). MGMT transfers the methyl group to a cysteine residue in the protein. In this process, each MGMT molecule can only be used once. Impairment of the DR pathway would allow the O^6-methylguanine to pair with thymine instead of cytosine, leading to G to A mutations (Kaina et al. 2007). When MGMT is unsuccessful in removing O^6-methylguanine during DR, the mismatch repair (MMR) pathway can recognize and fix the resulting O^6-methylguanine mispairs (Luo et al. 2010). Interestingly, glioma patients with *MGMT* gene inactivation, which would render the tumors incapable of repairing O^6-methylguanine, have better survival rates than patients with active *MGMT* following treatment with alkylating agents such as carmustine and temozolomide (Esteller et al. 2000, Hegi et al. 2005). As one would predict, lack of MMR has also been shown to render tumors resistant to alkylating agents, even in the absence of MGMT (Liu, Markowitz, and Gerson 1996).

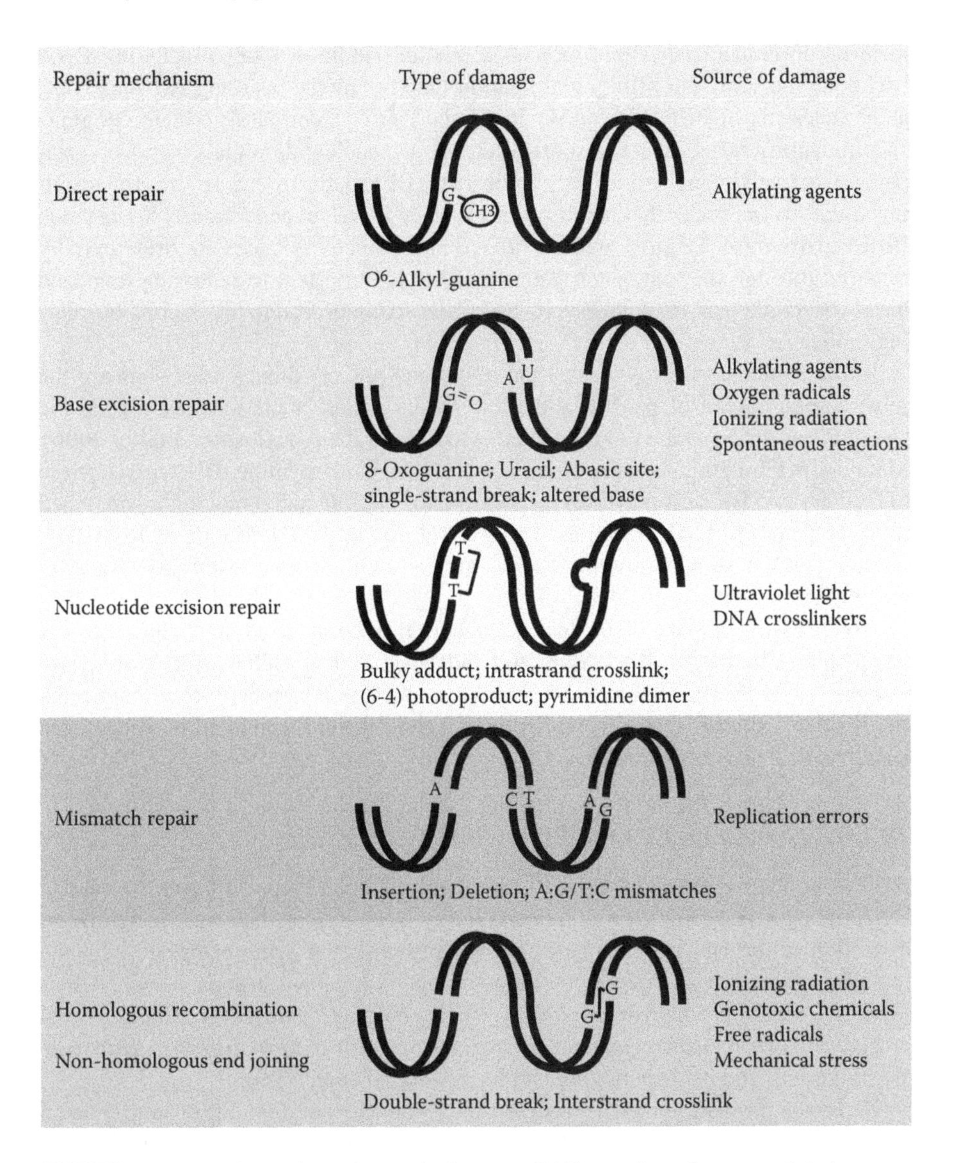

FIGURE 18.1 DNA repair pathways in humans. DNA repair pathways and their corresponding type of DNA damage and sources of endogenous and exogenous agents are summarized in this figure.

18.1.2 Base Excision Repair (BER)

BER is one of three excision repair pathways that happen to repair single stranded DNA damage. BER removes small, non-bulky lesions (do not distort the DNA helix) produced by alkylation, oxidation or deamination of bases. In this DNA repair mechanism, a DNA glycosylase-type enzyme removes a single damaged DNA base, forming an apurinic/apyrimidinic site (AP site). Additional steps including DNA backbone incision, gap filling, and ligation then repair the resulting AP site. Thus, a characteristic of BER is the diversity of the DNA glycosylases, which recognize specific substrates. Either the short-patch (single nucleotide replacement) or long-patch (two to eight nucleotides are synthesized) BER pathway can process the resulting single-strand break that results after cleavage by AP endonucleases. Short-patch BER repairs most AP sites, while oxidized and reduced AP sites are preferentially repaired through the long-patch pathway. BER is important to removing damaged bases that could lead to mutations by base mispairing or lead to breaks in DNA during replication.

In human cancer, C to T transition mutations at CpG dinucleotide sites are the most common kind of genetic alteration. In part, these mutations arise from the spontaneous deamination of methylated cytosines (5-methylcytosine) (Pfeifer 2006). Methyl-CpG binding domain protein 4 (MBD4) and thymine DNA glycosylase (TDG) are two BER glycosylases responsible for binding and removing mismatched thymine from DNA (Hendrich et al. 1999, Yoon et al. 2003). Mutations in MBD4, but not TDG, have been observed with cancers with genomic instability (Bader et al. 1999).

Another BER enzyme which, when mutated, may be involved in carcinogenesis is OGG1 (Chevillard et al. 1998, Shinmura and Yokota 2001). OGG1 is the glycosylase responsible for the excision of 8-oxoguanine, a mutagenic base byproduct that occurs as a result of exposure to reactive oxygen (Arai et al. 1997). Unrepaired 8-oxoguanines lead to G to T or G to C transversions.

18.1.3 Nucleotide Excision Repair (NER)

NER is another excision repair pathway involved in the repair of single stranded DNA damage. In NER, large adduct and bulky DNA lesions that cause a significant distortion of the DNA double helix are excised within a string of nucleotides and replaced with DNA as directed by the undamaged template strands. Thus, NER is the DNA repair mechanism used only when one of the two DNA strands is disturbed. This type of damage usually occurs as a result of cross-linking agents (e.g. UV radiation) and base-damaging carcinogens (Luo et al. 2010).

NER is a multi-step repair process that involves more than 30 proteins, listed in Table 18.1. There are two NER sub-pathways: global genomic repair (GGR) and transcription coupled repair (TCR). GGR acts throughout the genome, regardless of whether the specific sequence is the transcribed or non-transcribed strand of a gene (Sugasawa et al. 2001, Riedl, Hanaoka, and Egly 2003). As the name indicates, the TCR repair machinery removes lesions only from the transcribed strand of active genes, removing distorting lesions that block transcriptional elongation by RNA polymerases (Fousteri and Mullenders 2008, Hanawalt and Spivak 2008). The protein complexes that recognize the DNA damage site and initiates DNA repair determine the NER sub-pathway selection (Luo et al. 2010).

TABLE 18.1
NER Associated Genes

Human Gene	Protein	Subpathway	Function in NER
CCNH	Cyclin H	Both	CDK Activator Kinase (CAK) subunit
CDK7	Cyclin Dependent Kinase (CDK)7	Both	CAK subunit
CETN2	Centrin-2	GGR	Damage recognition; forms complex with XPC
DDB1	DDB1	GGR	Damage recognition; forms complex with DDB2
DDB2	DDB2	GGR	Damage recognition; recruits XPC
ERCC1	ERCC1	Both	Involved in incision on 3' side of damage; forms complex with XPF
ERCC2	XPD	Both	ATPase and helicase activity; transcription factor II H (TFIIH) subunit
ERCC3	XPB	Both	ATPase and helicase activity; transcription factor II H (TFIIH) subunit
ERCC4	XPF	Both	Involved in incision on 3' side of damage; structure specific endonuclease
ERCC5	XPG	Both	Involved in incision on 5' side of damage; stabilizes TFIIH; structure specific endonuclease
ERCC6	CSB	TCR	Transcription elongation factor; involved in transcription coupling and chromatin remodeling
ERCC8	CSA	TCR	Ubiquitin ligase complex; interacts with CSB and p44 of TFIIH
LIG1	DNA Ligase I	Both	Final ligation
MNAT1	MNAT1	Both	Stabilizes CAK complex
MMS19	MMS19	Both	Interacts with XPD and XPB subunits of TFIIH helicases
RAD23A	RAD23A	GGR	Damage recognition; forms complex with XPC
RAD23B	RAD23B	GGR	Damage recognition, forms complex with XPC
RPA1	RPA1	Both	Subunit of RFA complex
RPA2	RPA2	Both	Subunit of RFA complex
TFIIH	Transcription factor II H	Both	Involved in incision, forms complex around lesion
XAB2	XAB2	TCR	Damage recognition; interacts with XPA, CSA, and CSB
XPA	XPA	Both	Damage recognition
XPC	XPC	GGR	Damage recognition

The xeroderma pigmentosum group C protein, encoded by the XPC gene is a subunit of these damage recognition complexes and is essential for GGR (Friedberg 2001, Riedl, Hanaoka, and Egly 2003). For the TCR pathway, recognition of the DNA damage-blocked RNA polymerase by transcription-repair coupling factors is important. After damage recognition, both NER subclasses have the same or similar subsequent steps involved in nucleotide excision and gap filling by DNA polymerases.

18.1.4 Mismatch Repair (MMR)

The last excision repair pathway involved in the repair of single stranded DNA damage is MMR. During DNA replication mistakes can occur that escape the proofreading activity of DNA polymerase as it copies the two strands. The MMR pathway is responsible for recognizing and repairing single-base insertions, deletions, and mismatches that arise during normal DNA replication process (Luo et al. 2010, Fleck and Nielsen 2004). These errors that escape the proofreading activity of DNA polymerases happen with a frequency of about 1 in 10^9–10^{10} base pairs per cell division (Iyer et al. 2006). Furthermore, exposure to exogenous agents or endogenous reactive species may cause base modifications that lead to nucleotide mispairing (Li 2008). Loss of MMR affects genome stability (including microsatellite instability), which causes cancer predisposition (Jiricny 2006). In this pathway, PMS2, MLH1, LSH6, and MSH2 are proteins that recruit EXO1 to excise the segment of mutant DNA strand. Then DNA polymerases replace the missing section of the strand with a new section and the damage is repaired. The vast majority of hereditary non-polyposis colorectal cancers (HNPCC) are attributed to mutations in the *MSH2* and *MLH1* genes (Bronner et al. 1994).

18.1.5 Homologous Recombination (HR)

HR is one of two mechanisms through which DNA double-strand breaks can be repaired. DNA damage that has not been repaired before replication can cause DNA polymerase blockage, resulting in DNA double-strand breaks. HR is the repair pathway used to fix double-strand breaks detected during the S/G2 phases of the cell cycle, when a homologous template via the sister chromatid is available. Since HR requires a long homologous sequence to guide the repair, it is highly accurate in its repair (Fleck and Nielsen 2004). The DNA checkpoint responses are responsible for the regulation of double-strand break ends processing, which will determine which DNA double-strand break repair mechanism will perform the repair. This is a crucial stage in the recombination process (Lazzaro et al. 2009).

Two of the most studied genes and proteins that are involved in this repair pathway are BRCA1 and BRCA2. These tumor suppressor proteins form a complex along with RAD51 to repair DNA double-strand breaks (Duncan, Reeves, and Cooke 1998, Yoshida and Miki 2004). Cells missing BRCA1 and BRCA2 have a decreased rate of HR. Mutations in the *BRCA1* and *BRCA2* genes have been associated with considerably increased risk for breast and ovarian cancer (Miki et al. 1994, Wooster et al. 1994).

18.1.6 Non-Homologous End Joining (NHEJ)

NHEJ is the other pathway that repairs double-strand breaks in DNA. Unlike HR, NHEJ has the potential to relegate any type of DNA ends, without the need for a homologous template. Since NHEJ does not require an identical copy of DNA as a template, it is not restricted to a certain phase of the cell cycle, and it is prone to imprecise repair leading to loss or addition of bases in the ligation process (Fleck and Nielsen 2004). Inactivation of CDK1 increases NHEJ events in the G2 phase of the cell cycle (Lazzaro et al. 2009). DNA strands that are not repaired completely by NHEJ are subject to repair by HR (Essers et al. 2000).

The initial step in NHEJ is the recognition and binding of the Ku heterodimer at the DNA double-strand break (Mari et al. 2006). The Ku heterodimer is composed of Ku70 and Ku80, encoded by the *XRCC6* and *XRCC5* genes, respectively. Once the Ku heterodimer is bound the DNA double-strand break ends, it serves as a scaffold to recruit the other NHEJ factors to the damage site. No spontaneous Ku mutations have been found in humans, suggesting that both Ku70 and Ku80 are likely required for viability.

18.2 EPIGENETICS IN CANCER

It was long thought that tumorigenesis was mostly driven by genetic mutations and genomic instability. With the advent of whole-genome sequencing, cancers with a low rate of mutations have been identified and epigenetics has gained an ever-increasing role in the process of tumor progression (Zhang et al. 2012, Feinberg, Koldobskiy, and Gondor 2016). Epigenetics is defined as the inheritable changes in gene expression with no alterations in DNA sequences. During the past few years several studies showed the connection between disruptions of the epigenome, defined as the combination of changes in gene expression, and tumor progression. In the eukaryotic nucleus, DNA is compacted into a chromatin structure with the nucleosome as the basic unit, in which 147 bases of DNA surround each histone octamer. The histone octamer includes two elements of the core histone (H3, H4, H2A, and H2B) (Luger et al. 1997). Unlike the other histones H1, the "linker "histone, is not a component of the nucleosome. It interacts at the DNA entrance and exit site of the nucleosome and the linker DNA that connects adjacent nucleosome. There are three main epigenetic modifications that regulate chromatin structure and gene expression: DNA methylation, histone covalent modification and microRNAs (miRNAs) (Figure 18.2). All together, they constitute the "epigenetic code," that is capable of modulating the expression of the different cell types. Disruption of epigenetic processes can lead to altered gene function and malignant cellular transformation (Sharma, Kelly, and Jones 2010).

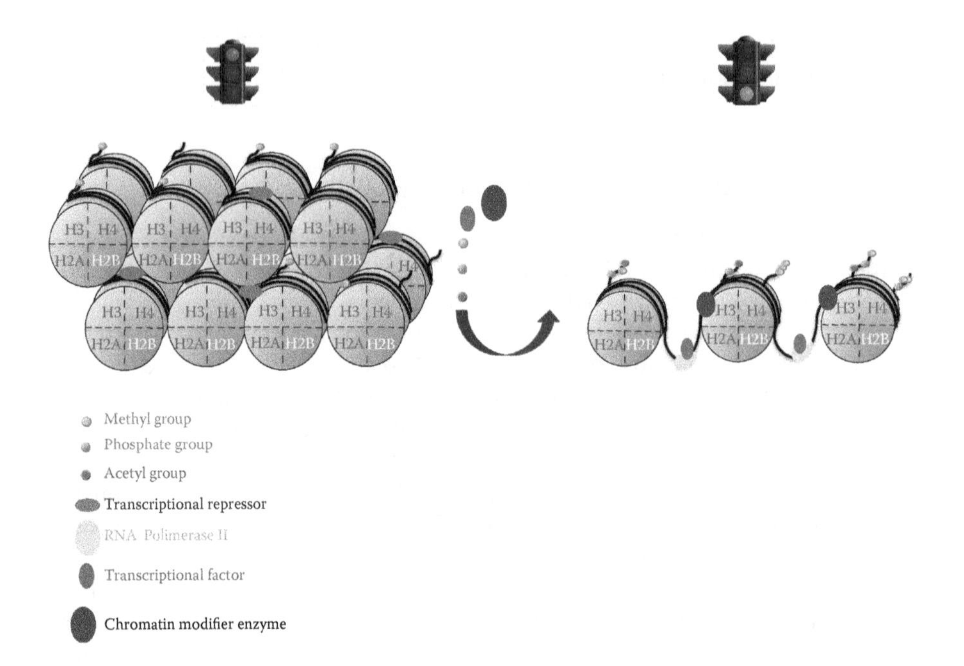

FIGURE 18.2 Euchromatin and heterochromatin landscape. Schematic representation of the nucleosome octamers. Condensed chromatin is not accessible to transcription factors (repressed chromatin, left). Epigenetics modifiers can render chromatin accessible to transcription factors and RNA polymerase II (active chromatin, right).

18.2.1 DNA Methylation

DNA methylation is a covalent modification of DNA that has been described in bacteria, plants, and mammals. It can occur following DNA replication, in order to re-establish the preexisting DNA methylation pattern or *de novo*, and in both situations acts to repress gene transcription (Chen et al. 2014). In eukaryotic cells, the 5' methyl group is added to the cytosine base, and this modification is most frequently found in the context of CpG dinucleotides. S-adenosyl-methionine is the methyl donor in a reaction catalyzed by the DNA methyltransferase (DNMT) family, including DNMT1, DNMT3A and DNMT3B. DNMT1 is responsible for the methylation of hemi-methylated DNA and thus DNA methylation maintenance, whereas DNMT3A and DNMT3B are involved in *de novo* DNA methylation, but they can also participate in methylation maintenance (Castillo-Aguilera et al. 2017). It can be speculated that DNA methylation is capable of preventing gene transcription either by blocking the combination of a transcription factor and its binding sites, or through the recruitment of methylated binding domain proteins that mediate inhibition of gene expression.

In some areas of the genome, CpG sites are concentrated in short CpG-rich DNA fragments or DNA fragments in the long repeat so-called 'CpG islands'. CpG island-containing gene promoters are usually un-methylated in normal cells to maintain euchromatic structure, which is the transcriptional active conformation allowing gene expression (Chen et al. 2014).

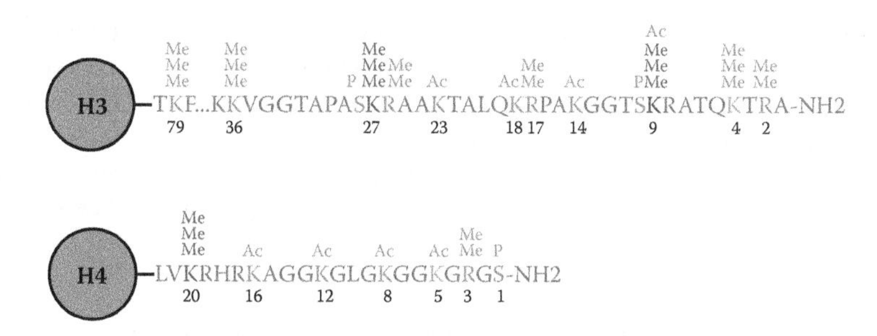

FIGURE 18.3 Predominant post-translational modifications of histones H3 and H4. Partial view of histones H3 and H4 N-terminal tail with the main post-translational modifications known to lead gene transcriptional activation or repression. The corresponding residue number is shown under each modified amino acid. Marked in green are the amino acids and post-translational modifications that lead to transcriptional activation. In red are the amino acid residues and post-translational modifications that lead to transcriptional repression. In purple is the amino acid that can be modified to both active and repressive transcription. Me = methyl group, P = phosphate group, Ac = acetyl group.

18.2.2 Histone Modifications

The histone octamer, the basic element of the nucleosome core particle, consists of two copies of each core histone proteins (H2A, H2B, H3, and H4). The N-terminals of histones protrude out of the nucleosome core, and amino acids of N-terminals easily undergo a series of covalent modifications, such as methylation, acetylation, phosphorylation, ubiquitination and sumoylation (Figure 18.3) (Tessarz and Kouzarides 2014, Cheung, Allis, and Sassone-Corsi 2000). These post-translational modifications can regulate important processes such as gene transcription, X-chromosome inactivation, mitosis, heterochromatin formation, DNA repair, and replication (Kouzarides 2007). Regarding gene transcription, histone H3 lysine 9 acetylation (H3K9ac), H3 serine 10 phosphorylation (H3S10ph), and H3 lysine 4 tri-methylation (H3K4me3) are reported to be associated with transcriptional activation. Conversely, H3K27me3 and hypoacetylation of H3 and H4 are correlated with transcriptional repression. Importantly, a large body of evidence supports a scenario in which combinatorial modifications correspond to specific functional chromatin states. Individual post-translational modifications can favor or inhibit consequent modifications on nearby residues of the same tail (Fischle, Wang, and Allis 2003, Latham and Dent 2007). For example phosphorylation of Ser-10 on H3, is a positive signal for subsequent acetylation at K14 on the same tail (Lo et al. 2000, Cheung, Allis, and Sassone-Corsi 2000) whereas histone deacetylation and methylation of H3-K9 lysine represses transcription (Fuks 2005).

18.2.3 microRNAs

miRNA encode small noncoding RNA molecules (19–25 nucleotides in length) that are complementary to the 3' untranslated regions of target mRNAs. This results in gene silencing through translational repression or target mRNA degradation (Bartel 2009). Studies on miRNAs have demonstrated how they regulate gene expression also at transcriptional level, and not only at post-transcriptional level as described above. Promoter-associated RNA (paRNA) can also regulate transcription of genes by targeting the promoter (Costa 2010). PaRNA can also modify the recruitment of the epigenetic machinery to enhance or silencing transcription of mRNA (Yan and Ma 2012).

18.2.4 The Epigenetic Classification System for Cancer Genes

The old classification of cancer genes into dominant oncogenes (*MYC, KRAS, PIK3CA, ABL1, BRAF*) and recessive tumor suppressor genes (*RB1, TP53, WT1, NF2, VHL, APC, CDKN2A*), has been replaced by the more functional epigenetics classification of the cancer genes, which includes the epigenetic modifiers, the mediators and the modulators (Feinberg, Koldobskiy, and Gondor 2016). Epigenetic modifiers are gene products capable of directly modifying the epigenome through DNA methylation, post-translational modification of chromatin, or the alteration of the structure of the chromatin. The epigenetic mediators are often the targets of epigenetic modification, although they are rarely mutated themselves; importantly, they appear to be responsible for the emergence of cancer stem cells (Feinberg, Koldobskiy, and Gondor 2016). Finally, the epigenetics modulators are defined as genes lying upstream of the modifiers and mediators in signaling and metabolic pathways, and serving as the mechanism by which environmental agents, injury, inflammation, and other forms of stress push tissues towards a neoplastic propensity and/or increase the likelihood that cancer will arise when a key mutation occurs by chance.

18.2.5 Epigenetic Modifier Mutations and Cancer

Epigenetic modifier mutations are a common occurrence in a wide range of cancers (Table 18.2). These occur in components at every level of the epigenetic machinery including DNA methylation and histone modification.

18.2.5.1 DNA Methylation

Hematological malignances are highly related to mutations in the DNA methylation machinery. These events clearly underline how epigenetics and genetics can cooperate in cancer initiation and progression. DNA methyltransferase3α (DNMT3A) mutations have been described in human acute myeloid leukemia, acute monocytic leukemia and T-cell lymphoma (Ley et al. 2010, Yan et al. 2011, Couronne, Bastard, and Bernard 2012). Moreover, *DNMT3A* mutations are considered a marker of poor prognosis both in acute myeloid leukemia and T-cell acute lymphoblastic leukemia (Grossmann et al. 2013, Ribeiro et al. 2012). *DNMT1* mutations have been described in colon cancer (Kanai et al. 2003). Mutations in *DNMT3B* have been associated with a rare autosomal recessive immunoglobulin deficiency, sometimes combined with defective cellular immunity called immunodeficiency-centromeric instability-facial anomalies (ICF) syndrome (Wijmenga et al. 2000). In addition, a single nucleotide polymorphism (SNPs) that involves a C to T transition on the promoter of this gene

TABLE 18.2
Epigenetic Modifier Mutations in Cancer

Gene	Function	Cancer
DNMT3A	DNA methylation (*de novo*)	Acute myeloid leukemia (Ley et al. 2010); T-cell lymphoma (Couronne, Bastard, and Bernard 2012)
DNMT1	DNA methylation	Colon cancer (Kanai et al. 2003)
DNMT3B	DNA methylation (*de novo*)	Lung adenocarcinoma (Shen et al. 2002); breast cancer (Roll et al. 2008)
Tet2	DNA demethylation	Myelodysplastic syndrome; myeloproliferative neoplasms; acute myeloid leukemia (Gaidzik et al. 2012)
IDH1/2	DNA demethylation	Glioma (Turcan et al. 2012); acute myeloid leukemia (Figueroa et al. 2010)
EP300	Histone acetylation	Acute myeloid leukemia (Wang, Gural et al. 2011)
HDAC1	Histone deacetylase	Prostate cancer (Halkidou et al. 2004); gastric cancer (Choi et al. 2001)
HDAC2	Histone deacetylase	Colorectal cancer (Ozdag et al. 2006)
HDAC4	Histone deacetylase	Breast cancer (Sjoblom et al. 2006)
HDAC7A	Histone deacetylase	Colorectal tumors (Ozdag et al. 2006)
KMT2A	Histone methyltransferase	Acute myeloid leukemia (Thirman et al. 1993)
KMT2B	Histone methyltransferase	Endometrial; large intestine; lung; glioma; liver carcinoma (Rao and Dou 2015)
KMT2C	Histone methyltransferase	Endometrial; large intestine; lung; bladder carcinoma (Rao and Dou 2015)
KMT2D	Histone methyltransferase	Acute myeloid leukemia; lung large intestine carcinoma; bladder carcinoma (Rao and Dou 2015)
EZH2	Histone methyltransferase	Non-Hodgkin lymphoma; solid tumors; T-cell leukemia (Feinberg, Koldobskiy, and Gondor 2016)
NSD1	Histone methyltransferase	Acute myeloid leukemia (Varier and Timmers 2011)
SMYD3	Histone methyltransferase	Colon; breast; hepatocellular carcinoma (Varier and Timmers 2011)
G9A	Histone methyltransferase	Hepatocellular carcinoma (Varier and Timmers 2011)
PRMT1	Arginine methyltransfrase	Breast cancer (Gao et al. 2016)
PRMT5	Arginine methyltransfrase	Hematologic and solid tumors (Tarighat et al. 2016)
LSD1	Histone demethylase	Bladder; colorectal cancer (Rotili and Mai 2011)
KDM6A	Histone demethylase	Myeloma; renal cell carcinoma (Rotili and Mai 2011)
BRCA1	Ubiquitin ligase	Breast; ovarian cancer (Zhu et al. 2011)
USP22	Ubiquitin hydrolase	Breast cancer (Zhang et al. 2011)

has been associated with increased *DNMT3B* promoter activity and increased risk of lung cancer, while *DNMT3B* overexpression can lead to hypermethylation and silencing of key genes in human breast cancer cell lines (Shen et al. 2002, Roll et al. 2008). DNA methylation can be reversed by erasers, which are part of the epigenetic machinery. Among these we can find TET (Ten-eleven-translocation) and AID (Activation-induced cytidine deaminase) demethylases (De Carvalho et al. 2012, Ko et al. 2010, Wu and Zhang 2010). Active DNA demethylation is currently thought of as being a stepwise process. First 5-methylcytosine (5mC) is converted into 5-hydroxymethylcytosine (5hmC) by the TET family of enzymes. Subsequently 5hmC can be deaminated by the AID/APOBEC family members to form 5-hydroxymethyluracil (5hmU). The DNA excision repair system can finally replace the cytosine without the methyl group (Bhutani, Burns, and Blau 2011). Three TET family members (TET1, TET2, and TET3) have been reported so far and each protein seems to have a distinct function in different cellular contexts (Cimmino et al. 2011). Mutations in TET2 including frame shift, nonsense and missense mutations, have been found in various myeloid neoplasms and gliomas (Gaidzik et al. 2012). Cytosolic isocitrate dehydrogenase 1/2 (*IDH*) mutants display global DNA hypermethylation along with the accumulation of 2-hydroxyglutarate, and they are also capable of impairing TET2 function. *IDH1/2* mutations were mutually exclusive with mutations in the α-ketoglutarate-dependent enzyme *TET2*, while *TET2* loss-of function mutations (a type of mutation in which the altered gene product lacks the molecular function of the wild-type gene mutations) are associated with similar epigenetic defects as *IDH1/2* mutants (Figueroa et al. 2010). *IDH1/2* mutations are described in different kind of gliomas and leukemias (Turcan et al. 2012). These mutants show impaired hematopoietic differentiation (Figueroa et al. 2010), and impaired cell differentiation (Lu et al. 2012).

18.2.5.2 Histone Acetylation and Deacetylation

The acetylation of lysine on histone is generally associated with active gene transcription. Acetyltransferase (HATs) can be grouped into three main categories according to their sequence similarity: Gcn5-related–N-acetyltransferase (GNAT), MYST (acronym for the founding members MOZ, Ybf2, Sas2, TIP60), and orphan (p300/CBP and nuclear receptors) (Yang 2004). Several mutations such as amplifications, point mutations or translocations of HATs have been described. Several publications connected acetyltransferase mutation to different types of cancer. Mutations, translocations or deletions of these genes are observed in colon, uterine, lung tumors, and leukemia (Esteller 2007). HATs can also modulate the activity of fusion proteins. It has been previously described how AML1-ETO, the most frequent fusion protein in acute myeloid leukemia needs p300-mediated site-specific acetylation to drive leukemogenesis (Wang, Gural, et al. 2011).

Histone deacetylases (HDAC) are erasers that can remove acetyl. There are five classes of histone deacetylases:

-class I: HDAC1, HDAC2, HDAC3, and HDAC8
-class IIa: HDAC4, HDAC5, HDAC7, HDAC9
-class IIb: HDAC6 and HDAC10
-class III: Sirtuins (SIRT1-7)
-class IV: HDAC11

Somatic changes in HDAC genes are implicated in cancer progression. *HDAC1, HDAC5* and *SIRT1* are downregulated in some renal, bladder, and colorectal tumors (Ozdag et al. 2006). *HDAC1* is overexpressed in prostate and gastric cancer (Halkidou et al. 2004, Choi et al. 2001). *HDAC2* mutations correlate with colorectal cancer (Ozdag et al. 2006), gastric (Ropero et al. 2006), and endometrial primary tumors (Ropero et al. 2006). The loss of HDAC2 protein expression renders those cells more resistant to the usual anti-proliferative and pro-apoptotic effect of histone deacetylase inhibitors. *HDAC4* mutations have been identified in breast cancer (Sjoblom et al. 2006). *HDAC5* is overexpressed in some colorectal tumors (Ozdag et al. 2006). And, higher expression of *HDAC7A* was observed in most colorectal tumors (Ozdag et al. 2006). Understanding the role of epigenetics modifiers in cancer can open new avenues for medical treatment through the identification of new drugs that specifically target these factors.

18.2.5.3 Histone Methylation

Methylation of arginine and lysine residues on histone protein tails can regulate chromatin structure and gene expression. One well-known example for alterations in histone methylation is mixed lineage leukemia (MLL). *MLL1* (also known as *KMT2A*) is the human homolog of the *trithorax* (*trx*) in Drosophila. The trithorax group of proteins typically function in large complexes formed with other proteins and are most commonly associated with gene activation. MLL regulates H3K4me3, an active mark for transcription. Translocations of *MLL1* with multiple different partners can originate fusion proteins that have abnormal patterns of H3K4me3 and/or abnormal patterns of histone modifier recruitment resulting in tumorigenesis. Rearrangement of the *MLL1* gene has been described in acute lymphoblastic and acute myeloid leukemia (Thirman et al. 1993). Mutations in *MLL1* have also been described in a large spectrum of solid tumors (colon, lung, bladder, endometrial, and breast cancers) (Rao and Dou 2015). Mutations in the coding region of *MLL2* (also known as *KMT2B*), another member of the MLL family of methylases, occur in endometrial, large intestine, lung, glioma, and liver carcinomas (Rao and Dou 2015). To date, hundreds of *MLL3* (known as *KMT2C*) and *MLL4* (known as *KMT2D*) mutations have been identified, making them among the most frequently mutated genes in human cancer. *MLL3* mutations are prevalent in lung, large intestine, breast, endometrial, and bladder carcinomas. All these together account for 60% of the total *KMT2C* mutations identified (Rao and Dou 2015). Nonsense, missense and frameshift mutation of *MLL4* (known as *KMT2D*) have been related to acute myeloid leukemia, lung, large intestine, endometrial carcinomas, and medulloblastoma (Rao and Dou 2015). *EZH2* (enhancer of zeste homologue 2) is a member the Polycomb repressive complex 2 (PRC2). EZH2 regulates H3K27me3 and there are two different classes of mutations that affect its function. Gain-of-function hotspot mutations (a type of mutation in which the altered gene product possesses a new molecular function or a new pattern of gene expression) and amplification have been reported in non-Hodgkin lymphomas and solid tumors. These events suggest how these tumors require an increased level of H3K27 tri-methylation (Feinberg, Koldobskiy, and Gondor 2016). On the contrary, loss-of-function mutations (a type of mutation in which the altered gene product lacks the molecular function of the wild-type gene) of *EZH2* have been described in myeloid malignancies, head and neck squamous carcinomas, and T-cell leukemia (Feinberg, Koldobskiy, and Gondor 2016). Other lysine histone methyl-transferases (HMTs) are aberrantly expressed in several cancers. NSD1 (Nuclear Receptor Binding SET Domain Protein 1) methylates Lys-36 of histone H3 and Lys-20 of histone H4 (*in vitro*). This transcriptional intermediary factor is capable of both negatively or positively influencing transcription, depending on the cellular context. NSD1 translocations have been described in acute myeloid leukemia (Varier and Timmers 2011). SMYD3 (SET and MYND Domain Containing 3) specifically methylates Lys-4 of histone H3, inducing di- and tri-methylation, but not mono-methylation. It also methylates Lys-5 of histone H4. SMYD3 overexpression has been described in colon, breast, and hepatocellular carcinoma (Varier and Timmers 2011). G9a regulates mono- and di-methylation of Lys-9 of histone H3 (H3K9me1 and H3K9me2, respectively) in euchromatin. H3K9me represents a specific tag for epigenetic transcriptional repression. G9a overexpression has been detected in hepatocellular carcinoma (Varier and Timmers 2011). Evidence for the role of arginine HMTs in tumorigenesis has not been as well established as that of lysine HMTs. PRMT1 (Protein Arginine Methyltransferase 1) constitutes the main enzyme that mediates mono-methylation and asymmetric di-methylation of histone H4 Arg-3 (H4R3me1 and H4R3me2a, respectively), a specific tag for epigenetic transcriptional activation. Upregulation of PRMT1 expression has been described in breast cancer (Gao et al. 2016). PRMT5 (Protein Arginine Methyltransferase 5) mediates the formation of omega-N mono-methyl-arginine (MMA) and symmetrical di-methylarginine (sDMA). Overexpression of PRMT5 has been reported in hematologic and solid malignancies (mantle cell lymphoma, lung and bladder cancer, gastric cancer, germ cell tumors) (Tarighat et al. 2016). Histone demethylases (KDMs) are erasers responsible for removing histone methylation. Aberrant expression of LSD1 (Lysine-specific histone demethylase 1), that demethylates both Lys-4 (H3K4me) and Lys-9 (H3K9me) of histone H3, has been shown in many types of cancers (bladder, small cell lung, and colorectal cancer) (Rotili and Mai 2011). Downregulation or inactivation of KDM6A/UTX (lysine-specific demethylase 6A), specific for demethylation of H3K27me3/me2, have been reported in various type of cancers such multiple myeloma, esophageal squamous cell carcinomas, and renal cell carcinomas (Rotili and Mai 2011). Another large class of histone demethylases is composed of the Jumonji family of Lysine demethylases. These enzymes can demethylate all three lysine methylation states (tri-, di-, and mono-methylation) at H3K4, H3K9, H3K27, and H3K36, as well as H1K26. Six different subfamilies (JMJD1s, JMJD2s, JARID1s, UTX/Y-JMJD3, PHFs, and FBXLs) of JmjC histone demethylases have been identified, which have different histone sequence and methylation state selectivity. Misregulation of JmjC KDMs has significantly been implicated in cancer initiation and progression (Rotili and Mai 2011).

18.2.5.4 Histone Ubiquitination

Histone H2A was the first protein identified to be modified by ubiquitination (Goldknopf et al. 1975). H2A and H2B are two of the most abundant ubiquitinated proteins present in the nucleus (5–15% for H2A and 1–2% for H2B) (Goldknopf et al. 1975, West and Bonner 1980, Robzyk, Recht, and Osley 2000). The most frequent forms of histone ubiquitination are mono-ubiquitination of H2A (H2Aub) and H2B (H2Bub). The residues that are normally mono-ubiquitinated are: Lys-119 for H2A, and Lys-123 in yeast, or Lys-120 in vertebrate for H2B (Goldknopf et al. 1975, West and Bonner 1980). It has been reported that H3 and H4 and linker histone H1 could be ubiquitinated, but the biological function still has to be elucidated (Pham and Sauer 2000, Jason et al. 2002, Wang et al. 2006). The modifier enzymes responsible for ubiquitination are called histone ubiquitin ligases. The first to be identified was RING1B (E3 ubiquitin ligase), and it is responsible for Lys-119 H2A mono-ubiquitination. It belongs to the Polycomb group proteins so it is related with gene silencing, while on the contrary H2B ubiquitination has been related with gene transcription activation (Cao and Yan 2012). Different studies underline how histone ubiquitination and other histone modifications are interconnected. It has been described in literature how histone H2B mono-ubiquitination is required for subsequent H3K4 methylation and H3K79 methylation, all markers of active gene transcription (Dover et al. 2002, Sun and Allis 2002, Lee, Shukla et al. 2007). *BRCA1*, a known tumor suppressor gene, contains in its RING finger an E3 ubiquitin ligase and it can catalyze H2A mono-ubiquitination *in vivo* (Zhu et al. 2011). Inactivation of BRCA1 causes the development of breast and ovarian cancer. RNF20, the major H2B specific E3 ubiquitin ligase in mammals, is considered a putative tumor suppressor gene since its depletion can increase the expression of *c-myc* and *c-FOS*, two proto-oncogenes (Shema et al. 2008). Histone ubiquitination can be reversed by ubiquitin hydrolase. USP22 can remove ubiquitin from monoubiquitinated H2A and H2B (Zhang, Varthi et al. 2008, Zhao et al. 2008). Elevated expression of USP22 is related with poor prognosis in patients with breast cancer (Zhang et al. 2011). All of this evidence underlines the important role played by histone ligase/hydrolase for normal cell function and makes these enzymes "drug-able" targets for future cancer therapies.

The expression patterns of histone modifiers in human cancer suggest these genes are important in neoplastic transformation and have characteristic patterns of expression depending on tissue of origin, with implications for potential clinical application.

18.2.6 Epigenetic Readers

Readers typically provide an accessible surface (such as a cavity or surface groove) to accommodate a modified histone residue, and determine the modification (acetylation vs. methylation) or state specificity (such as mono- vs. tri-methylation of lysine) (Yun et al. 2011). The SRA (Set and Ring Associated) domain of Uhrf1 behaves likes a "hand" with two fingers, capable of flipping out the methylated cytosine with subsequent recruitment of DNMT1 (DNA methyltransferase 1) to methylate the cytosine of the newly synthetized DNA strand. Uhrf1 TTD (Tandem Tudor Domain) and PHD (Plant Homo Domain) are instead capable of recruiting respectively the histone methyltransferases Suv39H1 or G9a. Uhrf1 overexpression has been reported to be upregulated in various types of cancers, including breast, lung, pancreatic, astrocytomas, cervical, bladder cancer, retinoblastoma, and leukemia (Kofunato et al. 2012, Benavente et al. 2014, Alhosin et al. 2016). Hells (helicase, lymphoid specific, also known as lymphoid-specific helicase) is a putative chromatin remodeler belonging to the SWI/SNF subfamily that plays a central role at repetitive pericentromeric heterochromatin. Hells can remodel chromatin in order to render it accessible to DNA methyltransferase enzymes Dnmt3a or Dnmt3b, but not Dnmt1, and therefore supports *de novo* DNA methylation and stable gene silencing. Hells upregulation has been described in human retinoblastoma and human prostate cancer (Benavente et al. 2014, von Eyss et al. 2012). BRD4 is a member of the bromodomain proteins of epigenetics readers. A bromodomain is an approximately 110 amino acid protein domain that recognizes acetylated lysine residues, such as those on the N-terminal tails of histones. The fusion of BRD4 with nuclear protein in testis (NUT) results in the development of NUT midline carcinoma (French et al. 2001). Downregulation of BRD7 is observed in hepatocellular carcinoma, and lower level of BRD7 expression is also used as an indicator of poor prognosis in patients with osteosarcoma (Chen et al. 2016). BRD8 somatic mutations have been reported in whole genome sequencing of human hepatocarcinoma (Fujimoto et al. 2012, Fujisawa and Filippakopoulos 2017). Selective inhibition of these epigenetic readers may be a novel tool for cancer treatment.

18.2.7 Epigenetic Mediators and Cancer

Epigenetics mediators are normally regulated in cancer by epigenetics modulators in order to increase pluripotency or survival. These genes are capable of counteracting proper maturation programs when ectopically expressed or overactive, and in order to do that, the mediators are capable of influencing the epigenetic states that define differentiated cell types. Large blocks of repressive H3K9me2 and H3K9me3 modifications along with DNA methylation coordinate the cell-type-specific repression of developmentally regulated genes. These areas are called large organized chromatin K9 modifications (LOCKs) and are largely absent from embryonic stem cells and cancer cell lines (Wen et al. 2009). Well-known pluripotency factors such as *NANOG* or *OCT4* (also known as *POU5F1*) and some WNT signaling members belong to this category (Feinberg, Koldobskiy, and Gondor 2016). *NANOG* is a transcription factor required for maintaining the pluripotency of embryonic stem cells and is not expressed in most normal adult tissues. However, several studies described NANOG overexpression on several tumors, including breast cancer. NANOG is not capable of inducing the mammary tumor alone, but when co-expressed with Wnt-1, promotes cell migration and invasion (Lu et al. 2014). OCT4 is another factor that plays a pivotal role as key regulator in pluripotency. It is believed that OCT4 maintains the pluripotency of spermatogonial (the earliest stage in the spermatocitic ontogeny) stem cells and keeps them in an undifferentiated, self-renewing state. Its aberrant expression may contribute to the neoplastic process in cancer cells (Gidekel et al. 2003). Moreover, the sex-determining Y-box (*SOX2*) gene, another important pluripotency factor, is amplified in small-cell lung cancer and squamous cell carcinomas of the lung and esophagus (Rudin et al. 2012, Bass et al. 2009). SOX2 is highly expressed in nasopharyngeal carcinoma, and lung adenocarcinoma (Luo et al. 2013, Chiou et al. 2010). NANOG and OCT4 overexpression are associated with increased metastatic potential in breast cancer and lung adenocarcinoma (Lu et al. 2014, Wang et al. 2014, Chiou et al. 2010). In conclusion *OCT4*, *NANOG*, and *SOX2* altered expression is capable of preventing proper maturation of the stem cells, and contributing to the development of different tumors. Finally, these genes could be potential markers of prognosis and a novel target of therapy for these tumors.

18.2.8 Epigenetic Modulators and Cancer

Epigenetic modulators are factors capable of influencing the activity of the epigenetics modifiers causing the destabilization of differentiation-specific epigenetic states. An example of epigenetic modulators is oncogenic *RAS* signaling. All RAS protein family members belong to a class of protein called small GTPase and are involved in transmitting signals within cells (cellular signal transduction). *KRAS* transformation, a member of the *RAS* genes, can drive the downregulation of TET enzymes (histone lysine demethylases described above) and this event increases DNA methylation, that facilitates the silencing of tumor suppressor genes (Wu and Brenner 2014). Moreover, about 70% of colorectal cancers with a *KRAS* mutation show chemical marks that "switch off" the expression of genes, known to suppress the growth of tumors. In actively growing human diploid fibroblasts, the *INK4A-ARF* locus is silenced by histone H3 lysine 27 tri-methylation (H3K27me3) directed by Polycomb group proteins. When such cells are exposed to cellular stress, such as oncogenic signals, the H3K27me3 mark on the locus is decreased, resulting in expression of *INK4A-ARF* tumor suppressor genes. KRAS can increase the level of the ZNF304 transcription factor that binds to the repressor complex made by KAP1-SETDB1-DNMT1. This event causes the lack of activation of the *INK4A-ARF* locus (Serra et al. 2014). Another example is *STAT3* (Signal transducer and activator of transcription 3) gene. This gene is an important regulator of *NANOG*, *OCT4*, and *SOX2* expression (all epigenetic mediators described above) (Do et al. 2013). STAT3 is also capable of promoting the acquisition of stem cell features in pancreatic cancer (Tyagi et al. 2016), so it can be speculated that external signals may lead to the activation of the epigenetic mediators through STAT3 activation. Moreover STAT3 is capable of interacting with epigenetic modifiers such as p300 histone acetyltransferase (HAT), or DNMT1 influencing gene expression and cell-type specific transcription (Hutchins, Diez, and Miranda-Saavedra 2013).

TP53, a tumor suppressor gene, is also capable of acting as an epigenetic modulator. *TP53* gain of function mutations can induce the expression of *MLL*1 and *MLL2* (mixed-lineage leukemia) genes, and this results in genome-wide increase in H3K4 trimethylation and gene transcription activation (Zhu et al. 2015). Moreover, mutated p53 can help in maintaining an open chromatin conformation at the *VEGFR2* promoter through the recruitment of the SWI/SNF complex (Pfister et al. 2015). Bi-allelic inactivation of *RB1* (retinoblastoma 1) gene, another important tumor suppressor, drives the development of human retinoblastoma, a pediatric tumor of the retina. Whole-genome sequencing of human retinoblastomas identified no genetic lesions in known tumor suppressor genes or oncogenes, other than *RB1*. Furthermore, the epigenetic profile showed profound changes compared to that observed in normal retinoblasts (Zhang et al. 2012). As for epigenetic mediators, the modulators are important targets for the cancer predisposing environment, and their mutations can lead to the destabilization of the epigenome.

18.2.9 microRNA and Cancer

MicroRNA (miRNAs) play an important role in regulating gene expression. MiRNA*s* can be classified as oncogenic, tumor suppressor, or context dependent

(Kasinski and Slack 2011). For example, miR-21 and miR-155 are frequently overexpressed in cancer, while miR-15~16 belong to the family of the onco-suppressor. MiR-146 instead is considered a context-dependent miRNA, because it may have opposing roles in tumorigenesis depending on the cellular context (Kasinski and Slack 2011). Interestingly it has been described in literature how miRNAs can regulate the expression of the epigenetic modifiers. Downregulation of miR-101 can lead to EZH2 (H3K27me3) overexpression in bladder and prostate cancer (Friedman et al. 2009, Varambally et al. 2008). Among the reported downregulated miRNAs in lung cancer, the miRNA miR-29 family (29a, 29b, and 29c) can target the 3'-UTRs of DNA methyltransferase DNMT3A and DNMT3B (*de novo* methyltransferases), two key enzymes involved in DNA methylation, that are frequently upregulated in lung cancer and associated with poor prognosis (Fabbri et al. 2007). In lung cancer miR-449 can downregulate HDAC1 expression, and this results in tumor suppression (Jeon et al. 2012, Rusek et al. 2015). Co-treatment with miR-449a and HDAC inhibitors had a significant growth reduction compared with HDAC inhibitor monotreatment. These results suggest that miR-449a/b may have a tumor suppressor function and might be a potential therapeutic candidate in patients with primary lung cancer. MiRNAs expression can be altered by epigenetic changes given that around half of the miRNA genomic sequences are associated with CpG islands (Weber et al. 2007). A good example is miR-127, which is embedded in a CpG island within a miRNA cluster. The expression of the whole cluster is downregulated or completely silenced in primary tumors (prostate, bladder and colon) and various cancer cell lines (HCT1116, HeLa and MCF7). Interestingly, miR-127 downregulation can be reversed using DNMT inhibitors (5-Aza-CdR) (Saito et al. 2006). MiR-9-1 and miR-34a/b are DNA hypermethylated in breast and colon cancer respectively (Lehmann et al. 2007, Toyota et al. 2008). MicroRNA also play a pivotal role in the early phase of cancer metastasis called epithelial-to-mesenchymal transition (EMT), characterized by the repression of E-cadherin. The zinc finger transcriptional repressors ZEB1, ZEB2, Snail1, and Twist1 are involved in E-cadherin regulation; and thus, EMT. The miR-200 family can inhibit ZEB in several cancer types such as breast, bladder, and ovarian cancers (Bendoraite et al. 2010, Gregory et al. 2008, Park et al. 2008, Adam et al. 2009). MiR-335 is also capable of suppressing migration/invasion through a different pathway involving the progenitor cell transcription factor SOX4 and extracellular matrix component Tenascin C (Tavazoie et al. 2008). MiR-34A is normally induced by the epigenetic modulator p53, while it can be repressed by the cytokine IL-6 and the oncogenic transcription factor STAT3. This event can promote EMT-mediated colorectal cancer invasion and metastasis (Rokavec et al. 2014). MiRNA-466 can reduce prostate cancer tumor growth and bone metastasis (Colden et al. 2017). The let-7 family is downregulated in several human cancers, which is thought to increase tumorigenicity and metastatic ability in breast cancer (Yu, Yao et al. 2007). Interestingly, the mature let-7 can be inhibited by another miRNA, miR-107, an event that promotes tumor progression and metastasis (Chen et al. 2011). Remarkably, circulating miRNAs may also serve as biomarkers for cancer prognosis. A study performed in serum samples from patients with colorectal cancer identified increased levels of circulating miR-92a and mir-29a (Huang et al. 2010).

Similarly, increased serum levels of miR-141 are observed in prostate cancer patients when compared to healthy individuals (Mitchell et al. 2008). The striking involvement of miRNA*s* in several critical cancer-associated processes makes them highly interesting molecules for therapeutic applications. So far, two potential approaches for the regulation of miRNA expression have been evaluated for their use in cancer treatment. One approach is to introduce antisense RNA (Anti-miRs), which can block the function of oncogenic miRNAs, or the re-introduction of a synthetic miRNAs to mimic the action of the tumor suppressor miRNAs. The other approach is focused on inducing the expression of the miRNAs using drugs. The use of drugs implies the modification of the oligonucleotide structure in order to avoid filtration by the kidneys and their clearance through the urine (Chan and Wang 2015).

18.2.10 Epigenetic Inactivation of DNA Repair Genes

Efficient DNA repair is crucial for preventing cancer. Earlier, we discussed how mutations in DNA repair genes could cause inherited cancer syndromes. We also examined how mutations in the epigenetic machinery contribute to cancer. Additionally, DNA repair pathways may be inactivated or decreased in effectiveness by epigenetic inactivation mechanisms affecting DNA repair genes. While mutations in the DNA repair machinery are quite rare in sporadic cancers, those that present with DNA repair deficiencies have one or more epigenetic alterations that reduce or silence the expression of the DNA repair genes (Bernstein and Bernstein 2015). DNA methylation at the promoter region of genes participating in DNA repair pathways including DR, BER, NER, HR, NHEJ, and others has been reported in several cancers, summarized in Table 18.3 and discussed below. It can be assumed that the epigenetic inactivation of DNA repair genes can result in an increase in genetic instability that contributes to tumor progression. On the other hand, diminished DNA repair may

TABLE 18.3
DNA Repair Genes Methylated in Cancer

DNA Repair Pathway	Gene Methylated
BER	MBD4 (Howard et al. 2009, Peng et al. 2006); TDG (Peng et al. 2006); OGG1 (Guan et al. 2008); NEIL1 (Do et al. 2014)
DR	MGMT (Herfarth et al. 1999)
NER	XPC (Yang et al. 2010); RAD23A (Peng et al. 2005); RAD23B, ERCC1 (Chen et al. 2010); ERCC4
MMR	MLH1 (Guan et al. 2008, Esteller et al. 1998, Wang et al. 2003, Kim et al. 2010, Hinrichsen et al. 2014); MSH2 (Wang et al. 2003, Lawes et al. 2005, Hinrichsen et al. 2014); MSH3 (Kim et al. 2010), MSH6 (Lawes et al. 2005); PMS2 (Hinrichsen et al. 2014)
HR	BRCA1 (Dobrovic and Simpfendorfer 1997, Lee, Tseng et al. 2007)
NHEJ	XRCC5 (Lee, Tseng et al. 2007)

also lead to reduced cell survival in general, and additional events are likely occurring that enable a cell with reduced repair capacity to undergo uncontrolled proliferation instead of cell death (e.g. *TP53* pathway inactivation).

18.2.10.1 DR

Epigenetic silencing of the *MGMT* gene has been broadly reported in solid tumors including colon cancer (Herfarth et al. 1999), glioblastoma (Esteller et al. 2000), non-small cell lung cancer (Wolf et al. 2001), and gastric cancer (Oue et al. 2001), among others. Furthermore, in glioma patients, epigenetic silencing of the *MGMT* gene correlates with better response to alkylating agent treatments when compared to patients with tumors with active *MGMT* (Esteller et al. 2000, Hegi et al. 2005).

18.2.10.2 BER

As discussed previously, in BER, MBD4 and TDG are important enzymes for counteracting the hydrolytic deamination of 5-methylcytosine. Promoter methylation of these two DNA repair genes has been observed in various cancer cell lines (Peng et al. 2006, Howard et al. 2009). Furthermore, epigenetic silencing of OGG1, involved in the repair of 8-oxoguanine, has also been observed in some cancer cell lines (Guan et al. 2008).

18.2.10.3 NER

The *XPC* gene, which encodes for the essential subunit for the damage recognition complexes in GGR (Friedberg 2001, Riedl, Hanaoka, and Egly 2003), is silenced in bladder cancer through DNA hypermethylation (Yang et al. 2010). In addition, *RAD23A* and *ERCC1*, two genes that are involved in DNA damage recognition and incision, respectively, as part of the NER pathway are also inactivated through DNA methylation of their promoter region. The *RAD23A* gene is methylated in the multiple myeloma cell line KAS-6/1 (Peng et al. 2005) and *ERCC1* is epigenetically silenced through DNA methylation is associated with drug resistance in glioma cell lines and glioma tumors (Chen et al. 2010).

18.2.10.4 MMR

Approximately 13% of all colorectal cancers present deficiencies in MMR. Among the majority of these—particularly in sporadic disease—have loss of MMR due to silencing of *MLH1* through DNA methylation of the promoter region of the gene (Truninger et al. 2005, Kane et al. 1997). Additionally, this gene is epigenetically inactivated in other types of cancer, including sporadic endometrial carcinoma (Esteller et al. 1998), gastric cancers (Fleisher et al. 1999), ovarian tumors (Gras et al. 2001, Zhang, Zhang et al. 2008), oral squamous cell carcinoma (Czerninski et al. 2009), neck squamous cell carcinoma (Liu et al. 2002), and acute myeloid leukemia (AML) (Seedhouse, Das-Gupta, and Russell 2003). Beyond *MLH1*, other genes that belong to the MMR pathway are also controlled by promoter methylation, including *MSH2*, *MSH3*, and *MSH6*. Indeed, *MSH2*, *MSH3* and *MSH6* are also DNA methylated in colorectal cancer (Lawes et al. 2005, Benachenhou et al. 1998). MSH2 is also DNA methylated in primary non-small cell lung cancer (Wang and Qin 2003), oral squamous cell carcinoma (Czerninski et al. 2009), and ovarian cancer

(Zhang, Zhang et al. 2008). Interestingly, the methylation frequencies in *MLH1* and *MSH3* were significantly higher in elderly gastric carcinoma patients than in younger patients (Kim et al. 2010). Thus, DNA methylation of these genes may have considerable importance in cancer development and as a prognostic factor.

18.2.10.5 HR

Mutation of the *BRCA1* tumor suppressor gene is an important contributing factor in hereditary breast and ovarian cancer. However, *BRCA1* mutations have not been detected in the sporadic forms of these cancers. Still, BRCA1 mRNA and protein levels are reduced in some sporadic breast and ovarian cancers. Aberrant promoter methylation of the *BRCA1* promoter is correlated with the low mRNA and protein levels (Dobrovic and Simpfendorfer 1997). The *BRCA1* promoter is also methylated in gastric cancer (Bernal et al. 2008), non-small cell lung cancer (Lee, Tseng et al. 2007), uterine leiomyosarcoma (Xing et al. 2009), and bladder cancer (Yu, Zhu et al. 2007).

18.2.10.6 NHEJ

While genetic mutation of the genes that regulate NHEJ have not been reported, a deficiency in expression of the Ku80 protein has been observed in melanoma (Korabiowska et al. 2002). In addition, low expression of Ku80 was found in 15% of adenocarcinoma type and 32% of squamous cell type non-small cell lung cancers, which correlated with hypermethylation of the *XRCC5* promoter (Lee, Tseng et al. 2007).

18.2.11 Epigenetics in Cancer Stem Cells

Cancer stem cells (CSCs) are a rare subpopulation of cells present within tumors with the capacity to self-renew and regenerate the whole tumor (Reya et al. 2001). While CSCs were first described in myeloid leukemia, they have also been identified in solid tumors including breast, prostate, colon, brain, pancreas, liver, ovary, melanoma, skin, and head and neck. CSCs are thought to be responsible for sustaining tumor growth and metastases. Most importantly, CSCs are more resistant to therapeutic agents than the non-stem tumor cells, suggesting that CSCs are responsible for tumor relapse. Most of the epigenetic mechanisms with roles in promoting the acquisition of uncontrolled self-renewal previously described in this chapter may contribute to CSC formation and maintenance. One of the best examples of the relevance of DNA methylation in CSC regulation and tumor growth is portrayed in leukemia stem cells, where abrogation of Dnmt1 expression blocks leukemia development in a mouse model. Furthermore, haploinsufficiency of Dnmt1 results in tumor suppressor gene activation, impaired CSC self-renewal and delayed progression of leukemogenesis (Trowbridge et al. 2012). Histone methylation also appears to be involved in CSC formation and maintenance. The Polycomb group complexes target similar sets of CpG-containing genes in embryonic stem cells as in cancer cells, suggesting that these genes may be responsible for the emergence of the CSC phenotype during tumorigenesis (Schlesinger et al. 2007, Widschwendter et al. 2007, Ohm et al. 2007). Further, a knock-in mouse model inducing the expression of EZH2 in hematopoietic stem cells was shown to promote myeloid expansion, which indicates a stem cell-specific EZH2 oncogenic role in myeloid disorders (Herrera-Merchan et al. 2012).

Several other studies support the idea that increased EZH2 expression in some tumors contributes to the maintenance of a reversible and undifferentiated stem-like phenotype in cancer cells and the expansion of breast CSCs (Chang et al. 2011, Burdach et al. 2009). Inhibition of other Polycomb proteins including LSD1 and MLL1 has also been shown to decrease CSC proliferation potential and tumorigenicity (Wang, Lu et al. 2011, Heddleston et al. 2012). Several miRNAs cooperate with Polycomb complexes and DNA methylation to regulate the balance between self-renewal and differentiation in CSCs (Esquela-Kerscher and Slack 2006; Volinia et al. 2006). Let-7 is thought to play a critical role in the breast CSC maintenance (Viswanathan et al. 2009; Yang et al. 2010) and contributes to EZH2 overexpression in prostate cancer (Kong et al. 2012). The downregulation of the miR-200 family members is linked to proliferation of CSCs and their ability to form tumors (Iliopoulos et al. 2010; Lo et al. 2011; Shimono et al. 2009). Further, miR-34a, which is underexpressed in CSCs, negatively regulates the tumor initiating capacity of prostate (Liu et al. 2011), pancreatic (Ji et al. 2009b), and breast (Yu et al. 2012) CSCs.

18.3 CONCLUSION AND FUTURE PROSPECTS

The ongoing molecular characterization of DNA damage response and repair pathways, and how they influence chemo- and radio-resistance, guides the development of novel cancer therapeutics. This information is instrumental in developing DNA repair inhibitors, the latest effort in creating more targeted anticancer treatments that cause less toxicity to normal cells. DNA repair inhibitors are essential in the application of synthetic lethal combinations of drugs and genetic deficiencies.

It is interesting to point out that while in most repair pathways there are several proteins involved in the repair process, within each DNA repair pathway there are specific genes that are preferentially epigenetically silenced. It is yet to be determined whether this specificity is due to selection of particular repair gene silencing events in promoting tumorigenesis or is due to preferential targeting of the DNA methylation machinery to specific DNA repair gene promoters (Lahtz and Pfeifer 2011).

It is also interesting that epigenetic inactivation of DNA repair pathways can lead to different clinical outcomes. Reduced repair capacity for alkylated guanines by promoter methylation of the *MGMT* gene provides a therapeutic benefit in patients with glioma (Esteller et al. 2000). On the other hand, inactivation of the MMR system is associated with resistance of cells to cisplatin treatment (Fink, Zheng et al. 1997, Fink, Nebel et al. 1997). With the mounting knowledge regarding the epigenome of specific cancer types, there is now an opportunity to develop chemotherapy regimens tailored to a patient's DNA repair gene status by incorporating information on epigenetic silencing of the relevant genes in the tumor.

Studies during the last decade emphasize the importance of epigenetic mechanism at most stages during cancer development, which given the reversible nature of epigenetics, present novel opportunities for therapeutic intervention. The observation that genes involved in epigenetic regulation are among the most commonly mutated gene families, underscores the importance of further understanding the epigenetic mechanisms participating in tumor maintenance and progression. Advances in the field will likely require elucidating how different epigenetic proteins contribute to gene-specific gene expression modulation, which would refine our understanding of their roles in tumor maintenance. Considering the multiple roles of epigenetics in cancer, and the possible interference with normal homeostasis, epigenetic interventions in carcinogenesis still faces many challenges, but also offers groundbreaking opportunities to the treatment of these devastating malignancies.

ACKNOWLEDGMENTS

Financial support for this work was provided by the National Institutes of Health, National Cancer Institute CA178207.

REFERENCES

Adam, L., M. Zhong, W. Choi, et al. 2009. "miR-200 expression regulates epithelial-to-mesenchymal transition in bladder cancer cells and reverses resistance to epidermal growth factor receptor therapy." *Clin Cancer Res* 15 (16):5060–72. doi: 10.1158/1078-0432.CCR-08-2245.

Alhosin, M., Z. Omran, M. A. Zamzami, et al. 2016. "Signalling pathways in UHRF1-dependent regulation of tumor suppressor genes in cancer." *J Exp Clin Cancer Res* 35 (1):174. doi: 10.1186/s13046-016-0453-5.

Arai, K., K. Morishita, K. Shinmura, et al. 1997. "Cloning of a human homolog of the yeast OGG1 gene that is involved in the repair of oxidative DNA damage." *Oncogene* 14 (23):2857–61. doi: 10.1038/sj.onc.1201139.

Bader, S., M. Walker, B. Hendrich, et al. 1999. "Somatic frameshift mutations in the MBD4 gene of sporadic colon cancers with mismatch repair deficiency." *Oncogene* 18 (56):8044–7. doi: 10.1038/sj.onc.1203229.

Bartel, D. P. 2009. "MicroRNAs: target recognition and regulatory functions." *Cell* 136 (2):215–33. doi: 10.1016/j.cell.2009.01.002.

Bass, A. J., H. Watanabe, C. H. Mermel, et al. 2009. "SOX2 is an amplified lineage-survival oncogene in lung and esophageal squamous cell carcinomas." *Nat Genet* 41 (11):1238–42. doi: 10.1038/ng.465.

Benachenhou, N., S. Guiral, I. Gorska-Flipot, et al. 1998. "Allelic losses and DNA methylation at DNA mismatch repair loci in sporadic colorectal cancer." *Carcinogenesis* 19 (11):1925–9.

Benavente, C. A., D. Finkelstein, D. A. Johnson, et al. 2014. "Chromatin remodelers HELLS and UHRF1 mediate the epigenetic deregulation of genes that drive retinoblastoma tumor progression." *Oncotarget* 5 (20):9594–608.

Bendoraite, A., E. C. Knouf, K. S. Garg, et al. 2010. "Regulation of miR-200 family microRNAs and ZEB transcription factors in ovarian cancer: evidence supporting a mesothelial-to-epithelial transition." *Gynecol Oncol* 116 (1):117–25. doi: 10.1016/j.ygyno.2009.08.009.

Bernal, C., M. Vargas, F. Ossandon, et al. 2008. "DNA methylation profile in diffuse type gastric cancer: evidence for hypermethylation of the BRCA1 promoter region in early-onset gastric carcinogenesis." *Biol Res* 41 (3):303–15. doi: /S0716-97602008000300007.

Bernstein, C., and H. Bernstein. 2015. "Epigenetic reduction of DNA repair in progression to gastrointestinal cancer." *World J Gastrointest Oncol* 7 (5):30–46. doi: 10.4251/wjgo.v7.i5.30.

Bhutani, N., D. M. Burns, and H. M. Blau. 2011. "DNA demethylation dynamics." *Cell* 146 (6):866–72. doi: 10.1016/j.cell.2011.08.042.

Bronner, C. E., S. M. Baker, P. T. Morrison, et al. 1994. "Mutation in the DNA mismatch repair gene homologue hMLH1 is associated with hereditary non-polyposis colon cancer." *Nature* 368 (6468):258–61. doi: 10.1038/368258a0.

Burdach, S., S. Plehm, R. Unland, et al. 2009. "Epigenetic maintenance of stemness and malignancy in peripheral neuroectodermal tumors by EZH2." *Cell Cycle* 8 (13):1991–6. doi: 10.4161/cc.8.13.8929.

Cao, J., and Q. Yan. 2012. "Histone ubiquitination and deubiquitination in transcription, DNA damage response, and cancer." *Front Oncol* 2:26. doi: 10.3389/fonc.2012.00026.

Castillo-Aguilera, O., P. Depreux, L. Halby, et al. 2017. "DNA methylation targeting: the DNMT/HMT crosstalk challenge." *Biomolecules* 7 (1). doi: 10.3390/biom7010003.

Chan, S. H., and L. H. Wang. 2015. "Regulation of cancer metastasis by microRNAs." *J Biomed Sci* 22:9. doi: 10.1186/s12929-015-0113-7.

Chang, C. J., J. Y. Yang, W. Xia, et al. 2011. "EZH2 promotes expansion of breast tumor initiating cells through activation of RAF1-beta-catenin signaling." *Cancer Cell* 19 (1):86–100. doi: 10.1016/j.ccr.2010.10.035.

Chen, C. L., Y. Wang, Q. Z. Pan, et al. 2016. "Bromodomain-containing protein 7 (BRD7) as a potential tumor suppressor in hepatocellular carcinoma." *Oncotarget* 7 (13):16248–61. doi: 10.18632/oncotarget.7637.

Chen, H. Y., C. J. Shao, F. R. Chen, et al. 2010. "Role of ERCC1 promoter hypermethylation in drug resistance to cisplatin in human gliomas." *Int J Cancer* 126 (8):1944–54. doi: 10.1002/ijc.24772.

Chen, P. S., J. L. Su, S. T. Cha, et al. 2011. "miR-107 promotes tumor progression by targeting the let-7 microRNA in mice and humans." *J Clin Invest* 121 (9):3442–55. doi: 10.1172/JCI45390.

Chen, Q. W., X. Y. Zhu, Y. Y. Li, et al. 2014. "Epigenetic regulation and cancer (review)." *Oncol Rep* 31 (2):523–32. doi: 10.3892/or.2013.2913.

Cheung, P., C. D. Allis, and P. Sassone-Corsi. 2000. "Signaling to chromatin through histone modifications." *Cell* 103 (2):263–71.

Chevillard, S., J. P. Radicella, C. Levalois, et al. 1998. "Mutations in OGG1, a gene involved in the repair of oxidative DNA damage, are found in human lung and kidney tumours." *Oncogene* 16 (23):3083–6. doi: 10.1038/sj.onc.1202096.

Chiou, S. H., M. L. Wang, Y. T. Chou, et al. 2010. "Coexpression of Oct4 and Nanog enhances malignancy in lung adenocarcinoma by inducing cancer stem cell-like properties and epithelial-mesenchymal transdifferentiation." *Cancer Res* 70 (24):10433–44. doi: 10.1158/0008-5472.CAN-10-2638.

Choi, J. H., H. J. Kwon, B. I. Yoon, et al. 2001. "Expression profile of histone deacetylase 1 in gastric cancer tissues." *Jpn J Cancer Res* 92 (12):1300–4.

Cimmino, L., O. Abdel-Wahab, R. L. Levine, et al. 2011. "TET family proteins and their role in stem cell differentiation and transformation." *Cell Stem Cell* 9 (3):193–204. doi: 10.1016/j.stem.2011.08.007.

Cleaver, J. E. 1968. "Defective repair replication of DNA in xeroderma pigmentosum." *Nature* 218 (5142):652–6.

Colden, M., A. A. Dar, S. Saini, et al. 2017. "MicroRNA-466 inhibits tumor growth and bone metastasis in prostate cancer by direct regulation of osteogenic transcription factor RUNX2." *Cell Death Dis* 8 (1):e2572. doi: 10.1038/cddis.2017.15.

Costa, F. F. 2010. "Non-coding RNAs: Meet thy masters." *Bioessays* 32 (7):599–608. doi: 10.1002/bies.200900112.

Couronne, L., C. Bastard, and O. A. Bernard. 2012. "TET2 and DNMT3A mutations in human T-cell lymphoma." *N Engl J Med* 366 (1):95–6. doi: 10.1056/NEJMc1111708.

Czerninski, R., S. Krichevsky, Y. Ashhab, et al. 2009. "Promoter hypermethylation of mismatch repair genes, hMLH1 and hMSH2 in oral squamous cell carcinoma." *Oral Dis* 15 (3):206–13. doi: 10.1111/j.1601-0825.2008.01510.x.

De Carvalho, D. D., S. Sharma, J. S. You, et al. 2012. "DNA methylation screening identifies driver epigenetic events of cancer cell survival." *Cancer Cell* 21 (5):655–67. doi: 10.1016/j.ccr.2012.03.045.

Do, D. V., J. Ueda, D. M. Messerschmidt, et al. 2013. "A genetic and developmental pathway from STAT3 to the OCT4-NANOG circuit is essential for maintenance of ICM lineages in vivo." *Genes Dev* 27 (12):1378–90. doi: 10.1101/gad.221176.113.

Do, H., N. C. Wong, C. Murone, et al. 2014. "A critical re-assessment of DNA repair gene promoter methylation in non-small cell lung carcinoma." *Sci Rep* 4:4186. doi: 10.1038/srep04186.

Dobrovic, A., and D. Simpfendorfer. 1997. "Methylation of the BRCA1 gene in sporadic breast cancer." *Cancer Res* 57 (16):3347–50.

Dover, J., J. Schneider, M. A. Tawiah-Boateng, et al. 2002. "Methylation of histone H3 by COMPASS requires ubiquitination of histone H2B by Rad6." *J Biol Chem* 277 (32):28368–71. doi: 10.1074/jbc.C200348200.

Duncan, J. A., J. R. Reeves, and T. G. Cooke. 1998. "BRCA1 and BRCA2 proteins: roles in health and disease." *Mol Pathol* 51 (5):237–47.

Esquela-Kerscher, A., and F. J. Slack. 2006. "Oncomirs - microRNAs with a role in cancer." *Nat Rev Cancer* 6 (4):259–69.

Essers, J., H. van Steeg, J. de Wit, et al. 2000. "Homologous and non-homologous recombination differentially affect DNA damage repair in mice." *EMBO J* 19 (7):1703–10. doi: 10.1093/emboj/19.7.1703.

Esteller, M. 2007. "Cancer epigenomics: DNA methylomes and histone-modification maps." *Nat Rev Genet* 8 (4):286-98. doi: 10.1038/nrg2005.

Esteller, M., J. Garcia-Foncillas, E. Andion, et al. 2000. "Inactivation of the DNA-repair gene MGMT and the clinical response of gliomas to alkylating agents." *N Engl J Med* 343 (19):1350–4. doi: 10.1056/NEJM200011093431901.

Esteller, M., R. Levine, S. B. Baylin, et al. 1998. "MLH1 promoter hypermethylation is associated with the microsatellite instability phenotype in sporadic endometrial carcinomas." *Oncogene* 17 (18):2413–7. doi: 10.1038/sj.onc.1202178.

Fabbri, M., R. Garzon, A. Cimmino, et al. 2007. "MicroRNA-29 family reverts aberrant methylation in lung cancer by targeting DNA methyltransferases 3A and 3B." *Proc Natl Acad Sci U S A* 104 (40):15805–10. doi: 10.1073/pnas.0707628104.

Feinberg, A. P., M. A. Koldobskiy, and A. Gondor. 2016. "Epigenetic modulators, modifiers and mediators in cancer aetiology and progression." *Nat Rev Genet* 17 (5):284–99. doi: 10.1038/nrg.2016.13.

Figueroa, M. E., O. Abdel-Wahab, C. Lu, et al. 2010. "Leukemic IDH1 and IDH2 mutations result in a hypermethylation phenotype, disrupt TET2 function, and impair hematopoietic differentiation." *Cancer Cell* 18 (6):553–67. doi: 10.1016/j.ccr.2010.11.015.

Fink, D., S. Nebel, S. Aebi, et al. 1997. "Loss of DNA mismatch repair due to knockout of MSH2 or PMS2 results in resistance to cisplatin and carboplatin." *Int J Oncol* 11 (3):539–42.

Fink, D., H. Zheng, S. Nebel, et al. 1997. "In vitro and in vivo resistance to cisplatin in cells that have lost DNA mismatch repair." *Cancer Res* 57 (10):1841–5.

Fischle, W., Y. Wang, and C. D. Allis. 2003. "Histone and chromatin cross-talk." *Curr Opin Cell Biol* 15 (2):172–83.

Fleck, O., and O. Nielsen. 2004. "DNA repair." *J Cell Sci* 117 (Pt 4):515–7. doi: 10.1242/jcs.00952.

Fleisher, A. S., M. Esteller, S. Wang, et al. 1999. "Hypermethylation of the hMLH1 gene promoter in human gastric cancers with microsatellite instability." *Cancer Res* 59 (5):1090–5.

Fousteri, M., and L. H. Mullenders. 2008. "Transcription-coupled nucleotide excision repair in mammalian cells: molecular mechanisms and biological effects." *Cell Res* 18 (1):73–84. doi: 10.1038/cr.2008.6.

French, C. A., I. Miyoshi, J. C. Aster, et al. 2001. "BRD4 bromodomain gene rearrangement in aggressive carcinoma with translocation t(15;19)." *Am J Pathol* 159 (6):1987–92. doi: 10.1016/S0002-9440(10)63049-0.

Friedberg, E. C. 2001. "How nucleotide excision repair protects against cancer." *Nat Rev Cancer* 1 (1):22–33. doi: 10.1038/35094000.

Friedman, J. M., G. Liang, C. C. Liu, et al. 2009. "The putative tumor suppressor microRNA-101 modulates the cancer epigenome by repressing the polycomb group protein EZH2." *Cancer Res* 69 (6):2623–9. doi: 10.1158/0008-5472.CAN-08-3114.

Fujimoto, A., Y. Totoki, T. Abe, et al. 2012. "Whole-genome sequencing of liver cancers identifies etiological influences on mutation patterns and recurrent mutations in chromatin regulators." *Nat Genet* 44 (7):760–4. doi: 10.1038/ng.2291.

Fujisawa, T., and P. Filippakopoulos. 2017. "Functions of bromodomain-containing proteins and their roles in homeostasis and cancer." *Nat Rev Mol Cell Biol.* doi: 10.1038/nrm.2016.143.

Fuks, F. 2005. "DNA methylation and histone modifications: teaming up to silence genes." *Curr Opin Genet Dev* 15 (5):490–5. doi: 10.1016/j.gde.2005.08.002.

Gaidzik, V. I., P. Paschka, D. Spath, et al. 2012. "TET2 mutations in acute myeloid leukemia (AML): results from a comprehensive genetic and clinical analysis of the AML study group." *J Clin Oncol* 30 (12):1350–7. doi: 10.1200/JCO.2011.39.2886.

Gao, Y., Y. Zhao, J. Zhang, et al. 2016. "The dual function of PRMT1 in modulating epithelial-mesenchymal transition and cellular senescence in breast cancer cells through regulation of ZEB1." *Sci Rep* 6:19874. doi: 10.1038/srep19874.

Gidekel, S., G. Pizov, Y. Bergman, and E. Pikarsky. 2003. "Oct-3/4 is a dose-dependent oncogenic fate determinant." *Cancer Cell* 4 (5):361–70.

Goldknopf, I. L., C. W. Taylor, R. M. Baum, et al. 1975. "Isolation and characterization of protein A24, a "histone-like" non-histone chromosomal protein." *J Biol Chem* 250 (18):7182–7.

Gras, E., L. Catasus, R. Arguelles, et al. 2001. "Microsatellite instability, MLH-1 promoter hypermethylation, and frameshift mutations at coding mononucleotide repeat microsatellites in ovarian tumors." *Cancer* 92 (11):2829–36.

Gregory, P. A., A. G. Bert, E. L. Paterson, et al. 2008. "The miR-200 family and miR-205 regulate epithelial to mesenchymal transition by targeting ZEB1 and SIP1." *Nat Cell Biol* 10 (5):593–601. doi: 10.1038/ncb1722.

Grossmann, V., C. Haferlach, S. Weissmann, et al. 2013. "The molecular profile of adult T-cell acute lymphoblastic leukemia: mutations in RUNX1 and DNMT3A are associated with poor prognosis in T-ALL." *Genes Chromosomes Cancer* 52 (4):410–2. doi: 10.1002/gcc.22039.

Guan, H., M. Ji, P. Hou, et al. 2008. "Hypermethylation of the DNA mismatch repair gene hMLH1 and its association with lymph node metastasis and T1799A BRAF mutation in patients with papillary thyroid cancer." *Cancer* 113 (2):247–55. doi: 10.1002/cncr.23548.

Halkidou, K., L. Gaughan, S. Cook, et al. 2004. "Upregulation and nuclear recruitment of HDAC1 in hormone refractory prostate cancer." *Prostate* 59 (2):177–89. doi: 10.1002/pros.20022.

Hanawalt, P. C., and G. Spivak. 2008. "Transcription-coupled DNA repair: two decades of progress and surprises." *Nat Rev Mol Cell Biol* 9 (12):958-70. doi: 10.1038/nrm2549.

Heddleston, J. M., Q. Wu, M. Rivera, et al. 2012. "Hypoxia-induced mixed-lineage leukemia 1 regulates glioma stem cell tumorigenic potential." *Cell Death Differ* 19 (3):428–39. doi: 10.1038/cdd.2011.109.

Hegi, M. E., A. C. Diserens, T. Gorlia, et al. 2005. "MGMT gene silencing and benefit from temozolomide in glioblastoma." *N Engl J Med* 352 (10):997–1003. doi: 10.1056/NEJMoa043331.

Hendrich, B., U. Hardeland, H. H. Ng, et al. 1999. "The thymine glycosylase MBD4 can bind to the product of deamination at methylated CpG sites." *Nature* 401 (6750):301–4. doi: 10.1038/45843.

Herfarth, K. K., T. P. Brent, R. P. Danam, et al. 1999. "A specific CpG methylation pattern of the MGMT promoter region associated with reduced MGMT expression in primary colorectal cancers." *Mol Carcinog* 24 (2):90–8.

Herrera-Merchan, A., L. Arranz, J. M. Ligos, et al. 2012. "Ectopic expression of the histone methyltransferase Ezh2 in haematopoietic stem cells causes myeloproliferative disease." *Nat Commun* 3:623. doi: 10.1038/ncomms1623.

Hinrichsen, I., M. Kemp, J. Peveling-Oberhag, et al. 2014. "Promoter methylation of MLH1, PMS2, MSH2 and p16 is a phenomenon of advanced-stage HCCs." *PLoS One* 9 (1):e84453. doi: 10.1371/journal.pone.0084453.

Howard, J. H., A. Frolov, C. W. Tzeng, et al. 2009. "Epigenetic downregulation of the DNA repair gene MED1/MBD4 in colorectal and ovarian cancer." *Cancer Biol Ther* 8 (1):94–100.

Huang, Z., D. Huang, S. Ni, et al. 2010. "Plasma microRNAs are promising novel biomarkers for early detection of colorectal cancer." *Int J Cancer* 127 (1):118–26. doi: 10.1002/ijc.25007.

Hutchins, A. P., D. Diez, and D. Miranda-Saavedra. 2013. "Genomic and computational approaches to dissect the mechanisms of STAT3's universal and cell type-specific functions." *JAKSTAT* 2 (4):e25097. doi: 10.4161/jkst.25097.

Iliopoulos, D., M. Lindahl-Allen, C. Polytarchou, et al. 2010. "Loss of miR-200 inhibition of Suz12 leads to polycomb-mediated repression required for the formation and maintenance of cancer stem cells." *Mol Cell* 39(5):761–2.

Iyer, R. R., A. Pluciennik, V. Burdett, et al. 2006. "DNA mismatch repair: functions and mechanisms." *Chem Rev* 106 (2):302–23. doi: 10.1021/cr0404794.

Jackson, S. P., and J. Bartek. 2009. "The DNA-damage response in human biology and disease." *Nature* 461 (7267):1071–8. doi: 10.1038/nature08467.

Jason, L. J., S. C. Moore, J. D. Lewis, et al. 2002. "Histone ubiquitination: a tagging tail unfolds?" *Bioessays* 24 (2):166–74. doi: 10.1002/bies.10038.

Jeon, H. S., S. Y. Lee, E. J. Lee, et al. 2012. "Combining microRNA-449a/b with a HDAC inhibitor has a synergistic effect on growth arrest in lung cancer." *Lung Cancer* 76 (2):171–6. doi: 10.1016/j.lungcan.2011.10.012.

Jiricny, J. 2006. "The multifaceted mismatch-repair system." *Nat Rev Mol Cell Biol* 7 (5):335–46. doi: 10.1038/nrm1907.

Kaina, B., M. Christmann, S. Naumann, et al. 2007. "MGMT: key node in the battle against genotoxicity, carcinogenicity and apoptosis induced by alkylating agents." *DNA Repair (Amst)* 6 (8):1079–99. doi: 10.1016/j.dnarep.2007.03.008.

Kanai, Y., S. Ushijima, Y. Nakanishi, et al. 2003. "Mutation of the DNA methyltransferase (DNMT) 1 gene in human colorectal cancers." *Cancer Lett* 192 (1):75–82.

Kane, M. F., M. Loda, G. M. Gaida, et al. 1997. "Methylation of the hMLH1 promoter correlates with lack of expression of hMLH1 in sporadic colon tumors and mismatch repair-defective human tumor cell lines." *Cancer Res* 57 (5):808–11.

Kasinski, A. L., and F. J. Slack. 2011. "Epigenetics and genetics. MicroRNAs en route to the clinic: progress in validating and targeting microRNAs for cancer therapy." *Nat Rev Cancer* 11 (12):849–64. doi: 10.1038/nrc3166.

Kelley, M. R., and M. L. Fishel. 2008. "DNA repair proteins as molecular targets for cancer therapeutics." *Anticancer Agents Med Chem* 8 (4):417–25.

Kim, H. G., S. Lee, D. Y. Kim, et al. 2010. "Aberrant methylation of DNA mismatch repair genes in elderly patients with sporadic gastric carcinoma: A comparison with younger patients." *J Surg Oncol* 101 (1):28–35. doi: 10.1002/jso.21432.

Ko, M., Y. Huang, A. M. Jankowska, et al. 2010. "Impaired hydroxylation of 5-methylcytosine in myeloid cancers with mutant TET2." *Nature* 468 (7325):839–43. doi: 10.1038/nature09586.

Kofunato, Y., K. Kumamoto, K. Saitou, et al. 2012. "UHRF1 expression is upregulated and associated with cellular proliferation in colorectal cancer." *Oncol Rep* 28 (6):1997–2002. doi: 10.3892/or.2012.2064.

Korabiowska, M., M. Tscherny, J. Stachura, et al. 2002. "Differential expression of DNA nonhomologous end-joining proteins Ku70 and Ku80 in melanoma progression." *Mod Pathol* 15 (4):426–33. doi: 10.1038/modpathol.3880542.

Kouzarides, T. 2007. "Chromatin modifications and their function." *Cell* 128 (4):693–705. doi: 10.1016/j.cell.2007.02.005.

Lahtz, C., and G. P. Pfeifer. 2011. "Epigenetic changes of DNA repair genes in cancer." *J Mol Cell Biol* 3 (1):51–8. doi: 10.1093/jmcb/mjq053.

Latham, J. A., and S. Y. Dent. 2007. "Cross-regulation of histone modifications." *Nat Struct Mol Biol* 14 (11):1017–24. doi: 10.1038/nsmb1307.

Lawes, D. A., T. Pearson, S. Sengupta, et al. 2005. "The role of MLH1, MSH2 and MSH6 in the development of multiple colorectal cancers." *Br J Cancer* 93 (4):472–7. doi: 10.1038/sj.bjc.6602708.

Lazzaro, F., M. Giannattasio, F. Puddu, et al. 2009. "Checkpoint mechanisms at the intersection between DNA damage and repair." *DNA Repair (Amst)* 8 (9):1055–67. doi: 10.1016/j.dnarep.2009.04.022.

Lee, J. S., A. Shukla, J. Schneider, et al. 2007. "Histone crosstalk between H2B monoubiquitination and H3 methylation mediated by COMPASS." *Cell* 131 (6):1084–96. doi: 10.1016/j.cell.2007.09.046.

Lee, M. N., R. C. Tseng, H. S. Hsu, et al. 2007. "Epigenetic inactivation of the chromosomal stability control genes BRCA1, BRCA2, and XRCC5 in non-small cell lung cancer." *Clin Cancer Res* 13 (3):832–8. doi: 10.1158/1078-0432.CCR-05-2694.

Lehmann, U., B. Hasemeier, D. Romermann, et al. 2007. "Epigenetic inactivation of microRNA genes in mammary carcinoma." *Verh Dtsch Ges Pathol* 91:214–20.

Ley, T. J., L. Ding, M. J. Walter, et al. 2010. "DNMT3A mutations in acute myeloid leukemia." *N Engl J Med* 363 (25):2424–33. doi: 10.1056/NEJMoa1005143.

Li, G. M. 2008. "Mechanisms and functions of DNA mismatch repair." *Cell Res* 18 (1):85–98. doi: 10.1038/cr.2007.115.

Liu, K., H. Huang, P. Mukunyadzi, et al. 2002. "Promoter hypermethylation: an important epigenetic mechanism for hMLH1 gene inactivation in head and neck squamous cell carcinoma." *Otolaryngol Head Neck Surg* 126 (5):548–53. doi: 10.1067/mhn.2002.124934.

Liu, L., S. Markowitz, and S. L. Gerson. 1996. "Mismatch repair mutations override alkyltransferase in conferring resistance to temozolomide but not to 1,3-bis(2-chloroethyl) nitrosourea." *Cancer Res* 56 (23):5375–9.

Lo, W. L., C. C. Yu, G. Y. Chiou, et al. 2011. "MicroRNA-200c attenuates tumour growth and metastasis of presumptive head and neck squamous cell carcinoma stem cells." *J Pathol* 223 (4):482–9.

Lo, W. S., R. C. Trievel, J. R. Rojas, et al. 2000. "Phosphorylation of serine 10 in histone H3 is functionally linked in vitro and in vivo to Gcn5-mediated acetylation at lysine 14." *Mol Cell* 5 (6):917–26.

Lu, C., P. S. Ward, G. S. Kapoor, et al. 2012. "IDH mutation impairs histone demethylation and results in a block to cell differentiation." *Nature* 483 (7390):474–8. doi: 10.1038/nature10860.

Lu, X., S. J. Mazur, T. Lin, et al. 2014. "The pluripotency factor nanog promotes breast cancer tumorigenesis and metastasis." *Oncogene* 33 (20):2655–64. doi: 10.1038/onc.2013.209.

Luger, K., A. W. Mader, R. K. Richmond, et al. 1997. "Crystal structure of the nucleosome core particle at 2.8 A resolution." *Nature* 389 (6648):251–60. doi: 10.1038/38444.

Luo, M., H. He, M. R. Kelley, et al. 2010. "Redox regulation of DNA repair: implications for human health and cancer therapeutic development." *Antioxid Redox Signal* 12 (11):1247–69. doi: 10.1089/ars.2009.2698.

Luo, W., S. Li, B. Peng, et al. 2013. "Embryonic stem cells markers SOX2, OCT4 and Nanog expression and their correlations with epithelial-mesenchymal transition in nasopharyngeal carcinoma." *PLoS One* 8 (2):e56324. doi: 10.1371/journal.pone.0056324.

Mari, P. O., B. I. Florea, S. P. Persengiev, et al. 2006. "Dynamic assembly of end-joining complexes requires interaction between Ku70/80 and XRCC4." *Proc Natl Acad Sci U S A* 103 (49):18597–602. doi: 10.1073/pnas.0609061103.

Miki, Y., J. Swensen, D. Shattuck-Eidens, et al. 1994. "A strong candidate for the breast and ovarian cancer susceptibility gene BRCA1." *Science* 266 (5182):66–71.

Mitchell, P. S., R. K. Parkin, E. M. Kroh, et al. 2008. "Circulating microRNAs as stable blood-based markers for cancer detection." *Proc Natl Acad Sci U S A* 105 (30):10513–8. doi: 10.1073/pnas.0804549105.

Natarajan, A. T., S. Vermeulen, F. Darroudi, et al. 1992. "Chromosomal localization of human O6-methylguanine-DNA methyltransferase (MGMT) gene by in situ hybridization." *Mutagenesis* 7 (1):83–5.

Ohm, J. E., K. M. McGarvey, X. Yu, et al. 2007. "A stem cell-like chromatin pattern may predispose tumor suppressor genes to DNA hypermethylation and heritable silencing." *Nat Genet* 39 (2):237–42. doi: 10.1038/ng1972.

Oue, N., K. Sentani, H. Yokozaki, et al. 2001. "Promoter methylation status of the DNA repair genes hMLH1 and MGMT in gastric carcinoma and metaplastic mucosa." *Pathobiology* 69 (3):143–9. doi: 48769.

Ozdag, H., A. E. Teschendorff, A. A. Ahmed, et al. 2006. "Differential expression of selected histone modifier genes in human solid cancers." *BMC Genomics* 7:90. doi: 10.1186/1471-2164-7-90.

Park, S. M., A. B. Gaur, E. Lengyel, et al. 2008. "The miR-200 family determines the epithelial phenotype of cancer cells by targeting the E-cadherin repressors ZEB1 and ZEB2." *Genes Dev* 22 (7):894–907. doi: 10.1101/gad.1640608.

Peng, B., D. R. Hodge, S. B. Thomas, et al. 2005. "Epigenetic silencing of the human nucleotide excision repair gene, hHR23B, in interleukin-6-responsive multiple myeloma KAS-6/1 cells." *J Biol Chem* 280 (6):4182–7. doi: 10.1074/jbc.M412566200.

Peng, B., E. M. Hurt, D. R. Hodge, et al. 2006. "DNA hypermethylation and partial gene silencing of human thymine- DNA glycosylase in multiple myeloma cell lines." *Epigenetics* 1 (3):138–45.

Pfeifer, G. P. 2006. "Mutagenesis at methylated CpG sequences." *Curr Top Microbiol Immunol* 301:259–81.

Pfister, N. T., V. Fomin, K. Regunath, et al. 2015. "Mutant p53 cooperates with the SWI/SNF chromatin remodeling complex to regulate VEGFR2 in breast cancer cells." *Genes Dev* 29 (12):1298–315. doi: 10.1101/gad.263202.115.

Pham, A. D., and F. Sauer. 2000. "Ubiquitin-activating/conjugating activity of TAFII250, a mediator of activation of gene expression in Drosophila." *Science* 289 (5488):2357–60.

Rao, R. C., and Y. Dou. 2015. "Hijacked in cancer: the KMT2 (MLL) family of methyltransferases." *Nat Rev Cancer* 15 (6):334–46. doi: 10.1038/nrc3929.

Reya, T., S. J. Morrison, M. F. Clarke, et al. 2001. "Stem cells, cancer, and cancer stem cells." *Nature* 414 (6859):105–11. doi: 10.1038/35102167.

Ribeiro, A. F., M. Pratcorona, C. Erpelinck-Verschueren, et al. 2012. "Mutant DNMT3A: a marker of poor prognosis in acute myeloid leukemia." *Blood* 119 (24):5824–31. doi: 10.1182/blood-2011-07-367961.

Riedl, T., F. Hanaoka, and J. M. Egly. 2003. "The comings and goings of nucleotide excision repair factors on damaged DNA." *EMBO J* 22 (19):5293–303. doi: 10.1093/emboj/cdg489.

Robzyk, K., J. Recht, and M. A. Osley. 2000. "Rad6-dependent ubiquitination of histone H2B in yeast." *Science* 287 (5452):501–4.

Rokavec, M., H. Li, L. Jiang, et al. 2014. "The p53/microRNA connection in gastrointestinal cancer." *Clin Exp Gastroenterol* 7:395–413. doi: 10.2147/CEG.S43738.

Roll, J. D., A. G. Rivenbark, W. D. Jones, et al. 2008. "DNMT3b overexpression contributes to a hypermethylator phenotype in human breast cancer cell lines." *Mol Cancer* 7:15. doi: 10.1186/1476-4598-7-15.

Ropero, S., M. F. Fraga, E. Ballestar, et al. 2006. "A truncating mutation of HDAC2 in human cancers confers resistance to histone deacetylase inhibition." *Nat Genet* 38 (5):566–9. doi: 10.1038/ng1773.

Rotili, D., and A. Mai. 2011. "Targeting Histone Demethylases: A New Avenue for the Fight against Cancer." *Genes Cancer* 2 (6):663–79. doi: 10.1177/1947601911417976.

Rudin, C. M., S. Durinck, E. W. Stawiski, et al. 2012. "Comprehensive genomic analysis identifies SOX2 as a frequently amplified gene in small-cell lung cancer." *Nat Genet* 44 (10):1111–6. doi: 10.1038/ng.2405.

Rusek, A. M., M. Abba, A. Eljaszewicz, et al. 2015. "MicroRNA modulators of epigenetic regulation, the tumor microenvironment and the immune system in lung cancer." *Mol Cancer* 14:34. doi: 10.1186/s12943-015-0302-8.

Saito, Y., G. Liang, G. Egger, et al. 2006. "Specific activation of microRNA-127 with down-regulation of the proto-oncogene BCL6 by chromatin-modifying drugs in human cancer cells." *Cancer Cell* 9 (6):435–43. doi: 10.1016/j.ccr.2006.04.020.

Schlesinger, Y., R. Straussman, I. Keshet, et al. 2007. "Polycomb-mediated methylation on Lys27 of histone H3 pre-marks genes for de novo methylation in cancer." *Nat Genet* 39 (2):232–6. doi: 10.1038/ng1950.

Seedhouse, C. H., E. P. Das-Gupta, and N. H. Russell. 2003. "Methylation of the hMLH1 promoter and its association with microsatellite instability in acute myeloid leukemia." *Leukemia* 17 (1):83–8. doi: 10.1038/sj.leu.2402747.

Serra, R. W., M. Fang, S. M. Park, et al. 2014. "A KRAS-directed transcriptional silencing pathway that mediates the CpG island methylator phenotype." *Elife* 3:e02313. doi: 10.7554/eLife.02313.

Setlow, R. B., J. D. Regan, J. German, et al. 1969. "Evidence that xeroderma pigmentosum cells do not perform the first step in the repair of ultraviolet damage to their DNA." *Proc Natl Acad Sci U S A* 64 (3):1035–41.

Sharma, S., T. K. Kelly, and P. A. Jones. 2010. "Epigenetics in cancer." *Carcinogenesis* 31 (1):27–36. doi: 10.1093/carcin/bgp220.

Shema, E., I. Tirosh, Y. Aylon, et al. 2008. "The histone H2B-specific ubiquitin ligase RNF20/hBRE1 acts as a putative tumor suppressor through selective regulation of gene expression." *Genes Dev* 22 (19):2664–76. doi: 10.1101/gad.1703008.

Shen, H., L. Wang, M. R. Spitz, et al. 2002. "A novel polymorphism in human cytosine DNA-methyltransferase-3B promoter is associated with an increased risk of lung cancer." *Cancer Res* 62 (17):4992–5.

Shimono, Y., M. Zabala, R. W. Cho, et al. 2009. "Downregulation of miRNA-200c links breast cancer stem cells with normal stem cells." *Cell* 138 (3):592–603.

Shinmura, K., and J. Yokota. 2001. "The OGG1 gene encodes a repair enzyme for oxidatively damaged DNA and is involved in human carcinogenesis." *Antioxid Redox Signal* 3 (4):597–609. doi: 10.1089/15230860152542952.

Sjoblom, T., S. Jones, L. D. Wood, et al. 2006. "The consensus coding sequences of human breast and colorectal cancers." *Science* 314 (5797):268–74. doi: 10.1126/science.1133427.

Sugasawa, K., T. Okamoto, Y. Shimizu, et al. 2001. "A multistep damage recognition mechanism for global genomic nucleotide excision repair." *Genes Dev* 15 (5):507–21. doi: 10.1101/gad.866301.

Sun, Z. W., and C. D. Allis. 2002. "Ubiquitination of histone H2B regulates H3 methylation and gene silencing in yeast." *Nature* 418 (6893):104-8. doi: 10.1038/nature00883.

Tano, K., S. Shiota, J. Collier, et al. 1990. "Isolation and structural characterization of a cDNA clone encoding the human DNA repair protein for O6-alkylguanine." *Proc Natl Acad Sci U S A* 87 (2):686–90.

Tarighat, S. S., R. Santhanam, D. Frankhouser, et al. 2016. "The dual epigenetic role of PRMT5 in acute myeloid leukemia: gene activation and repression via histone arginine methylation." *Leukemia* 30 (4):789–99. doi: 10.1038/leu.2015.308.

Tavazoie, S. F., C. Alarcon, T. Oskarsson, et al. 2008. "Endogenous human microRNAs that suppress breast cancer metastasis." *Nature* 451 (7175):147–52. doi: 10.1038/nature06487.

Tessarz, P., and T. Kouzarides. 2014. "Histone core modifications regulating nucleosome structure and dynamics." *Nat Rev Mol Cell Biol* 15 (11):703–8. doi: 10.1038/nrm3890.

Thirman, M. J., H. J. Gill, R. C. Burnett, et al. 1993. "Rearrangement of the MLL gene in acute lymphoblastic and acute myeloid leukemias with 11q23 chromosomal translocations." *N Engl J Med* 329 (13):909–14. doi: 10.1056/NEJM199309233291302.

Toyota, M., H. Suzuki, Y. Sasaki, et al. 2008. "Epigenetic silencing of microRNA-34b/c and B-cell translocation gene 4 is associated with CpG island methylation in colorectal cancer." *Cancer Res* 68 (11):4123–32. doi: 10.1158/0008-5472.CAN-08-0325.

Trowbridge, J. J., A. U. Sinha, N. Zhu, et al. 2012. "Haploinsufficiency of Dnmt1 impairs leukemia stem cell function through derepression of bivalent chromatin domains." *Genes Dev* 26 (4):344–9. doi: 10.1101/gad.184341.111.

Truninger, K., M. Menigatti, J. Luz, et al. 2005. "Immunohistochemical analysis reveals high frequency of PMS2 defects in colorectal cancer." *Gastroenterology* 128 (5):1160–71.

Turcan, S., D. Rohle, A. Goenka, et al. 2012. "IDH1 mutation is sufficient to establish the glioma hypermethylator phenotype." *Nature* 483 (7390):479–83. doi: 10.1038/nature10866.

Tyagi, N., S. Marimuthu, A. Bhardwaj, et al. 2016. "p-21 activated kinase 4 (PAK4) maintains stem cell-like phenotypes in pancreatic cancer cells through activation of STAT3 signaling." *Cancer Lett* 370 (2):260–7. doi: 10.1016/j.canlet.2015.10.028.

Varambally, S., Q. Cao, R. S. Mani, et al. 2008. "Genomic loss of microRNA-101 leads to overexpression of histone methyltransferase EZH2 in cancer." *Science* 322 (5908):1695–9. doi: 10.1126/science.1165395.

Varier, R. A., and H. T. Timmers. 2011. "Histone lysine methylation and demethylation pathways in cancer." *Biochim Biophys Acta* 1815 (1):75–89. doi: 10.1016/j.bbcan.2010.10.002.

Volinia, S., G. A. Calin, C. G. Liu, et al. 2006. "A microRNA expression signature of human solid tumors defines cancer gene targets." *Proc Natl Acad Sci U S A* 103 (7):2257–61.

von Eyss, B., J. Maaskola, S. Memczak, et al. 2012. "The SNF2-like helicase HELLS mediates E2F3-dependent transcription and cellular transformation." *EMBO J* 31 (4):972–85. doi: 10.1038/emboj.2011.451.

Wang, D., P. Lu, H. Zhang, M. et al. 2014. "Oct-4 and Nanog promote the epithelial-mesenchymal transition of breast cancer stem cells and are associated with poor prognosis in breast cancer patients." *Oncotarget* 5 (21):10803–15. doi: 10.18632/oncotarget.2506.

Wang, H., L. Zhai, J. Xu, et al. 2006. "Histone H3 and H4 ubiquitylation by the CUL4-DDB-ROC1 ubiquitin ligase facilitates cellular response to DNA damage." *Mol Cell* 22 (3):383–94. doi: 10.1016/j.molcel.2006.03.035.

Wang, J., F. Lu, Q. Ren, et al. 2011. "Novel histone demethylase LSD1 inhibitors selectively target cancer cells with pluripotent stem cell properties." *Cancer Res* 71 (23):7238–49. doi: 10.1158/0008-5472.CAN-11-0896.

Wang, L., A. Gural, X. J. Sun, et al. 2011. "The leukemogenicity of AML1-ETO is dependent on site-specific lysine acetylation." *Science* 333 (6043):765–9. doi: 10.1126/science.1201662.

Wang, Y. C., Y. P. Lu, R. C. Tseng, et al. 2003. "Inactivation of hMLH1 and hMSH2 by promoter methylation in primary non-small cell lung tumors and matched sputum samples." *J Clin Invest* 111 (6):887–95. doi: 10.1172/JCI15475.

Wang, Y., and J. Qin. 2003. "MSH2 and ATR form a signaling module and regulate two branches of the damage response to DNA methylation." *Proc Natl Acad Sci U S A* 100 (26):15387–92. doi: 10.1073/pnas.2536810100.

Weber, B., C. Stresemann, B. Brueckner, et al. 2007. "Methylation of human microRNA genes in normal and neoplastic cells." *Cell Cycle* 6 (9):1001–5. doi: 10.4161/cc.6.9.4209.

Wen, B., H. Wu, Y. Shinkai, et al. 2009. "Large histone H3 lysine 9 dimethylated chromatin blocks distinguish differentiated from embryonic stem cells." *Nat Genet* 41 (2):246–50. doi: 10.1038/ng.297.

West, M. H., and W. M. Bonner. 1980. "Histone 2B can be modified by the attachment of ubiquitin." *Nucleic Acids Res* 8 (20):4671–80.

Widschwendter, M., H. Fiegl, D. Egle, et al. 2007. "Epigenetic stem cell signature in cancer." *Nat Genet* 39 (2):157–8. doi: 10.1038/ng1941.

Wijmenga, C., R. S. Hansen, G. Gimelli, et al. 2000. "Genetic variation in ICF syndrome: evidence for genetic heterogeneity." *Hum Mutat* 16 (6):509–17. doi: 10.1002/1098-1004(200012)16:6<509::AID-HUMU8>3.0.CO;2-V.

Wolf, P., Y. C. Hu, K. Doffek, et al. 2001. "O(6)-Methylguanine-DNA methyltransferase promoter hypermethylation shifts the p53 mutational spectrum in non-small cell lung cancer." *Cancer Res* 61 (22):8113–7.

Wooster, R., S. L. Neuhausen, J. Mangion, et al. 1994. "Localization of a breast cancer susceptibility gene, BRCA2, to chromosome 13q12–13." *Science* 265 (5181):2088-90.

Wu, B. K., and C. Brenner. 2014. "Suppression of TET1-dependent DNA demethylation is essential for KRAS-mediated transformation." *Cell Rep* 9 (5):1827–40. doi: 10.1016/j.celrep.2014.10.063.

Wu, S. C., and Y. Zhang. 2010. "Active DNA demethylation: many roads lead to Rome." *Nat Rev Mol Cell Biol* 11 (9):607–20. doi: 10.1038/nrm2950.

Xing, D., G. Scangas, M. Nitta, et al. 2009. "A role for BRCA1 in uterine leiomyosarcoma." *Cancer Res* 69 (21):8231–5. doi: 10.1158/0008-5472.CAN-09-2543.

Yan, B. X., and J. X. Ma. 2012. "Promoter-associated RNAs and promoter-targeted RNAs." *Cell Mol Life Sci* 69 (17):2833–42. doi: 10.1007/s00018-012-0953-1.

Yan, X. J., J. Xu, Z. H. Gu, et al. 2011. "Exome sequencing identifies somatic mutations of DNA methyltransferase gene DNMT3A in acute monocytic leukemia." *Nat Genet* 43 (4):309–15. doi: 10.1038/ng.788.

Yang, J., Z. Xu, J. Li, et al. 2010. "XPC epigenetic silence coupled with p53 alteration has a significant impact on bladder cancer outcome." *J Urol* 184 (1):336–43. doi: 10.1016/j.juro.2010.03.044.

Yang, X. J. 2004. "The diverse superfamily of lysine acetyltransferases and their roles in leukemia and other diseases." *Nucleic Acids Res* 32 (3):959–76. doi: 10.1093/nar/gkh252.

Yoon, J. H., S. Iwai, T. R. O'Connor, et al. 2003. "Human thymine DNA glycosylase (TDG) and methyl-CpG-binding protein 4 (MBD4) excise thymine glycol (Tg) from a Tg:G mispair." *Nucleic Acids Res* 31 (18):5399–404.

Yoshida, K., and Y. Miki. 2004. "Role of BRCA1 and BRCA2 as regulators of DNA repair, transcription, and cell cycle in response to DNA damage." *Cancer Sci* 95 (11):866–71.

Yu, F., Y. Jiao, Y. Zhu, et al. 2012. "MicroRNA 34c gene down-regulation via DNA methylation promotes self-renewal and epithelial-mesenchymal transition in breast tumor-initiating cells." *J Biol Chem* 287 (1):465–73.

Yu, F., H. Yao, P. Zhu, et al. 2007. "let-7 regulates self renewal and tumorigenicity of breast cancer cells." *Cell* 131 (6):1109–23. doi: 10.1016/j.cell.2007.10.054.

Yu, J., T. Zhu, Z. Wang, et al. 2007. "A novel set of DNA methylation markers in urine sediments for sensitive/specific detection of bladder cancer." *Clin Cancer Res* 13 (24):7296–304. doi: 10.1158/1078-0432.CCR-07-0861.

Yun, M., J. Wu, J. L. Workman, et al. 2011. "Readers of histone modifications." *Cell Res* 21 (4):564–78. doi: 10.1038/cr.2011.42.

Zhang, H., S. Zhang, J. Cui, et al. 2008. "Expression and promoter methylation status of mismatch repair gene hMLH1 and hMSH2 in epithelial ovarian cancer." *Aust N Z J Obstet Gynaecol* 48 (5):505–9. doi: 10.1111/j.1479-828X.2008.00892.x.

Zhang, J., C. A. Benavente, J. McEvoy, et al. 2012. "A novel retinoblastoma therapy from genomic and epigenetic analyses." *Nature* 481 (7381):329–34. doi: 10.1038/nature10733.

Zhang, X. Y., M. Varthi, S. M. Sykes, et al. 2008. "The putative cancer stem cell marker USP22 is a subunit of the human SAGA complex required for activated transcription and cell-cycle progression." *Mol Cell* 29 (1):102–11. doi: 10.1016/j.molcel.2007.12.015.

Zhang, Y., L. Yao, X. Zhang, et al. 2011. "Elevated expression of USP22 in correlation with poor prognosis in patients with invasive breast cancer." *J Cancer Res Clin Oncol* 137 (8):1245–53. doi: 10.1007/s00432-011-0998-9.

Zhao, Y., G. Lang, S. Ito, et al. 2008. "A TFTC/STAGA module mediates histone H2A and H2B deubiquitination, coactivates nuclear receptors, and counteracts heterochromatin silencing." *Mol Cell* 29 (1):92–101. doi: 10.1016/j.molcel.2007.12.011.

Zhu, J., M. A. Sammons, G. Donahue, et al. 2015. "Gain-of-function p53 mutants co-opt chromatin pathways to drive cancer growth." *Nature* 525 (7568):206–11. doi: 10.1038/nature15251.

Zhu, Q., G. M. Pao, A. M. Huynh, et al. 2011. "BRCA1 tumour suppression occurs via heterochromatin-mediated silencing." *Nature* 477 (7363):179–84. doi: 10.1038/nature10371.

19 Genomic Engineering Utilizing the CRISPR/Cas System and Its Application in Cancer

Amit Kumar Chaudhary, Rajan Sharma Bhattarai, Chalet Tan, Channabasavaiah B. Gurumurthy, and Ram I. Mahato

CONTENTS

19.1 INTRODUCTION

Genome engineering is a process of making necessary changes to the genetic material by deleting, inserting or by modifying the DNA sequences to dissect the function of specific genes and regulatory elements. For decades, genome engineering has been an increasingly challenging task. The evolution of recombinant DNA technology marked the beginning of a new era for molecular biology. Since then, molecular biologists could manipulate genomic molecules. However, the development of easy and efficient engineering tools to edit genomic material in eukaryotic cells holds many obstacles with a promise to transform basic science, biotechnology and medicine. Recent progress in genome engineering is sparking a new revolution in life science, making it possible to study biological systems and applications in cell biology research, plant breeding, disease modeling, gene therapy, drug development, genetic variation, metabolic engineering of platform chemicals producing microbes, and so on. Today, our understanding of the genomic material and genetic engineering has progressed through the *de novo* assemblies of larger genomic material with a decrease in manufacturing cost.

The eukaryotic DNA molecule consists of two long polynucleotide chains composed of four types of nucleotide subunits, which make them difficult to manipulate. Utilization of a homologous recombination (HR) mechanism to integrate exogenous repair templates containing sequence homology at a specific site in the genome has been a game-changing revolution in genomics (Capecchi 1989a). HR-mediated gene targeting has generated transgenic animal models by germline competent stem cell manipulation and advanced the biological research arena. Although HR-mediated genome engineering results in highly precise alterations, the desired recombination events occur extremely infrequently (Capecchi 1989a). This frequency of low recombination events presents enormous challenges for large-scale applications of genome engineering tools. Therefore, numbers of genome editing technologies and tools have emerged, enabling targeted and efficient modification of a variety of eukaryotic systems.

19.2 GENOME ENGINEERING APPROACHES

The human genome is constantly under the influence of various intrinsic and extrinsic agents. These agents sometimes lead to the generation of thousands of DNA lesions including DNA double-strand break (DSB) (Khanna and Jackson 2001). DSB is one such type of damage, which recruits endogenous repair machinery for its repair by non-homologous end joining (NHEJ) or homology-directed repair (HDR) (Figure 19.1). During NHEJ-mediated repair, the process leads to the introduction of insertion or deletion mutations (indels) disrupting the protein coding sequence of the gene. The process is so active that no repair template or extensive DNA synthesis is needed for higher repair capacity in a short time. On the other hand, during HDR-mediated repair, the process leads to the introduction of site-specific mutations or insertions of desired sequences through recombination of exogenously supplied donor DNA template to the target locus. Targeted genome editing via HDR is a powerful approach in molecular biology (Capecchi 2005). However, the lower efficiency of engineered donor DNA template to correctly integrate into the desired target locus, lengthy selection process and adverse mutagenic effects hamper the use of the HDR system. DSBs are among the most lethal forms of DNA damage, which if unrepaired, result in the apoptosis or senescence of the injured cells. In addition, misprocessing of DSBs can lead to genomic instability and carcinogenesis. The technical advances in genome sequencing have advanced genome manipulating approaches. The wealth of biological data accumulated during the process has presented challenges of converting this information into clinically relevant knowledge. However, introducing site-specific changes in the genome of cells and organisms is difficult to achieve. Early approaches relied on the development of RNA interference (RNAi) for targeted gene knock-down, offering a

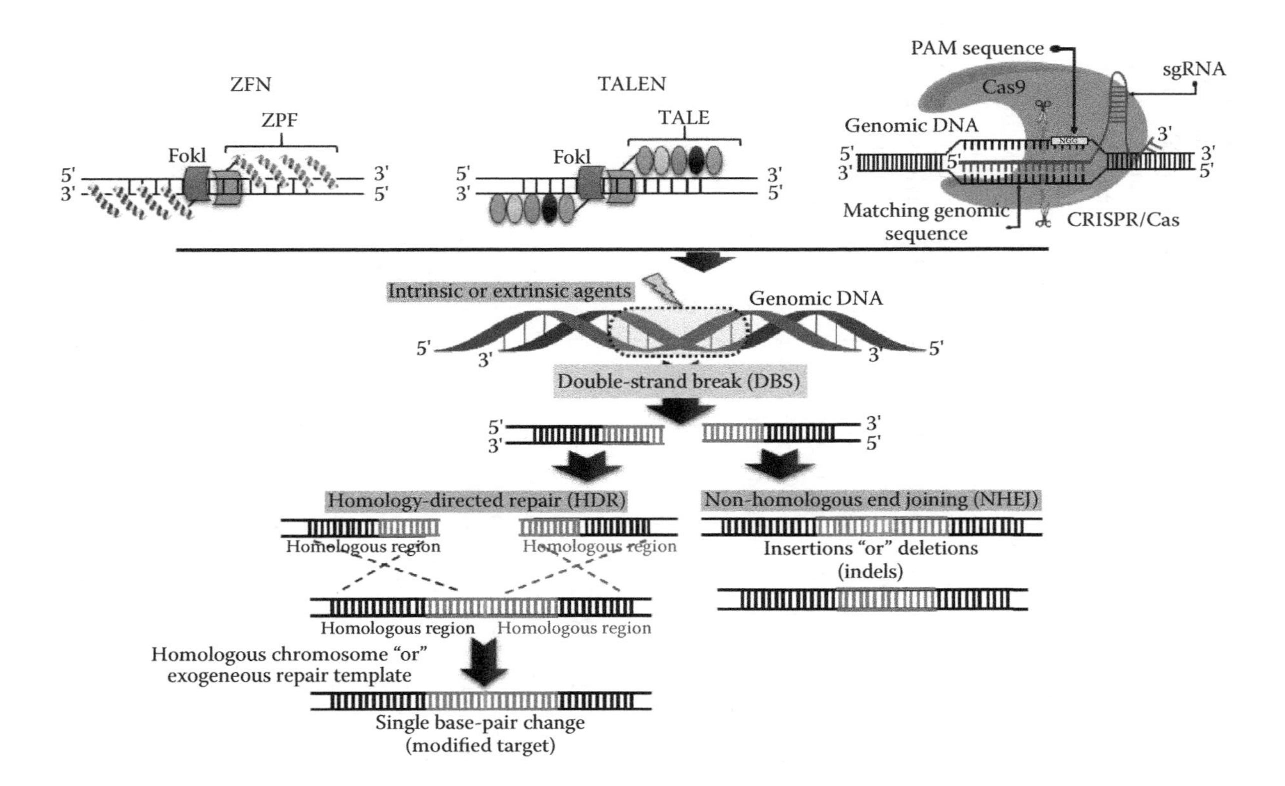

FIGURE 19.1 The nucleases and various intrinsic and extrinsic agents mediated DNA DSBs repair mechanisms by HDR and NHEJ.

cost-effective and high-throughput alternative to HDR. However, RNAi-mediated gene knock-down is incomplete and provides only temporary inhibition of the gene function. On the other hand, the rise of a nuclease-based method composed of site-specific DNA binding domains and nonspecific DNA cleavage module has allowed influencing genes of interest. Using DNA-protein recognition principles, zinc finger nucleases (ZFNs) and transcription activator-like effector nucleases (TALENs) were developed (Kim and Kim 2014; Gaj, Gersbach, and Barbas 2013; Joung and Sander 2013; Kim and Chandrasegaran 1994; Kim, Cha, and Chandrasegaran 1996; Boch et al. 2009; Christian et al. 2010). These approaches have enabled multifunctional DNA engineering, including the introduction of targeted DNA-DSBs and stimulating cellular DNA-repair mechanisms in an efficient and precise manner. Although these systems are effective, the difficulty in designing specific nucleases remained a barrier to adopt these engineered nucleases for routine use. On the other hand, the recently explored genome editing approach, which uses the prokaryotic immune defense system, has shown advantages over the previous regimes. The system is based on the clustered regularly interspaced short palindromic repeats (CRISPR) and CRISPR-associated (*Cas*) genes. The CRISPR/Cas system depends on the complementary base pairing between RNA from the CRISPR locus and the target DNA sequence (Jinek et al. 2012). A distinct feature of the CRISPR/Cas system is the use of an RNA molecule to guide the system's Cas nuclease to a specific DNA target. In addition, the targeting RNA molecule can easily be designed with cost-effective synthesis process. The comparison of three different types of genome editing approaches is presented in Table 19.1.

19.2.1 The CRISPR/Cas System

Discovery of the CRISPR/Cas system uncovered the presence of an adaptive immune system in bacteria and archaea, which was thought to be present only in eukaryotes. This system is utilized by the microorganisms to gain immunity against viruses and plasmids in a sequence-specific manner. CRISPR/Cas systems are highly diverse, of which Type II is mostly studied from *Streptococcus pyogenes* (Makarova et al. 2015; Heler et al. 2015). This diversification of CRISPR/Cas systems is based on the content and organization of *Cas* genes (Makarova et al. 2011). The enormous mechanistic difference between various types of CRISPR/Cas system has been reported. The general mode of action is structured into adaptation (spacer acquisition), maturation (expression and processing), and interference (silencing) (Figure 19.2). In a Type II CRISPR/Cas system, the CRISPR locus is composed of a cluster of *Cas* genes and the CRISPR array. The *Cas* genes are often located adjacent to the CRISPR array consisting of a series of repetitive sequences interspaced by short stretches of non-repetitive DNA called spacers. The spacers are short variable sequences of ~20 base pairs long acquired from invading exogenous viruses and plasmids, which are stored as a record against future encounters. Preceding the *Cas* genes, the trans-activating CRISPR RNA (tracrRNA) gene is located encoding for non-coding RNA homologous to repeats (Jiang and Doudna 2017).

TABLE 19.1
The Comparison between ZNF, TALEN, and CRISPR/Cas Genome-Editing Approaches

Properties	ZNF	TALEN	CRISPR/Cas
Molecular target	DNA	DNA	DNA
Generating target specificity	Difficult: Cloning and protein engineering steps are required	Moderate: Cloning steps are required	Easy: Simple oligo synthesis and cloning steps are required
Effect on target	Irreversible knock-out	Irreversible knock-out	Irreversible knock-out
Overall cost of target engineering	High	Moderate	Low
Generating large-scale libraries	Low: Protein engineering is required for each gene	Moderate: Challenging cloning steps are required	High: Simple oligo synthesis and cloning steps are required
Epigenetic and transcriptional control	DNA-binding zinc finger domains can be fused to new functional domains	DNA-binding domains can be fused to new functional domains	Catalytically inactive Cas9 can be fused to new functional domains
Off-target effect	Moderate	Low	Variable
Main nucleases	FokI	FokI	Cas9
Restriction in target site	GC-rich	Start with T and end with A	End with NGG or NAG sequence (PAM)
Cytotoxicity	Variable to high	Low	Low
In vivo delivery efficiency	Moderate: Viral vectors	Moderate: Viral vectors	Moderate: Viral vectors and nanoparticles
Mechanism of action	Induce DSB: Repair maybe NHEJ or HDR depending on tool design	Induce DSB: Repair maybe NHEJ or HDR depending on tool design	Induce DSB: Repair maybe NHEJ or HDR depending on tool design
Ease of design	Customized protein component for each gene sequence is required	Technical challenge due to extensive identical repeat sequences	Easy to design due to the requirement of simple cloning of 20-nucleotide long oligonucleotides targeting each gene

19.2.2 CRISPR/Cas Mechanism

In a Type II CRISPR/Cas system, the process begins at adaptation stage where Casl, Cas2, and Csn2 proteins identify exogenous genetic elements from invading pathogens and integrate them into the CRISPR array to form a new spacer sequence as an immunological memory (Heler et al. 2015). After integration, the CRISPR array is cotranscribed with the new spacer into a long precursor RNA (pre-crRNA) transcript containing repeats and spacers. On the other hand, the tracrRNA is also transcribed,

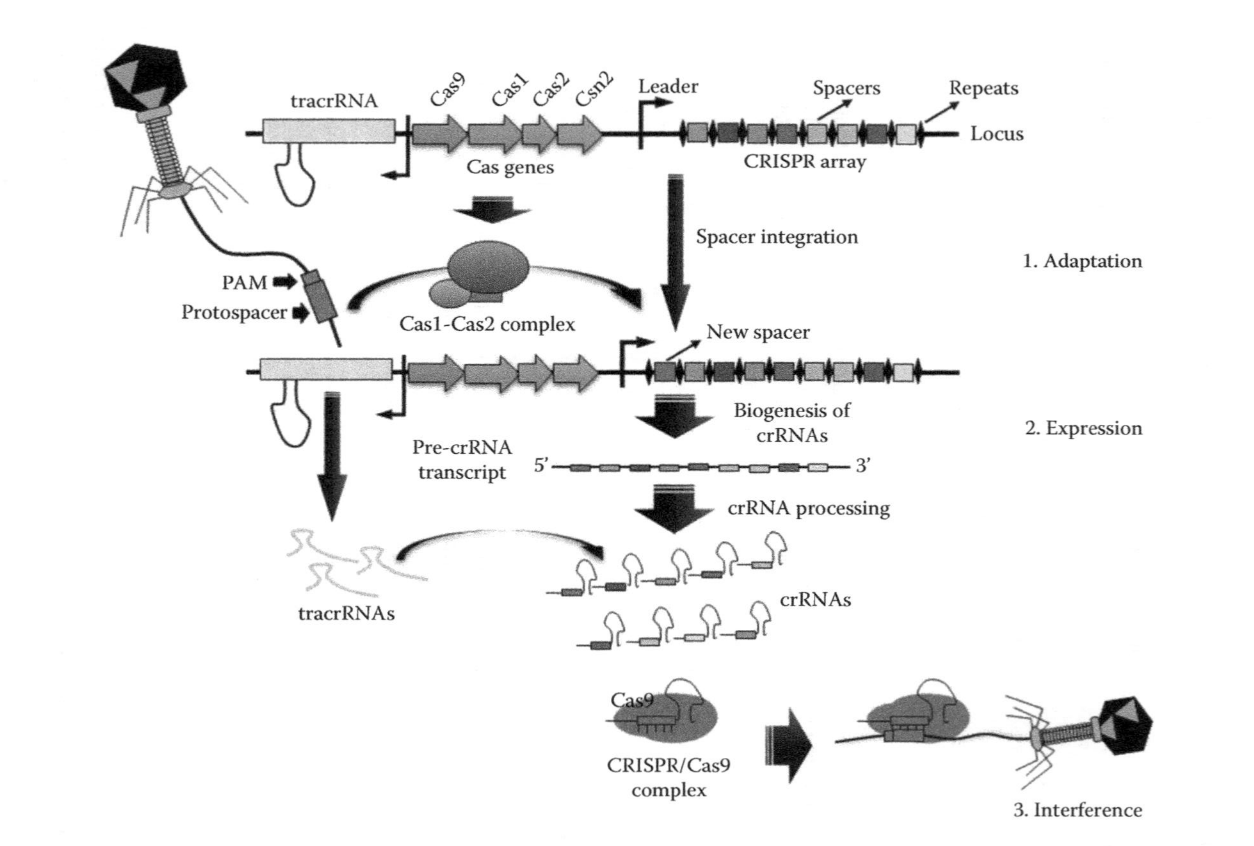

FIGURE 19.2 The Type II CRISPR/Cas9 system of *Streptococcus pyogenes*. The *Cas* genes are represented by multicolored arrows. The short spacers are represented by multicolored rectangular boxes separated by repeats represented in black.

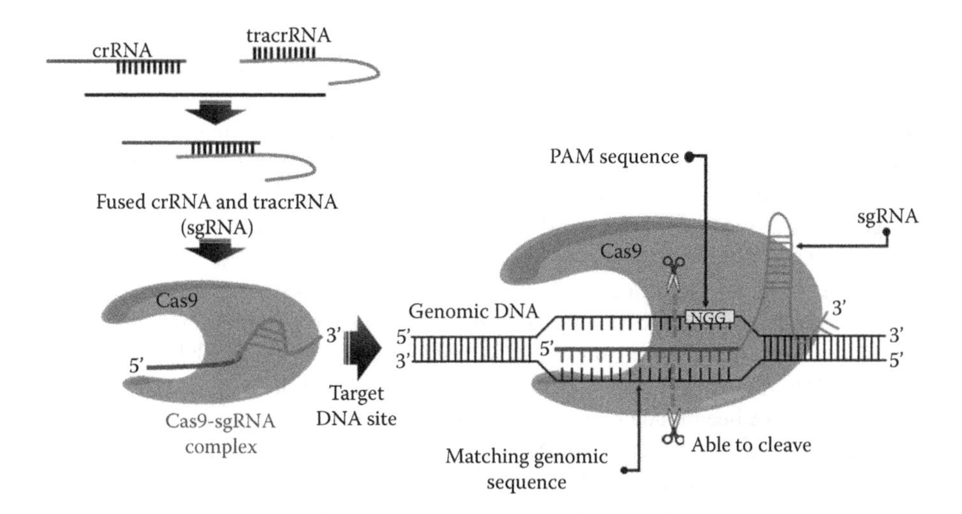

FIGURE 19.3 A representation of sgRNA/Cas9 complex recognizing the target DNA, resulting in the formation of DNA DSB break. The DNA DSB break is detected by the cellular DNA repair machinery and subsequently repaired via HDR and NHEJ.

which hybridizes to the repeats of pre-crRNA for its processing into the short crRNA unit in the maturation stage by the bacterial, double-stranded RNA-specific ribonuclease, RNase III. Further processing of the crRNA 5' end reduces its length to 20 nucleotide-long crRNA containing a single spacer flanked by a part of the repeat sequence. Finally, in the interference phase, the mature crRNA and tracrRNA form a complex with Cas-associated nuclease 9 (Cas9). The Cas9, which introduces a blunt DSB contains two cleavage domains: (i) a RuvC-like nuclease near the N-terminus which initiates cleavage of the DNA strand not complementary to the guide RNA and (ii) a McrA-like (HNH) nuclease in the center of Cas9 which cleaves the DNA strand complementary to the RNA guide (Jinek et al. 2012; Kennedy and Cullen 2015). On the other hand, a tracrRNA stabilizes the complex and activates Cas9. As the protospacer is always associated with a protospacer adjacent motif (PAM) within the target DNA, the crRNA-tracrRNA-Cas9 complex scans for the foreign DNA target inside the cells containing crRNA complementary sequence preceding PAM leading to degradation of target nucleic acid via the Watson-Crick base pairing method (Figure 19.3) (Marraffini and Sontheimer 2008; Brouns et al. 2008; Barrangou et al. 2007).

19.2.3 Off-Target Effects of the CRISPR/Cas System

Engineering biological systems holds great potential for applications across basic science including medicine. Precise editing of genomic material using programmable endonucleases has enabled engineering a broad range of species. The ability of the crRNA-tracrRNA-Cas9 complex to cleave sequence-specific DNA inspired the use of the CRISPR/Cas system to edit the genome. Since targeted nucleases have worked as a powerful tool for genome engineering in the past with high precision, the CRISPR/Cas system offers a new and efficient prospect for genome editing using a single guide

RNA (sgRNA) to guide Cas9 to the DNA target (Ran et al. 2013b). Therefore, the Type II CRISPR/Cas system has become the basis for the current genome engineering technology. The use of a genome editing system provides a tool to create model cells and transgenic animals, engineer metabolic pathways, and treat human diseases that are difficult to tackle by traditional medications. Although the CRISPR/Cas system has become a promising tool to successfully use in a variety of cells and organisms for genome editing, there are several aspects that affect its efficiency and specificity, including Cas9 activity, target site selection, sgRNA design, and delivery methods including off-target side effects. Several studies have demonstrated the introduction of off-target effects using CRISPR/Cas system, which generates unwanted mutations at random sites, thus affecting precise gene modification (Fu et al. 2013; Pattanayak et al. 2013; Ma, Zhang, and Huang 2014). However, over the past three years, considerable efforts have been made to diminish off-target effects by: (i) proper site selection and designing of highly active sgRNAs (Lin et al. 2014), (ii) pairing Cas9 mutant, which carries a mutation in the 10th amino acid position (D10A) converting aspartic acid (D) to alanine (A) for disabling the cleavage efficiency of the RuvC-like domain, converting Cas9 activity from catalyzing DSBs (nuclease) to catalyzing single-strand breaks (nickase) with sgRNAs (Jinek et al. 2012; Shen et al. 2014), and (iii) using truncated sgRNAs with a length of 17–18 nucleotides, which decreases undesired off-target effects and retains on-target efficiency (Fu et al. 2014).

19.3 APPLICATION OF CRISPR/Cas IN CANCER

Cancer is a complex genetic disorder that arises from multiple genetic and epigenetic alterations in onco- or tumor suppressor genes (Hanahan and Weinberg 2011). Improvements in whole-genome-sequencing have generated sufficient information and specific genetic alterations of human cancer cells. Over the past few decades, much has been learned about how point mutations, copy number alterations, and chromosomal rearrangements play causal roles in tumorigenesis. Oncogenes are usually hyperactivated via gain-of-function mutations, whereas tumor suppressor genes are typically inactivated via loss-of-function mutations. Epigenetic differences at the level of DNA methylation and histone marks present another layer of complexity in tumorigenesis (Lawrence et al. 2014; Cancer Genome Atlas Research Network 2014; ENCODE Project Consortium 2012). Experimental approaches to systematically manipulate the genomes of normal and cancer cells are therefore vital for the generation of animal tumor models as well as for the search of novel therapeutic modalities. The biggest hurdle with the treatment or management of cancer is the involvement of our own uncontrollably proliferating cells, against which our body's defense mechanisms do not work. Different treatment options are available for cancer, but over time, the cancerous cells develop resistance making traditional chemotherapy useless (Chaudhary et al. 2017; Kumar et al. 2017; Mondal et al. 2017). Therefore, genetic modification of the cancerous cells, controlling the expression of onco- or tumor suppressor genes could be a long-term solution for cancer treatment. The CRISPR/Cas system for genome editing has emerged as one of the most efficient, affordable, and flexible methods for modifying the genome of the cancerous cells.

19.3.1 Generation of Cancer Models

Many human disease-related genes have been mapped after the development of high-throughput sequencing technology in the post-genomic-era. It is now recognized that most diseases are genetically complex and require thousands of individuals to achieve adequate statistical power to ensure the validity of their association. Modeling the pathogenesis of illness by reproducing the patient's complex genetic alterations in specific cell types or organisms can ease the overall drug development process and the study of human diseases (Hutmacher et al. 2009; Das Thakur, Pryer, and Singh 2014). The laboratory mouse (*Mus musculus*) is an advantageous model due to its small size and similar genomic architecture as that of human, encoding for 23,000 genes with 3×10^9 nucleotides. Mice with specific gene modifications are a valuable system for understanding the function of a gene in diseases and discovering improved methods to prevent, diagnose, and treat diseases. Until recently, manipulating mouse genes to study their function in health and illness was hard to achieve. However, the generation of specific disease models in mammalian cells and animals that recapitulate the biological and clinical characteristics of humans by targeting specific locus in the genome has made CRISPR/Cas system an important subject for studying cancers (Van Dyke and Jacks 2002). The CRISPR/Cas system has been used to create germline and somatic mouse models with point mutations, deletions, and complex chromosomal rearrangements. Since genetic modifications in onco- or tumor suppressor genes are the major cause of cancer, the CRISPR/Cas system can be used to model the dynamics of genetic elements to generate cancer models. Useful mouse models for translational research need to replicate the genetics and genomics, the context and the heterogeneity of human tumors. A conventional chimeric animal generation has relied on genetically altered embryonic stem (ES) cells by HR techniques (Capecchi 1989b). By injecting such genetically modified ES cells into wild-type blastocysts, chimeric animals with altered germ lines are generated (Capecchi 2005). Although suitable for generating sophisticated genetic modifications, the rate of HR in ES cells is remarkably low. This is a time-consuming, laborious, and costly process, which involves tedious selection of ES cell clones followed by injection into blastocysts to generate chimeric animals. Also, the absence of ES cells in other mammalian species limits the studies that can be undertaken (Van Dyke and Jacks 2002; Dow and Lowe 2012). The construction of animals carrying many alterations comparatively adds substantially more time and effort. However, non-germline genetically engineered animals can simplify this process by bypassing the need for complex genetic crosses through a series of retargeting ES cells (Heyer et al. 2010). Since genetic modifications including gain-of-function mutations or loss-of-function mutations in proto-, onco-, and tumor suppressor genes respectively are the primary cause of cancer, the CRISPR/Cas system can be used to model the dynamics of genetic elements to generate disease models including footprint free point mutations (Gurumurthy et al. 2016), knock-in, and conditional knock-out models (Quadros et al. 2017). It is challenging to mimic the complex changes in genetic combinations due to the difficulty associated with large numbers of genetic crosses. However, some studies have used the CRISPR/Cas system to knock-out multiple genes of a pathway (Huang et al. 2016; O'Neill et al. 2016). Although CRISPR/Cas system has evolved as a promising tool for targeted genome editing, the *in vivo* delivery of Cas9 nuclease in somatic tissue remain challenging due to its large size (~4 kb) (Eyquem et al. 2017; Ran et al. 2015; Wang et al. 2016). Therefore, a step further is the development of Cas9 expressing animal, where any combination of desired genes can be inactivated by delivering sgRNAs for target genes to generate the disease models (Harms et al. 2014; Quadros et al. 2015). The generation of a Cas9 expressing mouse indeed offers a versatile tool for cancer research. As Rosa26 locus has become a standard 'safe harbor' located on chromosome 6 for transgenesis in mouse ES cells, Platt et al. generated a Cre-recombinase inducible Cas9 knock-in mouse by inserting *Cas9* transgene expression cassette into Rosa26 locus by HR in R1 ES cells and implanted them in C57BL/6N blastocysts to generate chimeric mice (Platt et al. 2014). These Cre-inducible Cas9 expressing mice were used to model most frequently mutated genes, such as *p53* (46%), *KRAS* (33%) and *LKB1* (17%) in lung adenocarcinoma (Cancer Genome Atlas Research Network 2014). Since mutations in *KRAS* are often missense resulting in a gain-of-function, the CRISPR/Cas system can be used to introduce an HDR-mediated missense mutation in *KRAS*. Delivery of a vector containing a donor template homologous to the first exon of *KRAS*, encoding a glycine (G) to aspartate (D) mutation in the 12th amino acid position (G12D) of *KRAS*, results in a gain-of-function oncogenic $KRAS^{G12D}$ mutation (Cancer Genome Atlas Research Network 2014). On the other hand, generating CRISPR/Cas-mediated indels in *p53* and *LKB1* at the predetermined cutting site can result in their loss-of-function mutation. Therefore, simultaneous induction of loss-of-function mutations in *p53* and *LKB1* followed by HDR-mediated gain-of-function mutations in *KRAS* ($KRAS^{G12D}$) leads to the generation of lung adenocarcinoma in mice (Platt et al. 2014). Hydrodynamic injection is a well-studied method to deliver foreign DNA molecule to the liver in animals. This method uses high-volume and high-pressure tail vein injection to express DNA in 20–30% of mouse hepatocytes transiently. Since the *in vivo* delivery of Cas9 nuclease is challenging, Xue et al. used hydrodynamic injection to deliver *Cas9* and sgRNAs containing plasmid to target tumor suppressor genes in mouse liver (Xue et al. 2014). Loss-of-functional mutation of tumor suppressor gene *PTEN*, a negative regulator of the phosphatidylinositol-3-kinase (PI3K)/Akt pathway (Song, Salmena, and Pandolfi 2012) has been associated with several cancers including liver cancer (Peyrou, Bourgoin, and Foti 2010). On the other hand, activation of *p53* signaling pathway in response to cascade of signaling networks provide *p53* tumor suppression properties. However, mutations or disturbance in signaling pathways related the loss of *p53* function leads to most of the human cancers. More than 75% of the mutations in *p53* lost their wild-type properties and exerted a dominant-negative regulation over any remaining wild-type *p53* (Petitjean et al. 2007). The mutant p53 proteins acquire oncogenic properties that enable them to promote invasion, metastasis, proliferation, and cell survival. Therefore, the delivery of plasmid encoding *Cas9* and sgRNAs for tumor suppressor genes *PTEN* and *p53* in combination induces liver tumors in mice (Xue et al. 2014).

19.3.2 Gene Therapy

The insertion, alteration, or removal of genetic material to treat or to improve and manage the clinical status of a patient is commonly known as gene therapy. It is a promising approach to treat diseases that conventional drugs fail to do. Gene therapy has potential to control a broad range of diseases, including cystic fibrosis, heart disease, diabetes, cancer, and blood diseases. The faulty genes can be corrected either by inserting a normal gene into a nonspecific location within the genome or by repairing to its normal function. Cellular activities are governed by basic signaling pathways, which coordinate cellular properties through a complex coordination of responses to the cell microenvironment. The obstruction within these pathways is one of the leading cause of various diseases including cancers. Targeting cellular activities that cause cancerous cells to grow is the principal therapeutic approach for cancer management. However, the cancerous cells adjust to the new environment, mutate to bypass the signaling blockade and develop new pathways for growth. These alternate mechanisms are the reasons for drug tolerance and drug resistance commonly observed in the cancerous cells. This prompts the use of the second-line therapy, which is not as effective in managing tumors. However, in contrast to traditional approaches, gene therapy provides cancer-specific treatment. Structural rearrangements of DNA molecules result in the exchange of coding or regulatory sequences between genes. Many such fusion genes may or may not function as wild-type genes, and thus acquire additional features that drive cancer progression or destroy oncogenic genes (Mertens et al. 2015). These hybrid genes serve as an attractive target as both therapeutic and diagnostic tools due to their inherent expression in cancerous tissue alone. Ninety-one percent of prostate cancer patients are positive for one of the *TRMT11-GRIK2*, *SLC45A2-AMACR*, *MTOR-TP53BP1*, *LRRC59-FLJ60017*, *TMEM135-CCDC67*, *KDM4-AC011523.2*, *MAN2A1-FER*, and *CCNH-C5orf30* fusion genes. They are either separated widely on a single chromosome (*TRMT11-GRIK2, TMEM135-CCDC67, CCNH-C5orf30, and KDM4B- AC011523.2*) or located on separate chromosomes (*MTOR-TP53BP1 and LRRC59-FLJ60017*). Among them, *CCNH-C5orf30*, *TMEM135-CCDC67*, and *LRRC59-FLJ60017* harbors genomic-breakpoints, where *TMEM135-CCDC67* is formed by the deletion of the 6-Mb genomic region between the genes coding for transmembrane protein 135 (TMEM135) and coiled-coil domain containing 67 (CCDC67) (Yu et al. 2014). In *TMEM135-CCDC67*, this phenomenon disturbs the gene encoding for CCDC67 (putative cancer suppressor) and truncates 65 amino acids from the C-terminus of TMEM135, thus making them a unique genotype-specific target for cancer therapy (Chen et al. 2017). Due to the precise specificity, the CRISPR/Cas system can be applied to target genomic-breakpoints (Chen et al. 2017). However, the off-target issues related to CRISPR/Cas system have been a concern, when using it as a therapeutic tool. Out of 20 nucleotide-long sgRNA, up to 3–5 mismatches are tolerated, translating into a substantial amount of potential off-target sites (Fu et al. 2013; Shen et al. 2014). Mismatches near the 5' end of the sgRNA are easy to tolerate. However, higher sensitivity is observed for mismatches in the 8–10 bases near the 3' end (Mali et al. 2013). To address this challenge, Cas9 specificity can be altered using two sgRNAs binding targets in combination with Cas9 nickase (Mali et al. 2013; Ran et al. 2013a; Cho et al. 2014). The introduction of two nicks in target DNA increases the efficiency of introducing sequences by 50–1,500-fold and decreases the off-target rate by 1/10,000 (Ran et al. 2013a). Since such specificity makes somatic genomic-targeting a viable approach in treating human diseases, Chen et al. used this method to create two nicks on complementary DNA strands (Chen et al. 2017) to introduce a suicide enzyme-encoding gene at the breakpoints of *TMEM135–CCDC67*. The prodrug-converting enzymes, such as thymidine kinases (TKs) from herpes simplex virus type 1 (HSV1-TK) convert non-toxic compounds (prodrugs) into toxic products. HSV1-TK phosphorylates the synthetic nucleoside ganciclovir (prodrug) into ganciclovir monophosphate in mammalian cells expressing HSV1-TK (Smith et al. 1982), which later is converted to triphosphate and blocks DNA synthesis through elongation termination (Van Rompay, Johansson, and Karlsson 2000). By simple delivery of two complementary sgRNAs designed from the flanking region of the breakpoint on the opposite strand of *TMEM135–CCDC67*, they incorporated a chimeric enhanced green fluorescence protein-HSV1-TK (*EGFP-TK*) gene by HDR into the *TMEM135–CCDC67* breakpoint of mouse genome. Which, after translation into protein, phosphorylate simultaneously delivered ganciclovir and incorporated into DNA, thus terminating DNA synthesis leading to cell death. Although, the HR rate reached up to 20–30%, this procedure has inherent drawback of not being able to achieve complete remission, owing to its inability to direct the system to every cancerous cell in the tissue and to random off-target effects. Therefore, a method which could increase the efficiency of CRISPR/Cas system instead of directly targeting cancerous cell, would be a clinically relevant approach for cancer treatment.

19.3.3 Cell Therapy

Understanding the role of the immune system in cancer has grown tremendously over the past few decades. Genetically engineered immune cells have been studied as a treatment option for multiple diseases including *human immunodeficiency virus* (*HIV*) and cancer. The T-cells of an adaptive immune system has potential to kill antigen-expressing cells after identification by a T-cell receptor protein (TCR). The genes encoding for TCRs are rearranged during T-cell development providing each T-cell a unique TCR for binding a specific antigen. However, the killing efficiency of a T-cell is usually weak over cancerous cells due to their nature of expressing self-antigens. This weak efficiency of T-cells can be increased by developing chimeric antigen receptor (CAR) expressing T-cells. CARs are genetically engineered fusion proteins containing an extracellular antigen-binding domain composed of a single-chain variable fragment derived from an antibody and intracellular signaling domains involved in T-cell functions (Finney et al. 1998). An engineered CAR to recognize a cluster of differentiation 19 (CD19) or any other antigens expressed in a cancer cell can be designed and genetically introduced into the T-cells for CAR T-cell generation. CAR T-cells are applicable for many types of cancer (Jensen and Riddell 2015; Brentjens et al. 2003) because they have the potential to overcome immunosuppressive tumor microenvironments, reduce toxicities, and prevent antigen escape.

On the other hand, cell therapy is the administration of live cells in a patient for the treatment of a disease (Haworth et al. 2017; Delhove and Qasim 2017). Gene and cell therapy are intended to cure disease related to the certain genetic modification. Blood transfusions and bone marrow transplantation are well-established treatment approach for blood disorders, and immunodeficiency diseases after identifying a suitable immunological match. Since both gene and cell therapies complement each other for effective therapy cell therapy can be used to overcome the issues related to gene therapy and increase the efficiency of the CRISPR/Cas system for clinical translation. More specifically, the drawbacks of the CRISPR/Cas system to target individual cells in bulk tumor and preventing the insertion of CAR at random locations in the genome can be eliminated, and more personalized therapy can be designed using CAR T-cells. The CRISPR/Cas system can advance immunotherapy by modulating T-cells into CAR T-cells, making them specific to various type of cancers. As the TCR locus is rearranged during T-cell development and expressed in the order of delta, gamma, beta, and alpha, Eyquem et al. used a CRISPR/Cas system to integrate gene encoding for *CAR* into a TCR alpha constant (*TRAC*) locus for targeting CD19 found on B-lymphocytes surface as treatment module for chemorefractory or relapsed B-cell malignancies (Eyquem et al. 2017; Sadelain 2015). The easy introduction of *CAR* coding sequence into the T-cell's genome, which is under the control of endogenous regulatory elements, not only results in more uniform CAR expression in human peripheral blood T-cells, but also increases the therapeutic potency of the engineered T-cells in the mouse with acute lymphoblastic leukemia.

19.4 DELIVERY OF THE CRISPR/Cas SYSTEM

The CRISPR/Cas system has revolutionized targeted genome editing and holds potential to give rise to an entirely new class of therapeutics. Achieving this potential requires exploration of its delivery to the target cells and to patients. The ~4 kb Cas9 from *Streptococcus pyogenes* can be integrated into an adeno-associated virus (AAV) with cargo size of ~4.5 kb (Eyquem et al. 2017; Ran et al. 2015; Wang et al. 2016). To accommodate larger cargo, adenoviral and lentiviral vectors with larger cargo capacities can also be used (Cheng et al. 2014). On the other hand, some non-viral vectors, such as electroporation, hydrodynamic delivery, cationic lipid-based vectors (Zuris et al. 2015), cationic polymer-based vectors (Platt et al. 2014), conjugated vectors (Ramakrishna et al. 2014), and the combination of viral and non-viral systems have also been studied. Depending on the necessity, Cas9 and sgRNAs can be formulated as DNA, RNA, or a complex of Cas9 and sgRNA. Most of these methods support delivery of the CRISPR/Cas components for transient expression, providing a safety advantage over viral delivery methods (Gori et al. 2015). These non-viral vectors need further refinement to make them more efficient in delivering through different routes other than direct injection into the target site. In addition, DNA nanoparticles reflect a promising approach for safe and effective delivery of CRISPR/Cas system (Sun et al. 2015). On the other hand, hydrodynamic injection of nucleic acids could result in the *in vivo* delivery of foreign DNA primarily in the liver (Xue et al. 2014).

19.5 REGULATORY AND ETHICAL ISSUES

The introduction of the CRISPR/Cas system as a new and efficient genome editing technology has not yet significantly altered any regulatory parameters set by most of the countries. In the United States, the Food and Drug Administration (FDA), US Department of Agriculture (USDA) and the Environmental Protection Agency (EPA) regulate genetically modified animals or insects through their guidelines. Most of the regulations, with few amendments, are based on the standards set in 1980s, and

90s making them incapable of addressing the recent concerns. The FDA issued a draft revised guidance in January 2017 on the "regulation of intentionally altered genomic-DNA in animals," a continuation to the one in 2009, which addresses animals with intentionally altered genomic-DNA developed using genome editing (ZFN, TALEN, and CRISPR) and recombinant DNA technologies (www.fda.gov/downloads/AnimalVeterinary/GuidanceComplianceEnforcement/GuidanceforIndustry/UCM113903.pdf). The ethical concerns related to CRISPR/Cas technology are not only limited to human germline editing but also to other organisms and the ecosystem. The scientific community has been bifurcated in their opinion about using this technology to propel the research on human germline editing forward or to be cautious and play safely without violating the ethical lines. The use of CRISPR/Cas technology to modify human embryos either to get rid of genetic defects or to develop immunity against other fatal diseases poses significant ethical concerns (Araki and Ishii 2014). Some scientists are pro CRISPR/Cas, while others caution about the limited understanding of gene edition in the long term. So, this debate is not going to end anytime soon, or not until we have sufficient evidence in animals to justify the safety of human beings. Apart from humans, the use of CRISPR/Cas technology to modify plants, animals, insects, and microorganisms to produce food and drugs for human use, must be considered as the affordability and efficiency of the technology will pose new threats and an ethical dilemma (Sugarman 2015). On the other hand, animals can be bred, without any ethical concerns, for a specific mutation in research to obtain less variability and more human-like animals. So, to address these types of ethical dilemmas, public reviews are conducted and strict biosafety measures are issued. This field is moving very quickly, and a 'sit back, relax, and think' approach might be helpful in identifying the favorable future for CRISPR/Cas technology (http://nationalacademies.org/cs/groups/genesite/documents/webpage/gene_177260.pdf).

19.6 CONCLUSION

Although the CRISPR/Cas system has been fruitfully used to model cancer, gene, and cell therapy, the system is still subject to technical restrictions. The most important challenges are concerned with the delivery of the necessary components to the target cell either *in vitro* or *in vivo* and off-target effects. Off-target cleavage of DNA can cause unwelcome genetic variations with unpredictable consequences. Many approaches have been used to increase the delivery efficiency while lowering the off-target effects of CRISPR/Cas components. The CRISPR/Cas system is becoming a superior technology over traditional methods, one that could be applied in all fields of life sciences. With the rapid advancement in CRISPR/Cas-based genome-engineering system, the application of this technology has changed the picture of cancer and genetic disorder research, providing new therapeutic approaches to personalized medicine, contributing to gene therapy and immunotherapy.

CONFLICT OF INTEREST

The authors declare no competing financial interests.

ACKNOWLEDGMENTS

We gratefully acknowledge the financial support by the National Institutes of Health (1R01EB017853 and 1R01GM113166) and the Faculty Start-up fund to RIM.

REFERENCES

Araki, M. and T. Ishii. 2014. "International Regulatory Landscape and Integration of Corrective Genome Editing into in Vitro Fertilization." *Reproductive Biology and Endocrinology : RB&E* 12: 108-7827-12-108.

Barrangou, R., C. Fremaux, H. Deveau, et al. 2007. "CRISPR Provides Acquired Resistance Against Viruses in Prokaryotes." *Science (New York, N.Y.)* 315 (5819): 1709–1712.

Boch, J., H. Scholze, S. Schornack, et al. 2009. "Breaking the Code of DNA Binding Specificity of TAL-Type III Effectors." *Science (New York, N.Y.)* 326 (5959): 1509–1512.

Brentjens, R. J., J. B. Latouche, E. Santos, et al. 2003. "Eradication of Systemic B-Cell Tumors by Genetically Targeted Human T Lymphocytes Co-Stimulated by CD80 and Interleukin-15." *Nature Medicine* 9 (3): 279–286.

Brouns, S. J., M. M. Jore, M. Lundgren, et al. 2008. "Small CRISPR RNAs Guide Antiviral Defense in Prokaryotes." *Science (New York, N.Y.)* 321 (5891): 960–964.

Cancer Genome Atlas Research Network. 2014. "Comprehensive Molecular Profiling of Lung Adenocarcinoma." *Nature* 511 (7511): 543–550.

Capecchi, M. R. 1989a. "Altering the Genome by Homologous Recombination." *Science (New York, N.Y.)* 244 (4910): 1288–1292.

Capecchi, M. R. 2005. "Gene Targeting in Mice: Functional Analysis of the Mammalian Genome for the Twenty-First Century." *Nature Reviews.Genetics* 6 (6): 507–512.

Capecchi, M. R. 1989b. "The New Mouse Genetics: Altering the Genome by Gene Targeting." *Trends in Genetics: TIG* 5 (3): 70–76.

Chaudhary, A. K., G. Mondal, V. Kumar, et al. 2017. "Chemosensitization and Inhibition of Pancreatic Cancer Stem Cell Proliferation by Overexpression of microRNA-205." *Cancer Letters* 402: 1–8.

Chen, Z. H., Y. P. Yu, Z. H. Zuo, et al. 2017. "Targeting Genomic Rearrangements in Tumor Cells through Cas9-Mediated Insertion of a Suicide Gene." *Nature Biotechnology* 35 (6): 543–550.

Cheng, R., J. Peng, Y. Yan, et al. 2014. "Efficient Gene Editing in Adult Mouse Livers Via Adenoviral Delivery of CRISPR/Cas9." *FEBS Letters* 588 (21): 3954–3958.

Cho, S. W., S. Kim, Y. Kim, et al. 2014. "Analysis of Off-Target Effects of CRISPR/Cas-Derived RNA-Guided Endonucleases and Nickases." *Genome Research* 24 (1): 132–141.

Christian, M., T. Cermak, E. L. Doyle, et al. 2010. "Targeting DNA Double-Strand Breaks with TAL Effector Nucleases." *Genetics* 186 (2): 757–761.

Das Thakur, M., N. K. Pryer, and M. Singh. 2014. "Mouse Tumour Models to Guide Drug Development and Identify Resistance Mechanisms." *The Journal of Pathology* 232 (2): 103–111.

Delhove, J. M. K. M. and W. Qasim. 2017. "Genome-Edited T Cell Therapies." *Current Stem Cell Reports* 3 (2): 124–136.

Dow, L. E. and S. W. Lowe. 2012. "Life in the Fast Lane: Mammalian Disease Models in the Genomics Era." *Cell* 148 (6): 1099–1109.

ENCODE Project Consortium. 2012. "An Integrated Encyclopedia of DNA Elements in the Human Genome." *Nature* 489 (7414): 57–74.

Eyquem, J., J. Mansilla-Soto, T. Giavridis, et al. 2017. "Targeting a CAR to the TRAC Locus with CRISPR/Cas9 Enhances Tumour Rejection." *Nature* 543 (7643): 113–117.

Finney, H. M., A. D. Lawson, C. R. Bebbington, et al. 1998. "Chimeric Receptors Providing both Primary and Costimulatory Signaling in T Cells from a Single Gene Product." *Journal of Immunology (Baltimore, Md.: 1950)* 161 (6): 2791–2797.

Fu, Y., J. A. Foden, C. Khayter, et al. 2013. "High-Frequency Off-Target Mutagenesis Induced by CRISPR-Cas Nucleases in Human Cells." *Nature Biotechnology* 31 (9): 822–826.

Fu, Y., J. D. Sander, D. Reyon, et al. 2014. "Improving CRISPR-Cas Nuclease Specificity using Truncated Guide RNAs." *Nature Biotechnology* 32 (3): 279–284.

Gaj, T., C. A. Gersbach, and C. F. Barbas 3rd. 2013. "ZFN, TALEN, and CRISPR/Cas-Based Methods for Genome Engineering." *Trends in Biotechnology* 31 (7): 397–405.

Gori, J. L., P. D. Hsu, M. L. Maeder, et al. 2015. "Delivery and Specificity of CRISPR-Cas9 Genome Editing Technologies for Human Gene Therapy." *Human Gene Therapy* 26 (7): 443–451.

Gurumurthy, C. B., M. Grati, M. Ohtsuka, et al. 2016. "CRISPR: A Versatile Tool for both Forward and Reverse Genetics Research." *Human Genetics* 135 (9): 971–976.

Hanahan, D. and R. A. Weinberg. 2011. "Hallmarks of Cancer: The Next Generation." *Cell* 144 (5): 646–674.

Harms, D. W., R. M. Quadros, D. Seruggia, et al. 2014. "Mouse Genome Editing using the CRISPR/Cas System." *Current Protocols in Human Genetics* 83: 15.7.1–27.

Haworth, K. G., C. Ironside, Z. K. Norgaard, et al. 2017. "In Vivo Murine-Matured Human CD3+ Cells as a Preclinical Model for T Cell-Based Immunotherapies." *Molecular Therapy. Methods & Clinical Development* 6: 17–30.

Heler, R., P. Samai, J. W. Modell, et al. 2015. "Cas9 Specifies Functional Viral Targets during CRISPR-Cas Adaptation." *Nature* 519 (7542): 199–202.

Heyer, J., L. N. Kwong, S. W. Lowe, et al. 2010. "Non-Germline Genetically Engineered Mouse Models for Translational Cancer Research." *Nature Reviews. Cancer* 10 (7): 470–480.

Huang, K., J. Zhang, K. L. O'Neill, et al. 2016. "Cleavage by Caspase 8 and Mitochondrial Membrane Association Activate the BH3-Only Protein Bid during TRAIL-Induced Apoptosis." *The Journal of Biological Chemistry* 291 (22): 11843–11851.

Hutmacher, D. W., R. E. Horch, D. Loessner, et al. 2009. "Translating Tissue Engineering Technology Platforms into Cancer Research." *Journal of Cellular and Molecular Medicine* 13 (8A): 1417–1427.

Jensen, M. C. and S. R. Riddell. 2015. "Designing Chimeric Antigen Receptors to Effectively and Safely Target Tumors." *Current Opinion in Immunology* 33: 9–15.

Jiang, F. and J. A. Doudna. 2017. "CRISPR-Cas9 Structures and Mechanisms." *Annual Review of Biophysics* 46: 505–529.

Jinek, M., K. Chylinski, I. Fonfara, et al. 2012. "A Programmable Dual-RNA-Guided DNA Endonuclease in Adaptive Bacterial Immunity." *Science (New York, N.Y.)* 337 (6096): 816–821.

Joung, J. K. and J. D. Sander. 2013. "TALENs: A Widely Applicable Technology for Targeted Genome Editing." *Nature Reviews. Molecular Cell Biology* 14 (1): 49–55.

Kennedy, E. M. and B. R. Cullen. 2015. "Bacterial CRISPR/Cas DNA Endonucleases: A Revolutionary Technology that could Dramatically Impact Viral Research and Treatment." *Virology* 479–480: 213–220.

Khanna, K. K. and S. P. Jackson. 2001. "DNA Double-Strand Breaks: Signaling, Repair and the Cancer Connection." *Nature Genetics* 27 (3): 247–254.

Kim, H. and J. S. Kim. 2014. "A Guide to Genome Engineering with Programmable Nucleases." *Nature Reviews. Genetics* 15 (5): 321–334.

Kim, Y. G., J. Cha, and S. Chandrasegaran. 1996. "Hybrid Restriction Enzymes: Zinc Finger Fusions to Fok I Cleavage Domain." *Proceedings of the National Academy of Sciences of the United States of America* 93 (3): 1156–1160.

Kim, Y. G. and S. Chandrasegaran. 1994. "Chimeric Restriction Endonuclease." *Proceedings of the National Academy of Sciences of the United States of America* 91 (3): 883–887.

Kumar, V., A. K. Chaudhary, Y. Dong, et al. 2017. "Design, Synthesis and Biological Evaluation of Novel Hedgehog Inhibitors for Treating Pancreatic Cancer." *Scientific Reports* 7 (1): 1665-017-01942-7.

Lawrence, M. S., P. Stojanov, C. H. Mermel, et al. 2014. "Discovery and Saturation Analysis of Cancer Genes Across 21 Tumour Types." *Nature* 505 (7484): 495–501.

Lin, Y., T. J. Cradick, M. T. Brown, et al. 2014. "CRISPR/Cas9 Systems have Off-Target Activity with Insertions Or Deletions between Target DNA and Guide RNA Sequences." *Nucleic Acids Research* 42 (11): 7473–7485.

Ma, Y., L. Zhang, and X. Huang. 2014. "Genome Modification by CRISPR/Cas9." *The FEBS Journal* 281 (23): 5186–5193.

Makarova, K. S., D. H. Haft, R. Barrangou, et al. 2011. "Evolution and Classification of the CRISPR-Cas Systems." *Nature Reviews.Microbiology* 9 (6): 467–477.

Makarova, K. S., Y. I. Wolf, O. S. Alkhnbashi, et al. 2015. "An Updated Evolutionary Classification of CRISPR-Cas Systems." *Nature Reviews.Microbiology* 13 (11): 722–736.

Mali, P., J. Aach, P. B. Stranges, et al. 2013. "CAS9 Transcriptional Activators for Target Specificity Screening and Paired Nickases for Cooperative Genome Engineering." *Nature Biotechnology* 31 (9): 833–838.

Marraffini, L. A. and E. J. Sontheimer. 2008. "CRISPR Interference Limits Horizontal Gene Transfer in Staphylococci by Targeting DNA." *Science (New York, N.Y.)* 322 (5909): 1843–1845.

Mertens, F., B. Johansson, T. Fioretos, et al. 2015. "The Emerging Complexity of Gene Fusions in Cancer." *Nature Reviews. Cancer* 15 (6): 371–381.

Mondal, G., S. Almawash, A. K. Chaudhary, et al. 2017. "EGFR-Targeted Cationic Polymeric Mixed Micelles for Codelivery of Gemcitabine and miR-205 for Treating Advanced Pancreatic Cancer." *Molecular Pharmaceutics* 14 (9): 3121–3133.

O'Neill, K. L., K. Huang, J. Zhang, et al. 2016. "Inactivation of Prosurvival Bcl-2 Proteins Activates Bax/Bak through the Outer Mitochondrial Membrane." *Genes & Development* 30 (8): 973–988.

Pattanayak, V., S. Lin, J. P. Guilinger, et al. 2013. "High-Throughput Profiling of Off-Target DNA Cleavage Reveals RNA-Programmed Cas9 Nuclease Specificity." *Nature Biotechnology* 31 (9): 839–843.

Petitjean, A., E. Mathe, S. Kato, et al. 2007. "Impact of Mutant p53 Functional Properties on TP53 Mutation Patterns and Tumor Phenotype: Lessons from Recent Developments in the IARC TP53 Database." *Human Mutation* 28 (6): 622–629.

Peyrou, M., L. Bourgoin, and M. Foti. 2010. "PTEN in Liver Diseases and Cancer." *World Journal of Gastroenterology* 16 (37): 4627–4633.

Platt, R. J., S. Chen, Y. Zhou, et al. 2014. "CRISPR-Cas9 Knockin Mice for Genome Editing and Cancer Modeling." *Cell* 159 (2): 440–455.

Quadros, R. M., D. W. Harms, M. Ohtsuka, et al. 2015. "Insertion of Sequences at the Original Provirus Integration Site of Mouse ROSA26 Locus using the CRISPR/Cas9 System." *FEBS Open Bio* 5: 191–197.

Quadros, R. M., H. Miura, D. W. Harms, et al. 2017. "Easi-CRISPR: A Robust Method for One-Step Generation of Mice Carrying Conditional and Insertion Alleles using Long ssDNA Donors and CRISPR Ribonucleoproteins." *Genome Biology* 18 (1): 92-017-1220-4.

Ramakrishna, S., A. B. Kwaku Dad, J. Beloor, et al. 2014. "Gene Disruption by Cell-Penetrating Peptide-Mediated Delivery of Cas9 Protein and Guide RNA." *Genome Research* 24 (6): 1020–1027.

Ran, F. A., L. Cong, W. X. Yan, et al. 2015. "In Vivo Genome Editing using Staphylococcus Aureus Cas9." *Nature* 520 (7546): 186–191.

Ran, F. A., P. D. Hsu, C. Y. Lin, et al. 2013a. "Double Nicking by RNA-Guided CRISPR Cas9 for Enhanced Genome Editing Specificity." *Cell* 154 (6): 1380–1389.

Ran, F. A., P. D. Hsu, J. Wright, et al. 2013b. "Genome Engineering using the CRISPR-Cas9 System." *Nature Protocols* 8 (11): 2281–2308.

Sadelain, M. 2015. "CAR Therapy: The CD19 Paradigm." *The Journal of Clinical Investigation* 125 (9): 3392–3400.

Shen, B., W. Zhang, J. Zhang, et al. 2014. "Efficient Genome Modification by CRISPR-Cas9 Nickase with Minimal Off-Target Effects." *Nature Methods* 11 (4): 399–402.

Smith, K. O., K. S. Galloway, W. L. Kennell, et al. 1982. "A New Nucleoside Analog, 9-[[2-Hydroxy-1-(Hydroxymethyl)Ethoxyl]Methyl]Guanine, Highly Active in Vitro Against Herpes Simplex Virus Types 1 and 2." *Antimicrobial Agents and Chemotherapy* 22 (1): 55–61.

Song, M. S., L. Salmena, and P. P. Pandolfi. 2012. "The Functions and Regulation of the PTEN Tumour Suppressor." *Nature Reviews.Molecular Cell Biology* 13 (5): 283–296.

Sugarman, J. 2015. "Ethics and Germline Gene Editing." *EMBO Reports* 16 (8): 879–880.

Sun, W., W. Ji, J. M. Hall, et al. 2015. "Self-Assembled DNA Nanoclews for the Efficient Delivery of CRISPR-Cas9 for Genome Editing." *Angewandte Chemie (International Ed.in English)* 54 (41): 12029–12033.

Van Dyke, T. and T. Jacks. 2002. "Cancer Modeling in the Modern Era: Progress and Challenges." *Cell* 108 (2): 135–144.

Van Rompay, A. R., M. Johansson, and A. Karlsson. 2000. "Phosphorylation of Nucleosides and Nucleoside Analogs by Mammalian Nucleoside Monophosphate Kinases." *Pharmacology & Therapeutics* 87 (2–3): 189–198.

Wang, L., F. Li, L. Dang, et al. 2016. "In Vivo Delivery Systems for Therapeutic Genome Editing." *International Journal of Molecular Sciences* 17 (5): 10.3390/ijms17050626.

Xue, W., S. Chen, H. Yin, et al. 2014. "CRISPR-Mediated Direct Mutation of Cancer Genes in the Mouse Liver." *Nature* 514 (7522): 380–384.

Yu, Y. P., Y. Ding, Z. Chen, et al. 2014. "Novel Fusion Transcripts Associate with Progressive Prostate Cancer." *The American Journal of Pathology* 184 (10): 2840–2849.

Zuris, J. A., D. B. Thompson, Y. Shu, et al. 2015. "Cationic Lipid-Mediated Delivery of Proteins Enables Efficient Protein-Based Genome Editing in Vitro and in Vivo." *Nature Biotechnology* 33 (1): 73–80.

Index

Page numbers followed by f and t indicate figures and tables, respectively.

B

F

I

K

L

N

O

P

Q

R

S

U

V

W

X

Y

Z